Basic Mathematics

TEACHER'S EDITION

6TH EDITION

Basic Mathematics

TEACHER'S EDITION

6TH EDITION

Mervin L. Keedy

Purdue University

Marvin L. Bittinger

Indiana University—Purdue University at Indianapolis

ADDISON-WESLEY PUBLISHING COMPANY

Reading, Massachusetts • Menlo Park, California • New York
Don Mills, Ontario • Wokingham, England • Amsterdam • Bonn
Sydney • Singapore • Tokyo • Madrid • San Juan

Sponsoring Editor	Elizabeth Burr
Managing Editor	Karen Guardino
Production Supervisor	Jack Casteel
Design, Editorial, and Production Services	Quadrata, Inc.
Illustrator	ST Associates, Inc., and Scientific Illustrators
Art Consultant	Loretta Bailey
Manufacturing Supervisor	Roy Logan
Cover Design and Photograph	Marshall Henrichs

PHOTO CREDITS

1, Elliott Erwitt, Magnum Photos, Inc. **73,** Bjorn Bolstad, Photo Researchers, Inc. **125,** Ken Karp **175,** Arthur Tress, Photo Researchers, Inc. **207,** Jean-Claude Lejeune, Stock/Boston **251,** Jeff Albertson, Stock/Boston **277,** Teri Leigh Stratford, Photo Researchers, Inc. **303,** FBI **335,** AP/Wide World Photos **381,** Cheryl Traendly **435,** The Image Bank **469,** Hirojui Kubota, Magnum Photos, Inc. **505,** Bill Bachman, Photo Researchers, Inc. **551,** Georg Gerster, Comstock

ABCDEFGHIJ-DO-94321

Contents

3 Addition and Subtraction: Fractional Notation 125

4 Addition and Subtraction: Decimal Notation 175

5 Multiplication and Division: Decimal Notation 207

6 Ratio and Proportion 251

7 Percent Notation 277

8 Descriptive Statistics 335

9 Geometry and Measures: Length and Area 381

10 More on Measures 435

11 The Real-Number System 469

12 Algebra: Solving Equations and Problems 505

Developmental Units

Preface

Intended for students who do not have basic arithmetic skills, this text is appropriate for a one-term course in arithmetic or prealgebra. It is the first in a series of texts that includes the following:

Keedy/Bittinger: *Basic Mathematics,* Sixth Edition,
Keedy/Bittinger: *Introductory Algebra,* Sixth Edition,
Keedy/Bittinger: *Intermediate Algebra,* Sixth Edition.

Basic Mathematics, Sixth Edition, is a significant revision of the Fifth Edition, with respect to content, pedagogy, and an expanded supplements package. Its unique approach, which has been developed over many years, is designed to help today's students both learn *and* retain mathematical concepts. The Sixth Edition is accompanied by a comprehensive supplements package that has been integrated with the text to provide maximum support for both instructor and student.

Following are some distinctive features of the approach and pedagogy that we feel will help meet some of the challenges all instructors face teaching developmental mathematics.

APPROACH

CAREFUL DEVELOPMENT OF CONCEPTS We have divided each section into discrete and manageable learning objectives. Within the presentation of each objective, there is a careful buildup of difficulty through a series of developmental and followup examples. These enable students to thoroughly understand the mathematical concepts involved at each step. Each objective is constructed in a similar way, which gives students a high level of comfort with both the text and their learning process.

FOCUS ON "WHY" Throughout the text, we present the appropriate mathematical rationale for a topic, rather than mathematical "shortcuts." For example, when manipulating rational expressions, we remove factors of 1 rather than cancel, although cancellation is mentioned with appropriate cautions. This helps prevent student errors in cancellation and other incorrectly remembered shortcuts in later courses.

PROBLEM SOLVING We include real-life applications and problem-solving techniques throughout the text to motivate students and encourage them to think about how mathematics can be used. We also introduce a five-step problem-solving process early in the text and use the basic steps of this process (Familiarize, Translate, Solve, Check, and State the Answer) whenever a problem is solved.

PEDAGOGY

INTERACTIVE WORKTEXT APPROACH The pedagogy of this text is designed to provide students with a clear set of learning objectives, and involve them with the development of the material, providing immediate and continual reinforcement.

Section objectives are keyed to appropriate sections of the text, exercises, and answers, so that students can easily find appropriate review material if they are unable to do an exercise.

Numerous *margin exercises* throughout the text provide immediate reinforcement of concepts covered in each section.

STUDY AID REFERENCES Many valuable study aids accompany this text. Each section is referenced to appropriate videotape, audiotape, and software diskette numbers to make it easy for students to find and use the correct support materials.

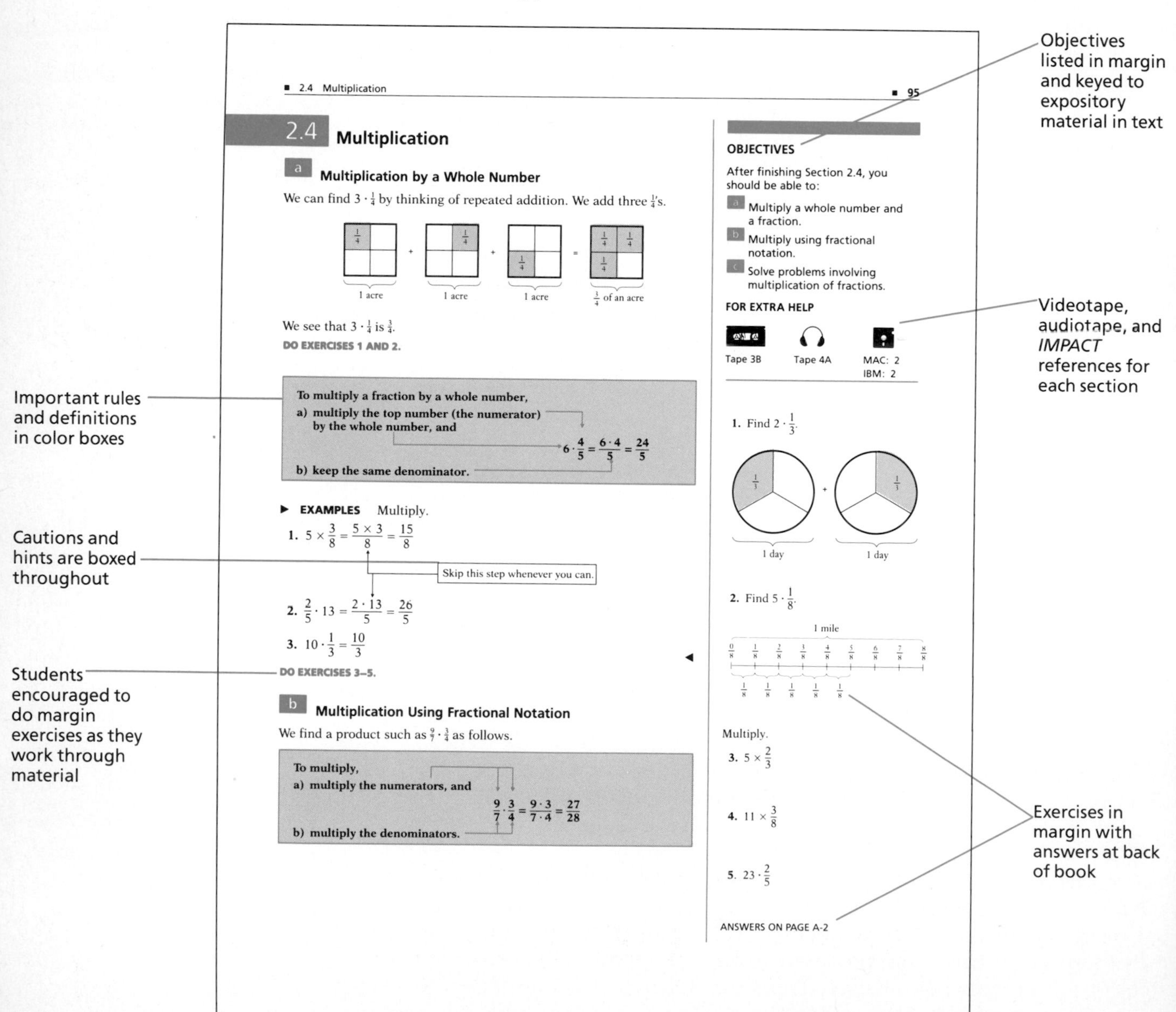
■ 2.4 Multiplication ■ 95

2.4 Multiplication

a Multiplication by a Whole Number

We can find $3 \cdot \frac{1}{4}$ by thinking of repeated addition. We add three $\frac{1}{4}$'s.

We see that $3 \cdot \frac{1}{4}$ is $\frac{3}{4}$.

DO EXERCISES 1 AND 2.

To multiply a fraction by a whole number,
a) multiply the top number (the numerator) by the whole number, and
b) keep the same denominator.

$6 \cdot \frac{4}{5} = \frac{6 \cdot 4}{5} = \frac{24}{5}$

▶ EXAMPLES Multiply.

1. $5 \times \frac{3}{8} = \frac{5 \times 3}{8} = \frac{15}{8}$

Skip this step whenever you can.

2. $\frac{2}{5} \cdot 13 = \frac{2 \cdot 13}{5} = \frac{26}{5}$

3. $10 \cdot \frac{1}{3} = \frac{10}{3}$

DO EXERCISES 3–5.

b Multiplication Using Fractional Notation

We find a product such as $\frac{9}{7} \cdot \frac{3}{4}$ as follows.

To multiply,
a) multiply the numerators, and
b) multiply the denominators.

$\frac{9}{7} \cdot \frac{3}{4} = \frac{9 \cdot 3}{7 \cdot 4} = \frac{27}{28}$

OBJECTIVES

After finishing Section 2.4, you should be able to:

a Multiply a whole number and a fraction.

b Multiply using fractional notation.

c Solve problems involving multiplication of fractions.

FOR EXTRA HELP

Tape 3B Tape 4A MAC: 2 IBM: 2

1. Find $2 \cdot \frac{1}{3}$.

2. Find $5 \cdot \frac{1}{8}$.

Multiply.

3. $5 \times \frac{2}{3}$

4. $11 \times \frac{3}{8}$

5. $23 \cdot \frac{2}{5}$

ANSWERS ON PAGE A-2

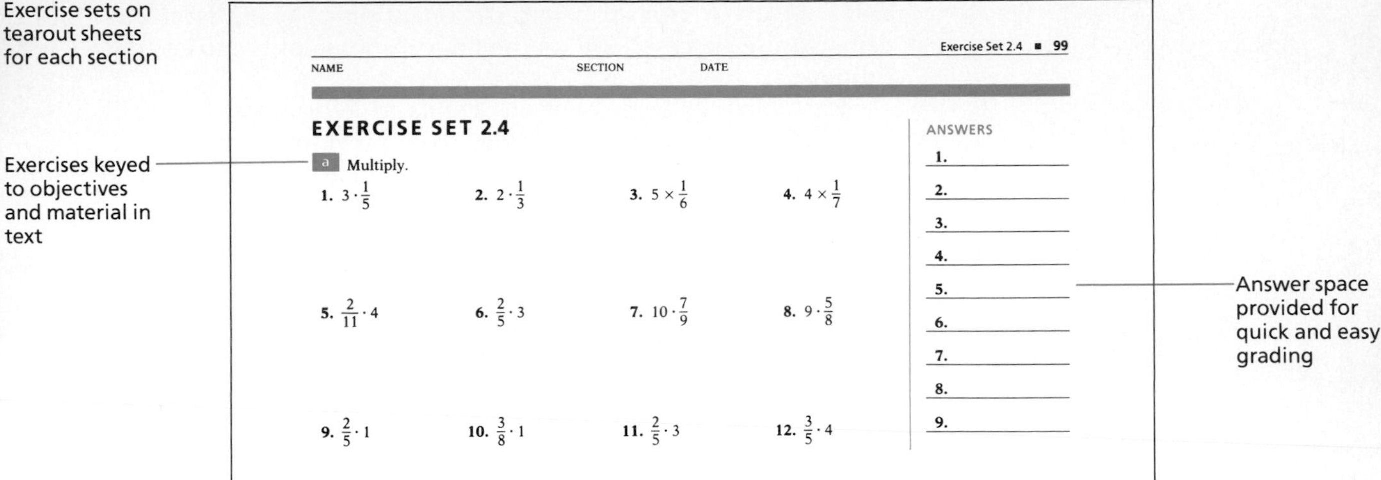

Exercise Set 2.4 ■ 99

NAME SECTION DATE

EXERCISE SET 2.4

a Multiply.

1. $3 \cdot \frac{1}{5}$	**2.** $2 \cdot \frac{1}{3}$	**3.** $5 \times \frac{1}{6}$	**4.** $4 \times \frac{1}{7}$
5. $\frac{2}{11} \cdot 4$	**6.** $\frac{2}{5} \cdot 3$	**7.** $10 \cdot \frac{7}{9}$	**8.** $9 \cdot \frac{5}{8}$
9. $\frac{2}{5} \cdot 1$	**10.** $\frac{3}{8} \cdot 1$	**11.** $\frac{2}{5} \cdot 3$	**12.** $\frac{3}{5} \cdot 4$

ANSWERS

1.
2.
3.
4.
5.
6.
7.
8.
9.

VERBALIZATION SKILLS AND "THINKING IT THROUGH" Students' perception that mathematics is a foreign language is a significant barrier to their ability to think mathematically and is a major cause of math anxiety. In the Sixth Edition we have encouraged students to think through mathematical situations, synthesize concepts, and verbalize mathematics whenever possible.

"Thinking It Through" exercises at the end of each chapter encourage students to both think and write about key mathematical ideas that they have encountered in the chapter.

"Synthesis Exercises" at the end of most exercise sets require students to synthesize several learning objectives or to think through and provide insight into the present material.

In addition, many important definitions, such as the laws of exponents, are presented verbally as well as symbolically, to help students learn to read mathematical notation.

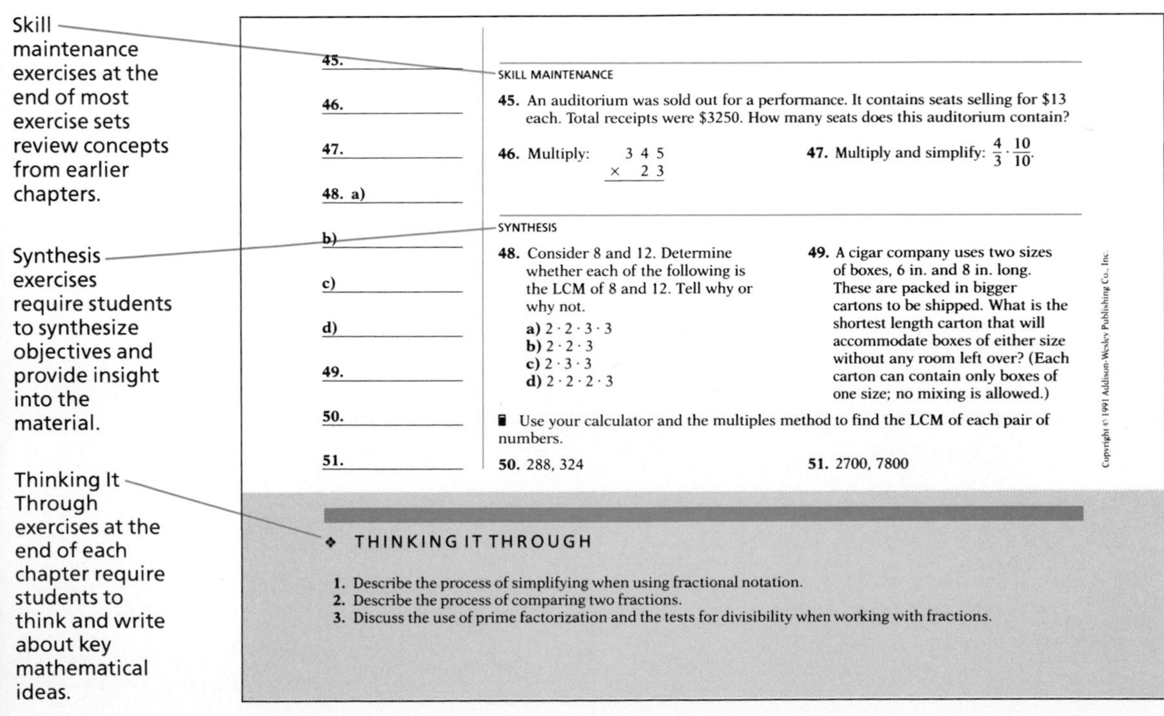

45.
46.
47.
48. a)
b)
c)
d)
49.
50.
51.

SKILL MAINTENANCE

45. An auditorium was sold out for a performance. It contains seats selling for $13 each. Total receipts were $3250. How many seats does this auditorium contain?

46. Multiply: $\begin{array}{r} 345 \\ \times\ 23 \\ \hline \end{array}$

47. Multiply and simplify: $\frac{4}{3} \cdot \frac{10}{10}$.

SYNTHESIS

48. Consider 8 and 12. Determine whether each of the following is the LCM of 8 and 12. Tell why or why not.
a) $2 \cdot 2 \cdot 3 \cdot 3$
b) $2 \cdot 2 \cdot 3$
c) $2 \cdot 3 \cdot 3$
d) $2 \cdot 2 \cdot 2 \cdot 3$

49. A cigar company uses two sizes of boxes, 6 in. and 8 in. long. These are packed in bigger cartons to be shipped. What is the shortest length carton that will accommodate boxes of either size without any room left over? (Each carton can contain only boxes of one size; no mixing is allowed.)

■ Use your calculator and the multiples method to find the LCM of each pair of numbers.

50. 288, 324

51. 2700, 7800

◆ THINKING IT THROUGH

1. Describe the process of simplifying when using fractional notation.
2. Describe the process of comparing two fractions.
3. Discuss the use of prime factorization and the tests for divisibility when working with fractions.

SKILL MAINTENANCE Because retention of skills is critical to students' future success, skill maintenance is a major emphasis of the Sixth Edition.

Each chapter begins with a *"Points to Remember"* box, which highlights key formulas and definitions from previous chapters.

In addition, we include *Skill Maintenance Exercises* at the end of most exercise sets. These review skills and concepts from earlier sections of the text.

At the end of each chapter, the *Summary and Review* summarizes important properties and formulas and includes extensive review exercises.

Each *Chapter Test* tests four review objectives from preceding chapters as well as the chapter objectives.

We also include a *Cumulative Review* at the end of each chapter; this reviews material from all preceding chapters.

At the back of the text are answers to all review exercises, together with section and objective references, so that students know exactly what material to restudy if they miss a review exercise.

TESTING AND SKILL ASSESSMENT Accurate assessment of student comprehension is an important factor in a student's long-term success. In the Sixth Edition, we have provided many assessment opportunities.

A *Diagnostic Pretest* at the beginning of the text can place students in the appropriate chapter for their skill level, and identifies both familiar material and specific trouble areas later in the text.

Chapter Pretests diagnose student skills and place the students appropriately within each chapter, allowing them to concentrate on topics with which they have particular difficulty.

Chapter Tests at the end of each chapter allow students to review and test comprehension of chapter skills.

Answers to each question on all tests are at the back of the text.

For additional testing options, we have developed a printed test bank with many alternative forms of each chapter test in both open-ended and multiple-choice formats. For a greater degree of flexibility in creating chapter tests, the text is also accompanied by extensive computerized testing programs for IBM, MAC, and Apple computers.

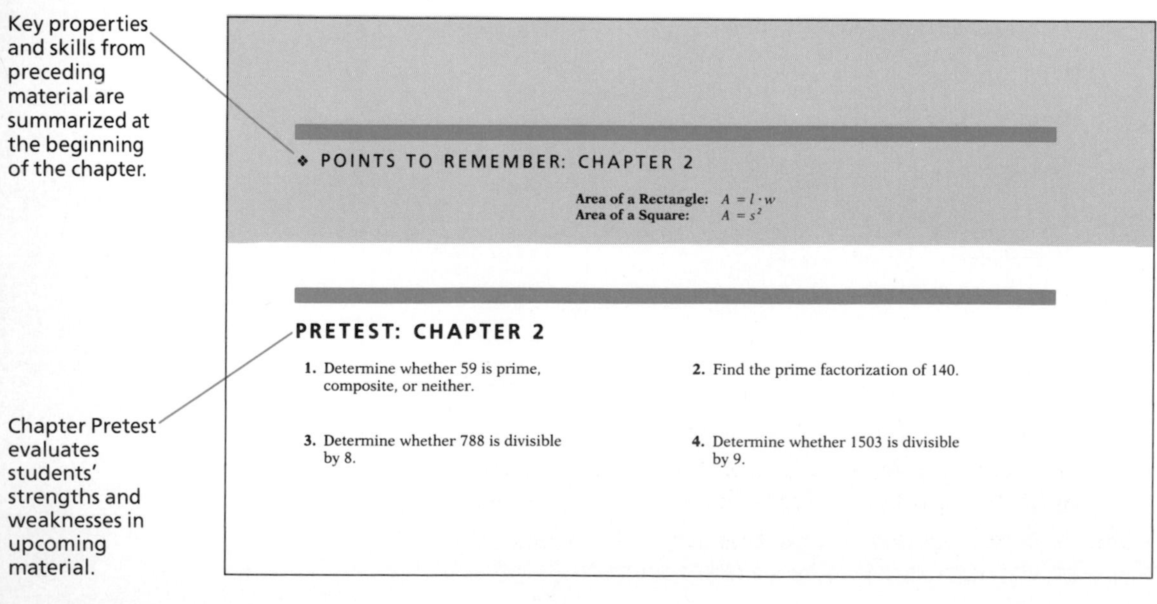

FLEXIBILITY OF TEACHING MODES

The flexible worktext format of *Basic Mathematics* allows the book to be used in many ways.

- **In a standard lecture.** To use the book in a lecture format, the instructor lectures in a conventional manner and encourages students to do the margin exercises while studying on their own. This greatly enhances the readability of the text.
- **For a modified lecture.** To bring student-centered activity into the class, the instructor stops lecturing and has the students do margin exercises.
- **For a no-lecture class.** The instructor makes assignments that students do on their own, including working the margin exercises. During the class period following the assignment, the instructor answers questions, and students have an extra day or two to polish their work before handing it in. In the meantime, they are working on the next assignment. This method provides individualization while keeping a class together. It also minimizes the number of instructor hours required and has been found to work well with large classes.
- **In a learning laboratory.** Because this text is highly readable and easy to understand, it can be used in a learning laboratory or any other self-study situation.

KEY CONTENT CHANGES

In response to both extensive user comments and reviewer feedback, there have been many organizational changes and revisions to the Sixth Edition. Detailed information about the changes made in this material is available in the form of a Conversion Guide. Please ask your local Addison-Wesley sales representative for more information. Following is a list of the major organizational changes for this revision:

- Word problems have been updated and added throughout. A five-step process for problem solving is introduced in Chapter 1, and the five steps are used throughout the text whenever an applied problem is solved.
- The concept of estimating and reasonable answers is integrated more extensively throughout the material.
- The introduction to exponents has been moved to Chapter 1, where it is presented very intuitively with the material on order of operations.
- Material on using fractions and decimals together in the same exercise or word problem has been added to Chapter 5, and examples and exercises of this type have been integrated into the rest of the text where appropriate.
- A new chapter on descriptive statistics has been added. This includes existing material on averages, medians, and new material on modes, tables, charts, bar graphs, line graphs, pictographs, and circle graphs.
- The algebra chapters have been revised so that the discussions of operations on integers and rational numbers are presented together.

SUPPLEMENTS

This text is accompanied by a comprehensive supplements package. Below is a brief list of these supplements, followed by a detailed description of each one.

For the Instructor	For the Student
Teacher's Edition	Student's Solutions Manual
Instructor's Solutions Manual	Videotapes
Instructor's Resource Guide	Audiotapes
Printed Test Bank	The Math Hotline
Lab Resource Manual	Comprehensive Tutorial Software
Answer Book	Drill and Practice Software
Computerized Testing	

SUPPLEMENTS FOR THE INSTRUCTOR

All supplements for the instructor are free upon adoption of this text.

Teacher's Edition

This is a specially bound version of the student text with exercise answers printed in a third color. It also includes additional information on the skill maintenance exercises, suggested syllabi for different length courses, and some information about the teaching aids that accompany the text.

Instructor's Solutions Manual

This manual by Judith A. Penna contains worked-out solutions to all even-numbered exercises and discussions of the "Thinking It Through" sections.

Instructor's Resource Guide

This guide contains the following:

- Additional "Thinking It Through" exercises.
- Extra practice problems for some of the most challenging topics in the text.
- Teaching essays on math anxiety and study skills.
- Indexes to the videotapes, the audiotapes, and the software that accompany the text.
- Number lines and grids for test preparation.
- Conversion guide that cross-references the Fifth Edition to the Sixth Edition.
- Black-line transparency masters including a selection of key definitions, procedures, problem-solving strategies, graphs, and figures to use in class.

Printed Test Bank

This is an extensive collection of alternative chapter test forms, including the following:

- 5 alternative test forms for each chapter with questions in the same topic order as the objectives presented in the chapter.
- 5 alternative test forms for each chapter with the questions in a different order.
- 3 multiple-choice test forms for each chapter.
- 2 cumulative review tests for each chapter.
- 9 alternative forms of the final examination, 3 with questions organized by chapter, 3 with questions scrambled, and 3 with multiple-choice questions.

Lab Resource Manual

This manual contains a selection of essays on setting up learning labs, including information on running large testing centers and setting up mastery learning programs. It also includes a directory of learning lab coordinators who are available to answer questions.

Answer Book

The Answer Book contains answers to all the exercises in the text for you to make available to your students.

Computerized Testing

OmniTest (IBM PC), AWTest (Apple II series)

This text is accompanied by algorithm-driven testing systems for both IBM and Apple. With both machine versions, it is easy to create up to 99 variations of a customized test with just a few keystrokes, choosing from over 300 open-ended and multiple-choice test items. Instructors can also print out tests in chapter-test format.

The IBM testing program, OmniTest, also allows users to enter their own test items and edit existing items in an easy-to-use WYSIWYG format.

LXR·TEST™ (MACINTOSH)

This is a versatile and flexible test item bank of more than 1200 multiple-choice and open-ended test items with complete math graphics and full editing capabilities. Tests can be created by selecting specific test items or by requesting the computer to select items randomly from designated objectives. LXR·TEST can create multiple test versions by scrambling the order of multiple-choice distractors or the order of the questions themselves.

SUPPLEMENTS FOR THE STUDENT

Student's Solutions Manual

This manual by Judith A. Penna contains completely worked out solutions with step-by-step annotations for all the odd-numbered exercises in the text. It is free to adopting instructors and may be purchased by your students from Addison-Wesley Publishing Company.

Videotapes

Using the chalkboard and manipulative aids, Donna DeSpain lectures in detail, works out exercises, and solves problems from most sections in the text on 15 70-minute videotapes. These tapes are ideal for students who have missed a lecture or who need extra help. Each section in the text is referenced to the appropriate tape number and section, underneath the icon . A complete set of videotapes is free to qualifying adopters.

Audiotapes

The audiotapes are designed to lead students through the material in each text section. Bill Saler explains solution steps to examples, cautions students about common errors, and instructs them to stop the tape and do exercises in the margin. He then reviews the margin-exercise solutions, pointing out potential errors. Each section in the text is referenced to the appropriate tape number and section, underneath the icon Audiotapes are free to qualifying adopters.

The Math Hotline

This telephone hotline is open 24 hours a day for students to receive detailed hints for exercises that have been developed by Larry Bittinger. Exercises covered include all the odd-numbered exercises in the exercise sets with the exception of the skill maintenance and synthesis exercises.

Tutorial Software

A variety of tutorial software packages is available to accompany this text. Please contact your Addison-Wesley representative for a software sampler that contains demonstration disks for these packages and a summary of our distribution policy.

Comprehensive Tutorials

IMPACT: An Interactive Mathematics Tutorial

by Wayne Mackey and Doug Proffer, Collin County Community College (IBM PC or MACINTOSH).

This software was developed exclusively for Addison-Wesley and is keyed section by section to this text. Icons at the beginning of each section reference the appropriate disk number. The disk menus correspond to the text's section numbers.

IMPACT is designed to generate practice exercises based on the exercise sets in this book. If students are having trouble with a particular exercise, they can ask to see an example or a step-by-step solution to the problem they are working on. Each step of the step-by-step solutions is treated interactively to keep students involved in the solution of the problem, and help them identify precisely where they are having trouble. *IMPACT* also keeps detailed records of students' scores.

Arithmetic Skillbuilder

by Virginia Hamilton and Dennis Owen (IBM PC or Apple II series).

This is an interactive software package covering 38 topics in basic mathematics, featuring a brief review of each topic with examples and over 300 skillbuilding exercises.

Drill and Practice Packages

The Math Lab
by Chris Avery and Chris Barker, DeAnza College (Apple II series, IBM PC, or Macintosh).

Students choose the topic, level of difficulty, and number of exercises. If they get a wrong answer, *The Math Lab* will prompt them with the first step of the solution. This software also keeps detailed records of student scores.

ACKNOWLEDGMENTS

Many of you who teach developmental mathematics have helped to shape the Sixth Edition of this text by reviewing, answering surveys, participating in focus groups, filling out questionnaires, and spending time with us on your campuses. Our heartfelt thanks to all of you, and many apologies to anyone we have missed on the following list.

TEXTBOOK REVIEWERS

Doloris Anenson, *Merced College;* Julia Brown, *Atlantic Community College;* Richard J. Burns, *Springfield Community College;* Joan Capps, *Raritan Valley Community College;* Joanne M. Kelly, *Palm Beach Community College;* Sue L. Korsak, *New Mexico State University;* Ellen Milosheff, *Triton College;* Charles A. Smith, *St. Petersburg Junior College;* Steve Sworder, *Saddleback College;* Richard Watkins, *Tidewater Community College.*

FORMAL AND INFORMAL FOCUS GROUP PARTICIPANTS

Geoff Akst, *Borough of Manhattan Community College;* Betty Jo Baker, *Lansing Community College;* Gene Beuthin, *Saginaw Valley State University;* Rheta Beaver, *Valencia Community College;* Roy Boersema, *Front Range Community College;* Dale Boye, *Schoolcraft College;* Jim Brenner, *Black Hawk College;* Ben Cheatham, *Valencia Community College;* Karen Clark, *Tacoma Community College;* Tom Clark, *Lane Community College;* Sally Copeland, *Johnson County Community College;* Ernie Danforth, *Corning Community College;* Sarah Evangelista, *Temple University;* Bill Freed, *Concordia College;* Sally Glover-Richard, *Pierce Community College;* Valerie Hayward, *Orange Coast College;* Eric Heinz, *Catonsville Community College;* Bruce Hoelter, *Raritan Valley Community College;* Lou Hoezle, *Bucks County Community College;* Linda Horner, *Broward Community College;* Mary Indelicato, *Normandale Community College;* Tom Jebson, *Pierce Community College;* Jeff Jones, *County College of Morris;* Judith Jones, *Valencia Community College;* Virginia Keen, *West Michigan University;* Roxanne King, *Prince Georges Community College;* Lee Marva Lacy, *Glendale Community College;* Ginny Licata, *Camden County College;* Randy Liefson, *Pierce Community College;* Charlie Luttrell, *Frederick Community College;* Marilyn MacDonald, *Red Deer College;* Sharon MacKendrick, *New Mexico State University;* Annette Magyar, *Southwestern Michigan College;* Bob Malena, *Community College of Allegheny County;* Marilyn Masterson, *Lansing Community College;* Don McNair, *Lane Community College;* John Pazdar, *Greater Hartford Community College;* Donald Perry, *Lee College;* Jeanne Romeo,

Delta College; Jack Rotman, *Lansing Community College;* Winona Sathre, *Valencia Community College;* Billie Stacey, *Sinclair Community College;* John Steele, *Lane Community College;* Dave Steinfort, *Grand Rapids Junior College;* Betty Swift, *Cerritos College;* Bill Wittinfeld, *Tacoma Community College; Faculty of St. Petersburg Junior College.*

QUESTIONNAIRE RESPONDEES

Albert Beron, *Moorpark College;* Linda Long, *Ricks College;* Mark Mays, *Indiana Vocational Technical College;* Cornelius McKenna, *Kishwaukee College;* Elizabeth Polen, *County College of Morris;* Jan Roy, *Montcalm Community College;* Carol Russell, *Indiana University Southeast;* Dan Snook, *Montcalm Community College;* Steve Sworder, *Saddleback College.*

We also wish to thank the many people without whose committed efforts our work could not have been completed. In particular, we would like to thank Judy Beecher, Barbara Johnson, and Judy Penna for their work on proofreading the manuscript and overseeing the production process. We would also like to thank Pat Pasternak, who did a marvelous job typing the text manuscript and answer section, and Bill Saler, Nancy Woods, Lauren Page, and Gary Hiday, who did a thorough and conscientious job of checking the manuscript.

M.L.K.
M.L.B.

To The Student

This text has many features that can help you succeed in basic mathematics. To familiarize yourself with these, you might read the preface that starts on page ix, and study the annotated pages that are included. Following are a few suggestions on how to use these features to enhance your learning process.

BEFORE YOU START THE TEXT

If you are in a classroom setting, your instructor might ask you to take the diagnostic pretest at the beginning of the text, checking your answers at the back of the text, to find out what material you already know and what material you need to spend time on. You can also use this pretest to skip material that you already know from an independent learning situation.

BEFORE YOU START A CHAPTER

The chapter opening page gives you an idea of the material that you are about to study and how it can be used. The chapter opening introduction also tells you what sections you will need to review in order to do the skill maintenance exercises on the chapter test. It's a good idea to restudy these sections to keep the material fresh in your mind for the midterm or final examination.

The first page of each chapter lists "Points to Remember" that will be needed to work certain examples and exercises in the chapter. You should try to review any skills listed here before beginning the chapter and learn any formulas or definitions.

This same page also includes a chapter pretest. You can work through this and check your answers at the back of the text to identify sections that you might skip or sections that give you particular difficulty and need extra concentration.

WORKING THROUGH A SECTION

First you should read the learning objectives for the section. The symbol next to an objective (a, b, c) appears next to the text, exercises, and answers that correspond to that objective, so you can always refer back to the appropriate material when you need to review a topic.

You will also notice that there are references to the audiotapes, videotapes, and software that are available for extra help for the section underneath the objective listing. The software referenced is a program called *IMPACT: An Interactive Mathematics Tutorial.*

As you work through a section, you will see an instruction to "Do Exercises x–xx." This refers to the exercises in the margin of the page. You should always stop and do these to practice what you have just studied because they greatly enhance the readability of the text. Answers to the margin exercises are at the back of the text.

After you have completed a section, you should do the assigned exercises in the exercise set. The exercises are keyed to the section objectives, so that if you get an incorrect answer, you know that you should restudy the text section that follows the corresponding symbol.

Answers to all the odd-numbered exercises are at the back of the text. A solutions manual with complete worked-out solutions to all the odd-numbered exercises is available from Addison-Wesley Publishing Company.

PREPARING FOR A CHAPTER TEST

To prepare for a chapter test, you can review your homework and restudy sections that were particularly difficult. You should also learn the "Important Properties and Formulas" that begin the chapter's summary and review and study the review sections that are listed at the beginning of the review exercises.

After studying, you might set aside a block of time to work through the summary and review as if it were a test. You can check your answers at the back of the text after you are done. The answers are coded to sections and objectives, so you can restudy any areas in which you are having trouble. You can also take the chapter test as practice, again checking your answers at the back of the text.

If you are still having difficulties with a topic, you might try either going to see your instructor or working with the videotapes, audiotapes, or tutorial software that are referenced at the beginning of the text sections. Be sure to start studying in time to get extra help before you must take the test.

PREPARING FOR A MIDTERM OR FINAL EXAMINATION

To keep material fresh in your mind for a midterm or a final examination, you can work through the cumulative reviews at the end of each chapter. You can also use these as practice midterms or finals. In addition, there is a final examination at the end of the text that has answers to all of its exercises at the back of the text.

OTHER STUDY TIPS

There is a saying in the real-estate business: "The three most important things to consider when buying a house are *location, location, location.*" When trying to learn mathematics, the three most important things are *time, time, time.* Try to carefully analyze your situation. Be sure to allow yourself *time* to do the lesson. Are you taking too many courses? Are you working so much that you do not have *time* to study? Are you taking *time* to maintain daily preparation? Other study tips are provided on pages marked "Sidelights" in the text.

Basic Mathematics

TEACHER'S EDITION

6TH EDITION

DIAGNOSTIC PRETEST

Chapter 1

1. Add: 1425 + 382.

2. Solve: $32 + x = 61$.

3. Multiply: $\begin{array}{r} 321 \\ \times\ 47 \\ \hline \end{array}$

4. A jar contains 128 oz of juice. How many 6-oz glasses can be filled from the jar? How many ounces will be left over?

Chapter 2

5. Solve: $\frac{5}{8} \cdot x = \frac{3}{16}$.

6. Multiply and simplify: $4 \cdot \frac{3}{8}$.

7. Find the prime factorization of 144.

8. A recipe calls for $\frac{3}{4}$ cup of flour. How much is needed to make $\frac{2}{3}$ of a recipe?

Chapter 3

9. Add: $\frac{3}{8} + \frac{1}{6}$.

10. Multiply and simplify: $4\frac{1}{5} \cdot 3\frac{2}{3}$.

11. A 3-m pole was set $1\frac{2}{5}$ m in the ground. How much was above the ground?

12. A car travels 249 mi on $8\frac{3}{10}$ gal of gas. How many miles per gallon did it get?

Chapter 4

13. Which number is larger, 0.00009 or 0.0001?

14. Round to the nearest tenth: 25.562.

15. Add: $\begin{array}{r} 12.035 \\ 0.08 \\ +\ 27.7 \\ \hline \end{array}$

16. A driver bought gasoline when the odometer read 68,123.2. At the next gasoline purchase the odometer read 68,310.1. How many miles were driven?

Chapter 5

17. Multiply. $\begin{array}{r} 0.012 \\ \times\ 2.5 \\ \hline \end{array}$

18. Find decimal notation: $\frac{7}{3}$.

19. Solve: $1.5 \times t = 3.6$.

20. What is the cost of 5 shirts at $23.99 each?

Chapter 6

21. Solve: $\frac{1.2}{x} = \frac{0.4}{1.5}$.

22. If 3 cans of green beans cost $1.19, how many cans of green beans can you buy for $4.76?

23. It costs $2.19 for a 22-oz box of cereal. Find the unit price in cents per ounce. Round to the nearest tenth of a cent.

Chapter 7

24. Find percent notation: $\frac{1}{8}$.

25. Find decimal notation: 1.35%.

26. The price of a pair of shoes was reduced from \$25 to \$19. Find the percent of decrease in price.

27. What is the simple interest on \$230 principal at the interest rate of 8.5% for one year?

Chapter 8

28. Find the average, median, and mode of the following set of numbers:

22, 25, 27, 25, 22, 25.

29. A car traveled 296 mi on 16 gal of gasoline. What was the average number of miles per gallon?

30. In order to get a B in math, a student must average 80 on four tests. Scores on the first three tests were 85, 72, and 78. What is the lowest score the student can get on the last test and still get a B?

Chapter 9

Complete.

31. 2 yd = ______ in.

32. 4 cm = ______ km

33. Find the area and the circumference of a circle with a diameter of 12 cm. Leave answers in terms of π.

34. Find the area and the perimeter of a rectangle with length 3 ft and width 2.5 ft.

Chapter 10

Complete.

35. 2 min = ______ sec

36. 5 mg = ______ g

37. $2\ \text{yd}^2$ = ______ ft^2

38. 10 qt = ______ gal

39. The diameter of a ball is 18 cm. Find the volume. Use 3.14 for π.

Chapter 11

40. Find the absolute value: $|-4.2|$.

41. Find decimal notation: $-\frac{4}{9}$.

Compute and simplify.

42. $-2-(-1.9)$

43. $\frac{5}{6}\left(-\frac{1}{10}\right)$

Chapter 12

Solve.

44. $2x-1=4x+5$

45. $\frac{1}{3}x+\frac{1}{5}=\frac{2}{3}-\frac{1}{4}x$

46. A student bought a sweater and a pair of jeans. The sweater cost \$33.95. This was \$8.39 more than the cost of the jeans. How much did the jeans cost?

47. A 22-oz box of cereal costs \$2.45. How many boxes of cereal can you buy for \$19.60?

Teacher's Resource Material

The information in this section is designed specifically for the instructor—to help him or her maximize the effectiveness of this text. It includes the following:

- A list of the review sections and objectives covered in the Skill Maintenance Exercises of each chapter test.
- A list of topics for which Extra Practice Problem sheets have been developed. The Extra Practice Problem sheets can be found in the *Instructor's Resource Guide.*
- Suggested generic syllabi for a one-semester 4-hour course, a one-semester 3-hour course, and a quarter course.
- A flowchart showing a path through the pedagogical features of the text.

This material is not included in the student edition of the text.

Additional resource material for instructors can be found in the *Instructor's Resource Guide.* See page xiv in the Preface for a detailed description of this supplement.

ADDISON-WESLEY SALES OFFICES

If you have any questions about this material, or about any of the supplements that accompany this text, please contact your Addison-Wesley sales representative or your regional sales office for more information. A list of regional sales office telephone numbers and addresses follows.

Eastern Regional Sales Office

Mike Simpson, Regional Manager
Jacob Way
Reading, MA 01867
(617) 944-3700 #2796

ME, VT, NH, NY, MA, RI, CT, PA, MD, NJ, DE, DC, VA, WV

Western Regional Sales Office

Ron Taylor, Regional Manager
390 Bridge Parkway
Redwood City, CA 94065
(415) 594-4410

CO, UT, NM, AZ, CA, NV, OR, WA, ID, MT, WY, AK, HI

In-House Sales Office

Nancy Kralowetz, Regional Manager
390 Bridge Parkway
Redwood City, CA 94065
(415) 594-4400

Midwestern Regional Sales Office

Susan Renwick, Regional Manager
1843 Hicks Road
Rolling Meadows, IL 60008
(708) 991-7878

OH, IN, IL, KY, KS, MO, WI, MI, MN, IA, ND, SD, NE

Southern Regional Sales Office

Walter Dinteman, Regional Manager
1100 Ashwood Drive, Suite #145
Atlanta, GA 30338
(404) 394-7268

NC, SC, GA, FL, AL, TN, AR, MS, LA, TX, OK

SKILL MAINTENANCE TOPICS

Each chapter test after Chapter 1 has skill maintenance questions that test students' comprehension of material from four specific sections in earlier chapters. The test questions are designed to help students retain fundamental skills in the key areas of manipulation, equation solving, and problem solving. The introduction to each chapter lists the review sections that will be tested on the chapter test. Following is a more detailed list of review sections for each chapter, which includes section and objective, as well as skill type.

CHAPTER 1

No skill maintenance exercises related to preceding chapters

CHAPTER 2

MANIPULATIVE SKILLS
[1.3d] Subtracting whole numbers
[1.6c] Dividing whole numbers

EQUATION-SOLVING SKILLS
[1.7b] Solving equations like $t + 26 = 54$, $26 \cdot x = 52$, and $58 \div 2 = m$

PROBLEM-SOLVING SKILLS
[1.8a] Solving problems involving addition, subtraction, multiplication, and division of whole numbers

CHAPTER 3

MANIPULATIVE SKILLS
[1.5b] Multiplying whole numbers
[2.6a] Multiplying and simplifying, using fractional notation
[2.7b] Dividing and simplifying, using fractional notation

PROBLEM-SOLVING SKILLS
[1.8a] Solving problems involving whole numbers

CHAPTER 4

MANIPULATIVE SKILLS
[1.2b] Adding whole numbers
[1.3d] Subtracting whole numbers
[3.2a, b] Adding using fractional notation
[3.3a] Subtracting using fractional notation

CHAPTER 5

MANIPULATIVE SKILLS
[2.1d] Finding the prime factorization of a composite number
[2.5b] Simplifying fractional notation
[3.5a, b] Adding and subtracting using mixed numerals
[3.6a, b] Multiplying and dividing using mixed numerals

CHAPTER 6

MANIPULATIVE SKILLS
[2.5c] Testing fractions for equality
[5.1a] Multiplying using decimal notation
[5.2a] Dividing using decimal notation

PROBLEM-SOLVING SKILLS
[4.4a] Solving problems involving addition and subtraction with decimals

CHAPTER 7

MANIPULATIVE SKILLS
[3.4b] Converting between mixed numerals and fractional notation
[5.3a] Converting from fractional notation to decimal notation

EQUATION-SOLVING SKILLS
[5.2b] Solving equations of the type $a \cdot x = b$, where a and b are in decimal notation
[6.1c] Solving proportions

CHAPTER 8

MANIPULATIVE SKILLS
[2.7b] Dividing and simplifying using fractional notation

PROBLEM-SOLVING SKILLS
[6.3a] Solving problems involving proportions
[7.3a] Solving basic percent problems using equations
[7.4a] Solving basic percent problems using proportions
[7.5a] Solving applied percent problems

CHAPTER 9

MANIPULATIVE SKILLS
[7.1b] Converting from percent notation to decimal notation
[7.1c] Converting from decimal notation to percent notation
[7.2a] Converting from fractional notation to percent notation
[7.2b] Converting from percent notation to fractional notation

CHAPTER 10

MANIPULATIVE SKILLS
[1.9b] Evaluating exponential notation
[9.1a] Converting from one American unit of length to another
[9.2a] Converting from one metric unit of length to another

PROBLEM-SOLVING SKILLS
[7.8a] Solving problems involving simple interest and percent

CHAPTER 11

MANIPULATIVE SKILLS

[2.1d] Finding the prime factorization of a composite number
[3.1a] Finding the LCM of two or more natural numbers
[7.3b] Solving basic percent problems using equations
[9.4a] Finding the area of a rectangle or square

CHAPTER 12

MANIPULATIVE SKILLS

[9.6a, b, c] Finding the radius or diameter, the circumference, and the area of a circle
[10.1a] Finding the volume of a rectangular solid
[11.2a] Adding real numbers
[11.4a] Multiplying real numbers

EXTRA PRACTICE PROBLEM TOPICS

The Extra Practice Sheets can be found in the *Instructor's Resource Guide*. They are designed to provide extra drill on the hardest topics in the text. Each practice sheet begins with a few examples, which are followed by a number of practice problems of an average level of difficulty. Answers are provided separately. The Extra Practice Sheets are an excellent source of practice and reteaching for students who have done poorly on a test and who are going to try again.

Following is a list of the sections and topics for which there are Extra Practice Sheets in the *Instructor's Resource Guide*.

Section	Topic
Sections 1.2–1.6	Operations on the Whole Numbers
Section 1.8	Solving Problems with Whole Numbers
Section 1.9	Order of Operations
Sections 2.1–2.2	Factorizations and Divisibility
Sections 2.6–3.6	Solving Problems with Fractional Notation and Mixed Numerals
Sections 2.7 and 3.3	Solving Equations with Fractional Notation
Sections 4.4 and 5.3	Solving Problems with Decimal Notation
Section 6.3	Solving Problems Using Proportions
Section 7.4–7.7	Solving Problems Involving Percents
Section 8.1	Averages, Medians, and Modes
Sections 8.2–8.4	Reading Tables, Charts, and Graphs
Section 9.7	Finding Squares and Square Roots
Sections 9.1 and 9.2	Converting Units of Length
Sections 9.4–9.6	Finding Areas
Sections 10.1–10.3	Converting Units of Mass and Volume
Sections 11.2 and 11.3	Addition and Subtraction of Real Numbers
Sections 11.4 and 11.5	Multiplication and Division of Real Numbers
Section 12.1	Using the Distributive Law
Sections 12.2–12.4	Solving Equations Using the Addition and Multiplication Principles
Section 12.5	Solving Problems Using Equations

SAMPLE SYLLABI

Following is a sample catalog description, suggested course guidelines, and three sample syllabi for *Basic Mathematics*, Sixth Edition. The first two syllabi are for a 4 semester-hour and 3 semester-hour course modeled after a similar course at Indiana University—Purdue University at Indianapolis. The third syllabus is for a quarter course, modeled after a similar course at Salt Lake City Community College.

CATALOG DESCRIPTION

MA 010. Basic Mathematics. Covers whole numbers and their operations, arithmetic numbers using fractional notation, arithmetic numbers using decimal notation, ratio and proportion, percent notation, averages, medians, and modes, geometry and measures, the real number system, and an introduction to algebra by solving equations and problems.

SUGGESTED COURSE GUIDELINES

1. A daily time schedule is given later in this syllabus. Follow it to the letter. Students in this course are procrastinators. They cannot be allowed to set the pace.
2. A typical daily class would consist of a period to answer questions about the preceding assignment, followed by a lecture about the new material. We have found a quiz at the end of each class day (other than a test day) to be quite helpful, but this is optional. Such a quiz would usually be for only five minutes, but has reduced the number of grades C or lower. You might use all the quizzes to make up an extra test.
3. Expect at least 45% of the students to withdraw or fail. This has been the average in the past.
4. The instructor *must* give eight tests and a two-hour comprehensive final examination. *All* students are to take all the tests *and* the final exam. Grade guidelines are as follows:

   90: A
   80: B
   65: C
   F

   No grades of D are to be given.

Alternative forms of the tests can be found in the *Test Bank*. *Please do not allow students to keep the tests.* Then future instructors can have the advantage of their use. If you wish to have students keep tests, you can prepare them yourself or obtain them using the *Computerized Testing*, which the math department secretaries can prepare for you.

The Testing Center is available for makeup tests. Check with the math department for the time and place. You can give the student a card and send him or her to the testing center. Send a makeup test through the math department and it will be proctered for you. Please mark the amount of time to be allowed for the test.

Audiotape and Videotape Supplements are available in the science study center on the first floor of the Krannert building at 38th St and in Room 425 of the Cavanaugh Building.

Skill Maintenance Exercises are included near the end of most exercise sets. These should always be part of the assignment. You will note that each chapter begins with a list of four sections that are to be specifically reviewed on each test, unless they are not part of the syllabus. Please include such questions on each of your tests. These have been found to increase scores significantly on the final exam.

TIME AND ASSIGNMENT SCHEDULES

4 Semester-Hour Course

This syllabus assumes 3 classes per week for 15 weeks except holidays and exam week. Each line represents a 1-hour and 15-minute class period.

Day	Text Section	Exercises
1	Getting started	
2	1.1	Odds
	1.2	Odds, 58
3	1.3	Odds, 70
	1.4	Odds, 58
4	1.5	Odds, 64
	1.6	Odds, 69
5	1.7	Odds, 58, 60
	1.8	Odds, 42
6	1.9	Odds
7	Test 1	Chapter 1
8	2.1	Odds, 68, 70
9	2.2	Odds, 18, 20
	2.3	Odds, 42, 44, 46
10	2.4	Odds, 46, 48
	2.5	Odds, 54, 56, 58
11	2.6	Odds, 58, 60
	2.7	Odds
12	Test 2	Chapter 2 plus skill maintenance exercises
13	3.1	Odds, 46, 48, 50
	3.2	Odds
14	3.3	Odds, 60
	3.4	Odds, 56, 58
15	3.5	Odds, 56, 58, 59
	3.6	Odds, 48, 50
16	Test 3	Chapter 3 plus skill maintenance exercises
17	4.1	Odds, 56, 58
	4.2	Odds, 68, 70, 72, 74
18	4.3	Odds, 70
	4.4	Odds, 38, 40
19	5.1	Odds, 64, 66, 68
	5.2	Odds, 70, 72
20	5.3	Odds, 56, 58, 60, All 61–75
	5.4	Odds, 28, 30

4 Semester-Hour Course *(Continued)*

Day	Text Section	Exercises
21	5.5	Odds, 44, 46
22	Test 4	Chapters 4 and 5 plus skill maintenance exercises
23	6.1	Odds, 46, 48, 50
	6.2	Odds, 36, 38
24	6.3	Odds
	7.1	Odds, 56, 58, 60
25	7.2	Odds, 42, 44, 46
	7.3	Odds, 36, 38, 40
26	7.5 (use equations, not proportions)	Odds
	7.8	Odds, 20, 22, 24
27	Test 5	Chapters 6 and 7 plus skill maintenance exercises
28	8.1	Odds, 22, 24, 26
	9.1	Odds, 38, 40
29	9.2	Odds, 46, 48, 50
	9.3	Odds, 22, 24
30	9.4	Odds, 26, 28
	9.5	Odds, 26, 28
31	9.6	Odds, 48, 50
	9.7	Odds, 54, 56
32	10.1	Odds, 32, 34
	10.2	Odds
33	10.3	Odds 1–53, 52, 54
34	Test 6	Chapters 8, 9, and 10 plus skill maintenance exercises
35	11.1	Odds, 54, 56
	11.2	Odds, 72, 74, 76
36	11.3	Odds, 76
	11.4	Odds 1–63, 62, 64
37	11.5	Odds
38	Test 7	Chapter 11 plus skill maintenance exercises
39	12.1	Odds
40	12.2	Odds 1–39, 38, 40
41	12.3	Odds, 38, 40
42	12.4	Odds 1–59, 58, 60
43	12.5	Odds 1–37, 40, 42, 43, 44
44	Test 8	Chapter 12 plus skill maintenance exercises
45	Review and catchup	
46	Final Examination: Comprehensive	

3 Semester-Hour Course

This syllabus assumes 3 classes per week for 15 weeks except holidays and exam week. Each line represents a 1-hour class period.

Day	Text Section	Exercises
1	Getting started	
2	1.1	Odds
	1.2	Odds, 58
3	1.3	Odds, 70
4	1.4	Odds, 58
5	1.5	Odds, 64
6	1.6	Odds, 69
7	1.7	Odds, 58, 60
	1.8	Odds, 42
8	1.9	Odds
9	Test 1	Chapter 1
10	2.1	Odds, 68, 70
11	2.2	Odds, 18, 20
	2.3	Odds, 42, 44, 46
12	2.4	Odds, 46, 48
	2.5	Odds, 54, 56, 58
13	2.6	Odds, 58, 60
	2.7	Odds
14	Test 2	Chapter 2 plus skill maintenance exercises
15	3.1	Odds, 46, 48, 50
	3.2	Odds
16	3.3	Odds, 60
	3.4	Odds, 56, 58
17	3.5	Odds, 56, 58, 59
	3.6	Odds, 48, 50
18	Test 3	Chapter 3 plus skill maintenance exercises
19	4.1	Odds, 56, 58
	4.2	Odds, 68, 70, 72, 74
20	4.3	Odds, 70
	4.4	Odds, 38, 40
21	5.1	Odds, 64, 66, 68
	5.2	Odds, 70, 72
22	5.3	Odds, 56, 58, 60, All 61–75
	5.4	Odds, 28, 30
23	5.5	Odds, 44, 46
24	Test 4	Chapters 4 and 5 plus skill maintenance exercises
25	6.1	Odds, 46, 48, 50
	6.2	Odds, 36, 38
26	6.3	Odds
	7.1	Odds, 56, 58, 60
27	7.2	Odds, 42, 44, 46
	7.3	Odds, 36, 38, 40

3 Semester-Hour Course *(Continued)*

Day	Text Section	Exercises
28	7.5 (Use equations, not proportions)	Odds
	7.8	Odds, 20, 22, 24
29	Test 5	Chapters 6 and 7 plus skill maintenance exercises
30	8.1	Odds, 22, 24, 26
	9.1	Odds, 38, 40
31	9.2	Odds, 46, 48, 50
	9.3	Odds, 22, 24
32	9.4	Odds, 26, 28
	9.5	Odds, 26, 28
33	9.6	Odds, 48, 50
	9.7	Odds, 54, 56
34	Test 6	Chapters 8 and 9 plus skill maintenance exercises
35	11.1	Odds, 54, 56
	11.2	Odds, 72, 74, 76
36	11.3	Odds, 76
	11.4	Odds 1–63, 62, 64
37	11.5	Odds
38	Test 7	Chapter 11 plus skill maintenance exercises
39	12.1	Odds
40	12.2	Odds 1–39, 38, 40
41	12.3	Odds, 38, 40
42	12.4	Odds 1–59, 58, 60
43	12.5	Odds 1–37, 40, 42, 43, 44
44	Test 8	Chapter 12 plus skill maintenance exercises
45	Review and catchup	
46	Final Examination: Comprehensive	

Quarter Course

This syllabus assumes 5 classes per week for 11 weeks except holidays and exam week. Each line represents a 1-hour class period.

Day	Text Section	Exercises
1	Getting started	
2	1.1	Odds
	1.2	Odds, 58
3	1.3	Odds, 70
4	1.4	Odds, 58
5	1.5	Odds, 64
6	1.6	Odds, 69
7	1.7	Odds, 58, 60
	1.8	Odds, 42
8	1.9, Review	Odds
9	Test 1	Chapter 1
10	2.1	Odds, 68, 70
11	2.2	Odds, 18, 20
	2.3	Odds, 42, 44, 46
12	2.4	Odds, 46, 48
	2.5	Odds, 54, 56, 58
13	2.6	Odds, 58, 60
	2.7	Odds
14	Review	
15	Test 2	Chapter 2 plus skill maintenance exercises
16	3.1	Odds, 46, 48, 50
	3.2	Odds
17	3.3	Odds, 60
	3.4	Odds, 56, 58
18	3.5	Odds, 56, 58, 59
	3.6	Odds, 48, 50
19	Review	
20	Test 3	Chapter 3 plus skill maintenance exercises
21	4.1	Odds, 56, 58
	4.2	Odds, 68, 70, 72, 74
22	4.3	Odds, 70
	4.4	Odds, 38, 40
23	5.1	Odds, 64, 66, 68
	5.2	Odds, 70, 72
24	5.3	Odds, 56, 58, 60, All 61–75
	5.4	Odds, 28, 30
25	5.5, Review	Odds, 44, 46
26	Test 4	Chapters 4 and 5 plus skill maintenance exercises
27	6.1	Odds, 46, 48, 50
	6.2	Odds, 36, 38
28	6.3	Odds
	7.1	Odds, 56, 58, 60
29	7.2	Odds, 42, 44, 46
	7.3	Odds, 36, 38, 40

Quarter Course *(Continued)*

Day	Text Section	Exercises
30	7.5 (Use equations, not proportions)	Odds
31	7.8	Odds, 20, 22, 24
32	Review	
33	Test 5	Chapters 6 and 7 plus skill maintenance exercises
34	8.1	Odds, 22, 24, 26
	9.1	Odds, 38, 40
35	9.2	Odds, 46, 48, 50
	9.3	Odds, 22, 24
36	9.4	Odds, 26, 28
	9.5	Odds, 26, 28
37	9.6	Odds, 48, 50
	9.7	Odds, 54, 56
38	Review	
39	Test 6	Chapters 8 and 9 plus skill maintenance exercises
40	11.1	Odds, 54, 56
	11.2	Odds, 72, 74, 76
41	11.3	Odds, 76
	11.4	Odds 1–63, 62, 64
42	11.5, Review	Odds
43	Test 7	Chapter 11 plus skill maintenance exercises
44	12.1	Odds
45	12.2	Odds 1–39, 38, 40
46	12.3	Odds, 38, 40
47	12.4	Odds 1–59, 58, 60
48	12.5	Odds 1–37, 40, 42, 43, 44
49	Review	
50	Test 8	Chapter 12 plus skill maintenance exercises
51	Review and catchup	
52	Final Examination: Comprehensive	

PEDAGOGICAL FLOWCHART

Following is a flowchart designed to show you how your students might use this text to enhance their learning process.

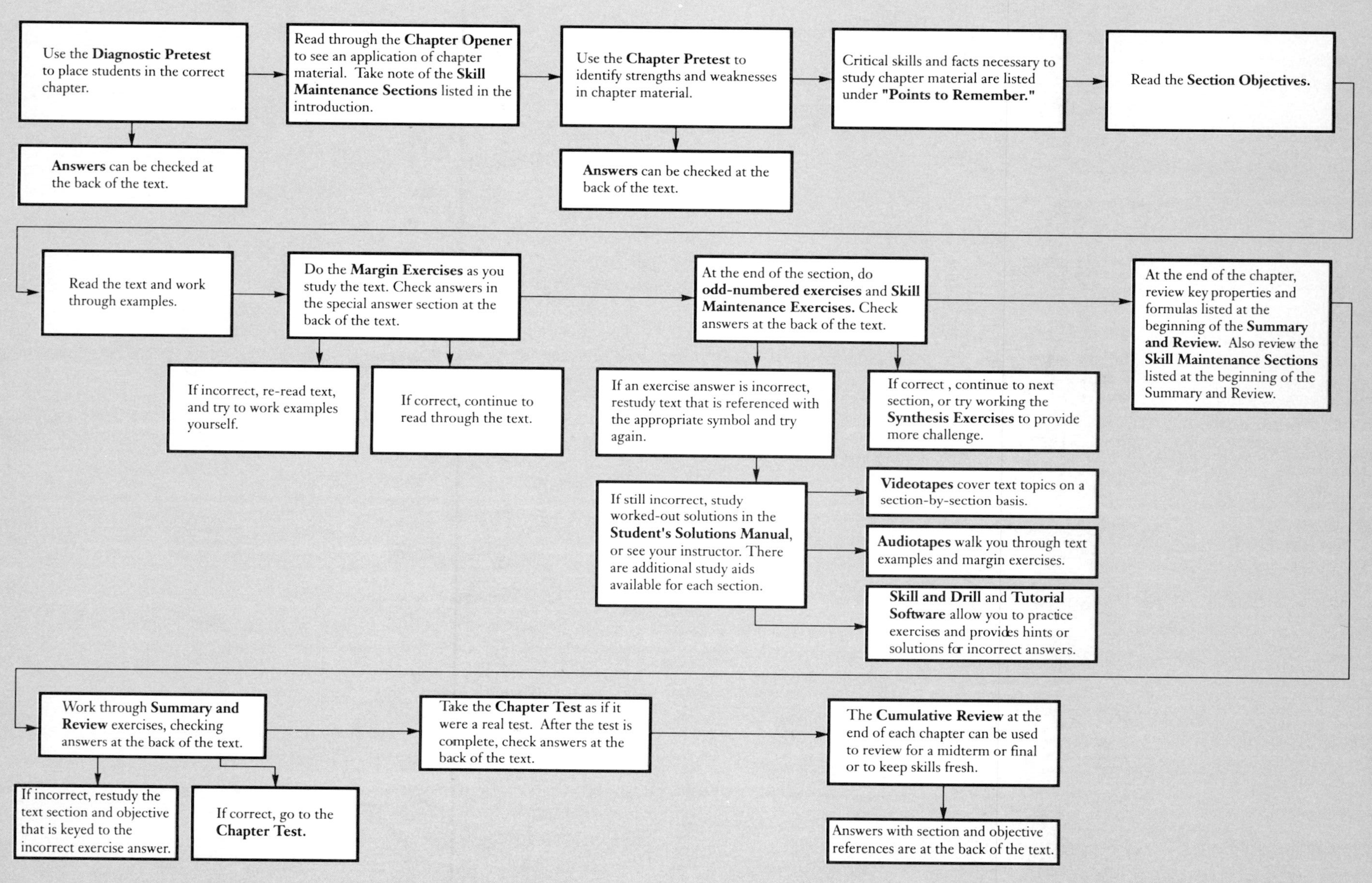

INTRODUCTION This chapter considers addition, subtraction, multiplication, and division of whole numbers. Then we study the solving of simple equations and apply our skills to the solving of problems. ❖

Operations on the Whole Numbers

AN APPLICATION

The John Hancock Building in Chicago is 1107 ft tall. It has two 342-ft antennas on top. How far are the tops of the antennas from the ground?

THE MATHEMATICS

Let $h =$ the height in question. Since we are combining lengths, addition can be used. We translate the problem to this equation:

$$\underbrace{h = 1107 + 342.}$$

Here is how addition can occur in problem solving.

❖ POINTS TO REMEMBER: CHAPTER 1

Area of a Rectangle: $A = l \cdot w$
Area of a Square: $A = s \cdot s$, or s^2

PRETEST: CHAPTER 1

1. Write a word name: 3,078,059.

2. Write expanded notation: 6987.

3. Write standard notation: Two billion, forty-seven million, three hundred ninety-eight thousand, five hundred eighty-nine.

4. What does the digit 6 mean in 2,967,342?

5. Round 956,449 to the nearest thousand.

6. Estimate the product $594 \cdot 126$ by first rounding the numbers to the nearest hundred.

7. Add.

$$\begin{array}{r} 7\ 3\ 1\ 2 \\ +\ 2\ 9\ 0\ 4 \\ \hline \end{array}$$

8. Subtract.

$$\begin{array}{r} 7\ 0\ 1\ 2 \\ -\ 2\ 9\ 0\ 4 \\ \hline \end{array}$$

9. Multiply: $359 \cdot 64$.

10. Divide: $23{,}149 \div 46$.

Use either $<$ or $>$ for $\square$ to write a true sentence.

11. $346 \ \square \ 364$

12. $54 \ \square \ 45$

Solve.

13. $326 \cdot 17 = m$

14. $y = 924 \div 42$

15. $19 + x = 53$

16. $34 \cdot n = 850$

Solve.

17. Betsy weighs 121 lb and Jennifer weighs 109 lb. How much more does Betsy weigh?

18. How many 12-jar cases can be filled with 1512 jars of spaghetti sauce?

19. The population of Illinois is 11,418,500. The population of Ohio is 10,797,600. What is the total population of Illinois and Ohio?

20. A lot measures 48 ft by 54 ft. A pool that is 15 ft by 20 ft is put on the lot. How much area is left over?

Evaluate.

21. 5^2

22. 4^3

Simplify.

23. $8^2 \div 8 \cdot 2 - (2 + 2 \cdot 7)$

24. $108 \div 9 - \{4 \cdot [18 - (5 \cdot 3)]\}$

1.1 Standard Notation

We study mathematics in order to be able to solve problems. In this chapter, we learn how to use operations on the whole numbers to solve various kinds of problems. We begin by studying how numbers are named.

a From Standard Notation to Expanded Notation

To answer questions such as "How many?", "How much?" and "How tall?" we use whole numbers.* The set of whole numbers is

0, 1, 2, 3, 4, 5, 6, 7, 8, 9, 10, 11, 12, ...

The set goes on indefinitely. There is no largest whole number, and the smallest whole number is 0. Each number can be named using various notations. For example, the height of the John Hancock Building (excluding antennas) is 1107 ft. **Standard notation** for this is 1107. We find **expanded notation** for 1107 as follows:

1107

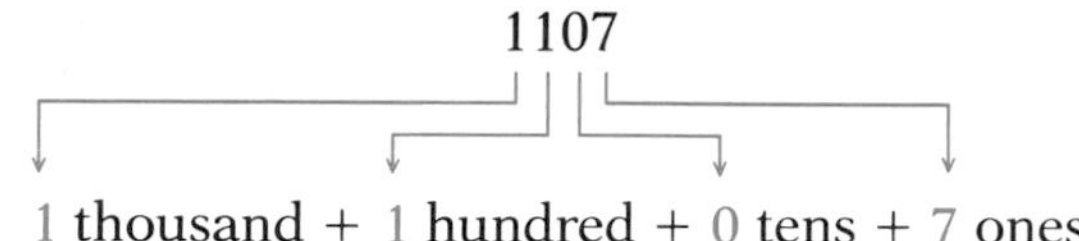

1 thousand + 1 hundred + 0 tens + 7 ones

▶ **EXAMPLE 1** Write expanded notation for 3742.

3742 = 3 thousands + 7 hundreds + 4 tens + 2 ones ◀

▶ **EXAMPLE 2** Write expanded notation for 54,567.

54,567 = 5 ten thousands + 4 thousands + 5 hundreds + 6 tens + 7 ones ◀

DO EXERCISES 1 AND 2 (IN THE MARGIN AT THE RIGHT).

▶ **EXAMPLE 3** Write expanded notation for 7091.

7091 = 7 thousands + 0 hundreds + 9 tens + 1 one, or 7 thousands + 9 tens + 1 one ◀

▶ **EXAMPLE 4** Write expanded notation for 3400.

3400 = 3 thousands + 4 hundreds + 0 tens + 0 ones, or 3 thousands + 4 hundreds ◀

DO EXERCISES 3–5.

*The set 1, 2, 3, 4, 5, . . . , without 0, is called the set of **natural numbers.**

OBJECTIVES

After finishing Section 1.1, you should be able to:

- a Convert from standard notation to expanded notation.
- b Convert from expanded notation to standard notation.
- c Write a word name for a number given standard notation.
- d Write standard notation for a number given a word name.
- e Given a standard notation like 278,342, tell what 8 means, 3 means, and so on; identify the hundreds digit, the thousands digit, and so on.

FOR EXTRA HELP

Tape 1A

Tape 1A

MAC: 1
IBM: 1

Write expanded notation.

1. 3728

3 thousands + 7 hundreds + 2 tens + 8 ones

2. 36,223

3 ten thousands + 6 thousands + 2 hundreds + 2 tens + 3 ones

Write expanded notation.

3. 3021

3 thousands + 2 tens + 1 one

4. 2009 2 thousands + 9 ones

5. 5700 5 thousands + 7 hundreds

ANSWERS ON PAGE A-1

Write standard notation.

6. 5 thousands + 6 hundreds + 8 tens + 9 ones 5689

7. 8 ten thousands + 7 thousands + 1 hundred + 2 tens + 8 ones

87,128

8. 9 thousands + 0 hundreds + 0 tens + 3 ones 9003

Write a word name.

9. 57 Fifty-seven

10. 29 Twenty-nine

11. 88 Eighty-eight

ANSWERS ON PAGE A-1

b From Expanded Notation to Standard Notation

► **EXAMPLE 5** Write standard notation for 2 thousands + 5 hundreds + 7 tens + 5 ones.

Standard notation is 2575. ◄

► **EXAMPLE 6** Write standard notation for 9 ten thousands + 6 thousands + 7 hundreds + 1 ten + 8 ones.

Standard notation is 96,718. ◄

► **EXAMPLE 7** Write standard notation for 2 thousands + 3 tens.

Standard notation is 2030. ◄

DO EXERCISES 6–8.

c Word Names

"Three," "two hundred one," and "forty-two" are **word names** for numbers. When we write word names for two-digit numbers like 42, 76, and 91, we use hyphens.

► **EXAMPLES** Write word names.

8. 42 Forty-two

9. 76 Seventy-six

10. 91 Ninety-one ◄

DO EXERCISES 9–11.

For large numbers, digits are separated into groups of three, called **periods.** Each period has a name like *ones, thousands, millions, billions,* and so on. When we write or read a large number, we start at the left with the largest period. The number named in the period is followed by the name of the period, then a comma is written and the next period is named. Recently, the U.S. national debt was $2,830,127,000,000. We can use a **place-value** chart to illustrate how to use periods to read the number 2,830,127,000,000.

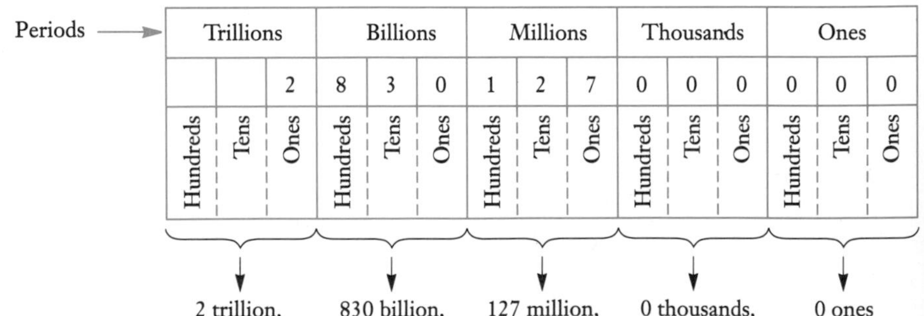

▶ **EXAMPLE 11** Write a word name for 46,625,314,732.

Forty-six billion
six-hundred twenty-five million,
three hundred fourteen thousand,
seven-hundred thirty-two ◀

The word "and" *does not* appear in word names for whole numbers. Although we commonly hear such expressions as "two hundred *and* one," the use of "and" is not, strictly speaking, correct in word names for whole numbers. For decimal notation like 317.4, it is appropriate to use "and" for the decimal point and read this as "three hundred seventeen *and* four tenths."

DO EXERCISES 12–15.

d Standard Notation

▶ **EXAMPLE 12** Write standard notation.

Five hundred six million,
three hundred forty-five thousand,
two hundred twelve

Standard notation is 506,345,212. ◀

DO EXERCISE 16.

e Digits

A **digit** is a number 0, 1, 2, 3, 4, 5, 6, 7, 8, or 9 that names a place-value location.

▶ **EXAMPLES** What does the digit 8 mean in each case?

13. 278,342 8 thousands
14. 872,342 8 hundred thousands
15. 28,343,399,223 8 billions ◀

DO EXERCISES 17–20.

▶ **EXAMPLE 16** In 278,346, what digit tells the number of:

a) Hundred thousands? 2
b) Thousands? 8 ◀

DO EXERCISES 21–24.

Write a word name.

12. 204 Two hundred four

13. 19,204

Nineteen thousand, two hundred four

14. 1,719,204

million, seven hundred nineteen thousand, two hundred four

15. 22,301,719,204

Twenty-two billion, three hundred one million, seven hundred nineteen thousand, two hundred four

16. Write standard notation.

Two hundred thirteen million, one hundred five thousand, three hundred twenty-nine

213,105,329

What does the digit 2 mean in each case?

17. 526,555 2 ten thousands

18. 265,789 2 hundred thousands

19. 42,789,654 2 millions

20. 24,789,654 2 ten millions

In 7,890,432, what digit tells the number of:

21. Hundreds? 4

22. Millions? 7

23. Ten thousands? 9

24. Thousands? 0

ANSWERS ON PAGE A-1

From time to time you will find a *"Sidelights"* like the one below. These are optional, but you may find them helpful and of interest. They will include such topics as study tips, career opportunities involving mathematics, applications, computer-calculator exercises, or other mathematical topics.

❖ SIDELIGHTS

Study Tips

Many students begin the study of a text by opening to the first section assigned by an instructor. There are many ways in which you can enhance your use of this book, and they have been outlined carefully in a page in the preface titled *To the student.* If you have not read that page, do so now before you start the exercise set on the next page.

There are some points on that page that bear repeating here.

- *Be sure to note the special symbols,* a, b, c, *and so on, that correspond to the objectives you are to learn.* They appear many places throughout the text. The first time you will see them is in the headings for the section. The second time you will see them is in the exercise set. You will also see them in the answers to the Review Exercises, the Chapter Tests, and the Cumulative Reviews. These allow you to reference back when you need to review a topic.
- *Be sure to note also the symbols in the margin under the list of objectives at the beginning of the section.* These refer to the many distinctive study aids that accompany the book.
- *Be sure to stop and do the margin exercises as you study a section.* When our students come to us troubled about how they are doing in the course, the first question we ask them is "Are you doing the margin exercises when directed to do so?" This is one of the most effective ways to enhance your ability to learn mathematics from this text. Don't deprive yourself of its benefits!
- *When you study the book, don't mark points you think are important, but mark the points you do not understand!* The book is written with all kinds of processes that highlight important points. Use your efforts to mark where you are having trouble. Then when you go to class or a math lab or a tutoring session, you are prepared to ask questions that close in on your difficultics.
- *Try to keep one section ahead of your syllabus.* We have tried to write a book that is readable for students. If you study ahead of your lectures, you can concentrate on just the lectures, rather than trying to write everything down. You can then take notes only of special points or of questions related to what is happening in class.

NAME SECTION DATE

EXERCISE SET 1.1

Always review the objectives before doing an exercise set. See page 3. Note how the objectives are keyed to the exercises.

a Write expanded notation.

1. 5742
5 thousands + 7 hundreds + 4 tens + 2 ones

2. 3897
3 thousands + 8 hundreds + 9 tens + 7 ones

3. 27,342
2 ten thousands + 7 thousands + 3 hundreds + 4 tens + 2 ones

4. 93,986
9 ten thousands + 3 thousands + 9 hundreds + 8 tens + 6 ones

5. 9010

6. 9990

7. 2300

8. 7020

b Write standard notation.

9. 2 thousands + 4 hundreds + 7 tens + 5 ones

10. 7 thousands + 9 hundreds + 8 tens + 3 ones

11. 6 ten thousands + 8 thousands + 9 hundreds + 3 tens + 9 ones

12. 1 ten thousand + 8 thousands + 4 hundreds + 6 tens + 1 one

13. 7 thousands + 3 hundreds + 0 tens + 4 ones

14. 8 thousands + 0 hundreds + 2 tens + 0 ones

15. 1 thousand + 0 hundreds + 0 tens + 9 ones

16. 2 thousands + 4 hundreds + 5 tens + 0 ones

c Write a word name.

17. 77

18. 48

19. 88,000

20. 45,987
Forty-five thousand, nine hundred eighty-seven

21. 123,765
One hundred twenty-three thousand, seven hundred sixty-five

22. 111,013

23. 7,754,211
Seven million, seven hundred fifty-four thousand, two hundred eleven

24. 43,550,651
Forty-three million, five hundred fifty thousand, six hundred fifty-one

Write a word name for the number in the sentence.

25. The population of the United States is 244,839,772.
Two hundred forty-four million, eight hundred thirty-nine thousand, seven hundred seventy-two

26. The diameter of the sun is 865,400 miles.
Eight hundred sixty-five thousand, four hundred

ANSWERS

1. ______
2. ______
3. ______
4. ______
5. 9 thousands + 1 ten
6. 9 thousands + 9 hundreds + 9 tens
7. 2 thousands + 3 hundreds
8. 7 thousands + 2 tens
9. 2475
10. 7983
11. 68,939
12. 18,461
13. 7304
14. 8020
15. 1009
16. 2450
17. Seventy-seven
18. Forty-eight
19. Eighty-eight thousand
20. ______
21. ______
22. One hundred eleven thousand, thirteen
23. ______
24. ______
25. ______
26. ______

ANSWERS

27.

28.

29. 2,233,812

30. 354,702

31. 8,000,000,000

32. 700,000,000

33. 217,503

34. 230,043,951,617

35. 2,173,638

36. 262,436,000

37. 206,658,000

38. 749,578,653

39. 5 thousands

40. 5 ten thousands

41. 5 hundreds

42. 5 hundred thousands

43. 3

44. 9

45. 0

46. 2

47.

27. There are 1,954,116 students in junior colleges.
One million, nine hundred fifty-four thousand, one hundred sixteen

28. The Harvard University library contains 10,707,266 books, more than any university in the country.
Ten million, seven hundred seven thousand, two hundred sixty-six

d Write standard notation.

29. Two million, two hundred thirty-three thousand, eight hundred twelve

30. Three hundred fifty-four thousand, seven hundred two

31. Eight billion

32. Seven hundred million

33. Two hundred seventeen thousand, five hundred three

34. Two hundred thirty billion, forty-three million, nine hundred fifty-one thousand, six hundred seventeen

Write standard notation for the number in the sentence.

35. In a recent year, two million, one hundred seventy-three thousand, six hundred thirty-eight people visited the Grand Canyon.

36. The population of Russia is two hundred sixty-two million, four hundred thirty-six thousand.

37. In one year, Americans use two hundred six million, six hundred fifty-eight thousand pounds of toothpaste.

38. The people of the United States burn seven hundred forty-nine million, five hundred seventy-eight thousand, six hundred fifty-three gallons of fuel annually driving to see motion pictures.

e What does the digit 5 mean in each case?

39. 235,888 **40.** 253,888 **41.** 488,526 **42.** 500,346

In 89,302 what digit tells the number of:

43. Hundreds?

44. Thousands?

45. Tens

46. Ones?

SYNTHESIS

Synthesis exercises are extra and optional, and usually more challenging, requiring you to put together objectives of this section or preceding sections of the text. Any exercises marked with a ▦ are to be worked with a calculator.

47. ▦ What is the largest number you can name on your calculator? How many digits does that number have? All 9's as digits. Answers may vary. For an 8-digit readout, it would be 99,999,999.

1.2 Addition

a Addition and the Real World

Addition of whole numbers corresponds to combining or putting things together. Let us look at various situations in which addition applies.

Combining Sets of Objects

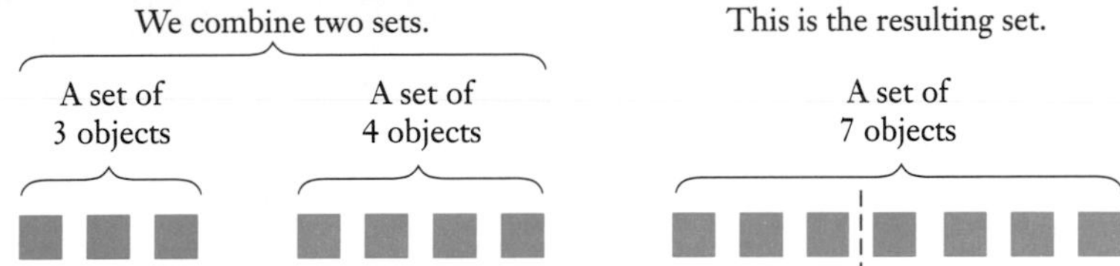

The addition that corresponds is

$$3 + 4 = 7.$$

We can find the number of objects in a set by counting. We count and find that the two sets have 3 members and 4 members, respectively. We count after combining and find that there are 7 objects. We say that the **sum** of 3 and 4 is 7. The numbers added are called **addends.**

$$3 + 4 = 7$$

Addend Addend Sum

▶ **EXAMPLE 1** Write an addition that corresponds to this situation.

A student has \$3 and earns \$10 more. How much money does the student have?

The addition that corresponds is

$$\$3 + \$10 = \$13.$$ ◀

DO EXERCISES 1 AND 2.

Addition also corresponds to combining distances or lengths.

▶ **EXAMPLE 2** Write an addition that corresponds to this situation.

A car is driven 3 mi (miles) from Dustville to Rainville. It is then driven 5 mi from Rainville to Mudville. How far is it from Dustville to Mudville along the same route?

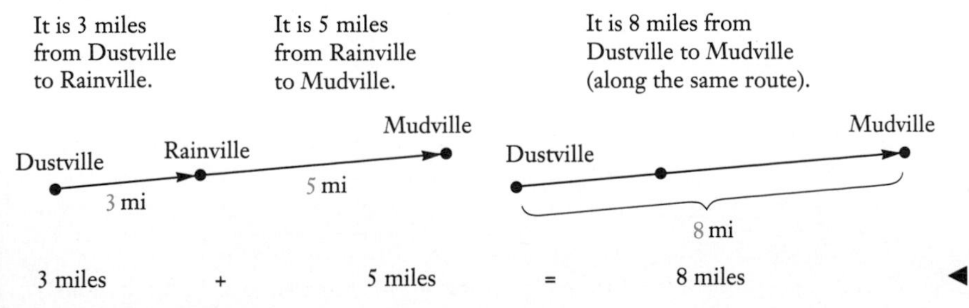

◀

DO EXERCISES 3 AND 4.

OBJECTIVES

After finishing Section 1.2, you should be able to:

a Determine what addition corresponds to a situation.

b Add whole numbers

FOR EXTRA HELP

Tape 1B Tape 1A MAC: 1 IBM: 1

Write an addition that corresponds to the situation.

1. John has 4 marbles. Then he wins 6 more. How many does he have in all? $4 + 6 = 10$

2. Sue earns \$15 on Thursday and \$13 on Friday. How much does she earn altogether on the two days? \$15 + \$13 = \$28

Write an addition that corresponds to the situation.

3. A car is driven 40 mi from Lafayette to Kokomo. Then it is driven 50 mi from Kokomo to Indianapolis. How far is it from Lafayette to Indianapolis along the same route?
40 mi + 50 mi = 90 mi

4. A rope 5 ft long is tied to a rope 7 ft long. How long is the resulting rope (ignoring the amount of rope it takes to tie the two ropes together)?
5 ft + 7 ft = 12 ft

ANSWERS ON PAGE A-1

Write an addition that corresponds to the situation.

5. One piece of paper has an area of 50 in^2 (square inches). Another piece of paper has an area of 60 in^2. What is the total area of the two pieces of paper?

$50\ in^2 + 60\ in^2 = 110\ in^2$

6. One plot of land contains 200 mi^2 (square miles). Another plot of land contains 400 mi^2. What is the total area of the two plots of land?

$200\ mi^2 + 400\ mi^2 = 600\ mi^2$

7. A motorist purchases 10 gal (gallons) of gasoline one day and 18 gal the next. How many gallons were bought in all?

10 gal + 18 gal = 28 gal

8. Ship A carries 3000 tons of sand and ship B carries 7000 tons. How many tons do they carry in all?

3000 tons + 7000 tons = 10,000 tons

9. Add.

$$\begin{array}{r} 6\,2\,0\,3 \\ +\,3\,5\,4\,2 \\ \hline \end{array}$$

9745

ANSWERS ON PAGE A-1

Addition also corresponds to combining areas.

▶ **EXAMPLE 3** Write an addition that corresponds to the following situation.

You have 5 square yards of nylon. You buy 7 more square yards. How much do you have in all?

You have 5 square yards of nylon.		You buy 7 more square yards.		You then have 12 square yards of nylon.
5 square yards	+	7 square yards	=	12 square yards ◀

Addition corresponds to combining volumes as well.

▶ **EXAMPLE 4** Write an addition that corresponds to the following situation.

Two trucks haul dirt to a construction site. One hauls 5 cubic yards and the other hauls 7 cubic yards. Altogether, how many yards of dirt have they hauled to the site?

Truck A hauls 5 cubic yards of dirt to a construction site.		Truck B hauls 7 cubic yards of dirt.		Altogether, they haul 12 cubic yards of dirt.

5 cubic yards	+	7 cubic yards	=	12 cubic yards ◀

DO EXERCISES 5–8.

b Addition of Whole Numbers

To add numbers, we can add the ones first, then the tens, then the hundreds, and so on.

▶ **EXAMPLE 5** Add: 7312 + 2504.

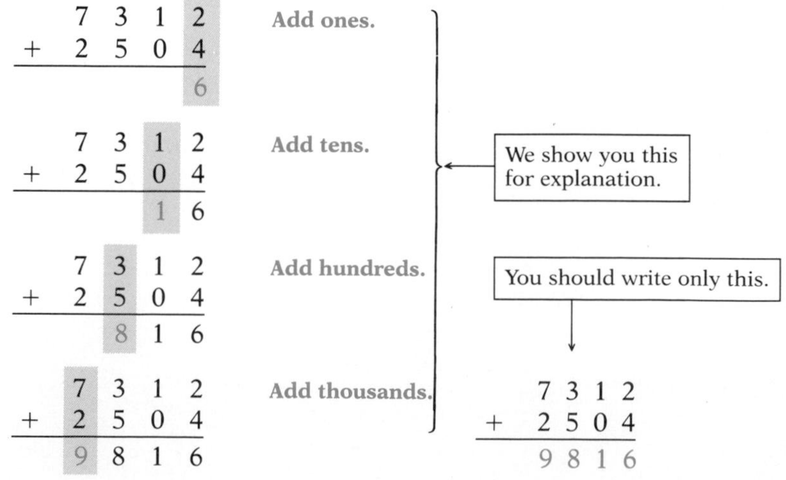

◀

DO EXERCISE 9.

▶ EXAMPLE 6 Add: 6878 + 4995.

$$\begin{array}{r} \;\;\;\;\;\;\;^{1}\;\; \\ 6\;8\;7\;8 \\ +\;4\;9\;9\;5 \\ \hline 3 \end{array}$$

Add ones. We get 13 ones, or 1 ten + 3 ones. Write 3 in the ones column and 1 above the tens. This is called *carrying*.

$$\begin{array}{r} ^{1}\;^{1}\;\;\; \\ 6\;8\;7\;8 \\ +\;4\;9\;9\;5 \\ \hline 7\;3 \end{array}$$

Add tens. We get 17 tens, or 1 hundred + 7 tens. Write 7 in the tens column and 1 above the hundreds.

$$\begin{array}{r} ^{1}\;^{1}\;^{1}\;\;\; \\ 6\;8\;7\;8 \\ +\;4\;9\;9\;5 \\ \hline 8\;7\;3 \end{array}$$

Add hundreds. We get 18 hundreds, or 1 thousand + 8 hundreds. Write 8 in the hundreds column and 1 above the thousands.

$$\begin{array}{r} ^{1}\;^{1}\;^{1}\;\;\; \\ 6\;8\;7\;8 \\ +\;4\;9\;9\;5 \\ \hline 1\;1\;8\;7\;3 \end{array}$$

Add thousands. We get 11 thousands. ◀

DO EXERCISE 10.

How do we do an addition of three numbers, like 2 + 3 + 6? We do so by adding 3 and 6, and then 2. We can show this with parentheses:

$$2 + (3 + 6) = 2 + 9 = 11.$$ Parentheses tell what to do first.

We could also add 2 and 3, and then 6:

$$(2 + 3) + 6 = 5 + 6 = 11.$$

Either way we get 11. It does not matter how we group the numbers. This illustrates the **associative law of addition,** $a + (b + c) = (a + b) + c$. We can also add whole numbers in any order. That is, 2 + 3 = 3 + 2. This illustrates the **commutative law of addition,** $a + b = b + a$. Together the commutative and associative laws tell us that to add more than two numbers, we can use any order and grouping we wish.

▶ EXAMPLE 7 Add from the top mentally.

$$\begin{array}{r} 8 \\ 9 \\ 7 \\ +\;6 \\ \hline \end{array}$$

We first add 8 and 9, getting 17; then 17 and 7, getting 24; then 24 and 6, getting 30.

$$\begin{array}{r} 8 \\ 9 \\ 7 \\ +\;6 \\ \hline 30 \end{array} \quad 8 + 9 \longrightarrow 17;\; 17 + 7 \longrightarrow 24;\; 24 + 6 \longrightarrow 30$$

30 ← You write only this. ◀

DO EXERCISE 11.

10. Add.

$$\begin{array}{r} 7\;9\;6\;8 \\ +\;5\;4\;9\;7 \\ \hline \end{array}$$

13,465

11. Add mentally from the top.

$$\begin{array}{r} 9 \\ 9 \\ 4 \\ +\;5 \\ \hline \end{array}$$

27

ANSWERS ON PAGE A-1

12. Add mentally from the bottom.

$$\begin{array}{r} 9 \\ 9 \\ 4 \\ +\ 5 \\ \hline 27 \end{array}$$

Add. Look for pairs of numbers whose sums are 10, 20, 30, and so on.

13.
$$\begin{array}{r} 15 \\ 7 \\ 5 \\ 3 \\ +\ 8 \\ \hline 38 \end{array}$$

14. 27 + 8 + 13 + 2 + 11 61

15. Add.

$$\begin{array}{r} 1932 \\ 6723 \\ 9878 \\ +\ 8941 \\ \hline 27{,}474 \end{array}$$

To the instructor and the student: This section presented a review of addition of whole numbers. Students who are successful should go on to Section 1.3. Those who have trouble should study developmental units A.1 and A.2 and then repeat Section 1.2. Students who still have trouble might study developmental unit A.3.

ANSWERS ON PAGE A-1

▶ **EXAMPLE 8** Add mentally from the bottom.

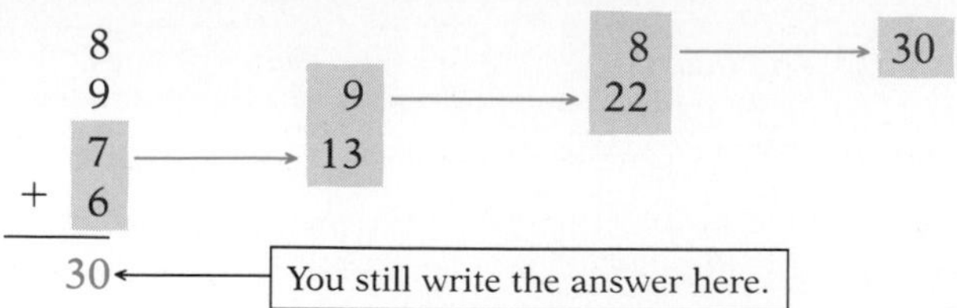

◀

DO EXERCISE 12.

Sometimes it is easier to look for pairs of numbers whose sums are 10 or 20 or 30, and so on.

▶ **EXAMPLES** Add.

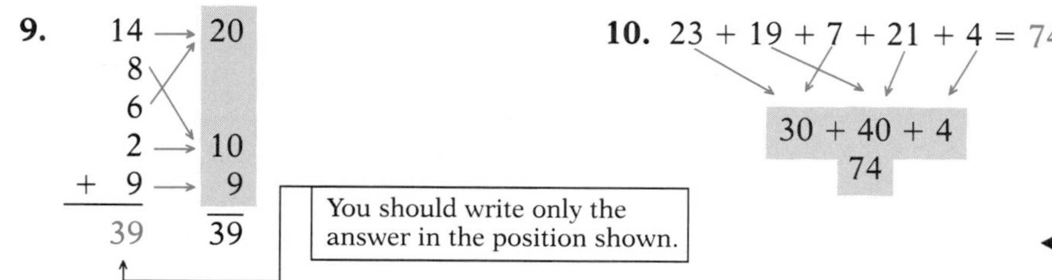

◀

DO EXERCISES 13 AND 14.

▶ **EXAMPLE 11** Add: 2391 + 3276 + 8789 + 1498.

$$\begin{array}{r} {}^{2}\ \ \\ 2391 \\ 3276 \\ 8789 \\ +\ 1498 \\ \hline 4 \end{array}$$

Add ones: We get 24, so we have 2 tens + 4 ones. Write 4 in the ones column and 2 above the tens.

$$\begin{array}{r} {}^{3\,2}\ \ \\ 2391 \\ 3276 \\ 8789 \\ +\ 1498 \\ \hline 54 \end{array}$$

Add tens: We get 35 tens, so we have 30 tens + 5 tens. This is also 3 hundreds + 5 tens. Write 5 in the tens column and 3 above the hundreds.

$$\begin{array}{r} {}^{1\,3\,2}\ \ \\ 2391 \\ 3276 \\ 8789 \\ +\ 1498 \\ \hline 954 \end{array}$$

Add hundreds: We get 19 hundreds, or 1 thousand + 9 hundreds. Write 9 in the hundreds column and 1 above the thousands.

$$\begin{array}{r} {}^{1\,3\,2}\ \ \\ 2391 \\ 3276 \\ 8789 \\ +\ 1498 \\ \hline 15954 \end{array}$$

Add thousands: We get 15 thousands.

◀

DO EXERCISE 15.

NAME SECTION DATE

EXERCISE SET 1.2

a Write an addition that corresponds to the situation.

1. A seamstress buys 3 yd of fabric on one day and 6 yd the next day. How many yards of fabric did the seamstress buy in all?

2. A jogger runs 4 mi one day and 5 mi the next. What total distance was run in the two days?

3. A student earns $23 one day and $31 the next. How much did the student earn in all?

4. You own a 40-acre farm. Then you buy an adjoining 80-acre farm. You now own a farm of how many acres?

40 acres + 80 acres = 120 acres

b Add.

5. $364 + 23$
6. $1721 + 348$
7. $1716 + 3282$
8. $7503 + 2683$
9. $86 + 78$
10. $99 + 1$
11. $999 + 1$
12. $999 + 11$
13. 999 + 111
14. 839 + 386
15. 909 + 101
16. 707 + 909
17. 811 + 390
18. 271 + 333
19. 356 + 491
20. 280 + 347
21. $8719 + 1420$
22. $3654 + 2700$
23. $4825 + 1783$
24. $6775 + 1432$
25. $9999 + 6785$
26. $45{,}879 + 21{,}786$
27. $23{,}443 + 10{,}989$
28. $67{,}654 + 98{,}786$
29. $77{,}543 + 23{,}767$
30. $44{,}654 + 4{,}765$
31. $99{,}999 + 112$
32. $127{,}556 + 68{,}766$

Add from the top. Then check by adding from the bottom.

33. 7, 9, 4, + 8
34. 5, 6, 5, + 4
35. 4, 3, 9, 1, + 8
36. 8, 6, 2, 3, + 7
37. 9, 4, 7, 8, + 7

ANSWERS

1. 3 yd + 6 yd = 9 yd
2. 4 mi + 5 mi = 9 mi
3. $23 + $31 = $54
4.
5. 387
6. 2069
7. 4998
8. 10,186
9. 164
10. 100
11. 1000
12. 1010
13. 1110
14. 1225
15. 1010
16. 1616
17. 1201
18. 604
19. 847
20. 627
21. 10,139
22. 6354
23. 6608
24. 8207
25. 16,784
26. 67,665
27. 34,432
28. 166,440
29. 101,310
30. 49,419
31. 100,111
32. 196,322
33. 28
34. 20
35. 25
36. 26
37. 35

ANSWERS

38. 67

39. 87

40. 57

41. 230

42. 187

43. 130

44. 91

45. 149

46. 108

47. 169

48. 932

49. 842

50. 1233

51. 11,679

52. 14,094

53. 22,654

54. 5636

55. 12,765,097

56. 97,326,211

57. 7992

58.

59.

Add. Look for pairs of numbers whose sums are 10, 20, 30 , and so on.

38. $7 + 18 + 3 + 37 + 2$

39. $23 + 16 + 11 + 18 + 19$

40. $7 + 24 + 15 + 6 + 5$

41. $45 + 25 + 36 + 44 + 80$

42. $38 + 27 + 32 + 14 + 76$

43. $23 + 62 + 45$

44. $43 + 11 + 37$

45. $51 + 36 + 62$

46. $31 + 53 + 24$

47. $26 + 82 + 61$

48. $324 + 126 + 482$

49. $207 + 295 + 340$

50. $248 + 314 + 671$

51. $2037 + 4923 + 3471 + 1248$

52. $4567 + 1023 + 4821 + 3683$

53. $3420 + 8719 + 4312 + 6203$

54. $2003 + 149 + 58 + 3426$

55. $5,678,987 + 1,409,312 + 898,888 + 4,777,910$

56. $78,899,311 + 6,784,170 + 11,541,913 + 100,817$

SKILL MAINTENANCE

The exercises that follow are *skill maintenance exercises,* which review any skill previously studied in the text. You can expect such exercises in almost every exercise set.

57. Write standard notation:
$7000 + 900 + 90 + 2.$

58. Write a word name for the number in the following sentence:
Each year Americans drink 8,395,000,000 gallons of soft drinks.

Eight billion, three hundred ninety-five million

SYNTHESIS

59. Try to discover a faster way of adding all the numbers from 1 to 100 inclusive.

$1 + 99 = 100$, $2 + 98 = 100$, and so on, . . . , $49 + 51 = 100$.
Then $49 \cdot 100 = 4900$ and $4900 + 50 + 100 = 5050$.

1.3 Subtraction

a Subtraction and the Real World: Take Away

Subtraction of whole numbers corresponds to two kinds of situations. The first one is called "take away."

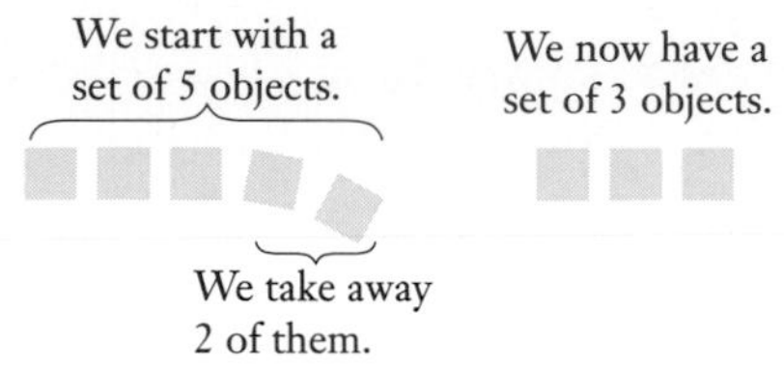

The subtraction that corresponds is as follows.

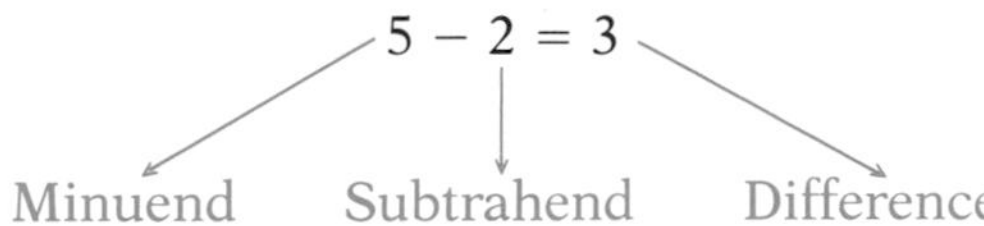

▶ **EXAMPLES** Write a subtraction that corresponds to each situation.

1. A bowler starts with 10 pins and knocks down 8 of them. How many pins are left?

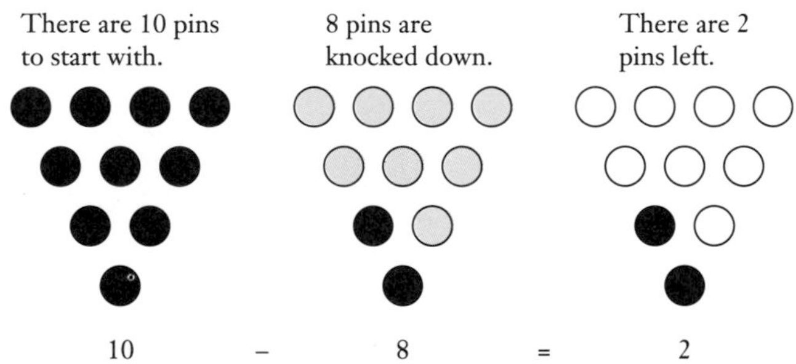

10 − 8 = 2

2. A shopper starts with $100 and spends $65 for groceries. How much money is left?

$100 − $65 = $35 ◀

DO EXERCISES 1 AND 2.

OBJECTIVES

After finishing Section 1.3, you should be able to:

- **a** Determine what subtraction corresponds to a situation involving "take away."
- **b** Given a subtraction sentence, write a related addition sentence; and given an addition sentence, write two related subtraction sentences.
- **c** Determine what subtraction corresponds to a situation involving "how much more."
- **d** Subtract whole numbers.

FOR EXTRA HELP

Tape 1C Tape 1B MAC: 1 IBM: 1

Write a subtraction that corresponds to the situation. You need not carry out the subtraction.

1. A cook pours 5 oz (ounces) of cooking oil out of a pitcher containing 16 oz. How many ounces are left?

16 oz − 5 oz = 11 oz

2. A farm contains 400 acres. The owner sells 100 acres of it. How many acres are left?

400 acres − 100 acres = 300 acres

ANSWERS ON PAGE A-1

b Related Sentences

Subtraction is defined in terms of addition. For example, $5 - 2$ is that number which when added to 2 gives 5. Thus for the subtraction sentence

$$5 - 2 = 3,$$ **Taking away 2 from 5 gives 3.**

there is a related addition sentence

$$5 = 3 + 2.$$ **Putting back the 2 gives 5 again.**

In fact, we know answers to subtractions are correct only because of the related addition, which provides a handy way to check a subtraction.

▶ **EXAMPLE 3** Write a related addition sentence: $8 - 5 = 3$.

$8 - 5 = 3$

This number gets added (after 3).

$8 = 3 + 5$

By the commutative law of addition, there is also another addition sentence:

$8 = 5 + 3.$

The related sentence is $8 = 3 + 5$. ◀

DO EXERCISES 3 AND 4.

Write a related addition sentence.

3. $7 - 5 = 2$ $7 = 2 + 5$

4. $17 - 8 = 9$ $17 = 9 + 8$

▶ **EXAMPLE 4** Write two related subtraction sentences: $4 + 3 = 7$.

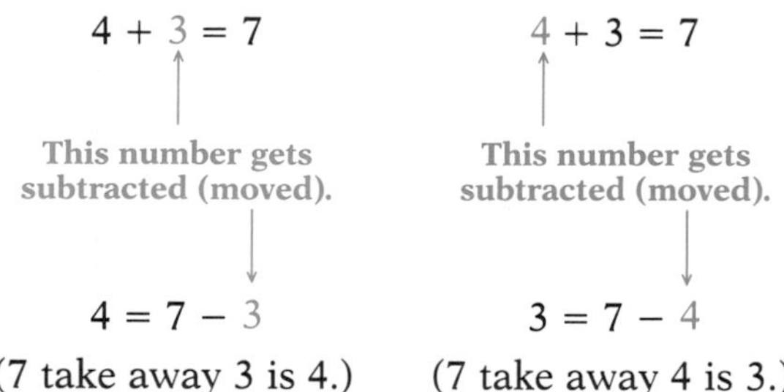

The related sentences are $4 = 7 - 3$ and $3 = 7 - 4$. ◀

DO EXERCISES 5 AND 6.

Write two related subtraction sentences.

5. $5 + 8 = 13$

$5 = 13 - 8$ and $8 = 13 - 5$

6. $11 + 3 = 14$

$11 = 14 - 3$ and $3 = 14 - 11$

ANSWERS ON PAGE A-1

c How Much More?

The second kind of situation for which subtraction corresponds is called "how much more"? From the related sentences, we see that finding a *missing addend* is the same as finding a *difference*. We need the concept of a missing addend for the "how-much-more" problems.

Missing addend: $12 = 3 + \square$ Difference: $12 - 3 = \square$

▶ **EXAMPLES** Write a subtraction that corresponds to each situation. You need not carry out the subtraction.

5. A student has \$17 and wants to buy a book that costs \$23. How much more is needed to buy the book?

To find the subtraction, we first consider addition.

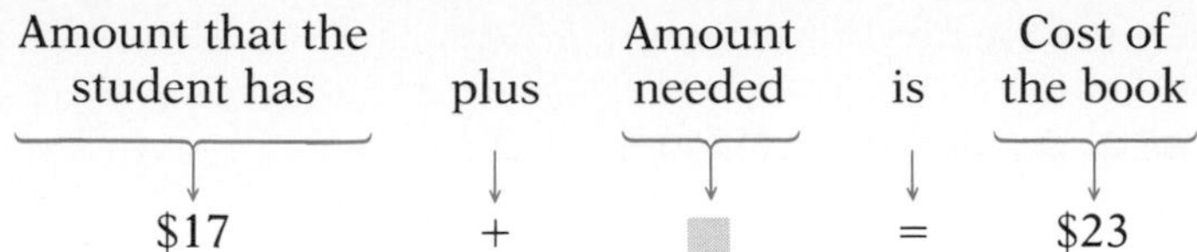

Now we write a related subtraction:

$17 + \square = 23$

$\square = 23 - 17.$ **17 gets subtracted (moved).**

6. It is 134 mi from Los Angeles to San Diego. A driver has gone 90 mi of the trip. How much farther does the driver have to go?

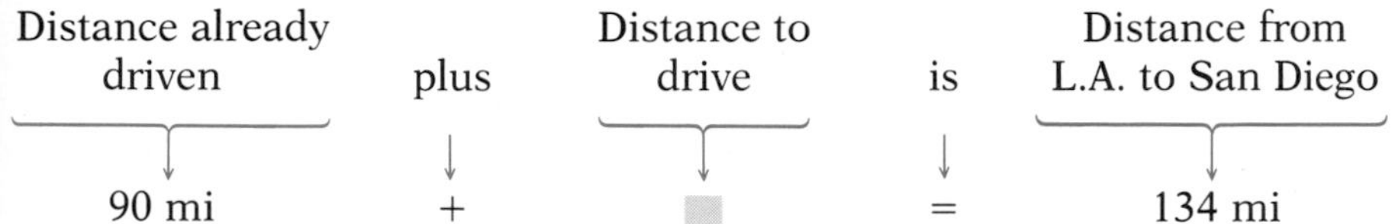

Now we write a related subtraction:

$90 + \square = 134$

$\square = 134 - 90.$ **90 gets subtracted (moved).** ◀

DO EXERCISES 7 AND 8.

d Subtraction of Whole Numbers

To subtract numbers, we can subtract ones first, then tens, then hundreds, and so on.

▶ **EXAMPLE 7** Subtract: 9768 − 4320.

$$\begin{array}{r} 9\ 7\ 6\ 8 \\ -\ 4\ 3\ 2\ 0 \\ \hline 8 \end{array}$$ **Subtract ones.**

$$\begin{array}{r} 9\ 7\ 6\ 8 \\ -\ 4\ 3\ 2\ 0 \\ \hline 4\ 8 \end{array}$$ **Subtract tens.**

$$\begin{array}{r} 9\ 7\ 6\ 8 \\ -\ 4\ 3\ 2\ 0 \\ \hline 4\ 4\ 8 \end{array}$$ **Subtract hundreds.**

$$\begin{array}{r} 9\ 7\ 6\ 8 \\ -\ 4\ 3\ 2\ 0 \\ \hline 5\ 4\ 4\ 8 \end{array}$$ **Subtract thousands.**

This is for explanation.

$$\begin{array}{r} 9\ 7\ 6\ 8 \\ -\ 4\ 3\ 2\ 0 \\ \hline 5\ 4\ 4\ 8 \end{array}$$

You should write only this. ◀

DO EXERCISE 9.

Write an addition sentence and a related subtraction sentence corresponding to the situation. You need not carry out the subtraction.

7. There are 32 million kangaroos and 15 million people in Australia. How many more kangaroos are there than people?

$15 + \square = 32;\ \square = 32 - 15$

8. A set of drapes requires 23 yd of material. The drapemaker has 10 yd of material. How much more is needed?

$10 + \square = 23;\ \square = 23 - 10$

9. Subtract.

$$\begin{array}{r} 7\ 8\ 9\ 3 \\ -\ 4\ 0\ 9\ 2 \\ \hline 3801 \end{array}$$

ANSWERS ON PAGE A-1

Subtract. Check by adding.

10. $$\begin{array}{r} 8686 \\ -\,2358 \\ \hline \end{array}$$
6328

11. $$\begin{array}{r} 7145 \\ -\,2398 \\ \hline \end{array}$$
4747

Subtract.

12. $$\begin{array}{r} 70 \\ -\,14 \\ \hline \end{array}$$
56

13. $$\begin{array}{r} 503 \\ -\,298 \\ \hline \end{array}$$
205

Subtract.

14. $$\begin{array}{r} 7007 \\ -\,6349 \\ \hline \end{array}$$
658

15. $$\begin{array}{r} 6000 \\ -\,3149 \\ \hline \end{array}$$
2851

16. $$\begin{array}{r} 9035 \\ -\,7489 \\ \hline \end{array}$$
1546

To the instructor and the student: This section presented a review of subtraction of whole numbers. Students who are successful should go on to Section 1.4. Those who have trouble should study developmental units S.1 and S.2 and then repeat Section 1.3. Students who still have trouble might study developmental unit S.3.

ANSWERS ON PAGE A-1

Sometimes we need to borrow.

▶ **EXAMPLE 8** Subtract: 6246 − 1879.

$$\begin{array}{r} \scriptstyle 3\ 16 \\ 6\,2\,\not4\,\not6 \\ -\,1\,8\,7\,9 \\ \hline 7 \end{array}$$

We cannot subtract 9 ones from 6 ones, but we can subtract 9 ones from 16 ones. We borrow 1 ten to get 16 ones.

$$\begin{array}{r} \scriptstyle 13 \\ \scriptstyle 1\ \not3\ 16 \\ 6\,\not2\,\not4\,\not6 \\ -\,1\,8\,7\,9 \\ \hline 6\,7 \end{array}$$

We cannot subtract 7 tens from 3 tens, but we can subtract 7 tens from 13 tens. We borrow 1 hundred to get 13 tens.

$$\begin{array}{r} \scriptstyle 11\ 13 \\ \scriptstyle 5\ \not1\ \not3\ 16 \\ \not6\,\not2\,\not4\,\not6 \\ -\,1\,8\,7\,9 \\ \hline 4\,3\,6\,7 \end{array}$$

We cannot subtract 8 hundreds from 1 hundred, but we can subtract 8 hundreds from 11 hundreds. We borrow 1 thousand to get 11 hundreds.

We can always check the answer by adding it to the number being subtracted.

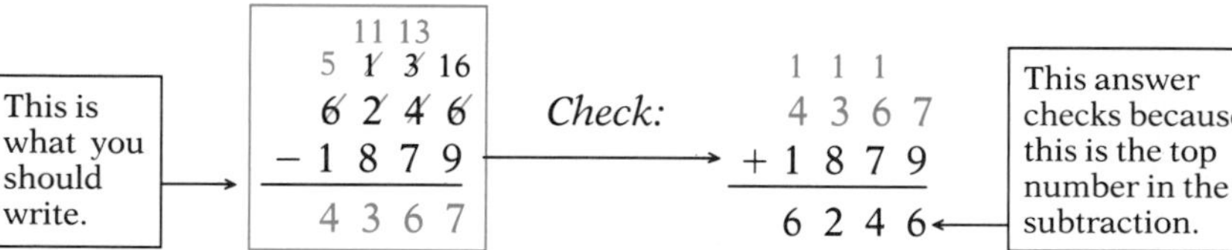

◀

DO EXERCISES 10 AND 11.

▶ **EXAMPLE 9** Subtract: 902 − 477.

$$\begin{array}{r} \scriptstyle 8\ 9\ 12 \\ \not9\,\not0\,\not2 \\ -\,4\,7\,7 \\ \hline 4\,2\,5 \end{array}$$

We cannot subtract 7 ones from 2 ones. We have 9 hundreds, or 90 tens. We borrow 1 ten to get 12 ones. We then have 89 tens. ◀

DO EXERCISES 12 AND 13.

▶ **EXAMPLE 10** Subtract: 8003 − 3667.

$$\begin{array}{r} \scriptstyle 7\ 9\ 9\ 13 \\ \not8\,\not0\,\not0\,\not3 \\ -\,3\,6\,6\,7 \\ \hline 4\,3\,3\,6 \end{array}$$

We have 8 thousands, or 800 tens. We borrow 1 ten to get 13 ones. We then have 799 tens. ◀

▶ **EXAMPLES**

11. Subtract: 6000 − 3762.

$$\begin{array}{r} \scriptstyle 5\ 9\ 9\ 10 \\ \not6\,\not0\,\not0\,\not0 \\ -\,3\,7\,6\,2 \\ \hline 2\,2\,3\,8 \end{array}$$

12. Subtract: 6024 − 2968.

$$\begin{array}{r} \scriptstyle 11 \\ \scriptstyle 5\ 9\ \not1\ 14 \\ \not6\,\not0\,\not2\,\not4 \\ -\,2\,9\,6\,8 \\ \hline 3\,0\,5\,6 \end{array}$$

◀

DO EXERCISES 14–16.

NAME SECTION DATE

EXERCISE SET 1.3

a Write a subtraction that corresponds to each situation. You need not carry out the subtraction.

1. A gasoline station has 2400 gal of lead-free gasoline. One day it sells 800 gal. How many gallons are left in the tank?

2. A consumer has $650 in a checking account and writes a check for $100. How much is left in the account?

b Write a related addition sentence.

3. 10 − 7 = 3
4. 12 − 5 = 7
5. 13 − 8 = 5
6. 9 − 9 = 0
7. 23 − 9 = 14
8. 20 − 8 = 12
9. 43 − 16 = 27
10. 51 − 18 = 33

Write two related subtraction sentences.

11. 6 + 9 = 15
 6 = 15 − 9; 9 = 15 − 6
12. 7 + 9 = 16
 7 = 16 − 9; 9 = 16 − 7
13. 8 + 7 = 15
 8 = 15 − 7; 7 = 15 − 8
14. 8 + 0 = 8
 8 = 8 − 0; 0 = 8 − 8
15. 17 + 6 = 23
 17 = 23 − 6; 6 = 23 − 17
16. 11 + 8 = 19
 11 = 19 − 8; 8 = 19 − 11
17. 23 + 9 = 32
 23 = 32 − 9; 9 = 32 − 23
18. 42 + 10 = 52
 42 = 52 − 10; 10 = 52 − 42

c Write an addition sentence and a related subtraction sentence corresponding to each situation. You need not carry out the subtraction.

19. One week a car dealer sets a goal of selling 220 cars. By Thursday it sells 190 cars. How many more does it have to sell in order to meet its goal?
 190 + ▒ = 220; ▒ = 220 − 190

20. Tuition will cost a student $3000. The student has $1250. How much more money is needed?
 $1250 + ▒ = $3000; ▒ = $3000 − $1250

d Subtract.

21. 16 − 4
22. 86 − 13
23. 65 − 21
24. 87 − 34
25. 866 − 333
26. 526 − 323
27. 4547 − 3421
28. 6875 − 2111
29. 86 − 47
30. 73 − 28
31. 625 − 327
32. 726 − 509
33. 835 − 609
34. 953 − 246
35. 981 − 747
36. 887 − 698

ANSWERS

1. 2400 − 800 = ▒
2. $650 − $100 = ▒
3. 10 = 3 + 7
4. 12 = 7 + 5
5. 13 = 5 + 8
6. 9 = 0 + 9
7. 23 = 14 + 9
8. 20 = 12 + 8
9. 43 = 27 + 16
10. 51 = 33 + 18
11.
12.
13.
14.
15.
16.
17.
18.
19.
20.
21. 12
22. 73
23. 44
24. 53
25. 533
26. 203
27. 1126
28. 4764
29. 39
30. 45
31. 298
32. 217
33. 226
34. 707
35. 234
36. 189

ANSWERS

37. 5382
38. 3535
39. 1493
40. 5885
41. 2187
42. 3629
43. 3831
44. 7404
45. 7748
46. 10,334
47. 33,794
48. 46,669
49. 56
50. 3
51. 36
52. 12
53. 84
54. 282
55. 454
56. 385
57. 771
58. 6150
59. 2191
60. 2213
61. 3749
62. 4813
63. 1053
64. 4418
65. 4206
66. 1458
67. 10,305
68. 5445
69. 7 ten thousands
70.

37. $7769 - 2387$

38. $6431 - 2896$

39. $3982 - 2489$

40. $7650 - 1765$

41. $5046 - 2859$

42. $6308 - 2679$

43. $7640 - 3809$

44. $8003 - 599$

45. $12{,}647 - 4{,}899$

46. $16{,}222 - 5{,}888$

47. $46{,}771 - 12{,}977$

48. $95{,}654 - 48{,}985$

49. $80 - 24$

50. $40 - 37$

51. $90 - 54$

52. $90 - 78$

53. $140 - 56$

54. $470 - 188$

55. $690 - 236$

56. $803 - 418$

57. $903 - 132$

58. $6408 - 258$

59. $2300 - 109$

60. $3506 - 1293$

61. $6808 - 3059$

62. $7840 - 3027$

63. $4027 - 2974$

64. $6007 - 1589$

65. $7000 - 2794$

66. $8001 - 6543$

67. $48{,}000 - 37{,}695$

68. $17{,}043 - 11{,}598$

SKILL MAINTENANCE

69. What does the digit 7 mean in 6,375,602?

70. Write a word name for 6,375,602.

Six million, three hundred seventy-five thousand, six hundred two

1.4 Rounding and Estimating; Order

a Rounding

We round numbers in various situations if we do not need an exact answer. We might round to check if an answer to a problem is reasonable or to check a calculation done by hand or on a calculator. We might also round to see if we are being charged the correct amount in a store.

To understand how to round, we first look at some examples using number lines, even though this is not the way we normally do rounding.

▶ **EXAMPLE 1** Round 47 to the nearest ten.

Here is part of a number line; 47 is between 40 and 50.

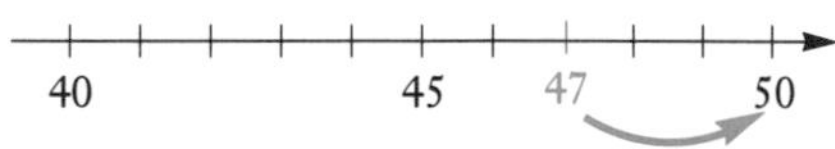

Since 47 is closer to 50, we round up to 50. ◀

▶ **EXAMPLE 2** Round 42 to the nearest ten.

42 is between 40 and 50.

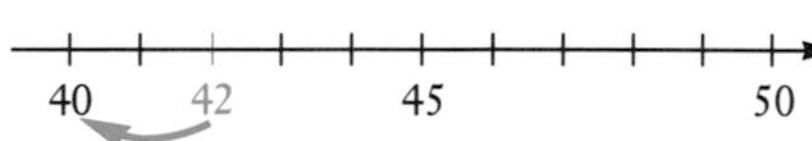

Since 42 is closer to 40, we round down to 40. ◀

DO EXERCISES 1–4.

▶ **EXAMPLE 3** Round 45 to the nearest ten.

45 is halfway between 40 and 50.

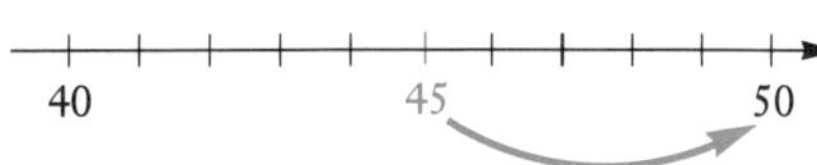

We could round 45 down to 40 or up to 50. We agree to round up to 50. ◀

When a number is halfway between rounding numbers, we agree to round up.

DO EXERCISES 5–7.

Here is a rule for rounding.

To round to a certain place:

a) Locate the digit in that place.

b) Then consider the next digit to the right.

c) If the digit to the right is 5 or higher, round up; if the digit to the right is 4 or lower, round down.

d) Change all digits to the right of the rounding location to zeros.

OBJECTIVES

After finishing Section 1.4, you should be able to:

a Round to the nearest ten, hundred, or thousand.

b Estimate sums and differences by rounding.

c Use < or > for ▭ to write a true sentence in a situation like 6 ▭ 10.

FOR EXTRA HELP

Tape 1D

Tape 1B

MAC: 1
IBM: 1

Round to the nearest ten.

1. 37 40

2. 52 50

3. 73 70

4. 98 100

Round to the nearest ten.

5. 35 40

6. 75 80

7. 85 90

ANSWERS ON PAGE A-1

Round to the nearest ten.

8. 137 140

9. 473 470

10. 235 240

11. 285 290

Round to the nearest hundred.

12. 641 600

13. 759 800

14. 750 800

15. 9325 9300

Round to the nearest thousand.

16. 7896 8000

17. 8459 8000

18. 19,343 19,000

19. 68,500 69,000

ANSWERS ON PAGE A-1

▶ **EXAMPLE 4** Round 6485 to the nearest ten.

a) Locate the digit in the tens place.

6 4 8 5
↑

b) Then consider the next digit to the right.

6 4 8 5
↑

c) Since that digit is 5 or higher, round 8 tens up to 9 tens.

d) Change all digits to the right of the tens digit to zeros.

6 4 9 0 ← This is the answer. ◀

▶ **EXAMPLE 5** Round 6485 to the nearest hundred.

a) Locate the digit in the hundreds place.

6 4 8 5
↑

b) Then consider the next digit to the right.

6 4 8 5
↑

c) Since that digit is 5 or higher, round 4 hundreds up to 5 hundreds.

d) Change all digits to the right of hundreds to zeros.

6 5 0 0 ← This is the answer. ◀

▶ **EXAMPLE 6** Round 6485 to the nearest thousand.

a) Locate the digit in the thousands place.

6 4 8 5
↑

b) Then consider the next digit to the right.

6 4 8 5
↑

c) Since that digit is 4 or lower, round down, meaning that 6 thousands stays as 6 thousands.

d) Change all digits to the right of thousands to zeros.

6 0 0 0 ← This is the answer. ◀

CAUTION! 7000 is not a correct answer to Example 6. It is incorrect to round from the ones digit over, as follows:

6485, 6490, 6500, 7000.

DO EXERCISES 8–19.

There are many methods of rounding. For example, in rounding 8563 to the nearest hundred, a different rule would call for us to **truncate,** meaning that we would simply change all digits to the right of the rounding location to zeros. Thus, 8563 would round to 8500, which is not the same answer that we would get using the rule in this section.

b Estimating

Estimating is done to make a problem simpler so that it can be done easily or mentally. Rounding is used when estimating. There are many ways to estimate.

▶ **EXAMPLE 7** Dick and Tom Van Arsdale are twin brothers who played professional basketball. Tom scored a total of 14,232 points in his career, and Dick scored a total of 15,079. Estimate how many points they scored in all.

There are many ways to get an answer, but there is no one perfect answer based on how the problem is worded. Let's consider a couple of methods.

Method 1. Round each number to the nearest thousand and then add.

$$\begin{array}{r} 14{,}232 \\ +\,15{,}079 \\ \hline \end{array} \qquad \begin{array}{r} 14{,}000 \\ +\,15{,}000 \\ \hline 29{,}000 \end{array} \leftarrow \text{Estimated answer}$$

Method 2. We might use a less formal approach, depending on how specific we want the answer to be. We note that both numbers are close to 15,000, and so the total is close to 30,000. In some contexts, such as sports commentary, this would be sufficient. ◀

The point to be made is that estimating can be done in many ways and can have many answers, even though in the problems that follow we ask you to round in a specific way.

▶ **EXAMPLE 8** Estimate this sum by rounding to the nearest ten:

$$45 + 53 + 32 + 88.$$

We round each number to the nearest ten. Then we add.

$$\begin{array}{r} 45 \\ 53 \\ 32 \\ +\,88 \\ \hline \end{array} \qquad \begin{array}{r} 50 \\ 50 \\ 30 \\ +\,90 \\ \hline 220 \end{array} \leftarrow \text{Estimated answer}$$

◀

DO EXERCISE 20.

▶ **EXAMPLE 9** Estimate this sum by rounding to the nearest hundred:

$$350 + 474 + 986 + 839.$$

We have

$$\begin{array}{r} 350 \\ 474 \\ 986 \\ +\,839 \\ \hline \end{array} \qquad \begin{array}{r} 400 \\ 500 \\ 1000 \\ +\,800 \\ \hline 2700 \end{array}$$

◀

DO EXERCISE 21.

20. Estimate the sum by rounding to the nearest ten. Show your work.

$$\begin{array}{r} 74 \\ 23 \\ 35 \\ +\,66 \\ \hline \end{array}$$

200

21. Estimate the sum by rounding to the nearest hundred. Show your work.

$$\begin{array}{r} 650 \\ 695 \\ 248 \\ +\,178 \\ \hline \end{array}$$

1800

ANSWERS ON PAGE A-1

22. Estimate the difference by rounding to the nearest hundred. Show your work.

$$\begin{array}{r} 8\,7\,2\,3 \\ -\,3\,6\,8\,5 \\ \hline \end{array}$$

5000

23. Estimate the difference by rounding to the nearest thousand. Show your work.

$$\begin{array}{r} 2\,3{,}2\,7\,8 \\ -\,1\,1{,}6\,9\,8 \\ \hline \end{array}$$

11,000

Use < or > for □ to write a true sentence. Draw a number line if necessary.

24. 8 □ 12   <

25. 12 □ 8   >

26. 76 □ 64   >

27. 64 □ 76   <

28. 217 □ 345   <

29. 345 □ 217   >

ANSWERS ON PAGE A-1

▶ **EXAMPLE 10** Estimate the difference by rounding to the nearest thousand: 9324 − 2849.

We have

$$\begin{array}{r} 9\,3\,2\,4 \\ -\,2\,8\,4\,9 \\ \hline \end{array} \qquad \begin{array}{r} 9\,0\,0\,0 \\ -\,3\,0\,0\,0 \\ \hline 6\,0\,0\,0 \end{array}$$

◀

DO EXERCISES 22 AND 23.

Later we will use the symbol "≈" when rounding. This symbol means **"approximately equal to."** Thus when 687 is rounded to the nearest ten, we may write

$$687 \approx 690.$$

C Order

We know that 2 is less than 5. We can see this order on a number line: 2 is to the left of 5.

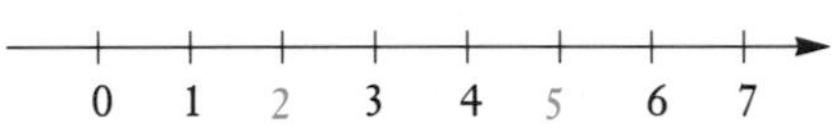

> **For any whole numbers a and b:**
>
> **1. $a < b$ (read "a less than b") is true when a is to the left of b on a number line.**
>
> **2. $a > b$ (read "a is greater than b") is true when a is to the right of b on a number line.**
>
> **We call "<" and ">" *inequality* symbols.**

▶ **EXAMPLE 11** Use < or > for □ to write a true sentence: 7 □ 11.

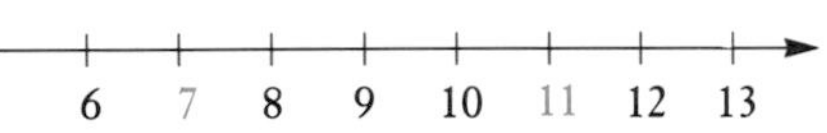

Since 7 is to the left of 11, $7 < 11$. ◀

▶ **EXAMPLE 12** Use < or > for □ to write a true sentence: 92 □ 87.

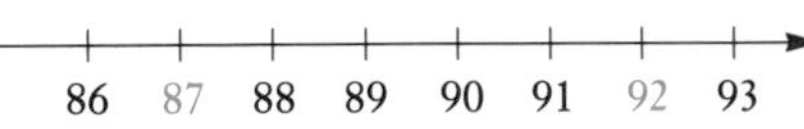

Since 92 is to the right of 87, $92 > 87$. ◀

▶ **EXAMPLE 13** Use < or > for □ to write a true sentence: 241 □ 200.

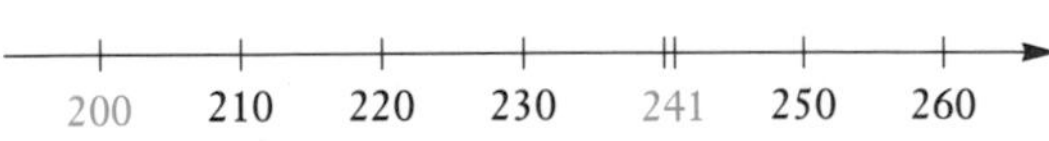

Since 241 is to the right of 200, $241 > 200$. ◀

A sentence like $7 < 11$ is called an **inequality.** The sentence $7 < 11$ is a true inequality. The sentence $23 > 69$ is a false inequality.

DO EXERCISES 24–29.

NAME SECTION DATE

EXERCISE SET 1.4

a Round to the nearest ten.

1. 48 **2.** 17 **3.** 67 **4.** 99

5. 731 **6.** 532 **7.** 895 **8.** 765

Round to the nearest hundred.

9. 146 **10.** 874 **11.** 957 **12.** 650

13. 3583 **14.** 4645 **15.** 2850 **16.** 4402

Round to the nearest thousand.

17. 5932 **18.** 4500 **19.** 7500 **20.** 13,855

21. 45,340 **22.** 735,562 **23.** 373,405 **24.** 2001

b Estimate the sum or difference by rounding to the nearest ten. Show your work.

25.
$$\begin{array}{r} 7848 \\ +9747 \\ \hline \end{array}$$

26.
$$\begin{array}{r} 31 \\ 91 \\ 87 \\ +58 \\ \hline \end{array}$$

27.
$$\begin{array}{r} 6882 \\ -1748 \\ \hline \end{array}$$

18.
$$\begin{array}{r} 670 \\ -\ \ 23 \\ \hline \end{array}$$

Estimate the sum by rounding to the nearest ten. Do any of the sums seem to be incorrect? Which ones?

29.
$$\begin{array}{r} 45 \\ 77 \\ 25 \\ +56 \\ \hline 343 \end{array}$$

30.
$$\begin{array}{r} 41 \\ 21 \\ 55 \\ +60 \\ \hline 177 \end{array}$$

31.
$$\begin{array}{r} 622 \\ 78 \\ 81 \\ +111 \\ \hline 932 \end{array}$$

32.
$$\begin{array}{r} 836 \\ 374 \\ 794 \\ +938 \\ \hline 3947 \end{array}$$

ANSWERS

1. 50
2. 20
3. 70
4. 100
5. 730
6. 530
7. 900
8. 770
9. 100
10. 900
11. 1000
12. 700
13. 3600
14. 4600
15. 2900
16. 4400
17. 6000
18. 5000
19. 8000
20. 14,000
21. 45,000
22. 736,000
23. 373,000
24. 2000
25. 17,600
26. 270
27. 5130
28. 650
29. 220; incorrect
30. 180
31. 890; incorrect
32. 2940; incorrect

ANSWERS

33. 17,500
34. 2900
35. 5200
36. 6600
37. 1600
38. 2800; incorrect
39. 1500
40. 2300
41. 31,000
42. 27,000
43. 69,000
44. 74,000
45. <
46. >
47. >
48. >
49. <
50. <
51. >
52. >
53. >
54. <
55. >
56. <
57. 86,754
58. 4415
59. 30,411
60. 27,060
61. 69,594
62. 73,780

Estimate the sum or difference by rounding to the nearest hundred. Show your work.

33. $\begin{array}{r} 7848 \\ +9747 \\ \hline \end{array}$ **34.** $\begin{array}{r} 836 \\ 374 \\ 794 \\ +938 \\ \hline \end{array}$ **35.** $\begin{array}{r} 6852 \\ 1748 \\ \hline \end{array}$ **36.** $\begin{array}{r} 9438 \\ 2787 \\ \hline \end{array}$

Estimate the sum by rounding to the nearest hundred. Do any of the sums seem to be incorrect? Which ones?

37. $\begin{array}{r} 216 \\ 84 \\ 745 \\ +595 \\ \hline 1640 \end{array}$ **38.** $\begin{array}{r} 481 \\ 702 \\ 623 \\ +1043 \\ \hline 1849 \end{array}$ **39.** $\begin{array}{r} 750 \\ 428 \\ 63 \\ +205 \\ \hline 1446 \end{array}$ **40.** $\begin{array}{r} 326 \\ 275 \\ 758 \\ +943 \\ \hline 2302 \end{array}$

Estimate the sum or difference by rounding to the nearest thousand. Show your work.

41. $\begin{array}{r} 9643 \\ 4821 \\ 8943 \\ +7004 \\ \hline \end{array}$ **42.** $\begin{array}{r} 7648 \\ 9348 \\ 7842 \\ +2222 \\ \hline \end{array}$ **43.** $\begin{array}{r} 92,149 \\ -22,555 \\ \hline \end{array}$ **44.** $\begin{array}{r} 84,890 \\ -11,110 \\ \hline \end{array}$

C Use < or > for □ to write a true sentence. Draw a number line if necessary.

45. 0 □ 17 **46.** 32 □ 0 **47.** 34 □ 12 **48.** 28 □ 18

49. 1000 □ 1001 **50.** 77 □ 117 **51.** 133 □ 132 **52.** 999 □ 997

53. 460 □ 17 **54.** 345 □ 456 **55.** 37 □ 11 **56.** 12 □ 32

SKILL MAINTENANCE

57. Add.

$$\begin{array}{r} 67,789 \\ +18,965 \\ \hline \end{array}$$

58. Subtract.

$$\begin{array}{r} 9002 \\ -4587 \\ \hline \end{array}$$

SYNTHESIS

59–62. Use a calculator to find the sums or differences in Exercises 41–44. Since you can still make errors on a calculator—say, by pressing the wrong buttons—you can check your answers by estimating.

1.5 Multiplication

a Multiplication and the Real World

Multiplication of whole numbers corresponds to two kinds of situations.

Repeated Addition

The multiplication 3 × 5 corresponds to this repeated addition:

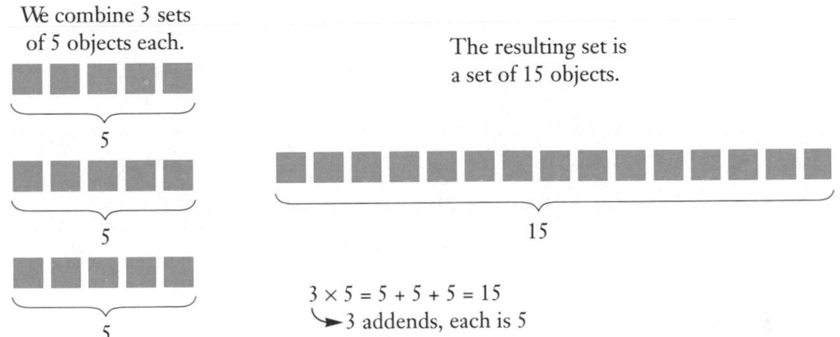

We say that the **product** of 3 and 5 is 15. The numbers 3 and 5 are called **factors.**

$$3 \times 5 = 15$$

Factors (3, 5) — Product (15)

Rectangular Arrays

The multiplication 3 × 5 corresponds to this rectangular array:

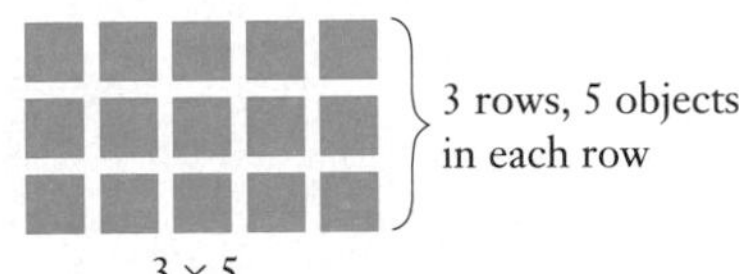

When you write a multiplication corresponding to a real-world situation, you should think of either a rectangular array or repeated addition. In some cases, it may help to think both ways.

We have used an " × " to denote multiplication. A dot " · " is also commonly used. It was invented by the German mathematician Leibniz in 1698. Parentheses are also used to denote multiplication. For example, (3)(5) = 15.

OBJECTIVES

After finishing Section 1.5, you should be able to:

a Determine what multiplication corresponds to a situation.

b Multiply whole numbers.

c Estimate products by rounding.

FOR EXTRA HELP

Tape 1E

Tape 2A

MAC: 1
IBM: 1

Write a multiplication that corresponds to the situation.

1. A jogger runs 4 mi on each of 8 days. How many miles are run in all? $8 \times 4 = 32$ mi

2. A lab technician pours 75 mL (milliliters) of acid into each of 10 beakers. How much acid is poured in all? $10 \cdot 75 = 750$ mL

3. A band is arranged rectangularly in 12 rows with 20 members in each row. How many people are in the band?

$12 \cdot 20 = 240$

4. What is the area of this region?

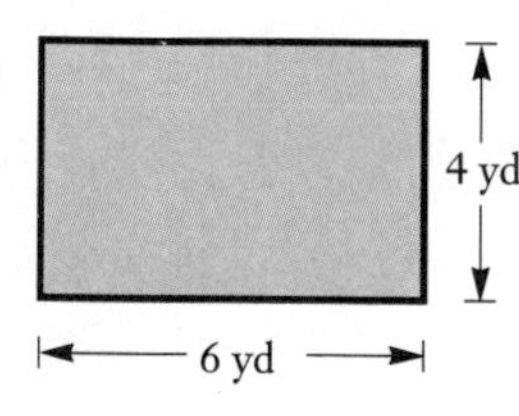

$4 \cdot 6 = 24$ yd^2

ANSWERS ON PAGE A-2

▶ **EXAMPLES** Write a multiplication that corresponds to each situation.

1. It is known that Americans drink 23 million gal of soft drinks per day (*per day* means *each day*). What quantity of soft drinks is consumed every 5 days?

   We draw a picture or at least visualize the situation. Repeated addition fits best in this case.

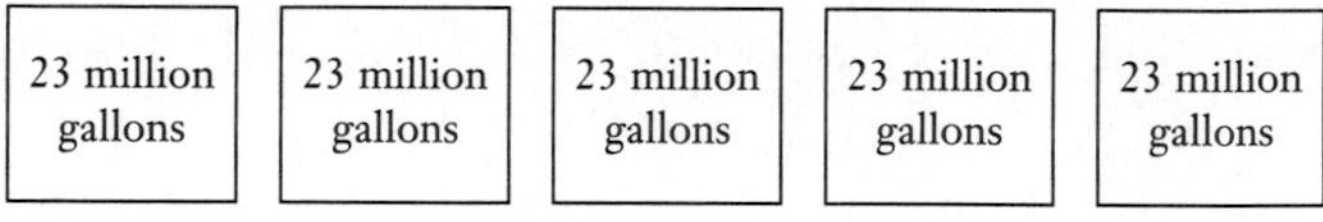

$5 \cdot 23$ million gallons $= 115$ million gallons

2. One side of a building has 6 floors with 7 windows on each floor. How many windows are there on that side of the building?

   We have a rectangular array and can easily draw a sketch.

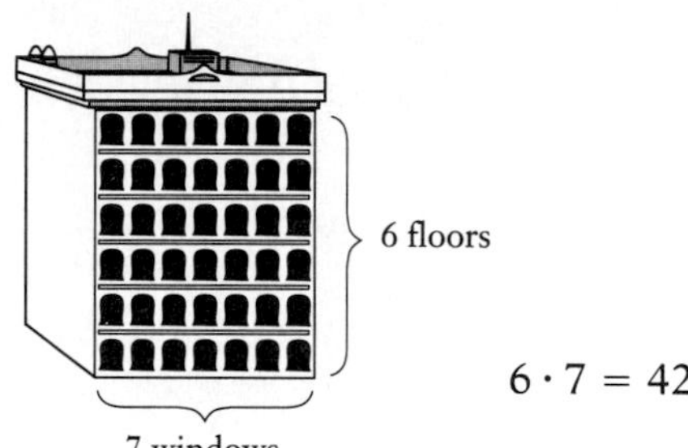

$6 \cdot 7 = 42$ ◀

Area

The area of a rectangular region is often considered to be the number of square units needed to fill it. Here is a rectangle 4 cm long and 3 cm wide. It takes 12 square centimeters (cm^2) to fill it.

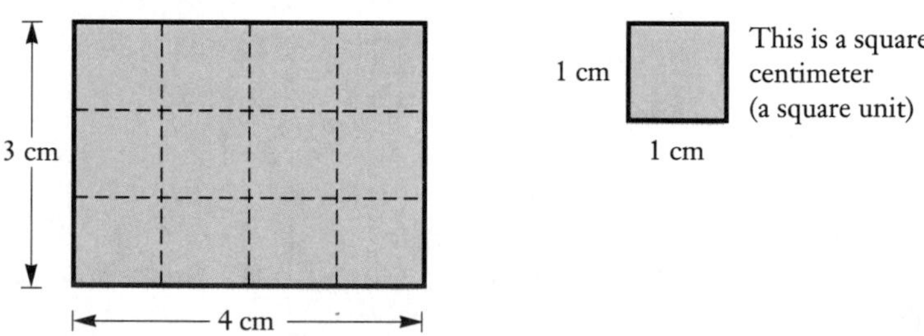

In this case, we have a rectangular array. The number of square units is $3 \cdot 4$, or 12.

▶ **EXAMPLE 3** Write a multiplication that corresponds to the following situation: A rectangular floor is 10 ft long and 8 ft wide. Find its area.

We draw a picture.

If we think of filling the rectangle with square feet, we have a rectangular array. The length $l = 10$ ft, and the width $w = 8$ ft. The area A is given by

$$A = l \cdot w = 10 \times 8 = 80 \text{ square feet (ft}^2\text{)}.$$ ◀

DO EXERCISES 1–4 ON THE PRECEDING PAGE.

b Multiplication of Whole Numbers

Let's find the product

$$\begin{array}{r} 54 \\ \times\ 32 \\ \hline \end{array}$$

To do this, we multiply 54 by 2, then 54 by 30, and then add.

$$\begin{array}{r} 54 \\ \times\quad 2 \\ \hline 108 \end{array} \qquad \begin{array}{r} {}^{1} \\ 54 \\ \times\ \ 30 \\ \hline 1620 \end{array}$$

Since we are going to add the results, let's write the work this way.

$$\begin{array}{rl} 54 & \\ \times\ 32 & \\ \hline 108 & \text{Multiplying by 2} \\ 1620 & \text{Multiplying by 30} \\ \hline 1728 & \text{Adding} \end{array}$$

The fact that we can do this is based on a property called the **distributive law.** It says that to multiply a number by a sum, $a \cdot (b + c)$, we can multiply the parts by a and then add like this: $(a \cdot b) + (a \cdot c)$. Thus, $a \cdot (b + c) = (a \cdot b) + (a \cdot c)$. Applied to the above example, the distributive law gives us

$$54 \cdot 32 = 54 \cdot (30 + 2) = (54 \cdot 30) + (54 \cdot 2).$$

Multiply.

5. $\begin{array}{r} 45 \\ \times\ 23 \\ \hline \end{array}$

1035

6. $\begin{array}{r} 63 \\ \times\ 48 \\ \hline \end{array}$

3024

Multiply.

7. $\begin{array}{r} 746 \\ \times\quad 62 \\ \hline \end{array}$

46,252

8. $\begin{array}{r} 837 \\ \times\ 245 \\ \hline \end{array}$

205,065

ANSWERS ON PAGE A-2

Multiply.

9.
```
    4 7 2
  × 3 0 6
```
144,432

10.
```
    7 0 4
  × 4 0 8
```
287,232

11.
```
    2 3 4 4
  × 6 0 0 5
```
14,075,720

Multiply.

12.
```
    4 7 2
  × 3 6 0
```
169,920

13.
```
    2 3 4 4
  × 7 4 0 0
```
17,345,600

ANSWERS ON PAGE A-2

▶ **EXAMPLE 4** Multiply 43 × 57.

```
      2
      5 7
  ×   4 3
  -------
    1 7 1      Multiplying by 3
```

```
      2
      2
      5 7
  ×   4 3
  -------
    1 7 1      Multiplying by 40. (We write a 0
  2 2 8 0      and then multiply 57 by 4.)
```

```
      2
      2
      5 7
  ×   4 3
  -------
    1 7 1
  2 2 8 0
  -------
  2 4 5 1      Adding
```

You may have learned that such a 0 does not have to be written. You may omit it if you wish. If you do omit it, remember, when multiplying by tens, to put the answer in the tens place.

DO EXERCISES 5 AND 6 ON THE PRECEDING PAGE.

▶ **EXAMPLE 5** Multiply: 457 × 683.

```
      5 2
      6 8 3
  ×   4 5 7
  ---------
      4 7 8 1      Multiplying 683 by 7
```

```
      4 1
      5 2
      6 8 3
  ×   4 5 7
  ---------
      4 7 8 1
    3 4 1 5 0      Multiplying 683 by 50
```

```
        3 1
        4 1
        5 2
        6 8 3
  ×     4 5 7
  -----------
        4 7 8 1
      3 4 1 5 0
    2 7 3 2 0 0    Multiplying 683 by 400
  -------------
    3 1 2, 1 3 1   Adding
```

DO EXERCISES 7 AND 8 ON THE PRECEDING PAGE.

Zeros in Multiplication

▶ **EXAMPLE 6** Multiply: 306×274.

Note that $306 = 3$ hundreds $+ 6$ ones.

$$\begin{array}{r} 2\ 7\ 4 \\ \times\ \ 3\ 0\ 6 \\ \hline 1\ 6\ 4\ 4 \\ 8\ 2\ 2\ 0\ 0 \\ \hline 8\ 3{,}8\ 4\ 4 \end{array}$$

Multiplying by 6

Multiplying by 3 hundreds. (We write 00 and then multiply 274 by 3.)

Adding ◀

DO EXERCISES 9–11 ON THE PRECEDING PAGE.

▶ **EXAMPLE 7** Multiply: 360×274.

Note that $360 = 3$ hundreds $+ 6$ tens.

$$\begin{array}{r} 2\ 7\ 4 \\ \times\ \ 3\ 6\ 0 \\ \hline 1\ 6\ 4\ 4\ 0 \\ 8\ 2\ 2\ 0\ 0 \\ \hline 9\ 8{,}6\ 4\ 0 \end{array}$$

Multiplying by 6 tens. (We write 0 and then multiply 274 by 6.)

Multiplying by 3 hundreds. (We write 00 and then multiply 274 by 3.)

Adding ◀

DO EXERCISES 12 AND 13 ON THE PRECEDING PAGE.

Note the following.

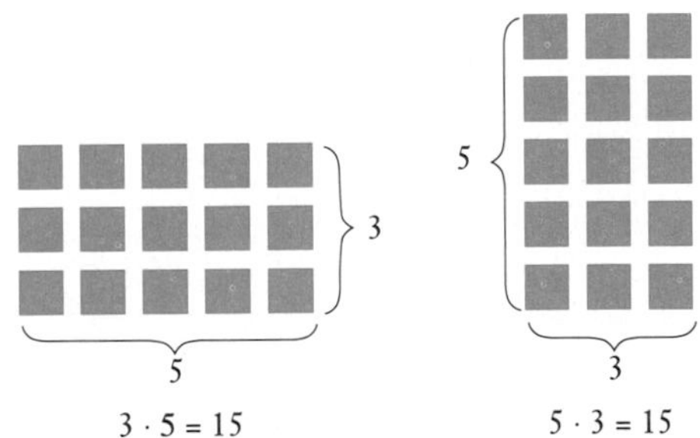

If we rotate the array on the left, we get the array on the right. The answers are the same. This illustrates the **commutative law of multiplication.** It says that we can multiply numbers in any order: $a \cdot b = b \cdot a$.

DO EXERCISE 14.

To multiply three or more numbers, we usually group them so that we multiply two at a time. Consider $2 \cdot (3 \cdot 4)$ and $(2 \cdot 3) \cdot 4$. The parentheses tell what to do first:

$$2 \cdot (3 \cdot 4) = 2 \cdot (12) = 24.$$ We multiply 3 and 4, then 2.

We can also multiply 2 and 3, then 4:

$$(2 \cdot 3) \cdot 4 = (6) \cdot 4 = 24.$$

Either way we get 24. It does not matter how we group the numbers. This illustrates that **multiplication is associative**: $a \cdot (b \cdot c) = (a \cdot b) \cdot c$. Together the commutative and associative laws tell us that to multiply more than two numbers, we can use any order and grouping we wish.

DO EXERCISES 15 AND 16.

14. a) Find $23 \cdot 47$. 1081

b) Find $47 \cdot 23$. 1081

c) Compare your answers to (a) and (b). Same

Multiply.

15. $5 \cdot 2 \cdot 4$ 40

16. $5 \cdot 1 \cdot 3$ 15

To the instructor and the student: This section presented a review of multiplication of whole numbers. Students who are successful should go on to Section 1.6. Those who have trouble should study developmental units M.1 and M.2 and then repeat Section 1.5. Students who still have trouble might study developmental unit M.3.

ANSWERS ON PAGE A-2

17. Estimate this product by rounding to the nearest ten and nearest hundred. Show your work.

$$\begin{array}{r} 837 \\ \times 245 \\ \hline \end{array}$$

210,000; 160,000

ANSWER ON PAGE A-2

C Rounding and Estimating

▶ **EXAMPLE 8** Estimate the following product by rounding to the nearest ten and to the nearest hundred.

Exact	*Nearest ten*	*Nearest hundred*
$\begin{array}{r} 683 \\ \times \quad 457 \\ \hline 4781 \\ 34150 \\ 273200 \\ \hline 312131 \end{array}$	$\begin{array}{r} 680 \\ \times \quad 460 \\ \hline 40800 \\ 272000 \\ \hline 312800 \end{array}$	$\begin{array}{r} 700 \\ \times \quad 500 \\ \hline 350000 \end{array}$

◀

Why does the rounding in Example 8 give a larger answer?

DO EXERCISE 17.

❖ SIDELIGHTS

Number Patterns: Magic Squares

The following is a *magic square*. The sum along any row, column, or diagonal is the same—in this case, 60. Check this on your calculator.

35	10	15
0	20	40
25	30	5

EXERCISES

1. Place the numbers 1 through 9 to form a magic square.

8	3	4
1	5	9
6	7	2

2. Place the numbers 1 through 16 to form a magic square. The sums will all be 34.

1	12	14	7
4	15	9	6
13	2	8	11
16	5	3	10

Each of the following is a magic square, but one number is incorrect. Find it.

3.

11	77	62	29
69	22	17	71
27	61	78	12
72	19	21	67

78 should be 79

4.

70	25	67	9
18	59	20	75
19	77	15	60
65	10	69	27

18 should be 17

NAME SECTION DATE

EXERCISE SET 1.5

a Write a multiplication that corresponds to each situation.

1. A store sold 32 calculators at $10 each. How much money did the store receive for the calculators?

2. There are 7 days in a week. How many days are there in 16 weeks?

3. A checkerboard contains 8 rows with 8 squares in each row. How many squares in all are there on a checkerboard?

4. A beverage carton contains 8 bottles, each of which holds 16 oz. How many ounces are there in the carton?

What is the area of each region?

5.

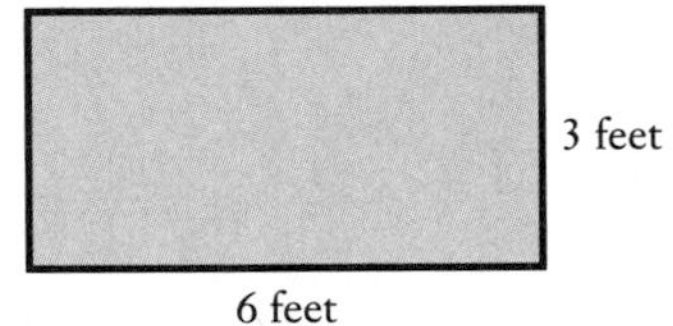

6.

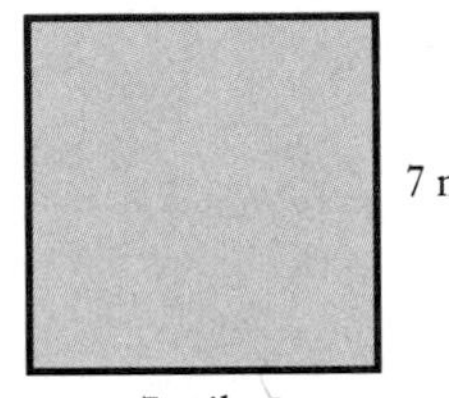

b Multiply.

7. $\begin{array}{r} 87 \\ \times\ 10 \\ \hline \end{array}$

8. $\begin{array}{r} 100 \\ \times\ 96 \\ \hline \end{array}$

9. $\begin{array}{r} 2340 \\ \times\ 1000 \\ \hline \end{array}$

10. $\begin{array}{r} 800 \\ \times\ 70 \\ \hline \end{array}$

11. $\begin{array}{r} 65 \\ \times\ 8 \\ \hline \end{array}$

12. $\begin{array}{r} 87 \\ \times\ 4 \\ \hline \end{array}$

13. $\begin{array}{r} 94 \\ \times\ 6 \\ \hline \end{array}$

14. $\begin{array}{r} 76 \\ \times\ 9 \\ \hline \end{array}$

15. $3 \cdot 509$

16. $7 \cdot 806$

17. $7 \cdot 9229$

18. $4 \cdot 7867$

19. $90 \cdot 53$

20. $60 \cdot 78$

21. $48 \cdot 65$

22. $34 \cdot 87$

23. $\begin{array}{r} 640 \\ \times\ 72 \\ \hline \end{array}$

24. $\begin{array}{r} 666 \\ \times\ 66 \\ \hline \end{array}$

25. $\begin{array}{r} 444 \\ \times\ 33 \\ \hline \end{array}$

26. $\begin{array}{r} 509 \\ \times\ 88 \\ \hline \end{array}$

27. $\begin{array}{r} 509 \\ \times\ 408 \\ \hline \end{array}$

28. $\begin{array}{r} 432 \\ \times\ 375 \\ \hline \end{array}$

29. $\begin{array}{r} 678 \\ \times\ 742 \\ \hline \end{array}$

30. $\begin{array}{r} 346 \\ \times\ 650 \\ \hline \end{array}$

ANSWERS

1. $32 \cdot \$10 = \$320$
2. $7 \cdot 16 = 112$
3. $8 \cdot 8 = 64$
4. $8 \cdot 16 = 128$ oz
5. $3 \cdot 6 = 18\ \text{ft}^2$
6. $7 \cdot 7 = 49\ \text{mi}^2$
7. 870
8. 9600
9. 2,340,000
10. 56,000
11. 520
12. 348
13. 564
14. 684
15. 1527
16. 5642
17. 64,603
18. 31,468
19. 4770
20. 4680
21. 3120
22. 2958
23. 46,080
24. 43,956
25. 14,652
26. 44,792
27. 207,672
28. 162,000
29. 503,076
30. 224,900

ANSWERS

31. 166,260
32. 1,132,800
33. 11,794,332
34. 3,843,360
35. 20,723,872
36. 28,319,616
37. 362,128
38. 38,858,112
39. 20,064,048
40. 15,818,370
41. 25,236,000
42. 35,978,400
43. 302,220
44. 106,145
45. 49,101,136
46. 22,421,949
47. 30,525
48. 78,144
49. 298,738
50. 1,534,160
51. $50 \cdot 70 = 3500$
52. $50 \cdot 80 = 4000$
53. $30 \cdot 30 = 900$
54. $60 \cdot 50 = 3000$
55. $900 \cdot 300 = 270,000$
56. $400 \cdot 300 = 120,000$
57. $400 \cdot 200 = 80,000$
58. $800 \cdot 400 = 320,000$
59.
60.
61.
62.
63. 4370
64. 3109
65. 2350; 2300; 2000

31. $\begin{array}{r} 489 \\ \times 340 \\ \hline \end{array}$ **32.** $\begin{array}{r} 7080 \\ \times 160 \\ \hline \end{array}$ **33.** $\begin{array}{r} 4378 \\ \times 2694 \\ \hline \end{array}$ **34.** $\begin{array}{r} 8007 \\ \times 480 \\ \hline \end{array}$

35. $\begin{array}{r} 6428 \\ \times 3224 \\ \hline \end{array}$ **36.** $\begin{array}{r} 8928 \\ \times 3172 \\ \hline \end{array}$ **37.** $\begin{array}{r} 3482 \\ \times 104 \\ \hline \end{array}$ **38.** $\begin{array}{r} 6408 \\ \times 6064 \\ \hline \end{array}$

39. $\begin{array}{r} 5006 \\ \times 4008 \\ \hline \end{array}$ **40.** $\begin{array}{r} 6789 \\ \times 2330 \\ \hline \end{array}$ **41.** $\begin{array}{r} 5608 \\ \times 4500 \\ \hline \end{array}$ **42.** $\begin{array}{r} 4560 \\ \times 7890 \\ \hline \end{array}$

43. $\begin{array}{r} 876 \\ \times 345 \\ \hline \end{array}$ **44.** $\begin{array}{r} 355 \\ \times 299 \\ \hline \end{array}$ **45.** $\begin{array}{r} 7889 \\ \times 6224 \\ \hline \end{array}$ **46.** $\begin{array}{r} 6501 \\ \times 3449 \\ \hline \end{array}$

47. $\begin{array}{r} 555 \\ \times 55 \\ \hline \end{array}$ **48.** $\begin{array}{r} 888 \\ \times 88 \\ \hline \end{array}$ **49.** $\begin{array}{r} 734 \\ \times 407 \\ \hline \end{array}$ **50.** $\begin{array}{r} 5080 \\ \times 302 \\ \hline \end{array}$

c Estimate the product by rounding to the nearest ten. Show your work.

51. $\begin{array}{r} 45 \\ \times 67 \\ \hline \end{array}$ **52.** $\begin{array}{r} 51 \\ \times 78 \\ \hline \end{array}$ **53.** $\begin{array}{r} 34 \\ \times 29 \\ \hline \end{array}$ **54.** $\begin{array}{r} 63 \\ \times 54 \\ \hline \end{array}$

Estimate the product by rounding to the nearest hundred. Show your work.

55. $\begin{array}{r} 876 \\ \times 345 \\ \hline \end{array}$ **56.** $\begin{array}{r} 355 \\ \times 299 \\ \hline \end{array}$ **57.** $\begin{array}{r} 432 \\ \times 199 \\ \hline \end{array}$ **58.** $\begin{array}{r} 789 \\ \times 434 \\ \hline \end{array}$

Estimate the product by rounding to the nearest thousand. Show your work.

59. $\begin{array}{r} 5608 \\ \times 4576 \\ \hline \end{array}$ $6000 \cdot 5000 = 30,000,000$

60. $\begin{array}{r} 2344 \\ \times 6123 \\ \hline \end{array}$ $2000 \cdot 6000 = 12,000,000$

61. $\begin{array}{r} 7888 \\ \times 6224 \\ \hline \end{array}$ $8000 \cdot 6000 = 48,000,000$

62. $\begin{array}{r} 6501 \\ \times 3449 \\ \hline \end{array}$ $7000 \cdot 3000 = 21,000,000$

SKILL MAINTENANCE

63. Add: $\begin{array}{r} 20 \\ 850 \\ +3500 \\ \hline \end{array}$

64. Subtract: $\begin{array}{r} 6003 \\ -2894 \\ \hline \end{array}$

65. Round 2345 to the nearest ten, hundred, and thousand.

1.6 Division

a Division and the Real World

Division of whole numbers corresponds to two kinds of situations. Consider the division $20 \div 5$, read "20 divided by 5." We can think of 20 objects arranged in a rectangular array. We ask "How many rows, each with 5 objects, are there?

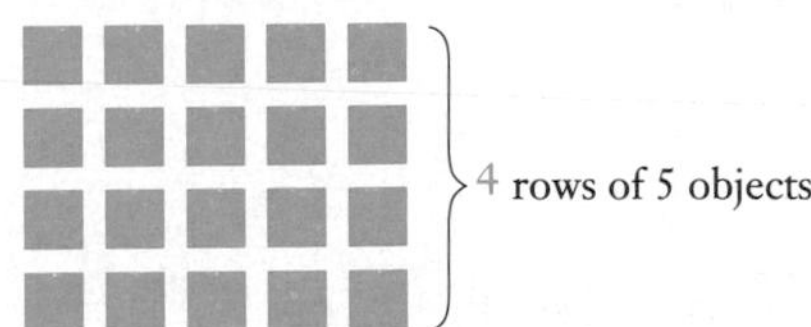

Since there are 4 rows of 5 objects each, we have

$$20 \div 5 = 4.$$

We can also ask, "If we make 5 rows, how many objects will there be in each row?

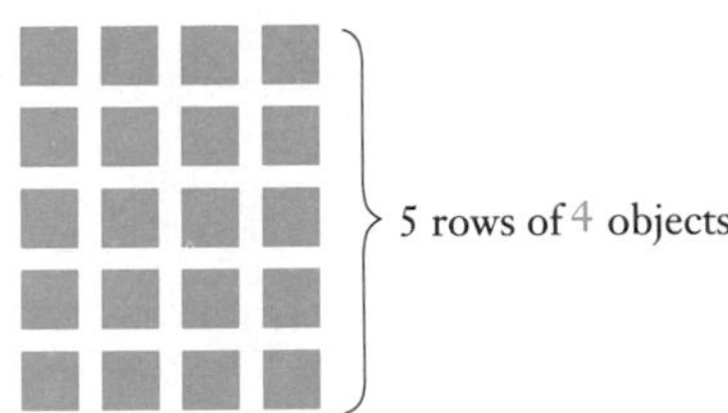

Since there are 4 objects in each of the 5 rows, we have

$$20 \div 5 = 4.$$

We also write a division such as $20 \div 5$ as

$$20/5 \quad \text{or} \quad \frac{20}{5}.$$

▶ **EXAMPLE 1** Write a division that corresponds to the following.

A parent gives \$24 to 3 children, giving the same amount to each. How much does each child get?

We think of an array with 3 rows. Each row will go to a child. How many dollars will be in each row?

$24 \div 3 = 8$ ◀

OBJECTIVES

After finishing Section 1.6, you should be able to:

- a Write a division that corresponds to a situation.
- b Given a division sentence, write a related multiplication sentence; and given a multiplication sentence, write two related division sentences.
- c Divide whole numbers.

FOR EXTRA HELP

 Tape 1F

 Tape 2A

 MAC: 1 IBM: 1

Write a division that corresponds to the situation. You need not carry out the division.

1. There are 112 students in a band and they are marching with 14 in each row. How many rows are there? $112 \div 14 = \square$

2. A marching band is in a rectangular array. There are 112 students in the band, and they are marching in 8 rows. How many students are there in each row? $112 \div 8 = \square$

ANSWERS ON PAGE A-2

▶ **EXAMPLE 2** Write a division that corresponds to this situation. You need not carry out the division.

How many radios at \$45 each can be purchased for \$495?

We think of an array with \$45 in each row. The money in each row will buy a radio. How many rows will there be?

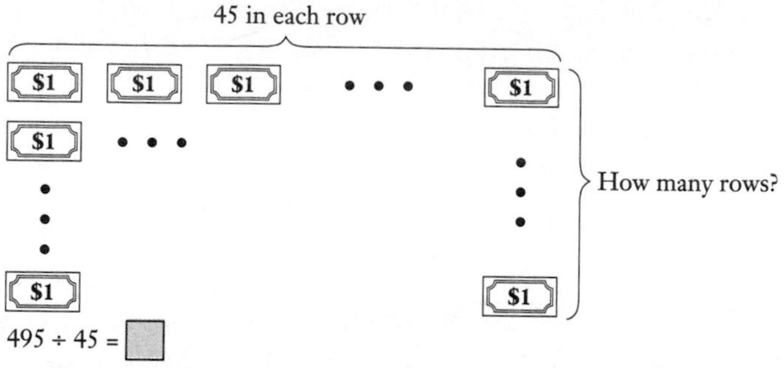

$495 \div 45 = \square$ ◀

Whenever we have a rectangular array, we know the following:

The total number) ÷ (The number of rows) = The number in each row).

Also:

(The total number) ÷ (The number in each row) = (The number of rows).

DO EXERCISES 1 AND 2.

b Related Sentences

By looking at rectangular arrays, we can see how multiplication and division are related. The following array shows that $4 \cdot 5 = 20$.

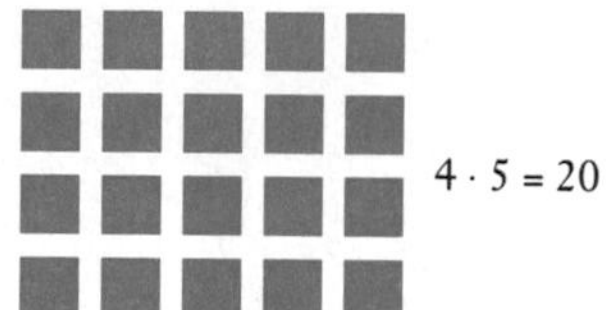

The array also shows the following:

$$20 \div 5 = 4 \text{ and } 20 \div 4 = 5.$$

Division is actually defined in terms of multiplication. For example, $20 \div 5$ is defined to be the number which when multiplied by 5 gives 20. Thus, for every division sentence, there is a related multiplication sentence.

$20 \div 5 = 4$ **Division sentence**

$20 = 4 \cdot 5$ **Related multiplication sentence**

To get the related multiplication sentence, we move the 5 to the other side and then write a multiplication.

▶ **EXAMPLE 3** Write a related multiplication sentence: $12 \div 6 = 2$.

We have

$$12 \div 6 = 2$$ This number moves to the right.

The related multiplication sentence is $12 = 2 \cdot 6$.

By the commutative law of multiplication, there is also another multiplication sentence: $12 = 6 \cdot 2$. ◀

DO EXERCISES 3 AND 4.

For every multiplication sentence, we can write related divisions, as we can see from the preceding array. We move one of the factors to the opposite side and then write a division.

▶ **EXAMPLE 4** Write two related division sentences: $7 \cdot 8 = 56$.

We move a factor to the other side and then write a division:

$$7 \cdot 8 = 56 \qquad 7 \cdot 8 = 56$$

$$7 = 56 \div 8 \qquad 8 = 56 \div 7$$ ◀

DO EXERCISES 5 AND 6.

With multiplication and division, we use the following words.

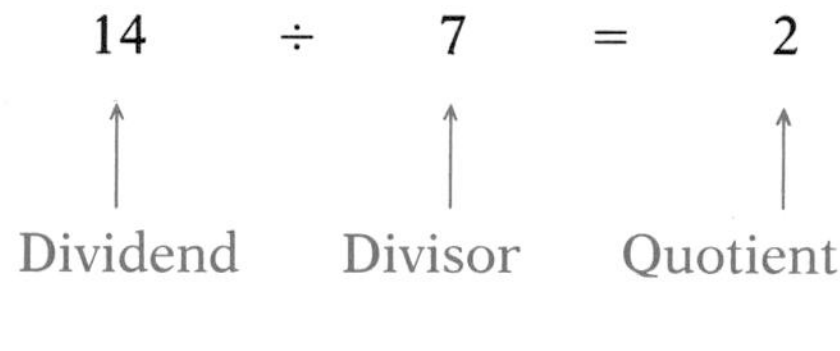

14 = 7 · 2

Product Factor Factor

C Division of Whole Numbers

Multiplication can be thought of as repeated addition. Division can be thought of as repeated subtraction. Compare.

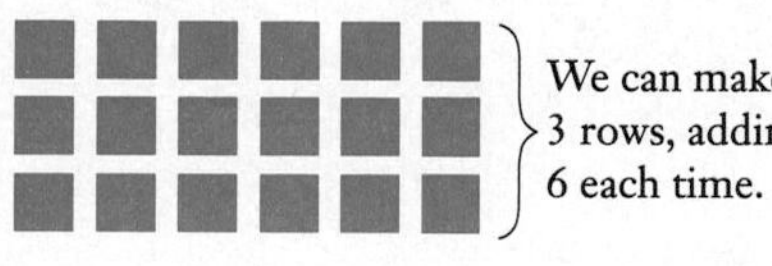

We can make 3 rows, adding 6 each time.

If we take away 6 objects at a time, we can do so 3 times.

$$18 = 6 + 6 + 6$$
$$= 3 \cdot 6$$

$$18 - 6 - 6 - 6 = 0$$

3 times

$$18 \div 6 = 3$$

Write a related multiplication sentence.

3. $15 \div 3 = 5$ $15 = 5 \cdot 3$

4. $72 \div 8 = 9$ $72 = 9 \cdot 8$

Write two related division sentences.

5. $6 \cdot 2 = 12$

$6 = 12 \div 2$; $2 = 12 \div 6$

6. $7 \cdot 6 = 42$

$6 = 42 \div 7$; $7 = 42 \div 6$

ANSWERS ON PAGE A-2

Divide by repeated subtraction. Then check.

7. $54 \div 9$

6; $9 \cdot 6 = 54$

8. $61 \div 9$

6 R 7; $6 \cdot 9 = 54$, $54 + 7 = 61$

9. $53 \div 12$

4 R 5; $4 \cdot 12 = 48$, $48 + 5 = 53$

10. $157 \div 24$

6 R 13; $24 \cdot 6 = 144$, $144 + 13 = 157$

ANSWERS ON PAGE A-2

To divide by repeated subtraction, we keep track of the number of times we subtract.

▶ **EXAMPLE 5** Divide by repeated subtraction: $20 \div 4$.

$$\begin{array}{r} 20 \\ -\;4 \\ \hline 16 \\ -\;4 \\ \hline 12 \\ -\;4 \\ \hline 8 \\ -\;4 \\ \hline 4 \\ -\;4 \\ \hline 0 \end{array}$$

We subtracted 5 times, so $20 \div 4 = 5$. ◀

▶ **EXAMPLE 6** Divide by repeated subtraction: $23 \div 5$.

$$\begin{array}{r} 23 \\ -\;5 \\ \hline 18 \\ -\;5 \\ \hline 13 \\ -\;5 \\ \hline 8 \\ -\;5 \\ \hline 3 \end{array}$$

We subtracted 4 times.

3 → We have 3 left. This number is called the *remainder*. ◀

We write

$$23 \div 5 = 4 \text{ R } 3$$

Dividend (23), Divisor (5), Quotient (4), Remainder (3)

Checking divisions. To check a division, we multiply. Suppose we divide 98 by 2 and get 49:

$$98 \div 2 = 49.$$

To check, we think of the related sentence $49 \cdot 2 = \square$. We multiply 49 by 2 and see if we get 98.

If there is a remainder, we add it after multiplying.

▶ **EXAMPLE 7** Check the division in Example 6.

We found that $23 \div 5 = 4$ R 3. To check, we multiply 5 by 4. This gives us 20. Then we add 3 to get 23. The dividend is 23, so the answer checks. ◀

DO EXERCISES 7–10.

When we use the general division process, we are doing repeated subtraction, even though we are going about it in a different way.

To divide, we start from the digit of highest place value in the dividend and work down to the lowest through the remainders. At each step we ask if there are multiples of the divisor in the quotient.

▶ **EXAMPLE 8** Divide and check: 3642 ÷ 5.

```
     ?
5 ) 3 6 4 2
```

1. We start with the thousands digit in the dividend. Are there any thousands in the quotient? No; 5 · 1000 = 5000, and 5000 is larger than 3000.

```
       7
5 ) 3 6 4 2
    3 5 0 0
    -------
      1 4 2
```

2. Now we go to the hundreds place in the dividend. Are there any hundreds in the quotient? Think of the dividend as 36 hundreds. Estimate 7 hundreds. Write 7 in the hundreds place, multiply 700 by 5, write the answer below 3642, and subtract.

```
       7 3   ← No!
5 ) 3 6 4 2
    3 5 0 0
    -------
      1 4 2   Can't
      1 5 0 ← subtract
```

3. a) We go to the tens place of the first remainder. Are there any tens in the tens place of the quotient? To answer the question, think of the first remainder as 14 tens. Estimate 3 tens. When we multiply, we get 150, which is too large.
b) We lower our estimate to 2 tens. Write 2 in the tens place, multiply 20 by 5, and subtract.

```
       7 2
5 ) 3 6 4 2
    3 5 0 0
    -------
      1 4 2
      1 0 0
      -----
        4 2
```

4. We go to the ones place of the second remainder. Are there any ones in the ones place of the quotient? To answer the question, think of the second remainder as 42 ones. Estimate 8 ones. Write 8 in the ones place, multiply 8 by 5, and subtract.

```
       7 2 8
5 ) 3 6 4 2
    3 5 0 0
    -------
      1 4 2
      1 0 0
      -----
        4 2
        4 0
        ---
          2
```

You may have learned to divide like this, not writing the extra zeros. You may omit them if desired. →

```
     7 2 8
5)3 6 4 2
  3 5
  ---
    1 4
    1 0
    ---
      4 2
      4 0
      ---
        2
```

The answer is 728 R 2. To check, we multiply 728 by 5. This gives us 3640. Then we add 2 to get 3642. The dividend is 3642, so the answer checks. ◀

DO EXERCISES 11–13.

Divide and check.

11. $4\overline{)239}$ 59 R 3

12. $6\overline{)8855}$ 1475 R 5

13. $5\overline{)5075}$ 1015

ANSWERS ON PAGE A-2

Divide.

14. $45\overline{)6030}$ 134

15. $52\overline{)3288}$ 63 R 12

ANSWERS ON PAGE A-2

Sometimes rounding the divisor helps us find estimates.

▶ **EXAMPLE 9** Divide: $8904 \div 42$.

We mentally round 42 to 40.

```
       2
42 ) 8904   ← Think: 89 hundreds ÷ 40.
     8400     Estimate 2 hundreds.
      504
```

```
       21
42 ) 8904
     8400
      504   ← Think: 50 tens ÷ 40.
      420     Estimate 1 ten.
       84
```

```
       212
42 ) 8904
     8400
      504
      420
       84   ← Think: 84 ones ÷ 40.
       84     Estimate 2 ones.
        0
```

CAUTION! Be careful to keep the digits lined up correctly.

The answer is 212. *Remember:* if after estimating and multiplying you get a number that is larger than the divisor, you cannot subtract, so lower your estimate. ◀

DO EXERCISES 14 AND 15.

Zeros in Quotients

▶ **EXAMPLE 10** Divide: 6341 ÷ 7.

```
       9
  7 ) 6 3 4 1   ← Think: 63 hundreds ÷ 7.
      6 3 0 0     Estimate 9 hundreds.
          4 1
```

```
       9 0
  7 ) 6 3 4 1
      6 3 0 0
          4 1   ← Think: 4 tens ÷ 7. There are no tens
                  in the quotient (other than the tens in 900).
                  We write a 0 to show this.
```

```
       9 0 5
  7 ) 6 3 4 1
      6 3 0 0
          4 1   ← Think: 41 ones ÷ 7.
          3 5     Estimate 5 ones.
            6
```

The answer is 905 R 6. ◀

DO EXERCISES 16 AND 17.

▶ **EXAMPLE 11** Divide: 8889 ÷ 37.

We round 37 to 40.

```
         2
  3 7 ) 8 8 8 9 ← Think: 37 ≈ 40; 88 hundreds ÷ 40.
        7 4 0 0   Estimate 2 hundreds.
        1 4 8 9
```

```
         2 4
  3 7 ) 8 8 8 9
        7 4 0 0
        1 4 8 9 ← Think: 148 tens ÷ 40.
        1 4 8 0   Estimate 4 tens.
              9
```

```
         2 4 0
  3 7 ) 8 8 8 9
        7 4 0 0
        1 4 8 9
        1 4 8 0
              9 ← Think: 9 ones ÷ 40.
                  There are no ones in the quotient.
```

The answer is 240 R 9. ◀

DO EXERCISES 18 AND 19.

Divide.

16. 6)4 8 4 6 807 R 4

17. 7)7 6 1 6 1088

Divide.

18. 2 7)9 7 2 4 360 R 4

19. 5 6)4 4,8 4 7 800 R 47

To the instructor and the student: This section presented a review of division of whole numbers. Students who are successful should go on to Section 1.7. Those who have trouble should study developmental units D.1, D.2, and D.3, and then repeat Section 1.6.

ANSWERS ON PAGE A-2

❖ SIDELIGHTS

Careers and Their Uses of Mathematics

Students typically ask the question "Why do we have to study mathematics?" This is a question with a complex set of answers. Certainly, one answer is that you will use this mathematics in the next course. While it is a correct answer, it sometimes frustrates students, because this answer can be given in the next mathematics course, and the next one, and so on. Sometimes an answer can be given by applications like those you have seen or will see in this book. Another answer is that you are living in a society in which mathematics becomes more and more critical with each passing day. Evidence of this was provided recently by a nationwide symposium sponsored by the National Research Council's Mathematical Sciences Education Board. Results showed that "Other than demographic factors, the *strongest* predictor of earnings nine years after high school is the number of mathematics courses taken." This is a significant testimony to the need for you to take as many mathematics courses as possible.

We try to provide other answers to "Why do we have to study mathematics?" in what follows. We have listed several occupations that are attractive and popular to students. Below each occupation are listed various kinds of mathematics that are useful in that occupation.

Doctor	Lawyer
Equations	Equations
Percent notation	Percent notation
Graphing	Graphing
Statistics	Probability
Geometry	Statistics
Measurement	Ratio and proportion
Estimation	Area and volume
Exponents	Negative numbers
Logic	Formulas
	Calculator skills

Pilot	Firefighter
Equations	Percent notation
Percent notation	Graphing
Graphing	Estimation
Trigonometry	Formulas
Angles and geometry	Angles and geometry
Calculator skills	Probability
Computer skills	Statistics
Ratio and proportion	Area and geometry
Vectors	Square roots
	Exponents
	Pythagorean theorem

Accountant and Business	Travel Agent
Computer skills	Whole-number skills
Calculator skills	Fraction/decimal skills
Equations	Estimation
Systems of equations	Percent notation
Formulas	Equations
Probability	Calculator skills
Statistics	Computer skills
Ratio and proportion	
Percent notation	
Estimation	

Librarian	Machinist
Whole-number skills	Whole-number skills
Fraction/decimal skills	Fraction/decimal skills
Estimation	Estimation
Percent notation	Percent notation
Ratio and proportion	Length, area, volume, and perimeter
Area and perimeter	Angle measures
Formulas	Geometry
Calculator skills	Pythagorean theorem
Computer skills	Square roots
	Equations
	Formulas
	Graphing
	Calculator skills
	Computer skills

Nurse	Police Officer
Whole-number skills	Whole-number skills
Fraction/decimal skills	Fraction/decimal skills
Estimation	Estimation
Percent notation	Percent notation
Ratio and proportion	Ratio and proportion
Estimation	Geometry
Equations	Negative numbers
English/Metric measurement	Probability
Probability	Statistics
Statistics	Calculator skills
Formulas	
Exponents and scientific notation	
Calculator skills	
Computer skills	

NAME SECTION DATE

EXERCISE SET 1.6

a Write a division that corresponds to each situation. You need not carry out the division.

1. A candy factory made 176 lb (pounds) of chocolates. They put them in 4-lb boxes. How many boxes did they fill?

2. A beverage company put 222 bottles of soda into 6-bottle cartons. How many cartons did they fill?

3. A family divides an inheritance of \$184,000 among its children, giving each of them \$23,000. How many children are there?

$184,000 ÷ $23,000 = ■

4. A lab technician pours 455 mL of acid into 5 beakers, putting the same amount in each. How much acid is in each beaker?

b Write a related multiplication sentence.

5. $24 \div 8 = 3$ **6.** $72 \div 9 = 8$ **7.** $22 \div 22 = 1$ **8.** $32 \div 1 = 32$

9. $54 \div 6 = 9$ **10.** $72 \div 8 = 9$ **11.** $37 \div 1 = 37$ **12.** $28 \div 28 = 1$

Write two related division sentences.

13. $9 \times 5 = 45$

$9 = 45 \div 5$; $5 = 45 \div 9$

14. $2 \cdot 7 = 14$

$2 = 14 \div 7$; $7 = 14 \div 2$

15. $37 \cdot 1 = 37$

$37 = 37 \div 1$; $1 = 37 \div 37$

16. $4 \cdot 12 = 48$

$4 = 48 \div 12$; $12 = 48 \div 4$

17. $8 \times 8 = 64$

18. $9 \cdot 7 = 63$

$9 = 63 \div 7$; $7 = 63 \div 9$

19. $11 \cdot 6 = 66$

$11 = 66 \div 6$; $6 = 66 \div 11$

20. $1 \cdot 43 = 43$

$1 = 43 \div 43$; $43 = 43 \div 1$

c Divide.

21. $277 \div 5$ **22.** $699 \div 3$ **23.** $864 \div 8$ **24.** $869 \div 8$

25. $4\overline{)1228}$ **26.** $3\overline{)2124}$ **27.** $7\overline{)6345}$ **28.** $9\overline{)9110}$

29. $4\overline{)297}$ **30.** $2\overline{)389}$ **31.** $8\overline{)738}$ **32.** $6\overline{)881}$

33. $5\overline{)8515}$ **34.** $3\overline{)6027}$ **35.** $9\overline{)8888}$ **36.** $8\overline{)4139}$

37. $70\overline{)3692}$ **38.** $20\overline{)5798}$ **39.** $30\overline{)875}$ **40.** $40\overline{)987}$

ANSWERS

1. 176 ÷ 4 = ■
2. 222 ÷ 6 = ■
3.
4. 455 ÷ 5 = ■
5. $24 = 3 \cdot 8$
6. $72 = 8 \cdot 9$
7. $22 = 1 \cdot 22$
8. $32 = 32 \cdot 1$
9. $54 = 9 \cdot 6$
10. $72 = 9 \cdot 8$
11. $37 = 37 \cdot 1$
12. $28 = 1 \cdot 28$
13.
14.
15.
16.
17. $8 = 64 \div 8$
18.
19.
20.
21. 55 R 2
22. 233
23. 108
24. 108 R 5
25. 307
26. 708
27. 906 R 3
28. 1012 R 2
29. 74 R 1
30. 194 R 1
31. 92 R 2
32. 146 R 5
33. 1703
34. 2009
35. 987 R 5
36. 517 R 3
37. 52 R 52
38. 289 R 18
39. 29 R 5
40. 24 R 27

ANSWERS

41. 40 R 12
42. 40 R 22
43. 90 R 22
44. 50 R 29
45. 29
46. 55 R 2
47. 105 R 3
48. 107
49. 507 R 1
50. 808 R 1
51. 1007 R 1
52. 1010 R 4
53. 23
54. 301 R 18
55. 107 R 1
56. 102 R 3
57. 370
58. 210 R 3
59. 609 R 15
60. 803 R 21
61. 304
62. 984
63. 3508 R 219
64. 2904 R 264
65. 8070
66. 7002
67.
68. >
69.

41. $852 \div 21$

42. $942 \div 23$

43. $85\overline{)7672}$

44. $54\overline{)2729}$

45. $111\overline{)3219}$

46. $102\overline{)5612}$

47. $8\overline{)843}$

48. $7\overline{)749}$

49. $6\overline{)3043}$

50. $9\overline{)7273}$

51. $5\overline{)5036}$

52. $7\overline{)7074}$

53. $46\overline{)1058}$

54. $24\overline{)7242}$

55. $32\overline{)3425}$

56. $48\overline{)4899}$

57. $24\overline{)8880}$

58. $36\overline{)7563}$

59. $28\overline{)17{,}067}$

60. $36\overline{)28{,}929}$

61. $80\overline{)24{,}320}$

62. $90\overline{)88{,}560}$

63. $285\overline{)999{,}999}$

64. $306\overline{)888{,}888}$

65. $456\overline{)3{,}679{,}920}$

66. $803\overline{)5{,}622{,}606}$

SKILL MAINTENANCE

67. Write expanded notation for 7882.
7 thousands + 8 hundreds + 8 tens + 2 ones

68. Use $<$ or $>$ for ▢ to write a true sentence:

888 ▢ 788.

SYNTHESIS

69. A person addicted to smoking but without any cash figures out that there is enough tobacco in four cigarette butts to make one new cigarette. How many cigarettes can the person make from 29 cigarette butts? How many butts are left over? 9, with 2 left over (The butts from the first seven cigarettes are recycled.)

1.7 Solving Equations

a Solutions

Let's find a number that we can put in the blank to make this sentence true:

$$9 = 3 + \square\ .$$

We are asking "9 is 3 plus what number?" The answer is 6.

$$9 = 3 + \boxed{6}$$

DO EXERCISES 1 AND 2.

A sentence with = is called an **equation.** A **solution** of an equation is a number that makes the sentence true. Thus, 6 is a solution of

$9 = 3 + \square$ because $9 = 3 + \boxed{6}$ is true.

But 7 is not a solution of

$9 = 3 + \square$ because $9 = 3 + \boxed{7}$ is false.

DO EXERCISES 3 AND 4.

We can use a letter instead of a blank. For example,

$$x + 8 = 11.$$

We call x a **variable** because it can represent any number.

> **A *solution* is a replacement for the letter that makes the equation true. When we find the solutions, we say that we have *solved* the equation.**

▶ **EXAMPLE 1** Solve $x + 12 = 27$ by trial.

We replace x by several numbers.

If we replace x by 13, we get a false equation: $13 + 12 = 27$.
If we replace x by 14, we get a false equation: $14 + 12 = 27$.
If we replace x by 15, we get a true equation: $15 + 12 = 27$.

No other replacement makes the equation true, so the solution is 15. ◀

▶ **EXAMPLES** Solve.

2. $7 + n = 22$
(7 plus what number is 22?)
The solution is 15.
3. $8 \cdot 23 = y$
(8 times 23 is what?)
The solution is 184. ◀

Note, as in Example 3, that when the letter is alone on one side of the equation, the other side shows us what calculations to do in order to find the solution.

DO EXERCISES 5–8.

OBJECTIVES

After finishing Section 1.7, you should be able to:

a Solve simple equations by trial.

b Solve equations like $t + 28 = 54$, $28 \cdot x = 168$, and $98 \div 2 = y$.

FOR EXTRA HELP

Tape 2A

Tape 2B

MAC: 1
IBM: 1

Find a number that makes the sentence true.

1. $8 = 1 + \square$ 7

2. $\square + 2 = 7$ 5

3. Determine whether 7 is a solution of $\square + 5 = 9$.
No

4. Determine whether 4 is a solution of $\square + 5 = 9$.
Yes

Solve by trial.

5. $n + 3 = 8$ 5

6. $x - 2 = 8$ 10

7. $45 \div 9 = y$ 5

8. $10 + t = 32$ 22

ANSWERS ON PAGE A-2

Solve.

9. $346 \times 65 = y$ 22,490

10. $x = 2347 + 6675$ 9022

11. $4560 \div 8 = t$ 570

12. $x = 6007 - 2346$ 3661

Solve.

13. $x + 9 = 17$ 8

14. $77 = m + 32$ 45

ANSWERS ON PAGE A-2

b Solving Equations

We now begin to develop some more efficient ways to solve certain equations. When an equation has a variable alone on one side, it is easy to see the solution or to compute it. For example, the solution of

$$x = 12$$

is 12.

When a calculation is on one side and the variable is alone on the other, we can find the solution by carrying out the calculation.

▶ **EXAMPLE 4** Solve: $x = 245 \times 34$.

To solve the equation, we carry out the calculation.

$$\begin{array}{r} 2\,4\,5 \\ \times \quad 3\,4 \\ \hline 9\,8\,0 \\ 7\,3\,5\,0 \\ \hline 8\,3\,3\,0 \end{array}$$

The solution is 8330. ◀

DO EXERCISES 9–12.

If we can get an equation in a form with the letter alone on one side, we can "see" the solution.

Look at

$$x + 12 = 27.$$

We can get x alone by writing a related sentence:

$x = 27 - 12$ **12 gets subtracted to find the related subtraction.**

$x = 15.$ **Doing the subtraction**

It is useful to think of this as "subtracting 12 *on both sides.*" Thus,

$x + 12 - 12 = 27 - 12$ **Subtracting 12 on both sides**

$x + 0 = 15$ **Carrying out the subtraction**

$x = 15.$ $x + 0 = x$

> **To solve $x + a = b$, subtract a on both sides.**

▶ **EXAMPLE 5** Solve: $t + 28 = 54$.

We have

$t + 28 = 54$

$t + 28 - 28 = 54 - 28$ **Subtracting 28 on both sides**

$t + 0 = 26$

$t = 26.$

The solution is 26. ◀

DO EXERCISES 13 AND 14.

▶ **EXAMPLE 6** Solve: $182 = 65 + n$.

We have

$$182 = 65 + n$$
$$182 - 65 = 65 + n - 65 \quad \text{Subtracting 65 on both sides}$$
$$117 = 0 + n \quad \text{65 plus } n \text{ minus 65 is } 0 + n$$
$$117 = n$$

The solution is 117. ◀

DO EXERCISE 15.

▶ **EXAMPLE 7** Solve: $7381 + x = 8067$.

We have

$$7381 + x = 8067$$
$$7381 + x - 7381 = 8067 - 7381 \quad \text{Subtracting 7381 on both sides}$$
$$x = 686.$$

The solution is 686. ◀

DO EXERCISES 16 AND 17.

We now learn to solve equations like $8 \cdot n = 96$. Look at

$$8 \cdot n = 96.$$

We can get n alone by writing a related division sentence:

$$n = 96 \div 8 = \frac{96}{8} \quad \text{We move 8 to the other side and write a division.}$$
$$n = 12. \quad \text{Doing the division}$$

Note that $n = 12$ is easier to solve than $8 \cdot n = 96$. This is because we see easily that if we replace n on the left side by 12, we get a true sentence: $12 = 12$. The solution of $n = 12$ is 12, which is also the solution of $8 \cdot n = 96$.

It is useful to think of the preceding as "dividing by 8 *on both sides.*" Thus,

$$\frac{8 \cdot n}{8} = \frac{96}{8} \quad \text{Dividing by 8 on both sides}$$
$$n = 12. \quad \text{8 times } n \text{ divided by 8 is } n.$$

To solve $a \cdot x = b$, divide by a on both sides.

▶ **EXAMPLE 8** Solve: $10 \cdot x = 240$.

We have

$$10 \cdot x = 240$$
$$\frac{10 \cdot x}{10} = \frac{240}{10} \quad \text{Dividing by 10 on both sides}$$
$$x = 24.$$

The solution is 24. ◀

DO EXERCISES 18 AND 19.

15. Solve: $155 = t + 78$. 77

Solve.

16. $4566 + x = 7877$ 3311

17. $8172 = h + 2058$ 6114

Solve.

18. $8 \cdot x = 64$ 8

19. $144 = 9 \cdot n$ 16

ANSWERS ON PAGE A-2

20. Solve: $5152 = 8 \cdot t$. 644

21. Solve: $18 \cdot y = 1728$. 96

22. Solve: $n \cdot 48 = 4512$. 94

ANSWERS ON PAGE A-2

▶ **EXAMPLE 9** Solve: $5202 = 9 \cdot t$.

We have

$$5202 = 9 \cdot t$$

$$\frac{5202}{9} = \frac{9 \cdot t}{9} \quad \text{Dividing by 9 on both sides}$$

$$578 = t.$$

The solution is 578. ◀

DO EXERCISE 20.

▶ **EXAMPLE 10** Solve: $14 \cdot y = 1092$.

We have

$$14 \cdot y = 1092$$

$$\frac{14 \cdot y}{14} = \frac{1092}{14} \quad \text{Dividing by 14 on both sides}$$

$$y = 78.$$

The solution is 78. ◀

DO EXERCISE 21.

▶ **EXAMPLE 11** Solve: $n \cdot 56 = 4648$.

We have

$$n \cdot 56 = 4648$$

$$\frac{n \cdot 56}{56} = \frac{4648}{56} \quad \text{Dividing by 56 on both sides}$$

$$n = 83.$$

The solution is 83. ◀

DO EXERCISE 22.

NAME SECTION DATE

EXERCISE SET 1.7

a Solve by trial.

1. $x + 0 = 14$ **2.** $x - 7 = 18$ **3.** $y \cdot 17 = 0$ **4.** $56 \div m = 7$

b Solve.

5. $13 + x = 42$ **6.** $15 + t = 22$ **7.** $12 = 12 + m$

8. $16 = t + 16$ **9.** $3 \cdot x = 24$ **10.** $6 \cdot x = 42$

11. $112 = n \cdot 8$ **12.** $162 = 9 \cdot m$ **13.** $45 \times 23 = x$

14. $23 \times 78 = y$ **15.** $t = 125 \div 5$ **16.** $w = 256 \div 16$

17. $p = 908 - 458$ **18.** $9007 - 5667 = m$ **19.** $x = 12{,}345 + 78{,}555$

20. $5678 + 9034 = t$ **21.** $3 \cdot m = 96$ **22.** $4 \cdot y = 96$

23. $715 = 5 \cdot z$ **24.** $741 = 3 \cdot t$ **25.** $10 + x = 89$

26. $20 + x = 57$ **27.** $61 = 16 + y$ **28.** $53 = 17 + w$

29. $6 \cdot p = 1944$ **30.** $4 \cdot w = 3404$ **31.** $5 \cdot x = 3715$

32. $9 \cdot x = 1269$ **33.** $47 + n = 84$ **34.** $56 + p = 92$

ANSWERS

1. 14
2. 25
3. 0
4. 8
5. 29
6. 7
7. 0
8. 0
9. 8
10. 7
11. 14
12. 18
13. 1035
14. 1794
15. 25
16. 16
17. 450
18. 3340
19. 90,900
20. 14,712
21. 32
22. 24
23. 143
24. 247
25. 79
26. 37
27. 45
28. 36
29. 324
30. 851
31. 743
32. 141
33. 37
34. 36

ANSWERS

35. 66
36. 66
37. 15
38. 55
39. 48
40. 49
41. 175
42. 112
43. 335
44. 369
45. 104
46. 320
47. 45
48. 75
49. 4056
50. 959
51. 17,603
52. 959
53. 18,252
54. 23
55. 205
56. 95
57. 55
58. $7 = 15 - 8$; $8 = 15 - 7$
59. $6 = 48 \div 8$; $8 = 48 \div 6$
60. $<$
61. $>$

35. $x + 78 = 144$

36. $z + 67 = 133$

37. $165 = 11 \cdot n$

38. $660 = 12 \cdot n$

39. $624 = t \cdot 13$

40. $784 = y \cdot 16$

41. $x + 214 = 389$

42. $x + 221 = 333$

43. $567 + x = 902$

44. $438 + x = 807$

45. $18 \cdot x = 1872$

46. $19 \cdot x = 6080$

47. $40 \cdot x = 1800$

48. $20 \cdot x = 1500$

49. $2344 + y = 6400$

50. $9281 = 8322 + t$

51. $8322 + 9281 = x$

52. $9281 - 8322 = y$

53. $234 \times 78 = y$

54. $10{,}534 \div 458 = q$

55. $58 \cdot m = 11{,}890$

56. $233 \cdot x = 22{,}135$

57. $x \cdot 198 = 10{,}890$

SKILL MAINTENANCE

58. Write two related subtraction sentences: $7 + 8 = 15$.

59. Write two related division sentences: $6 \cdot 8 = 48$.

Use $>$ or $<$ for ▒ to write a true sentence.

60. 123 ▒ 789

61. 342 ▒ 339

1.8 Solving Problems

OBJECTIVE

After finishing Section 1.8, you should be able to:

a Solve problems involving addition, subtraction, multiplication, or division of whole numbers.

FOR EXTRA HELP

Tape 2B

Tape 2B

MAC: 1
IBM: 1

a To solve a problem using the operations on the whole numbers, we first look at the situation. We try to translate the problem to an equation. Then we solve the equation. We check to see if the solution of the equation is a solution to the original problem. We are using the following five-step strategy.

> Problem Solving Tips
>
> 1. ***Familiarize* yourself with the situation. If it is described in words, as in a textbook, *read carefully*. In any case, think about the situation. Draw a picture whenever it makes sense to do so. Choose a letter, or *variable*, to represent the unknown quantity to be solved for.**
> 2. ***Translate* the problem to an equation.**
> 3. ***Solve* the equation.**
> 4. ***Check* the answer in the original wording of the problem.**
> 5. ***State* the answer to the problem clearly with appropriate units.**

▶ **EXAMPLE 1** There are 87 boxcars on a freight train. A train behind it has 112 boxcars, and a third train has 98 boxcars. The trains are put together to make one long train. How many boxcars are there on the long train?

1. *Familiarize.* We can make a drawing or at least visualize the situation.

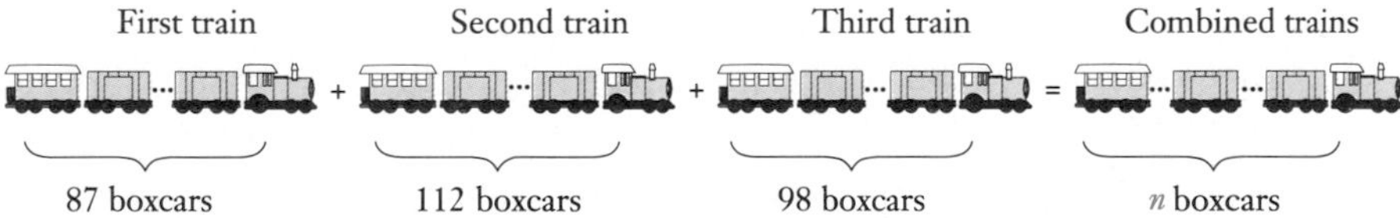

Since we are combining objects, addition can be used. To define the unknown, we let n = the total number of boxcars on the train.

2. *Translate.* We translate to an equation by writing a number sentence that corresponds to the situation:

$$98 + 112 + 87 = n.$$

3. *Solve.* We solve the equation by carrying out the addition.

$$\begin{array}{r} {}^{11} \\ 98 \\ 112 \\ +\ 87 \\ \hline 297 \end{array}$$

Note that even if we did not make one long train, the total number of boxcars would still be 297.

Thus, $297 = n$, or $n = 297$.

4. *Check.* We check 297 boxcars in the original problem. There are many ways to check. We can repeat the calculation. (We leave this to the student.) We can also check the reasonableness of the answer. We would expect our answer to be larger than any of the separate trains, which it is. We can also find an estimated answer by rounding:

$$87 + 112 + 98 \approx 90 + 100 + 100 = 290 \approx 297.$$

1. In a tournament, a professional bowler rolled games of 212, 198, and 249. What was the total? 659

2. On a long four-day trip a family bought the following amounts of gasoline:

   23 gallons, 24 gallons,
   26 gallons, 25 gallons.

   How much gasoline did they buy in all? 98 gal

3. The area of the state of Kansas is 82,056 sq mi. The area of the state of Nebraska is 76,522 sq mi. What is the total area of the two states?

   158,578 sq mi

4. It takes 109 kilowatt-hours (kWh) to operate a record player for a year. It takes 440 kWh to operate a TV for a year. How much energy is needed for both?

   549 kWh

ANSWERS ON PAGE A-2

If we had gotten an estimate like 1290 or 850, we might be suspicious that our calculated answer is incorrect. Since our estimated answer is close to our calculation, we are further convinced that our answer checks.

5. *State.* The answer is that there are 297 boxcars altogether. ◀

DO EXERCISES 1 AND 2.

> In the real world, problems are not usually given in words. You must still become familiar with the situation before you can solve the problem.

▶ **EXAMPLE 2** The John Hancock Building in Chicago is 1107 ft tall. It has two 342-ft antennas on top. How far are the tops of the antennas from the ground?

1. *Familiarize.* We first make a drawing.

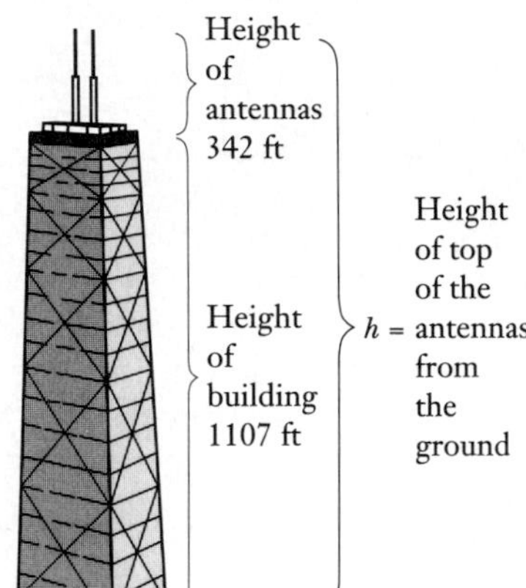

Since we are combining lengths, addition can be used. To define the unknown, we let h = the height of the top of the antennas from the ground.

2. *Translate.* We translate the problem to the following addition sentence:

$$1107 + 342 = h.$$

3. *Solve.* To solve the equation, we carry out the addition.

$$\begin{array}{r} 1\,1\,0\,7 \\ +\quad 3\,4\,2 \\ \hline 1\,4\,4\,9 \end{array}$$

Thus, $1449 = h$, or $h = 1449$.

4. *Check.* We check the height of 1449 ft in the original problem. We can repeat the calculation. We can also check the reasonableness of the answer. We would expect our answer to be larger than either of the heights, which it is. We can also find an estimated answer by rounding:

$$1107 + 342 \approx 1100 + 300 = 1400 \approx 1449.$$

The answer checks.

5. *State.* The height of the top of the antennas from the ground is 1449 ft. ◀

DO EXERCISES 3 AND 4.

▶ **EXAMPLE 3** A farm contains 2679 acres. If the owner sells 1884 acres, how many acres are left?

1. *Familiarize.* We first draw a picture or at least visualize the situation. We let A = the number of acres left.

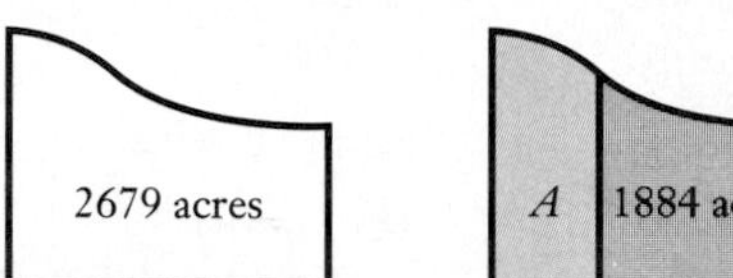

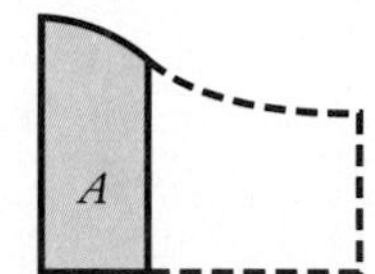

2. *Translate.* We see that this is a "take-away" situation. We translate to an equation.

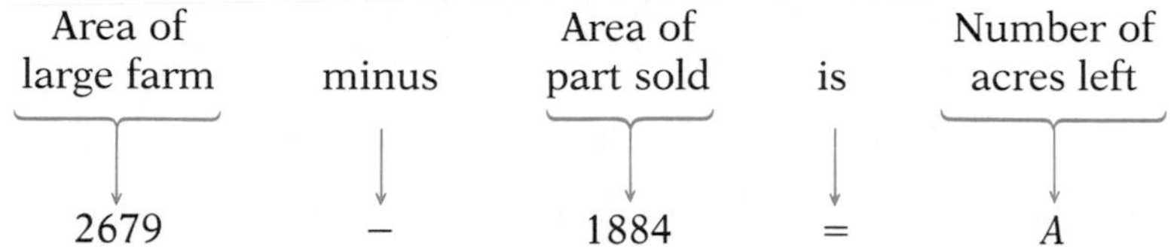

3. *Solve.* This sentence tells us what to do. We subtract.

$$\begin{array}{r} \overset{1}{\cancel{2}}\,\overset{\overset{15}{\cancel{5}}}{\cancel{6}}\,\overset{17}{\cancel{7}}\,9 \\ -\ 1\ 8\ 8\ 4 \\ \hline 7\ 9\ 5 \end{array}$$

Thus, $795 = A$, or $A = 795$.

4. *Check.* We check 795 acres. We can repeat the calculation. We note that the answer should be less than the original acreage, 2679 acres, which it is. We can add the answer, 795, to the number being subtracted, 1884: $1884 + 795 = 2679$. We can also estimate:

$$2679 - 1884 \approx 2700 - 1900 = 800 \approx 795.$$

The answer checks.

5. *State.* There are 795 acres left. ◀

DO EXERCISES 5 AND 6.

5. A person has \$948 in a checking account. A check is written for \$427. How much is left in the checking account? \$521

6. An oil company has 7890 gal of gasoline in a tank. It drains out 5630 gal into another tank. How much is left in the original tank? 2260 gal

ANSWERS ON PAGE A-2

▶ **EXAMPLE 4** It is 1154 mi from Indianapolis to Denver. A driver has traveled 685 mi of that distance. How much farther is it to Denver?

1. *Familiarize.* We first make a drawing or at least visualize the situation. We let x = the remaining distance to Denver.

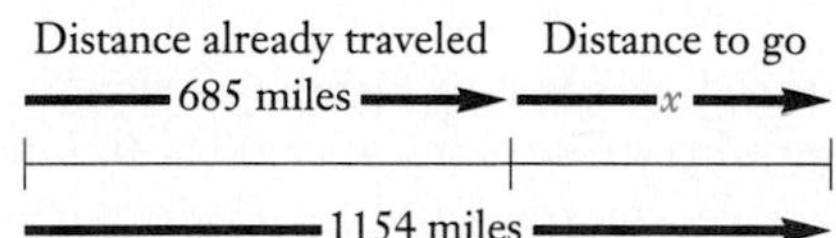

2. *Translate.* We see that this is a "how-much-more" situation. We translate to an equation.

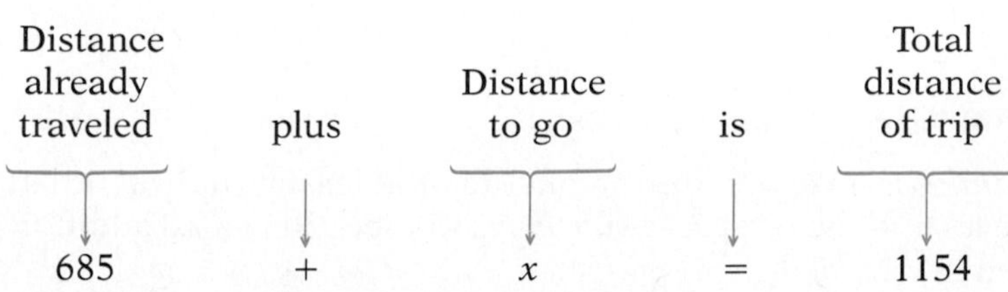

7. Gold has a melting point of 1063° C. Silver has a melting point of 960° C. How much higher is the melting point of gold? 103° C

3. *Solve.* We solve the equation.

$$685 + x = 1154$$

$$685 + x - 685 = 1154 - 685 \qquad \text{Subtracting 685 on both sides}$$

$$x = 469$$

$$\begin{array}{r} \scriptstyle 10\ 14 \\ \scriptstyle 0\ \ \cancel{0}\ \ \cancel{4}\ 14 \\ \cancel{1}\ \cancel{1}\ \cancel{5}\ \cancel{4} \\ -\quad 6\ 8\ 5 \\ \hline 4\ 6\ 9 \end{array}$$

4. *Check.* We check 469 mi in the original problem. This number should be less than the total distance, 1154 mi, which it is. We can repeat the calculation. We add the result, 469, to the number being subtracted, 685: $685 + 469 = 1154$. We can also estimate as follows:

$$1154 - 685 \approx 1200 - 700 = 500 \approx 469.$$

This can be handy if you were using a calculator. The answer checks.

5. *State.* It is 469 mi to Denver. ◀

DO EXERCISES 7 AND 8.

8. Annual income in Washington, D.C., is $16,845 per person. In New Jersey it is $15,285. How much greater is the income in Washington, D.C.? $1560

▶ **EXAMPLE 5** A ream of paper contains 500 sheets. How many sheets are in 9 reams?

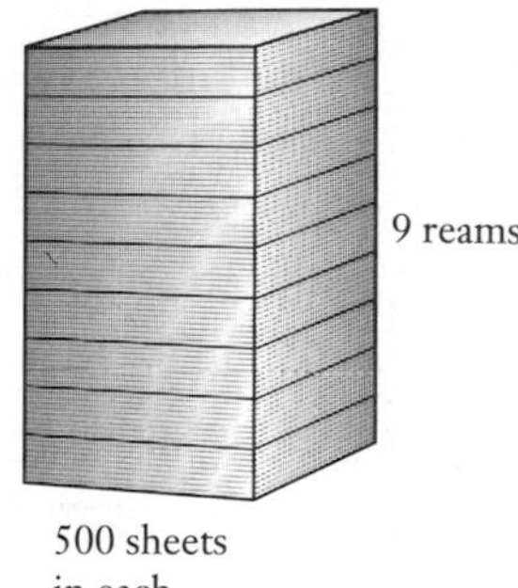

1. *Familiarize.* We first draw a picture or at least visualize the situation. We can think of this situation as a stack of reams. We let n = the total number of sheets in 9 reams.

2. *Translate.* Then we translate and solve as follows.

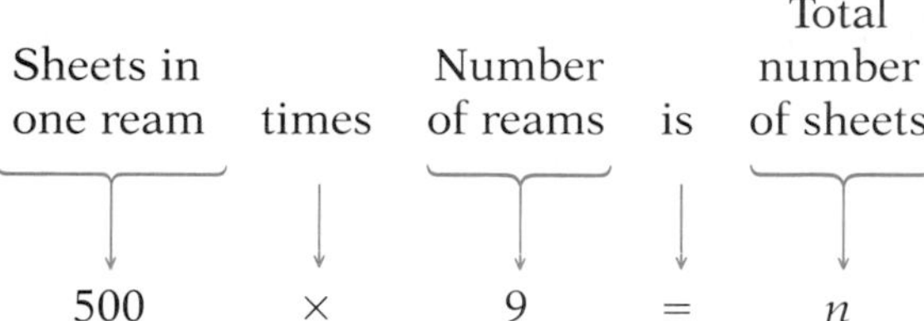

9. A certain type of shelf can hold 40 books. How many books can 70 shelves hold? 2800

3. *Solve.* To solve the equation, we multiply: $500 \times 9 = 4500$. Thus, $4500 = n$, or $n = 4500$.

4. *Check.* An estimated answer is almost a repeated calculation. Certainly our answer should be larger than 500, since we are multiplying, so the answer seems reasonable. The answer checks.

5. *State.* There are 4500 sheets in 9 reams. ◀

DO EXERCISE 9.

▶ **EXAMPLE 6** What is the cost of 5 television sets at $145 each?

1. *Familiarize.* We first draw a picture or at least visualize the situation. We let n = the total cost of 5 television sets. Repeated addition works well here.

ANSWERS ON PAGE A-2

2. *Translate.* We translate to an equation and solve.

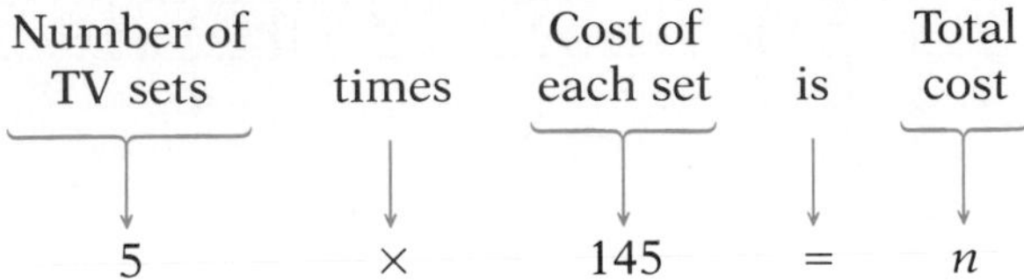

$$5 \times 145 = n$$

3. *Solve.* This sentence tells us what to do. We multiply.

$$\begin{array}{r} {\scriptstyle 2\,2} \\ 1\,4\,5 \\ \times \quad 5 \\ \hline 7\,2\,5 \end{array} \qquad \text{Thus, } n = 725.$$

4. *Check.* We have an answer that is much larger than the cost of any individual television, which is reasonable. We can repeat our calculation. We can also check by estimating as follows:

$$5 \times 145 \approx 5 \times 150 = 750 \approx 725.$$

5. *State.* The cost of 5 television sets is $725. ◀

DO EXERCISE 10.

▶ **EXAMPLE 7** The state of Colorado is 270 mi by 380 mi. What is its area?

1. *Familiarize.* We first make a drawing. We let A = the area.

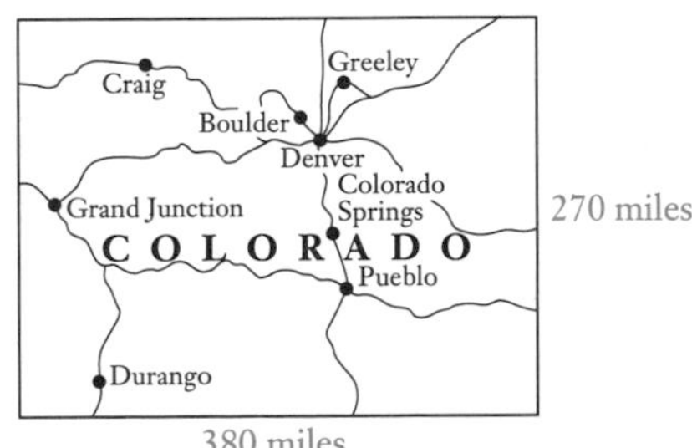

2. *Translate.* Using a formula for area, we have $A = l \cdot w = 380 \cdot 270$.

3. *Solve.* We carry out the multiplication.

$$\begin{array}{r} 3\,8\,0 \\ \times \quad 2\,7\,0 \\ \hline 2\,6\,6\,0\,0 \\ 7\,6\,0\,0\,0 \\ \hline 1\,0\,2\,6\,0\,0 \end{array} \qquad \text{Thus, } A = 102{,}600.$$

4. *Check.* We repeat our calculation. We also note that the answer is larger than either the length or the width, which it should be. (This might not be the case, if we were using decimals.) The answer checks.

5. *State.* The area is 102,600 sq mi. ◀

10. An electronics firm sells 324 calculators one month, each at a price of $16. How much money did it receive from the calculators? $5184

ANSWER ON PAGE A-2

11. The state of Wyoming is 275 mi by 365 mi. What is its area?

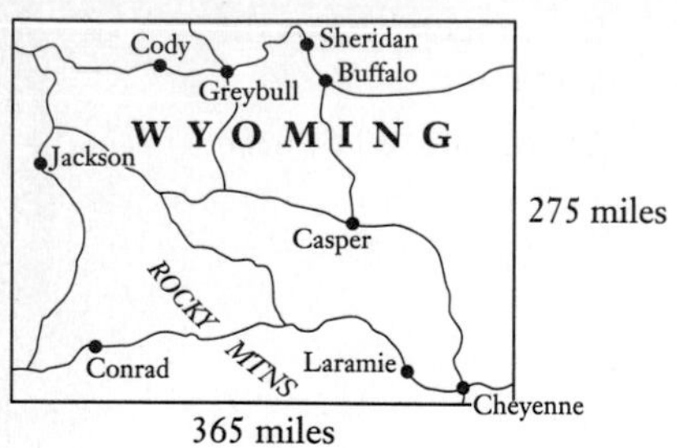

100,375 sq mi

12. There are 60 minutes in an hour. How many minutes are there in 72 hours? 4320 min

ANSWERS ON PAGE A-2

DO EXERCISE 11.

▶ **EXAMPLE 8** There are 24 hours in a day and 7 days in a week. How many hours are there in a week?

1. *Familiarize.* We first make a drawing. We let y = the number of hours in a week. Repeated addition works well here.

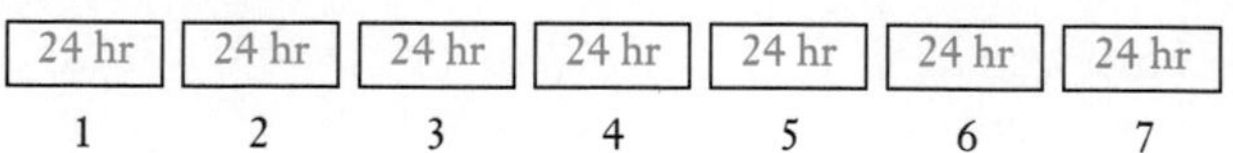

2. *Translate.* We translate to a number sentence.

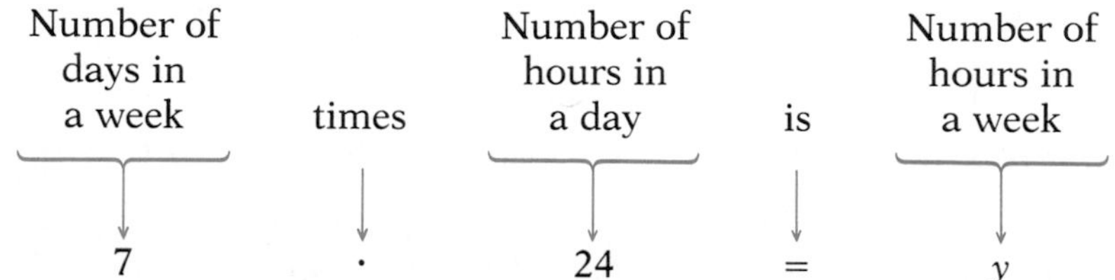

3. *Solve.* To solve the equation, we carry out the multiplication.

$$\begin{array}{r} {}^{2}\\ 24 \\ \times\ \ 7 \\ \hline 168 \end{array} \qquad \text{Thus, } y = 168.$$

4. *Check.* We check our answer by estimating:

$$7 \times 24 \approx 7 \times 30 = 210 \approx 168.$$

We can also check by repeating our calculation. We note that there are more hours in a week than in a day, which we would expect.

5. *State.* There are 168 hours in a week. ◀

DO EXERCISE 12.

▶ **EXAMPLE 9** A beverage company produces 2269 bottles of soda. How many 6-bottle cartons can be filled? How many bottles will be left over?

1. *Familiarize.* We first draw a picture. We let n = the number of 6-bottle cartons to be filled.

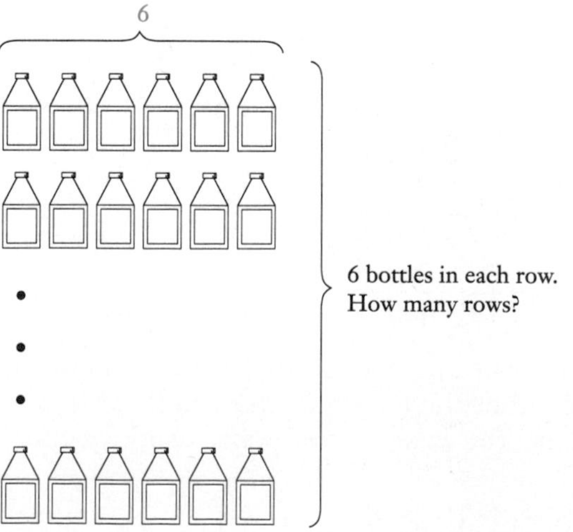

2.,3. *Translate* and *Solve.* We translate to an equation and solve as follows:

$$2269 \div 6 = n.$$

$$\begin{array}{r} 378 \\ 6\overline{)2269} \\ 1800 \\ \hline 469 \\ 420 \\ \hline 49 \\ 48 \\ \hline 1 \end{array}$$

4. *Check.* We can check by multiplying the number of cartons by 6 and adding the remainder of 1:

$$6 \cdot 378 = 2268, \qquad 2268 + 1 = 2269.$$

5. *State.* Thus, 378 six-bottle cartons can be filled. There will be 1 bottle left over. ◀

DO EXERCISE 13.

13. A beverage company produces 2205 bottles of soda. How many 8-bottle cartons can be filled? How many bottles will be left over?

275 cartons; 5 left over

▶ **EXAMPLE 10** An automobile with a 5-speed transmission gets 27 mi to the gallon in city driving. How many gallons will it take to travel 7020 mi of city driving?

1. *Familiarize.* We first draw a picture. It is often helpful to be descriptive about how you define a variable. In this example, we let g = the number of gallons (g comes from "gallons").

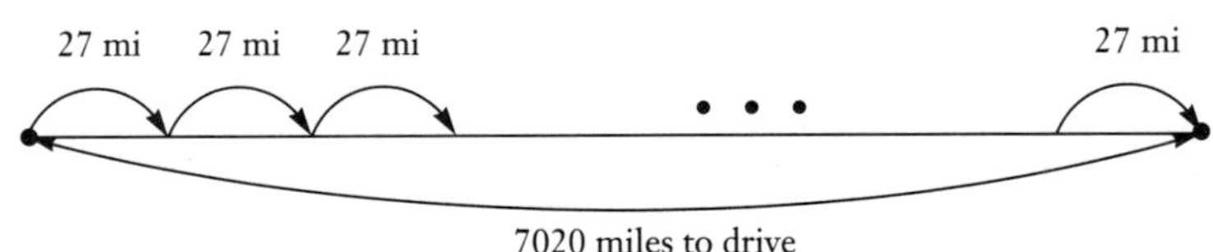

2. *Translate.* Repeated addition applies here. Thus the following multiplication corresponds to the situation.

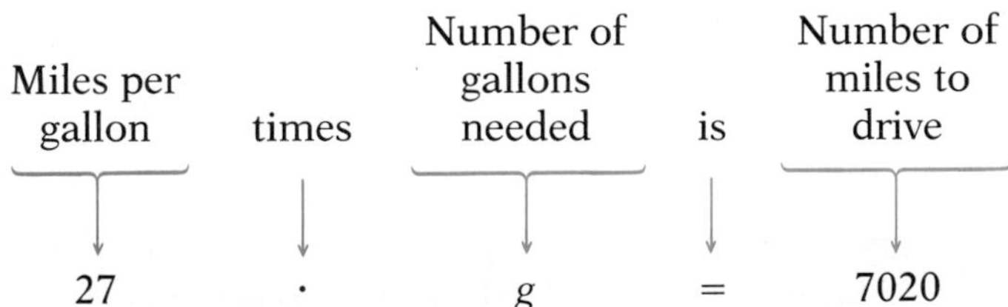

3. *Solve.* To solve the equation, we divide on both sides by 27.

$$27 \cdot g = 7020$$

$$\frac{27 \cdot g}{27} = \frac{7020}{27}$$

$$g = 260$$

$$\begin{array}{r} 260 \\ 27\overline{)7020} \\ 5400 \\ \hline 1620 \\ 1620 \\ \hline 0 \end{array}$$

4. *Check.* To check, we multiply 260 by 27: $27 \cdot 260 = 7020$.

5. *State.* Thus, 260 gal will be needed. ◀

DO EXERCISE 14.

14. An automobile with a 5-speed transmission gets 33 mi to the gallon in city driving. How many gallons will it take to drive 1485 mi?

45 gal

ANSWERS ON PAGE A-2

15. Use the information in the table in Example 11. How long do you have to swim in order to lose one pound? 70 min

16. There are 27 bones in each human hand and 26 bones in each human foot. How many bones are there in all in the hands and feet? 106

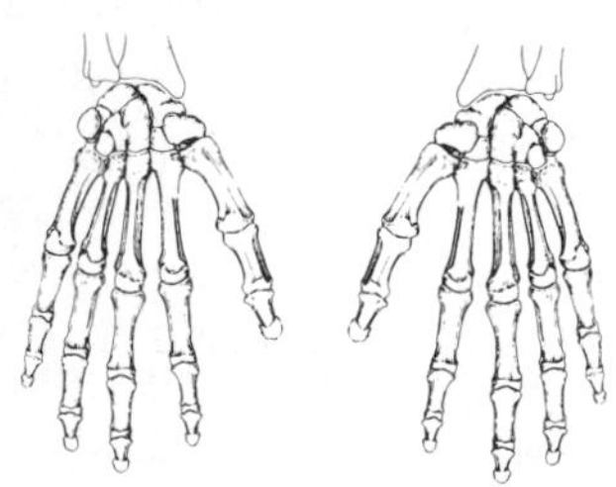

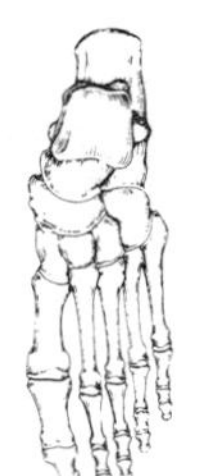

ANSWERS ON PAGE A-2

Multistep Problems

Sometimes we must use more than one operation to solve a problem. We do so in the following example.

▶ **EXAMPLE 11** To lose one pound, you must burn off about 3500 calories.

To burn off 100 calories, you have to:

* **Run for 8 min at a brisk pace, or**
* **Swim for 2 min at a brisk pace, or**
* **Bicycle for 15 min at 9 mph, or**
* **Do aerobic exercises for 15 min.**

How long do you have to run in order to lose one pound?

1. *Familiarize.* We first draw a picture.

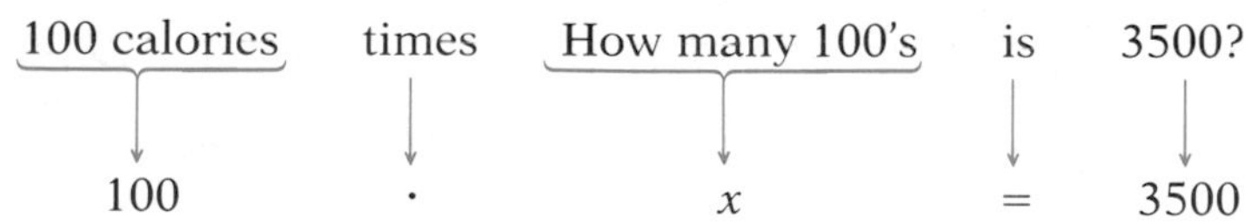

One pound			
3500 calories			
100 cal	100 cal	. . .	100 cal
8 min	8 min	. . .	8 min

2. *Translate.* Repeated addition applies here. Thus the following multiplication corresponds to the situation. We must find out how many 100's there are in 3500. We let x = the number of 100's in 3500.

100 calories ↓ 100   times ↓ $\cdot$ How many 100's ↓ x is ↓ $=$ 3500? ↓ 3500

$$100 \cdot x = 3500$$

3. *Solve.* To solve the equation, we divide on both sides by 100.

$$100 \cdot x = 3500$$
$$\frac{100 \cdot x}{100} = \frac{3500}{100}$$
$$x = 35$$

$$\begin{array}{r} 35 \\ 100\overline{)3500} \\ \underline{3000} \\ 500 \\ \underline{500} \\ 0 \end{array}$$

We know that running for 8 min will burn off 100 calories. To do this 35 times will burn off a pound, so we need to run for 35 times 8 minutes in order to burn off one pound. We let t = the time it takes to run off a pound.

$$35 \times 8 = t$$

$$\begin{array}{r} 35 \\ \times \quad 8 \\ \hline 280 \end{array}$$

4. *Check.* Suppose you run for 280 minutes. If we divide 280 by 8, we get 35, and 35 times 100 is 3500, the number of calories it takes to lose one pound.

5. *State.* It will take 280 min, or 4 hr 40 min, of running to lose one pound. ◀

DO EXERCISES 15 AND 16.

NAME SECTION DATE

EXERCISE SET 1.8

a Solve.

1. Ty Cobb hit 3052 singles in his career. Stan Musial hit 2641. How many did they hit together?

2. In the Revolutionary War, there were 4435 battle deaths. In the War of 1812, there were 2260. How many battle deaths were there in both wars?

3. The Empire State building is 381 m (meters) tall. It has a 68-m antenna on top. How far is the top of the antenna from the ground?

4. In a recent year, 1,580,000 people visited the United States from Japan. In addition, 1,200,000 came from England, 800,000 from Mexico, and 510,000 from West Germany. How many, in all, came from these countries?

5. A medical researcher poured first 2340 cubic centimeters of water and then 655 cubic centimeters of alcohol into a beaker. How much liquid was poured?

6. A person's water loss each day includes 400 cubic centimeters from the lungs and 500 cubic centimeters from the skin. How much is lost from both?

7. In Exercise 6, how much more water is lost from the skin than from the lungs?

8. It takes Venus 225 days to rotate about the sun. It takes the earth 365 days. How much longer does it take the earth?

9. In the Tokyo–Yokohama area, there are 17,317,000 people. In the New York–northeastern New Jersey area, there are 17,013,000 people. How many more people are there in the Tokyo–Yokohama area?

10. O'Hare International Airport is the busiest in the world, handling 45,700,000 passengers each year. Hartsfield Atlanta International is the second busiest, handling 39,000,000 passengers. How many more passengers does O'Hare handle than Hartsfield?

ANSWERS

1. 5693
2. 6695
3. 449 m
4. 4,090,000
5. 2995 cubic centimeters
6. 900 cubic centimeters
7. 100 cubic centimeters
8. 140 days
9. 304,000
10. 6,700,000

ANSWERS

11. $91

12. $12,380

13. 665 cal

14. 73,000

15. 3600 sec

16. 4380; 105,120; 38,368,800

17. 2808 sq ft

18. 4080 sq m

19. 7815 mi

20. 9 sec

11. You wrote checks of $45, $78, and $32. Your balance before that was $246. What is your new balance?

12. A family bought a house for $108,900. They spent $16,500 for an extra room. They then sold the house for $137,780. How much did they make on the house?

13. Playing golf for an hour burns 133 calories. How many calories would be burned in 5 hours?

14. Each day 200 people in the United States become millionaires. How many people become millionaires in one year? (Use 365 days for one year.)

15. There are 60 seconds in a minute and 60 minutes in an hour. How many seconds are there in an hour?

16. The average heartbeat is 73 beats per minute. How many beats are there in one hour? one day? one year?

17. A standard-sized tennis court is 36 ft by 78 ft. What is the area of a tennis court?

18. A rectangular field measures 48 m by 85 m. What is the area of the field?

19. The diameter of a circle is the length of a line through its center. The diameter of Jupiter is about 85,965 mi. The diameter of Jupiter is about 11 times the diameter of the earth. What is the diameter of the earth?

20. Sound travels at a speed of about 1087 ft per second (*per* means *for each*). How long does it take the sound of an airplane engine to reach your ear when the plane is 9783 ft overhead?

21. A customer buys 8 suits at $195 each and 3 shirts at $26 each. How much is spent?

22. What is the cost of 11 radios at $27 each and 6 television sets at $736 each?

23. A college has a vacant rectangular lot that is 324 yd by 25 yd. On the lot students dug a garden that was 165 yd by 18 yd. How much area was left over?

24. A college student pays $107 a month for rent and $88 for food during 9 months at college. What is the total cost?

25. How many 16-oz bottles can be filled with 608 oz of catsup?

26. How many 24-can cases can be filled with 768 cans of beans?

27. There are 225 members in a band, and they are marching with 15 in each row. How many rows are there?

28. There are 225 members in a band, and they are marching in 5 rows. How many band members are there in each row?

29. A loan of $324 will be paid off in 12 monthly payments. How much is each payment?

30. A loan of $1404 will be paid off in 36 monthly payments. How much is each payment?

ANSWERS

21. $1638

22. $4713

23. 5130 sq yd

24. $1755

25. 38

26. 32

27. 15

28. 45

29. $27

30. $39

ANSWERS

31. 38 bags; 11 kg left over

32. 16 injections; 2 cubic centimeters left over

33. 16

34. 32

35. 11 in.; 770 mi

36. 690 mi; 12 in.

37. 480

38. 60 in.

39. 525 min, or 8 hr 45 min

40. 525 min, or 8 hr 45 min

41. 186,000 mi

42. 35 lb

31. How many 23-kg (kilogram) bags can be filled by 885 kg of sand? How many kilograms of sand will be left over?

32. A vial contains 50 cubic centimeters of penicillin. How many 3-cubic-centimeter injections can be filled from the vial? How much will be left over?

33. A student bought 5 coats at $64 each and paid for them with $20 bills. How many $20 bills did it take?

34. How many $10 bills would it take to buy the 5 coats mentioned in Exercise 33?

35. A map has a scale of 55 mi to the inch. How far apart *on the map* are two cities that, in reality, are 605 mi apart? How far apart *in reality* are two cities that are 14 in. apart on the map?

36. A map has a scale of 46 mi to the inch. How far apart *in reality* are two cities that are 15 in. apart on the map? How far apart *on the map* are two cities that, in reality, are 552 mi apart?

37. A beverage company fills 640 12-oz bottles with soda. How many 16-oz bottles can be filled with the same amount of soda?

38. A rectangular piece of cardboard measures 64 in. by 15 in. Suppose the width were changed to 16 in. What would the length have to be in order to have the same area?

39. Use the information from the table in Example 11. How long do you have to bicycle at 9 mph in order to lose one pound?

40. Use the information from the table in Example 11. How long do you have to do aerobic exercises in order to lose one pound?

SYNTHESIS

41. 🖩 Light travels at a speed of 8,370,000 mi in 45 sec. How far does it travel in 1 sec?

42. The thickness of paper is often measured by weights. For example, 20-lb paper is that weight of the paper for which 1000 letter-size sheets weigh 20 lb. A ream of paper is 500 sheets. How much does a ream of 70-lb paper weigh?

1.9 Exponential Notation and Order of Operations

a Exponential Notation

Consider the product $3 \cdot 3 \cdot 3 \cdot 3$. Such products occur often enough that mathematicians have found it convenient to invent a shorter notation, called **exponential notation,** explained as follows.

$\underbrace{3 \cdot 3 \cdot 3 \cdot 3}_{\text{4 factors}}$ is shortened to $3^4 \leftarrow$ exponent

We read 3^4 as "three to the fourth power," 5^3 as "five cubed," and 5^2 as "five squared." The latter comes from the fact that a square of side s has area A given by $A = s^2$.

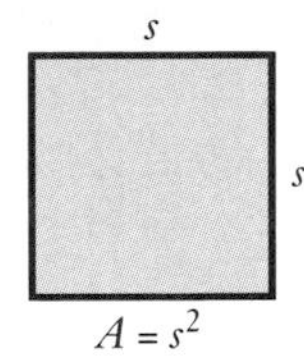

▶ **EXAMPLE 1** Write exponential notation for $10 \cdot 10 \cdot 10 \cdot 10 \cdot 10$.

Exponential notation is 10^5. **5 is the *exponent*. 10 is the *base*.** ◀

▶ **EXAMPLE 2** Write exponential notation for $2 \cdot 2 \cdot 2$.

Exponential notation is 2^3. ◀

DO EXERCISES 1–4.

b Evaluating Exponential Notation

▶ **EXAMPLE 3** Evaluate: 10^3.

$$10^3 = 10 \cdot 10 \cdot 10 = 1000$$ ◀

▶ **EXAMPLE 4** Evaluate: 5^4.

$$5^4 = 5 \cdot 5 \cdot 5 \cdot 5 = 625$$ ◀

DO EXERCISES 5–8.

OBJECTIVES

After finishing Section 1.9, you should be able to:

- a Write exponential notation for products such as $4 \cdot 4 \cdot 4$.
- b Evaluate exponential notation.
- c Simplify expressions using the rules for order of operations.
- d Remove parentheses within parentheses.

FOR EXTRA HELP

Tape 2C

Tape 3A

MAC: 1
IBM: 1

Write exponential notation.

1. $5 \cdot 5 \cdot 5 \cdot 5$ 5^4
2. $5 \cdot 5 \cdot 5 \cdot 5 \cdot 5$ 5^5
3. $10 \cdot 10$ 10^2
4. $10 \cdot 10 \cdot 10 \cdot 10$ 10^4

Evaluate.

5. 10^4 10,000
6. 10^2 100
7. 8^3 512
8. 2^5 32

ANSWERS ON PAGE A-2

Simplify.

9. $93 - 14 \cdot 3$ 51

10. $104 \div 4 + 4$ 30

11. $25 \cdot 26 - (56 + 10)$ 584

12. $75 \div 5 + (83 - 14)$ 84

Simplify and compare.

13. $64 \div (32 \div 2)$ and $(64 \div 32) \div 2$
4; 1

14. $(28 + 13) + 11$ and
$28 + (13 + 11)$ 52; 52

ANSWERS ON PAGE A-2

C Simplifying Expressions

Suppose we have a calculation like the following:

$$8 \cdot 6 - 1.$$

How do we find the answer? Do we subtract 1 from 6 and then multiply by 8, or do we multiply 8 by 6 and then subtract 1? In the first case, the answer is 40. In the second case, the answer is 47.

Consider the calculation

$$7 \cdot 14 - (12 + 18).$$

What do the parentheses mean? To deal with these questions, we must make some agreement regarding the order in which we perform operations. The rules are as follows.

Rules for Order of Operations

1. **Do all calculations within parentheses before operations outside.**
2. **Evaluate all exponential expressions.**
3. **Do all multiplications and divisions in order from left to right.**
4. **Do all additions and subtractions in order from left to right.**

It is worth noting that these are the rules that a computer uses to do computations. In order to program a computer, one must know these rules.

▶ **EXAMPLE 5** Simplify: $8 \cdot 6 - 1$.

There are no parentheses or exponents, so we start with the third step.

$8 \cdot 6 - 1 = 48 - 1$ **Doing all multiplications and divisions in order from left to right**

$= 47$ **Doing all additions and subtractions in order from left to right** ◀

▶ **EXAMPLE 6** Simplify: $7 \cdot 14 - (12 + 18)$.

$7 \cdot 14 - (12 + 18) = 7 \cdot 14 - 30$ **Carrying out operations inside parentheses**

$= 98 - 30$ **Doing all multiplications and divisions**

$= 68$ **Doing all additions and subtractions** ◀

DO EXERCISES 9–12.

▶ **EXAMPLE 7** Simplify and compare:

$$23 - (10 - 9) \quad \text{and} \quad (23 - 10) - 9.$$

We have

$$23 - (10 - 9) = 23 - 1 = 22;$$
$$(23 - 10) - 9 = 13 - 9 = 4.$$

We can see that $23 - (10 - 9)$ and $(23 - 10) - 9$ represent different numbers. ◀

DO EXERCISES 13 AND 14.

▶ **EXAMPLE 8** Simplify: $7 \cdot 2 - (12 + 0) \div 3 - (5 - 2)$.

$7 \cdot 2 - (12 + 0) \div 3 - (5 - 2) = 7 \cdot 2 - 12 \div 3 - 3$ Carrying out operations inside parentheses

$= 14 - 4 - 3$ Doing all multiplications and divisions in order from left to right

$= 7$ Doing all additions and subtractions in order from left to right ◀

DO EXERCISE 15.

▶ **EXAMPLE 9** Simplify: $15 \div 3 \cdot 2 \div (10 - 8)$.

$15 \div 3 \cdot 2 \div (10 - 8) = 15 \div 3 \cdot 2 \div 2$ Carrying out operations inside parentheses

$= 5 \cdot 2 \div 2$ Doing all multiplications and divisions in order from left to right

$= 10 \div 2$

$= 5$ ◀

DO EXERCISES 16–18.

▶ **EXAMPLE 10** Simplify and compare: $(3 + 5)^2$ and $3^2 + 5^2$.

We have

$$(3 + 5)^2 = 8^2 = 64;$$
$$3^2 + 5^2 = 9 + 25 = 34.$$

We see that $(3 + 5)^2$ and $3^2 + 5^2$ do not represent the same numbers. ◀

DO EXERCISE 19.

▶ **EXAMPLE 11** Simplify: $6^3 \div (10 - 8)^2$.

$6^3 \div (10 - 8)^2 = 6^3 \div 2^2$ Carrying out operations inside parentheses first

$= 216 \div 4$ Evaluating exponential expressions second

$= 54$ Dividing ◀

▶ **EXAMPLE 12** Simplify: $2^4 + 51 \cdot 4 - 2 \cdot (37 + 23 \cdot 2)$.

$2^4 + 51 \cdot 4 - 2 \cdot (37 + 23 \cdot 2)$

$= 2^4 + 51 \cdot 4 - 2 \cdot (37 + 46)$ Carrying out operations inside parentheses. To do this, we first multiply 23 by 2.

$= 2^4 + 51 \cdot 4 - 2 \cdot 83$ Completing the addition inside parentheses

$= 16 + 51 \cdot 4 - 2 \cdot 83$ Evaluating exponential expressions

$= 16 + 204 - 166$ Doing the multiplications

$= 220 - 166$
$= 54$ } Doing the additions and subtractions in order from left to right ◀

DO EXERCISES 20–22.

15. Simplify:
$9 \times 4 - (20 + 4) \div 8 - (6 - 2)$.

29

Simplify.

16. $5 \cdot 5 \cdot 5 + 26 \cdot 71 - (16 + 25 \cdot 3)$

1880

17. $4 \cdot 4 \cdot 4 + 10 \cdot 20 + 8 \cdot 8 - 23$

305

18. $95 - 2 \cdot 2 \cdot 2 \cdot 5 \div (24 - 4)$

93

19. Simplify and compare:
$(4 + 6)^2$ and $4^2 + 6^2$.

100; 52

Simplify.

20. $5^3 + 26 \cdot 71 - (16 + 25 \cdot 3)$

1880

21. $(1 + 3)^3 + 10 \cdot 20 + 8^2 - 23$

305

22. $95 - 2^3 \cdot 5 \div (24 - 4)$

93

ANSWERS ON PAGE A-2

Simplify.

23. $9 \times 5 + \{6 \div [14 - (5 + 3)]\}$

46

24. $[18 - (2 + 7) \div 3] - (31 - 10 \times 2)$

4

ANSWERS ON PAGE A-2

d Parentheses Within Parentheses

When parentheses occur within parentheses, we can make them different shapes, such as [ ] (also called "brackets") and { } (also called "braces"). All of these have the same meaning. When parentheses occur within parentheses, computations in the innermost ones are to be done first.

▶ **EXAMPLE 13** Simplify: $16 \div 2 + \{40 - [13 - (4 + 2)]\}$.

$16 \div 2 + \{40 - [13 - (4 + 2)]\}$

$= 16 \div 2 + \{40 - [13 - 6]\}$ **Doing the calculations in the innermost parentheses first**

$= 16 \div 2 + \{40 - 7\}$ **Again, doing the calculations in the innermost parentheses**

$= 16 \div 2 + 33$

$= 8 + 33$ **Doing all multiplications and divisions in order from left to right**

$= 41$ **Doing all additions and subtractions in order from left to right** ◀

▶ **EXAMPLE 14** Simplify: $[25 - (4 + 3) \times 3] \div (11 - 7)$.

$$\begin{aligned} [25 - (4 + 3) \times 3] \div (11 - 7) &= [25 - 7 \times 3] \div (11 - 7) \\ &= [25 - 21] \div (11 - 7) \\ &= 4 \div 4 \\ &= 1 \end{aligned}$$ ◀

DO EXERCISES 23 AND 24.

❖ SIDELIGHTS

Palindrome Numbers

Words like "radar" and "toot" read the same backward and forward. A number that reads the same backward and forward is called a *palindrome.* For example,

11, 121, 202, and 34543

are palindrome numbers. Many numbers can be transformed to palindromes by reversing the digits and adding, and so on. For example,

257	← Not palindrome
752	← Reverse digits
1009	← Add
9001	← Reverse digits
10010	← Add
01001	← Reverse digits and add
11011	← Palindrome

EXERCISES

To what palindrome can the number be transformed?

1. 356 11,011 **2.** 471 5115

NAME SECTION DATE

EXERCISE SET 1.9

a Write exponential notation.

1. $3 \cdot 3 \cdot 3 \cdot 3$ **2.** $2 \cdot 2 \cdot 2 \cdot 2 \cdot 2$ **3.** $5 \cdot 5$ **4.** $13 \cdot 13 \cdot 13$

5. $7 \cdot 7 \cdot 7 \cdot 7 \cdot 7$ **6.** $10 \cdot 10$ **7.** $10 \cdot 10 \cdot 10$ **8.** $1 \cdot 1 \cdot 1 \cdot 1$

b Evaluate.

9. 7^2 **10.** 5^3 **11.** 9^3 **12.** 10^2

13. 12^4 **14.** 10^5 **15.** 11^2 **16.** 6^3

c Simplify.

17. $12 + (6 + 4)$ **18.** $(12 + 6) + 18$ **19.** $52 - (40 - 8)$

20. $(52 - 40) - 8$ **21.** $1000 \div (100 \div 10)$ **22.** $(1000 \div 100) \div 10$

23. $(256 \div 64) \div 4$ **24.** $256 \div (64 \div 4)$ **25.** $(2 + 5)^2$

26. $2^2 + 5^2$ **27.** $2 + 5^2$ **28.** $2^2 + 5$

29. $16 \cdot 24 + 50$ **30.** $23 + 18 \cdot 20$ **31.** $83 - 7 \cdot 6$

32. $10 \cdot 7 - 4$ **33.** $10 \cdot 10 - 3 \cdot 4$ **34.** $90 - 5 \cdot 5 \cdot 2$

35. $4^3 \div 8 - 4$ **36.** $8^2 - 8 \cdot 2$ **37.** $17 \cdot 20 - (17 + 20)$

38. $1000 \div 25 - (15 + 5)$ **39.** $6 \cdot 10 - 4 \cdot 10$ **40.** $3 \cdot 8 + 5 \cdot 8$

ANSWERS

1. 3^4
2. 2^5
3. 5^2
4. 13^3
5. 7^5
6. 10^2
7. 10^3
8. 1^4
9. 49
10. 125
11. 729
12. 100
13. 20,736
14. 100,000
15. 121
16. 216
17. 22
18. 36
19. 20
20. 4
21. 100
22. 1
23. 1
24. 16
25. 49
26. 29
27. 27
28. 9
29. 434
30. 383
31. 41
32. 66
33. 88
34. 40
35. 4
36. 48
37. 303
38. 20
39. 20
40. 64

ANSWERS

41. 70

42. 34

43. 295

44. 311

45. 32

46. 33

47. 906

48. 83

49. 62

50. 68

51. 102

52. 68

53. 110

54. 10

55. 7

56. 58

57. 544

58. 9

59. 708

60. 27

61. 24; $1 + 5 \cdot (4 + 3) = 36$

62. 7; $12 \div (4 + 2) \cdot (3 - 2) = 2$

63. 7; $12 \div (4 + 2) \cdot 3 - 2 = 4$

64.

41. $300 \div 5 + 10$

42. $144 \div 4 - 2$

43. $3 \cdot (2 + 8)^2 - 5 \cdot (4 - 3)^2$

44. $7 \cdot (10 - 3)^2 - 2 \cdot (3 + 1)^2$

45. $4^2 + 8^2 \div 2^2$

46. $6^2 - 3^4 \div 3^3$

47. $10^3 - 10 \cdot 6 - (4 + 5 \cdot 6)$

48. $7^2 + 20 \cdot 4 - (28 + 9 \cdot 2)$

49. $6 \cdot 11 - (7 + 3) \div 5 - (6 - 4)$

50. $8 \times 9 - (12 - 8) \div 4 - (10 - 7)$

51. $120 - 3^3 \cdot 4 \div (30 - 24)$

52. $80 - 2^4 \cdot 15 \div (35 - 15)$

d Simplify.

53. $8 \times 13 + \{42 \div [18 - (6 + 5)]\}$

54. $72 \div 6 - \{2 \times [9 - (4 \times 2)]\}$

55. $[14 - (3 + 5) \div 2] - [18 \div (8 - 2)]$

56. $[92 \times (6 - 4) \div 8] + [7 \times (8 - 3)]$

57. $(82 - 14) \times [(10 + 45 \div 5) - (6 \cdot 6 - 5 \cdot 5)]$

58. $(18 \div 2) \cdot \{[(9 \cdot 9 - 1) \div 2] - [5 \cdot 20 - (7 \cdot 9 - 2)]\}$

59. $4 \times \{(200 - 50 \div 5) - [(35 \div 7) \cdot (35 \div 7) - 4 \times 3]\}$

60. $\{[18 - 2 \cdot 6] - [40 \div (17 - 9)]\} + \{48 - 13 \times 3 + [(50 - 7 \cdot 5) + 2]\}$

SYNTHESIS

Each of the expressions in Exercises 61–63 is incorrect. First find the correct answer. Then place as many parentheses as needed in the expression in order to make the incorrect answer correct.

61. $1 + 5 \cdot 4 + 3 = 36$

62. $12 \div 4 + 2 \cdot 3 - 2 = 2$

63. $12 \div 4 + 2 \cdot 3 - 2 = 4$

64. Use any grouping symbols and one occurrence each of 1, 2, 3, 4, 5, 6, 7, 8, and 9 to represent 100.

Answers may vary; $(1 \cdot 2 \cdot 3) + (6 \cdot 7) - (4 \cdot 5) + (8 \cdot 9) = 100$

SUMMARY AND REVIEW EXERCISES: CHAPTER 1

The review exercises that follow are for practice. Answers are at the back of the book. If you miss an exercise, restudy the section and objective indicated alongside the answer.

1. Write expanded notation: 2793.

2. Write a word name: 2,781,427.

3. What does the digit 7 mean in 4,678,952?

4. Write standard notation for the number in this sentence: The gross national product is two trillion, six hundred twenty-six billion, one hundred million dollars.

Add.

5. $\begin{array}{r} 3847 \\ +\,2132 \\ \hline \end{array}$

6. $\begin{array}{r} 27{,}609 \\ +\,38{,}415 \\ \hline \end{array}$

7. $\begin{array}{r} 2743 \\ 4125 \\ 6274 \\ +\,8956 \\ \hline \end{array}$

8. $\begin{array}{r} 91{,}426 \\ +\quad 7{,}495 \\ \hline \end{array}$

Subtract.

9. $\begin{array}{r} 8465 \\ -\,7312 \\ \hline \end{array}$

10. $\begin{array}{r} 3743 \\ -\,2596 \\ \hline \end{array}$

11. $\begin{array}{r} 6003 \\ -\,3729 \\ \hline \end{array}$

12. $\begin{array}{r} 37{,}405 \\ -\,19{,}648 \\ \hline \end{array}$

13. $678 - 234$

14. $6000 - 1234$

Multiply.

15. $\begin{array}{r} 700 \\ \times\,600 \\ \hline \end{array}$

16. $\begin{array}{r} 7846 \\ \times\quad 800 \\ \hline \end{array}$

17. $\begin{array}{r} 76 \\ \times\quad 9 \\ \hline \end{array}$

18. $\begin{array}{r} 6394 \\ \times\qquad 7 \\ \hline \end{array}$

19. $\begin{array}{r} 74 \\ \times\,46 \\ \hline \end{array}$

20. $\begin{array}{r} 726 \\ \times\,698 \\ \hline \end{array}$

21. $\begin{array}{r} 587 \\ \times\quad 47 \\ \hline \end{array}$

22. $\begin{array}{r} 3456 \\ \times\,1000 \\ \hline \end{array}$

Divide.

23. $80 \div 16$

24. $63 \div 5$

25. $7\overline{)560}$

26. $4\overline{)830}$

27. $8\overline{)3073}$

28. $60\overline{)286}$

29. $79\overline{)4266}$

30. $38\overline{)17{,}176}$

31. $14\overline{)70{,}112}$

32. $12\overline{)52{,}668}$

Solve.

33. $47 + x = 92$

34. $x = 782 - 236$

35. $46 \cdot n = 368$

Solve.

36. In 1909 the first "Lincoln-head" pennies were minted. Seventy-three years later, these pennies were first minted with a decreased copper content. In what year was the copper content reduced?

37. A farmer harvested 625 bu (bushels) of corn, 865 bu of wheat, 698 bu of soybeans, and 597 bu of potatoes. What was the farmer's total harvest?

38. A family budgets \$4950 yearly for food and clothing and an additional \$3585 for entertainment. The yearly income of the family was \$28,283. How much of this income remained after these two allotments?

39. A certain cottage cheese contains 113 calories per ounce. A bulk container of this cheese contains 25 ounces. What is the caloric content of this container?

40. A sweater costs \$28 and a coat costs \$37. Find the total cost of 6 sweaters and 9 coats.

41. A chemist has 2753 L (liters) of acid. How many 18-L beakers can be filled? How much will be left over?

42. A student buys 8 books at \$25 each and pays for them with \$20 bills. How many \$20 bills does it take?

Round 345,759 to the nearest:

43. Hundred.

44. Ten.

45. Thousand.

Estimate each sum, difference, or product by rounding to the nearest hundred. Show your work.

46. $\begin{array}{r} 41{,}348 \\ +\,19{,}749 \\ \hline \end{array}$

47. $\begin{array}{r} 38{,}652 \\ -\,24{,}549 \\ \hline \end{array}$

48. $\begin{array}{r} 396 \\ \times\,748 \\ \hline \end{array}$

Use $<$ or $>$ for ▒ to write a true sentence.

49. 67 ▒ 56

50. 1 ▒ 23

51. Write exponential notation: $8 \cdot 8 \cdot 8$.

Evaluate.

52. 2^4

53. 6^2

Simplify.

54. $8 \times 6 + 17$

55. $10 \times 24 - (18 + 2) \div 4 - (9 - 7)$

56. $7 + (4 + 3)^2$

57. $7 + 4^2 + 3^2$

58. $(80 \div 16) \times [(20 - 56 \div 8) + (8 \cdot 8 - 5 \cdot 5)]$

❖ THINKING IT THROUGH

1. Discuss the difference between a “take away” problem situation and a “how much more” situation in subtraction.
2. Describe at least three reasons for rounding and estimating.
3. Describe two ways in which multiplication can occur in the real world.
4. Describe two ways in which division can occur in the real world.

NAME SECTION DATE

TEST: CHAPTER 1

1. Write expanded notation: 8843.

[1.1a] 8 thousands + 8 hundreds + 4 tens + 3 ones

2. Write a word name: 38,403,277.

[1.1c] Thirty-eight million, four hundred three thousand, two hundred seventy-seven

3. In the number 546,789, which digit tells the number of hundred thousands?

Add.

4. $\begin{array}{r} 6811 \\ +3178 \\ \hline \end{array}$

5. $\begin{array}{r} 45{,}889 \\ +17{,}902 \\ \hline \end{array}$

6. $\begin{array}{r} 12 \\ 8 \\ 3 \\ 7 \\ +\ 4 \\ \hline \end{array}$

7. $\begin{array}{r} 6203 \\ +4312 \\ \hline \end{array}$

Subtract.

8. $\begin{array}{r} 7983 \\ -4353 \\ \hline \end{array}$

9. $\begin{array}{r} 2974 \\ -1935 \\ \hline \end{array}$

10. $\begin{array}{r} 8907 \\ -2059 \\ \hline \end{array}$

11. $\begin{array}{r} 23{,}067 \\ -17{,}892 \\ \hline \end{array}$

Multiply.

12. $\begin{array}{r} 4568 \\ \times\quad 9 \\ \hline \end{array}$

13. $\begin{array}{r} 8876 \\ \times\ 600 \\ \hline \end{array}$

14. $\begin{array}{r} 65 \\ \times 37 \\ \hline \end{array}$

15. $\begin{array}{r} 678 \\ \times 788 \\ \hline \end{array}$

Divide.

16. $15 \div 4$

17. $420 \div 6$

18. $89\overline{)8633}$

19. $44\overline{)35{,}428}$

Solve.

20. James Dean was 24 years old when he died. He was born in 1931. In what year did he die?

21. A beverage company produces 739 cans of soda. How many 8-can packages can be filled? How many cans will be left over?

22. A customer buys 15 pieces of lumber at $12 each and pays for them with $10 bills. How many $10 bills does it take?

23. A rectangular lot measures 200 m by 600 m. What is the area of the lot?

24. A sack of oranges weighs 27 lb. A sack of apples weighs 32 lb. Find the total weight of 16 bags of oranges and 43 bags of apples.

25. A box contains 5000 staples. How many staplers can be filled from the box if each stapler holds 250 staples?

ANSWERS

1.
2.
3. [1.1e] 5
4. [1.2b] 9989
5. [1.2b] 63,791
6. [1.2b] 34
7. [1.2b] 10,515
8. [1.3d] 3630
9. [1.3d] 1039
10. [1.3d] 6848
11. [1.3d] 5175
12. [1.5b] 41,112
13. [1.5b] 5,325,600
14. [1.5b] 2405
15. [1.5b] 534,264
16. [1.6c] 3 R 3
17. [1.6c] 70
18. [1.6c] 97
19. [1.6c] 805 R 8
20. [1.8a] 1955
21. [1.8a] 92 packages, 3 left over
22. [1.8a] 18
23. [1.8a] 120,000 sq m
24. [1.8a] 1808 lb
25. [1.8a] 20

ANSWERS

26. [1.8a] 305 sq mi

27. [1.8a] 56

28. [1.8a] 66,444 sq mi

29. [1.8a] $271

30. [1.7b] 46

31. [1.7b] 13

32. [1.7b] 14

33. [1.4a] 35,000

34. [1.4a] 34,580

35. [1.4a] 34,600

36. [1.4b] 23,600 + 54,700 = 78,300

37. [1.4b] 54,800 − 23,600 = 31,200

38. [1.5c] $800 \cdot 500 =$ 400,000

39. [1.4c] >

40. [1.4c] <

41. [1.9a] 12^4

42. [1.9b] 343

43. [1.9b] 8

44. [1.9c] 64

45. [1.9c] 96

46. [1.9c] 2

47. [1.9d] 216

48. [1.9c] 18

26. The area of Vermont is 9609 sq mi. The area of New Hampshire is 9304 sq mi. How much more area does Vermont have?

27. A professional bowler rolled a game of 245. Then the bowler rolled a game of 189. How much higher was the first game?

28. Listed below are the areas, in square miles, of the New England states. What is the total area of New England?

Maine	33,215
Massachusetts	8,093
New Hampshire	9,304
Vermont	9,609
Connecticut	5,009
Rhode Island	1,214

29. You have $345 in a checking account. You write checks for $45 and $29. How much money is left in the checking account?

Solve.

30. $28 + x = 74$

31. $169 \div 13 = n$

32. $38 \cdot y = 532$

Round 34,578 to the nearest:

33. Thousand.

34. Ten.

35. Hundred.

Estimate each sum, difference, or product by rounding to the nearest hundred. Show your work.

36. $\begin{array}{r} 23,649 \\ +\,54,746 \\ \hline \end{array}$

37. $\begin{array}{r} 54,751 \\ -\,23,649 \\ \hline \end{array}$

38. $\begin{array}{r} 824 \\ \times\,489 \\ \hline \end{array}$

Use < or > for ▒ to write a true sentence.

39. 34 ▒ 17

40. 117 ▒ 157

41. Write exponential notation: $12 \cdot 12 \cdot 12 \cdot 12$.

Evaluate.

42. 7^3

43. 2^3

Simplify.

44. $(10 - 2)^2$

45. $10^2 - 2^2$

46. $(25 - 15) \div 5$

47. $8 \times \{(20 - 11) \cdot [(12 + 48) \div 6 - (9 - 2)]\}$

48. $2^4 + 24 \div 12$

INTRODUCTION Multiplication and division using fractional notation is considered in this chapter. To aid such study, the chapter begins with factorizations and rules for divisibility. After multiplication and division are discussed, those skills are used to solve equations and problems.

The review sections to be tested in addition to the material in this chapter are 1.3, 1.6, 1.7, and 1.8. ❖

Multiplication and Division: Fractional Notation

AN APPLICATION

Business people have determined that $\frac{1}{4}$ of the items on a mailing list will change in one year. A business has a mailing list of 2500 people. After one year, how many addresses on that list will be incorrect?

THE MATHEMATICS

Let a = the number of addresses. Then the problem can be translated to this equation:

$$\underbrace{\frac{1}{4} \cdot 2500}_{\uparrow} = a.$$

This multiplication using fractional notation occurs in problem solving.

❖ POINTS TO REMEMBER: CHAPTER 2

Area of a Rectangle: $A = l \cdot w$
Area of a Square: $A = s^2$

PRETEST: CHAPTER 2

1. Determine whether 59 is prime, composite, or neither.

2. Find the prime factorization of 140.

3. Determine whether 788 is divisible by 8.

4. Determine whether 1503 is divisible by 9.

Simplify.

5. $\frac{57}{57}$

6. $\frac{68}{1}$

7. $\frac{0}{50}$

8. $\frac{8}{32}$

Multiply and simplify.

9. $\frac{1}{3} \cdot \frac{18}{5}$

10. $\frac{5}{6} \cdot 24$

11. $\frac{2}{5} \cdot \frac{25}{8}$

Find the reciprocal.

12. $\frac{7}{8}$

13. 11

Divide and simplify.

14. $15 \div \frac{5}{8}$

15. $\frac{2}{3} \div \frac{8}{9}$

16. Solve:

$$\frac{7}{10} \cdot x = 21.$$

17. Use = or ≠ for ▢ to write a true sentence:

$$\frac{5}{11} \;▢\; \frac{1}{2}.$$

Solve.

18. A person earns \$48 for working a full day. How much is earned for working $\frac{3}{4}$ of a day?

19. A piece of rope $\frac{5}{8}$ m long is to be cut into 15 pieces of the same length. What is the length of each piece?

2.1 Factorizations

In this chapter, we begin our work with fractions. Certain skills make such work easier. For example, in order to simplify

$$\frac{12}{32},$$

it is important that we be able to *factor* the 12 and the 32:

$$\frac{12}{32} = \frac{4 \cdot 3}{4 \cdot 8}.$$

Then we "remove" a factor of 1:

$$\frac{4 \cdot 3}{4 \cdot 8} = \frac{4}{4} \cdot \frac{3}{8} = 1 \cdot \frac{3}{8} = \frac{3}{8}.$$

Thus factoring is an important skill in working with fractions.

a Factors and Factorization

In Sections 2.1 and 2.2, we consider only the **natural numbers** 1, 2, 3, and so on.

Let's look at the product $3 \cdot 4 = 12$. We say that 3 and 4 are **factors** of 12. Since $12 = 12 \cdot 1$, we also know that 12 and 1 are factors of 12.

> **A *factor* of a given number is a number multiplied in a product.**
> **A *factorization* of a number is an equation that expresses the number as a product of natural numbers.**

For example, each of the following is a factorization of 12.

$12 = 4 \cdot 3$ ⟵ This factorization shows that 4 and 3 are factors of 12.

$12 = 12 \cdot 1$ ⟵ This factorization shows that 12 and 1 are factors of 12.

$12 = 6 \cdot 2$ ⟵ This factorization shows that 6 and 2 are factors of 12.

$12 = 2 \cdot 3 \cdot 2$ ⟵ This factorization shows that 2 and 3 are factors of 12.

Since $n = n \cdot 1$, every number has a factorization and every number has factors even if its only factors are itself and 1.

▶ **EXAMPLE 1** Find all the factors of 24.

We first find some factorizations.

$24 = 1 \cdot 24$ $24 = 3 \cdot 8$

$24 = 2 \cdot 12$ $24 = 4 \cdot 6$

Note that all but one of the factors of a natural number are *less* than the number.

Factors: 1, 2, 3, 4, 6, 8, 12, 24. ◀

DO EXERCISES 1–4.

OBJECTIVES

After finishing Section 2.1, you should be able to:

a Find the factors of a number.

b Find some multiples of a number, and determine whether a number is divisible by another.

c Given a number from 1 to 50, tell whether it is prime, composite, or neither.

d Find the prime factorization of a composite number.

FOR EXTRA HELP

Tape 3A

Tape 3A

MAC: 2
IBM: 2

Find all the factors of the number. (*Hint:* Find some factorizations of the number.)

1. 6 1, 2, 3, 6

2. 8 1, 2, 4, 8

3. 10 1, 2, 5, 10

4. 32 1, 2, 4, 8, 16, 32

ANSWERS ON PAGE A-2

b Multiples and Divisibility

A **multiple** of a natural number is a product of it and some natural number. For example, some multiples of 2 are:

2 (because $2 = 1 \cdot 2$);
4 (because $4 = 2 \cdot 2$);
6 (because $6 = 3 \cdot 2$);
8 (because $8 = 4 \cdot 2$);
10 (because $10 = 5 \cdot 2$).

Note that all but one of the multiples of a number are *larger* than the number.

We find multiples of 2 by counting by twos: 2, 4, 6, 8, and so on. We can find multiples of 3 by counting by threes: 3, 6, 9, 12, and so on.

▶ **EXAMPLE 2** Show that each of the numbers 3, 6, 9, and 15 is a multiple of 3.

$$3 = 1 \cdot 3 \qquad 9 = 3 \cdot 3$$
$$6 = 2 \cdot 3 \qquad 15 = 5 \cdot 3$$ ◀

DO EXERCISES 5 AND 6.

▶ **EXAMPLE 3** Multiply by 1, 2, 3, and so on, to find ten multiples of 7.

$$1 \cdot 7 = 7 \qquad 6 \cdot 7 = 42$$
$$2 \cdot 7 = 14 \qquad 7 \cdot 7 = 49$$
$$3 \cdot 7 = 21 \qquad 8 \cdot 7 = 56$$
$$4 \cdot 7 = 28 \qquad 9 \cdot 7 = 63$$
$$5 \cdot 7 = 35 \qquad 10 \cdot 7 = \ \ 0$$ ◀

DO EXERCISE 7.

> **A number *b* is said to be *divisible* by another number *a* if *b* is a multiple of *a*.**

Thus,

4 is divisible by 2 because 4 is a multiple of 2 ($4 = 2 \cdot 2$);
27 is divisible by 3 because 27 is a multiple of 3 ($27 = 9 \cdot 3$);
100 is divisible by 25 because 100 is a multiple of 25 ($100 = 4 \cdot 25$).

> ***A number b is divisible by another number a if division of b by a results in a remainder of zero. We sometimes say that a divides b "evenly."***

▶ **EXAMPLE 4** Determine whether 24 is divisible by 3.

We divide 24 by 3:

$$\begin{array}{r} 8 \\ 3\overline{)24} \\ \underline{24} \\ 0 \end{array}$$

Since the remainder is 0, 24 is divisible by 3. ◀

DO EXERCISES 8–10.

5. Show that each of the numbers 5, 45, and 100 is a multiple of 5.
$5 = 1 \cdot 5$, $45 = 9 \cdot 5$, $100 = 20 \cdot 5$

6. Show that each of the numbers 10, 60, and 110 is a multiple of 10. $10 = 1 \cdot 10$, $60 = 6 \cdot 10$, $110 = 11 \cdot 10$

7. Multiply by 1, 2, 3, and so on, to find ten multiples of 5.
5, 10, 15, 20, 25, 30, 35, 40, 45, 50

8. Determine whether 16 is divisible by 2. Yes

9. Determine whether 125 is divisible by 5. Yes

10. Determine whether 125 is divisible by 6. No

ANSWERS ON PAGE A-2

c Prime and Composite Numbers

A natural number that has exactly two different factors, itself and 1, is called a *prime number*.

▶ **EXAMPLE 5** Tell whether the numbers 2, 3, 5, 7, and 11 are prime.

The number 2 is prime. It has only the factors 1 and 2.

The number 5 is prime. It has only the factors 1 and 5.

The numbers 3, 7, and 11 are also prime. ◀

Some numbers are not prime.

▶ **EXAMPLE 6** Tell whether the numbers 4, 6, 8, 10, 63, and 1 are prime.

The number 4 is not prime. It has the factors 1, 2, and 4.

The numbers 6, 8, 10, and 63 are not prime. Each has more than two different factors.

The number 1 is not prime. It does not have two *different* factors. ◀

A natural number, other than 1, that is not prime is called *composite*.

In other words, if a number can be factored into a product of natural numbers, some of which are not the number itself or 1, it is composite. Thus,

2, 3, 5, 7, and 11 are prime;

4, 6, 8, 10, and 63 are composite;

1 is neither prime nor composite.

DO EXERCISE 11.

d Prime Factorizations

To factor a composite number into a product of primes is to find a **prime factorization** of the number. To do this, we consider the primes

2, 3, 5, 7, 11, 13, 17, 19, 23, and so on,

and determine whether a given number is divisible by the primes.

▶ **EXAMPLE 7** Find the prime factorization of 39.

a) We divide by the first prime, 2.

$$2\overline{)39} \quad 19 \quad R = 1$$

Since the remainder is not 0, 2 is not a factor of 39.

b) We divide by the next prime, 3.

$$3\overline{)39} \quad 13 \quad R = 0$$

Since 13 is prime, we are finished. The prime factorization is

$$39 = 3 \cdot 13.$$ ◀

11. Tell whether each number is prime, composite, or neither.

1, 4, 6, 8, 13, 19, 41

13, 19, 41 are prime; 4, 6, 8 are composite; 1 is neither

ANSWER ON PAGE A-2

▶ **EXAMPLE 8** Find the prime factorization of 76.

a) We divide by the first prime, 2.

$$\begin{array}{r} 38 \quad R = 0 \\ 2\overline{)76} \end{array}$$

b) Since 38 is composite, we start with 2 again:

$$\begin{array}{r} 19 \quad R = 0 \\ 2\overline{)38} \end{array}$$

Because 19 is a prime, we are finished. The prime factorization is

$$76 = 2 \cdot 2 \cdot 19.$$

We abbreviate our procedure as follows.

$$\begin{array}{r} 19 \\ 2\overline{)38} \\ 2\overline{)76} \end{array}$$

$$76 = 2 \cdot 2 \cdot 19$$ ◀

Multiplication is commutative so a factorization such as $2 \cdot 2 \cdot 19$ could also be expressed as $2 \cdot 19 \cdot 2$ or $19 \cdot 2 \cdot 2$, but the prime factors are still the same. For this reason, we agree that any of these is "the" prime factorization of 76.

Every number has just one (unique) prime factorization.

▶ **EXAMPLE 9** Find the prime factorization of 72.

$$\begin{array}{r} 3 \\ 3\overline{)9} \\ 2\overline{)18} \\ 2\overline{)36} \\ 2\overline{)72} \end{array}$$

$$72 = 2 \cdot 2 \cdot 2 \cdot 3 \cdot 3$$

Another way to find a prime factorization is by using a **factor tree** as follows:

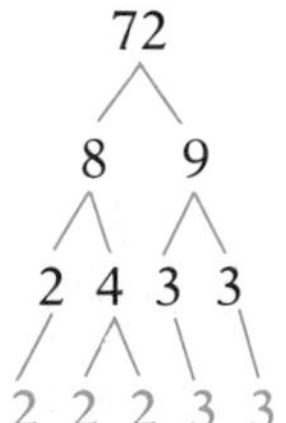

◀

▶ **EXAMPLE 10** Find the prime factorization of 189.

We can use a string of successive divisions:

$$\begin{array}{r} 7 \\ 3\overline{)21} \\ 3\overline{)63} \\ 3\overline{)189} \end{array}$$

189 is not divisible by 2. We move to 3.
63 is not divisible by 2. We move to 3.
21 is not divisible by 2. We move to 3.

$189 = 3 \cdot 3 \cdot 3 \cdot 7$

We can also use a factor tree.

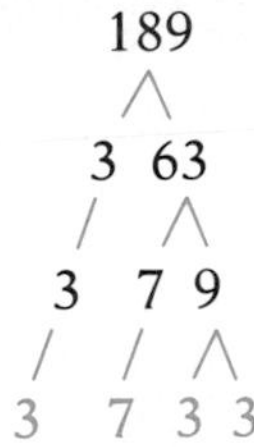

◀

▶ **EXAMPLE 11** Find the prime factorization of 65.

We can use a string of successive divisions.

$$\begin{array}{r} 13 \\ 5\overline{)65} \end{array}$$

65 is not divisible by 2 or 3. We move to 5.

$65 = 5 \cdot 13$ ◀

We can also use a factor tree.

◀

DO EXERCISES 12–17.

Find the prime factorization of the number.

12. 6 $2 \cdot 3$

13. 12 $2 \cdot 2 \cdot 3$

14. 45 $3 \cdot 3 \cdot 5$

15. 98 $2 \cdot 7 \cdot 7$

16. 126 $2 \cdot 3 \cdot 3 \cdot 7$

17. 144 $2 \cdot 2 \cdot 2 \cdot 2 \cdot 3 \cdot 3$

To the student and the instructor: Recall that the *Skill Maintenance Exercises,* which occur at the end of most exercise sets, review *any* skill that has been studied before in the text. Beginning with this chapter, however, certain objectives from four particular sections, along with the material of this chapter, will be tested on the chapter test. For this chapter, the review sections and objectives to be tested are Sections [1.3d], [1.6c], [1.7b], and [1.8a].

ANSWERS ON PAGE A-2

❖ SIDELIGHTS

Factors and Sums

To *factor* a number is to express it as a product. Since $15 = 5 \cdot 3$, we say that 15 is *factored* and that 5 and 3 are *factors* of 15. In the table below, the top number has been factored in such a way that the sum of the factors is the bottom number. For example, in the first column, 56 has been factored as $7 \cdot 8$, and $7 + 8 = 15$, the bottom number. Such thinking will be important in knowing the meaning of a factor and in algebra.

Product	56	63	36	72	140	96	48	168	110	90	432	63
Factor	7	7	18	36	14	12	6	21	11	9	24	3
Factor	8	9	2	2	10	8	8	8	10	10	18	21
Sum	15	16	20	38	24	20	14	29	21	19	42	24

EXERCISE

Find the missing numbers in the table.

NAME SECTION DATE

EXERCISE SET 2.1

a Find all the factors of the number.

1. 16

2. 18

3. 54

4. 48

1, 2, 3, 4, 6, 8, 12, 16, 24, 48

5. 4

6. 9

7. 7

8. 11

9. 1

10. 3

11. 98

12. 100

1, 2, 4, 5, 10, 20, 25, 50, 100

b Multiply by 1, 2, 3, and so on, to find ten multiples of the number.

13. 4

4, 8, 12, 16, 20, 24, 28, 32, 36, 40

14. 14

14, 28, 42, 56, 70, 84, 98, 112, 126, 140

15. 20

20, 40, 60, 80, 100, 120, 140, 160, 180, 200

16. 50

50, 100, 150, 200, 250, 300, 350, 400, 450, 500

17. 3

3, 6, 9, 12, 15, 18, 21, 24, 27, 30

18. 5

5, 10, 15, 20, 25, 30, 35, 40, 45, 50

19. 12

12, 24, 36, 48, 60, 72, 84, 96, 108, 120

20. 17

17, 34, 51, 68, 85, 102, 119, 136, 153, 170

21. 10

10, 20, 30, 40, 50, 60, 70, 80, 90, 100

22. 6

6, 12, 18, 24, 30, 36, 42, 48, 54, 60

23. 9

9, 18, 27, 36, 45, 54, 63, 72, 81, 90

24. 11

11, 22, 33, 44, 55, 66, 77, 88, 99, 110

25. Determine whether 26 is divisible by 6.

26. Determine whether 29 is divisible by 9.

27. Determine whether 1880 is divisible by 8.

28. Determine whether 4227 is divisible by 3.

29. Determine whether 256 is divisible by 16.

30. Determine whether 102 is divisible by 4.

31. Determine whether 4227 is divisible by 9.

32. Determine whether 200 is divisible by 25.

33. Determine whether 8650 is divisible by 16.

34. Determine whether 4143 is divisible by 7.

ANSWERS

1. 1, 2, 4, 8, 16
2. 1, 2, 3, 6, 9, 18
3. 1, 2, 3, 6, 9, 18, 27, 54
4.
5. 1, 2, 4
6. 1, 3, 9
7. 1, 7
8. 1, 11
9. 1
10. 1, 3
11. 1, 2, 7, 14, 49, 98
12.
13.
14.
15.
16.
17.
18.
19.
20.
21.
22.
23.
24.
25. No
26. No
27. Yes
28. Yes
29. Yes
30. No
31. No
32. Yes
33. No
34. No

ANSWERS

35. Neither
36. Prime
37. Composite
38. Prime
39. Prime
40. Composite
41. Prime
42. Composite
43. $2 \cdot 2 \cdot 2$
44. $2 \cdot 2 \cdot 2 \cdot 2$
45. $2 \cdot 7$
46. $3 \cdot 5$
47. $2 \cdot 11$
48. $2 \cdot 2 \cdot 2 \cdot 2 \cdot 2$
49. $5 \cdot 5$
50. $2 \cdot 2 \cdot 2 \cdot 5$
51. $2 \cdot 5 \cdot 5$
52. $2 \cdot 31$
53. $13 \cdot 13$
54. $2 \cdot 2 \cdot 5 \cdot 7$
55. $2 \cdot 2 \cdot 5 \cdot 5$
56. $2 \cdot 5 \cdot 11$
57. $5 \cdot 7$
58. $2 \cdot 5 \cdot 7$
59. $2 \cdot 2 \cdot 2 \cdot 3 \cdot 3$
60. $2 \cdot 43$
61. $7 \cdot 11$
62. $3 \cdot 3 \cdot 11$
63. $2 \cdot 2 \cdot 2 \cdot 2 \cdot 7$
64. $2 \cdot 71$
65. $2 \cdot 2 \cdot 3 \cdot 5 \cdot 5$
66. $5 \cdot 5 \cdot 7$
67. 26
68. 425
69. 0
70. 1
71.
72.

c Determine whether the number is prime, composite, or neither.

35. 1 **36.** 2 **37.** 9 **38.** 19

39. 11 **40.** 27 **41.** 29 **42.** 49

d Find the prime factorization of the number.

43. 8 **44.** 16 **45.** 14 **46.** 15

47. 22 **48.** 32 **49.** 25 **50.** 40

51. 50 **52.** 62 **53.** 169 **54.** 140

55. 100 **56.** 110 **57.** 35 **58.** 70

59. 72 **60.** 86 **61.** 77 **62.** 99

63. 112 **64.** 142 **65.** 300 **66.** 175

71. A rectangular array of 6 rows of 9 objects each, or 9 rows of 6 objects each

72. A 3-dimensional rectangular array: 2 tiers of 12 objects, each tier a rectangular array of 4 rows of 3 objects. Answers may vary.

SKILL MAINTENANCE

Multiply.

67. $2 \cdot 13$

68. $17 \cdot 25$

Divide.

69. $0 \div 22$

70. $22 \div 22$

SYNTHESIS

71. Describe an arrangement of 54 objects that corresponds to the factorization $54 = 6 \times 9$.

72. Describe an arrangement of 24 objects that corresponds to the factorization $24 = 2 \cdot 3 \cdot 4$.

2.2 Divisibility

Suppose you are asked to find the simplest fractional notation for

$$\frac{117}{225}.$$

Since the numbers are quite large, you might feel that the task is difficult. However, both the numerator and the denominator have 9 as a factor. If you knew this, you could factor and simplify quickly as follows:

$$\frac{117}{225} = \frac{9 \cdot 13}{9 \cdot 25} = \frac{9}{9} \cdot \frac{13}{25} = 1 \cdot \frac{13}{25} = \frac{13}{25}.$$

How did we know that both numbers have 9 as a factor? There are fast tests for such determinations. If the sum of the digits of a number is divisible by 9, then the number is divisible by 9; that is, it has 9 as a factor. Since $1 + 1 + 7 = 9$ and $2 + 2 + 5 = 9$, both numbers have 9 as a factor.

a Rules for Divisibility

In this section we learn fast ways of determining whether numbers are divisible by 2, 3, 4, 5, 6, 8, 9, and 10. This will make simplifying with fractional notation much easier.

Divisibility by 2

You may already know the test for divisibility by 2.

A number is divisible by 2 (is *even*) if it has a ones digit of 0, 2, 4, 6, or 8 (that is, it has an even ones digit).

Let's see why. Consider 354, which is

3 hundreds + 5 tens + 4.

Hundreds and tens are both multiples of 2. If the last digit is a multiple of 2, the entire number is.

▶ **EXAMPLES** Determine whether the number is divisible by 2.

1. 355 is not a multiple of 2; 5 is *not* even.
2. 4786 is a multiple of 2; 6 is even.
3. 8990 is a multiple of 2; 0 is even.
4. 4261 is not a multiple of 2; 1 is *not* even. ◀

DO EXERCISES 1–4.

OBJECTIVE

After finishing Section 2.2, you should be able to:

a Determine whether a number is divisible by 2, 3, 4, 5, 6, 8, 9, or 10.

FOR EXTRA HELP

Tape NC

Tape 3B

MAC: 2
IBM: 2

Determine whether the number is divisible by 2.

1. 84 Yes

2. 59 No

3. 998 Yes

4. 2225 No

ANSWERS ON PAGE A-2

Determine whether the number is divisible by 9.

5. 16 No

6. 117 Yes

7. 930 No

8. 29,223 Yes

Determine whether the number is divisible by 3.

9. 111 Yes

10. 1111 No

11. 309 Yes

12. 17,216 No

Determine whether the number is divisible by 4.

13. 216 Yes

14. 217 No

15. 5865 No

16. 23,524 Yes

ANSWERS ON PAGE A-2

Divisibility by 9

A number is divisible by 9 if the sum of the digits is divisible by 9.

▶ **EXAMPLE 5** The number 6984 is divisible by 9 because

$$6 + 9 + 8 + 4 = 27$$

and 27 is divisible by 9. ◀

▶ **EXAMPLE 6** The number 322 is *not* divisible by 9 because

$$3 + 2 + 2 = 7$$

and 7 is not divisible by 9. ◀

DO EXERCISES 5–8.

Divisibility by 3

The test for divisibility by 3 is similar to the test for divisibility by 9.

A number is divisible by 3 if the sum of the digits is divisible by 3.

▶ **EXAMPLES** Determine whether the number is divisible by 3.

7. 18 $1 + 8 = 9$

8. 93 $9 + 3 = 12$

9. 201 $2 + 0 + 1 = 3$

All divisible by 3 because the sums of their digits are divisible by 3.

10. 256 $2 + 5 + 6 = 13$ The sum is not divisible by 3, so 256 is not divisible by 3. ◀

DO EXERCISES 9–12.

Divisibility by 4

The test for divisibility by 4 is similar to the test for divisibility by 2.

A number is divisible by 4 if the number named by the last *two* digits is divisible by 4.

▶ **EXAMPLES** Determine whether the number is divisible by 4.

11. 8212 is divisible by 4 because 12 is divisible by 4.

12. 5216 is divisible by 4 because 16 is divisible by 4.

13. 8211 is *not* divisible by 4 because 11 is *not* divisible by 4.

14. 7515 is *not* divisible by 4 because 15 is *not* divisible by 4. ◀

DO EXERCISES 13–16.

To see why the test for divisibility by 4 works, consider 516:

$$516 = 5 \text{ hundreds} + 16.$$

Hundreds are multiples of 4. If the number named by the last two digits is a multiple of 4, then the entire number is a multiple of 4.

Divisibility by 8

The test for divisibility by 8 is an extension of the tests for divisibility by 2 and 4.

> **A number is divisible by 8 if the number named by the last *three* digits is divisible by 8.**

▶ **EXAMPLES** Determine whether each number is divisible by 8.

15. 5648 is divisible by 8 because 648 is divisible by 8.

16. 96,088 is divisible by 8 because 88 is divisible by 8.

17. 7324 is *not* divisible by 8 because 324 is *not* divisible by 8.

18. 13,420 is *not* divisible by 8 because 420 is *not* divisible by 8. ◀

DO EXERCISES 17–20.

Determine whether the number is divisible by 8.

17. 7564 No

18. 7864 Yes

19. 17,560 Yes

20. 25,716 No

Divisibility by 6

A number divisible by 6 is a multiple of 6. But $6 = 2 \cdot 3$, so the number is also a multiple of 2 and 3. Thus:

> **In order for a number to be divisible by 6, the sum of the digits must be divisible by 3 and the ones digit must be 0, 2, 4, 6, or 8 (even).**

▶ **EXAMPLES** Determine whether the number is divisible by 6.

19. 720

Since 720 is even, it is divisible by 2. Also, $7 + 2 + 0 = 9$, so 720 is divisible by 3. Thus, 720 is divisible by 6.

720 $\quad 7 + 2 + 0 = 9$

Even $\quad$ Divisible by 3

20. 73

73 is *not* divisible by 6 because it is *not* divisible by 2.

21. 256

256 is *not* divisible by 6 because the sum of the digits is *not* divisible by 3.

$$2 + 5 + 6 = 13$$

Not divisible by 3 ◀

DO EXERCISES 21–24.

Determine whether the number is divisible by 6.

21. 420 Yes

22. 106 No

23. 321 No

24. 444 Yes

ANSWERS ON PAGE A-2

Divisibility by 10

A number is divisible by 10 if the ones digit is 0.

We know that this test works because the product of 10 and *any* number has a ones digit of 0.

▶ **EXAMPLES** Determine whether the number is divisible by 10.

22. 3440 is divisible by 10 because the ones digit is 0.

23. 3447 is *not* divisible by 10 because the ones digit is not 0. ◀

DO EXERCISES 25–28.

Determine whether the number is divisible by 10.

25. 305 No

26. 300 Yes

27. 847 No

28. 8760 Yes

Divisibility by 5

A number is divisible by 5 if the ones digit is 0 or 5.

▶ **EXAMPLES** Determine whether the number is divisible by 5.

24. 220 is divisible by 5 because the ones digit is 0.

25. 475 is divisible by 5 because the ones digit is 5.

26. 6514 is *not* divisible by 5 because the ones digit is neither a 0 nor a 5. ◀

DO EXERCISES 29–32.

Determine whether the number is divisible by 5.

29. 5780 Yes

30. 3427 No

31. 34,678 No

32. 7775 Yes

Let's see why the test for 5 works. Consider 7830:

$$7830 = 10 \cdot 783 = 5 \cdot 2 \cdot 783.$$

Since 7830 is divisible by 10 and 5 is a factor of 10, 7830 is divisible by 5.

Consider 6734:

$$6734 = 673 \text{ tens} + 4.$$

Tens are multiples of 5, so the only number that must be checked is the ones digit. If the last digit is a multiple of 5, the entire number is: 4 is not a multiple of 5, so 6734 is not divisible by 5.

A Note About Divisibility by 7

There are several tests for divisibility by 7, but all of them are more complicated than simply dividing by 7. So if you want to test for divisibility by 7, divide by 7.

ANSWERS ON PAGE A-2

NAME SECTION DATE

EXERCISE SET 2.2

a To answer Exercises 1–8, consider the following numbers.

46	300	85
224	36	711
19	45,270	13,251
555	4444	254,765

1. Which of the above are divisible by 2?

2. Which of the above are divisible by 3?

3. Which of the above are divisible by 4?

4. Which of the above are divisible by 5?

5. Which of the above are divisible by 6?

6. Which of the above are divisible by 8?

7. Which of the above are divisible by 9?

8. Which of the above are divisible by 10?

To answer Exercises 9–16, consider the following numbers.

56	200	75
324	42	812
784	501	2345
55,555	3009	2001

9. Which of the above are divisible by 3?

10. Which of the above are divisible by 2?

ANSWERS

1. 46; 300; 224; 36; 45,270; 4444

2. 300; 36; 711; 45,270; 13,251; 555

3. 300; 224; 36; 4444

4. 300; 85; 45,270; 555; 254,765

5. 300; 36; 45,270

6. 224

7. 36; 711; 45,270

8. 300; 45,270

9. 75; 324; 42; 501; 3009; 2001

10. 56; 200; 324; 42; 812; 784

ANSWERS

11. 200; 75; 2345; 55,555

12. 56; 200; 324; 812; 784

13. 324

14. 324; 42

15. 200

16. 56; 200; 784

17. 138

18. 26

19. \$680

20. 4003

21. $2 \cdot 2 \cdot 2 \cdot 3 \cdot 5 \cdot 5 \cdot 13$

22. $2 \cdot 2 \cdot 2 \cdot 3 \cdot 3 \cdot 5 \cdot 7$

23. $2 \cdot 2 \cdot 3 \cdot 3 \cdot 7 \cdot 11$

24. $2 \cdot 3 \cdot 3 \cdot 3 \cdot 37$

11. Which of the above are divisible by 5?

12. Which of the above are divisible by 4?

13. Which of the above are divisible by 9?

14. Which of the above are divisible by 6?

15. Which of the above are divisible by 10?

16. Which of the above are divisible by 8?

SKILL MAINTENANCE

Solve.

17. $56 + x = 194$

18. $24 \cdot m = 624$

19. Find the total cost of 12 shirts at \$37 each and 4 pairs of trousers at \$59 each.

20. Divide: $45\overline{)180,135}$.

SYNTHESIS

Use the tests of divisibility to find the prime factorization of the number.

21. 7800

22. 2520

23. 2772

24. 1998

2.3 Fractions

The study of arithmetic begins with the set of whole numbers

0, 1, 2, 3, 4, 5, 6, 7, 8, 9, 10, 11, and so on.

The need soon arises for fractional parts of numbers such as halves, thirds, fourths, and so on. Here are some examples:

$\frac{1}{25}$ of the parking spaces in a commercial area in the state of Indiana are to be marked for the handicapped.

For $\frac{1}{10}$ of the people in the United States, English is not the primary language.

$\frac{1}{4}$ of the minimum daily requirement of calcium is provided by a cup of yogurt.

$\frac{16}{177}$ of the outdoor drive-in theaters in this country are in California.

a Identifying Numerators and Denominators

The following are some additional examples of fractions:

$$\frac{1}{2}, \quad \frac{3}{4}, \quad \frac{8}{5}, \quad \frac{11}{23}.$$

This way of writing number names is called **fractional notation.** The top number is called the **numerator** and the bottom number is called the **denominator.**

▶ **EXAMPLE 1** Identify the numerator and the denominator.

$7 \leftarrow$ Numerator
$8 \leftarrow$ Denominator ◀

DO EXERCISES 1–3.

b Fractions and the Real World

▶ **EXAMPLE 2** What part is shaded?

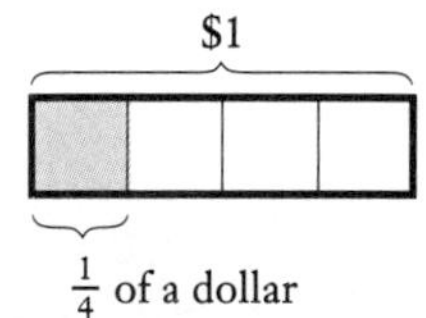

When an object is divided into 4 parts of the same size, each of these parts is $\frac{1}{4}$ of the object. Thus, $\frac{1}{4}$ (*one-fourth*) is shaded. ◀

DO EXERCISES 4–7.

OBJECTIVES

After finishing Section 2.3, you should be able to:

a Identify the numerator and the denominator of a fraction.

b Write fractional notation for part of an object or part of a set of objects.

c Simplify fractional notation like *n*/*n* to 1, 0/*n* to 0, and *n*/1 to *n*.

FOR EXTRA HELP

Tape NC

Tape 3B

MAC: 2
IBM: 2

Identify the numerator and the denominator.

1. $\frac{1}{6}$ **2.** $\frac{5}{7}$ **3.** $\frac{22}{3}$

1 numerator; 6 denominator 5 numerator 7 denominator 22 numerator; 3 denominator

What part is shaded?

4. $\frac{1}{2}$

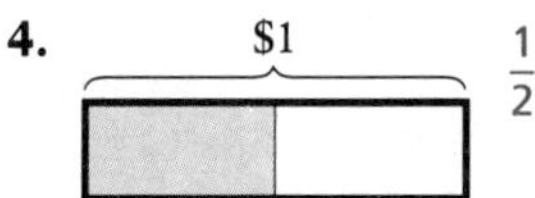

5. 1 mile $\frac{1}{3}$

6. $\frac{1}{3}$

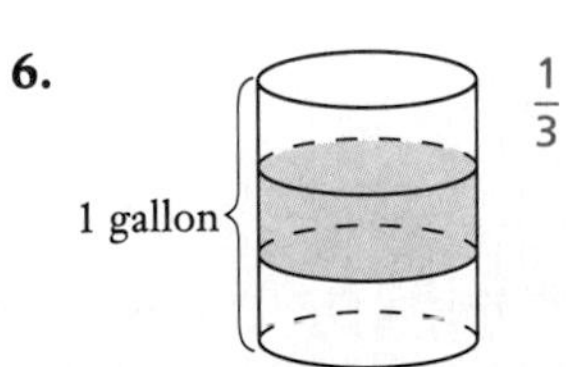

7. $\frac{1}{6}$

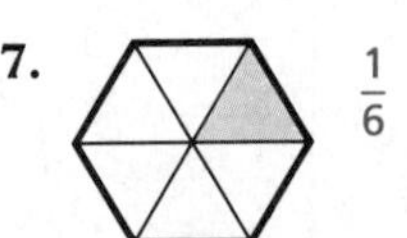

ANSWERS ON PAGE A-2

What part is shaded?

8. $\frac{5}{8}$

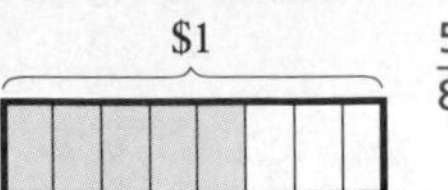

9. 1 mile $\frac{2}{3}$

10. $\frac{3}{4}$

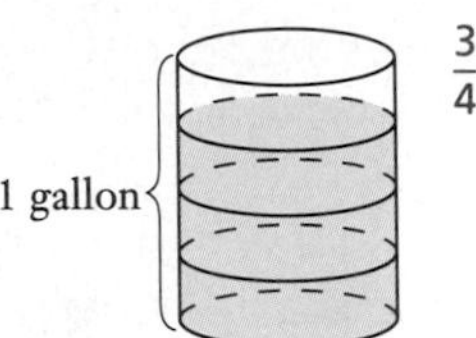

11. $\frac{4}{6}$

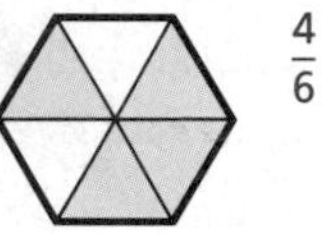

What part is shaded?

12. $\frac{4}{3}$

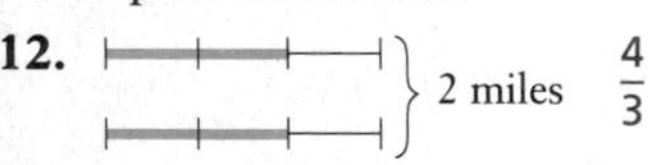

13. $\frac{5}{5}$

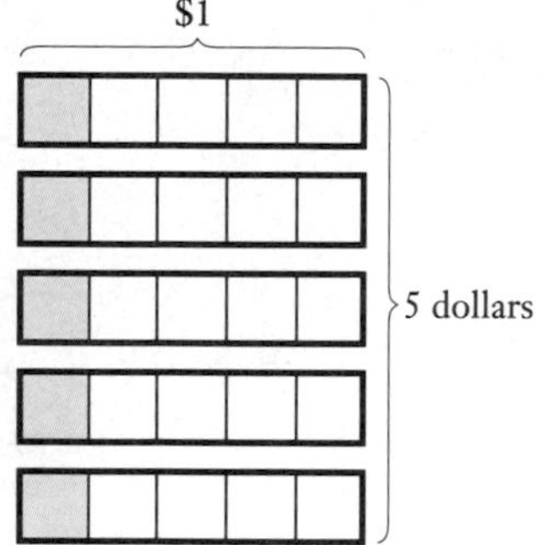

What part is shaded?

14. $\frac{5}{4}$

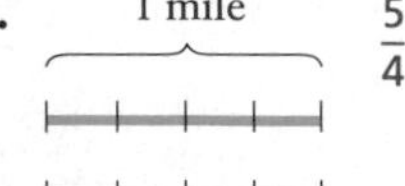

15. $\frac{7}{4}$

ANSWERS ON PAGE A-2

► **EXAMPLE 3** What part is shaded?

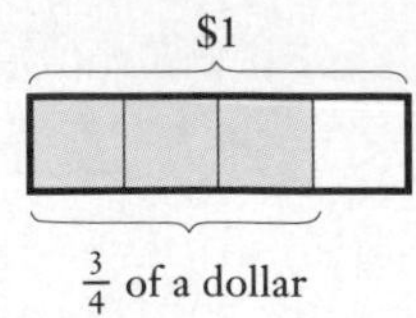

The object is divided into 4 parts of the same size, and 3 of them are shaded. This is $3 \cdot \frac{1}{4}$, or $\frac{3}{4}$. Thus, $\frac{3}{4}$ (*three-fourths*) of the object is shaded. ◄

DO EXERCISES 8–11.

The fraction $\frac{3}{4}$ corresponds to another situation. We take 3 objects, divide them into fourths, and take $\frac{1}{4}$ of the entire amount. This is $\frac{1}{4} \cdot 3$, or $\frac{3}{4}$, or $3 \div 4$.

► **EXAMPLE 4** What part is shaded?

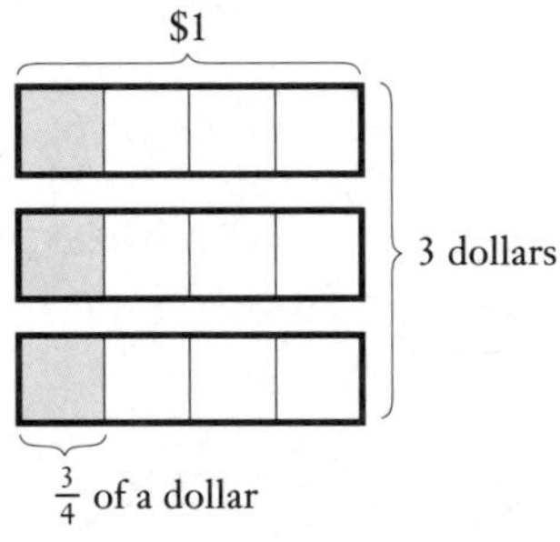

Thus, $\$\frac{3}{4}$ is shaded. ◄

DO EXERCISES 12 AND 13.

Fractions greater than 1 correspond to situations like the following.

► **EXAMPLE 5** What part is shaded?

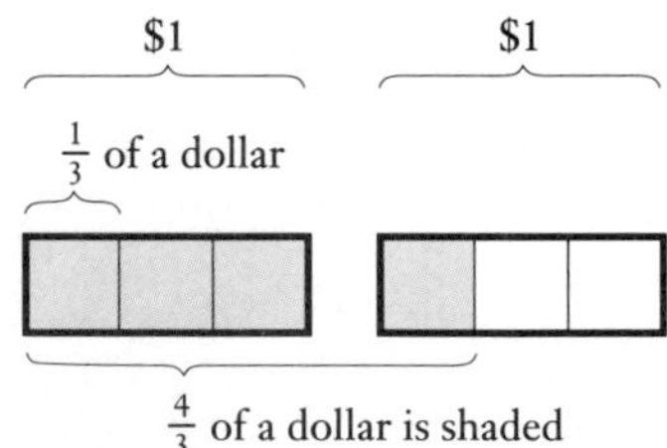

We divide the two objects into 3 parts each and take 4 of those parts. We have more than one whole object. In this case, it is $4 \cdot \frac{1}{3}$, or $\frac{4}{3}$. ◄

DO EXERCISES 14–15.

Fractional notation also corresponds to situations involving part of a set.

▶ **EXAMPLE 6** What part of this set of tools are wrenches?

There are 5 tools, and 3 are wrenches. We say that three-fifths of the tools are wrenches; that is, $\frac{3}{5}$ of the set consists of wrenches. ◀

DO EXERCISES 16–18.

16. What part of the set of tools in Example 6 are hammers? $\frac{2}{5}$

17. What part of this set is shaded? $\frac{2}{3}$

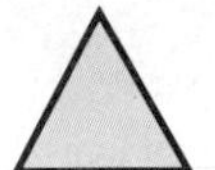

18. What part of this set are or were United States presidents? are recording stars?

Abraham Lincoln
Debbie Gibson
Elton John
George Bush
Linda Ronstadt
Gloria Estefan $\frac{2}{6}$; $\frac{4}{6}$

C Some Fractional Notation for Whole Numbers

Fractional Notation for 1

The number 1 corresponds to situations like the following.

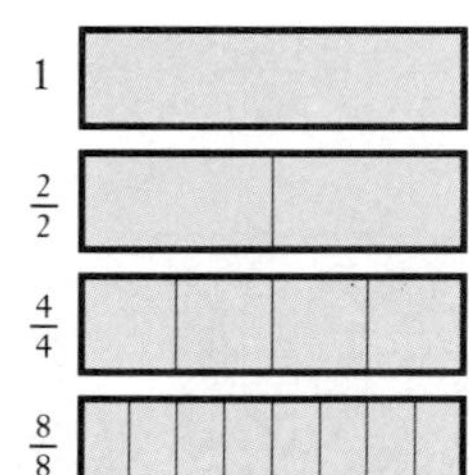

If we divide an object into n parts and take n of them, we get all of the object (1 whole object).

$\frac{n}{n} = 1$, for any whole number n that is not 0.

▶ **EXAMPLES** Simplify.

7. $\frac{5}{5} = 1$

8. $\frac{9}{9} = 1$

9. $\frac{23}{23} = 1$ ◀

DO EXERCISES 19–24.

Simplify.

19. $\frac{1}{1}$ 1

20. $\frac{4}{4}$ 1

21. $\frac{34}{34}$ 1

22. $\frac{100}{100}$ 1

23. $\frac{2347}{2347}$ 1

24. $\frac{103}{103}$ 1

ANSWERS ON PAGE A-2

Simplify.

25. $\frac{0}{1}$ 0

26. $\frac{0}{8}$ 0

27. $\frac{0}{107}$ 0

28. $\frac{4-4}{567}$ 0

Simplify.

29. $\frac{8}{1}$ 8

30. $\frac{10}{1}$ 10

31. $\frac{346}{1}$ 346

32. $\frac{24-1}{23}$ 1

ANSWERS ON PAGE A-2

Fractional Notation for 0

Consider $\frac{0}{4}$. This corresponds to dividing an object into 4 parts and taking none of them. We get 0.

> $\frac{0}{n} = 0,$ **for any whole number n that is not 0.**

▶ **EXAMPLES** Simplify.

10. $\frac{0}{1} = 0$

11. $\frac{0}{9} = 0$

12. $\frac{0}{23} = 0$ ◀

DO EXERCISES 25–28.

Fractional notation with a denominator of 0, such as $n/0$, is meaningless because we cannot speak of an object divided into *zero* parts. (If it is not divided at all, then we say that it is undivided and remains in one part.)

> $\frac{n}{0}$ **is not defined for any whole number n.**

Other Whole Numbers

Consider $\frac{4}{1}$. This corresponds to taking 4 objects and dividing them into 1 part. (We do not divide them.) We have 4 objects.

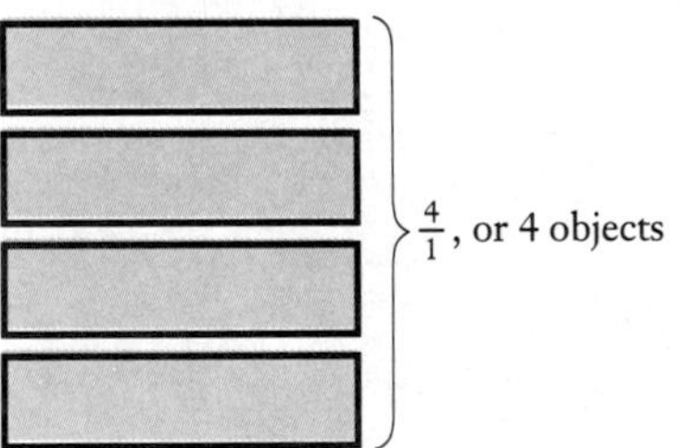

> **Any whole number divided by 1 is the whole number. That is,**
>
> $\frac{n}{1} = n,$ **for any whole number n.**

▶ **EXAMPLES** Simplify.

13. $\frac{2}{1} = 2$

14. $\frac{9}{1} = 9$

15. $\frac{34}{1} = 34$ ◀

DO EXERCISES 29–32.

NAME SECTION DATE

EXERCISE SET 2.3

a Identify the numerator and the denominator.

1. $\frac{3}{4}$ **2.** $\frac{9}{10}$ **3.** $\frac{11}{20}$ **4.** $\frac{18}{5}$

b What part of each object or set of objects is shaded?

5.

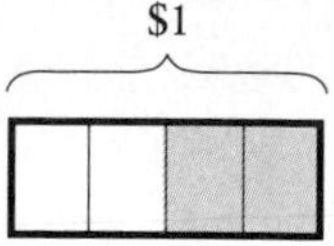

6.

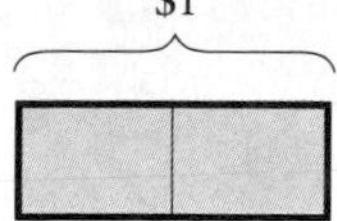

7.

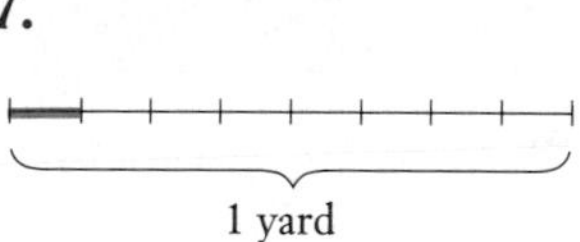

8.

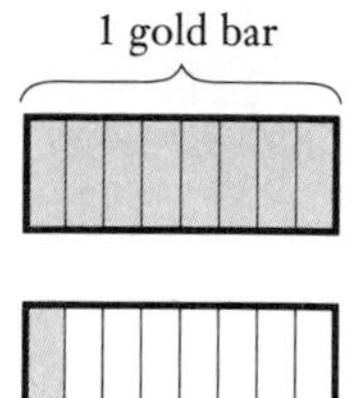

9.

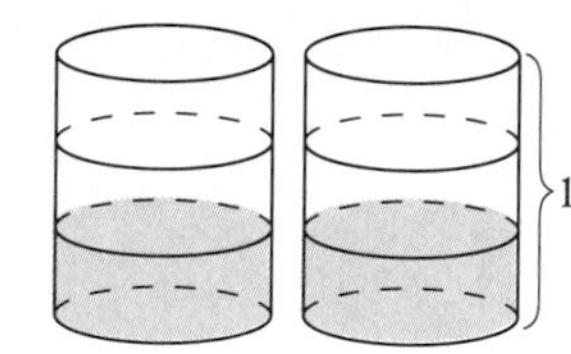

10.

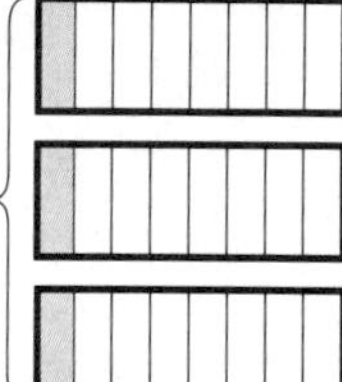

11.

12.

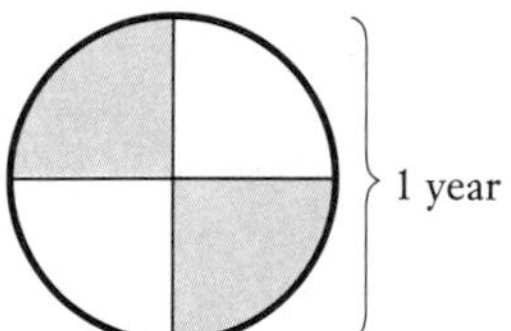

13.

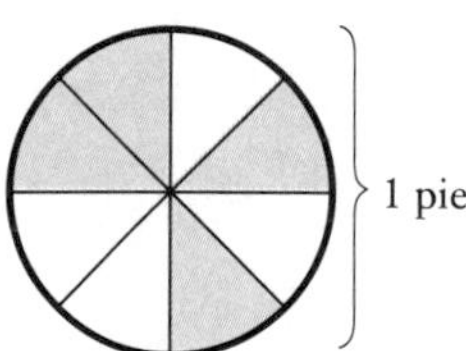

14.

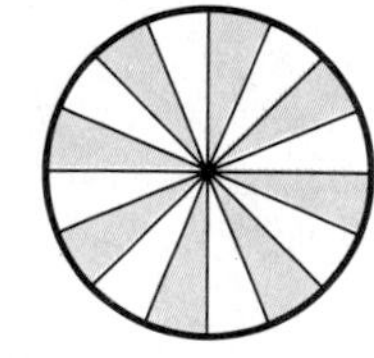

15.

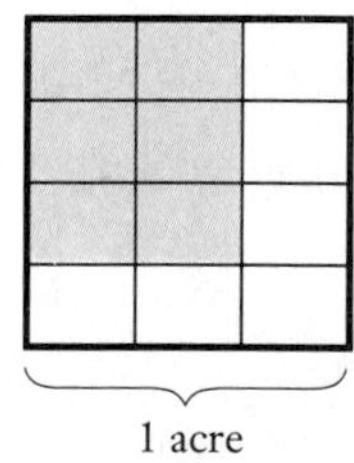

16.

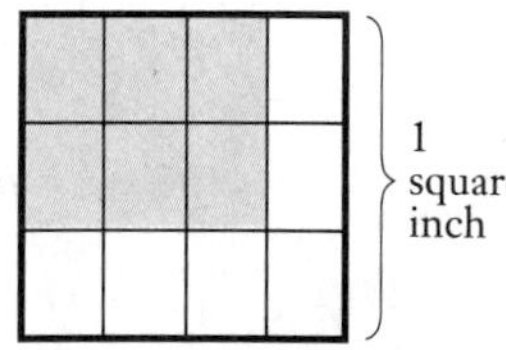

17.

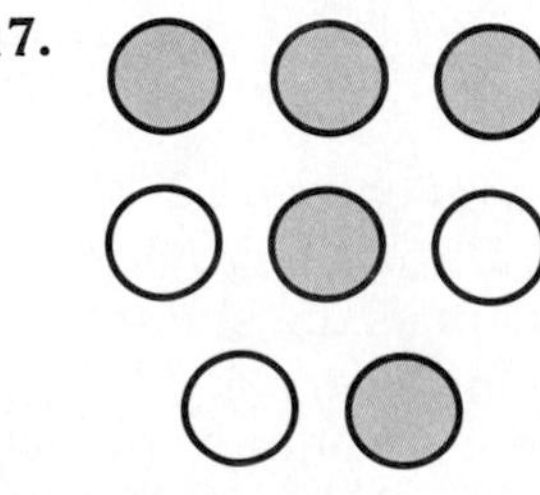

18.

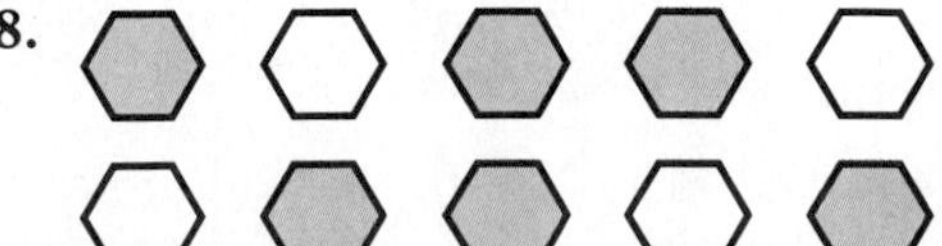

ANSWERS

1. 4 denominator
2. 9 numerator; 10 denominator
3. 11 numerator; 20 denominator
4. 5 denominator
5. $\frac{2}{4}$
6. $\frac{2}{2}$
7. $\frac{1}{8}$
8. $\frac{9}{8}$
9. $\frac{2}{3}$
10. $\frac{3}{8}$
11. $\frac{3}{4}$
12. $\frac{2}{4}$
13. $\frac{4}{8}$
14. $\frac{8}{16}$
15. $\frac{6}{12}$
16. $\frac{6}{12}$
17. $\frac{5}{8}$
18. $\frac{6}{10}$

ANSWERS

19. $\frac{3}{5}$
20. $\frac{4}{7}$
21. 0
22. 1
23. 5
24. 10
25. 1
26. 20
27. 1
28. 7
29. 0
30. 234
31. 1
32. 0
33. 1
34. 1
35. 1
36. 1
37. 1
38. 0
39. 8
40. 0
41. 34,560
42. 34,600
43. 35,000
44. $3413
45. 201 min
46. $\frac{3}{4}$; $\frac{1}{4}$
47. $\frac{1200}{2700}$; $\frac{540}{2700}$; $\frac{360}{2700}$; $\frac{600}{2700}$

19.

20.

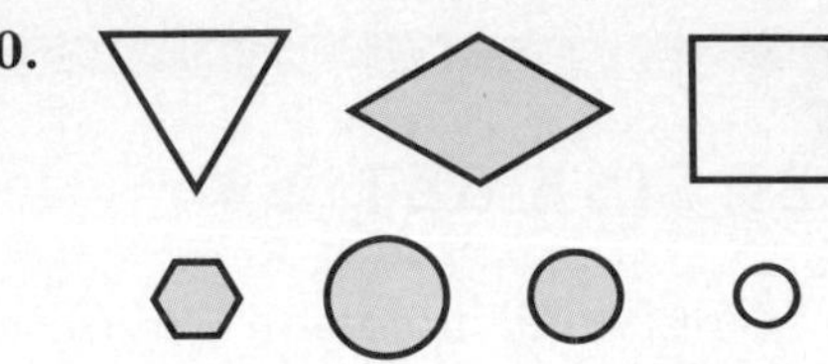

c Simplify.

21. $\frac{0}{5}$ **22.** $\frac{5}{5}$ **23.** $\frac{5}{1}$ **24.** $\frac{10}{1}$

25. $\frac{20}{20}$ **26.** $\frac{20}{1}$ **27.** $\frac{45}{45}$ **28.** $\frac{7}{1}$

29. $\frac{0}{234}$ **30.** $\frac{234}{1}$ **31.** $\frac{234}{234}$ **32.** $\frac{0}{1}$

33. $\frac{3}{3}$ **34.** $\frac{56}{56}$ **35.** $\frac{57}{57}$ **36.** $\frac{58}{58}$

37. $\frac{8}{8}$ **38.** $\frac{0}{8}$ **39.** $\frac{8}{1}$ **40.** $\frac{8-8}{1247}$

SKILL MAINTENANCE

Round 34,562 to the nearest:

41. Ten. **42.** Hundred. **43.** Thousand.

44. The annual income of people living in Alaska is $17,155 per person. This is the highest in the United States. In Colorado the annual income is $13,742. How much more do people in Alaska make than those living in Colorado?

45. A typist can type 62 words per minute. How long will it take to type 12,462 words?

SYNTHESIS

46. The surface of the earth is 3 parts water and 1 part land. What part of the earth is water? land?

47. A college student earned $2700 one summer. During the following year, the student spent $1200 for tuition, $540 for rent, and $360 for food. The rest went for miscellaneous expenses. What part of the income went for tuition? rent? food? miscellaneous expenses?

2.4 Multiplication

a Multiplication by a Whole Number

We can find $3 \cdot \frac{1}{4}$ by thinking of repeated addition. We add three $\frac{1}{4}$'s.

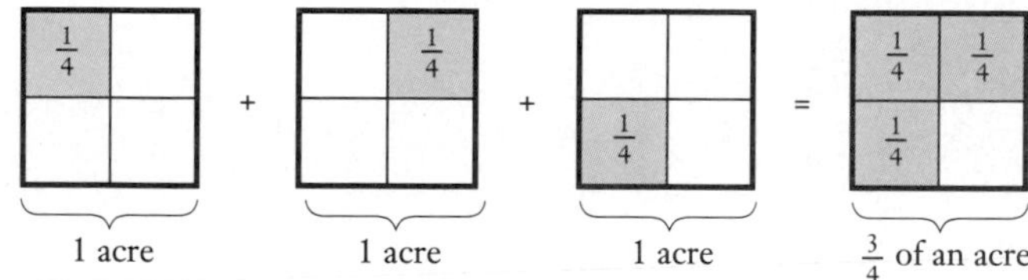

We see that $3 \cdot \frac{1}{4}$ is $\frac{3}{4}$.

DO EXERCISES 1 AND 2.

> **To multiply a fraction by a whole number,**
> **a) multiply the top number (the numerator) by the whole number, and**
> **b) keep the same denominator.**
>
> $$6 \cdot \frac{4}{5} = \frac{6 \cdot 4}{5} = \frac{24}{5}$$

▶ **EXAMPLES** Multiply.

1. $5 \times \frac{3}{8} = \frac{5 \times 3}{8} = \frac{15}{8}$

Skip this step whenever you can.

2. $\frac{2}{5} \cdot 13 = \frac{2 \cdot 13}{5} = \frac{26}{5}$

3. $10 \cdot \frac{1}{3} = \frac{10}{3}$ ◀

DO EXERCISES 3–5.

b Multiplication Using Fractional Notation

We find a product such as $\frac{9}{7} \cdot \frac{3}{4}$ as follows.

> **To multiply,**
> **a) multiply the numerators, and**
> **b) multiply the denominators.**
>
> $$\frac{9}{7} \cdot \frac{3}{4} = \frac{9 \cdot 3}{7 \cdot 4} = \frac{27}{28}$$

OBJECTIVES

After finishing Section 2.4, you should be able to:

a Multiply a whole number and a fraction.

b Multiply using fractional notation.

c Solve problems involving multiplication of fractions.

FOR EXTRA HELP

Tape 3B

Tape 4A

MAC: 2
IBM: 2

1. Find $2 \cdot \frac{1}{3}$. $\frac{2}{3}$

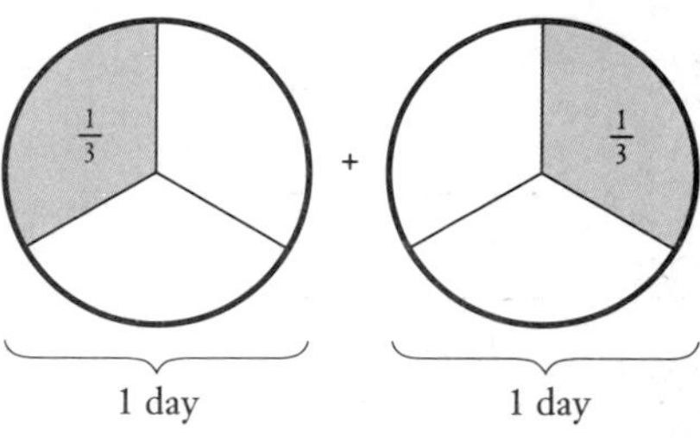

2. Find $5 \cdot \frac{1}{8}$. $\frac{5}{8}$

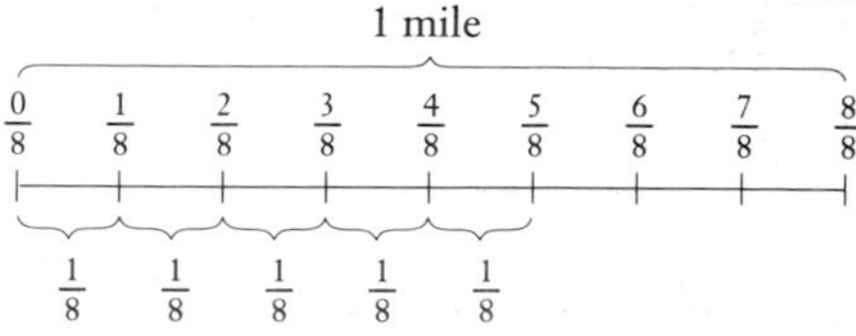

Multiply.

3. $5 \times \frac{2}{3}$ $\frac{10}{3}$

4. $11 \times \frac{3}{8}$ $\frac{33}{8}$

5. $23 \cdot \frac{2}{5}$ $\frac{46}{5}$

ANSWERS ON PAGE A-2

Multiply.

6. $\frac{3}{8} \cdot \frac{5}{7}$ $\frac{15}{56}$

7. $\frac{4}{3} \times \frac{8}{5}$ $\frac{32}{15}$

8. $\frac{3}{10} \cdot \frac{1}{10}$ $\frac{3}{100}$

9. $7 \cdot \frac{2}{3}$ $\frac{14}{3}$

10. Draw diagrams like those in the text to show how the multiplication $\frac{1}{3} \cdot \frac{4}{5}$ corresponds to a real-world situation.

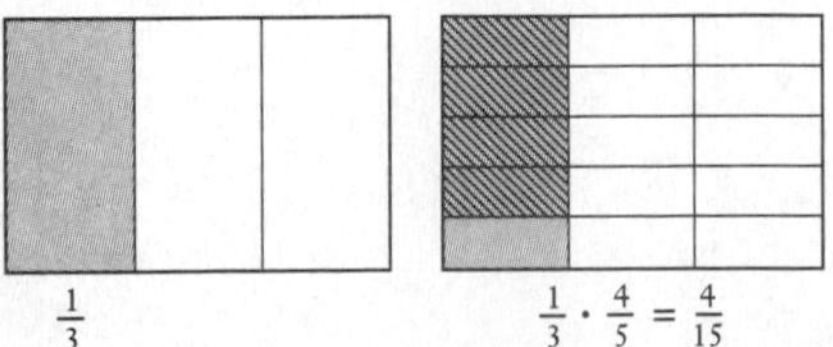

ANSWERS ON PAGE A-2

▶ **EXAMPLES** Multiply.

4. $\frac{5}{6} \times \frac{7}{4} = \frac{5 \times 7}{6 \times 4} = \frac{35}{24}$

Skip this step whenever you can.

5. $\frac{3}{5} \cdot \frac{7}{8} = \frac{3 \cdot 7}{5 \cdot 8} = \frac{21}{40}$

6. $\frac{3}{5} \cdot \frac{3}{4} = \frac{9}{20}$

7. $\frac{1}{4} \cdot \frac{1}{3} = \frac{1}{12}$

8. $6 \cdot \frac{4}{5} = \frac{6}{1} \cdot \frac{4}{5} = \frac{24}{5}$ ◀

DO EXERCISES 6–9.

Unless one of the factors is a whole number, multiplication does not correspond to repeated addition. Let us see how multiplication of fractions corresponds to situations in the real world. We consider the multiplication

$$\frac{3}{5} \cdot \frac{3}{4}.$$

We first consider some object and take $\frac{3}{4}$ of it. We divide it into 4 parts and take 3 of them. That is shown in the shading below.

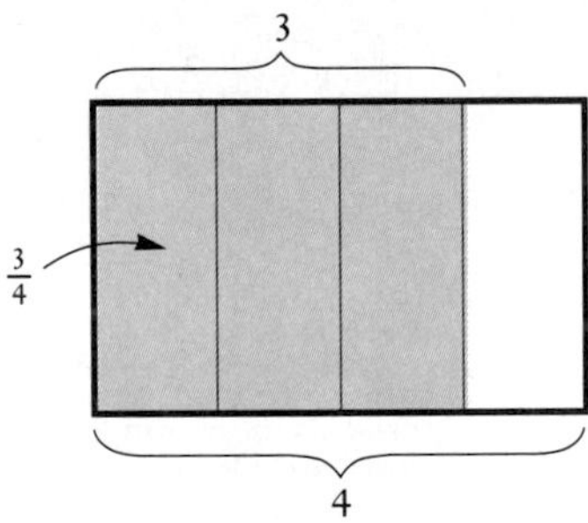

Next, we take $\frac{3}{5}$ of the result. We divide the shaded part into 5 parts and take 3 of them. That is shown below.

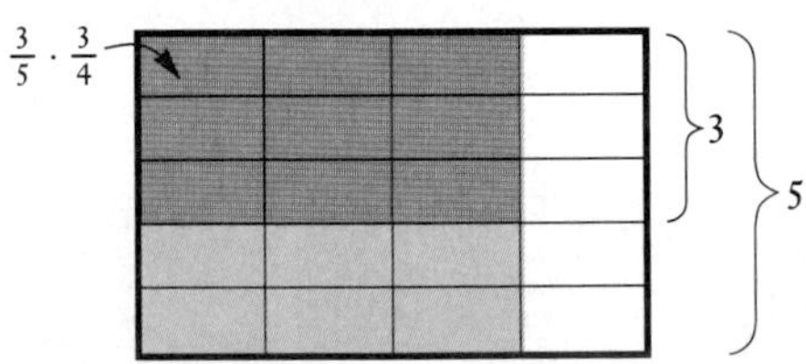

The entire object has been divided into 20 parts, and we have shaded 9 of them for a second time:

$$\frac{3}{5} \cdot \frac{3}{4} = \frac{3 \cdot 3}{5 \cdot 4} = \frac{9}{20}.$$

The figure above shows a rectangular array inside a rectangular array. The number of pieces in the entire array is $5 \cdot 4$ (the product of the denominators). The number of pieces doubly shaded is $3 \cdot 3$ (the product of the numerators). For the answer, we take 9 pieces out of a set of 20 to get $\frac{9}{20}$.

DO EXERCISE 10.

C Solving Problems

Most problems that can be solved by multiplying fractions can be thought of in terms of rectangular arrays.

▶ **EXAMPLE 9** A farmer owns a square mile of land. He gives $\frac{4}{5}$ of it to his daughter and she gives $\frac{2}{3}$ of her share to her son. How much land goes to the son?

1. *Familiarize.* We draw a picture to help solve the problem. The land may not be square. It could be in a shape like A or B below, or it could even be in more than one piece. But to think out the problem, we can think of it as a square, as shown by shape C.

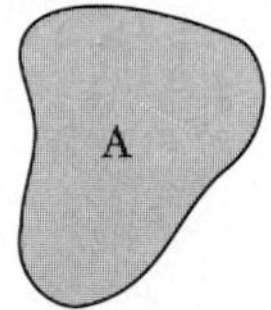

1 square mile

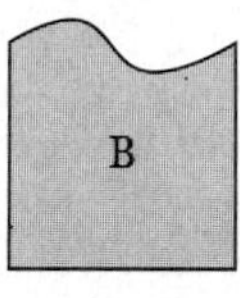

1 square mile

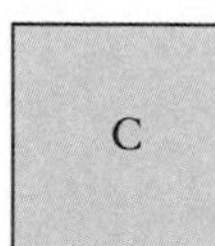

1 square mile

The daughter gets $\frac{4}{5}$ of the land. We shade $\frac{4}{5}$.

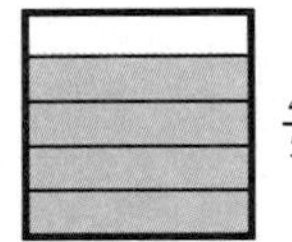

Her son gets $\frac{2}{3}$ of her part. We shade that.

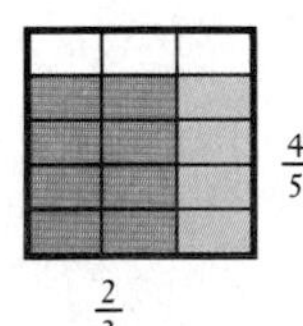

2. *Translate.* We let n = the part of the land that goes to the son. We are taking "two-thirds of four-fifths." The word "of" corresponds to multiplication. Thus the following multiplication sentence corresponds to the situation:

$$\frac{2}{3} \cdot \frac{4}{5} = n.$$

3. *Solve.* The number sentence tells us what to do. We multiply:

$$\frac{2}{3} \cdot \frac{4}{5} = \frac{8}{15}.$$

4. *Check.* We can check partially by noting that the answer is smaller than the original area, 1, which we expect since the farmer is giving parts of the land away. Thus, $\frac{8}{15}$ is a reasonable answer.
5. *State.* The son gets $\frac{8}{15}$ of a square mile of land. ◀

DO EXERCISE 11.

11. A family uses $\frac{3}{4}$ of its land for a play area. Of that, $\frac{1}{2}$ is used for a swimming pool. What part of the land is used for the swimming pool? $\frac{3}{8}$

ANSWER ON PAGE A-2

12. The length of a key on a calculator is $\frac{9}{10}$ of a centimeter. The width is $\frac{7}{10}$ of a centimeter. What is the area? $\frac{63}{100}$ sq cm

13. Of the students in the sophomore class, $\frac{1}{8}$ participate in sports and $\frac{3}{7}$ of these play football. What part of the number of students in the sophomore class play football? $\frac{3}{56}$

ANSWERS ON PAGE A-2

Example 9 and the preceding discussion indicate that the area of a rectangular region can be found by multiplying length by width. That is true whether length and width are whole numbers or not. Remember, the area of a rectangular region is given by the formula

$$A = l \cdot w.$$

▶ **EXAMPLE 10** The length of a rectangular button on a calculator is $\frac{7}{10}$ of a centimeter. The width is $\frac{3}{10}$ of a centimeter. What is the area?

1. *Familiarize.* Recall that area is length times width. We draw a picture, letting A = the area of the calculator button.

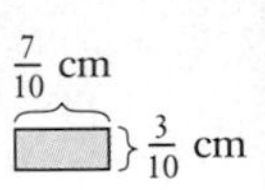

2. *Translate.* Then we translate.

Area	is	length	times	width
↓	↓	↓	↓	↓
A	$=$	$\frac{7}{10}$	$\times$	$\frac{3}{10}$

3. *Solve.* The sentence tells us what to do. We multiply:

$$\frac{7}{10} \cdot \frac{3}{10} = \frac{7 \cdot 3}{10 \cdot 10} = \frac{21}{100}.$$

4. *Check.* We check by repeating the calculation. This is left to the student.
5. *State.* The area is $\frac{21}{100}$ square centimeter. ◀

DO EXERCISE 12.

▶ **EXAMPLE 11** A recipe calls for $\frac{3}{4}$ of a cup of flour. A chef is making an amount that is $\frac{1}{2}$ of the recipe. How much flour should the chef use?

1. *Familiarize.* We draw a picture or at least visualize the situation. We let n = the amount of flour the chef should use.

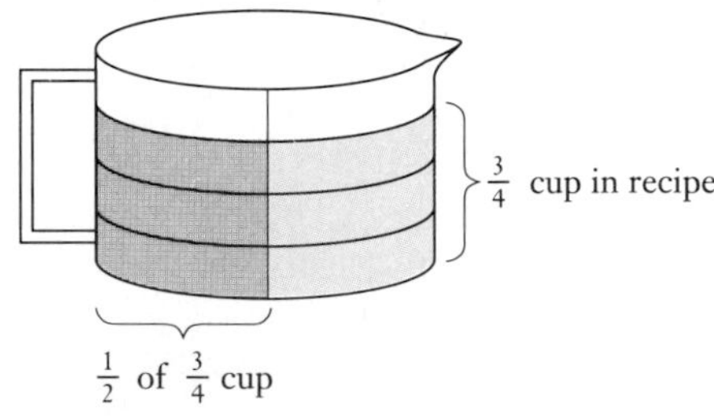

2. *Translate.* The multiplication sentence $\frac{1}{2} \cdot \frac{3}{4} = n$ corresponds to the situation.
3. *Solve.* We carry out the multiplication:

$$\frac{1}{2} \cdot \frac{3}{4} = \frac{1 \cdot 3}{2 \cdot 4} = \frac{3}{8}.$$

4. *Check.* We check by repeating the calculation. This is left to the student.
5. *State.* The chef should use $\frac{3}{8}$ of a cup of flour. ◀

DO EXERCISE 13.

NAME SECTION DATE

EXERCISE SET 2.4

a Multiply.

1. $3 \cdot \frac{1}{5}$

2. $2 \cdot \frac{1}{3}$

3. $5 \times \frac{1}{6}$

4. $4 \times \frac{1}{7}$

5. $\frac{2}{11} \cdot 4$

6. $\frac{2}{5} \cdot 3$

7. $10 \cdot \frac{7}{9}$

8. $9 \cdot \frac{5}{8}$

9. $\frac{2}{5} \cdot 1$

10. $\frac{3}{8} \cdot 1$

11. $\frac{2}{5} \cdot 3$

12. $\frac{3}{5} \cdot 4$

13. $7 \cdot \frac{3}{4}$

14. $7 \cdot \frac{2}{5}$

15. $17 \times \frac{5}{6}$

16. $\frac{3}{7} \cdot 40$

b Multiply.

17. $\frac{1}{2} \cdot \frac{1}{3}$

18. $\frac{1}{4} \cdot \frac{1}{5}$

19. $\frac{1}{4} \times \frac{1}{10}$

20. $\frac{1}{3} \times \frac{1}{10}$

21. $\frac{2}{3} \times \frac{1}{5}$

22. $\frac{3}{5} \times \frac{1}{5}$

23. $\frac{2}{5} \cdot \frac{2}{3}$

24. $\frac{3}{4} \cdot \frac{3}{5}$

25. $\frac{3}{4} \cdot \frac{3}{4}$

26. $\frac{3}{7} \cdot \frac{4}{5}$

27. $\frac{2}{3} \cdot \frac{7}{13}$

28. $\frac{3}{11} \cdot \frac{4}{5}$

ANSWERS

1. $\frac{3}{5}$
2. $\frac{2}{3}$
3. $\frac{5}{6}$
4. $\frac{4}{7}$
5. $\frac{8}{11}$
6. $\frac{6}{5}$
7. $\frac{70}{9}$
8. $\frac{45}{8}$
9. $\frac{2}{5}$
10. $\frac{3}{8}$
11. $\frac{6}{5}$
12. $\frac{12}{5}$
13. $\frac{21}{4}$
14. $\frac{14}{5}$
15. $\frac{85}{6}$
16. $\frac{120}{7}$
17. $\frac{1}{6}$
18. $\frac{1}{20}$
19. $\frac{1}{40}$
20. $\frac{1}{30}$
21. $\frac{2}{15}$
22. $\frac{3}{25}$
23. $\frac{4}{15}$
24. $\frac{9}{20}$
25. $\frac{9}{16}$
26. $\frac{12}{35}$
27. $\frac{14}{39}$
28. $\frac{12}{55}$

ANSWERS

29. $\frac{7}{100}$

30. $\frac{9}{100}$

31. $\frac{49}{64}$

32. $\frac{16}{25}$

33. $\frac{1}{1000}$

34. $\frac{21}{1000}$

35. $\frac{182}{285}$

36. $\frac{144}{169}$

37. $\frac{12}{25}$ sq m

38. $\frac{1}{12}$

39. $\frac{7}{16}$ L

40. $\frac{1}{1521}$

41. $\frac{3}{8}$ cup

42. $\frac{10}{3}$ yd

43. $\frac{230}{1000}$

44. $\frac{492}{1000}$

45. 204

46. 700

47. 3001

48. 6 hundred thousands

29. $\frac{1}{10} \cdot \frac{7}{10}$

30. $\frac{3}{10} \cdot \frac{3}{10}$

31. $\frac{7}{8} \cdot \frac{7}{8}$

32. $\frac{4}{5} \cdot \frac{4}{5}$

33. $\frac{1}{10} \cdot \frac{1}{100}$

34. $\frac{3}{10} \cdot \frac{7}{100}$

35. $\frac{14}{15} \cdot \frac{13}{19}$

36. $\frac{12}{13} \cdot \frac{12}{13}$

c Solve.

37. A rectangular table top measures $\frac{4}{5}$ of a meter long by $\frac{3}{5}$ of a meter wide. What is its area?

38. If each piece of pie is $\frac{1}{6}$ of a pie, how much of the pie is $\frac{1}{2}$ of a piece?

39. A can holds $\frac{7}{8}$ of a liter. How much will it hold when it is $\frac{1}{2}$ full?

40. One out of 39 high school football players plays college football. One out of 39 college players plays professional football. What fraction of the number of high school players plays professional football?

41. A cereal recipe calls for $\frac{3}{4}$ of a cup of honey. How much is needed to make $\frac{1}{2}$ of a recipe?

42. It takes $\frac{2}{3}$ of a yard of silk to make a bow. How much silk is needed for 5 bows?

43. Out of every 1000 people who attend movies, 230 are in the 16–20 age group. What fraction of the number of all moviegoers is in the 16–20 age group?

44. Out of every 1000 musical recordings that are purchased, 492 are cassettes. What fraction of the number of musical recordings that are purchased are cassettes?

SKILL MAINTENANCE

Divide.

45. $35\overline{)7140}$

46. $46\overline{)32{,}200}$

47. $9\overline{)27{,}009}$

48. What does the digit 6 mean in 4,678,952?

2.5 Simplifying

a Multiplying by 1

Recall the following:

$$1 = \frac{1}{1} = \frac{2}{2} = \frac{3}{3} = \frac{4}{4} = \frac{10}{10} = \frac{45}{45} = \frac{100}{100} = \frac{n}{n}.$$

Any nonzero number divided by itself is 1.

When we multiply a number by 1, we get the same number.

$$\frac{3}{5} \cdot 1 = \frac{3}{5} \cdot \frac{4}{4} = \frac{12}{20}$$

Since $\frac{3}{5} \cdot 1 = \frac{12}{20}$, we know that $\frac{3}{5}$ and $\frac{12}{20}$ are two names for the same number. We also say that $\frac{3}{5}$ and $\frac{12}{20}$ are **equivalent.**

DO EXERCISES 1–4.

Suppose we want to find a name for $\frac{2}{3}$, but one that has a denominator of 9. We can multiply by 1 to find equivalent fractions:

$$\frac{2}{3} = \frac{2}{3} \cdot \frac{3}{3} = \frac{2 \cdot 3}{3 \cdot 3} = \frac{6}{9}.$$

We chose $\frac{3}{3}$ for 1 in order to get a denominator of 9.

▶ **EXAMPLE 1** Find a name for $\frac{1}{4}$ with a denominator of 24.

Since $4 \cdot 6 = 24$, we multiply by $\frac{6}{6}$:

$$\frac{1}{4} = \frac{1}{4} \cdot \frac{6}{6} = \frac{1 \cdot 6}{4 \cdot 6} = \frac{6}{24}.$$ ◀

▶ **EXAMPLE 2** Find a name for $\frac{2}{5}$ with a denominator of 35.

Since $5 \cdot 7 = 35$, we multiply by $\frac{7}{7}$:

$$\frac{2}{5} = \frac{2}{5} \cdot \frac{7}{7} = \frac{2 \cdot 7}{5 \cdot 7} = \frac{14}{35}.$$ ◀

DO EXERCISES 5–9.

b Simplifying

All of the following are names for three-fourths:

$$\frac{3}{4}, \frac{6}{8}, \frac{9}{12}, \frac{12}{16}, \frac{15}{20}.$$

We say that $\frac{3}{4}$ is **simplest** because it has the smallest numerator and denominator.

OBJECTIVES

After finishing Section 2.5, you should be able to:

a Find another name for a number, but having a new denominator. Use multiplying by 1.

b Simplify fractional notation.

c Test fractions for equality.

FOR EXTRA HELP

Tape 3C | Tape 4A | MAC: 2 IBM: 2

Multiply.

1. $\frac{1}{2} \cdot \frac{8}{8}$ $\frac{8}{16}$

2. $\frac{3}{5} \cdot \frac{10}{10}$ $\frac{30}{50}$

3. $\frac{13}{25} \cdot \frac{4}{4}$ $\frac{52}{100}$

4. $\frac{8}{3} \cdot \frac{25}{25}$ $\frac{200}{75}$

Find another name for the number, but with the denominator indicated. Use multiplying by 1.

5. $\frac{4}{3} = \frac{?}{9}$ $\frac{12}{9}$

6. $\frac{3}{4} = \frac{?}{24}$ $\frac{18}{24}$

7. $\frac{9}{10} = \frac{?}{100}$ $\frac{90}{100}$

8. $\frac{3}{15} = \frac{?}{45}$ $\frac{9}{45}$

9. $\frac{8}{7} = \frac{?}{49}$ $\frac{56}{49}$

ANSWERS ON PAGE A-2

Simplify.

10. $\frac{2}{8}$ $\frac{1}{4}$

11. $\frac{10}{12}$ $\frac{5}{6}$

12. $\frac{40}{8}$ 5

13. $\frac{24}{18}$ $\frac{4}{3}$

ANSWERS ON PAGE A-2

To simplify, we reverse the process of "multiplying by 1."

Remove the greatest common factor of the numerator and the denominator.

$$\frac{12}{18} = \frac{2 \cdot 6}{3 \cdot 6} \quad \begin{matrix}\longleftarrow \textbf{Factor the numerator.} \\ \longleftarrow \textbf{Factor the denominator.}\end{matrix}$$

$$= \frac{2}{3} \cdot \frac{6}{6} \longleftarrow \textbf{Factor the fraction.}$$

$$= \frac{2}{3} \cdot 1 \longleftarrow \frac{6}{6} = 1$$

$$= \frac{2}{3} \quad \longleftarrow \textbf{Removing a factor of 1: } \frac{2}{3} \cdot 1 = \frac{2}{3}$$

▶ **EXAMPLES** Simplify.

3. $\frac{8}{20} = \frac{2 \cdot 4}{5 \cdot 4} = \frac{2}{5} \cdot \frac{4}{4} = \frac{2}{5}$

4. $\frac{2}{6} = \frac{1 \cdot 2}{3 \cdot 2} = \frac{1}{3} \cdot \frac{2}{2} = \frac{1}{3}$

The number 1 allows for pairing of factors in the numerator and the denominator.

5. $\frac{30}{6} = \frac{5 \cdot 6}{1 \cdot 6} = \frac{5}{1} \cdot \frac{6}{6} = \frac{5}{1} = 5$

We could also simplify $\frac{30}{6}$ by doing the division $30 \div 6$. That is, $\frac{30}{6} = 30 \div 6 = 5$. ◀

DO EXERCISES 10–13.

The use of prime factorizations can be helpful for larger numbers.

▶ **EXAMPLE 6** Simplify: $\frac{90}{84}$.

$$\frac{90}{84} = \frac{2 \cdot 3 \cdot 3 \cdot 5}{2 \cdot 2 \cdot 3 \cdot 7} \quad \textbf{Factoring the numerator and the denominator into primes}$$

$$= \frac{2 \cdot 3 \cdot 3 \cdot 5}{2 \cdot 3 \cdot 2 \cdot 7} \quad \textbf{Changing the order so that like primes are above and below each other}$$

$$= \frac{2}{2} \cdot \frac{3}{3} \cdot \frac{3 \cdot 5}{2 \cdot 7} \quad \textbf{Factoring the fraction}$$

$$= 1 \cdot 1 \cdot \frac{3 \cdot 5}{2 \cdot 7}$$

$$= \frac{3 \cdot 5}{2 \cdot 7} \quad \textbf{Removing factors of 1}$$

$$= \frac{15}{14}$$ ◀

We could have shortened the preceding example had we recalled our tests for divisibility (Section 2.2) and noted that 6 is a factor of both the numerator and the denominator. Then

$$\frac{90}{84} = \frac{6 \cdot 15}{6 \cdot 14} = \frac{6}{6} \cdot \frac{15}{14} = \frac{15}{14}.$$

The tests for divisibility are very helpful in simplifying.

▸ **EXAMPLE 7** Simplify: $\frac{603}{207}$.

At first glance this looks difficult. But note, using the test for divisibility by 9 (sum of digits divisible by 9), that both the numerator and the denominator are divisible by 9. Thus we can factor 9 from both numbers:

$$\frac{603}{207} = \frac{9 \cdot 67}{9 \cdot 23} = \frac{9}{9} \cdot \frac{67}{23} = \frac{67}{23}.$$ ◂

DO EXERCISES 14–17.

CANCELING. Canceling is a shortcut that you may have used for removing a factor of 1 when working with fractional notation. With *great* concern, we mention it as a possibility of speeding up your work. Canceling may be done only when removing common factors in numerators and denominators. Each common factor allows us to remove a factor of 1 in a product. Canceling may not be done in sums. Our concern is that canceling be done with care and understanding. In effect, slashes are used to indicate factors of 1 that have been removed. Example 6 might have been done faster as follows:

$$\frac{90}{84} = \frac{2 \cdot 3 \cdot 3 \cdot 5}{2 \cdot 2 \cdot 3 \cdot 7}$$ Factoring the numerator and the denominator

$$= \frac{\not{2} \cdot \not{3} \cdot 3 \cdot 5}{2 \cdot \not{2} \cdot \not{3} \cdot 7}$$ When a factor of 1 is noted, it is "canceled" as shown: $\frac{2 \cdot 3}{2 \cdot 3} = 1$.

$$= \frac{3 \cdot 5}{2 \cdot 7} = \frac{15}{14}.$$

CAUTION! The difficulty with canceling is that it is often applied incorrectly in situations like the following:

$$\frac{\not{2} + 3}{\not{2}} = 3; \qquad \frac{\not{4} + 1}{\not{4} + 2} = \frac{1}{2}; \qquad \frac{1\not{5}}{\not{5}4} = \frac{1}{4}.$$

Wrong! Wrong! Wrong!

The correct answers are

$$\frac{2 + 3}{2} = \frac{5}{2}; \qquad \frac{4 + 1}{4 + 2} = \frac{5}{6}; \qquad \frac{15}{54} = \frac{5}{18}.$$

In each situation, the number canceled was *not* a factor of 1. Factors are parts of products. For example, in $2 \cdot 3$, 2 and 3 are factors, but in $2 + 3$, 2 and 3 are *not* factors.

If you cannot factor, do not cancel! If in doubt, do not cancel!

c A Test for Equality

Suppose we want to compare $\frac{2}{4}$ and $\frac{3}{6}$. When denominators are the same, we say that the fractions have a **common denominator.** We find a common denominator and compare numerators. To do this, we multiply by 1 using symbols for 1 formed by looking at opposite denominators.

$$\frac{2}{4} = \frac{2}{4} \cdot \frac{6}{6} = \frac{2 \cdot 6}{4 \cdot 6} = \frac{12}{24}$$

$$\frac{3}{6} = \frac{3}{6} \cdot \frac{4}{4} = \frac{3 \cdot 4}{6 \cdot 4} = \frac{12}{24}$$

We see that $\frac{2}{4} = \frac{3}{6}$.

Simplify.

14. $\frac{35}{40}$ $\frac{7}{8}$

15. $\frac{801}{702}$ $\frac{89}{78}$

16. $\frac{24}{21}$ $\frac{8}{7}$

17. $\frac{75}{300}$ $\frac{1}{4}$

ANSWERS ON PAGE A-2

Use = or ≠ for $\square$ to write a true sentence.

18. $\frac{2}{6} \square \frac{3}{9}$ =

19. $\frac{2}{3} \square \frac{14}{20}$ ≠

ANSWERS ON PAGE A-2

Note in the preceding that if

$$\frac{2}{4} = \frac{3}{6}, \quad \text{then } 2 \cdot 6 = 4 \cdot 3.$$

We need only check the products $2 \cdot 6$ and $4 \cdot 3$ to compare the fractions.

A Test For Equality

We multiply these two numbers: 3 · 4.

We multiply these two numbers: 6 · 2.

$$\frac{3}{6} \square \frac{2}{4}$$

Since 3 · 4 = 6 · 2, we know that

$$\frac{3}{6} = \frac{2}{4}.$$

We call 3 · 4 and 6 · 2 *cross products*.

If a sentence $a = b$ is true, it means that a and b name the same number. If a sentence $a \neq b$ is true, it means that a and b do *not* name the same number.

▶ **EXAMPLE 8** Use = or ≠ for $\square$ to write a true sentence:

$$\frac{6}{7} \square \frac{7}{8}.$$

We multiply these two numbers: $6 \cdot 8 = 48$.

We multiply these two numbers: $7 \cdot 7 = 49$.

$$\frac{6}{7} \quad \frac{7}{8}$$

Since $48 \neq 49$ (read "48 is not the same as 49"), $\frac{6}{7} = \frac{7}{8}$ is not a true sentence. Thus,

$$\frac{6}{7} \neq \frac{7}{8}.$$ ◀

▶ **EXAMPLE 9** Use = or ≠ for $\square$ to write a true sentence:

$$\frac{6}{10} \square \frac{3}{5}.$$

We multiply these two numbers: $6 \cdot 5 = 30$.

We multiply these two numbers: $10 \cdot 3 = 30$.

$$\frac{6}{10} \quad \frac{3}{5}.$$

Since the cross products are the same, we have

$$\frac{6}{10} = \frac{3}{5}.$$ ◀

DO EXERCISES 18 AND 19.

NAME SECTION DATE

EXERCISE SET 2.5

a Find another name for the given number, but with the denominator indicated. Use multiplying by 1.

1. $\frac{1}{2} = \frac{?}{10}$ **2.** $\frac{1}{6} = \frac{?}{12}$ **3.** $\frac{3}{4} = \frac{?}{48}$ **4.** $\frac{2}{9} = \frac{?}{18}$

5. $\frac{9}{10} = \frac{?}{30}$ **6.** $\frac{3}{8} = \frac{?}{48}$ **7.** $\frac{7}{8} = \frac{?}{32}$ **8.** $\frac{2}{5} = \frac{?}{25}$

9. $\frac{5}{12} = \frac{?}{48}$ **10.** $\frac{7}{8} = \frac{?}{56}$ **11.** $\frac{17}{18} = \frac{?}{54}$ **12.** $\frac{11}{16} = \frac{?}{256}$

13. $\frac{5}{3} = \frac{?}{45}$ **14.** $\frac{11}{5} = \frac{?}{30}$ **15.** $\frac{7}{22} = \frac{?}{132}$ **16.** $\frac{10}{21} = \frac{?}{126}$

b Simplify.

17. $\frac{2}{4}$ **18.** $\frac{3}{6}$ **19.** $\frac{6}{8}$ **20.** $\frac{9}{12}$ **21.** $\frac{3}{15}$

22. $\frac{8}{10}$ **23.** $\frac{24}{8}$ **24.** $\frac{36}{4}$ **25.** $\frac{18}{24}$ **26.** $\frac{42}{48}$

27. $\frac{14}{16}$ **28.** $\frac{15}{25}$ **29.** $\frac{12}{10}$ **30.** $\frac{16}{14}$ **31.** $\frac{16}{48}$

32. $\frac{100}{20}$ **33.** $\frac{150}{25}$ **34.** $\frac{19}{76}$ **35.** $\frac{17}{51}$ **36.** $\frac{425}{525}$

ANSWERS

1. $\frac{5}{10}$
2. $\frac{2}{12}$
3. $\frac{36}{48}$
4. $\frac{4}{18}$
5. $\frac{27}{30}$
6. $\frac{18}{48}$
7. $\frac{28}{32}$
8. $\frac{10}{25}$
9. $\frac{20}{48}$
10. $\frac{49}{56}$
11. $\frac{51}{54}$
12. $\frac{176}{256}$
13. $\frac{75}{45}$
14. $\frac{66}{30}$
15. $\frac{42}{132}$
16. $\frac{60}{126}$
17. $\frac{1}{2}$
18. $\frac{1}{2}$
19. $\frac{3}{4}$
20. $\frac{3}{4}$
21. $\frac{1}{5}$
22. $\frac{4}{5}$
23. 3
24. 9
25. $\frac{3}{4}$
26. $\frac{7}{8}$
27. $\frac{7}{8}$
28. $\frac{3}{5}$
29. $\frac{6}{5}$
30. $\frac{8}{7}$
31. $\frac{1}{3}$
32. 5
33. 6
34. $\frac{1}{4}$
35. $\frac{1}{3}$
36. $\frac{17}{21}$

ANSWERS

37. =

38. =

39. ≠

40. ≠

41. =

42. =

43. ≠

44. ≠

45. =

46. =

47. ≠

48. ≠

49. =

50. =

51. ≠

52. ≠

53. 4992 sq ft

54. $928

55. $\frac{2}{5}$

56. No; $\frac{3}{4} \neq \frac{4}{5}$, because $3 \cdot 5 \neq 4 \cdot 4$

57. No; $\frac{92}{564} \neq \frac{84}{634}$, because $92 \cdot 634 \neq 564 \cdot 84$

58. No; $\frac{63}{78} \neq \frac{75}{100}$, because $63 \cdot 100 \neq 78 \cdot 75$

C Use = or ≠ for ■ to write a true sentence.

37. $\frac{3}{4}$ ■ $\frac{9}{12}$

38. $\frac{4}{8}$ ■ $\frac{3}{6}$

39. $\frac{1}{5}$ ■ $\frac{2}{9}$

40. $\frac{1}{4}$ ■ $\frac{2}{9}$

41. $\frac{3}{8}$ ■ $\frac{6}{16}$

42. $\frac{2}{6}$ ■ $\frac{6}{18}$

43. $\frac{2}{5}$ ■ $\frac{3}{7}$

44. $\frac{1}{3}$ ■ $\frac{1}{4}$

45. $\frac{12}{9}$ ■ $\frac{8}{6}$

46. $\frac{16}{14}$ ■ $\frac{8}{7}$

47. $\frac{5}{2}$ ■ $\frac{17}{7}$

48. $\frac{3}{10}$ ■ $\frac{7}{24}$

49. $\frac{3}{10}$ ■ $\frac{30}{100}$

50. $\frac{700}{1000}$ ■ $\frac{70}{100}$

51. $\frac{5}{10}$ ■ $\frac{520}{1000}$

52. $\frac{49}{100}$ ■ $\frac{50}{1000}$

SKILL MAINTENANCE

Solve.

53. A playing field is 78 ft long and 64 ft wide. What is its area?

54. A landscaper buys 13 maple trees and 17 oak trees for a project. A maple costs $23 and an oak costs $37. How much is spent for all the trees?

SYNTHESIS

55. Sociologists have found that 4 out of 10 people are shy. Write fractional notation for the part of the population that is shy. Simplify.

56. Is $\frac{3}{4}$ of a gallon the same as $\frac{4}{5}$ of a gallon? Why or why not?

57. ■ In a recent year, Jim Rice of the Boston Red Sox got 92 hits in 564 times at bat. George Brett of the Kansas City Royals got 84 hits in 634 times at bat. Did they have the same batting average? Why or why not?

58. ■ On a test of 78 questions, a student got 63 correct. On another test of 100 questions, the student got 75 correct. Did the student get the same portion of each test correct? Why or why not?

2.6 Multiplying and Simplifying

OBJECTIVES

After finishing Section 2.6, you should be able to:

a Multiply and simplify using fractional notation.

b Solve problems involving multiplication.

FOR EXTRA HELP

Tape 3D | Tape 4B | MAC: 2 IBM: 2

a Simplifying After Multiplying

We usually simplify after we multiply. To make such simplifying easier, it is generally best not to carry out the products in the numerator and the denominator, but to factor and simplify. Consider the product

$$\frac{3}{8} \cdot \frac{4}{9}.$$

We proceed as follows:

$$\begin{aligned}
\frac{3}{8} \cdot \frac{4}{9} &= \frac{3 \cdot 4}{8 \cdot 9} && \text{We write the products in the numerator and the denominator but we do not carry them out.} \\
&= \frac{3 \cdot 2 \cdot 2}{2 \cdot 2 \cdot 2 \cdot 3 \cdot 3} && \text{Factoring the numerator and the denominator} \\
&= \frac{3 \cdot 2 \cdot 2}{3 \cdot 2 \cdot 2} \cdot \frac{1}{2 \cdot 3} && \text{Factoring the fraction} \\
&= 1 \cdot \frac{1}{2 \cdot 3} \\
&= \frac{1}{2 \cdot 3} && \text{Removing a factor of 1} \\
&= \frac{1}{6}.
\end{aligned}$$

The procedure could have been shortened had we noticed that 4 is a factor of the 8 in the denominator:

$$\frac{3}{8} \cdot \frac{4}{9} = \frac{3 \cdot 4}{8 \cdot 9} = \frac{3 \cdot 4}{4 \cdot 2 \cdot 3 \cdot 3} = \frac{3 \cdot 4}{3 \cdot 4} \cdot \frac{1}{2 \cdot 3} = 1 \cdot \frac{1}{2 \cdot 3} = \frac{1}{2 \cdot 3} = \frac{1}{6}.$$

To multiply and simplify:

a) Write the products in the numerator and the denominator, but do not carry out the products.

b) Factor the numerator and the denominator.

c) Factor the fraction to remove factors of 1.

d) Carry out the remaining products.

▶ **EXAMPLES** Multiply and simplify.

1. $\frac{2}{3} \cdot \frac{9}{4} = \frac{2 \cdot 9}{3 \cdot 4} = \frac{2 \cdot 3 \cdot 3}{3 \cdot 2 \cdot 2} = \frac{2 \cdot 3}{2 \cdot 3} \cdot \frac{3}{2} = 1 \cdot \frac{3}{2} = \frac{3}{2}$

2. $\frac{6}{7} \cdot \frac{5}{3} = \frac{6 \cdot 5}{7 \cdot 3} = \frac{3 \cdot 2 \cdot 5}{7 \cdot 3} = \frac{3}{3} \cdot \frac{2 \cdot 5}{7} = 1 \cdot \frac{2 \cdot 5}{7} = \frac{2 \cdot 5}{7} = \frac{10}{7}$

3. $40 \cdot \frac{7}{8} = \frac{40 \cdot 7}{8} = \frac{8 \cdot 5 \cdot 7}{8 \cdot 1} = \frac{8}{8} \cdot \frac{5 \cdot 7}{1} = 1 \cdot \frac{5 \cdot 7}{1} = \frac{5 \cdot 7}{1} = 35$ ◀

Multiply and simplify.

1. $\frac{2}{3} \cdot \frac{7}{8}$ $\frac{7}{12}$

2. $\frac{4}{5} \cdot \frac{5}{12}$ $\frac{1}{3}$

3. $16 \cdot \frac{3}{8}$ 6

4. $\frac{5}{8} \cdot 4$ $\frac{5}{2}$

5. A florist uses $\frac{2}{5}$ of a pound of peat moss for a rosebush. How much will be needed for 25 rosebushes? 10 lb

ANSWERS ON PAGE A-2

CAUTION! Canceling can be used as follows for these examples.

1. $\frac{2}{3} \cdot \frac{9}{4} = \frac{2 \cdot 9}{3 \cdot 4} = \frac{\not{2} \cdot \not{3} \cdot 3}{\not{3} \cdot \not{2} \cdot 2} = \frac{3}{2}$ Removing a factor of 1: $\frac{2 \cdot 3}{2 \cdot 3} = 1$

2. $\frac{6}{7} \cdot \frac{5}{3} = \frac{6 \cdot 5}{7 \cdot 3} = \frac{\not{3} \cdot 2 \cdot 5}{7 \cdot \not{3}} = \frac{2 \cdot 5}{7} = \frac{10}{7}$ Removing a factor of 1: $\frac{3}{3} = 1$

3. $40 \cdot \frac{7}{8} = \frac{40 \cdot 7}{8} = \frac{\not{8} \cdot 5 \cdot 7}{\not{8} \cdot 1} = \frac{5 \cdot 7}{1} = 35$ Removing a factor of 1: $\frac{8}{8} = 1$

Remember, if you can't factor, you can't cancel!

DO EXERCISES 1–4.

b Solving Problems

▶ **EXAMPLE 4** How much steak will be needed to serve 30 people if each person gets $\frac{2}{3}$ of a pound?

1. *Familiarize.* We first draw a picture or at least visualize the situation. Repeated addition will work here.

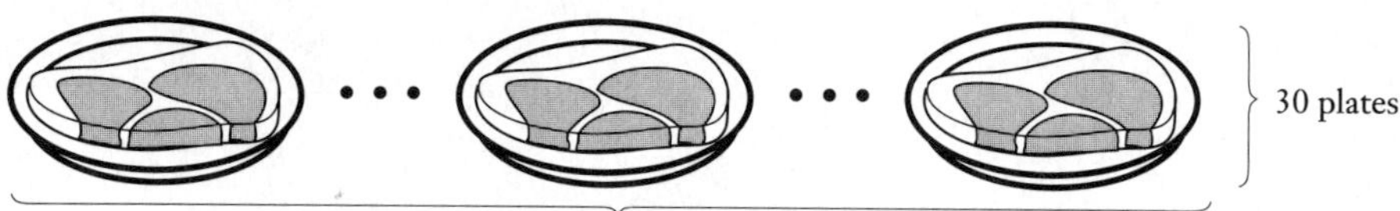

We let n = the number of pounds of steak needed.

2. *Translate.* The problem translates to the following equation:

$$n = 30 \cdot \frac{2}{3}.$$

3. *Solve.* To solve the equation, we carry out the multiplication:

$$\begin{aligned} n &= 30 \cdot \frac{2}{3} = \frac{30 \cdot 2}{3} \quad \text{Multiplying} \\ &= \frac{3 \cdot 10 \cdot 2}{3 \cdot 1} \\ &= \frac{3}{3} \cdot \frac{10 \cdot 2}{1} \\ &= 20. \quad \text{Simplifying} \end{aligned}$$

4. *Check.* We check by repeating the calculation. (We leave the check to the student.) We can also think about the reasonableness of the answer. We are multiplying 30 by a number less than 1, so the product will be less than 30. Since 20 is less than 30, we have a partial check of the reasonableness of the answer. The number 20 checks.

5. *State.* Thus, 20 lb of steak will be needed. ◀

DO EXERCISE 5.

NAME SECTION DATE

EXERCISE SET 2.6

a Multiply and simplify. Don't forget to simplify!

1. $\frac{2}{3} \cdot \frac{1}{2}$

2. $\frac{4}{5} \cdot \frac{1}{4}$

3. $\frac{7}{8} \cdot \frac{1}{7}$

4. $\frac{5}{6} \cdot \frac{1}{5}$

5. $\frac{1}{8} \cdot \frac{4}{5}$

6. $\frac{2}{5} \cdot \frac{1}{6}$

7. $\frac{1}{4} \cdot \frac{2}{3}$

8. $\frac{3}{6} \cdot \frac{1}{6}$

9. $\frac{12}{5} \cdot \frac{9}{8}$

10. $\frac{16}{15} \cdot \frac{5}{4}$

11. $\frac{10}{9} \cdot \frac{7}{5}$

12. $\frac{25}{12} \cdot \frac{4}{3}$

13. $9 \cdot \frac{1}{9}$

14. $4 \cdot \frac{1}{4}$

15. $\frac{1}{3} \cdot 3$

16. $\frac{1}{6} \cdot 6$

17. $\frac{7}{10} \cdot \frac{10}{7}$

18. $\frac{8}{9} \cdot \frac{9}{8}$

19. $\frac{7}{5} \cdot \frac{5}{7}$

20. $\frac{2}{11} \cdot \frac{11}{2}$

ANSWERS

1. $\frac{1}{3}$

2. $\frac{1}{5}$

3. $\frac{1}{8}$

4. $\frac{1}{6}$

5. $\frac{1}{10}$

6. $\frac{1}{15}$

7. $\frac{1}{6}$

8. $\frac{1}{12}$

9. $\frac{27}{10}$

10. $\frac{4}{3}$

11. $\frac{14}{9}$

12. $\frac{25}{9}$

13. 1

14. 1

15. 1

16. 1

17. 1

18. 1

19. 1

20. 1

ANSWERS

21. 2
22. 2
23. 5
24. 7
25. 9
26. 15
27. 9
28. 8
29. $\frac{26}{5}$
30. $\frac{5}{2}$
31. $\frac{98}{5}$
32. $\frac{85}{4}$
33. 60
34. 40
35. 30
36. 30
37. $\frac{1}{5}$
38. $\frac{119}{750}$
39. $\frac{9}{25}$
40. $\frac{6}{25}$

21. $\frac{1}{4} \cdot 8$

22. $\frac{1}{6} \cdot 12$

23. $15 \cdot \frac{1}{3}$

24. $14 \cdot \frac{1}{2}$

25. $12 \cdot \frac{3}{4}$

26. $18 \cdot \frac{5}{6}$

27. $\frac{3}{8} \cdot 24$

28. $\frac{2}{9} \cdot 36$

29. $13 \cdot \frac{2}{5}$

30. $15 \cdot \frac{1}{6}$

31. $\frac{7}{10} \cdot 28$

32. $\frac{5}{8} \cdot 34$

33. $\frac{1}{6} \cdot 360$

34. $\frac{1}{3} \cdot 120$

35. $240 \cdot \frac{1}{8}$

36. $150 \cdot \frac{1}{5}$

37. $\frac{4}{10} \cdot \frac{5}{10}$

38. $\frac{7}{10} \cdot \frac{34}{150}$

39. $\frac{8}{10} \cdot \frac{45}{100}$

40. $\frac{3}{10} \cdot \frac{8}{10}$

41. $\frac{11}{24} \cdot \frac{3}{5}$

42. $\frac{15}{22} \cdot \frac{4}{7}$

43. $\frac{10}{21} \cdot \frac{3}{4}$

44. $\frac{17}{18} \cdot \frac{3}{5}$

b Solve.

45. A person receives $36 for working a full day. How much is received for working $\frac{1}{4}$ of a day?

46. A person receives $45 for working a full day. How much is received for working $\frac{1}{5}$ of a day?

47. Business people have determined that $\frac{1}{4}$ of the items on a mailing list will change in one year. A business has a mailing list of 2500 people. After one year, how many addresses on that list will be incorrect?

48. Sociologists have determined that $\frac{2}{5}$ of the people in the world are shy. A sales manager is interviewing 650 people for an aggressive sales position. How many of these people might be shy?

49. A recipe calls for $\frac{2}{3}$ of a cup of flour. A chef is making an amount that is $\frac{1}{2}$ of the recipe. How much flour should the chef use?

50. Of the students in the freshman class, $\frac{2}{5}$ have cameras; $\frac{1}{4}$ of these students also join the college photography club. What fraction of the students in the freshman class join the photography club?

ANSWERS

41. $\frac{11}{40}$

42. $\frac{30}{77}$

43. $\frac{5}{14}$

44. $\frac{17}{30}$

45. $9

46. $9

47. 625

48. 260

49. $\frac{1}{3}$ cup

50. $\frac{1}{10}$

ANSWERS

51. $1600

52. $2100

53. 160 mi

54. 90 mi

55. $3375 for food; $2700 for housing; $1350 for clothing; $1500 for savings; $3375 for taxes; $1200 for other

56. $3150 for food; $2520 for housing; $1260 for clothing; $1400 for savings; $3150 for taxes; $1120 for other

57. 35

58. 8546

59. 4673

60. 5338

51. A student's tuition was $2400. A loan was obtained for $\frac{2}{3}$ of the tuition. How much was the loan?

52. A student's tuition was $2800. A loan was obtained for $\frac{3}{4}$ of the tuition. How much was the loan?

53. On a map, 1 in. represents 240 mi. How much does $\frac{2}{3}$ in. represent?

54. On a map, 1 in. represents 120 mi. How much does $\frac{3}{4}$ in. represent?

55. A family has an annual income of $13,500. Of this, $\frac{1}{4}$ is spent for food, $\frac{1}{5}$ for housing, $\frac{1}{10}$ for clothing, $\frac{1}{9}$ for savings, $\frac{1}{4}$ for taxes, and the rest for other expenses. How much is spent for each?

56. A family has an annual income of $12,600. Of this, $\frac{1}{4}$ is spent for food, $\frac{1}{5}$ for housing, $\frac{1}{10}$ for clothing, $\frac{1}{9}$ for savings, $\frac{1}{4}$ for taxes, and the rest for other expenses. How much is spent for each?

SKILL MAINTENANCE

Solve.

57. $48 \cdot t = 1680$

58. $456 + x = 9002$

Subtract.

59. $\begin{array}{r} 9\,0\,6\,0 \\ -\,4\,3\,8\,7 \\ \hline \end{array}$

60. $\begin{array}{r} 7\,8\,0\,0 \\ -\,2\,4\,6\,2 \\ \hline \end{array}$

2.7 Reciprocals and Division

a Reciprocals

Look at these products:

$$8 \cdot \frac{1}{8} = \frac{8 \cdot 1}{8} = \frac{8}{8} = 1; \qquad \frac{2}{3} \cdot \frac{3}{2} = \frac{2 \cdot 3}{3 \cdot 2} = \frac{6}{6} = 1.$$

If the product of two numbers is 1, we say that they are *reciprocals* of each other. To find a reciprocal, interchange the numerator and the denominator:

Number $\longrightarrow \frac{3}{4} \quad \frac{4}{3}. \longleftarrow$ **Reciprocal**

▶ **EXAMPLES** Find the reciprocal.

1. The reciprocal of $\frac{4}{5}$ is $\frac{5}{4}$.
2. The reciprocal of $\frac{8}{7}$ is $\frac{7}{8}$.
3. The reciprocal of 8 is $\frac{1}{8}$. Think of 8 as $\frac{8}{1}$.
4. The reciprocal of $\frac{1}{3}$ is 3. ◀

DO EXERCISES 1–4.

Does 0 have a reciprocal? If it did, it would have to be a number x such that

$$0 \cdot x = 1.$$

But 0 times any number is 0. Thus,

The number 0, or $\frac{0}{n}$, has no reciprocal! (Recall that $\frac{n}{0}$ is not defined.)

b Division

Recall that $a \div b$ is that number which when multiplied by b gives a. Consider the division $\frac{3}{4} \div \frac{1}{8}$. We are asking how many $\frac{1}{8}$'s are in $\frac{3}{4}$. We can answer this by looking at the figure below.

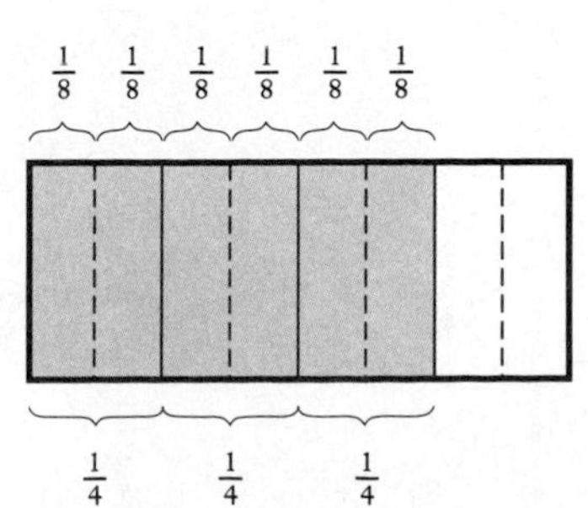

OBJECTIVES

After finishing Section 2.7, you should be able to:

- a Find the reciprocal of a number.
- b Divide and simplify, using fractional notation.
- c Solve equations of the type $a \cdot x = b$ and $x \cdot a = b$, where a and b may be fractions.
- d Solve problems involving division.

FOR EXTRA HELP

Tape 3E

Tape 4B

MAC: 2
IBM: 2

Find the reciprocal.

1. $\frac{2}{5}$ $\frac{5}{2}$
2. $\frac{10}{7}$ $\frac{7}{10}$
3. 9 $\frac{1}{9}$
4. $\frac{1}{5}$ 5

ANSWERS ON PAGE A-2

Divide and simplify.

5. $\frac{6}{7} \div \frac{3}{4}$ $\frac{8}{7}$

6. $\frac{2}{3} \div \frac{1}{4}$ $\frac{8}{3}$

7. $\frac{4}{5} \div 8$ $\frac{1}{10}$

8. $60 \div \frac{3}{5}$ 100

9. $\frac{3}{5} \div \frac{3}{5}$ 1

ANSWERS ON PAGE A-2

We see that there are six $\frac{1}{8}$'s in $\frac{3}{4}$. Thus,

$$\frac{3}{4} \div \frac{1}{8} = 6.$$

We can check this by multiplying:

$$6 \cdot \frac{1}{8} = \frac{6}{8} = \frac{3}{4}.$$

Here is a faster way to divide.

To divide, multiply the dividend by the reciprocal of the divisor:

$$\frac{2}{5} \div \frac{3}{4} = \frac{2}{5} \cdot \frac{4}{3} = \frac{2 \cdot 4}{5 \cdot 3} = \frac{8}{15}.$$

Multiply by the reciprocal of the divisor.

▶ **EXAMPLES** Divide and simplify.

5. $\frac{5}{6} \div \frac{2}{3} = \frac{5}{6} \cdot \frac{3}{2} = \frac{5 \cdot 3}{6 \cdot 2} = \frac{5 \cdot 3}{3 \cdot 2 \cdot 2} = \frac{3}{3} \cdot \frac{5}{2 \cdot 2} = \frac{5}{2 \cdot 2} = \frac{5}{4}$

6. $\frac{3}{4} \div \frac{1}{8} = \frac{3}{4} \cdot 8 = \frac{3 \cdot 8}{4} = \frac{3 \cdot 4 \cdot 2}{4 \cdot 1} = \frac{4}{4} \cdot \frac{3 \cdot 2}{1} = \frac{3 \cdot 2}{1} = 6$

7. $\frac{2}{5} \div 6 = \frac{2}{5} \cdot \frac{1}{6} = \frac{2 \cdot 1}{5 \cdot 6} = \frac{2 \cdot 1}{5 \cdot 2 \cdot 3} = \frac{2}{2} \cdot \frac{1}{5 \cdot 3} = \frac{1}{5 \cdot 3} = \frac{1}{15}$

8. $\frac{3}{5} \div \frac{1}{2} = \frac{3}{5} \cdot 2 = \frac{3 \cdot 2}{5} = \frac{6}{5}$ ◀

CAUTION! Canceling can be used as follows for Examples 5–7.

5. $\frac{5}{6} \div \frac{2}{3} = \frac{5}{6} \cdot \frac{3}{2} = \frac{5 \cdot 3}{6 \cdot 2} = \frac{5 \cdot \cancel{3}}{\cancel{3} \cdot 2 \cdot 2} = \frac{5}{2 \cdot 2} = \frac{5}{4}$ Removing a factor of 1: $\frac{3}{3} = 1$

6. $\frac{3}{4} \div \frac{1}{8} = \frac{3}{4} \cdot 8 = \frac{3 \cdot 8}{4} = \frac{3 \cdot \cancel{4} \cdot 2}{\cancel{4} \cdot 1} = \frac{3 \cdot 2}{1} = 6$ Removing a factor of 1: $\frac{4}{4} = 1$

7. $\frac{2}{5} \div 6 = \frac{2}{5} \cdot \frac{1}{6} = \frac{2 \cdot 1}{5 \cdot 6} = \frac{\cancel{2} \cdot 1}{5 \cdot \cancel{2} \cdot 3} = \frac{1}{5 \cdot 3} = \frac{1}{15}$ Removing a factor of 1: $\frac{2}{2} = 1$

Remember, if you can't factor, you can't cancel!

DO EXERCISES 5–9.

Why do we multiply by a reciprocal when dividing? To see this, let's consider $\frac{2}{3} \div \frac{7}{5}$. We will multiply by 1. The name for 1 that we will use is

$$\frac{\frac{5}{7}}{\frac{5}{7}}.$$

Then we multiply as follows:

$$\frac{2}{3} \div \frac{7}{5} = \frac{\frac{2}{3}}{\frac{7}{5}}$$ Writing fractional notation for the division

$$= \frac{\frac{2}{3}}{\frac{7}{5}} \cdot 1$$ Multiplying by 1

$$= \frac{\frac{2}{3}}{\frac{7}{5}} \cdot \frac{\frac{5}{7}}{\frac{5}{7}}$$ Multiplying by 1; $\frac{5}{7}$ is the reciprocal of $\frac{7}{5}$ and $\frac{\frac{5}{7}}{\frac{5}{7}} = 1$

$$= \frac{\frac{2}{3} \cdot \frac{5}{7}}{\frac{7}{5} \cdot \frac{5}{7}}$$ Multiplying the numerators and the denominators

$$= \frac{\frac{2}{3} \cdot \frac{5}{7}}{1} = \frac{2}{3} \cdot \frac{5}{7} = \frac{10}{21}$$ After we multiplied, we got 1 for the denominator. The numerator (in color) shows the multiplication by the reciprocal.

DO EXERCISE 10.

10. Divide by multiplying by 1:

$$\frac{\frac{4}{5}}{\frac{6}{7}} \quad \frac{14}{15}$$

c Solving Equations

Now let us solve equations $a \cdot x = b$ and $x \cdot a = b$, where a and b may be fractions. Proceeding as we have before, we divide on both sides by a.

▶ **EXAMPLE 9** Solve: $\frac{4}{3} \cdot x = \frac{6}{7}$.

$$\frac{4}{3} \cdot x = \frac{6}{7}$$

$$x = \frac{6}{7} \div \frac{4}{3}$$ Dividing on both sides by $\frac{4}{3}$

$$= \frac{6}{7} \cdot \frac{3}{4}$$ Multiplying by the reciprocal

$$= \frac{2 \cdot 3 \cdot 3}{7 \cdot 2 \cdot 2} = \frac{2}{2} \cdot \frac{3 \cdot 3}{7 \cdot 2} = \frac{3 \cdot 3}{7 \cdot 2} = \frac{9}{14}$$

The solution is $\frac{9}{14}$. ◀

▶ **EXAMPLE 10** Solve: $t \cdot \frac{4}{5} = 80$.

Dividing on both sides by $\frac{4}{5}$, we get

$$t = 80 \div \frac{4}{5} = 80 \cdot \frac{5}{4} = \frac{80 \cdot 5}{4} = \frac{4 \cdot 20 \cdot 5}{4 \cdot 1} = \frac{4}{4} \cdot \frac{20 \cdot 5}{1} = \frac{20 \cdot 5}{1} = 100.$$

The solution is 100. ◀

DO EXERCISES 11 AND 12.

Solve.

11. $\frac{5}{6} \cdot y = \frac{2}{3}$ $\quad \frac{4}{5}$

12. $\frac{3}{4} \cdot n = 24$ $\quad 32$

d Solving Problems

▶ **EXAMPLE 11** How many test tubes, each containing $\frac{3}{5}$ mL, can be filled from a container of 60 mL?

1. *Familiarize.* Repeated addition will apply here. We let n = the num-

ANSWERS ON PAGE A-2

13. Each loop in a spring takes $\frac{3}{8}$ in. of wire. How many loops can be made from 120 in. of wire? 320

14. A tank had 175 gal of oil when it was $\frac{7}{8}$ full. How much could it hold? 200 gal

ANSWERS ON PAGE A-2

ber of test tubes in all. We draw a picture.

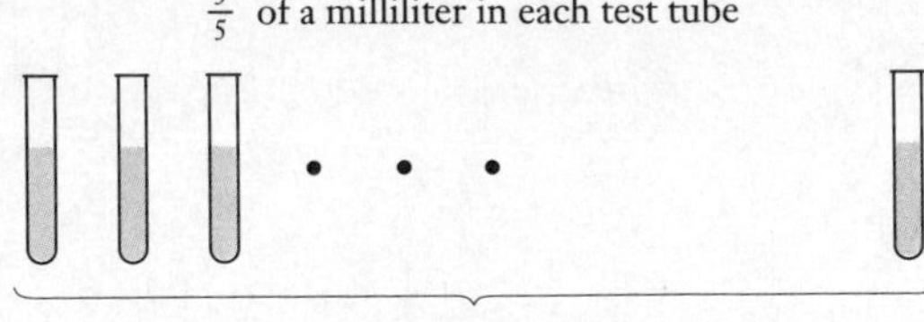

2. *Translate.* The multiplication that corresponds to the situation is

$$n \cdot \frac{3}{5} = 60.$$

3. *Solve.* We solve the equation by dividing on both sides by $\frac{3}{5}$ and carrying out the division:

$$n = 60 \div \frac{3}{5} = 60 \cdot \frac{5}{3} = \frac{60 \cdot 5}{3} = \frac{3 \cdot 20 \cdot 5}{3 \cdot 1} = \frac{3}{3} \cdot \frac{20 \cdot 5}{1} = 100.$$

4. *Check.* We check by repeating the calculation.
5. *State.* Thus, 100 test tubes can be filled. ◀

DO EXERCISE 13.

▶ **EXAMPLE 12** After driving 210 mi, $\frac{5}{6}$ of a trip was completed. How long was the total trip?

1. *Familiarize.* We first draw a picture or at least visualize the situation. We let n = the length of the trip.

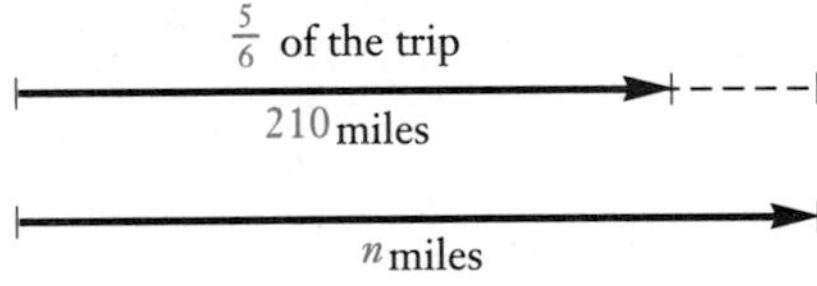

2. *Translate.* We translate to an equation.

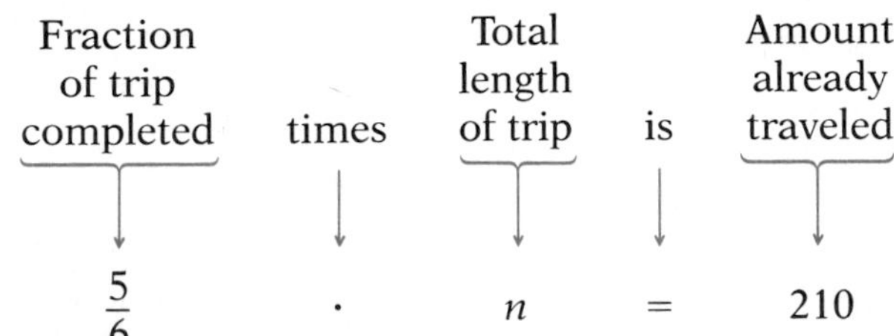

$$\frac{5}{6} \quad \cdot \quad n \quad = \quad 210$$

3. *Solve.* The equation that corresponds to the situation is

$$\frac{5}{6} \cdot n = 210.$$

We solve the equation by dividing on both sides by $\frac{5}{6}$ and carrying out the division:

$$n = 210 \div \frac{5}{6} = 210 \cdot \frac{6}{5} = \frac{210 \cdot 6}{5} = \frac{5 \cdot 42 \cdot 6}{5 \cdot 1} = \frac{5}{5} \cdot \frac{42 \cdot 6}{1} = 252.$$

4. *Check.* We check by repeating the calculation.
5. *State.* The total trip was 252 mi. ◀

DO EXERCISE 14.

NAME SECTION DATE

EXERCISE SET 2.7

a Find the reciprocal.

1. $\frac{5}{6}$ **2.** $\frac{3}{8}$ **3.** 6 **4.** 2

5. $\frac{1}{6}$ **6.** $\frac{1}{4}$ **7.** $\frac{10}{3}$ **8.** $\frac{12}{5}$

b Divide and simplify.

Don't forget to simplify!

9. $\frac{3}{5} \div \frac{3}{4}$ **10.** $\frac{2}{3} \div \frac{3}{4}$ **11.** $\frac{3}{5} \div \frac{9}{4}$ **12.** $\frac{6}{7} \div \frac{3}{5}$

13. $\frac{4}{3} \div \frac{1}{3}$ **14.** $\frac{10}{9} \div \frac{1}{2}$ **15.** $\frac{1}{3} \div \frac{1}{6}$ **16.** $\frac{1}{4} \div \frac{1}{5}$

17. $\frac{3}{8} \div 3$ **18.** $\frac{5}{6} \div 5$ **19.** $\frac{12}{7} \div 4$ **20.** $\frac{16}{5} \div 2$

21. $12 \div \frac{3}{2}$ **22.** $24 \div \frac{3}{8}$ **23.** $28 \div \frac{4}{5}$ **24.** $40 \div \frac{2}{3}$

25. $\frac{5}{8} \div \frac{5}{8}$ **26.** $\frac{2}{5} \div \frac{2}{5}$ **27.** $\frac{8}{15} \div \frac{4}{5}$ **28.** $\frac{6}{13} \div \frac{3}{26}$

29. $\frac{9}{5} \div \frac{4}{5}$ **30.** $\frac{5}{12} \div \frac{25}{36}$ **31.** $120 \div \frac{5}{6}$ **32.** $360 \div \frac{8}{7}$

c Solve.

33. $\frac{4}{5} \cdot x = 60$ **34.** $\frac{3}{2} \cdot t = 90$ **35.** $\frac{5}{3} \cdot y = \frac{10}{3}$ **36.** $\frac{4}{9} \cdot m = \frac{8}{3}$

ANSWERS

1. $\frac{6}{5}$
2. $\frac{8}{3}$
3. $\frac{1}{6}$
4. $\frac{1}{2}$
5. 6
6. 4
7. $\frac{3}{10}$
8. $\frac{5}{12}$
9. $\frac{4}{5}$
10. $\frac{8}{9}$
11. $\frac{4}{15}$
12. $\frac{10}{7}$
13. 4
14. $\frac{20}{9}$
15. 2
16. $\frac{5}{4}$
17. $\frac{1}{8}$
18. $\frac{1}{6}$
19. $\frac{3}{7}$
20. $\frac{8}{5}$
21. 8
22. 64
23. 35
24. 60
25. 1
26. 1
27. $\frac{2}{3}$
28. 4
29. $\frac{9}{4}$
30. $\frac{3}{5}$
31. 144
32. 315
33. 75
34. 60
35. 2
36. 6

ANSWERS

37. $\frac{3}{5}$

38. $\frac{2}{3}$

39. 315

40. 144

41. $\frac{1}{10}$ m

42. $\frac{1}{10}$ m

43. 32

44. 30

45. 24

46. 15

47. 16 L

48. 25 L

49. 288 km; 108 km

50. 400 km; 160 km

37. $x \cdot \frac{25}{36} = \frac{5}{12}$

38. $p \cdot \frac{4}{5} = \frac{8}{15}$

39. $n \cdot \frac{8}{7} = 360$

40. $y \cdot \frac{5}{6} = 120$

d Solve.

41. A piece of wire $\frac{3}{5}$ m long is to be cut into six pieces of the same length. What is the length of each piece?

42. A piece of wire $\frac{4}{5}$ m long is to be cut into eight pieces of the same length. What is the length of each piece?

43. A child's teeshirt requires $\frac{3}{4}$ yd of fabric. How many shirts can be made from 24 yd of fabric?

44. A child's shirt requires $\frac{5}{6}$ yd of fabric. How many shirts can be made from 25 yd of fabric?

45. How many $\frac{2}{3}$-cup sugar bowls can be filled from 16 cups of sugar?

46. How many $\frac{2}{3}$-cup sugar bowls can be filled from 10 cups of sugar?

47. A bucket had 12 L of water in it when it was $\frac{3}{4}$ full. How much could it hold?

48. A pail had 20 L of gasoline in it when it was $\frac{4}{5}$ full. How much could it hold?

49. After driving 180 km, $\frac{5}{8}$ of a trip was completed. How long was the total trip? How many kilometers were left to drive?

50. After driving 240 km, $\frac{3}{5}$ of a trip was completed. How long was the total trip? How many kilometers were left to drive?

SUMMARY AND REVIEW EXERCISES: CHAPTER 2

Beginning with this chapter, material from certain sections of preceding chapters will be covered on the chapter tests. Accordingly, the review exercises and the chapter tests will contain skill maintenance exercises. The review sections and objectives to be tested in addition to the material in this chapter are [1.3d], [1.6c], [1.7b], and [1.8a].

Find the prime factorization of the number.

1. 70

2. 30

3. 45

4. 150

5. Determine whether 2432 is divisible by 6.

6. Determine whether 182 is divisible by 4.

7. Determine whether 4344 is divisible by 8.

8. Determine whether 4344 is divisible by 9.

9. Determine whether 37 is prime, composite, or neither.

10. Identify the numerator and denominator of $\frac{2}{7}$.

What part is shaded?

11. 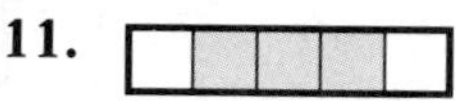

12. 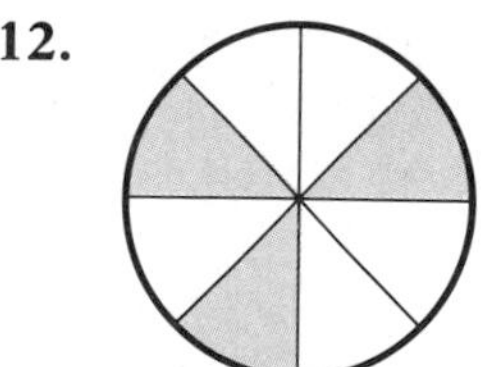

Simplify.

13. $\frac{0}{4}$

14. $\frac{23}{23}$

15. $\frac{48}{1}$

16. $\frac{48}{8}$

17. $\frac{10}{15}$

18. $\frac{7}{28}$

19. $\frac{21}{21}$

20. $\frac{0}{25}$

21. $\frac{12}{30}$

22. $\frac{18}{1}$

23. $\frac{32}{8}$

24. $\frac{9}{27}$

Use $=$ or $\neq$ for ▒ to write a true sentence.

25. $\frac{3}{5}$ ▒ $\frac{4}{6}$

26. $\frac{4}{7}$ ▒ $\frac{8}{14}$

27. $\frac{4}{5}$ ▒ $\frac{5}{6}$

28. $\frac{4}{3}$ ▒ $\frac{28}{21}$

Multiply and simplify.

29. $4 \cdot \frac{3}{8}$

30. $\frac{7}{3} \cdot 24$

31. $9 \cdot \frac{5}{18}$

32. $\frac{6}{5} \cdot 20$

33. $\frac{3}{4} \cdot \frac{8}{9}$

34. $\frac{5}{7} \cdot \frac{1}{10}$

35. $\frac{3}{7} \cdot \frac{14}{9}$

36. $\frac{1}{4} \cdot \frac{2}{11}$

Find the reciprocal.

37. $\frac{4}{5}$

38. 3

39. $\frac{1}{9}$

40. $\frac{47}{36}$

Divide and simplify.

41. $\frac{5}{9} \div \frac{5}{18}$

42. $6 \div \frac{4}{3}$

43. $\frac{1}{6} \div \frac{1}{11}$

44. $\frac{3}{14} \div \frac{6}{7}$

45. $180 \div \frac{3}{5}$

46. $\frac{1}{4} \div \frac{1}{9}$

47. $\frac{23}{25} \div \frac{23}{25}$

48. $\frac{2}{3} \div \frac{3}{2}$

Solve.

49. $\frac{5}{4} \cdot t = \frac{3}{8}$

50. $x \cdot \frac{2}{3} = 160$

Solve.

51. After driving 60 km, $\frac{3}{8}$ of a trip is complete. How long is the trip?

52. A recipe calls for $\frac{4}{5}$ of a cup of sugar. In making $\frac{1}{2}$ of this recipe, how much sugar should be used?

53. A person usually earns \$42 for working a full day. How much is received for working $\frac{1}{7}$ of a day?

54. How many $\frac{2}{3}$-cup sugar bowls can be filled from 12 cups of sugar?

SKILL MAINTENANCE

Solve.

55. $17 \cdot x = 408$

56. $765 + t = 1234$

57. An economy car gets 43 mi per gallon on the highway. How far can it go on a full tank of 26 gal?

58. You wrote checks for \$78, \$97, and \$102. Your balance before that was \$789. What is your new balance?

59. Divide: $36 \overline{)14{,}697}$.

60. Subtract: $\begin{array}{r} 5604 \\ -\,1997 \\ \hline \end{array}$

❖ THINKING IT THROUGH

1. Describe the process of simplifying when using fractional notation.
2. Describe the process of comparing two fractions.
3. Discuss the use of prime factorization and the tests for divisibility when working with fractions.

NAME SECTION DATE

TEST: CHAPTER 2

Find the prime factorization of the number.

1. 18

2. 60

3. Determine whether 1784 is divisible by 8.

4. Determine whether 784 is divisible by 9.

5. Identify the numerator and denominator of $\frac{4}{9}$.

6. What part is shaded?

Simplify.

7. $\frac{26}{1}$

8. $\frac{12}{12}$

9. $\frac{0}{16}$

10. $\frac{12}{24}$

11. $\frac{42}{7}$

12. $\frac{2}{28}$

Use = or ≠ for ▢ to write a true sentence.

13. $\frac{3}{4}$ ▢ $\frac{6}{8}$

14. $\frac{5}{4}$ ▢ $\frac{9}{7}$

Multiply and simplify.

15. $\frac{4}{3} \cdot 24$

16. $5 \cdot \frac{3}{10}$

17. $\frac{2}{3} \cdot \frac{15}{4}$

18. $\frac{3}{5} \cdot \frac{1}{6}$

ANSWERS

1. [2.1d] $2 \cdot 3 \cdot 3$

2. [2.1d] $2 \cdot 2 \cdot 3 \cdot 5$

3. [2.2a] Yes

4. [2.2a] No

5. [2.3a] 4 numerator; 9 denominator

6. [2.3b] $\frac{3}{4}$

7. [2.3c] 26

8. [2.3c] 1

9. [2.3c] 0

10. [2.5b] $\frac{1}{2}$

11. [2.5b] 6

12. [2.5b] $\frac{1}{14}$

13. [2.5c] =

14. [2.5c] ≠

15. [2.6a] 32

16. [2.6a] $\frac{3}{2}$

17. [2.6a] $\frac{5}{2}$

18. [2.6a] $\frac{1}{10}$

ANSWERS

19. [2.7a] $\frac{8}{5}$

20. [2.7a] 4

21. [2.7a] $\frac{1}{18}$

22. [2.7b] $\frac{3}{10}$

23. [2.7b] $\frac{8}{5}$

24. [2.7b] 18

25. [2.7c] 64

26. [2.7c] $\frac{7}{4}$

27. [2.6b] 28 lb

28. [2.7d] $\frac{3}{40}$ m

29. [1.7b] 1805

30. [1.7b] 101

31. [1.8a] 3635 mi

32. [1.6c] 380 R 7

33. [1.3d] 4434

Find the reciprocal.

19. $\frac{5}{8}$

20. $\frac{1}{4}$

21. 18

Divide and simplify.

22. $\frac{3}{8} \div \frac{5}{4}$

23. $\frac{1}{5} \div \frac{1}{8}$

24. $12 \div \frac{2}{3}$

Solve.

25. $\frac{7}{8} \cdot x = 56$

26. $\frac{2}{5} \cdot t = \frac{7}{10}$

Solve.

27. It takes $\frac{7}{8}$ lb of salt to use in the ice of one batch of homemade ice cream. How much salt would it take for 32 batches?

28. A board $\frac{9}{10}$ m long is cut into 12 equal pieces. What is the length of each piece?

SKILL MAINTENANCE

Solve.

29. $x + 198 = 2003$

30. $47 \cdot t = 4747$

31. It is 2060 mi from San Francisco to Winnipeg, Canada. It is 1575 mi from Winnipeg to Atlanta. What is the total length of a route from San Francisco to Winnipeg to Atlanta?

32. Divide: $24\overline{)9127}$

33. Subtract: $\begin{array}{r} 8001 \\ -\,3567 \\ \hline \end{array}$

CUMULATIVE REVIEW: CHAPTERS 1–2

1. Write standard notation for the number in the following sentence: The earth travels five hundred eighty-four million, seventeen thousand, eight hundred miles around the sun.

2. Write a word name: 5,380,621.

3. In the number 2,751,043, which digit tells the number of hundreds?

Add.

4. $\begin{array}{r} 14{,}862 \\ +\ 2{,}935 \\ \hline \end{array}$

5. $\begin{array}{r} 7989 \\ 798 \\ +\ \ 79 \\ \hline \end{array}$

Subtract.

6. $\begin{array}{r} 5376 \\ -\ 430 \\ \hline \end{array}$

7. $\begin{array}{r} 2004 \\ -\ 579 \\ \hline \end{array}$

Multiply and simplify.

8. $\begin{array}{r} 621 \\ \times\ 27 \\ \hline \end{array}$

9. $\begin{array}{r} 2505 \\ \times 3300 \\ \hline \end{array}$

10. $5 \times \frac{3}{100}$

11. $\frac{4}{9} \cdot \frac{3}{8}$

Divide and simplify.

12. $19\overline{)4580}$

13. $62\overline{)3844}$

14. $\frac{3}{10} \div 5$

15. $\frac{8}{9} \div \frac{15}{6}$

16. Round 427,931 to the nearest thousand.

17. Round 5309 to the nearest hundred.

Estimate the sum or product by rounding to the nearest hundred. Show your work.

18. $\begin{array}{r} 749{,}559 \\ +301{,}362 \\ \hline \end{array}$

19. $\begin{array}{r} 749 \\ \times 531 \\ \hline \end{array}$

20. Use $<$ or $>$ for ▢ to write a true sentence:

26 ▢ 17.

21. Use $=$ or $\neq$ for ▢ to write a true sentence:

$\frac{7}{10}$ ▢ $\frac{5}{7}$.

22. Evaluate: 3^4.

Simplify.

23. $35 - 25 \div 5 + 2 \times 3$

24. $\{17 - [8 - (5 - 2 \times 2)]\} \div (3 + 12 \div 6)$

25. Find all the factors of 28.

26. Find the prime factorization of 28.

27. Determine whether 39 is prime, composite, or neither.

28. Determine whether 32,712 is divisible by 3.

29. Determine whether 32,712 is divisible by 5.

Simplify.

30. $\frac{35}{1}$

31. $\frac{77}{11}$

32. $\frac{28}{98}$

33. $\frac{0}{47}$

Solve.

34. $x + 13 = 50$

35. $\frac{1}{5} \cdot t = \frac{3}{10}$

36. $13 \cdot y = 39$

37. $384 \div 16 = n$

Solve.

38. There were 750,619 books in print in 1988 and 718,500 in print in 1987. How many more books were in print in 1988 than in 1987?

39. Four of the largest hotels in the United States are in Las Vegas. One has 3174 rooms, the second has 2920 rooms, the third has 2832 rooms, and the fourth has 2793 rooms. What is the total number of rooms in these four hotels?

40. A student is offered a job paying \$3900 a year. How much is each weekly paycheck?

41. One $\frac{3}{4}$-cup serving of macaroni and cheese contains 290 calories. A box makes 4 servings. How many cups of macaroni and cheese does the box make?

42. It takes 6 hr to paint the trim on a certain house. If the painter can work only $\frac{3}{4}$ hr per day, how many days will it take to finish the job?

43. Eastside Appliance sells a refrigerator for \$600 and \$30 tax with no delivery charge. Westside Appliance sells the same model for \$560 and \$28 tax plus a \$25 delivery charge. Which is the better buy?

SYNTHESIS

44. A student works 25 hr a week at \$5/hr. The employer withholds $\frac{1}{4}$ of the total salary for taxes. Room and board expenses are \$25 a week and tuition is \$800 for a 16-week semester. Books cost \$150 for the semester. Will the student make enough to cover the listed expenses? If so, how much will be left over at the end of the semester?

45. A can of mixed nuts is 1 part cashews, 1 part almonds, 1 part pecans, and 3 parts peanuts. What part of the mixture is peanuts?

INTRODUCTION In this chapter, addition and subtraction using fractional notation are considered. Also discussed are addition, subtraction, multiplication, and division using mixed numerals. All these operations are then applied to problem solving.

The review sections to be used in addition to the material in this chapter are 1.5, 1.8, 2.6, and 2.7. ❖

Addition and Subtraction: Fractional Notation

AN APPLICATION

A long-playing record makes $33\frac{1}{3}$ revolutions per minute. It plays for 12 minutes. How many revolutions does it make?

THE MATHEMATICS

Let n = the total number of revolutions. The problem translates to the following equation:

$$\underbrace{33\frac{1}{3}} \cdot 12 = n.$$

This is a mixed numeral.

❖ POINTS TO REMEMBER: CHAPTER 3

Area of a Rectangle: $A = lw$
Multiplication and Division Using Fractional Notation

PRETEST: CHAPTER 3

1. Find the LCM of 15 and 24.

2. Use < or > for ▢ to write a true sentence:

$$\frac{7}{9} \;\square\; \frac{4}{5}.$$

3. Convert to fractional notation: $7\frac{5}{8}$.

4. Convert to a mixed numeral: $\frac{11}{2}$.

5. Divide. Write a mixed numeral for the answer.

$$12\overline{)4789}$$

6. Add. Write a mixed numeral for the answer.

$$8\frac{11}{12} + 2\frac{3}{5}$$

7. Subtract. Write a mixed numeral for the answer.

$$14 - 7\frac{5}{6}$$

8. Multiply. Write a mixed numeral for the answer.

$$3 \cdot 4\frac{8}{15}$$

9. Multiply. Write a mixed numeral for the answer.

$$6\frac{2}{3} \cdot 3\frac{1}{4}$$

10. Divide. Write a mixed numeral for the answer.

$$35 \div 5\frac{5}{6}$$

11. Divide. Write a mixed numeral for the answer.

$$5\frac{5}{12} \div 3\frac{1}{4}$$

12. Solve:

$$\frac{2}{3} + x = \frac{8}{9}.$$

13. A cook bought 100 lb of potatoes and used $78\frac{3}{4}$ lb. How many pounds were left?

14. The weight of water is $62\frac{1}{2}$ lb per cubic foot. How many cubic feet would be occupied by $265\frac{5}{8}$ lb of water?

15. A traveler drove $214\frac{3}{10}$ km one day and $136\frac{9}{10}$ km the next. How far did she travel in all?

16. A cake recipe calls for $3\frac{3}{4}$ cups of flour. How much flour would be used to make 6 cakes?

3.1 Least Common Multiples

In this chapter we study addition and subtraction using fractional notation. Suppose we want to add $\frac{2}{3}$ and $\frac{1}{2}$. To do so, we find the least common multiple of the denominators: $\frac{2}{3} + \frac{1}{2} = \frac{4}{6} + \frac{3}{6}$. Then we add the numerators and keep the common denominator, 6. Before we do this, though, we study finding the **least common denominator,** or **least common multiple,** of the denominators.

a Finding Least Common Multiples

The *least common multiple,* or LCM, of two natural numbers is the smallest number that is a multiple of both.

▶ **EXAMPLE 1** Find the LCM of 20 and 30.

a) First list some multiples of 20 by multiplying 20 by 1, 2, 3, and so on:

20, 40, 60, 80, 100, 120, 140, 160, 180, 200, 220, 240, ...

b) Then list some multiples of 30 by multiplying 30 by 1, 2, 3, and so on:

30, 60, 90, 120, 150, 180, 210, 240, ...

c) Now list the numbers *common* to both lists, the common multiples:

60, 120, 180, 240, ...

d) These are the common multiples of 20 and 30. What is the smallest? The LCM of 20 and 30 is 60. ◀

DO EXERCISE 1.

Below we develop two efficient methods for finding LCMs. You may choose to learn either method (consult with your instructor), or both, but if you are going on to a study of algebra, you should definitely learn method 2.

Method 1: Finding LCMs Using One List of Multiples

***Method 1:* To find the LCM of a set of numbers (9, 12):**

a) Determine whether the greatest number is a multiple of the others. If it is, it is the LCM. If one number is a factor of another, the LCM is the greater number.

(12 is not a multiple of 9)

b) If not, check multiples of the largest number until you get one that is a multiple of the others.

(2 · 12 = 24, not a multiple of 9)

(3 · 12 = 36, a multiple of 9)

c) That number is the LCM.

LCM = 36

OBJECTIVE

After finishing Section 3.1, you should be able to:

a Find the LCM of two or more numbers using a list of multiples or factorizations.

FOR EXTRA HELP

Tape 4A

Tape 5A

MAC: 3
IBM: 3

1. By examining lists of multiples, find the LCM of 9 and 15. 45

ANSWER ON PAGE A-2

2. By examining lists of multiples, find the LCM of 8 and 10. 40

Find the LCM.

3. 10, 15 30

4. 6, 8 24

Find the LCM.

5. 5, 10 10

6. 20, 40, 80 80

ANSWERS ON PAGE A-2

▶ **EXAMPLE 2** Find the LCM of 12 and 15.

a) 15 is not a multiple of 12.

b) Check multiples:

$$2 \cdot 15 = 30, \quad \text{Not a multiple of 12}$$
$$3 \cdot 15 = 45, \quad \text{Not a multiple of 12}$$
$$4 \cdot 15 = 60. \quad \text{A multiple of 12}$$

c) The LCM = 60. ◀

DO EXERCISE 2.

▶ **EXAMPLE 3** Find the LCM of 4 and 6.

a) 6 is not a multiple of 4.

b) Check multiples:

$$2 \cdot 6 = 12. \quad \text{A multiple of 4}$$

c) The LCM = 12. ◀

DO EXERCISES 3 AND 4.

▶ **EXAMPLE 4** Find the LCM of 4 and 8.

a) 8 is a multiple of 4, so it is the LCM.

c) The LCM = 8. ◀

▶ **EXAMPLE 5** Find the LCM of 10, 100, and 1000.

a) 1000 is a multiple of 10 and 100, so it is the LCM.

c) The LCM = 1000. ◀

DO EXERCISES 5 AND 6.

Method 2: Finding LCMs Using Factorizations

A second method for finding LCMs uses prime factorizations. Consider again 20 and 30. Their prime factorizations are

$$20 = 2 \cdot 2 \cdot 5 \quad \text{and} \quad 30 = 2 \cdot 3 \cdot 5.$$

Let's look at these prime factorizations in order to find the LCM. Any multiple of 20 will have to have *two* 2's as factors and *one* 5 as a factor. Any multiple of 30 will have to have *one* 2, *one* 3, and *one* 5 as factors. The smallest number satisfying these conditions is

Two 2's, one 5; makes 20 a factor

$$2 \cdot 2 \cdot 3 \cdot 5.$$

One 2, one 3, one 5; makes 30 a factor

The LCM must have all the factors of 20 and all the factors of 30, but the factors need not be repeated when they are common to both numbers.

The greatest number of times a 2 occurs as a factor of either 20 or 30 is two, and the LCM has 2 as a factor twice. The greatest number of times a 3 occurs as a factor of either 20 or 30 is one, and the LCM has 3 as a factor once. The greatest number of times 5 occurs as a factor of either 20 or 30 is one, and the LCM has 5 as a factor once.

***Method 2.* To find the LCM of a set of numbers using prime factorizations:**

a) Find the prime factorization of each number.

b) Create a product of factors, using each factor the greatest number of times it occurs in any one factorization.

▶ **EXAMPLE 6** Find the LCM of 6 and 8.

a) Find the prime factorization of each number.

$$6 = 2 \cdot 3, \qquad 8 = 2 \cdot 2 \cdot 2$$

b) Create a product by writing factors, using each the greatest number of times it occurs in any one factorization.

Consider the factor 2. The greatest number of times 2 occurs in any one factorization is three. We write 2 as a factor three times.

$$2 \cdot 2 \cdot 2 \cdot ?$$

Consider the factor 3. The greatest number of times 3 occurs in any one factorization is one. We write 3 as a factor one time.

$$2 \cdot 2 \cdot 2 \cdot 3 \cdot ?$$

Since there are no other prime factors in either factorization, the

LCM is $2 \cdot 2 \cdot 2 \cdot 3$, or 24. ◀

▶ **EXAMPLE 7** Find the LCM of 24 and 36.

a) Find the prime factorization of each number.

$$24 = 2 \cdot 2 \cdot 2 \cdot 3, \qquad 36 = 2 \cdot 2 \cdot 3 \cdot 3$$

b) Create a product by writing factors, using each the greatest number of times it occurs in any one factorization.

Consider the factor 2. The greatest number of times 2 occurs in any one factorization is three. We write 2 as a factor three times:

$$2 \cdot 2 \cdot 2 \cdot ?$$

Consider the factor 3. The greatest number of times 3 occurs in any one factorization is two. We write 3 as a factor two times:

$$2 \cdot 2 \cdot 2 \cdot 3 \cdot 3 \cdot ?$$

Since there are no other prime factors in either factorization, the

LCM is $2 \cdot 2 \cdot 2 \cdot 3 \cdot 3$, or 72. ◀

DO EXERCISES 7 AND 8.

Use prime factorizations to find the LCM.

7. 8, 10 40

8. 18, 40 360

ANSWERS ON PAGE A-2

9. Find the LCM of 24, 35, and 45.
2520

Find the LCM.

10. 3, 18 18

11. 12, 24 24

ANSWERS ON PAGE A-2

▶ **EXAMPLE 8** Find the LCM of 27, 90, and 84.

a) Find the prime factorization of each number.

$$27 = 3 \cdot 3 \cdot 3, \qquad 90 = 2 \cdot 3 \cdot 3 \cdot 5, \qquad 84 = 2 \cdot 2 \cdot 3 \cdot 7$$

b) Create a product by writing factors, using each the greatest number of times it occurs in any one factorization.

Consider the factor 2. The greatest number of times 2 occurs in any one factorization is two. We write 2 as a factor two times:

$$2 \cdot 2 \cdot ?$$

Consider the factor 3. The greatest number of times 3 occurs in any one factorization is three. We write 3 as a factor three times:

$$2 \cdot 2 \cdot 3 \cdot 3 \cdot 3 \cdot ?$$

Consider the factor 5. The greatest number of times 5 occurs in any one factorization is one. We write 5 as a factor one time:

$$2 \cdot 2 \cdot 3 \cdot 3 \cdot 3 \cdot 5 \cdot ?$$

Consider the factor 7. The greatest number of times 7 occurs in any one factorization is one. We write 7 as a factor one time:

$$2 \cdot 2 \cdot 3 \cdot 3 \cdot 3 \cdot 5 \cdot 7 \cdot ?$$

Since no other prime factors are possible in any of the factorizations, the

LCM is $2 \cdot 2 \cdot 3 \cdot 3 \cdot 3 \cdot 5 \cdot 7$, or 3780. ◀

DO EXERCISE 9.

▶ **EXAMPLE 9** Find the LCM of 7 and 21.

a) Find the prime factorization of each number. Since 7 is prime, it has no prime factorization. We think of $7 = 7$ as a "factorization" in order to carry out our procedure.

$$7 = 7, \qquad 21 = 3 \cdot 7$$

b) Create a product by writing factors, using each the greatest number of times it occurs in any one factorization.

Consider the factor 7. The greatest number of times 7 occurs in any one factorization is one. We write 7 as a factor one time:

$$7 \cdot ?$$

Consider the factor 3. The greatest number of times 3 occurs in any one factorization is one. We write 3 as a factor one time:

$$7 \cdot 3 \cdot ?$$

Since no other prime factors are possible in any of the factorizations, the

LCM is $7 \cdot 3$, or 21. ◀

Note, in Example 9, that 7 is a factor of 21. We stated earlier that if one number is a factor of another, the LCM is the larger of the numbers. Thus, if you notice this at the outset, you can find the LCM quickly without using factorizations.

DO EXERCISES 10 AND 11.

► **EXAMPLE 10** Find the LCM of 8 and 9.

a) Find the prime factorization of each number.

$$8 = 2 \cdot 2 \cdot 2, \qquad 9 = 3 \cdot 3$$

b) Create a product by writing factors, using each the greatest number of times it occurs in any one factorization.

Consider the factor 2. The greatest number of times 2 occurs in any one factorization is three. We write 2 as a factor three times.

$$2 \cdot 2 \cdot 2 \cdot ?$$

Consider the factor 3. The greatest number of times 3 occurs in any one factorization is two. We write 3 as a factor two times.

$$2 \cdot 2 \cdot 2 \cdot 3 \cdot 3 \cdot ?$$

Since no other prime factors are possible in any of the factorizations, the

LCM is $2 \cdot 2 \cdot 2 \cdot 3 \cdot 3$, or 72. ◄

Note, in Example 10, that the two numbers, 8 and 9, have no common prime factor. When this happens, the LCM is just the product of the two numbers. Thus, when you notice this at the outset, you can find the LCM quickly by multiplying the two numbers.

DO EXERCISES 12 AND 13.

Find the LCM.

12. 4, 9 36

13. 5, 6, 7 210

Let's compare the two methods considered for finding LCMs: the multiples method and the factorization method.

Method 1, the **multiples method,** can be longer than the factorization method when the LCM is large or when there are more than two numbers. But this method is faster and easier to use mentally for two numbers.

Method 2, the **factorization method,** works well for several numbers. It is just like a method used in algebra. If you are going to study algebra, you should definitely learn the factorization method.

Method 3: A Third Method for Finding LCMs (Optional)

Here is another method for finding LCMs that may work well for you. Suppose you want to find the LCM of 48, 72, and 80. Using only prime numbers, find a number that divides any two of these numbers with no remainder. Do the division and bring the third number down, unless the third number is divisible by the prime also. Repeat the process until you can divide no more. Multiply, as shown at the right, all the numbers at the side by all the numbers at the bottom. The LCM is

2	48	72	80
3	24	36	40
2	8	12	40
2	4	6	20
2	2	3	10
	1	3	5

$$2 \cdot 3 \cdot 2 \cdot 2 \cdot 2 \cdot 1 \cdot 3 \cdot 5, \text{ or } 720.$$

DO EXERCISES 14 AND 15.

Find the LCM using the optional method.

14. 24, 35, 45 2520

15. 27, 90, 84 3780

ANSWERS ON PAGE A-2

❖ SIDELIGHTS

Application of LCM's: Planet Orbits

The earth, Jupiter, Saturn, and Uranus all revolve around the sun. The earth takes 1 year, Jupiter 12 years, Saturn 30 years, and Uranus 84 years to make a complete revolution. On a certain night you look at all the planets and wonder how many years it will take before they have the same position again. (*Hint:* To find out, you find the LCM of 12, 30, and 84. It will be that number of years.)

EXERCISES

1. How often will Jupiter and Saturn appear in the same direction in the night sky as seen from the earth? Every 60 yr
2. How often will Saturn and Uranus appear in the same direction in the night sky as seen from the earth? Every 420 yr
3. How often will Jupiter, Saturn, and Uranus appear in the same direction in the night sky as seen from the earth? Every 420 yr

NAME SECTION DATE

EXERCISE SET 3.1

a Find the LCM of each set of numbers. Do so mentally, if possible.

1. 2, 4 **2.** 3, 10 **3.** 10, 25 **4.** 3, 15

5. 20, 40 **6.** 8, 12 **7.** 18, 27 **8.** 9, 11

9. 30, 50 **10.** 24, 36 **11.** 30, 40 **12.** 13, 23

13. 18, 24 **14.** 12, 18 **15.** 60, 70 **16.** 35, 45

17. 16, 36 **18.** 18, 20 **19.** 32, 36 **20.** 36, 48

21. 2, 3, 5 **22.** 7, 18, 3 **23.** 3, 5, 7 **24.** 6, 12, 18

25. 24, 36, 12 **26.** 8, 16, 22 **27.** 5, 12, 15 **28.** 12, 18, 40

29. 9, 12, 6 **30.** 8, 16, 12 **31.** 3, 6, 8 **32.** 12, 8, 4

ANSWERS

1. 4
2. 30
3. 50
4. 15
5. 40
6. 24
7. 54
8. 99
9. 150
10. 72
11. 120
12. 299
13. 72
14. 36
15. 420
16. 315
17. 144
18. 180
19. 288
20. 144
21. 30
22. 126
23. 105
24. 36
25. 72
26. 176
27. 60
28. 360
29. 36
30. 48
31. 24
32. 24

ANSWERS

33. 48
34. 32
35. 50
36. 72
37. 143
38. 182
39. 420
40. 575
41. 378
42. 504
43. 810
44. 300
45. 250
46. 7935
47. $\frac{4}{3}$
48. a)
b)
c)
d)
49. 24 in.
50. 2592
51. 70,200

33. 8, 48 **34.** 16, 32 **35.** 5, 50 **36.** 12, 72

37. 11, 13 **38.** 13, 14 **39.** 12, 35 **40.** 23, 25

41. 54, 63 **42.** 56, 72 **43.** 81, 90 **44.** 75, 100

SKILL MAINTENANCE

45. An auditorium was sold out for a performance. It contains seats selling for $13 each. Total receipts were $3250. How many seats does this auditorium contain?

46. Multiply: $\begin{array}{r} 345 \\ \times\ 23 \\ \hline \end{array}$

47. Multiply and simplify: $\frac{4}{3} \cdot \frac{10}{10}$.

SYNTHESIS

48. Consider 8 and 12. Determine whether each of the following is the LCM of 8 and 12. Tell why or why not.

a) $2 \cdot 2 \cdot 3 \cdot 3$ No; needs another 2, one less 3
b) $2 \cdot 2 \cdot 3$ No; needs another 2
c) $2 \cdot 3 \cdot 3$ No; needs two more 2's, one less 3
d) $2 \cdot 2 \cdot 2 \cdot 3$ Yes; has three 2's, one 3

49. A cigar company uses two sizes of boxes, 6 in. and 8 in. long. These are packed in bigger cartons to be shipped. What is the shortest length carton that will accommodate boxes of either size without any room left over? (Each carton can contain only boxes of one size; no mixing is allowed.)

▦ Use your calculator and the multiples method to find the LCM of each pair of numbers.

50. 288, 324 **51.** 2700, 7800

3.2 Addition

a Like Denominators

Addition using fractional notation still corresponds to combining or putting like things together, even though we may not be adding whole numbers.

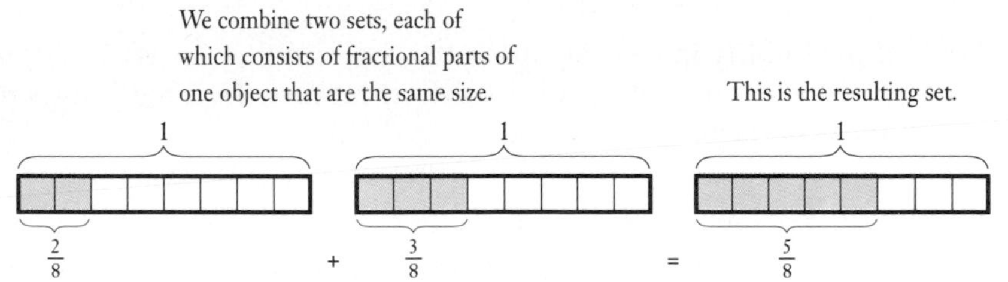

2 eighths + 3 eighths = 5 eighths,

or

$$2 \cdot \frac{1}{8} + 3 \cdot \frac{1}{8} = 5 \cdot \frac{1}{8},$$

or

$$\frac{2}{8} + \frac{3}{8} = \frac{5}{8}.$$

DO EXERCISE 1.

> **To add when denominators are the same,**
> **a) add the numerators,**
> **b) keep the denominator,**
> **and**
> **c) simplify, if possible.**
>
> $$\frac{2}{6} + \frac{5}{6} = \frac{2+5}{6} = \frac{7}{6}$$

► **EXAMPLES** Add and simplify.

1. $\frac{2}{4} + \frac{1}{4} = \frac{2+1}{4} = \frac{3}{4}$ **No simplifying is possible.**

2. $\frac{11}{6} + \frac{3}{6} = \frac{11+3}{6} = \frac{14}{6} = \frac{2 \cdot 7}{2 \cdot 3} = \frac{2}{2} \cdot \frac{7}{3} = 1 \cdot \frac{7}{3} = \frac{7}{3}$ **Here we simplified.**

3. $\frac{3}{12} + \frac{5}{12} = \frac{3+5}{12} = \frac{8}{12} = \frac{4 \cdot 2}{4 \cdot 3} = \frac{4}{4} \cdot \frac{2}{3} = 1 \cdot \frac{2}{3} = \frac{2}{3}$ ◄

DO EXERCISES 2–4.

b Addition Using the LCD: Different Denominators

What do we do when denominators are different? We try to find a common denominator. We can do this by multiplying by 1. Consider adding $\frac{1}{6}$ and $\frac{3}{4}$. There are several common denominators that can be obtained. Let's look at two possibilities.

OBJECTIVES

After finishing Section 3.2, you should be able to:

a Add with fractional notation when denominators are the same.

b Add with fractional notation when denominators are different, by multiplying by 1 to find the least common denominator.

c Solve problems involving addition with fractional notation.

FOR EXTRA HELP

Tape 4B

Tape 5A

MAC: 3
IBM: 3

1. Find $\frac{1}{5} + \frac{3}{5}$. $\frac{4}{5}$

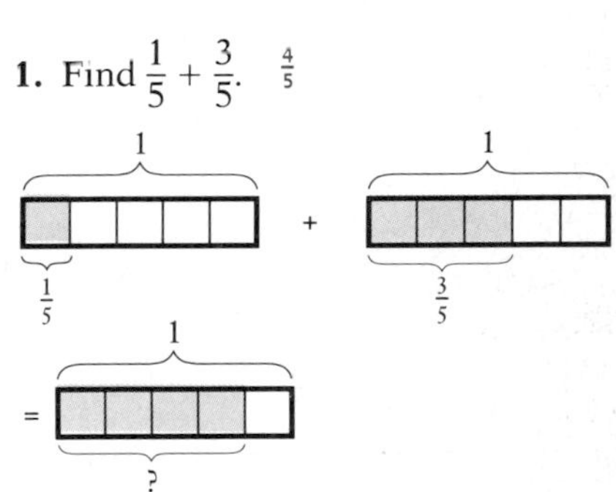

Add and simplify.

2. $\frac{1}{3} + \frac{2}{3}$ 1

3. $\frac{5}{12} + \frac{1}{12}$ $\frac{1}{2}$

4. $\frac{9}{16} + \frac{3}{16}$ $\frac{3}{4}$

ANSWERS ON PAGE A-2

5. Add. Find the least common denominator.

$\frac{2}{3} + \frac{1}{6}$ $\frac{5}{6}$

6. Add: $\frac{3}{8} + \frac{5}{6}$. $\frac{29}{24}$

ANSWERS ON PAGE A-2

A.
$$\frac{1}{6} + \frac{3}{4} = \frac{1}{6} \cdot 1 + \frac{3}{4} \cdot 1 = \frac{1}{6} \cdot \frac{4}{4} + \frac{3}{4} \cdot \frac{6}{6} = \frac{4}{24} + \frac{18}{24} = \frac{22}{24} = \frac{11}{12}$$

B.
$$\frac{1}{6} + \frac{3}{4} = \frac{1}{6} \cdot 1 + \frac{3}{4} \cdot 1 = \frac{1}{6} \cdot \frac{2}{2} + \frac{3}{4} \cdot \frac{3}{3} = \frac{2}{12} + \frac{9}{12} = \frac{11}{12}$$

We had to simplify in (A). We didn't have to simplify in (B). In (B) we used the least common multiple of the denominators, 12. That number is called the **least common denominator,** or LCD.

To add when denominators are different:

a) Find the least common multiple of the denominators. That number is the least common denominator, LCD.

b) Multiply by 1, using an appropriate notation, *n*/*n*, to obtain the LCD for each number.

c) Add and simplify, if appropriate.

▶ **EXAMPLE 4** Add: $\frac{3}{4} + \frac{1}{8}$.

The LCD is 8. 4 is a factor of 8 so the LCM of 4 and 8 is 8.

$$\frac{3}{4} + \frac{1}{8} = \frac{3}{4} \cdot 1 + \frac{1}{8}$$ ← This fraction already has the LCD as its denominator.

$$= \frac{3}{4} \cdot \frac{2}{2} + \frac{1}{8}$$ *Think*: 4 × ▒ = 8. The answer is 2, so we multiply by 1, using $\frac{2}{2}$.

$$= \frac{6}{8} + \frac{1}{8} = \frac{7}{8}$$ ◀

DO EXERCISE 5.

▶ **EXAMPLE 5** Add: $\frac{1}{9} + \frac{5}{6}$.

The LCD is 18. $9 = 3 \cdot 3$ and $6 = 2 \cdot 3$, so the LCM of 9 and 6 is $2 \cdot 3 \cdot 3$, or 18.

$$\frac{1}{9} + \frac{5}{6} = \frac{1}{9} \cdot 1 + \frac{5}{6} \cdot 1 = \frac{1}{9} \cdot \frac{2}{2} + \frac{5}{6} \cdot \frac{3}{3}$$

Think: 6 × ▒ = 18. The answer is 3, so we multiply by 1 using $\frac{3}{3}$.

Think: 9 × ▒ = 18. The answer is 2, so we multiply by 1, using $\frac{2}{2}$.

$$= \frac{2}{18} + \frac{15}{18} = \frac{17}{18}$$ ◀

DO EXERCISE 6.

▶ **EXAMPLE 6** Add: $\frac{5}{9} + \frac{11}{18}$.

The LCD is 18.

$$\frac{5}{9} + \frac{11}{18} = \frac{5}{9} \cdot \frac{2}{2} + \frac{11}{18}$$
$$= \frac{10}{18} + \frac{11}{18}$$
$$= \frac{21}{18}$$
$$= \frac{7}{6}$$

We may still have to simplify, but it is usually easier if we have used the LCD. ◀

DO EXERCISE 7.

▶ **EXAMPLE 7** Add: $\frac{1}{10} + \frac{3}{100} + \frac{7}{1000}$.

Since 10 and 100 are factors of 1000, the LCD is 1000. Then

$$\frac{1}{10} + \frac{3}{100} + \frac{7}{1000} = \frac{1}{10} \cdot \frac{100}{100} + \frac{3}{100} \cdot \frac{10}{10} + \frac{7}{1000}$$
$$= \frac{100}{1000} + \frac{30}{1000} + \frac{7}{1000}$$
$$= \frac{137}{1000}.$$

Look back over this example. Try to think it out so that you can do it mentally. ◀

▶ **EXAMPLE 8** Add: $\frac{13}{70} + \frac{11}{21} + \frac{6}{15}$.

We have

$$\frac{13}{70} + \frac{11}{21} + \frac{6}{15} = \frac{13}{2 \cdot 5 \cdot 7} + \frac{11}{3 \cdot 7} + \frac{6}{3 \cdot 5}. \quad \text{Factoring denominators}$$

The LCD is $2 \cdot 3 \cdot 5 \cdot 7$, or 210. Then

$$\frac{13}{70} + \frac{11}{21} + \frac{6}{15} = \frac{13}{2 \cdot 5 \cdot 7} \cdot \frac{3}{3} + \frac{11}{3 \cdot 7} \cdot \frac{2 \cdot 5}{2 \cdot 5} + \frac{6}{3 \cdot 5} \cdot \frac{7 \cdot 2}{7 \cdot 2}$$
$$= \frac{13 \cdot 3}{2 \cdot 5 \cdot 7 \cdot 3} + \frac{11 \cdot 2 \cdot 5}{3 \cdot 7 \cdot 2 \cdot 5} + \frac{6 \cdot 7 \cdot 2}{3 \cdot 5 \cdot 7 \cdot 2}$$
$$= \frac{39}{3 \cdot 5 \cdot 7 \cdot 2} + \frac{110}{3 \cdot 5 \cdot 7 \cdot 2} + \frac{84}{3 \cdot 5 \cdot 7 \cdot 2}$$
$$= \frac{233}{3 \cdot 5 \cdot 7 \cdot 2}$$
$$= \frac{233}{210} \quad \text{We left 210 factored until we knew we could not simplify.}$$

In each case, we multiply by 1 to obtain the LCD. In other words, look at the prime factorization of the LCD. Multiply each number by 1 to obtain what is missing in the LCD. ◀

DO EXERCISES 8–10.

7. Add: $\frac{1}{6} + \frac{7}{18}$. $\frac{5}{9}$

Add.

8. $\frac{4}{10} + \frac{1}{100} + \frac{3}{1000}$ $\frac{413}{1000}$

9. $\frac{7}{10} + \frac{5}{100} + \frac{9}{1000}$ $\frac{759}{1000}$

(Try to do this one mentally.)

10. $\frac{7}{10} + \frac{2}{21} + \frac{1}{7}$ $\frac{197}{210}$

ANSWERS ON PAGE A-2

11. A consumer bought $\frac{1}{2}$ lb of peanuts and $\frac{3}{5}$ lb of cashews. How many pounds of nuts were bought altogether? $\frac{11}{10}$ lb

ANSWER ON PAGE A-2

C Solving Problems

▶ **EXAMPLE 9** One jogger ran $\frac{4}{5}$ of a mile and another ran $\frac{1}{10}$ of a mile. How far did they run in all?

1. *Familiarize.* We first draw a picture. We let D = the distance run in all.

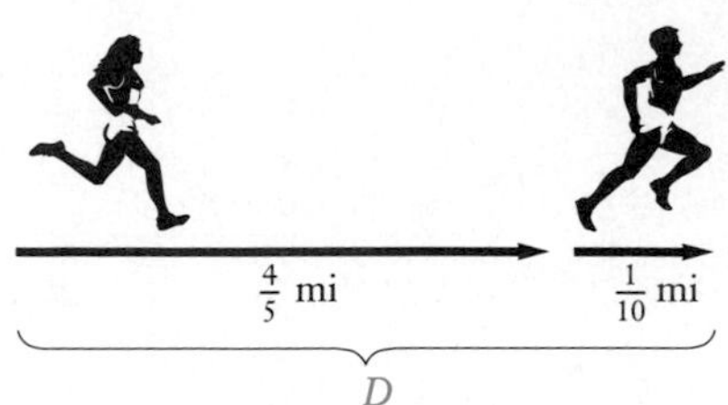

2. *Translate.* The problem can be translated to an equation as follows.

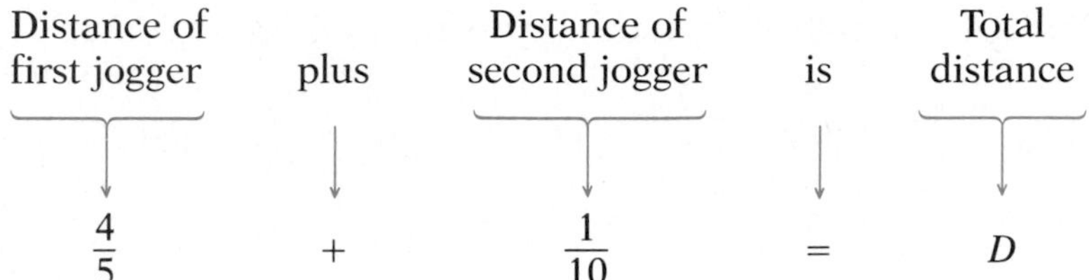

$$\frac{4}{5} + \frac{1}{10} = D$$

3. *Solve.* To solve the equation, we carry out the addition. The LCM of the denominators is 10 because 5 is a factor of 10. We multiply by 1 in order to obtain the LCD.

$$\frac{4}{5}\cdot\frac{2}{2} + \frac{1}{10} = D$$
$$\frac{8}{10} + \frac{1}{10} = D$$
$$\frac{9}{10} = D$$

4. *Check.* We check by repeating the calculation. We also note that the sum should be larger than either of the individual distances, which it is. This gives us a partial check on the reasonableness of the answer.
5. *State.* In all, the joggers ran $\frac{9}{10}$ of a mile. ◀

DO EXERCISE 11.

NAME SECTION DATE

EXERCISE SET 3.2

a, b Add and simplify.

1. $\frac{7}{8} + \frac{1}{8}$
2. $\frac{1}{4} + \frac{1}{4}$
3. $\frac{1}{8} + \frac{5}{8}$
4. $\frac{2}{9} + \frac{4}{9}$
5. $\frac{2}{3} + \frac{5}{6}$
6. $\frac{1}{4} + \frac{5}{6}$
7. $\frac{1}{8} + \frac{1}{6}$
8. $\frac{1}{9} + \frac{1}{6}$
9. $\frac{4}{5} + \frac{7}{10}$
10. $\frac{3}{4} + \frac{1}{12}$
11. $\frac{5}{12} + \frac{3}{8}$
12. $\frac{7}{8} + \frac{1}{16}$
13. $\frac{3}{20} + \frac{3}{4}$
14. $\frac{2}{15} + \frac{2}{5}$
15. $\frac{5}{6} + \frac{7}{9}$
16. $\frac{5}{8} + \frac{5}{6}$
17. $\frac{3}{10} + \frac{1}{100}$
18. $\frac{9}{10} + \frac{3}{100}$
19. $\frac{5}{12} + \frac{4}{15}$
20. $\frac{3}{16} + \frac{1}{12}$
21. $\frac{9}{10} + \frac{99}{100}$
22. $\frac{3}{10} + \frac{27}{100}$
23. $\frac{7}{8} + \frac{0}{1}$
24. $\frac{0}{1} + \frac{5}{6}$
25. $\frac{3}{8} + \frac{1}{6}$
26. $\frac{7}{8} + \frac{1}{6}$
27. $\frac{5}{12} + \frac{7}{24}$
28. $\frac{1}{18} + \frac{7}{12}$
29. $\frac{3}{16} + \frac{5}{16} + \frac{4}{16}$
30. $\frac{3}{8} + \frac{1}{8} + \frac{2}{8}$
31. $\frac{8}{10} + \frac{7}{100} + \frac{4}{1000}$

ANSWERS

1. 1
2. $\frac{1}{2}$
3. $\frac{3}{4}$
4. $\frac{2}{3}$
5. $\frac{3}{2}$
6. $\frac{13}{12}$
7. $\frac{7}{24}$
8. $\frac{5}{18}$
9. $\frac{3}{2}$
10. $\frac{5}{6}$
11. $\frac{19}{24}$
12. $\frac{15}{16}$
13. $\frac{9}{10}$
14. $\frac{8}{15}$
15. $\frac{29}{18}$
16. $\frac{35}{24}$
17. $\frac{31}{100}$
18. $\frac{93}{100}$
19. $\frac{41}{60}$
20. $\frac{13}{48}$
21. $\frac{189}{100}$
22. $\frac{57}{100}$
23. $\frac{7}{8}$
24. $\frac{5}{6}$
25. $\frac{13}{24}$
26. $\frac{25}{24}$
27. $\frac{17}{24}$
28. $\frac{23}{36}$
29. $\frac{3}{4}$
30. $\frac{3}{4}$
31. $\frac{437}{500}$

ANSWERS

32. $\frac{123}{1000}$

33. $\frac{53}{40}$

34. $\frac{9}{8}$

35. $\frac{391}{144}$

36. $\frac{268}{91}$

37. $\frac{3}{4}$ lb

38. $\frac{5}{6}$ lb

39. $\frac{23}{12}$ mi

40. $\frac{51}{40}$ mi

41. $\frac{4}{5}$ L; $\frac{8}{5}$ L; $\frac{2}{5}$ L

42. 690 kg; $\frac{14}{23}$ cement, $\frac{5}{23}$ stone, $\frac{4}{23}$ sand; 1

43. $\frac{173}{100}$ cm

44. $\frac{13}{12}$ lb

45. $\frac{4}{9}$

32. $\frac{1}{10} + \frac{2}{100} + \frac{3}{1000}$

33. $\frac{3}{8} + \frac{5}{12} + \frac{8}{15}$

34. $\frac{1}{2} + \frac{3}{8} + \frac{1}{4}$

35. $\frac{15}{24} + \frac{7}{36} + \frac{91}{48}$

36. $\frac{5}{7} + \frac{25}{52} + \frac{7}{4}$

C Solve.

37. A consumer bought $\frac{1}{4}$ lb of bonbons and $\frac{1}{2}$ lb of caramels. How many pounds of candy were bought?

38. A consumer bought $\frac{1}{3}$ lb of Turkish blend tobacco and $\frac{1}{2}$ lb of Virginia tobacco. How many pounds of tobacco were bought?

39. A student walked $\frac{7}{6}$ mi to a friend's house, and then $\frac{3}{4}$ mi to class. How far did the student walk?

40. A student walked $\frac{7}{8}$ mi to a friend's house, and then $\frac{2}{5}$ mi to class. How far did the student walk?

41. For strawberry punch, the recipe required $\frac{1}{5}$ L of ginger ale and $\frac{3}{5}$ L of strawberry cocktail juice. How much liquid was needed? If the recipe is doubled, how much liquid is needed? If the recipe is halved, how much liquid is needed?

42. A cubic meter of concrete mix contains 420 kg of cement, 150 kg of stone, and 120 kg of sand. What is the total weight of the cubic meter of concrete mix? What part is cement? stone? sand? Add these amounts. What is the result?

43. A board $\frac{9}{10}$ cm thick is glued to a board $\frac{8}{10}$ cm thick. The glue is $\frac{3}{100}$ cm thick. How thick is the result?

44. A baker used $\frac{1}{2}$ lb of flour for rolls, $\frac{1}{4}$ lb for donuts, and $\frac{1}{3}$ lb for cookies. How much flour was used?

SYNTHESIS

45. A student gets a part-time job on a weekend. The student does one-half of four-ninths of the job on Saturday and two-fifths of five-ninths of the job on Sunday. What part of the job is done in total on the two days?

3.3 Subtraction and Order

a Subtraction

Like Denominators

The difference $\frac{4}{8} - \frac{3}{8}$ can be considered as it was before, either "take away" or "how much more." Let us consider "take away."

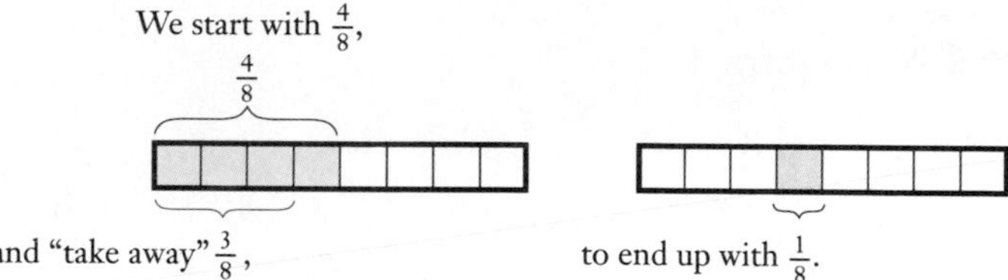

We start with 4 eighths, and take away 3 eighths:

$$4 \text{ eighths} - 3 \text{ eighths} = 1 \text{ eighth},$$

or

$$4 \cdot \frac{1}{8} - 3 \cdot \frac{1}{8} = \frac{1}{8},$$

or

$$\frac{4}{8} - \frac{3}{8} = \frac{1}{8}.$$

To subtract when denominators are the same,
a) subtract the numerators,
b) keep the denominator,
and
c) simplify, if possible.

$$\frac{7}{10} - \frac{4}{10} = \frac{7-4}{10} = \frac{3}{10}$$

Answers should be simplified, if possible.

▶ **EXAMPLES** Subtract and simplify.

1. $\frac{7}{10} - \frac{3}{10} = \frac{7-3}{10} = \frac{4}{10} = \frac{2 \cdot 2}{5 \cdot 2} = \frac{2}{5} \cdot \frac{2}{2} = \frac{2}{5} \cdot 1 = \frac{2}{5}$

2. $\frac{8}{9} - \frac{2}{9} = \frac{8-2}{9} = \frac{6}{9} = \frac{2 \cdot 3}{3 \cdot 3} = \frac{2}{3} \cdot \frac{3}{3} = \frac{2}{3} \cdot 1 = \frac{2}{3}$

3. $\frac{32}{12} - \frac{25}{12} = \frac{32-25}{12} = \frac{7}{12}$

DO EXERCISES 1–3.

OBJECTIVES

After finishing Section 3.3, you should be able to:

a Subtract using fractional notation.

b Use < or > to write a true sentence.

c Solve equations of the type $a + x = b$ and $x + a = b$, where a and b may be fractions.

d Solve problems involving subtraction.

FOR EXTRA HELP

Tape 4C

Tape 5B

MAC: 3
IBM: 3

Subtract and simplify.

1. $\frac{7}{8} - \frac{3}{8}$ $\frac{1}{2}$

2. $\frac{10}{16} - \frac{4}{16}$ $\frac{3}{8}$

3. $\frac{8}{10} - \frac{3}{10}$ $\frac{1}{2}$

ANSWERS ON PAGE A-3

4. Subtract: $\frac{3}{4} - \frac{2}{3}$. $\frac{1}{12}$

Subtract.

5. $\frac{5}{6} - \frac{1}{9}$ $\frac{13}{18}$

6. $\frac{4}{5} - \frac{3}{10}$ $\frac{1}{2}$

ANSWERS ON PAGE A-3

Different Denominators

To subtract when denominators are different:

a) Find the least common multiple of the denominators. That number is the least common denominator, LCD.

b) Multiply by 1, using an appropriate notation, *n/n*, to obtain the LCD for each number.

c) Subtract and simplify, if appropriate.

▶ **EXAMPLE 4** Subtract: $\frac{2}{5} - \frac{3}{8}$.

The LCM of 5 and 8 is 40. The LCD is 40.

$$\frac{2}{5} - \frac{3}{8} = \frac{2}{5} \cdot \frac{8}{8} - \frac{3}{8} \cdot \frac{5}{5}$$

Think: 8 × ▢ = 40. The answer is 5, so we multiply by 1, using $\frac{5}{5}$.

Think: 5 × ▢ = 40. The answer is 8, so we multiply by 1, using $\frac{8}{8}$.

$$= \frac{16}{40} - \frac{15}{40}$$

$$= \frac{16 - 15}{40} = \frac{1}{40}$$ ◀

DO EXERCISE 4.

▶ **EXAMPLE 5** Subtract: $\frac{5}{6} - \frac{7}{12}$.

Since 6 is a factor of 12, the LCM of 6 and 12 is 12. The LCD is 12.

$$\frac{5}{6} - \frac{7}{12} = \frac{5}{6} \cdot \frac{2}{2} - \frac{7}{12}$$

$$= \frac{10}{12} - \frac{7}{12}$$

$$= \frac{10 - 7}{12} = \frac{3}{12}$$

$$= \frac{3 \cdot 1}{3 \cdot 4} = \frac{3}{3} \cdot \frac{1}{4}$$

$$= \frac{1}{4}$$

We could have used 12 in factored form until we knew whether we could simplify. ◀

DO EXERCISES 5 AND 6.

b Order

We see that $\frac{4}{5} > \frac{3}{5}$. That is, $\frac{4}{5}$ is greater than $\frac{3}{5}$.

$\frac{4}{5}$ $\frac{3}{5}$

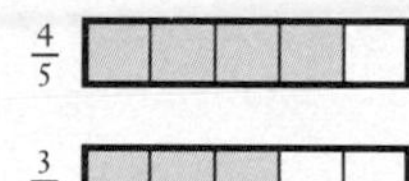

To determine which of two numbers is greater when there is a common denominator, compare the numerators:

$$\frac{4}{5}, \frac{3}{5} \qquad 4 > 3 \qquad \frac{4}{5} > \frac{3}{5}.$$

DO EXERCISES 7 AND 8.

When denominators are different, we multiply by 1 to make the denominators the same.

▶ **EXAMPLE 6** Use < or > for ■ to write a true sentence.

$$\frac{2}{5} \;■\; \frac{3}{4}$$

We have

$$\frac{2}{5} \cdot \frac{4}{4} = \frac{8}{20};$$ We choose $\frac{4}{4}$ by looking at the denominator of $\frac{3}{4}$.

$$\frac{3}{4} \cdot \frac{5}{5} = \frac{15}{20}.$$ We choose $\frac{5}{5}$ by looking at the denominator of $\frac{2}{5}$.

Since $8 < 15$, it follows that $\frac{8}{20} < \frac{15}{20}$, so

$$\frac{2}{5} < \frac{3}{4}.$$ ◀

▶ **EXAMPLE 7** Use < or > for ■ to write a true sentence.

$$\frac{9}{10} \;■\; \frac{89}{100}$$

The LCD is 100.

$$\frac{9}{10} \cdot \frac{10}{10} = \frac{90}{100}$$ We multiply by $\frac{10}{10}$ to get the LCD.

Since $90 > 89$, it follows that $\frac{90}{100} > \frac{89}{100}$, so

$$\frac{9}{10} > \frac{89}{100}.$$ ◀

DO EXERCISES 9–11.

c Solving Equations

Now let us solve equations of the form $x + a = b$ or $a + x = b$, where a and b may be fractions. Proceeding as we have before, we subtract a on both sides of the equation.

7. Use < or > for ■ to write a true sentence.

$\frac{3}{8}$ ■ $\frac{5}{8}$ <

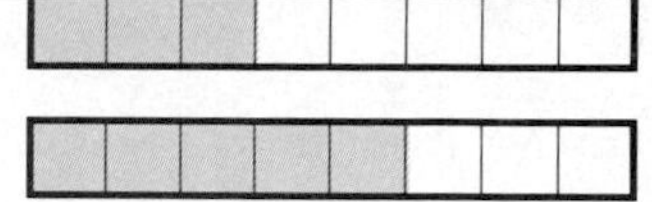

8. Use < or > for ■ to write a true sentence.

$\frac{7}{10}$ ■ $\frac{6}{10}$ >

Use < or > for ■ to write a true sentence.

9. $\frac{2}{3}$ ■ $\frac{5}{8}$ >

10. $\frac{3}{4}$ ■ $\frac{8}{12}$ >

11. $\frac{5}{6}$ ■ $\frac{7}{8}$ <

ANSWERS ON PAGE A-3

Solve.

12. $x + \frac{2}{3} = \frac{5}{6}$ $\frac{1}{6}$

13. $\frac{3}{5} + t = \frac{7}{8}$ $\frac{11}{40}$

14. There is $\frac{1}{4}$ cup of cooking oil in a pitcher. How much oil must be added so there will be $\frac{4}{5}$ cup of oil in the pitcher? $\frac{11}{20}$ cup

ANSWERS ON PAGE A-3

▶ **EXAMPLE 8** Solve: $x + \frac{1}{4} = \frac{3}{5}$.

$$x + \frac{1}{4} = \frac{3}{5}$$

$$x + \frac{1}{4} - \frac{1}{4} = \frac{3}{5} - \frac{1}{4} \qquad \text{Subtracting } \tfrac{1}{4} \text{ on both sides}$$

$$x + 0 = \frac{3}{5}\cdot\frac{4}{4} - \frac{1}{4}\cdot\frac{5}{5} \qquad \text{The LCD is 20. We multiply by 1 to get the LCD.}$$

$$x = \frac{12}{20} - \frac{5}{20} = \frac{7}{20}$$

The solution is $\frac{7}{20}$. ◀

DO EXERCISES 12 AND 13.

d Solving Problems

▶ **EXAMPLE 9** A jogger has run $\frac{2}{3}$ mi and will stop running when she has run $\frac{7}{8}$ mi. How much farther does the jogger have to go?

1. *Familiarize.* We first draw a picture or at least visualize the situation. We let d = the distance to go.

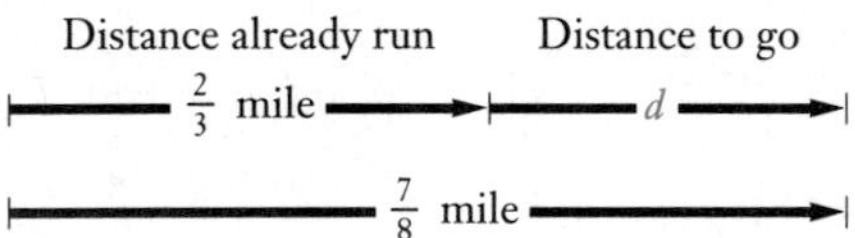

2. *Translate.* We see that this is a "how much more" situation. Now we translate to an equation.

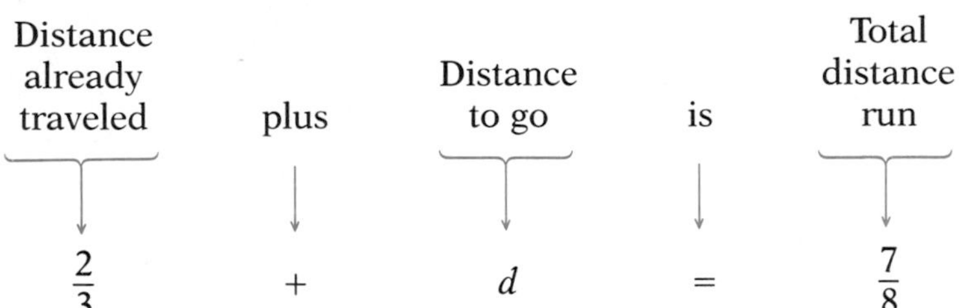

3. *Solve.* To solve the equation, we subtract $\frac{2}{3}$ on both sides.

$$\frac{2}{3} + d - \frac{2}{3} = \frac{7}{8} - \frac{2}{3} \qquad \text{Subtracting } \tfrac{2}{3} \text{ on both sides}$$

$$d + 0 = \frac{7}{8}\cdot\frac{3}{3} - \frac{2}{3}\cdot\frac{8}{8} \qquad \text{The LCD is 24. We multiply by 1 to obtain the LCD.}$$

$$d = \frac{21}{24} - \frac{16}{24} = \frac{5}{24}$$

4. *Check.* To check, we return to the original problem and add:

$$\frac{2}{3} + \frac{5}{24} = \frac{2}{3}\cdot\frac{8}{8} + \frac{5}{24} = \frac{16}{24} + \frac{5}{24} = \frac{21}{24} = \frac{7}{8}\cdot\frac{3}{3} = \frac{7}{8}.$$

   This checks.

5. *State.* The jogger has $\frac{5}{24}$ mi to go. ◀

DO EXERCISE 14.

NAME SECTION DATE

EXERCISE SET 3.3

a Subtract and simplify.

1. $\frac{5}{6} - \frac{1}{6}$ **2.** $\frac{7}{5} - \frac{2}{5}$ **3.** $\frac{11}{12} - \frac{2}{12}$ **4.** $\frac{15}{16} - \frac{11}{16}$

5. $\frac{3}{4} - \frac{1}{8}$ **6.** $\frac{2}{3} - \frac{1}{9}$ **7.** $\frac{1}{8} - \frac{1}{12}$ **8.** $\frac{1}{6} - \frac{1}{8}$

9. $\frac{4}{3} - \frac{5}{6}$ **10.** $\frac{7}{8} - \frac{1}{16}$ **11.** $\frac{3}{4} - \frac{3}{28}$ **12.** $\frac{2}{5} - \frac{2}{15}$

13. $\frac{3}{4} - \frac{3}{20}$ **14.** $\frac{5}{6} - \frac{1}{2}$ **15.** $\frac{3}{4} - \frac{1}{20}$ **16.** $\frac{3}{4} - \frac{4}{16}$

17. $\frac{5}{12} - \frac{2}{15}$ **18.** $\frac{9}{10} - \frac{11}{16}$ **19.** $\frac{6}{10} - \frac{7}{100}$ **20.** $\frac{9}{10} - \frac{3}{100}$

21. $\frac{7}{15} - \frac{3}{25}$ **22.** $\frac{18}{25} - \frac{4}{35}$ **23.** $\frac{99}{100} - \frac{9}{10}$ **24.** $\frac{78}{100} - \frac{11}{20}$

25. $\frac{2}{3} - \frac{1}{8}$ **26.** $\frac{3}{4} - \frac{1}{2}$ **27.** $\frac{3}{5} - \frac{1}{2}$ **28.** $\frac{5}{6} - \frac{2}{3}$

29. $\frac{5}{12} - \frac{3}{8}$ **30.** $\frac{7}{12} - \frac{2}{9}$ **31.** $\frac{7}{8} - \frac{1}{16}$ **32.** $\frac{5}{12} - \frac{5}{16}$

33. $\frac{17}{25} - \frac{4}{15}$ **34.** $\frac{11}{18} - \frac{7}{24}$ **35.** $\frac{23}{25} - \frac{112}{150}$ **36.** $\frac{89}{90} - \frac{53}{120}$

b Use $<$ or $>$ for $\square$ to write a true sentence.

37. $\frac{5}{8} \square \frac{6}{8}$ **38.** $\frac{7}{9} \square \frac{5}{9}$ **39.** $\frac{1}{3} \square \frac{1}{4}$ **40.** $\frac{1}{8} \square \frac{1}{6}$

ANSWERS

1. $\frac{2}{3}$
2. 1
3. $\frac{3}{4}$
4. $\frac{1}{4}$
5. $\frac{5}{8}$
6. $\frac{5}{9}$
7. $\frac{1}{24}$
8. $\frac{1}{24}$
9. $\frac{1}{2}$
10. $\frac{13}{16}$
11. $\frac{9}{14}$
12. $\frac{4}{15}$
13. $\frac{3}{5}$
14. $\frac{1}{3}$
15. $\frac{7}{10}$
16. $\frac{1}{2}$
17. $\frac{17}{60}$
18. $\frac{17}{80}$
19. $\frac{53}{100}$
20. $\frac{87}{100}$
21. $\frac{26}{75}$
22. $\frac{106}{175}$
23. $\frac{9}{100}$
24. $\frac{23}{100}$
25. $\frac{13}{24}$
26. $\frac{1}{4}$
27. $\frac{1}{10}$
28. $\frac{1}{6}$
29. $\frac{1}{24}$
30. $\frac{13}{36}$
31. $\frac{13}{16}$
32. $\frac{5}{48}$
33. $\frac{31}{75}$
34. $\frac{23}{72}$
35. $\frac{13}{75}$
36. $\frac{197}{360}$
37. $<$
38. $>$
39. $>$
40. $<$

ANSWERS

41. $<$

42. $>$

43. $<$

44. $>$

45. $>$

46. $>$

47. $>$

48. $>$

49. $<$

50. $<$

51. $\frac{1}{15}$

52. $\frac{1}{6}$

53. $\frac{2}{15}$

54. $\frac{5}{24}$

55. $\frac{1}{15}$

56. $\frac{1}{2}$

57. $\frac{1}{4}$

58. $\frac{5}{16}$

59. $\frac{3}{2}$

60. 10 cups

61. Rice

62. $\frac{41}{90}$

63. $\frac{19}{24}$

64. $\frac{71}{35}$

65. $\frac{145}{144}$

66. 3 hr

41. $\frac{2}{3} \square \frac{5}{7}$ **42.** $\frac{3}{5} \square \frac{4}{7}$ **43.** $\frac{4}{5} \square \frac{5}{6}$ **44.** $\frac{3}{2} \square \frac{7}{5}$

45. $\frac{19}{20} \square \frac{4}{5}$ **46.** $\frac{5}{6} \square \frac{13}{16}$ **47.** $\frac{19}{20} \square \frac{9}{10}$ **48.** $\frac{3}{4} \square \frac{11}{15}$

49. $\frac{31}{21} \square \frac{41}{13}$ **50.** $\frac{12}{7} \square \frac{132}{49}$

c Solve.

51. $x + \frac{1}{30} = \frac{1}{10}$ **52.** $y + \frac{9}{12} = \frac{11}{12}$ **53.** $\frac{2}{3} + t = \frac{4}{5}$

54. $\frac{2}{3} + p = \frac{7}{8}$ **55.** $m + \frac{5}{6} = \frac{9}{10}$ **56.** $x + \frac{1}{3} = \frac{5}{6}$

d Solve.

57. A business was owned by three people. One owned $\frac{7}{12}$ of the business and the second owned $\frac{1}{6}$. How much did the third person own?

58. A parent died and left an estate to four children. One got $\frac{1}{4}$ of the estate, the second got $\frac{1}{16}$, and the third got $\frac{3}{8}$. How much did the fourth get?

SKILL MAINTENANCE

59. Divide and simplify:

$$\frac{9}{10} \div \frac{3}{5}.$$

60. A chocolate cake requires $\frac{5}{6}$ cup of sugar. How much sugar is needed to make 12 cakes?

SYNTHESIS

61. ▦ In a recent year Jim Rice of the Boston Red Sox got 174 hits in 564 times at bat. George Brett of the Kansas City Royals got 195 hits in 634 times at bat. Who had the highest batting average?

Simplify. Use the rules for order of operations given in Section 1.9.

62. $\frac{2}{5} + \frac{1}{6} \div 3$ **63.** $\frac{7}{8} - \frac{1}{10} \times \frac{5}{6}$ **64.** $5 \times \frac{3}{7} - \frac{1}{7} \times \frac{4}{5}$ **65.** $\left(\frac{2}{3}\right)^2 + \left(\frac{3}{4}\right)^2$

66. A video cassette recorder is purchased that records tapes up to a maximum of 6 hr. It can also record a tape of the same length at a speed of 4 hr , or 2 hr, meaning that it will fill a 6-hr tape up in either 4 hr or 2 hr because it runs at faster speeds. A 6-hr tape is placed in the machine. It records for $\frac{1}{2}$ hr at the 4-hr speed and $\frac{3}{4}$ hr at the 2-hr speed. How much time is left on the tape to record at the 6-hr speed?

3.4 Mixed Numerals

a What Is a Mixed Numeral?

A symbol like $2\frac{3}{4}$ is called a **mixed numeral.**

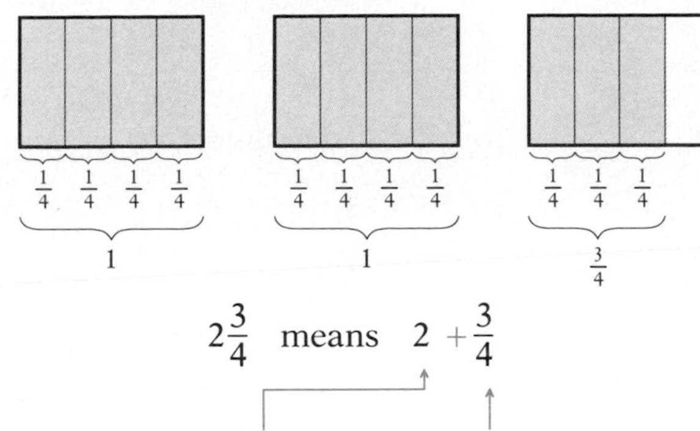

$$2\frac{3}{4} \quad \text{means} \quad 2 + \frac{3}{4}$$

This is a whole number. This is a number less than 1.

▶ **EXAMPLES** Convert to a mixed numeral.

1. $7 + \frac{2}{5} = 7\frac{2}{5}$

2. $4 + \frac{3}{10} = 4\frac{3}{10}$ ◀

DO EXERCISES 1–3.

The notation $2\frac{3}{4}$ has a plus sign left out. To aid in understanding, we sometimes write the missing plus sign.

▶ **EXAMPLES** Convert to fractional notation.

3. $2\frac{3}{4} = 2 + \frac{3}{4}$ **Inserting the missing plus sign**

$= \frac{2}{1} + \frac{3}{4}$ $\quad 2 = \frac{2}{1}$

$= \frac{2}{1} \cdot \frac{4}{4} + \frac{3}{4}$ **Finding a common denominator**

$= \frac{8}{4} + \frac{3}{4}$

$= \frac{11}{4}$

4. $4\frac{3}{10} = 4 + \frac{3}{10} = \frac{4}{1} + \frac{3}{10} = \frac{4}{1} \cdot \frac{10}{10} + \frac{3}{10} = \frac{40}{10} + \frac{3}{10} = \frac{43}{10}$ ◀

DO EXERCISES 4 AND 5.

OBJECTIVES

After finishing Section 3.4, you should be able to:

a Convert from mixed numerals to fractional notation.

b Convert from fractional notation to mixed numerals, and divide writing a mixed numeral for the quotient.

FOR EXTRA HELP

Tape 5A Tape 5B MAC: 3 IBM: 3

1. $1 + \frac{2}{3} = \square$ — Convert to a mixed numeral. $1\frac{2}{3}$

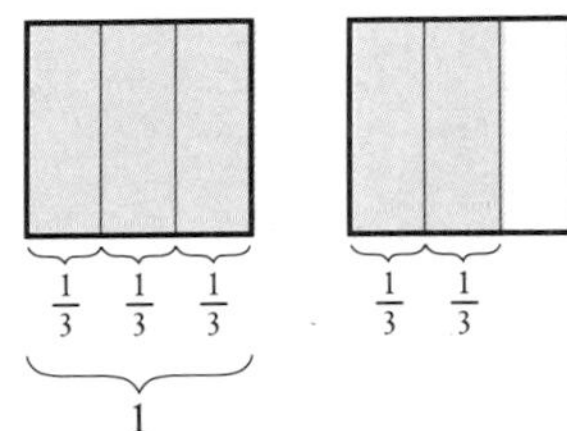

Convert to a mixed numeral.

2. $8 + \frac{3}{4}$ $8\frac{3}{4}$

3. $12 + \frac{2}{3}$ $12\frac{2}{3}$

Convert to fractional notation.

4. $4\frac{2}{5}$ $\frac{22}{5}$

5. $6\frac{1}{10}$ $\frac{61}{10}$

ANSWERS ON PAGE A-3

Convert to fractional notation. Use the faster way.

6. $4\frac{5}{6}$ $\frac{29}{6}$

7. $9\frac{1}{4}$ $\frac{37}{4}$

8. $20\frac{2}{3}$ $\frac{62}{3}$

ANSWERS ON PAGE A-3

To convert from a mixed numeral to fractional notation:

① Multiply: $4 \cdot 10 = 40$.
② Add: $40 + 3 = 43$.
③ Keep the denominator.

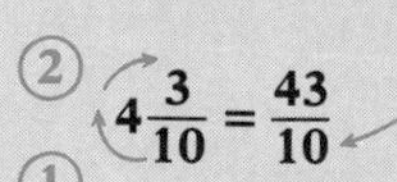

$$4\frac{3}{10} = \frac{43}{10}$$

▶ **EXAMPLES** Convert to fractional notation.

5. $6\frac{2}{3} = \frac{20}{3}$ $\quad 6 \cdot 3 = 18,\ 18 + 2 = 20$

6. $8\frac{2}{9} = \frac{74}{9}$

7. $10\frac{7}{8} = \frac{87}{8}$ ◀

DO EXERCISES 6–8.

b Writing Mixed Numerals

We can find a mixed numeral for $\frac{5}{3}$ as follows:

$$\frac{5}{3} = \frac{3}{3} + \frac{2}{3} = 1 + \frac{2}{3} = 1\frac{2}{3}.$$

Fractional symbols like $\frac{5}{3}$ also indicate division. Let's divide.

$$3\overline{)5}\ \text{quotient } 1\tfrac{2}{3}$$
$$\underline{3}$$
$$2 \longleftarrow 3\overline{)2}\ \text{gives } \tfrac{2}{3} \text{ or } 2 \div 3 = \tfrac{2}{3}$$

Thus, $\frac{5}{3} = 1\frac{2}{3}$.

In terms of objects, we can think of $\frac{5}{3}$ as 5 objects, each divided into 3 equal parts, as shown below.

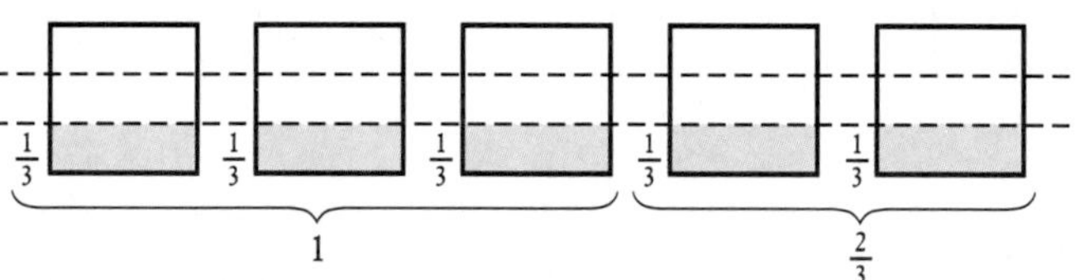

To convert from fractional notation to a mixed numeral, divide.

$\frac{13}{5}$ $\quad 5\overline{)13}$ with quotient 2 — The quotient; 10; 3 — The remainder $\quad \rightarrow 2\frac{3}{5}$

▶ **EXAMPLES** Convert to a mixed numeral.

8. $\frac{8}{5}$

$$\begin{array}{r} 1 \\ 5\overline{)8} \\ \underline{5} \\ 3 \end{array} \qquad \frac{8}{5} = 1\frac{3}{5}$$

A fraction larger than 1, such as $\frac{8}{5}$, is sometimes referred to as an "improper" fraction. We have intentionally avoided such terminology. The use of such notation, simplified, as $\frac{8}{5}$, $\frac{69}{10}$, and so on, is quite proper and very common in algebra.

9. $\frac{69}{10}$

$$\begin{array}{r} 6 \\ 10\overline{)69} \\ \underline{60} \\ 9 \end{array} \qquad \frac{69}{10} = 6\frac{9}{10}$$

10. $\frac{122}{8}$

$$\begin{array}{r} 15 \\ 8\overline{)122} \\ \underline{80} \\ 42 \\ \underline{40} \\ 2 \end{array} \qquad \frac{122}{8} = 15\frac{2}{8} = 15\frac{1}{4}$$

◀

DO EXERCISES 9–11.

It is quite common when dividing whole numbers to write the quotient using a mixed numeral. The remainder is the fractional part of the mixed numeral.

▶ **EXAMPLE 11** Divide. Write a mixed numeral for the quotient.

$$7\overline{)6\ 3\ 4\ 1}$$

We first divide as usual.

$$\begin{array}{r} 9\ 0\ 5 \\ 7\overline{)6\ 3\ 4\ 1} \\ \underline{6\ 3\ 0\ 0} \\ 4\ 1 \\ \underline{3\ 5} \\ 6 \end{array}$$

The answer is 905 R 6. We write a mixed numeral for the answer as follows:

$$905\frac{6}{7}.$$

The division 6341 ÷ 7 can be expressed using fractional notation as follows:

$$\frac{6341}{7} = 905\frac{6}{7}.$$

◀

Convert to a mixed numeral.

9. $\frac{7}{3}$ $2\frac{1}{3}$

10. $\frac{11}{10}$ $1\frac{1}{10}$

11. $\frac{110}{6}$ $18\frac{1}{3}$

ANSWERS ON PAGE A-3

Divide. Write a mixed numeral for the answer.

12. $6\overline{)4846}$ $807\frac{2}{3}$

13. $45\overline{)6053}$ $134\frac{23}{45}$

ANSWERS ON PAGE A-3

▶ **EXAMPLE 12** Divide. Write a mixed numeral for the answer.

$$42\overline{)8915}$$

We first divide as usual.

```
        2 1 2
 4 2)8 9 1 5
     8 4 0 0
       5 1 5
       4 2 0
         9 5
         8 4
         1 1
```

$$\frac{8915}{42} = 212\frac{11}{42}$$

The answer is $212\frac{11}{42}$. ◀

DO EXERCISES 12 AND 13.

NAME SECTION DATE

EXERCISE SET 3.4

a Convert to fractional notation.

1. $5\frac{2}{3}$ **2.** $4\frac{3}{5}$ **3.** $6\frac{1}{4}$ **4.** $8\frac{1}{2}$

5. $10\frac{1}{8}$ **6.** $10\frac{1}{3}$ **7.** $5\frac{1}{10}$ **8.** $8\frac{1}{10}$

9. $20\frac{3}{5}$ **10.** $30\frac{4}{5}$ **11.** $9\frac{5}{6}$ **12.** $8\frac{7}{8}$

13. $7\frac{3}{10}$ **14.** $6\frac{9}{10}$ **15.** $1\frac{5}{8}$ **16.** $1\frac{3}{5}$

17. $12\frac{3}{4}$ **18.** $15\frac{2}{3}$ **19.** $4\frac{3}{10}$ **20.** $5\frac{7}{10}$

21. $2\frac{3}{100}$ **22.** $5\frac{7}{100}$ **23.** $66\frac{2}{3}$ **24.** $33\frac{1}{3}$

25. $5\frac{29}{50}$ **26.** $84\frac{3}{8}$

b Convert to a mixed numeral.

27. $\frac{8}{5}$ **28.** $\frac{7}{4}$ **29.** $\frac{14}{3}$ **30.** $\frac{19}{8}$ **31.** $\frac{27}{6}$

ANSWERS

1. $\frac{17}{3}$
2. $\frac{23}{5}$
3. $\frac{25}{4}$
4. $\frac{17}{2}$
5. $\frac{81}{8}$
6. $\frac{31}{3}$
7. $\frac{51}{10}$
8. $\frac{81}{10}$
9. $\frac{103}{5}$
10. $\frac{154}{5}$
11. $\frac{59}{6}$
12. $\frac{71}{8}$
13. $\frac{73}{10}$
14. $\frac{69}{10}$
15. $\frac{13}{8}$
16. $\frac{8}{5}$
17. $\frac{51}{4}$
18. $\frac{47}{3}$
19. $\frac{43}{10}$
20. $\frac{57}{10}$
21. $\frac{203}{100}$
22. $\frac{507}{100}$
23. $\frac{200}{3}$
24. $\frac{100}{3}$
25. $\frac{279}{50}$
26. $\frac{675}{8}$
27. $1\frac{3}{5}$
28. $1\frac{3}{4}$
29. $4\frac{2}{3}$
30. $2\frac{3}{8}$
31. $4\frac{1}{2}$

ANSWERS

32. $3\frac{1}{3}$
33. $5\frac{7}{10}$
34. $6\frac{9}{10}$
35. $7\frac{4}{7}$
36. $6\frac{5}{8}$
37. $7\frac{1}{2}$
38. $6\frac{1}{4}$
39. $11\frac{1}{2}$
40. $4\frac{1}{3}$
41. $1\frac{1}{2}$
42. $4\frac{2}{3}$
43. $7\frac{57}{100}$
44. $4\frac{67}{100}$
45. $43\frac{1}{8}$
46. $55\frac{3}{4}$
47. $108\frac{5}{8}$
48. $708\frac{2}{3}$
49. $906\frac{3}{7}$
50. $1012\frac{2}{9}$
51. $40\frac{4}{7}$
52. $90\frac{22}{85}$
53. $55\frac{1}{51}$
54. $23\frac{1}{2}$
55. 18
56. $\frac{5}{2}$
57. $\frac{1}{4}$
58. $\frac{2}{5}$
59. $8\frac{2}{3}$
60. $6\frac{5}{6}$
61. $52\frac{2}{7}$
62. $52\frac{1}{7}$

32. $\frac{30}{9}$ **33.** $\frac{57}{10}$ **34.** $\frac{69}{10}$ **35.** $\frac{53}{7}$ **36.** $\frac{53}{8}$

37. $\frac{45}{6}$ **38.** $\frac{50}{8}$ **39.** $\frac{46}{4}$ **40.** $\frac{39}{9}$ **41.** $\frac{12}{8}$

42. $\frac{28}{6}$ **43.** $\frac{757}{100}$ **44.** $\frac{467}{100}$ **45.** $\frac{345}{8}$ **46.** $\frac{223}{4}$

Divide. Write a mixed numeral for the answer.

47. $8\overline{)869}$ **48.** $3\overline{)2126}$ **49.** $7\overline{)6345}$ **50.** $9\overline{)9110}$

51. $21\overline{)852}$ **52.** $85\overline{)7672}$ **53.** $102\overline{)5612}$ **54.** $46\overline{)1081}$

SKILL MAINTENANCE

Multiply and simplify.

55. $\frac{6}{5} \cdot 15$ **56.** $\frac{5}{12} \cdot 6$ **57.** $\frac{7}{10} \cdot \frac{5}{14}$ **58.** $\frac{1}{10} \cdot \frac{20}{5}$

SYNTHESIS

Write a mixed numeral.

59. $\frac{56}{7} + \frac{2}{3}$ **60.** $\frac{72}{12} + \frac{5}{6}$

61. There are $\frac{366}{7}$ weeks in a leap year. Write a mixed numeral.

62. There are $\frac{365}{7}$ weeks in a year. Write a mixed numeral.

3.5 Addition and Subtraction Using Mixed Numerals

a Addition

To find the sum $1\frac{5}{8} + 3\frac{1}{8}$, we add the fractions. Then we add the whole numbers.

$$\begin{array}{rl} 1\ \frac{5}{8} & = \\ +\,3\ \frac{1}{8} & = \\ \hline \frac{6}{8} & \end{array} \qquad \begin{array}{rl} 1\ \frac{5}{8} \\ +\ 3\ \frac{1}{8} \\ \hline 4\ \frac{6}{8} = 4\frac{3}{4} \end{array}$$

Add the fractions. Add the whole numbers.

DO EXERCISE 1.

▶ **EXAMPLE 1** Add: $5\frac{2}{3} + 3\frac{5}{6}$. Write a mixed numeral for the answer.

The LCD is 6.

$$\begin{array}{rcl} 5\ \frac{2}{3}\cdot\frac{2}{2} & = & 5\frac{4}{6} \\ +\ 3\ \frac{5}{6} & = & +\ 3\frac{5}{6} \\ \hline & & 8\frac{9}{6} = 8 + \frac{9}{6} \\ & & = 8 + 1\frac{1}{2} \\ & & = 9\frac{1}{2} \end{array}$$

To find a mixed numeral for $\frac{9}{6}$, we divide:

$$6\overline{)9}\ \text{with quotient } 1;\ 9 - 6 = 3 \qquad \frac{9}{6} = 1\frac{3}{6} = 1\frac{1}{2}$$

$\frac{19}{2}$ is also a correct answer, but it is not a mixed numeral, which is what we are working with in Sections 3.4, 3.5, and 3.6. ◀

DO EXERCISE 2.

▶ **EXAMPLE 2** Add: $10\frac{5}{6} + 7\frac{3}{8}$.

The LCD is 24.

$$\begin{array}{rcl} 10\ \frac{5}{6}\cdot\frac{4}{4} & = & 10\frac{20}{24} \\ +\ 7\ \frac{3}{8}\cdot\frac{3}{3} & = & +\ 7\frac{9}{24} \\ \hline & & 17\frac{29}{24} = 18\frac{5}{24} \end{array}$$ ◀

DO EXERCISE 3.

OBJECTIVES

After finishing Section 3.5, you should be able to:

- a Add using mixed numerals.
- b Subtract using mixed numerals.
- c Solve problems involving addition and subtraction with mixed numerals.

FOR EXTRA HELP

Tape 5B Tape 6A MAC: 3 IBM: 3

1. Add.

$$\begin{array}{r} 2\frac{3}{10} \\ +\ 5\frac{1}{10} \\ \hline \end{array}$$

$7\frac{2}{5}$

2. Add.

$$\begin{array}{r} 8\frac{2}{5} \\ +\ 3\frac{7}{10} \\ \hline \end{array}$$

$12\frac{1}{10}$

3. Add.

$$\begin{array}{r} 9\frac{3}{4} \\ +\ 3\frac{5}{6} \\ \hline \end{array}$$

$13\frac{7}{12}$

ANSWERS ON PAGE A-3

4. Subtract.

$$\begin{array}{r} 10\frac{7}{8} \\ -\ 9\frac{3}{8} \\ \hline \end{array}$$

$1\frac{1}{2}$

5. Subtract.

$$\begin{array}{r} 8\frac{2}{3} \\ -5\frac{1}{2} \\ \hline \end{array}$$

$3\frac{1}{6}$

6. Subtract.

$$\begin{array}{r} 5\frac{1}{12} \\ -1\frac{3}{4} \\ \hline \end{array}$$

$3\frac{1}{3}$

7. Subtract.

$$\begin{array}{r} 5 \\ -1\frac{1}{3} \\ \hline \end{array}$$

$3\frac{2}{3}$

b Subtraction

▶ **EXAMPLE 3** Subtract: $7\frac{3}{4} - 2\frac{1}{4}$.

$$\begin{array}{rl} 7\ \frac{3}{4} & = \\ -\ 2\ \frac{1}{4} & = \\ \hline \frac{2}{4} & \end{array} \qquad \begin{array}{r} 7\ \frac{3}{4} \\ -\ 2\ \frac{1}{4} \\ \hline 5\ \frac{2}{4} = 5\frac{1}{2} \end{array}$$

Subtract the fractions. Subtract the whole numbers.

DO EXERCISE 4.

▶ **EXAMPLE 4** Subtract: $9\frac{4}{5} - 3\frac{1}{2}$.

The LCD is 10.

$$\begin{array}{rcr} 9\ \frac{4}{5}\cdot\frac{2}{2} & = & 9\frac{8}{10} \\ -\ 3\ \frac{1}{2}\cdot\frac{5}{5} & = & -\ 3\frac{5}{10} \\ \hline & & 6\frac{3}{10} \end{array}$$

DO EXERCISE 5.

▶ **EXAMPLE 5** Subtract: $7\frac{1}{6} - 2\frac{1}{4}$.

The LCD is 12.

$$\begin{array}{rcr} 7\ \frac{1}{6}\cdot\frac{2}{2} & = & 7\frac{2}{12} \\ -\ 2\ \frac{1}{4}\cdot\frac{3}{3} & = & -2\frac{3}{12} \\ \hline \end{array}$$

We cannot subtract $\frac{3}{12}$ from $\frac{2}{12}$. We borrow 1, or $\frac{12}{12}$, from 7: $7\frac{2}{12} = 6 + 1 + \frac{2}{12} = 6 + \frac{12}{12} + \frac{2}{12} = 6\frac{14}{12}$.

We can write this as

$$\begin{array}{rcr} 7\frac{2}{12} & = & 6\frac{14}{12} \\ -\ 2\frac{3}{12} & = & -\ 2\frac{3}{12} \\ \hline & & 4\frac{11}{12} \end{array}$$

DO EXERCISE 6.

▶ **EXAMPLE 6** Subtract: $12 - 9\frac{3}{8}$.

$$\begin{array}{rcr} 12 & = & 11\frac{8}{8} \\ -\ 9\frac{3}{8} & = & -\ 9\frac{3}{8} \\ \hline & & 2\frac{5}{8} \end{array}$$

$12 = 11 + 1 = 11 + \frac{8}{8} = 11\frac{8}{8}$

DO EXERCISE 7.

C Solving Problems

▶ **EXAMPLE 7** On two business days, a salesperson drove $144\frac{9}{10}$ km and $87\frac{1}{4}$ km. What was the total distance driven?

1. *Familiarize.* We let d = the total distance driven.
2. *Translate.* We translate as follows.

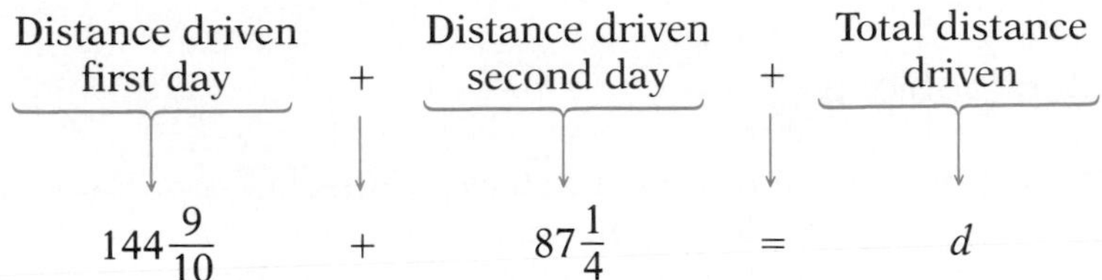

$$144\frac{9}{10} + 87\frac{1}{4} = d$$

3. *Solve.* The sentence tells us what to do. We add. The LCD is 20.

$$144\frac{9}{10} = 144\frac{9}{10}\cdot\frac{2}{2} = 144\frac{18}{20}$$

$$+\ 87\frac{1}{4} = +\ 87\frac{1}{4}\cdot\frac{5}{5} = +\ 87\frac{5}{20}$$

$$231\frac{23}{20} = 232\frac{3}{20}$$

Thus, $d = 232\frac{3}{20}$.

4. *Check.* We check by repeating the calculation. We also note that the answer is larger than any of the distances driven, which gives us a partial check of the reasonableness of the answer.
5. *State.* The total distance driven was $232\frac{3}{20}$ km. ◀

DO EXERCISE 8.

▶ **EXAMPLE 8** On a recent day, the stock of Ohio Edison opened at $\$19\frac{7}{8}$ and closed at $\$19\frac{1}{2}$. How much did it drop?

1. *Familiarize.* We let d = the amount of money the stock dropped.
2. *Translate.* We translate as follows.

Amount at opening	−	Amount at closing	=	Amount of drop
$19\frac{7}{8}$	−	$19\frac{1}{2}$	=	d

3. *Solve.* To solve the equation, we carry out the subtraction. The LCD is 8.

$$19\frac{7}{8} = 19\ \frac{7}{8} = 19\frac{7}{8}$$

$$-\ 19\frac{1}{2} = -\ 19\ \frac{1}{2}\cdot\frac{4}{4} = -\ 19\frac{4}{8}$$

$$\frac{3}{8}$$

Thus, $d = \frac{3}{8}$.

8. A fabric store sold two pieces of burlap $6\frac{1}{4}$ yd and $10\frac{5}{6}$ yd long. What was the total length of the burlap? $17\frac{1}{12}$ yd

ANSWER ON PAGE A-3

9. A $6\frac{1}{2}$-m pole was set $2\frac{3}{4}$ m in the ground. How much was above the ground? $3\frac{3}{4}$ m

10. There are $20\frac{1}{3}$ gal of water in a barrel; $5\frac{3}{4}$ gal are poured out and $8\frac{2}{3}$ gal are poured back in. How many gallons of water are then in the tank? $23\frac{1}{4}$ gal

ANSWERS ON PAGE A-3

4. *Check.* To check, we add the amount that the stock dropped to the closing price:

$$\frac{3}{8} + 19\frac{1}{2} = \frac{3}{8} + 19\frac{4}{8} = 19\frac{7}{8}. \qquad \text{This checks.}$$

5. *State.* The stock dropped $\$\frac{3}{8}$. ◀

DO EXERCISE 9.

Multistep Problems

▶ **EXAMPLE 9** One morning, the stock of Hitachi Corporation opened at a price of $\$100\frac{3}{8}$ per share. By noon, the price had risen $\$4\frac{7}{8}$. At the end of the day, it had fallen $\$10\frac{3}{4}$ from the price at noon. What was the closing price?

1. *Familiarize.* We first draw a picture or at least visualize the situation. We let p = the price at noon, after the rise, and c = the price after the drop, at the close.

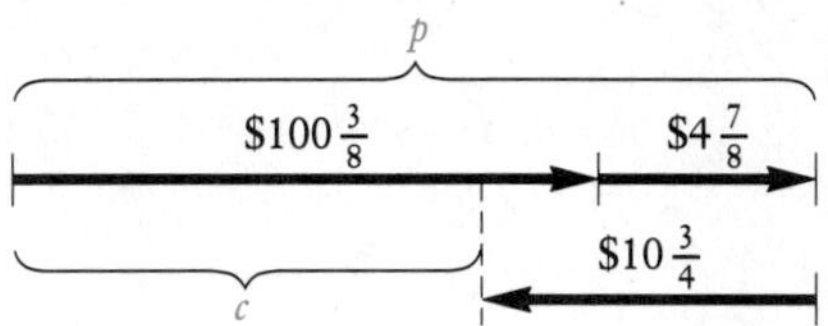

2. *Translate.* From the figure, we see that the price after the drop is the price at noon minus the amount of the drop. Thus,

$$c = p - \$10\frac{3}{4} = \left(\$100\frac{3}{8} + \$4\frac{7}{8}\right) - \$10\frac{3}{4}.$$

3. *Solve.* This is a two-step problem.

a) We first add $\$4\frac{7}{8}$ to $\$100\frac{3}{8}$ to find the price p of the stock at noon.

$$\begin{array}{r} 100\frac{3}{8} \\ +\ \ 4\frac{7}{8} \\ \hline 104\frac{10}{8} \end{array} = 105\frac{1}{4} = p$$

b) Next we subtract $\$10\frac{3}{4}$ from $\$105\frac{1}{4}$ to find the price c of the stock at closing.

$$\begin{array}{rcr} 105\frac{1}{4} & = & 104\frac{5}{4} \\ -\ 10\frac{3}{4} & = & -\ \ 10\frac{3}{4} \\ \hline & & 94\frac{2}{4} \end{array} = 94\frac{1}{2} = c$$

4. *Check.* We check by repeating the calculation.

5. *State.* The price of the stock at closing is $\$94\frac{1}{2}$. ◀

DO EXERCISE 10.

NAME SECTION DATE

EXERCISE SET 3.5

a Add. Write mixed numerals for the answers.

1. $2\frac{7}{8} + 3\frac{5}{8}$

2. $4\frac{5}{6} + 3\frac{5}{6}$

3. $1\frac{1}{4} + 1\frac{2}{3}$

4. $4\frac{1}{3} + 5\frac{2}{9}$

5. $8\frac{3}{4} + 5\frac{5}{6}$

6. $4\frac{3}{8} + 6\frac{5}{12}$

7. $3\frac{2}{5} + 8\frac{7}{10}$

8. $5\frac{1}{2} + 3\frac{7}{10}$

9. $5\frac{3}{8} + 10\frac{5}{6}$

10. $\frac{5}{8} + 1\frac{5}{6}$

11. $12\frac{4}{5} + 8\frac{7}{10}$

12. $15\frac{5}{8} + 11\frac{3}{4}$

13. $14\frac{5}{8} + 13\frac{1}{4}$

14. $16\frac{1}{4} + 15\frac{7}{8}$

15. $7\frac{1}{8} + 9\frac{2}{3} + 10\frac{3}{4}$

16. $45\frac{2}{3} + 31\frac{3}{5} + 12\frac{1}{4}$

b Subtract. Write mixed numerals for the answers.

17. $4\frac{1}{5} - 2\frac{3}{5}$

18. $5\frac{1}{8} - 2\frac{3}{8}$

19. $6\frac{3}{5} - 2\frac{1}{2}$

20. $7\frac{2}{3} - 6\frac{1}{2}$

ANSWERS

1. $6\frac{1}{2}$

2. $8\frac{2}{3}$

3. $2\frac{11}{12}$

4. $9\frac{5}{9}$

5. $14\frac{7}{12}$

6. $10\frac{19}{24}$

7. $12\frac{1}{10}$

8. $9\frac{1}{5}$

9. $16\frac{5}{24}$

10. $2\frac{11}{24}$

11. $21\frac{1}{2}$

12. $27\frac{3}{8}$

13. $27\frac{7}{8}$

14. $32\frac{1}{8}$

15. $27\frac{13}{24}$

16. $89\frac{31}{60}$

17. $1\frac{3}{5}$

18. $2\frac{3}{4}$

19. $4\frac{1}{10}$

20. $1\frac{1}{6}$

ANSWERS

21. $21\frac{17}{24}$

22. $6\frac{9}{16}$

23. $12\frac{1}{4}$

24. $38\frac{1}{8}$

25. $15\frac{3}{8}$

26. $3\frac{1}{4}$

27. $7\frac{5}{12}$

28. $18\frac{31}{60}$

29. $13\frac{3}{8}$

30. $27\frac{1}{2}$

31. $11\frac{5}{18}$

32. $8\frac{35}{48}$

33. $5\frac{14}{15}$ lb

34. $7\frac{5}{12}$ lb

35. $17\frac{11}{20}$ cm

36. $6\frac{7}{20}$ cm

37. $18\frac{4}{5}$ cm

38. $19\frac{1}{16}$ ft

21. $34\frac{1}{3} - 12\frac{5}{8}$

22. $23\frac{5}{16} - 16\frac{3}{4}$

23. $21 - 8\frac{3}{4}$

24. $42 - 3\frac{7}{8}$

25. $34 - 18\frac{5}{8}$

26. $23 - 19\frac{3}{4}$

27. $21\frac{1}{6} - 13\frac{3}{4}$

28. $42\frac{1}{10} - 23\frac{7}{12}$

29. $14\frac{1}{8} - \frac{3}{4}$

30. $28\frac{1}{6} - \frac{2}{3}$

31. $25\frac{1}{9} - 13\frac{5}{6}$

32. $23\frac{5}{16} - 14\frac{7}{12}$

c Solve.

33. Chuck Stake, a butcher, sold packages of meat weighing $1\frac{1}{3}$ lb and $4\frac{3}{5}$ lb. What was the total weight of the meat?

34. R.U. Ground, a butcher, sold packages of hamburger weighing $1\frac{2}{3}$ lb and $5\frac{3}{4}$ lb. What was the total weight of the meat?

35. A woman is $168\frac{1}{4}$ cm tall and her son is $150\frac{7}{10}$ cm tall. How much taller is the woman?

36. A man is $187\frac{1}{10}$ cm tall and his daughter is $180\frac{3}{4}$ cm tall. How much taller is the man?

37. The standard pencil is $16\frac{9}{10}$ cm wood and $1\frac{9}{10}$ cm eraser. What is the length of the standard pencil?

38. A plumber uses pipes of length $10\frac{5}{16}$ ft and $8\frac{3}{4}$ ft in the installation of a sink. How much pipe was used?

39. A businessperson drove $180\frac{7}{10}$ km away from Los Angeles one day. The next day that person drove $85\frac{1}{2}$ km back toward Los Angeles. How far was the person from Los Angeles then?

40. A woman is $4\frac{1}{2}$ cm taller than her daughter. The daughter is $169\frac{3}{10}$ cm tall. How tall is the woman?

41. A standard sheet of paper is $8\frac{1}{2}$ in. by 11 in. What is the total distance around the paper?

42. One standard book size is $8\frac{1}{2}$ in. by $9\frac{3}{4}$ in. What is the total distance around the front cover of such a book?

43. On a recent day, the stock of International Business Machines Corporation opened at $\$104\frac{5}{8}$ and dropped $\$1\frac{1}{4}$ during the course of the day. What was the closing price?

44. On a recent day, the stock of Shearson-Lehman-Hutton Corporation opened at $\$26\frac{7}{8}$ and closed at $\$27\frac{1}{4}$. How much did it gain that day?

45. A family room took $1\frac{2}{3}$ gal of paint and a bedroom took $1\frac{1}{2}$ gal. How much paint was used in all?

46. A living room took $1\frac{3}{4}$ gal of paint and a family room took $1\frac{1}{3}$ gal. How much paint was used in all?

47. A man is $5\frac{1}{4}$ cm taller than his son. The son is $182\frac{9}{10}$ cm tall. How tall is the man?

48. A plane flew 640 km on a nonstop flight. On the return flight, it landed after it had flown $320\frac{3}{10}$ km. How far was it from its original destination?

49. A painter had $3\frac{1}{2}$ gal of paint. It took $2\frac{3}{4}$ gal for a family room. It was estimated that it would take $2\frac{1}{4}$ gal to paint the living room. How much more paint was needed?

50. A person worked $10\frac{1}{2}$ hr over a three-day period. If the person worked $2\frac{1}{2}$ hr the first day and $4\frac{1}{5}$ hr the second, how many hours were worked the third day?

ANSWERS

39. $95\frac{1}{5}$ km

40. $173\frac{4}{5}$ cm

41. 39 in.

42. $36\frac{1}{2}$ in.

43. $\$103\frac{3}{8}$

44. $\$\frac{3}{8}$

45. $3\frac{1}{6}$ gal

46. $3\frac{1}{12}$ gal

47. $188\frac{3}{20}$ cm

48. $319\frac{7}{10}$ km

49. $1\frac{1}{2}$ gal

50. $3\frac{4}{5}$ hr

ANSWERS

51. $28\frac{3}{4}$ yd

52. $27\frac{1}{2}$ in.

53. $7\frac{3}{8}$ ft

54. $6\frac{1}{10}$ in.

55. $1\frac{9}{16}$ in.

56. 286 bottles; 2 oz left over

57. $\frac{1}{10}$

58. 33 ft

59. $8\frac{7}{12}$

51. Find the distance around this figure.

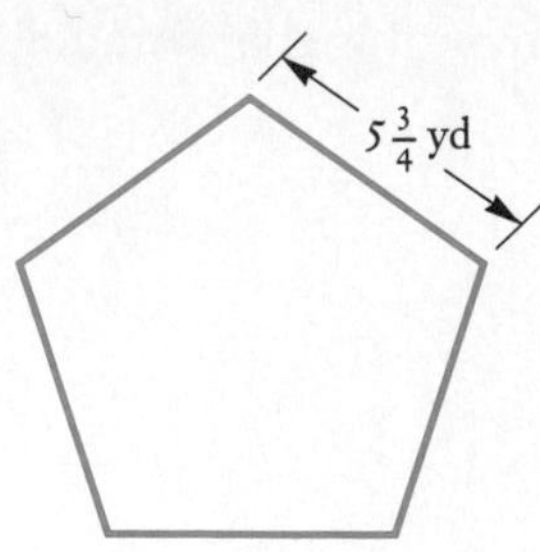

52. Find the distance around this figure.

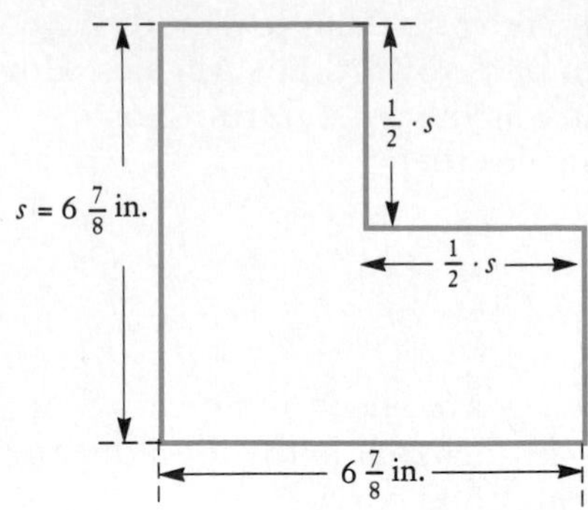

53. Find the length d in this figure.

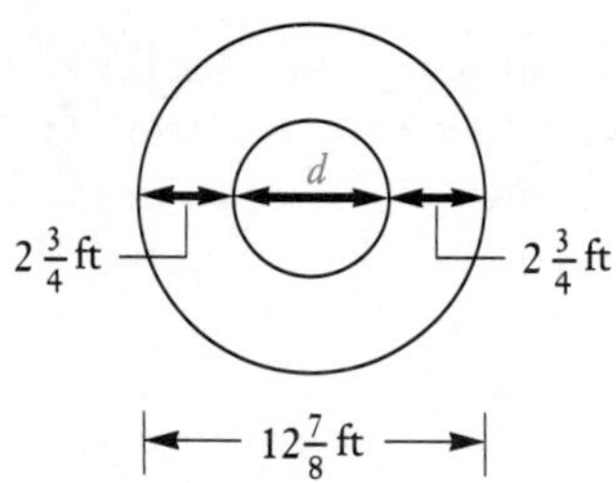

54. Find the length d in this figure.

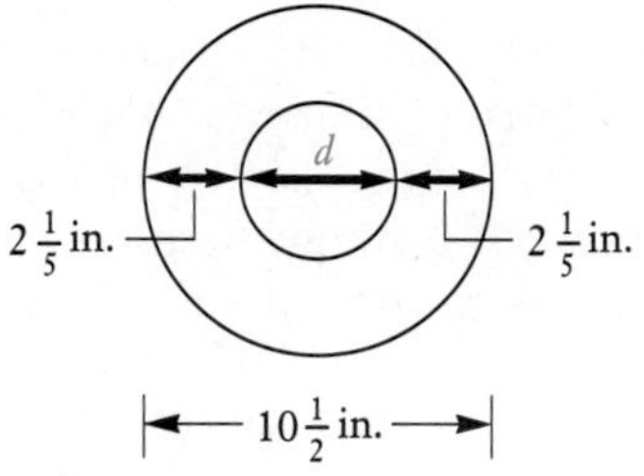

55. Find the smallest length of a bolt that will pass through a piece of tubing with an outside diameter of $\frac{1}{2}$ in., a washer $\frac{1}{16}$ in. thick, a piece of tubing with a $\frac{3}{4}$-in. outside diameter, another washer, and a nut $\frac{3}{16}$ in. thick.

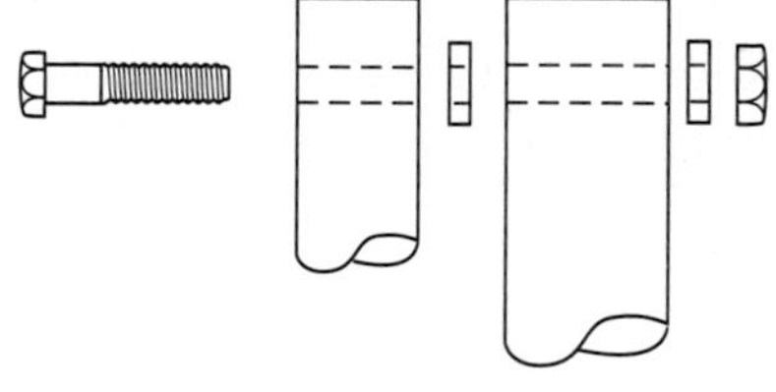

SKILL MAINTENANCE

56. A dairy produced 4578 oz of milk one week. How many 16-oz bottles were filled? How much milk was left over?

57. Divide and simplify: $\frac{12}{25} \div \frac{24}{5}$.

SYNTHESIS

58. A post is placed through some water into the mud at the bottom of the lake. Half of the post is in the mud, $\frac{1}{3}$ is in the water, and the part above water is $5\frac{1}{2}$ ft long. How long is the post?

59. Solve: $47\frac{2}{3} + n = 56\frac{1}{4}$.

3.6 Multiplication and Division Using Mixed Numerals

a Multiplication

To carry out addition and subtraction with mixed numerals, it is easiest to leave them as mixed numerals. With multiplication and division, however, it is easiest to convert them first to fractional notation.

To multiply using mixed numerals, first convert to fractional notation. Then multiply with fractional notation and convert the answer back to a mixed numeral, if appropriate.

▶ **EXAMPLE 1** Multiply: $6 \cdot 2\frac{1}{2}$.

$$6 \cdot 2\frac{1}{2} = \frac{6}{1} \cdot \frac{5}{2} = \frac{6 \cdot 5}{1 \cdot 2} = \frac{2 \cdot 3 \cdot 5}{2 \cdot 1} = \frac{2}{2} \cdot \frac{3 \cdot 5}{1} = 15$$

Here we write fractional notation. ◀

DO EXERCISE 1.

▶ **EXAMPLE 2** Multiply: $3\frac{1}{2} \cdot \frac{3}{4}$.

$$3\frac{1}{2} \cdot \frac{3}{4} = \frac{7}{2} \cdot \frac{3}{4} = \frac{21}{8} = 2\frac{5}{8}$$

Note here that we need fractional notation to carry out the multiplication. ◀

DO EXERCISE 2.

▶ **EXAMPLE 3** Multiply: $8 \cdot 4\frac{2}{3}$.

$$8 \cdot 4\frac{2}{3} = \frac{8}{1} \cdot \frac{14}{3} = \frac{112}{3} = 37\frac{1}{3}$$ ◀

DO EXERCISE 3.

▶ **EXAMPLE 4** Multiply: $2\frac{1}{4} \cdot 3\frac{2}{5}$.

$$2\frac{1}{4} \cdot 3\frac{2}{5} = \frac{9}{4} \cdot \frac{17}{5} = \frac{153}{20} = 7\frac{13}{20}$$ ◀

CAUTION! $2\frac{1}{4} \cdot 3\frac{2}{5} \neq 6\frac{2}{20}$. A common error is to just multiply the whole numbers and then the fractions. This does not give the correct answer, $7\frac{13}{20}$, which is found by converting first to fractional notation.

DO EXERCISE 4.

OBJECTIVES

After finishing Section 3.6, you should be able to:

a Multiply using mixed numerals.

b Divide using mixed numerals.

c Solve problems involving multiplication and division with mixed numerals.

FOR EXTRA HELP

Tape 5C | Tape 6A | MAC: 3 IBM: 3

1. Multiply: $6 \cdot 3\frac{1}{3}$. 20

2. Multiply: $2\frac{1}{2} \cdot \frac{3}{4}$. $1\frac{7}{8}$

3. Multiply: $2 \cdot 6\frac{2}{5}$. $12\frac{4}{5}$

4. Multiply: $3\frac{1}{3} \cdot 2\frac{1}{2}$. $8\frac{1}{3}$

ANSWERS ON PAGE A-3

5. Divide: $84 \div 5\frac{1}{4}$. 16

6. Divide: $26 \div 3\frac{1}{2}$. $7\frac{3}{7}$

Divide.

7. $2\frac{1}{4} \div 1\frac{1}{5}$ $1\frac{7}{8}$

8. $1\frac{3}{4} \div 2\frac{1}{2}$ $\frac{7}{10}$

ANSWERS ON PAGE A-3

b Division

The division $1\frac{1}{2} \div \frac{1}{6}$ is shown here.

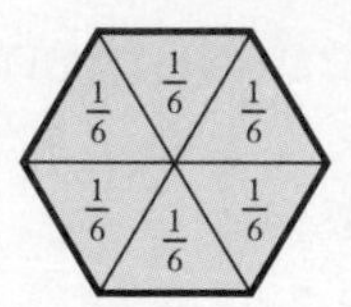

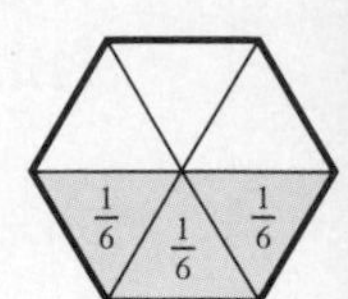

$$1\frac{1}{2} \div \frac{1}{6} = \frac{3}{2} \div \frac{1}{6}$$
$$= \frac{3}{2} \cdot 6 = \frac{3 \cdot 6}{2} = \frac{3 \cdot 3 \cdot 2}{2 \cdot 1} = \frac{3 \cdot 3}{1} \cdot \frac{2}{2} = \frac{3 \cdot 3}{1} \cdot 1 = 9$$

To divide using mixed numerals, first write fractional notation. Then divide with fractional notation and convert the answer back to a mixed numeral, if appropriate.

▶ **EXAMPLE 5** Divide: $32 \div 3\frac{1}{5}$.

$$32 \div 3\frac{1}{5} = \frac{32}{1} \div \frac{16}{5}$$
$$= \frac{32}{1} \cdot \frac{5}{16} = \frac{32 \cdot 5}{1 \cdot 16} = \frac{2 \cdot 16 \cdot 5}{1 \cdot 16} = \frac{16}{16} \cdot \frac{2 \cdot 5}{1} = 10$$ ◀

Remember to multiply by the reciprocal.

DO EXERCISE 5.

▶ **EXAMPLE 6** Divide: $35 \div 4\frac{1}{3}$.

$$35 \div 4\frac{1}{3} = \frac{35}{1} \div \frac{13}{3} = \frac{35}{1} \cdot \frac{3}{13} = \frac{105}{13} = 8\frac{1}{13}$$ ◀

CAUTION! The reciprocal of $4\frac{1}{3}$ is *not* $3\frac{1}{4}$!

DO EXERCISE 6.

▶ **EXAMPLE 7** Divide: $2\frac{1}{3} \div 1\frac{3}{4}$.

$$2\frac{1}{3} \div 1\frac{3}{4} = \frac{7}{3} \div \frac{7}{4} = \frac{7}{3} \cdot \frac{4}{7} = \frac{7 \cdot 4}{7 \cdot 3} = \frac{7}{7} \cdot \frac{4}{3} = 1 \cdot \frac{4}{3} = \frac{4}{3} = 1\frac{1}{3}$$ ◀

▶ **EXAMPLE 8** Divide: $1\frac{3}{5} \div 3\frac{1}{3}$.

$$1\frac{3}{5} \div 3\frac{1}{3} = \frac{8}{5} \div \frac{10}{3} = \frac{8}{5} \cdot \frac{3}{10} = \frac{2 \cdot 4 \cdot 3}{5 \cdot 2 \cdot 5} = \frac{2}{2} \cdot \frac{4 \cdot 3}{5 \cdot 5} = \frac{4 \cdot 3}{5 \cdot 5} = 1 \cdot \frac{12}{25} = \frac{12}{25}$$ ◀

DO EXERCISES 7 AND 8.

c Solving Problems

▶ **EXAMPLE 9** A long-playing record makes $33\frac{1}{3}$ revolutions per minute. It plays for 12 minutes. How many revolutions does it make?

1. *Familiarize.* We first draw a picture. A rectangular array works well here. We draw a circle for each revolution and a row for each minute. We let n = the number of revolutions.

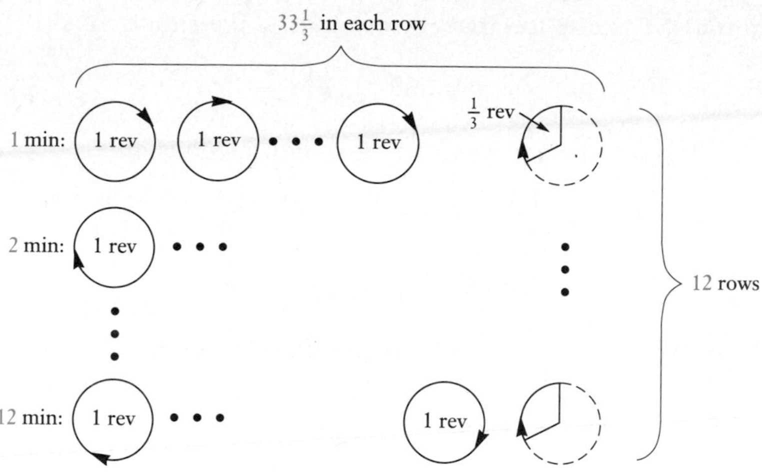

2. *Translate.* We then translate as follows.

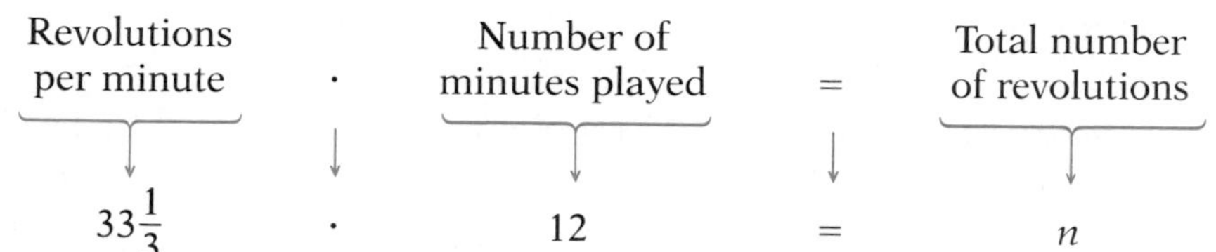

$$33\frac{1}{3} \cdot 12 = n$$

3. *Solve.* To solve the equation, we carry out the multiplication:

$$n = 33\frac{1}{3} \cdot 12 = \frac{100}{3} \cdot \frac{12}{1} = \frac{100 \cdot 3 \cdot 4}{3 \cdot 1} = \frac{3}{3} \cdot \frac{100 \cdot 4}{1} = 1 \cdot 400 = 400.$$

4. *Check.* We check by repeating the calculation. We can do a partial check by noting that $33\frac{1}{3} \approx 33$ and $12 \approx 10$. Then the product is about 330. Thus our answer is reasonable.

5. *State.* The long-playing record makes 400 revolutions in 12 minutes. ◀

DO EXERCISE 9.

▶ **EXAMPLE 10** A long-playing record makes $33\frac{1}{3}$ revolutions per minute. It makes 500 revolutions. How long does it play?

1. *Familiarize.* We first draw a picture. We draw a circle for each revolution and a row for each minute. Let t = the time the record plays. The last row may be incomplete.

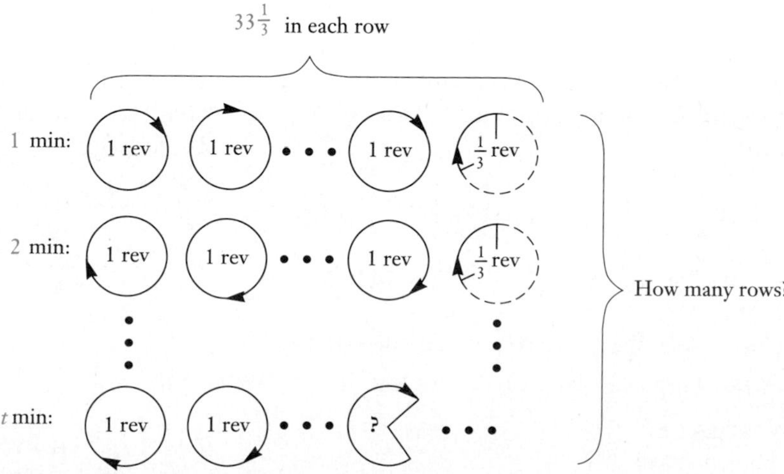

9. A car travels on an interstate highway at 65 mph for $3\frac{1}{2}$ hr. How far does it travel? $227\frac{1}{2}$ mi

ANSWER ON PAGE A-3

10. A car travels 302 mi on $15\frac{1}{10}$ gal of gas. How many miles per gallon did it get? 20

2. *Translate.* The division that corresponds to the situation is

$$t = 500 \div 33\frac{1}{3}.$$

3. *Solve.* To solve the equation, we carry out the division:

$$t = 500 \div 33\frac{1}{3} = \frac{500}{1} \div \frac{100}{3} = \frac{500}{1} \cdot \frac{3}{100} = \frac{100 \cdot 5 \cdot 3}{1 \cdot 100}$$
$$= \frac{100}{100} \cdot \frac{5 \cdot 3}{1} = 15.$$

4. *Check.* We check by multiplying the time by the number of revolutions per minute:

$$33\frac{1}{3} \cdot 15 = \frac{100}{3} \cdot \frac{15}{1} = \frac{100 \cdot 15}{3 \cdot 1} = \frac{100 \cdot 5 \cdot 3}{1 \cdot 3} = \frac{100 \cdot 5}{1} \cdot \frac{3}{3} = 500.$$

5. *State.* The record plays 15 minutes. ◀

DO EXERCISE 10.

▶ **EXAMPLE 11** Find the total area of a rectangle that is $8\frac{1}{2}$ by 11 in. and one that is $6\frac{1}{2}$ by $7\frac{1}{2}$ in.

1. *Familiarize.* We draw a picture of the situation. We let a = the total area.

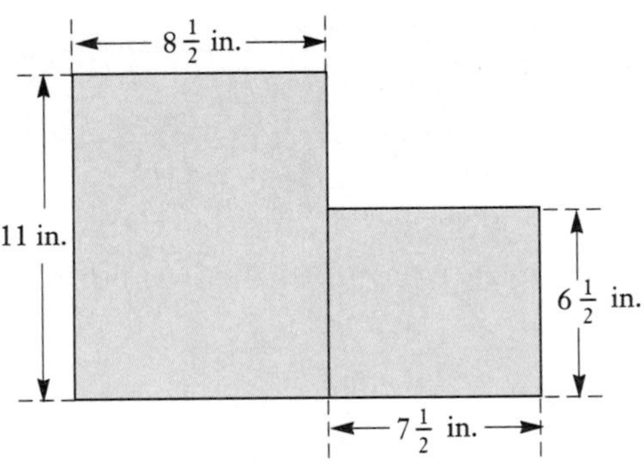

11. A room is $22\frac{1}{2}$ ft by $15\frac{1}{2}$ ft. A 9-ft by 12-ft rug is placed on the floor. How much area is not covered by the rug? $240\frac{3}{4}$ sq ft

2. *Translate.* The total area is the sum of the areas of the two rectangles. This gives us the following equation:

$$a = \left(8\frac{1}{2}\right) \cdot (11) + \left(7\frac{1}{2}\right) \cdot \left(6\frac{1}{2}\right).$$

3. *Solve.* This is a multistep problem. We can carry it out doing each multiplication and then add. This follows our rules for order of operations:

$$a = \left(8\frac{1}{2}\right) \cdot (11) + \left(7\frac{1}{2}\right) \cdot \left(6\frac{1}{2}\right) = \frac{17}{2} \cdot 11 + \frac{15}{2} \cdot \frac{13}{2} = \frac{17 \cdot 11}{2} + \frac{15 \cdot 13}{2 \cdot 2}$$
$$= 93\frac{1}{2} + 48\frac{3}{4} = 93\frac{2}{4} + 48\frac{3}{4} = 141\frac{5}{4} = 141 + 1 + \frac{1}{4} + 142\frac{1}{4}.$$

4. *Check.* We check by repeating the calculation.

5. *State.* The total area of the rectangles is $142\frac{1}{4}$ sq in. ◀

DO EXERCISE 11.

ANSWERS ON PAGE A-3

NAME SECTION DATE

EXERCISE SET 3.6

a Multiply. Write a mixed numeral for the answer.

1. $8 \cdot 2\frac{5}{6}$ **2.** $5 \cdot 3\frac{3}{4}$ **3.** $3\frac{5}{8} \cdot \frac{2}{3}$ **4.** $6\frac{2}{3} \cdot \frac{1}{4}$

5. $3\frac{1}{2} \cdot 2\frac{1}{3}$ **6.** $4\frac{1}{5} \cdot 5\frac{1}{4}$ **7.** $3\frac{2}{5} \cdot 2\frac{7}{8}$ **8.** $2\frac{3}{10} \cdot 4\frac{2}{5}$

9. $4\frac{7}{10} \cdot 5\frac{3}{10}$ **10.** $6\frac{3}{10} \cdot 5\frac{7}{10}$ **11.** $20\frac{1}{2} \cdot 10\frac{1}{5}$ **12.** $21\frac{1}{3} \cdot 11\frac{1}{3}$

b Divide. Write a mixed numeral for the answer.

13. $20 \div 3\frac{1}{5}$ **14.** $18 \div 2\frac{1}{4}$ **15.** $8\frac{2}{5} \div 7$ **16.** $3\frac{3}{8} \div 3$

17. $4\frac{3}{4} \div 1\frac{1}{3}$ **18.** $5\frac{4}{5} \div 2\frac{1}{2}$ **19.** $1\frac{7}{8} \div 1\frac{2}{3}$ **20.** $4\frac{3}{8} \div 2\frac{5}{6}$

21. $5\frac{1}{10} \div 4\frac{3}{10}$ **22.** $4\frac{1}{10} \div 2\frac{1}{10}$ **23.** $20\frac{1}{4} \div 90$ **24.** $12\frac{1}{2} \div 50$

ANSWERS

1. $22\frac{2}{3}$
2. $18\frac{3}{4}$
3. $2\frac{5}{12}$
4. $1\frac{2}{3}$
5. $8\frac{1}{6}$
6. $22\frac{1}{20}$
7. $9\frac{31}{40}$
8. $10\frac{3}{25}$
9. $24\frac{91}{100}$
10. $35\frac{91}{100}$
11. $209\frac{1}{10}$
12. $241\frac{7}{9}$
13. $6\frac{1}{4}$
14. 8
15. $1\frac{1}{5}$
16. $1\frac{1}{8}$
17. $3\frac{9}{16}$
18. $2\frac{8}{25}$
19. $1\frac{1}{8}$
20. $1\frac{37}{68}$
21. $1\frac{8}{43}$
22. $1\frac{20}{21}$
23. $\frac{9}{40}$
24. $\frac{1}{4}$

ANSWERS

c Solve.

25. A long-playing record makes $33\frac{1}{3}$ revolutions per minute. If it plays for 21 min, how many revolutions does it make?

26. A long-playing record makes $33\frac{1}{3}$ revolutions per minute. If it plays for 27 min, how many revolutions does it make?

27. One serving of meat is about $3\frac{1}{2}$ oz. A person needs 2 servings a day for proper nutrition. How many ounces of meat is this?

28. Round steak contains $3\frac{1}{2}$ servings per pound. How many servings are there in 10 lb of round steak?

29. The weight of water is $62\frac{1}{2}$ lb per cubic foot. What is the weight of $5\frac{1}{2}$ cubic feet of water?

30. The weight of water is $62\frac{1}{2}$ lb per cubic foot. What is the weight of $2\frac{1}{4}$ cubic feet of water?

31. Listed below are the ingredients for *opossum and sweet potatoes*. What are the ingredients for $\frac{1}{2}$ the recipe? for 3 recipes?

Opossum and sweet potatoes
$2\frac{1}{2}$ lb opossum meat
$2\frac{1}{2}$ teaspoons salt
$1\frac{1}{3}$ teaspoons black pepper
$\frac{2}{3}$ cup flour
$\frac{1}{2}$ cup water
4 medium sweet potatoes
2 tablespoons sugar

25. 700

26. 900

27. 7 oz

28. 35

29. $43\frac{3}{4}$ lb

30. $140\frac{5}{8}$ lb

31. $1\frac{1}{4}$ lb opossum meat, $1\frac{1}{4}$ tsp salt, $\frac{2}{3}$ tsp black pepper, $\frac{1}{3}$ cup flour, $\frac{1}{4}$ cup water, 2 medium sweet potatoes, 1 tbsp sugar; $7\frac{1}{2}$ lb opossum meat, $7\frac{1}{2}$ tsp salt, 4 tsp black pepper, 2 cups flour, $1\frac{1}{2}$ cups water, 12 medium sweet potatoes, 6 tbsp sugar

32. Listed below are the ingredients for *venison stew*. What are the ingredients for $\frac{1}{2}$ the recipe? for 3 recipes?

Venison stew
$1\frac{1}{2}$ lb venison flank
$\frac{1}{4}$ cup flour
$1\frac{1}{2}$ teaspoons salt
$\frac{1}{4}$ teaspoon pepper
1 small onion
$\frac{1}{3}$ cup cubed turnips
$\frac{1}{3}$ cup cubed carrots
4 cups potatoes

$\frac{3}{4}$ lb venison flank, $\frac{1}{8}$ cup flour, $\frac{3}{4}$ tsp salt, $\frac{1}{8}$ tsp pepper, $\frac{1}{2}$ small onion, $\frac{1}{6}$ cup cubed turnips, $\frac{1}{6}$ cup cubed carrots, 2 cups potatoes; $4\frac{1}{2}$ lb venison flank, $\frac{3}{4}$ cup flour, $4\frac{1}{2}$ tsp salt, $\frac{3}{4}$ tsp pepper, 3 small onions, 1 cup cubed turnips, 1 cup cubed carrots, 12 cups potatoes

33. Fahrenheit temperature can be obtained from Celsius (centigrade) temperature by multiplying by $1\frac{4}{5}$ and adding 32°. What Fahrenheit temperature corresponds to a Celsius temperature of 20°?

34. Fahrenheit temperature can be obtained from Celsius (centigrade) temperature by multiplying by $1\frac{4}{5}$ and adding 32°. What Fahrenheit temperature corresponds to the Celsius temperature of boiling water, which is 100°?

35. If a record made 480 revolutions at $33\frac{1}{3}$ revolutions per minute, how long did it play?

36. If a record made 420 revolutions at $33\frac{1}{3}$ revolutions per minute, how long did it play?

37. A car traveled 213 mi on $14\frac{2}{10}$ gal of gas. How many miles per gallon did it get?

38. A car traveled 385 mi on $15\frac{4}{10}$ gal of gas. How many miles per gallon did it get?

39. The weight of water is $62\frac{1}{2}$ lb per cubic foot. How many cubic feet would be occupied by 250 lb of water?

40. The weight of water is $62\frac{1}{2}$ lb per cubic foot. How many cubic feet would be occupied by 375 lb of water?

ANSWERS

32. ______

33. 68°F

34. 212°F

35. $14\frac{2}{5}$ min

36. $12\frac{3}{5}$ min

37. 15 mpg

38. 25 mpg

39. 4 cu ft

40. 6 cu ft

ANSWERS

41. 24 lb

42. 16

43. $35\frac{115}{256}$ sq m

44. $76\frac{1}{4}$ sq ft

45. $59{,}538\frac{1}{8}$ sq m

46. $96\frac{1}{3}$ sq yd

47. 1,429,017

48. 45,800

49. 588

50. $\frac{2}{3}$

51. $35\frac{57}{64}$

52. $14\frac{11}{48}$

53. $\frac{4}{9}$

54. $\frac{13}{8}$ or $1\frac{5}{8}$

55. $\frac{9}{5}$ or $1\frac{4}{5}$

56. $5\frac{3}{16}$

41. Turkey contains $1\frac{1}{3}$ servings per pound. How many pounds would be needed for 32 servings?

42. Most space shuttles orbit the earth once every $1\frac{1}{2}$ hr. How many orbits are made every 24 hours?

43. Find the area of the shaded region.

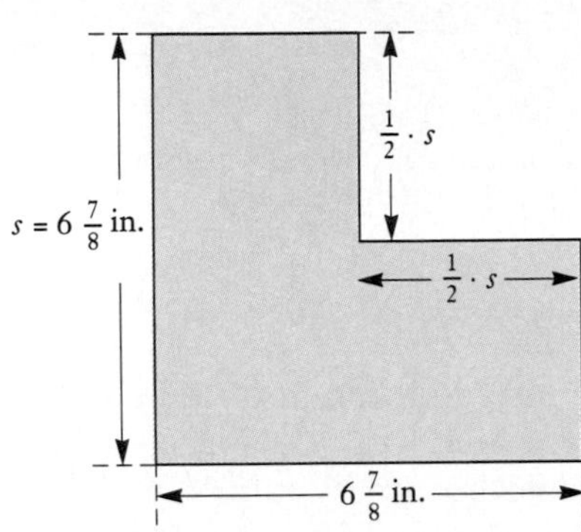

44. Find the area of the shaded region.

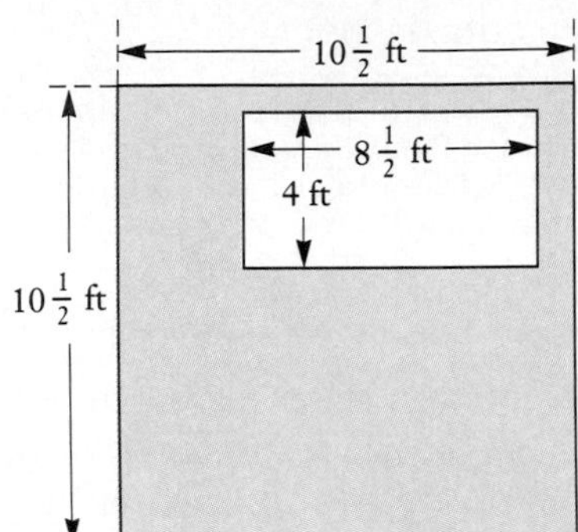

45. A rectangular lot has dimensions of $302\frac{1}{2}$ by $205\frac{1}{4}$ m. A building with dimensions of 100 by $25\frac{1}{2}$ m is built on the lot. How much area is left over?

46. Find the total area of 3 squares, each of which is $5\frac{2}{3}$ yd on a side.

SKILL MAINTENANCE

47. $\begin{array}{r} 6\,7\,0\,9 \\ \times \quad 2\,1\,3 \\ \hline \end{array}$

48. Round to the nearest hundred: 45,765.

49. Solve: $\frac{5}{7}\cdot t = 420$.

50. Divide and simplify: $\frac{4}{5} \div \frac{6}{5}$.

SYNTHESIS

Simplify.

51. $8 \div \frac{1}{2} + \frac{3}{4} + \left(5 - \frac{5}{8}\right)^2$

52. $\left(\frac{5}{9} - \frac{1}{4}\right) \times 12 + \left(4 - \frac{3}{4}\right)^2$

53. $\frac{1}{3} \div \left(\frac{1}{2} - \frac{1}{5}\right) \times \frac{1}{4} + \frac{1}{6}$

54. $\frac{7}{8} - 1\frac{1}{8} \times \frac{2}{3} + \frac{9}{10} \div \frac{3}{5}$

55. $4\frac{1}{2} \div 2\frac{1}{2} + 8 - 4 \div \frac{1}{2}$

56. $6 - 2\frac{1}{3} \times \frac{3}{4} + \frac{5}{8} \div \frac{2}{3}$

SUMMARY AND REVIEW EXERCISES: CHAPTER 3

The review sections and objectives to be tested in addition to the material in this chapter are [1.5b], [1.8a], [2.6a], and [2.7b].

Find the LCM.

1. 12 and 18

2. 18 and 45

3. 3, 6, and 30

Add and simplify.

4. $\frac{6}{5} + \frac{3}{8}$

5. $\frac{5}{16} + \frac{1}{12}$

6. $\frac{6}{5} + \frac{11}{15}$

7. $\frac{5}{16} + \frac{3}{24}$

Subtract and simplify.

8. $\frac{5}{9} - \frac{2}{9}$

9. $\frac{7}{8} - \frac{3}{4}$

10. $\frac{11}{27} - \frac{2}{9}$

11. $\frac{5}{6} - \frac{2}{9}$

Use < or > for ■ to write a true sentence.

12. $\frac{4}{7}$ ■ $\frac{5}{9}$

13. $\frac{8}{9}$ ■ $\frac{11}{13}$

Solve.

14. $x + \frac{2}{5} = \frac{7}{8}$

15. $\frac{1}{2} + y = \frac{9}{10}$

Convert to fractional notation.

16. $7\frac{1}{2}$

17. $8\frac{3}{8}$

18. $4\frac{1}{3}$

19. $10\frac{5}{7}$

Convert to a mixed numeral.

20. $\frac{7}{3}$

21. $\frac{27}{4}$

22. $\frac{63}{5}$

23. $\frac{7}{2}$

Divide. Write a mixed numeral for the answer.

24. $9\overline{)7896}$

25. $23\overline{)10{,}493}$

Add. Write a mixed numeral for the answer.

26. $5\frac{3}{5} + 4\frac{4}{5}$

27. $8\frac{1}{3} + 3\frac{2}{5}$

28. $5\frac{5}{6} + 4\frac{5}{6}$

29. $2\frac{3}{4} + 5\frac{1}{2}$

Subtract. Write a mixed numeral for the answer.

30. $12 - 4\frac{2}{9}$

31. $9\frac{3}{5} - 4\frac{13}{15}$

32. $10\frac{1}{4} - 6\frac{1}{10}$

33. $24 - 10\frac{5}{8}$

Multiply. Write a mixed numeral for the answer.

34. $6 \cdot 2\frac{2}{3}$

35. $5\frac{1}{4} \cdot \frac{2}{3}$

36. $2\frac{1}{5} \cdot 1\frac{1}{10}$

37. $2\frac{2}{5} \cdot 2\frac{1}{2}$

Divide. Write a mixed numeral for the answer.

38. $27 \div 2\frac{1}{4}$

39. $2\frac{2}{5} \div 1\frac{7}{10}$

40. $3\frac{1}{4} \div 26$

41. $4\frac{1}{5} \div 4\frac{2}{3}$

Solve.

42. A curtain requires $2\frac{3}{5}$ m of material. How many of these curtains can be made from 39 m of material?

43. On the first day of trading on the stock market, stock in General Mills opened at $\$67\frac{3}{4}$ and rose by $\$2\frac{5}{8}$ at the close of trading. What was the stock's closing price?

44. A recipe calls for $1\frac{2}{3}$ cups of sugar. How much is needed for 18 recipes?

45. A wedding cake recipe requires 12 cups of shortening. Being calorie conscious, the wedding couple decides to reduce the shortening by $3\frac{5}{8}$ cups. How many cups of shortening are used in their new recipe?

SKILL MAINTENANCE

46. Multiply and simplify: $\frac{9}{10} \cdot \frac{4}{3}$.

47. Divide and simplify: $\frac{5}{4} \div \frac{5}{6}$.

48. Multiply: 4023×176

49. A factory produces 85 radios per day. How long will it take to fill an order for 1445 radios?

❖ THINKING IT THROUGH

1. Explain why $2\frac{1}{4} \cdot 3\frac{2}{5} \neq 6\frac{2}{20}$.
2. Explain why $5 \cdot 3\frac{2}{5} \neq (5 \cdot 3) \cdot (5 \cdot \frac{2}{5})$.
3. Discuss the role of least common multiples in adding and subtracting with fractional notation.

NAME SECTION DATE

TEST: CHAPTER 3

1. Find the LCM of 12 and 16.

Add and simplify.

2. $\frac{1}{2} + \frac{5}{2}$

3. $\frac{7}{8} + \frac{2}{3}$

4. $\frac{7}{10} + \frac{9}{100}$

Subtract and simplify.

5. $\frac{5}{6} - \frac{3}{6}$

6. $\frac{5}{6} - \frac{3}{4}$

7. $\frac{17}{24} - \frac{5}{8}$

8. Use < or > for ▢ to write a true sentence.

$\frac{6}{7}$ ▢ $\frac{21}{25}$

9. Solve: $x + \frac{2}{3} = \frac{11}{12}$.

Convert to fractional notation.

10. $3\frac{1}{2}$

11. $9\frac{7}{8}$

Convert to a mixed numeral.

12. $\frac{9}{2}$

13. $\frac{74}{9}$

Divide. Write a mixed numeral for the answer.

14. $11\overline{)1789}$

Add. Write a mixed numeral for the answer.

15. $6\frac{2}{5} + 7\frac{4}{5}$

16. $9\frac{1}{4} + 5\frac{1}{6}$

ANSWERS

1. [3.1a] 48

2. [3.2a] 3

3. [3.2b] $\frac{37}{24}$

4. [3.2b] $\frac{79}{100}$

5. [3.3a] $\frac{1}{3}$

6. [3.3a] $\frac{1}{12}$

7. [3.3a] $\frac{1}{12}$

8. [3.3b] >

9. [3.3c] $\frac{1}{4}$

10. [3.4a] $\frac{7}{2}$

11. [3.4a] $\frac{79}{8}$

12. [3.4b] $4\frac{1}{2}$

13. [3.4b] $8\frac{2}{9}$

14. [3.4b] $162\frac{7}{11}$

15. [3.5a] $14\frac{1}{5}$

16. [3.5a] $14\frac{5}{12}$

ANSWERS

17. [3.5b] $4\frac{7}{24}$

18. [3.5b] $6\frac{1}{6}$

19. [3.6a] 39

20. [3.6a] $4\frac{1}{2}$

21. [3.6a] $5\frac{5}{6}$

22. [3.6b] 6

23. [3.6b] 2

24. [3.6b] $\frac{1}{36}$

25. [3.6c] $17\frac{1}{2}$ cups

26. [3.6c] 80

27. [3.5c] $160\frac{5}{12}$ kg

28. [3.5c] $2\frac{1}{2}$ in.

29. [1.5b] 346,636

30. [2.7b] $\frac{8}{5}$

31. [2.6a] $\frac{10}{9}$

32. [1.8a] 535 bottles; 10 oz left over

Subtract. Write a mixed numeral for the answer.

17. $10\frac{1}{6} - 5\frac{7}{8}$

18. $14 - 7\frac{5}{6}$

Multiply. Write a mixed numeral for the answer.

19. $9 \cdot 4\frac{1}{3}$

20. $6\frac{3}{4} \cdot \frac{2}{3}$

21. $3\frac{1}{3} \cdot 1\frac{3}{4}$

Divide. Write a mixed numeral for the answer.

22. $33 \div 5\frac{1}{2}$

23. $2\frac{1}{3} \div 1\frac{1}{6}$

24. $2\frac{1}{12} \div 75$

Solve.

25. A fruitcake recipe calls for $3\frac{1}{2}$ cups of diced fruit. How much diced fruit is needed for 5 recipes?

26. An order of books for a math course weighs 220 lb. Each book weighs $2\frac{3}{4}$ lb. How many books are in the order?

27. The weights of two students are $83\frac{2}{3}$ kg and $76\frac{3}{4}$ kg. What is their total weight?

28. A standard piece of paper is $8\frac{1}{2}$ in. by 11 in. By how much does the length exceed the width?

SKILL MAINTENANCE

29. Multiply: 4561×76

30. Divide and simplify: $\frac{4}{3} \div \frac{5}{6}$.

31. Multiply and simplify: $\frac{4}{3} \cdot \frac{5}{6}$.

32. A container has 8570 oz of beverage with which to fill 16-oz bottles. How many of these bottles can be filled? How much beverage will be left over?

CUMULATIVE REVIEW: CHAPTERS 1–3

1. In the number 2753, what digit names tens?

2. Write expanded notation for 6075.

3. Write a word name for the number in the following sentence: The diameter of Uranus is 29,500 miles.

Add and simplify.

4. $\begin{array}{r} 628 \\ +271 \\ \hline \end{array}$

5. $\begin{array}{r} 3704 \\ +5278 \\ \hline \end{array}$

6. $\frac{3}{8}+\frac{1}{24}$

7. $\begin{array}{r} 2\frac{3}{4} \\ +5\frac{1}{2} \\ \hline \end{array}$

Subtract and simplify.

8. $\begin{array}{r} 7469 \\ -2345 \\ \hline \end{array}$

9. $\begin{array}{r} 7605 \\ -3087 \\ \hline \end{array}$

10. $\frac{3}{4}-\frac{1}{3}$

11. $\begin{array}{r} 2\frac{1}{3} \\ -1\frac{1}{6} \\ \hline \end{array}$

Multiply and simplify.

12. $\begin{array}{r} 278 \\ \times\ 18 \\ \hline \end{array}$

13. $\begin{array}{r} 894 \\ \times 328 \\ \hline \end{array}$

14. $\frac{9}{10}\cdot\frac{5}{3}$

15. $18\cdot\frac{5}{6}$

16. $2\frac{1}{3}\cdot 3\frac{1}{7}$

Divide. Write the answers with remainders in the form 34 R 7.

17. $6\overline{)4290}$

18. $45\overline{)2531}$

19. In Exercise 18, write a mixed numeral for the answer.

Divide and simplify, where appropriate.

20. $\frac{2}{5}\div\frac{7}{10}$

21. $2\frac{1}{5}\div\frac{3}{10}$

22. Round 38,478 to the nearest hundred.

23. Find the LCM of 18 and 24.

24. Determine whether 3718 is divisible by 8.

25. Find all factors of 16.

26. What part is shaded?

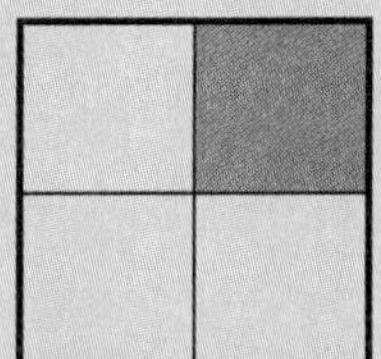

Use $<, >$, or $=$ for ▒ to write a true sentence.

27. $\frac{4}{5}$ ▒ $\frac{4}{6}$

28. $\frac{5}{12}$ ▒ $\frac{3}{7}$

Simplify.

29. $\frac{36}{45}$

30. $\frac{320}{10}$

31. Convert to fractional notation: $4\frac{5}{8}$.

32. Convert to a mixed numeral: $\frac{17}{3}$.

Solve.

33. $x + 24 = 117$

34. $x + \frac{7}{9} = \frac{4}{3}$

35. $\frac{7}{9} \cdot t = \frac{4}{3}$

36. $y = 32{,}580 \div 36$

Solve.

37. A jacket costs $87 and a coat costs $148. How much does it cost to buy both?

38. A fund contains $423. From this fund, $148 and $167 are withdrawn for expenses. How much is left in the fund?

39. A lot measures 27 ft by 11 ft. What is its area?

40. How many people can get equal $16 shares from a total of $496?

41. A recipe calls for $\frac{4}{5}$ teaspoon of salt. How much salt should be used in $\frac{1}{2}$ of the recipe?

42. A book weighs $2\frac{3}{5}$ lb. How much do 15 books weigh?

43. How many pieces, each $2\frac{3}{8}$ cm long, can be cut from a piece of wire 38 cm long?

44. In a walkathon, one person walked $\frac{9}{10}$ km and another walked $\frac{75}{100}$ km. What was the total distance walked?

INTRODUCTION In this chapter, the meaning of decimal notation is considered. Also discussed are rounding, addition and subtraction, equation solving, and problem solving involving decimal notation.

The review sections to be tested in addition to the material in this chapter are 1.2, 1.3, 3.2, and 3.3. ❖

Addition and Subtraction: Decimal Notation

AN APPLICATION

Each year, each of us drinks 43.7 gal of soft drinks, 37.3 gal of water, 27.3 gal of coffee, 21.1 gal of milk, and 8.1 gal of fruit juice. What is the total amount that each of us drinks?

THE MATHEMATICS

Let t = the total amount of these liquids. The problem translates to the following equation:

$$\underbrace{43.7}_{\uparrow} + 37.3 + 27.3 + 21.1 + 8.1 = t.$$

This is decimal notation. (↑ points to 43.7)

❖ POINTS TO REMEMBER: CHAPTER 4

Addition and subtraction using fractional notation
Skill of comparing numbers using fractional notation

PRETEST: CHAPTER 4

1. Write a word name for 2.347.

2. Write a word name, as on a check, for \$3264.78.

Write fractional notation.

3. 0.21

4. 5.408

Write decimal notation.

5. $\frac{379}{100}$

6. $\frac{539}{10,000}$

Which number is larger?

7. 3.2, 0.321

8. 0.099, 0.091

9. 0.54, 0.562

Round 21.0448 to the nearest:

10. Tenth.

11. Hundredth.

12. Thousandth.

Add.

13.
$$\begin{array}{r} 601.3 \\ 5.81 \\ +\ \ 0.109 \\ \hline \end{array}$$

14.
$$\begin{array}{r} 0.8 \\ 0.06 \\ 0.007 \\ +\,0.0014 \\ \hline \end{array}$$

15. 102.4 + 10.24 + 1.024

16. 4.127 + 0.5 + 2.1167

Subtract.

17.
$$\begin{array}{r} 94.061 \\ -\ \ 2.329 \\ \hline \end{array}$$

18.
$$\begin{array}{r} 40.0 \\ -\ \ 0.9099 \\ \hline \end{array}$$

19. 8 − 0.0049

20. 344.6788 − 91.6851

Solve.

21. $x + 2.33 = 5.6$

22. $54.906 + q = 6400.1177$

Solve.

23. A checking account contained \$434.19. After a \$148.24 check was drawn, how much was left in the account?

24. On a three-day trip, a traveler drove the following distances: 432.6 mi, 179.2 mi, and 469.8 mi. What is the total number of miles driven?

4.1 Decimal Notation

The set of **arithmetic numbers, or nonnegative rational numbers,** consists of the whole numbers

$$0,\ 1,\ 2,\ 3,\ 4,\ 5,\ 6,\ 7,\ 8,\ 9,\ 10, \text{ and so on,}$$

and fractions like

$$\frac{1}{2}, \frac{2}{3}, \frac{7}{8}, \frac{17}{10}, \text{ and so on.}$$

We studied the use of fractional notation for arithmetic numbers in Chapters 2 and 3. In Chapters 4 and 5, we will study the use of *decimal notation.* We are still considering the same set of numbers, but we are using different notation. For example, instead of using fractional notation for $\frac{7}{8}$, we use decimal notation, 0.875.

In this chapter, we will learn the meaning of decimal notation, rounding, and addition, subtraction, and problem solving involving decimal notation.

a Decimal Notation and Word Names

Decimal notation for the women's shotput record is

69.675 ft.

To understand what 69.675 means, we use a **place-value chart.** The value of each place is $\frac{1}{10}$ as large as the one to its left.

Place-Value Chart							
Hundreds	Tens	Ones	Ten*ths*	Hundred*ths*	Thousand*ths*	Ten Thousand*ths*	Hundred Thousand*ths*
100	10	1	$\frac{1}{10}$	$\frac{1}{100}$	$\frac{1}{1000}$	$\frac{1}{10,000}$	$\frac{1}{100,000}$
	6	9 •	6	7	5		

The decimal notation 69.675 means

$$6 \text{ tens} + 9 \text{ ones} + 6 \text{ tenths} + 7 \text{ hundredths} + 5 \text{ thousandths},$$

or

$$6 \cdot 10 + 9 \cdot 1 + 6 \cdot \frac{1}{10} + 7 \cdot \frac{1}{100} + 5 \cdot \frac{1}{1000},$$

or

$$60 + 9 + \frac{6}{10} + \frac{7}{100} + \frac{5}{1000}.$$

A mixed numeral for 69.675 is $69\frac{675}{1000}$. We read 69.675 as "sixty-nine and six hundred seventy-five thousandths." When we come to the decimal point, we read "and." We can also read 69.675 as "six nine *point* six seven five."

OBJECTIVES

After finishing Section 4.1, you should be able to:

a Given decimal notation, write a word name, and write a word name for an amount of money.

b Convert from decimal notation to fractional notation.

c Convert from fractional notation to decimal notation.

FOR EXTRA HELP

Tape 6A

Tape 6B

MAC: 4
IBM: 4

To write a word name from decimal notation:

a) **write a word name for the whole number (the number named to the left of the decimal point),** — 397.685 → Three hundred ninety-seven

b) **write the word "and" for the decimal point, and** — 397.685 → Three hundred ninety-seven and

c) **write a word name for the number named to the right of the decimal point, followed by the place value of the last digit.** — 397.685 → Three hundred ninety-seven and six hundred eighty-five *thousandths.*

▶ **EXAMPLE 1** Write a word name for the number in this sentence: Each person in this country consumes an average of 43.7 gallons of soft drinks per year.

Forty-three and seven tenths ◀

▶ **EXAMPLE 2** Write a word name for 413.87.

Four hundred thirteen and eighty-seven hundredths ◀

▶ **EXAMPLE 3** Write a word name for the number in this sentence: The world record in the men's marathon is 2.2525 hours.

Two and two thousand five hundred twenty-five ten thousandths ◀

▶ **EXAMPLE 4** Write a word name for 1788.405.

One thousand, seven hundred eighty-eight and four hundred five thousandths. ◀

DO EXERCISES 1–4.

Decimal notation is also used with money. It is common on a check to write "and ninety-five cents" as "and $\frac{95}{100}$ dollars."

▶ **EXAMPLE 5** Write a word name for the amount on the check, $5876.95.

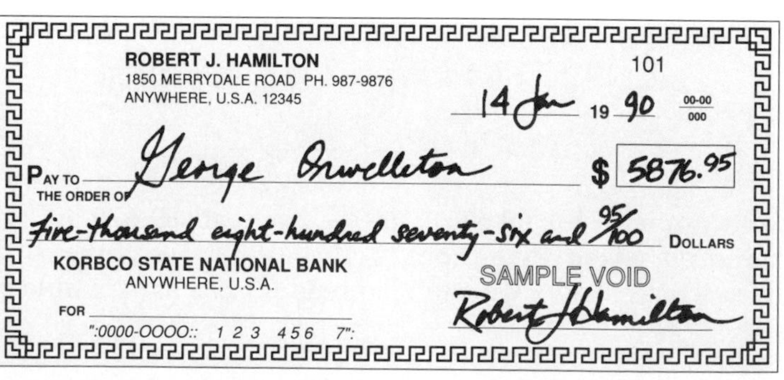

Five thousand, eight hundred seventy-six and $\frac{95}{100}$ dollars ◀

DO EXERCISES 5 AND 6.

Write a word name for the number.

1. Each person in this country consumes an average of 27.3 gallons of coffee per year.

   Twenty-seven and three tenths

2. The racehorse *Swale* won the Belmont Stakes in a time of 2.4533 minutes.

   Two and four thousand five hundred thirty-three ten thousandths

3. 245.89

   Two hundred forty-five and eighty-nine hundredths

4. 31,479.764

   Thirty-one thousand, four hundred seventy-nine and seven hundred sixty-four thousandths

Write a word name as on a check.

5. $4217.56

   Four thousand, two hundred seventeen and $\frac{56}{100}$ dollars

6. $13.98

   Thirteen and $\frac{98}{100}$ dollars

ANSWERS ON PAGE A-3

b Converting from Decimal Notation to Fractional Notation

We can find fractional notation as follows:

$$9.875 = 9 + \frac{8}{10} + \frac{7}{100} + \frac{5}{1000}$$
$$= 9 \cdot \frac{1000}{1000} + \frac{8}{10} \cdot \frac{100}{100} + \frac{7}{100} \cdot \frac{10}{10} + \frac{5}{1000}$$
$$= \frac{9000}{1000} + \frac{800}{1000} + \frac{70}{1000} + \frac{5}{1000} = \frac{9875}{1000}.$$

Note the following:

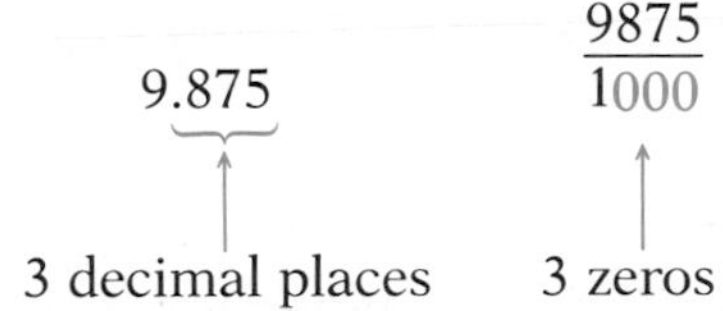

To convert from decimal to fractional notation:

a) count the number of decimal places, — **4.98** — **2 places**

b) move the decimal point that many places to the right, and — **4.98.** — **Move 2 places**

c) write the answer over a denominator with that number of zeros. — $\frac{498}{100}$ — **2 zeros**

▶ **EXAMPLE 6** Write fractional notation for 0.876. Do not simplify.

0.876 0.876. 0.876 $= \frac{876}{1000}$ ◀

3 places

CAUTION! For a number like 0.876, we normally write a 0 before the decimal to avoid forgetting or omitting the decimal point.

▶ **EXAMPLE 7** Write fractional notation for 56.23. Do not simplify.

56.23 56.23. 56.23 $= \frac{5623}{100}$ ◀

2 places

▶ **EXAMPLE 8** Write fractional notation for 1.5018. Do not simplify.

1.5018 1.5018. 1.5018 $= \frac{15{,}018}{10{,}000}$ ◀

4 places

DO EXERCISES 7–10.

Write fractional notation.

7. 0.896 $\frac{896}{1000}$

8. 23.78 $\frac{2378}{100}$

9. 5.6789 $\frac{56{,}789}{10{,}000}$

10. 1.9 $\frac{19}{10}$

ANSWERS ON PAGE A-3

Write decimal notation.

11. $\frac{743}{100}$ 7.43

12. $\frac{406}{1000}$ 0.406

13. $\frac{67{,}089}{10{,}000}$ 6.7089

14. $\frac{9}{10}$ 0.9

ANSWERS ON PAGE A-3

c Converting from Fractional Notation to Decimal Notation

If fractional notation has a denominator that is a power of ten, such as 10, 100, 1000, and so on, we reverse the procedure we used before.

To convert from fractional notation to decimal notation when the denominator is 10, 100, 1000, and so on,

a) count the number of zeros, and $\frac{8679}{1000}$ — **3 zeros**

b) move the decimal point that number of places to the left. Leave off the denominator. **8.679.** **Move 3 places**

▶ **EXAMPLE 9** Write decimal notation for $\frac{47}{10}$.

$\frac{47}{10}$ — 1 zero 4.7. $\frac{47}{10} = 4.7$ ◀

▶ **EXAMPLE 10** Write decimal notation for $\frac{123{,}067}{10{,}000}$.

$\frac{123{,}067}{10{,}000}$ — 4 zeros 12.3067. $\frac{123{,}067}{10{,}000} = 12.3067$ ◀

DO EXERCISES 11–14.

When denominators are numbers other than 10, 100, and so on, we will use another method for conversion. It will be considered in Section 5.3.

NAME SECTION DATE

EXERCISE SET 4.1

a Write a word name for the number.

1. The average age of a bride is 23.2 years.

Twenty-three and two tenths

2. The world record in the woman's marathon was 2.6693 hours.

Two and six thousand six hundred ninety-three ten thousandths

3. Recently one dollar was worth 135.87 Japanese yen.

One hundred thirty-five and eighty-seven hundredths

4. Each day the average person spends $3.50 on health care.

Three and fifty hundredths

5. 34.891

Thirty-four and eight hundred ninety-one thousandths

6. 12.345

Twelve and three hundred forty-five thousandths

Write a word name as on a check.

7. $326.48

Three hundred twenty-six and $\frac{48}{100}$ dollars

8. $125.99

One hundred twenty-five and $\frac{99}{100}$ dollars

9. $0.67

Zero and $\frac{67}{100}$ dollars

10. $3.25

Three and $\frac{25}{100}$ dollars

b Write fractional notation.

11. 6.8 **12.** 7.2 **13.** 0.17 **14.** 0.89

15. 1.46 **16.** 2.78 **17.** 204.6 **18.** 314.8

19. 3.142 **20.** 1.732 **21.** 46.03 **22.** 53.81

23. 0.00013 **24.** 0.0109 **25.** 20.003 **26.** 1000.3

27. 1.0008 **28.** 2.0114 **29.** 4567.2 **30.** 0.1104

ANSWERS

1.

2.

3.

4.

5.

6.

7.

8.

9.

10.

11. $\frac{68}{10}$

12. $\frac{72}{10}$

13. $\frac{17}{100}$

14. $\frac{89}{100}$

15. $\frac{146}{100}$

16. $\frac{278}{100}$

17. $\frac{2046}{10}$

18. $\frac{3148}{10}$

19. $\frac{3142}{1000}$

20. $\frac{1732}{1000}$

21. $\frac{4603}{100}$

22. $\frac{5381}{100}$

23. $\frac{13}{100,000}$

24. $\frac{109}{10,000}$

25. $\frac{20,003}{1000}$

26. $\frac{10,003}{10}$

27. $\frac{10,008}{10,000}$

28. $\frac{20,114}{10,000}$

29. $\frac{45,672}{10}$

30. $\frac{1104}{10,000}$

ANSWERS

31. 0.8
32. 0.1
33. 0.92
34. 0.04
35. 9.3
36. 6.7
37. 8.89
38. 6.94
39. 250.8
40. 670.1
41. 3.798
42. 0.078
43. 0.0078
44. 0.0904
45. 0.56788
46. 0.00019
47. 21.73
48. 67.43
49. 0.66
50. 1.78
51. 34.17
52. 95.63
53. 0.376193
54. 8.953073
55. 6170
56. 6200
57. 6000
58. 99.44
59. 4.909

c Write decimal notation.

31. $\frac{8}{10}$ **32.** $\frac{1}{10}$ **33.** $\frac{92}{100}$ **34.** $\frac{4}{100}$

35. $\frac{93}{10}$ **36.** $\frac{67}{10}$ **37.** $\frac{889}{100}$ **38.** $\frac{694}{100}$

39. $\frac{2508}{10}$ **40.** $\frac{6701}{10}$ **41.** $\frac{3798}{1000}$ **42.** $\frac{78}{1000}$

43. $\frac{78}{10,000}$ **44.** $\frac{904}{10,000}$ **45.** $\frac{56,788}{100,000}$ **46.** $\frac{19}{100,000}$

47. $\frac{2173}{100}$ **48.** $\frac{6743}{100}$ **49.** $\frac{66}{100}$ **50.** $\frac{178}{100}$

51. $\frac{3417}{100}$ **52.** $\frac{9563}{100}$ **53.** $\frac{376,193}{1,000,000}$ **54.** $\frac{8,953,073}{1,000,000}$

SKILL MAINTENANCE

Round 6172 to the nearest:

55. Ten. **56.** Hundred. **57.** Thousand.

SYNTHESIS

Write decimal notation.

58. $99\frac{44}{100}$ **59.** $4\frac{909}{1000}$

4.2 Order and Rounding

a Order

To understand how to compare numbers in decimal notation, consider 0.85 and 0.9. First note that 0.9 = 0.90 because $\frac{9}{10} = \frac{90}{100}$. Then $0.85 = \frac{85}{100}$ and $0.90 = \frac{90}{100}$. Since $\frac{85}{100} < \frac{90}{100}$, it follows that 0.85 < 0.90. This leads us to a quick way to compare two numbers named in decimal notation.

To compare two numbers in decimal notation, start at the left and compare corresponding digits. When two digits differ, the number with the larger digit is the larger of the two numbers. To ease the comparison, extra zeros can be written to the right of the decimal point, if necessary, so the number of decimal places is the same.

▶ **EXAMPLE 1** Which of 2.109 and 2.1 is larger?

2.109		2.109		2.109		2.109	
↕	The same	↕	The same	↕	The same	↕	These digits differ, and 9 is larger than 0.
2.1		2.1		2.10		2.100	

Thus, 2.109 is larger. ◀

▶ **EXAMPLE 2** Which of 0.09 and 0.108 is larger?

0.09		0.09	
↕	The same	↕	These digits differ, and 1 is larger than 0.
0.108		0.108	

Thus, 0.108 is larger. ◀

DO EXERCISES 1–6.

b Rounding

Rounding is done as for whole numbers. To understand, we first consider an example using a number line.

▶ **EXAMPLE 3** Round 0.37 to the nearest tenth.

Here is part of a number line.

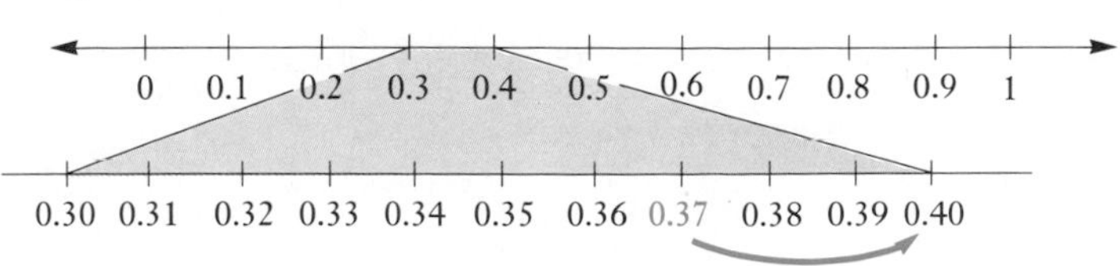

We see that 0.37 is closer to 0.40 than to 0.30. Thus, 0.37 rounded to the nearest tenth is 0.4. ◀

OBJECTIVES

After finishing Section 4.2, you should be able to:

a Given a pair of numbers named by decimal notation, tell which is larger.

b Round to the nearest thousandth, hundredth, tenth, one, ten, hundred, or thousand.

FOR EXTRA HELP

Tape 6B Tape 6B MAC: 4 IBM: 4

Which number is larger?

1. 2.04, 2.039 2.04

2. 0.06, 0.008 0.06

3. 0.5, 0.58 0.58

4. 1, 0.9999 1

5. 0.8989, 0.09898 0.8989

6. 21.006, 21.05 21.05

ANSWERS ON PAGE A-3

Round to the nearest tenth.

7. 2.76 2.8

8. 13.85 13.9

9. 234.448 234.4

10. 7.009 7.0

Round to the nearest hundredth.

11. 0.636 0.64

12. 7.834 7.83

13. 34.675 34.68

14. 0.025 0.03

Round to the nearest thousandth.

15. 0.9434 0.943

16. 8.0038 8.004

17. 43.1119 43.112

18. 37.4005 37.401

Round 7459.3548 to the nearest:

19. Thousandth. 7459.355

20. Hundredth. 7459.35

21. Tenth. 7459.4

22. One. 7459

23. Ten. (*Caution:* "Tens" are not "tenths.") 7460

24. Hundred. 7500

25. Thousand. 7000

ANSWERS ON PAGE A-3

To round to a certain place:

a) Locate the digit in that place.

b) Consider the next digit to the right.

c) If the digit to the right is 5 or higher, round up; if the digit to the right is less than 5, round down.

▶ **EXAMPLE 4** Round 3872.2459 to the nearest tenth.

a) Locate the digit in the tenths place.

3 8 7 2.2 4 5 9
↑

b) Consider the next digit to the right.

3 8 7 2.2 4 5 9
↑

c) Since that digit is less than 5, round down.

3 8 7 2.2 ← **This is the answer.** ◀

CAUTION! 3872.3 is not a correct answer to Example 4. It is incorrect to round from the ten thousandths digit over as follows:

3872.246, 3872.25, 3872.3.

▶ **EXAMPLE 5** Round 3872.2459 to the nearest hundredth.

a) Locate the digit in the hundredths place. 3 8 7 2.2 4 5 9 ↑

b) Consider the next digit to the right. 3 8 7 2.2 4 5 9 ↑

c) Since that digit is 5 or higher, round up.

3 8 7 2.2 5 ← **This is the answer.** ◀

▶ **EXAMPLE 6** Round 3872.2459 to the nearest thousandth, hundredth, tenth, one, ten, hundred, and thousand.

Thousandth:	3872.246	Ten:	3870
Hundredth:	3872.25	Hundred:	3900
Tenth:	3872.2	Thousand:	4000
One:	3872		

◀

▶ **EXAMPLE 7** Round 0.008 to the nearest tenth.

a) Locate the digit in the tenths place. 0.0 0 8 ↑

b) Consider the next digit to the right. 0.0 0 8 ↑

c) Since that digit is less than 5, round down.

The answer is 0.0, or 0. ◀

DO EXERCISES 7–25.

NAME SECTION DATE

EXERCISE SET 4.2

a Which number is larger?

1. 0.06, 0.58 **2.** 0.003, 0.3 **3.** 0.1, 0.111

4. 31.08, 31.2 **5.** 0.0009, 0.001 **6.** 4.056, 4.043

7. 234.07 235.07 **8.** 0.99999, 1.0 **9.** 0.4545, 0.05454

10. 0.6, 0.05 **11.** 0.004, $\frac{4}{100}$ **12.** $\frac{43}{10}$, 0.43

13. 0.54, 0.78 **14.** 0.432, 0.4325 **15.** 0.8437, 0.84384

16. 0.872, 0.873 **17.** 0.19, 1.9 **18.** 0.22, 0.2367

b Round to the nearest tenth.

19. 0.11 **20.** 0.15 **21.** 0.16 **22.** 0.29

23. 0.5794 **24.** 0.88 **25.** 2.7449 **26.** 4.78

27. 13.41 **28.** 41.23 **29.** 123.65 **30.** 36.049

Round to the nearest hundredth.

31. 0.893 **32.** 0.675 **33.** 0.666 **34.** 0.6666

35. 0.4246 **36.** 6.529 **37.** 1.435 **38.** 0.406

ANSWERS

1. 0.58
2. 0.3
3. 0.111
4. 31.2
5. 0.001
6. 4.056
7. 235.07
8. 1.0
9. 0.4545
10. 0.6
11. $\frac{4}{100}$
12. $\frac{43}{10}$
13. 0.78
14. 0.4325
15. 0.84384
16. 0.873
17. 1.9
18. 0.2367
19. 0.1
20. 0.2
21. 0.2
22. 0.3
23. 0.6
24. 0.9
25. 2.7
26. 4.8
27. 13.4
28. 41.2
29. 123.7
30. 36.0
31. 0.89
32. 0.68
33. 0.67
34. 0.67
35. 0.42
36. 6.53
37. 1.44
38. 0.41

ANSWERS

39. 3.58
40. 283.14
41. 0.01
42. 4.89
43. 0.325
44. 0.428
45. 0.667
46. 7.429
47. 17.002
48. 2.678
49. 0.001
50. 123.456
51. 10.101
52. 67.101
53. 0.116
54. 9.999
55. 300
56. 283.1
57. 283.136
58. 283.14
59. 283
60. 280
61. 34.5439
62. 34.544
63. 34.54
64. 34.5
65. 35
66. 30
67. 830
68. $\frac{83}{100}$
69. 182
70. $\frac{91}{50}$
71. 6.78346
72. 6.78346
73. 99.99999
74. 0.03030

39. 3.581 **40.** 283.1379 **41.** 0.007 **42.** 4.889

Round to the nearest thousandth.

43. 0.3246 **44.** 0.4278 **45.** 0.6666 **46.** 7.4294

47. 17.0015 **48.** 2.6776 **49.** 0.0009 **50.** 123.4562

51. 10.1011 **52.** 67.1006 **53.** 0.1161 **54.** 9.9989

Round 283.1359 to the nearest:

55. Hundred. **56.** Tenth. **57.** Thousandth.

58. Hundredth. **59.** One. **60.** Ten.

Round 34.54389 to the nearest:

61. Ten thousandth. **62.** Thousandth. **63.** Hundredth.

64. Tenth. **65.** One. **66.** Ten.

SKILL MAINTENANCE

Add.

67. $\begin{array}{r} 681 \\ +149 \\ \hline \end{array}$

68. $\frac{681}{1000} + \frac{149}{1000}$

Subtract.

69. $\begin{array}{r} 267 \\ -85 \\ \hline \end{array}$

70. $\frac{267}{100} - \frac{85}{100}$

SYNTHESIS

There are other methods of rounding decimal notation. A computer or calculator oftens uses a method called **truncating.** To round using truncating, we simply drop off all decimal places past the rounding place, which is the same as changing all digits to the right to zeros. For example, rounding 6.78093456285102 to the ninth decimal place, using truncating, we get 6.780934562. Use truncating to round each of the following to the fifth decimal place, that is, the nearest hundred thousandth.

71. 6.78346123 **72.** 6.783461902

73. 99.999999999 **74.** 0.030303030303

4.3 Addition and Subtraction with Decimals

a Addition

Adding with decimal notation is similar to adding whole numbers. First we line up the decimal points. Then we add digits from the right. For example, we add the thousandths, and then the hundredths, carrying if necessary. Then we go on to the tenths, then the ones, and so on. If desired, we can add extra zeros to the right of the decimal point so that the number of places is the same.

▶ **EXAMPLE 1** Add: 56.314 + 17.78.

$$\begin{array}{r} 56.314 \\ +17.780 \\ \hline \end{array}$$

Lining up the decimal points in order to add
Adding an extra zero to the right of the decimal point

$$\begin{array}{r} 56.314 \\ +17.780 \\ \hline 4 \end{array}$$

Adding thousandths

$$\begin{array}{r} 56.314 \\ +17.780 \\ \hline 94 \end{array}$$

Adding hundredths

$$\begin{array}{r} {}^{1}\quad\;\; \\ 56.314 \\ +17.780 \\ \hline .094 \end{array}$$

Adding tenths
Write a decimal point in the answer.

We get 10 tenths = 1 one + 0 tenths, so we carry the 1 to the ones column.

$$\begin{array}{r} {}^{1\,1}\quad\;\;\; \\ 56.314 \\ +17.780 \\ \hline 4.094 \end{array}$$

Adding ones

We get 14 = 1 ten + 4 ones, so we carry the 1 to the tens column.

$$\begin{array}{r} {}^{1\,1}\quad\;\;\; \\ 56.314 \\ +17.780 \\ \hline 74.094 \end{array}$$

Adding tens ◀

DO EXERCISES 1 AND 2.

If we want, we can write extra zeros to the right of the decimal point to get the same number of decimal places.

▶ **EXAMPLE 2** Add: 3.42 + 0.237 + 14.1.

$$\begin{array}{r} 3.420 \\ 0.237 \\ +14.100 \\ \hline 17.757 \end{array}$$

Writing extra zeros

Adding ◀

DO EXERCISES 3–5.

OBJECTIVES

After finishing Section 4.3, you should be able to:

- a Add using decimal notation.
- b Subtract using decimal notation.
- c Solve equations of the type $x + a = b$ and $a + x = b$, where a and b may be in decimal notation.

FOR EXTRA HELP

Tape 6C Tape 7A MAC: 4 IBM: 4

Add.

1. $\begin{array}{r} 0.847 \\ +10.07 \\ \hline \end{array}$ 10.917

2. $\begin{array}{r} 2.1 \\ 0.739 \\ +31.3689 \\ \hline \end{array}$ 34.2079

Add.

3. 0.02 + 4.3 + 0.649 4.969

4. 0.12 + 3.006 + 0.4357 3.5617

5. 0.4591 + 0.2374 + 8.70894 9.40544

ANSWERS ON PAGE A-3

Add.

6. 789 + 123.67 912.67

7. 45.78 + 2467 + 1.993 2514.773

Subtract.

8.
```
  37.428
- 26.674
  10.754
```

9.
```
  0.347
- 0.008
  0.339
```

ANSWERS ON PAGE A-3

Consider the addition 3456 + 19.347. Keep in mind that a whole number, such as 3456, has an "unwritten" decimal point at the right with 0 fractional parts. When adding, we can always write in that decimal point and extra zeros if desired.

▶ **EXAMPLE 3** Add: 3456 + 19.347.

```
  3456.000   Writing in the decimal point and extra zeros
+   19.347   Lining up the decimal points in order to add
  3475.347   Adding
```
◀

DO EXERCISES 6 AND 7.

b Subtraction

Subtracting with decimal notation is similar to subtracting whole numbers. First we line up the decimal points. Then we subtract digits from the right. For example, we subtract the thousandths, and then the hundredths, the tenths, and so on, borrowing if necessary.

▶ **EXAMPLE 4** Subtract: 56.314 − 17.78.

```
  56.314   Lining up the decimal points in order to subtract
- 17.780   Writing an extra 0

  56.314   Subtracting thousandths
- 17.780
       4

     2 11
  56.3 1 4   Borrowing tenths to subtract hundredths
- 17.7 8 0
       3 4

     12
   5 2 11
  56.3 1 4   Borrowing ones to subtract tenths
- 17.7 8 0
    .5 3 4   Writing a decimal point

  15 12
  4 5 2 11
  5 6.3 1 4   Borrowing tens to subtract ones
- 1 7.7 8 0
    8.5 3 4

  15 12
  4 5 2 11
  5 6.3 1 4   Subtracting tens
- 1 7.7 8 0
  3 8.5 3 4
```
◀

DO EXERCISES 8 AND 9.

▶ **EXAMPLE 5** Subtract: 13.07 − 9.205.

$$\begin{array}{r} 13.070 \\ -\ 9.205 \\ \hline 3.865 \end{array}$$

Writing an extra zero

Subtracting ◀

▶ **EXAMPLE 6** Subtract: 23.08 − 5.0053.

$$\begin{array}{r} 23.0800 \\ -\ 5.0053 \\ \hline 18.0747 \end{array}$$

Writing two extra zeros

Subtracting ◀

DO EXERCISES 10–12.

When subtraction involves a whole number, again keep in mind that there is an "unwritten" decimal point that can be written in if desired. Extra zeros can also be written in to the right of the decimal point.

▶ **EXAMPLE 7** Subtract: 456 − 2.467.

$$\begin{array}{r} 456.000 \\ -\ 2.467 \\ \hline 453.533 \end{array}$$

Writing in the decimal point and extra zeros

Subtracting

DO EXERCISES 13 AND 14.

C Solving Equations

Now let us solve equations $x + a = b$ and $a + x = b$, where a and b may be in decimal notation. Proceeding as we have before, we subtract a on both sides.

▶ **EXAMPLE 8** Solve: $x + 28.89 = 74.567$.

We have

$$x + 28.89 - 28.89 = 74.567 - 28.89 \quad \text{Subtracting 28.89 on both sides}$$
$$x = 45.677.$$

$$\begin{array}{r} 74.567 \\ -\ 28.890 \\ \hline 45.677 \end{array}$$

The solution is 45.677. ◀

▶ **EXAMPLE 9** Solve: $0.8879 + y = 9.0026$.

We have

$$0.8879 + y - 0.8879 = 9.0026 - 0.8879 \quad \text{Subtracting 0.8879 on both sides}$$
$$y = 8.1147.$$

$$\begin{array}{r} 9.0026 \\ -\ 0.8879 \\ \hline 8.1147 \end{array}$$

The solution is 8.1147. ◀

DO EXERCISES 15 AND 16.

Subtract.

10. 1.2345 − 0.7 0.5345

11. 0.9564 − 0.4392 0.5172

12. 7.37 − 0.00008 7.36992

Subtract.

13. 1277 − 82.78 1194.22

14. 5 − 0.0089 4.9911

Solve.

15. $x + 17.78 = 56.314$ 38.534

16. $8.906 + t = 23.07$ 14.164

ANSWERS ON PAGE A-3

❖ SIDELIGHTS

A Number Pattern

Show that each of the following is true by simplifying each side of the equation. Look for a pattern.

$$1 = \frac{1 \cdot 2}{2}$$

$$1 + 2 = \frac{2 \cdot 3}{2}$$

$$1 + 2 + 3 = \frac{3 \cdot 4}{2}$$

$$1 + 2 + 3 + 4 = \frac{4 \cdot 5}{2}$$

$$1 + 2 + 3 + 4 + 5 = \frac{5 \cdot 6}{2}$$

EXERCISES

Use the pattern of the above to find these sums without adding.

1. $1 + 2 + 3 + 4 + 5 + 6$ 21

2. $1 + 2 + 3 + 4 + 5 + 6 + 7 + 8$ 36

3. $1 + 2 + 3 + 4 + 5 + 6 + 7 + 8 + 9 + 10 + 11 + 12$ 78

4. $1 + 2 + 3 + \cdots + 100$ (The dots stand for the symbols we did not write.) 5050

NAME SECTION DATE

EXERCISE SET 4.3

a Add.

1. $\begin{array}{r} 316.25 \\ +\ 18.12 \\ \hline \end{array}$

2. $\begin{array}{r} 41.823 \\ +614.915 \\ \hline \end{array}$

3. $\begin{array}{r} 659.403 \\ +916.812 \\ \hline \end{array}$

4. $\begin{array}{r} 3.25 \\ +1123.39 \\ \hline \end{array}$

5. $\begin{array}{r} 9.104 \\ +123.456 \\ \hline \end{array}$

6. $\begin{array}{r} 4.1523 \\ +3.2778 \\ \hline \end{array}$

7. $\begin{array}{r} 61.006 \\ +\ 3.407 \\ \hline \end{array}$

8. 0.8096 + 0.7856

9. 20.0124 + 30.0124

10. 0.263 + 0.8

11. 0.83 + 0.005

12. 0.347 + 10.04

13. 0.34 + 3.5 + 0.127 + 768

14. 2.3 + 0.729 + 23

15. 17 + 3.24 + 0.256 + 0.3689

16. $\begin{array}{r} 47.8 \\ 219.852 \\ 43.59 \\ +666.713 \\ \hline \end{array}$

17. $\begin{array}{r} 2.703 \\ 78.33 \\ 28.0009 \\ +118.4341 \\ \hline \end{array}$

18. $\begin{array}{r} 13.72 \\ 9.112 \\ 6542.7908 \\ +\ 23.901 \\ \hline \end{array}$

19. 99.6001 + 7285.18 + 500.042 + 870

20. 65.987 + 9.4703 + 6744.02 + 1.0003 + 200.895

ANSWERS

1. 334.37

2. 656.738

3. 1576.215

4. 1126.64

5. 132.560

6. 7.4301

7. 64.413

8. 1.5952

9. 50.0248

10. 1.063

11. 0.835

12. 10.387

13. 771.967

14. 26.029

15. 20.8649

16. 977.955

17. 227.4680

18. 6589.5238

19. 8754.8221

20. 7021.3726

ANSWERS

21. 1.3
22. 9.240
23. 49.02
24. 31.13
25. 45.61
26. 44.06
27. 85.921
28. 0.3114
29. 2.4975
30. 27.72
31. 3.397
32. 0.45
33. 8.85
34. 100.008
35. 3.37
36. 19.92
37. 1.045
38. 0.991
39. 3.703
40. 0.28
41. 0.9902
42. 0.9092

b Subtract.

21. $\begin{array}{r} 5.2 \\ -3.9 \\ \hline \end{array}$

22. $\begin{array}{r} 11.345 \\ -\ 2.105 \\ \hline \end{array}$

23. $\begin{array}{r} 51.31 \\ -\ 2.29 \\ \hline \end{array}$

24. $\begin{array}{r} 37.45 \\ -\ 6.32 \\ \hline \end{array}$

25. $\begin{array}{r} 48.76 \\ -\ 3.15 \\ \hline \end{array}$

26. $\begin{array}{r} 47.21 \\ -\ 3.15 \\ \hline \end{array}$

27. $\begin{array}{r} 92.341 \\ -\ 6.42 \\ \hline \end{array}$

28. $\begin{array}{r} 0.346 \\ -0.0346 \\ \hline \end{array}$

29. $\begin{array}{r} 2.5 \\ -0.0025 \\ \hline \end{array}$

30. $\begin{array}{r} 28.0 \\ -\ 0.28 \\ \hline \end{array}$

31. $\begin{array}{r} 3.4 \\ -0.003 \\ \hline \end{array}$

32. $\begin{array}{r} 1.5 \\ -1.05 \\ \hline \end{array}$

33. 28.2 − 19.35

34. 100.12 − 0.112

35. 34.07 − 30.7

36. 36.2 − 16.28

37. 8.45 − 7.405

38. 3.801 − 2.81

39. 6.003 − 2.3

40. 9.087 − 8.807

41. 1 − 0.0098

42. 2 − 1.0908

43. $100 - 0.34$

44. $624 - 18.79$

45. $7.48 - 2.6$

46. $3 - 2.006$

47. $25.008 - 12.4$

48. $263.7 - 102.08$

49. $2548.98 - 2.007$

50. $19 - 1.198$

51. $45 - 0.999$

52. $10.056 - 0.392$

53. $3.907 - 1.416$

54. $70.0009 - 23.0567$

55. $\begin{array}{r} 32.7978 \\ -\ 0.0592 \\ \hline \end{array}$

56. $\begin{array}{r} 0.49634 \\ -0.12678 \\ \hline \end{array}$

57. $\begin{array}{r} 3.0074 \\ -1.3408 \\ \hline \end{array}$

58. $\begin{array}{r} 6.07 \\ -2.0078 \\ \hline \end{array}$

59. $\begin{array}{r} 2345.90786 \\ -\ 0.999 \\ \hline \end{array}$

60. $\begin{array}{r} 1.0 \\ -0.9999 \\ \hline \end{array}$

ANSWERS

43. 99.66

44. 605.21

45. 4.88

46. 0.994

47. 12.608

48. 161.62

49. 2546.973

50. 17.802

51. 44.001

52. 9.664

53. 2.491

54. 46.9442

55. 32.7386

56. 0.36956

57. 1.6666

58. 4.0622

59. 2344.90886

60. 0.0001

ANSWERS

61. 199.897

62. 3.37

63. 19.251

64. 54.06

65. 384.68

66. 46.4996

67. 582.97

68. 32.43

69. 35,000

70. 34,000

71. 345.8

C Solve.

61. $x + 0.223 = 200.12$

62. $t + 50.7 = 54.07$

63. $3.205 + m = 22.456$

64. $4.26 + q = 58.32$

65. $17.95 + p = 402.63$

66. $w + 1.3004 = 47.8$

67. $13{,}083.3 = x + 12{,}500.33$

68. $100.23 = 67.8 + z$

SKILL MAINTENANCE

69. Round 34,567 to the nearest thousand.

70. Round 34,496 to the nearest thousand.

SYNTHESIS

71. A student presses the wrong button when using a calculator and adds 235.7 instead of subtracting it. The incorrect answer is 817.2. What is the correct answer?

4.4 Solving Problems

a Solving problems using decimals is like solving problems with whole numbers. We translate to an equation that corresponds to the situation. Then we solve the equation.

▶ **EXAMPLE 1** A patient was given injections of 3.68 milligrams (mg), 2.7 mg, 3.65 mg, and 5.0 mg over a 24-hr period. What was the total amount of the injections?

1. *Familiarize.* We draw a picture or at least visualize the situation. We let $t =$ the amount of the injections.

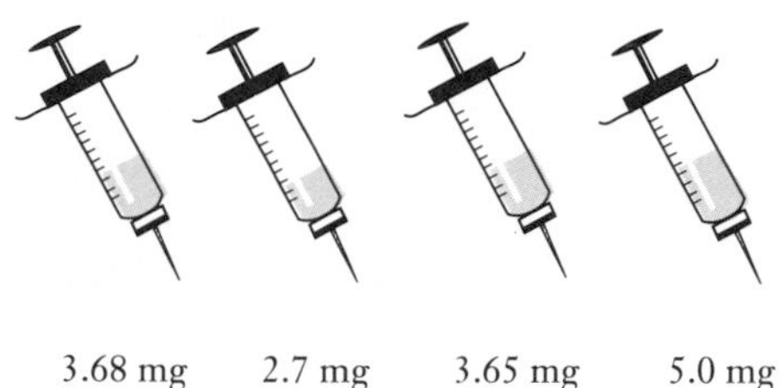

2. *Translate.* Amounts are being combined. We translate to an equation:

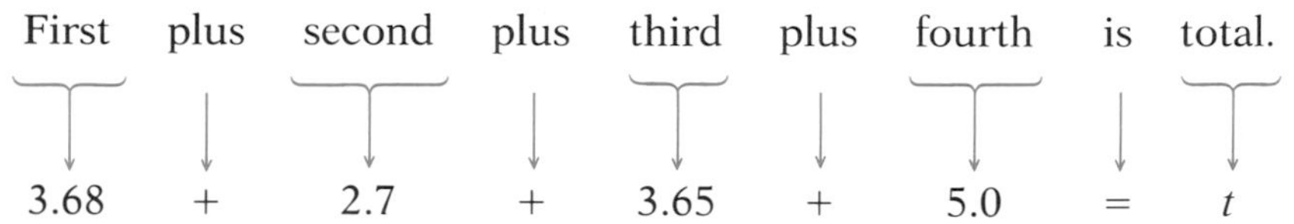

3. *Solve.* To solve, we carry out the addition.

$$\begin{array}{r} {\scriptstyle 2\ 1} \\ 3.68 \\ 2.70 \\ 3.65 \\ +\ 5.00 \\ \hline 15.03 \end{array}$$

Thus, $t = 15.03$.

4. *Check.* We can check by repeating our addition. We can also see whether our answer is reasonable by first noting that it is indeed larger than any of the numbers being added. We can also check by rounding:

$$3.68 + 2.7 + 3.65 + 5.0 \approx 4.0 + 3.0 + 4.0 + 5.0 = 16 \approx 15.03.$$

If we had gotten an answer like 150.3 or 0.1503, then our estimate, 16, would have told us that we did something wrong, like not lining up the decimal points.

5. *State.* The total of the injections was 15.03 mg. ◀

DO EXERCISE 1.

OBJECTIVE

After finishing Section 4.4, you should be able to:

a Solve problems involving addition and subtraction with decimals.

FOR EXTRA HELP

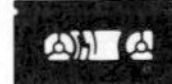  

Tape 6D | Tape 7A | MAC: 4 IBM: 4

1. Each year, each of us drinks 43.7 gal of soft drinks, 37.3 gal of water, 27.3 gal of coffee, 21.1 gal of milk, and 8.1 gal of fruit juice. What is the total amount that each of us drinks?

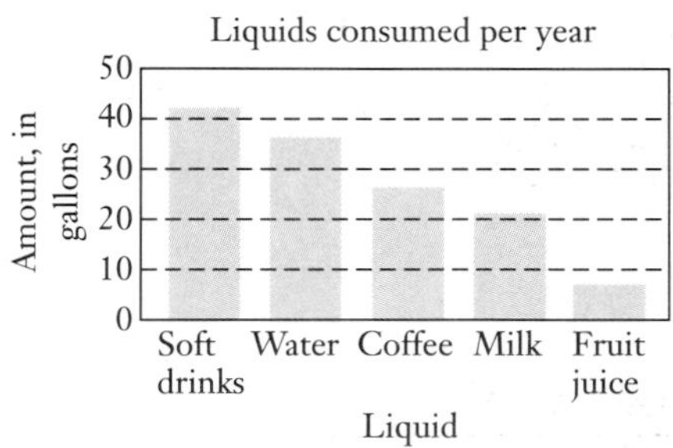

137.5 gal

ANSWER ON PAGE A-3

2. Coffee prices recently made a dramatic increase. The price was \$2.73 per pound in May of 1988. In May of 1989, the price was \$3.16. How much was the increase? \$0.43 per pound

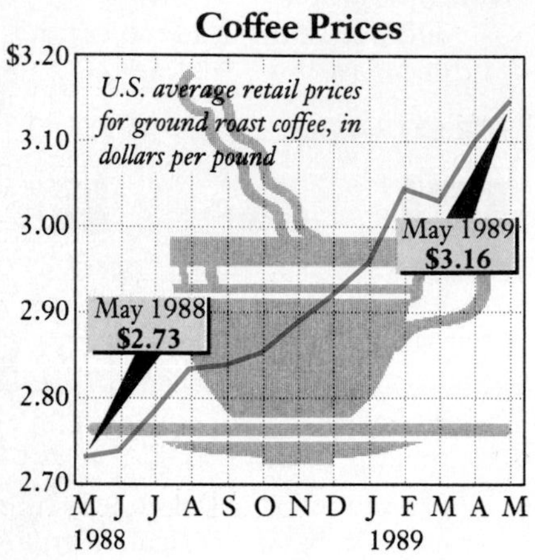

ANSWER ON PAGE A-3

▶ **EXAMPLE 2** Normal body temperature is 98.6°F. When fevered, most people will die if their bodies reach 107°F. This is a rise of how many degrees?

1. *Familiarize.* We draw a picture or at least visualize the situation. We let n = the number of degrees of rise in temperature.

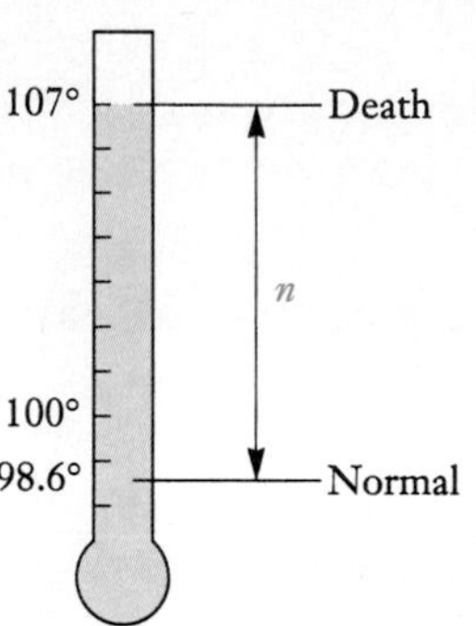

2. *Translate.* This is a "how-much-more" situation. We translate as follows.

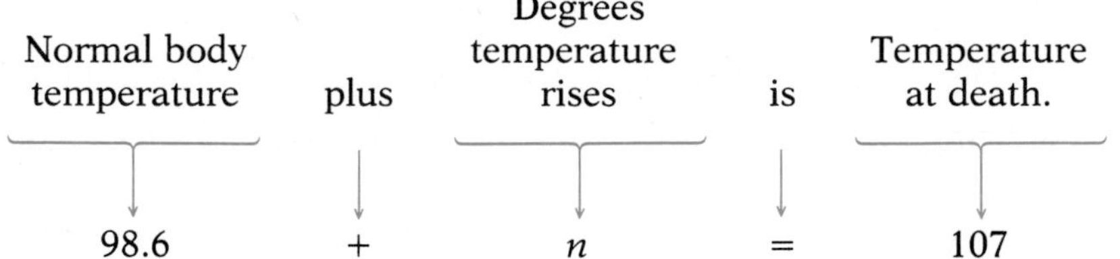

3. *Solve.* We solve the equation. We subtract 98.6 on both sides.

$$98.6 + n - 98.6 = 107 - 98.6$$
$$n = 8.4$$

$$\begin{array}{r} \overset{9\ \overset{16}{\not 6}\ 10}{\not 1 \not 0\ 7.\not 0} \\ -\quad 9\ 8.6 \\ \hline 8.4 \end{array}$$

4. *Check.* We can check by adding 8.4 to 98.6, to get 107. This checks.
5. *State.* A rise of 8.4°F will cause death. ◀

DO EXERCISE 2.

NAME SECTION DATE

EXERCISE SET 4.4

ANSWERS

a Solve.

1. On a six-day trip, a driver bought the following amounts of gasoline: 23.6 gal, 17.7 gal, 20.8 gal, 17.2 gal, 25.4 gal, and 13.8 gal. How many gallons of gasoline were purchased?

2. One week a pilot flew a plane the following distances: 247.6 mi, 80.5 mi, 536.8 mi, 198.2 mi, and 360 mi. What was the total number of miles that the pilot flew?

3. A consumer bought a record for \$6.99 and paid with a \$10 bill. How much change was there?

4. A student bought a book for \$14.68 and paid with a \$20 bill. How much change was there?

5. Normal body temperature is 98.6°F. During an illness, it rose 4.2°. What was the new temperature?

6. A medical assistant draws 17.85 mg of blood and uses 9.68 mg in a blood test. How much is left?

7. A family read the odometer before starting a trip. It read 22,456.8 and they drove 234.7 mi. What did it read at the end of the trip?

8. A driver bought gasoline when the odometer read 14,296.3. At the next gasoline purchase, the odometer read 14,515.8. How many miles had been driven?

9. The average cost of lunch for one person in a hotel in New York is \$20.27. The cost of a lunch in Los Angeles is \$13.68. How much higher is a lunch in New York than in Los Angeles?

10. In 1911 Ray Harroun won the first Indianapolis 500-mile race with an average speed of 74.59 mph. Bobby Rahal set a record with the highest average speed ever in the 1986 race with a speed of 170.722 mph. How much faster was Rahal?

ANSWERS

1. 118.5 gal
2. 1423.1 mi
3. \$3.01
4. \$5.32
5. 102.8°F
6. 8.17 mg
7. 22,691.5
8. 219.5 mi
9. \$6.59
10. 96.132 mph

ANSWERS

11. $984.89

12. $110.40

13. 3.4 yr

14. 93.9 km

15. 18.09 min

16. 42.7 sec

17. a) 2438 mi

b) 3900.8 km

18. a) 36.85 in.

b) 93.599 cm

19. 225.8 mi

20. 19,181.8

11. One day the following checks were written: $26.79, $268.10, and $690.00. How much was spent?

12. A student bought a sweater for $17.95, a pair of slacks for $23.95, and a coat for $68.50. How much was spent?

13. In 1961 the average age of a bride was 19.8. In 1984 the average age was 23.2. How much older was the average bride in 1984 than in 1961?

14. The length of the Panama Canal is 81.6 kilometers (km). The length of the Suez Canal is 175.5 km. How much longer is the Suez Canal?

15. On a 4-mile relay, the times of each member were 4.25 min, 4.86 min, 3.98 min, and 5.0 min. What was their combined time?

16. On a 400-meter relay, the times were 10.8 sec, 10.6 sec, 11.1 sec, and 10.2 sec. What was their combined time?

17. The distance, by air, from New York to St. Louis is 876 mi (1401.6 kilometers); from St. Louis to Los Angeles, it is 1562 mi (2499.2 km).

a) How far in miles is it from New York to Los Angeles?

b) How far in kilometers is it from New York to Los Angeles?

18. Dallas, Texas, receives an average of 34.55 in. (87.757 cm) of rain and 2.3 in. (5.842 cm) of snow each year.

a) What is the total average amount of precipitation in inches?

b) What is the total average amount of precipitation in centimeters?

19. A driver bought gasoline when the odometer read 28,576.8. At the next gasoline purchase, the odometer read 28,802.6. How many miles had been driven?

20. A family read the odometer before starting a trip. It read 18,788.9 and they drove 356.4 mi one day and 36.5 mi the next. What did it read at the end of the trip?

21. A businesswoman has $1123.56 in her checking account. She writes checks of $23.82, $507.88, and $98.32 to pay some bills. She then deposits a paycheck of $678.20. How much is in her account after these changes?

22. A student had $185.00 to spend for apparel: $44.95 was spent for shoes, $71.95 for a sport coat, and $55.35 for slacks. How much was left?

The table below shows the number of passengers per year who travel through the country's busiest airports. (Use the table for Exercises 23–26.)

Airport	**Passengers (in millions)**
O'Hare International (Chicago)	45.7
Hartsfield (Atlanta)	39.0
Los Angeles International	34.4
Dallas/Ft. Worth	32.3
Kennedy International	29.9

23. How many more passengers does O'Hare handle than Kennedy?

24. How many more passengers does Hartsfield handle than Los Angeles?

25. How many passengers do Dallas/Ft. Worth and Kennedy handle together?

26. How many passengers do all these airports handle together?

27. A pair of cotton slacks costs $39.95. A student bought two pairs. The tax on the purchase was $4.80. How much did the student pay for the slacks, with tax?

28. A pair of sports shoes costs $79.95. A jogger buys two pairs. The tax on the purchase was $9.40. How much did the jogger pay for the shoes, with tax?

ANSWERS

21. $1171.74

22. $12.75

23. 15.8 million

24. 4.6 million

25. 62.2 million

26. 181.3 million

27. $84.70

28. $169.30

ANSWERS

29. 78.1 cm

30. 391.5 yd

31. 2.31 cm

32. 2.72 cm

33. 1.4°F

34. 29.6°F

35. No

36. Yes

37. 6335

38. $\frac{31}{24}$

39. 2803

40. $\frac{1}{24}$

29. Find the distance around this figure.

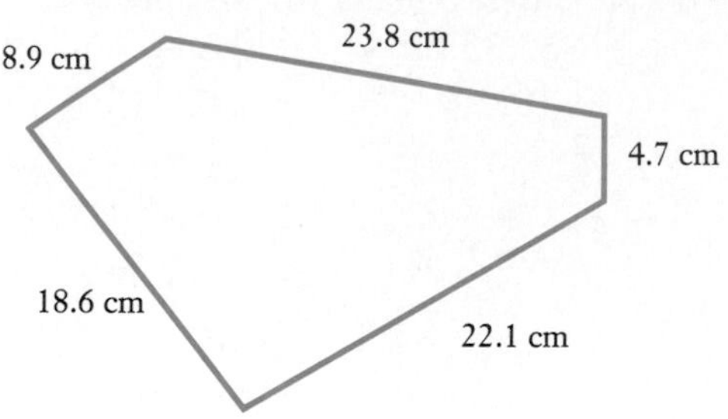

30. Find the distance around this figure.

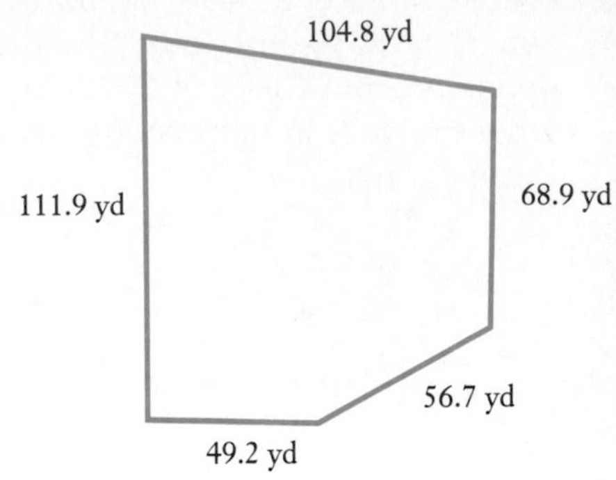

31. Find the length d in this figure.

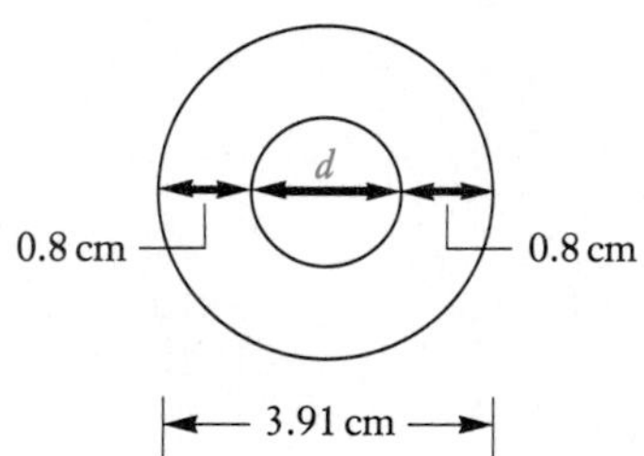

32. Find the length d in this figure.

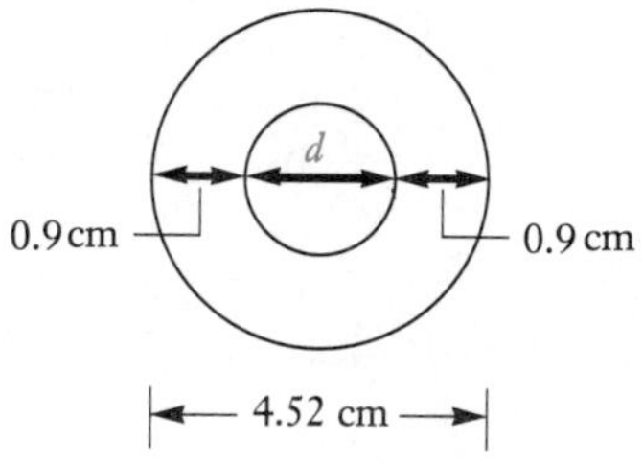

33. Normal body temperature is 98.6°F. A baby's bath water should be 100°F. How many degrees above normal body temperature is this?

34. Normal body temperature is 98.6°F. The lowest temperature at which a patient has survived is 69°F. How many degrees below normal is this?

35. A student used a \$20 bill to buy a book for \$10.75. The change was a five-dollar bill, three one-dollar bills, a dime, and two nickles. Was the change correct?

36. A customer used a \$20 bill to buy two albums for \$13.88. The change was a five-dollar bill, a one-dollar bill, one dime, and two pennies. Was the change correct?

SKILL MAINTENANCE

37. 4569 + 1766

38. $\frac{2}{3} + \frac{5}{8}$

Subtract.

39. 4569 − 1766

40. $\frac{2}{3} - \frac{5}{8}$

SUMMARY AND REVIEW EXERCISES: CHAPTER 4

The review sections and objectives to be tested in addition to the material in this chapter are [1.2b]; [1.3d]; [3.2a, b]; and [3.3a].

Write a word name.

1. 3.47

2. 0.031

Write a word name as on a check.

3. \$597.25

4. \$0.98

Write fractional notation.

5. 0.09

6. 4.561

7. 0.089

8. 3.0227

Write decimal notation.

9. $\frac{34}{1000}$

10. $\frac{42,603}{10,000}$

11. $\frac{2791}{100}$

12. $\frac{6}{1000}$

Which number is larger?

13. 0.034, 0.0185

14. 0.91, 0.19

15. 0.741, 0.6943

16. 1.038, 1.041

Round 17.4287 to the nearest:

17. Tenth.

18. Hundredth.

19. Thousandth.

Round 4.2716 to the nearest:

20. Thousandth.

21. Hundredth.

22. Tenth.

Add.

23.
$$\begin{array}{r} 2.048 \\ 65.371 \\ +\,507.1 \\ \hline \end{array}$$

24.
$$\begin{array}{l} 0.6 \\ 0.004 \\ 0.07 \\ 0.0098 \\ \hline \end{array}$$

25.
$$\begin{array}{r} 236.231 \\ 263.4 \\ +\quad 0.198 \\ \hline \end{array}$$

26.
$$\begin{array}{r} 14.6 \\ 15.08 \\ 16.004 \\ +\,17.0092 \\ \hline \end{array}$$

27. $219.3 + 2.8 + 7$

28. $0.41 + 4.1 + 41 + 0.091$

Subtract.

29. $\begin{array}{r} 30.0 \\ -\ 0.7908 \\ \hline \end{array}$

30. $\begin{array}{r} 845.08 \\ -\ 54.79 \\ \hline \end{array}$

31. $\begin{array}{r} 37.645 \\ -\ 8.497 \\ \hline \end{array}$

32. $\begin{array}{r} 70.8 \\ -\ 0.0109 \\ \hline \end{array}$

33. 745.0109 − 59.959

34. 8 − 0.0047

Solve.

35. $x + 51.748 = 548.0275$

36. $0.0089 + y = 5$

Solve.

37. In the United States, there are 51.81 telephone poles for every hundred people. In Canada, there are 40.65. How many more telephone poles for every hundred people are there in the United States?

38. In 1961 the average age of a groom was 22.0. In 1984 the average age was 25.5. How much older was the average groom in 1984 than in 1961?

39. A farmer has 4 corn fields. One year the harvest in each field was 1419.3 bushels, 1761.8 bushels, 1095.2 bushels, and 2088.8 bushels. What was the year's total harvest?

40. A checking account contained $6274.53. After a $385.79 check was drawn, what was left in the account?

SKILL MAINTENANCE

Add.

41. 5608 + 8997

42. $\frac{4}{5} + \frac{5}{6}$

43. 8997 − 5608

44. $\frac{5}{6} - \frac{4}{5}$

❖ THINKING IT THROUGH

Find the error in each of the following.

1. Add:

$$\begin{array}{r} 13.07 \\ +\ 9.205 \\ \hline 10.512 \end{array}$$

2. Subtract:

$$\begin{array}{r} 73.089 \\ -\ 5.0061 \\ \hline 2.3028 \end{array}$$

3. Discuss how decimal notation is defined in terms of fractional notation.

NAME SECTION DATE

TEST: CHAPTER 4

1. Write a word name for 2.34.

2. Write a word name, as on a check, for $1234.78.

Write fractional notation.

3. 0.91

✓4. 2.769

Write decimal notation.

5. $\frac{74}{100}$

6. $\frac{37,047}{10,000}$

Which number is larger?

7. 0.07, 0.162

✓8. 0.09, 0.9

9. 0.078, 0.06

Round 5.6783 to the nearest:

10. Tenth.

✓ 11. Hundredth.

12. Thousandth.

Add.

13.
$$\begin{array}{r} 402.3 \\ 2.81 \\ +\quad 0.109 \\ \hline \end{array}$$

14.
$$\begin{array}{r} 0.7 \\ 0.08 \\ 0.009 \\ +0.0012 \\ \hline \end{array}$$

ANSWERS

1. [4.1a] Two and thirty-four hundredths

2. [4.1a] One thousand, two hundred thirty-four and $\frac{78}{100}$ dollars

3. [4.1b] $\frac{91}{100}$

4. [4.1b] $\frac{2769}{1000}$

5. [4.1c] 0.74

6. [4.1c] 3.7047

7. [4.2a] 0.162

8. [4.2a] 0.9

9. [4.2a] 0.078

10. [4.2b] 5.7

11. [4.2b] 5.68

12. [4.2b] 5.678

13. [4.3a] 405.219

14. [4.3a] 0.7902

15. 102.4 + 6.1 + 78

16. 0.93 + 9.3 + 93 + 930

Subtract.

17.
$$\begin{array}{r} 52.678 \\ -\ 4.321 \\ \hline \end{array}$$

18.
$$\begin{array}{r} 20.0 \\ -\ 0.9099 \\ \hline \end{array}$$

19. 2 − 0.0054

20. 234.6788 − 81.7854

Solve.

21. $x + 0.018 = 9$

22. $34.908 + q = 3400.5677$

Solve.

23. A student wrote checks of $123.89, $56.78, and $3446.98. How much was written in checks altogether?

24. In the 1896 Olympics, Alfred Hajos won the 100-meter freestyle in 60.37 sec. In the 1984 Olympics, Rowdy Gaines won in 49.80 sec. How much faster was Gaines?

SKILL MAINTENANCE

Add.

25. $\frac{5}{6} + \frac{2}{9}$

26.
$$\begin{array}{r} 4509 \\ 6789 \\ +\ 2354 \\ \hline \end{array}$$

Subtract.

27. $\frac{5}{6} - \frac{2}{9}$

28.
$$\begin{array}{r} 4509 \\ -\ 2354 \\ \hline \end{array}$$

ANSWERS

15. [4.3a] 186.5

16. [4.3a] 1033.23

17. [4.3b] 48.357

18. [4.3b] 19.0901

19. [4.3b] 1.9946

20. [4.3b] 152.8934

21. [4.3c] 8.982

22. [4.3c] 3365.6597

23. [4.4a] $3627.65

24. [4.4a] 10.57 sec

25. [3.2b] $\frac{19}{18}$

26. [1.2b] 13,652

27. [3.3a] $\frac{11}{18}$

28. [1.3d] 2155

CUMULATIVE REVIEW: CHAPTERS 1–4

1. Write expanded notation: 12,758.

2. Write a word name as on a check: \$802.53.

3. Write fractional notation: 10.09.

4. Convert to fractional notation: $3\frac{3}{8}$.

5. Write decimal notation: $\frac{35}{1000}$.

6. Find all the factors of 66.

7. Find the prime factorization of 66.

8. Find the LCM of 28 and 35.

9. Round 6962.4721 to the nearest hundred.

10. Round 6962.4721 to the nearest hundredth.

Add and simplify.

11. $3\frac{2}{3} + 2\frac{5}{9}$

12. $110.863 + 0.73 + 121.9 + 1.904$

13. $5249 + 215 + 31$

14. $\frac{4}{15} + \frac{7}{30}$

Subtract.

15. $6813 - 5987$

16. $9010 - 563.47$

17. $\frac{8}{9} - \frac{7}{8}$

18. $7\frac{1}{5} - 3\frac{4}{5}$

Multiply and simplify.

19. 831×220

20. $\frac{3}{5} \times \frac{10}{21}$

21. $3\frac{2}{11} \cdot 4\frac{2}{7}$

22. $5 \cdot \frac{3}{10}$

Divide and simplify.

23. $2\frac{4}{5} \div 1\frac{13}{15}$

24. $\frac{6}{5} \div \frac{7}{8}$

Divide. Write the answers with remainders as mixed numerals.

25. $16\overline{)608}$

26. $21\overline{)4325}$

Use $<$, $>$, or $=$ for ■ to write a true sentence.

27. $\frac{5}{9}$ ■ $\frac{8}{11}$

28. 10 ■ 9.999

29. $\frac{4}{6}$ ■ $\frac{6}{9}$

30. 2.222 ■ 2.3

Solve.

31. $2.305 + x = 3.1$

32. $\frac{2}{5} \cdot t = \frac{5}{4}$

33. $45 \cdot x = 5445$

34. $y + \frac{1}{15} = \frac{2}{9}$

35. $38.015 + t = 43$

36. $x \cdot \frac{11}{15} = \frac{22}{25}$

Solve.

37. There are 411,293 students taking courses in Spanish in colleges in the United States. There are 275,328 taking French, and 121,022 studying German. What is the total enrollment in Spanish, French, and German classes?

38. The Cleveland Stadium seats 74,383 fans. Fenway Park in Boston seats 33,583. How many more does Cleveland Stadium seat?

39. There were 16.7 billion pennies minted in 1982 and 9.6 billion minted in 1987. How many more were made in 1982 than in 1987?

40. A family spent \$55.56 on groceries one week, \$43.19 the next week, \$86.02 the third week, and \$19.77 the fourth week. How much was spent on groceries for the four weeks?

41. A cookie recipe calls for $2\frac{1}{4}$ cups of flour. How much flour should be used to make half a recipe?

42. A piece of fabric $2\frac{1}{8}$ yd long is cut from a longer piece originally measuring 4 yd. How much fabric is left?

43. One pie is cut into 6 equal pieces and another is cut into 8 equal pieces. One piece from each pie is left over after a meal. What part of a pie is left over?

44. How many stamps are in 128 books of stamps if each book contains 12 stamps?

SYNTHESIS

45. Simplify:

$$\left(\frac{3}{4}\right)^2 - \frac{1}{8} \cdot \left(3 - 1\frac{1}{2}\right)^2.$$

46. Add, writing the answer in decimal notation:

$$5.42 + \frac{355}{100} + \frac{89}{10} + \frac{17}{1000}.$$

INTRODUCTION In this chapter, we first consider multiplication and division using decimal notation. This will allow us to solve problems like the one below. Also studied are equation and problem solving, as well as estimating sums, differences, products, and quotients. We also consider conversion from fractional to decimal notation where the decimal notation may be repeating.

The review sections to be tested in addition to the material in this chapter are 2.1, 2.5, 3.5, and 3.6. ❖

Multiplication and Division: Decimal Notation

AN APPLICATION

A driver filled the gasoline tank and noted that the odometer read 67,507.8. After the next filling, the odometer read 68,006.1. It took 16.5 gal to fill the tank the second time. How many miles per gallon did the driver get?

THE MATHEMATICS

This is a two-step problem, First, we let n = the number of miles driven between fillups. We find n by solving the following equation:

$$\begin{aligned} 67{,}507.8 + n &= 68{,}006.1 \\ n &= 68{,}006.1 - 67{,}507.8 \\ n &= 498.3. \end{aligned}$$

We let m = the number of miles per gallon. We find m as follows:

$$\underbrace{498.3 \div 16.5}_{\uparrow} = m.$$

This is division using decimal notation.

❖ POINTS TO REMEMBER: CHAPTER 5

Skills of Multiplying and Dividing with Whole Numbers	
Order of Operations:	Section 1.9
Rounding and Estimating Skills:	Section 1.4
Area of a Rectangle:	$A = l \cdot w$

PRETEST: CHAPTER 5

Multiply.

1. 47×0.82

2. 0.835×0.74

3. 0.463×100

4. 57.299×7.6

5. 6.8×0.54

6. 4.07×0.105

7. 324.56×0.001

8. 73.962×10

Divide.

9. $8\overline{)29}$

10. $25\overline{)33}$

11. $42\overline{)20.16}$

12. $6.6\overline{)200.64}$

13. $82\overline{)31.16}$

14. $\dfrac{576.98}{1000}$

15. $\dfrac{756.89}{0.01}$

16. $\dfrac{0.004653}{100}$

17. Solve: $9.6 \cdot y = 808.896$.

18. Estimate the product 6.92×32.458 by rounding to the nearest one.

19. Estimate the quotient $74.882209 \div 15.03$ by rounding to the nearest ten.

Find decimal notation. Use multiplying by 1.

20. $\dfrac{7}{5}$

21. $\dfrac{23}{16}$

22. $\dfrac{53}{4}$

Find decimal notation. Use division.

23. $\dfrac{11}{4}$

24. $\dfrac{7}{9}$

25. $\dfrac{29}{7}$

Round the answer to Exercise 25 to the nearest:

26. Tenth.

27. Hundredth.

28. Thousandth.

Solve.

29. What is the cost of 6 compact discs at \$14.95 each?

30. A person walked 10.85 km in 5 hr. How far did the person walk in 1 hr?

31. A developer paid \$47,567.89 for 14 acres of land. How much did it cost for 1 acre? Round to the nearest cent.

32. Convert from dollars to cents: \$74.96.

33. Convert from cents to dollars: 13,549 cents.

34. Convert to standard notation: 48.6 trillion.

Calculate.

35. $256 \div 3.2 \div 2 - 3.685 + 78.325 \times 0.03$

36. $(1 - 0.06)^2 + \{8[5(12.1 - 7.8) + 20(17.3 - 8.7)]\}$

37. $\dfrac{5}{8} \times 78.95$

38. $\dfrac{2}{3} \times 89.95 - \dfrac{5}{9} \times 3.234$

5.1 Multiplication with Decimal Notation

a Multiplication

Look at these products.

$$0.1 \times 38 = \frac{1}{10} \times 38 = \frac{38}{10} = 3.8$$

$$0.01 \times 38 = \frac{1}{100} \times 38 = \frac{38}{100} = 0.38$$

$$0.001 \times 38 = \frac{1}{1000} \times 38 = \frac{38}{1000} = 0.038$$

$$0.0001 \times 38 = \frac{1}{10{,}000} \times 38 = \frac{38}{10{,}000} = 0.0038$$

To multiply a whole number by a tenth, hundredth, or thousandth:

a) count the number of decimal places, and — **0.01 × 765** → **2 places**

b) move the decimal point of the whole number that many places to the left. — **0.01 × 765 = 7.65.** Move 2 places to the left.

▶ **EXAMPLES** Multiply.

1. 0.1 × 45 = 4.5
2. 0.01 × 2346 = 23.46
3. 0.001 × 347 = 0.347
4. 0.0001 × 2346 = 0.2346 ◀

DO EXERCISES 1–4.

Let's find the product

$$2.3 \times 1.12.$$

To understand how we find such products, we convert to fractional notation:

$$2.3 \times 1.12 = \frac{23}{10} \times \frac{112}{100} = \frac{23 \times 112}{10 \times 100} = \frac{2576}{1000} = 2.576.$$

We multiply the whole numbers 23 and 112, and then divide by 1000. Note the number of decimal places.

```
  1.1 2    (2 decimal place)
×   2.3    (1 decimal place)
-------
2.5 7 6    (3 decimal places)
```

OBJECTIVES

After finishing Section 5.1, you should be able to:

a Multiply using decimal notation.

b Convert from dollars to cents and cents to dollars, and from notation like 45.7 million to standard notation.

FOR EXTRA HELP

Tape 7A   Tape 7B

MAC: 5
IBM: 5

Multiply.

1. 0.1 × 3482   348.2

2. 0.01 × 3482   34.82

3. 0.001 × 3482   3.482

4. 0.0001 × 3482   0.3482

ANSWERS ON PAGE A-3

5. Multiply.

$$\begin{array}{r} 85.4 \\ \times\ 6.2 \\ \hline \end{array}$$

529.48

To multiply using decimals: **0.8 × 0.43**

a) Ignore the decimal points and multiply as though both factors were whole numbers.

$$\begin{array}{r} {}^{2} \\ 0.43 \\ \times\ 0.8 \\ \hline 344 \end{array}$$

Ignore the decimal points for now.

b) Then place the decimal point in the result. The number of decimal places in the product is the sum of the numbers of places in the factors (count places from the right).

$$\begin{array}{rl} 0.43 & \text{(2 decimal places)} \\ \times\ 0.8 & \text{(1 decimal place)} \\ \hline 0.344 & \text{(3 decimal places)} \end{array}$$

▶ **EXAMPLE 5** Multiply: 8.3 × 74.6.

a) Ignore the decimal points and multiply as though factors were whole numbers:

$$\begin{array}{r} {}^{3\,4} \\ {}^{1\,1} \\ 74.6 \\ \times\ 8.3 \\ \hline 2238 \\ 59680 \\ \hline 61918 \end{array}$$

b) Place the decimal point in the result. The number of decimal places in the product is the sum, 1 + 1, of the number of places in the factors.

$$\begin{array}{rl} 74.6 & \text{(1 decimal place)} \\ \times\ 8.3 & \text{(1 decimal place)} \\ \hline 2238 & \\ 59680 & \downarrow \\ \hline 619.18 & \text{(2 decimal places)} \end{array}$$ ◀

DO EXERCISE 5.

▶ **EXAMPLE 6** Multiply: 0.0032 × 2148.

As we catch on to the skill, we can combine the two steps.

$$\begin{array}{rl} 2148 & \text{(0 decimal places)} \\ \times\ 0.0032 & \text{(4 decimal places)} \\ \hline 4296 & \\ 64440 & \downarrow \\ \hline 6.8736 & \text{(4 decimal places)} \end{array}$$ ◀

ANSWER ON PAGE A-3

▶ **EXAMPLE 7** Multiply: 0.14×0.867.

$$\begin{array}{rl} 0.8\,6\,7 & \text{(3 decimal places)} \\ \times \quad 0.1\,4 & \text{(2 decimal places)} \\ \hline 3\,4\,6\,8 & \\ 8\,6\,7\,0 & \downarrow \\ \hline 0.1\,2\,1\,3\,8 & \text{(5 decimal places)} \end{array}$$

◀

DO EXERCISES 6 AND 7.

▶ **EXAMPLE 8** Multiply: 100×8.415.

$$\begin{array}{rl} 8.4\,1\,5 & \text{(3 decimal places)} \\ \times \quad 1\,0\,0 & \text{(0 decimal places)} \\ \hline 8\,4\,1.5\,0\,0 & \text{(3 decimal places)} \end{array}$$

◀

▶ **EXAMPLE 9** Multiply: 0.001×97.04.

$$\begin{array}{rl} 9\,7.0\,4 & \text{(2 decimal places)} \\ \times \quad 0.0\,0\,1 & \text{(3 decimal places)} \\ \hline 0.0\,9\,7\,0\,4 & \text{(5 decimal places)} \end{array}$$

◀

DO EXERCISES 8 AND 9.

To multiply any number by a power of ten such as 10, 100, 1000, and so on:

a) count the number of zeros, and

1000 × 34.45678
(1000 → 3 zeros)

b) move the decimal point that many places to the right.

1000 × 34.45678 = 34.456.78
Move 3 places to the right.

To multiply any number by a tenth, hundredth, or thousandth:

a) count the number of decimal places in the tenth, hundredth, or thousandth, and

0.001 × 34.45678
(0.001 → 3 places)

b) move the decimal point that many places to the left.

0.001 × 34.45678 = 0.034.45678
Move 3 places to the left.

Multiply.

6. 1234×0.0041 5.0594

7. 42.65×0.804 34.2906

Multiply.

8. 1000×8.415 8415

9. 0.01×5.6 0.056

ANSWERS ON PAGE A-3

Multiply.

10. 100×345.906 34,590.6

11. 0.01×345.906 3.45906

12. 0.001×0.73 0.00073

13. 1000×0.73 730

Convert from dollars to cents.

14. \$15.69 1569¢

15. \$0.17 17¢

ANSWERS ON PAGE A-3

▶ **EXAMPLE 10** Multiply: 10×0.037.

10×0.037, 0.0.37, $10 \times 0.037 = 0.37$

Move 1 place to the right. ◀

▶ **EXAMPLE 11** Multiply: 0.01×0.037.

0.01×0.037, 0.00.037, $0.01 \times 0.037 = 0.00037$

Move 2 places to the left. ◀

DO EXERCISES 10–13.

b Applications

Converting from dollars to cents is like multiplying by 100. To see why, consider \$19.43.

\$19.43 = 19.43 × \$1   **We think of \$19.43 as 19.43 × 1 dollar, or 19.43 × \$1.**

= 19.43 × 100¢   **Substituting 100¢ for \$1: \$1 = 100¢**

= 1943¢   **Multiplying**

To convert from dollars to cents, move the decimal point two places to the right and change from the \$ sign in front to the ¢ sign at the end.

▶ **EXAMPLES** Convert from dollars to cents.

12. \$189.64 = 18,964¢

13. \$0.75 = 75¢ ◀

DO EXERCISES 14 AND 15.

Converting from cents to dollars is like multiplying by 0.01. To see why, consider 65¢.

65¢ = 65 × 1¢   **We think of 65¢ as 65 × 1 cent, or 65 × 1¢.**

= 65 × \$0.01   **Substituting \$0.01 for 1¢: 1¢ = \$0.01**

= \$0.65   **Multiplying**

To convert from cents to dollars, move the decimal point two places to the left and change from the ¢ sign at the end to the \$ sign at the front.

▶ **EXAMPLES** Convert from cents to dollars.

14. 395¢ = \$3.95

15. 8503¢ = \$85.03 ◀

DO EXERCISES 16 AND 17.

We often see notation like the following in newspapers and magazines and on television.

O'Hare International Airport handles 45.7 million passengers per year.

Americans drink 17 million gallons of coffee each day.

The population of the world is 5.1 billion.

To understand such notation, it helps to consider the following table.

1 hundred = 100 = 10^2 → 2 zeros

1 thousand = 1000 = 10^3 → 3 zeros

1 million = 1,000,000 = 10^6 → 6 zeros

1 billion = 1,000,000,000 = 10^9 → 9 zeros

1 trillion = 1,000,000,000,000 = 10^{12} → 12 zeros

To convert to standard notation, we proceed as follows.

▶ **EXAMPLE 16** Convert the number in this sentence to standard notation: O'Hare handles 45.7 million passengers per year.

$$45.7 \text{ million} = 45.7 \times 1 \text{ million}$$
$$= 45.7 \times 1{,}000{,}000 = 45{,}700{,}000$$ ◀

DO EXERCISES 18 AND 19.

Convert from cents to dollars.

16. 35¢ \$0.35

17. 577¢ \$5.77

Convert the number in the sentence to standard notation.

18. Americans drink 17 million gallons of coffee each day.
17,000,000

19. The population of the world is 5.1 billion. 5,100,000,000

ANSWERS ON PAGE A-3

❖ SIDELIGHTS

Calculator Corner: Number Patterns

There are many interesting number patterns in mathematics. Look for a pattern in the following. We can use a calculator for the computations.

$$6^2 = 36$$
$$66^2 = 4356$$
$$666^2 = 443556$$
$$6666^2 = 44435556$$

Do you see a pattern? If so, find 66666^2 without the use of your calculator.

EXERCISES

In each of the following, do the first four calculations using your calculator. Look for a pattern. Use the pattern to do the last calculation without the use of your calculator.

1. 3^2 9
33^2 1089
333^2 110889
3333^2 11108889
33333^2 1111088889

2. $9 \cdot 6$ 54
$99 \cdot 66$ 6534
$999 \cdot 666$ 665334
$9999 \cdot 6666$ 66653334
$99999 \cdot 66666$ 6666533334

3. $37 \cdot 3$ 111
$37 \cdot 33$ 1221
$37 \cdot 333$ 12321
$37 \cdot 3333$ 123321
$37 \cdot 33333$ 1233321

4. $37 \cdot 3$ 111
$37 \cdot 6$ 222
$37 \cdot 9$ 333
$37 \cdot 12$ 444
$37 \cdot 15$ 555

5. $(9 \cdot 9) + 7$ 88
$(98 \cdot 9) + 6$ 888
$(987 \cdot 9) + 5$ 8888
$(9876 \cdot 9) + 4$ 88888
$(98765 \cdot 9) + 3$ 888888

6. $(1 \cdot 8) + 1$ 9
$(12 \cdot 8) + 2$ 98
$(123 \cdot 8) + 3$ 987
$(1234 \cdot 8) + 4$ 9876
$(12345 \cdot 8) + 5$ 98765

7. $8 \cdot 6$ 48
$68 \cdot 6$ 408
$668 \cdot 6$ 4008
$6668 \cdot 6$ 40008
$66668 \cdot 6$ 400008

8. $6 \cdot 4$ 24
$66 \cdot 44$ 2904
$666 \cdot 444$ 295704
$6666 \cdot 4444$ 29623704
$66666 \cdot 44444$ 2962903704

9. 1^2 1
11^2 121
111^2 12321
1111^2 1234321
11111^2 123454321

10. 9^2 81
99^2 9801
999^2 998001
9999^2 99980001
99999^2 9999800001

11. $77 \cdot 78$ 6006
$777 \cdot 78$ 60606
$7777 \cdot 78$ 606606
$77777 \cdot 78$ 6066606
$777777 \cdot 78$ 60666606

12. $1 \cdot 13 \cdot 76{,}923$ 999999
$2 \cdot 13 \cdot 76{,}923$ 1999998
$3 \cdot 13 \cdot 76{,}923$ 2999997
$4 \cdot 13 \cdot 76{,}923$ 3999996
$5 \cdot 13 \cdot 76{,}923$ 4999995

NAME SECTION DATE

EXERCISE SET 5.1

a Multiply.

1. 8.6×7
2. 47×0.9
3. 0.84×8
4. 7.3×0.6
5. 6.3×0.04
6. 7.8×0.09
7. 87×0.006
8. 8.7×0.06
9. 10×23.76
10. 100×2.8793
11. 1000×783.686852
12. 0.34×1000
13. 7.8×100
14. 0.00238×10
15. 0.1×89.23
16. 0.01×789.235
17. 0.001×97.68
18. 8976.23×0.001
19. 78.2×0.01
20. 0.0235×0.1
21. 32.6×16
22. 7.28×5.4
23. 0.984×3.3
24. 7.489×8.2
25. 374×2.4
26. 569×1.05
27. 749×0.43
28. 876×20.4
29. 0.87×64
30. 7.25×60
31. 46.50×75
32. 8.24×703
33. 81.7×0.612
34. 31.82×7.15
35. 10.105×11.324
36. 151.2×4.555

ANSWERS

1. 60.2
2. 42.3
3. 6.72
4. 4.38
5. 0.252
6. 0.702
7. 0.522
8. 0.522
9. 237.6
10. 287.93
11. 783,686.852
12. 340
13. 780
14. 0.0238
15. 8.923
16. 7.89235
17. 0.09768
18. 8.97623
19. 0.782
20. 0.00235
21. 521.6
22. 39.312
23. 3.2472
24. 61.4098
25. 897.6
26. 597.45
27. 322.07
28. 17,870.4
29. 55.68
30. 435
31. 3487.5
32. 5792.72
33. 50.0004
34. 227.513
35. 114.42902
36. 688.716

ANSWERS

37. 789
38. 0.1155
39. 13.284
40. 0.18768
41. 90.72
42. 0.060675
43. 0.0028728
44. 7.62128
45. 0.72523
46. 2.701644
47. 1.872115
48. 0.00836608
49. 45,678
50. 0.045678
51. 4567.8
52. 0.45678
53. 2888¢
54. 6743¢
55. 66¢
56. 178¢
57. $0.34
58. $0.95
59. $34.45
60. $9.33
61. 2,830,000,000,000
62. 5,800,000
63. $11\frac{1}{5}$
64. $\frac{35}{72}$
65. 342
66. 87
67. 10^{21}
68. 10^{15}

37. 0.789×1000

38. 0.231×0.5

39. 12.3×1.08

40. 7.82×0.024

41. 32.4×2.8

42. 8.09×0.0075

43. 0.00342×0.84

44. 2.0056×3.8

45. 0.347×2.09

46. 2.532×1.067

47. 3.005×0.623

48. 16.34×0.000512

49. 45.678×1000

50. 0.001×45.678

51. 100×45.678

52. 0.01×45.678

b Convert from dollars to cents.

53. $28.88 **54.** $67.43 **55.** $0.66 **56.** $1.78

Convert from cents to dollars.

57. 34¢ **58.** 95¢ **59.** 3445¢ **60.** 933¢

Convert the number in the sentence to standard notation.

61. The national debt was $2.83 trillion.

62. We eat 5.8 million pounds of chocolate each day.

SKILL MAINTENANCE

63. Multiply: $2\frac{1}{3} \cdot 4\frac{4}{5}$.

64. Divide: $2\frac{1}{3} \div 4\frac{4}{5}$.

Divide.

65. $24\overline{)8208}$

66. $4\overline{)348}$

SYNTHESIS

Express as a power of 10.

67. (1 trillion) · (1 billion)

68. (1 million) · (1 billion)

5.2 Division with Decimal Notation

a Division

Whole-Number Divisors

Compare these divisions.

$$\frac{588}{7} = 84$$

$$\frac{58.8}{7} = 8.4$$

$$\frac{5.88}{7} = 0.84$$

$$\frac{0.588}{7} = 0.084$$

The number of decimal places in the quotient is the same as the number of decimal places in the dividend.

These lead us to the following method for dividing by a whole number.

> **To divide by a whole number:**
>
> **a) place the decimal point directly above the decimal point in the dividend, and**
>
> **b) divide as though dividing whole numbers.**

```
    0.8 4
7 )5.8 8
   5 6 0
     2 8
     2 8
       0
```

▶ **EXAMPLE 1** Divide: 82.08 ÷ 24.

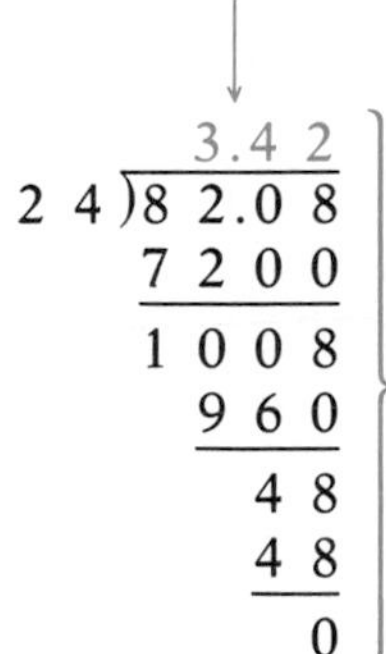

Place the decimal point.

Divide as though dividing whole numbers. ◀

DO EXERCISES 1–3.

OBJECTIVES

After finishing Section 5.2, you should be able to:

a Divide using decimal notation.

b Solve equations of the type $a \cdot x = b$, where a and b may be in decimal notation.

c Simplify expressions using the rules for order of operations.

FOR EXTRA HELP

Tape 7B

Tape 7B

MAC: 5
IBM: 5

Divide.

1. $9\overline{)5.4}$ 0.6

2. $15\overline{)22.5}$ 1.5

3. $82\overline{)38.54}$ 0.47

ANSWERS ON PAGE A-3

Divide.

4. $25\overline{)8}$ 0.32

5. $4\overline{)15}$ 3.75

6. $86\overline{)21.5}$ 0.25

ANSWERS ON PAGE A-3

Extra Zeros

Sometimes it helps to write some extra zeros to the right of the decimal point. They don't change the number.

▶ **EXAMPLE 2** Divide: 30 ÷ 8.

```
      3.
 8 )3 0.
    2 4
      6
```

Place the decimal point and divide to find how many ones.

```
      3.
 8 )3 0.0
    2 4 ↓
      6 0
```

Write an extra zero.

```
      3.7
 8 )3 0.0
    2 4
      6 0
      5 6
        4
```

Divide to find how many tenths.

```
      3.7
 8 )3 0.0 0
    2 4   |
      6 0 |
      5 6 ↓
        4 0
```

Write an extra zero.

```
      3.7 5
 8 )3 0.0 0
    2 4
      6 0
      5 6
        4 0
        4 0
          0
```

Divide to find how many hundredths. ◀

▶ **EXAMPLE 3** Divide: 4 ÷ 25.

```
       0.1 6
 2 5 )4.0 0
       2 5
       1 5 0
       1 5 0
           0
```

◀

DO EXERCISES 4–6.

Divisors That Are Not Whole Numbers

Consider the division

$$0.24\overline{)8.208}$$

We write the division as $\frac{8.208}{0.24}$. Then we multiply by 1 to change to a whole-number divisor:

$$\frac{8.208}{0.24} = \frac{8.208}{0.24} \times \frac{100}{100} = \frac{820.8}{24}.$$

The divisor is now a whole number. The division

$$0.24\overline{)8.208}$$

is the same as

$$24\overline{)820.8}$$

To divide when the divisor is not a whole number:

a) move the decimal point (multiply by 10, 100, and so on) to make the divisor a whole number;

$0.24\overline{)8.208}$ **Move 2 places to the right.**

b) move the decimal point (multiply the same way) in the dividend the same number of places; and

$0.24\overline{)8.208}$ **Move 2 places to the right.**

c) place the decimal point directly above the decimal point in the dividend and divide as though dividing whole numbers.

$$\begin{array}{r} 34.2 \\ 0.24_\wedge\overline{)8.20_\wedge 8} \\ 7200 \\ \hline 1008 \\ 960 \\ \hline 48 \\ 48 \\ \hline 0 \end{array}$$

(The new decimal point in the dividend is indicated by a caret.)

▶ **EXAMPLE 4** Divide: 5.848 ÷ 8.6.

$8.6\overline{)5.848}$ **Multiply the divisor by 10 (move the decimal point 1 place). Multiply the same way in the dividend (move 1 place).**

$$\begin{array}{r} 0.68 \\ 8.6_\wedge\overline{)5.8_\wedge 48} \\ 5160 \\ \hline 688 \\ 688 \\ \hline 0 \end{array}$$

Then divide. ◀

DO EXERCISES 7–9.

7. a) Complete.

$$\frac{3.75}{0.25} = \frac{3.75}{0.25} \times \frac{100}{100} = \frac{(\quad)}{25} \quad 375$$

b) Divide.

$0.25\overline{)3.75}$ 15

Divide.

8. $0.83\overline{)4.067}$ 4.9

9. $3.5\overline{)44.8}$ 12.8

ANSWERS ON PAGE A-3

10. Divide.

$1.6\overline{)25}$ 15.625

ANSWER ON PAGE A-3

EXAMPLE 5 Divide: $12 \div 0.64$.

$0.64\overline{)12.}$ Put a decimal point at the end of the whole number.

$0.64\overline{)12.00}$ Multiply the divisor by 100 (move the decimal point 2 places). Multiply the same way in the dividend (move 2 places).

$$
\begin{array}{r}
18.75 \\
0.64\overline{)12.00_{\wedge}00} \\
\underline{640} \\
560 \\
\underline{512} \\
480 \\
\underline{448} \\
320 \\
\underline{320} \\
0
\end{array}
$$

Then divide.

DO EXERCISE 10.

It is often helpful to be able to divide quickly by a ten, hundred, or thousand, or by a tenth, hundredth, or thousandth. The procedure we use is based on multiplying by 1. Consider the following examples:

$$\frac{23.789}{1000} = \frac{23.789}{1000} \cdot \frac{1000}{1000} = \frac{23{,}789}{1{,}000{,}000} = 0.023789;$$

$$\frac{23.789}{0.01} = \frac{23.789}{0.01} \cdot \frac{100}{100} = \frac{2378.9}{1} = 2378.9.$$

We use the following procedure.

To divide by a power of ten, such as 10, 100, or 1000, and so on:

a) count the number of zeros in the divisor, and $\frac{713.49}{100}$ → **2 zeros**

b) move the decimal point that number of places to the left.

$\frac{713.49}{100}$, 7.13.49 $\frac{713.49}{100} = 7.1349$

2 places to the left

To divide by a tenth, hundredth, or thousandth:

a) count the number of decimal places in the divisor, and $\frac{713.49}{0.001}$ → **3 places**

b) move the decimal point that number of places to the right.

$\frac{713.49}{0.001}$, 713.490. $\frac{713.49}{0.001} = 713{,}490$

3 places to the right

▶ **EXAMPLE 6** Divide: $\frac{0.0104}{10}$.

$\frac{0.0104}{10}$, 0.0.0104, $\frac{0.0104}{10} = 0.00104$

1 place to the left to change 10 to 1. ◀

▶ **EXAMPLE 7** Divide: $\frac{23.738}{0.001}$.

$\frac{23.738}{0.001}$, 23.738. $\frac{23.738}{0.001} = 23{,}738$

3 places to the right to change 0.001 to 1. ◀

DO EXERCISES 11–14.

b Solving Equations

Now let us solve equations of the type $a \cdot x = b$, where a and b may be in decimal notation. Proceeding as we have before, we divide on both sides by a.

▶ **EXAMPLE 8** Solve: $2.9 \times t = 0.14616$.

We have

$$\frac{2.9 \times t}{2.9} = \frac{0.14616}{2.9} \quad \text{Dividing on both sides by 2.9}$$

$$t = 0.0504.$$

```
       0.0 5 0 4
2.9 )0.1 4 6 1 6
       1 4 5 0 0
       ---------
           1 1 6
           1 1 6
           -----
               0
```

The solution is 0.0504. ◀

DO EXERCISES 15 AND 16.

c Order of Operations: Decimal Notation

The same rules for order of operations used with whole numbers apply when simplifying expressions involving decimal notation. For review, we list these rules.

> **Rules for Order of Operations**
>
> 1. **Do all calculations within parentheses before operations outside.**
> 2. **Evaluate all exponential expressions.**
> 3. **Do all multiplications and divisions in order from left to right.**
> 4. **Do all additions and subtractions in order from left to right.**

Divide.

11. $\frac{0.1278}{0.01}$ 12.78

12. $\frac{0.1278}{100}$ 0.001278

13. $\frac{98.47}{1000}$ 0.09847

14. $\frac{6.7832}{0.1}$ 67.832

Solve.

15. $100 \cdot x = 78.314$ 0.78314

16. $0.25 \cdot y = 276.4$ 1105.6

ANSWERS ON PAGE A-3

Simplify.

17. $0.25 \cdot (1 + 0.08) - 0.0274$

0.2426

18. $20^2 - 3.4^2 + \{0.25[100(9.2 - 5.6)] + 5(100 - 50)\}$ 728.44

ANSWERS ON PAGE A-3

▶ **EXAMPLE 9** Simplify: $(5 - 0.06) \div 2 + 3.42 \times 0.1$.

$(5 - 0.06) \div 2 + 3.42 \times 0.1 = 4.94 \div 2 + 3.42 \times 0.1$ **Carrying out operations inside parentheses**

$= 2.47 + 0.342$ **Doing all multiplications and divisions in order from left to right**

$= 2.812$ ◀

▶ **EXAMPLE 10** Simplify: $10^2 \times \{[(3 - 0.24) \div 2.4] - (0.21 - 0.092)\}$.

$10^2 \times \{[(3 - 0.24) \div 2.4] - (0.21 - 0.092)\}$

$= 10^2 \times \{[2.76 \div 2.4] - 0.118\}$ **Doing the calculations in the innermost parentheses first**

$= 10^2 \times \{1.15 - 0.118\}$ **Again, doing the calculations in the innermost parentheses**

$= 10^2 \times 1.032$ **Subtracting inside the parentheses**

$= 100 \times 1.032$ **Evaluating the exponential expression**

$= 103.2$ ◀

DO EXERCISES 17 AND 18.

NAME SECTION DATE

EXERCISE SET 5.2

a Divide.

1. $2\overline{)5.98}$
2. $5\overline{)16}$
3. $4\overline{)95.12}$
4. $8\overline{)25.92}$
5. $12\overline{)89.76}$
6. $21\overline{)22.89}$
7. $33\overline{)237.6}$
8. $9.3 \div 3$
9. $9.144 \div 8$
10. $3.6 \div 4$
11. $12.123 \div 3$
12. $6\overline{)5.4}$
13. $5\overline{)0.35}$
14. $0.04\overline{)1.68}$
15. $0.12\overline{)8.4}$
16. $0.36\overline{)2.88}$
17. $3.4\overline{)68}$
18. $0.25\overline{)5}$
19. $15\overline{)6}$
20. $12\overline{)1.8}$
21. $36\overline{)14.76}$

ANSWERS

1. 2.99
2. 3.2
3. 23.78
4. 3.24
5. 7.48
6. 1.09
7. 7.2
8. 3.1
9. 1.143
10. 0.9
11. 4.041
12. 0.9
13. 0.07
14. 42
15. 70
16. 8
17. 20
18. 20
19. 0.4
20. 0.15
21. 0.41

ANSWERS

22. 2.3
23. 8.5
24. 3.2
25. 9.3
26. 0.023
27. 0.625
28. 0.875
29. 0.26
30. 0.5
31. 15.625
32. 225
33. 2.34
34. 93
35. 0.47
36. 0.0043
37. 0.2134567
38. 2.134567
39. 21.34567

22. $52\overline{)119.6}$

23. $3.2\overline{)27.2}$

24. $8.5\overline{)27.2}$

25. $4.2\overline{)39.06}$

26. $3.6\overline{)0.0828}$

27. $8\overline{)5}$

28. $8\overline{)7}$

29. $0.47\overline{)0.1222}$

30. $0.54\overline{)0.27}$

31. $4.8\overline{)75}$

32. $0.28\overline{)63}$

33. $0.032\overline{)0.07488}$

34. $0.017\overline{)1.581}$

35. $82\overline{)38.54}$

36. $34\overline{)0.1462}$

37. $\frac{213.4567}{1000}$

38. $\frac{213.4567}{100}$

39. $\frac{213.4567}{10}$

40. $\frac{100.7604}{0.1}$

41. $\frac{1.0237}{0.001}$

42. $\frac{1.0237}{0.01}$

43. $\frac{56.78}{0.001}$

44. $\frac{0.5668}{1000}$

b Solve.

45. $4.2 \cdot x = 39.06$

46. $36 \cdot y = 14.76$

47. $1000 \cdot y = 9.0678$

48. $789.23 = 0.25 \cdot q$

c Simplify.

49. $14 \times (82.6 + 67.9)$

50. $(26.2 - 14.8) \times 12$

51. $0.003 + 3.03 \div 0.01$

52. $9.94 + 4.26 \div (6.02 - 4.6) - 0.9$

53. $42 \times (10.6 + 0.024)$

54. $(18.6 - 4.9) \times 13$

55. $4.2 \times 5.7 + 0.7 \div 3.5$

56. $123.3 - 4.24 \times 1.01$

ANSWERS

40. 1007.604

41. 1023.7

42. 102.37

43. 56,780

44. 0.0005668

45. 9.3

46. 0.41

47. 0.0090678

48. 3156.92

49. 2107

50. 136.8

51. 303.003

52. 12.04

53. 446.208

54. 178.1

55. 24.14

56. 119.0176

ANSWERS

57. 13.0072

58. 399.9892

59. 19.3204

60. 0.7125

61. 96.13

62. 7.3

63. 10.49

64. 5.417

65. 911.13

66. 9.5

67. 205

68. 1.206

69. $15\frac{1}{8}$

70. $5\frac{7}{8}$

71. $\frac{6}{7}$

72. $2 \cdot 3 \cdot 3 \cdot 3 \cdot 3$

57. $9.0072 + 0.04 \div 0.1^2$

58. $12 \div 0.03 - 12 \times 0.03^2$

59. $(8 - 0.04)^2 \div 4 + 8.7 \times 0.4$

60. $(5 - 2.5)^2 \div 100 + 0.1 \times 6.5$

61. $86.13 + 95.7 \div (9.6 - 0.03)$

62. $2.48 \div (1 - 0.504) + 24.3 - 11 \times 2$

63. $4 \div 0.4 + 0.1 \times 5 - 0.1^2$

64. $6 \times 0.9 + 0.1 \div 4 - 0.2^3$

65. $5.5^2 \times [(6 - 4.2) \div 0.06 + 0.12]$

66. $12^2 \div (12 + 2.4) - [(2 - 1.6) \div 0.8]$

67. $200 \times \{[(4 - 0.25) \div 2.5] - (4.5 - 4.025)\}$

68. $0.03 \times \{1 \times 50.2 - [(8 - 7.5) \div 0.05]\}$

SKILL MAINTENANCE

69. Add: $10\frac{1}{2} + 4\frac{5}{8}$.

70. Subtract: $10\frac{1}{2} - 4\frac{5}{8}$.

71. Simplify: $\frac{36}{42}$.

72. Find the prime factorization of 162.

5.3 Converting from Fractional Notation to Decimal Notation

a Fractional Notation to Decimal Notation

When a denominator has no prime factors other than 2's and 5's, we can find decimal notation by multiplying by 1. We multiply to get a denominator that is a power of ten like 10, 100, or 1000.

▶ **EXAMPLE 1** Find decimal notation for $\frac{3}{5}$.

$$\frac{3}{5} = \frac{3}{5}\cdot\frac{2}{2} = \frac{6}{10} = 0.6$$ We use $\frac{2}{2}$ for 1 to get a denominator of 10. ◀

▶ **EXAMPLE 2** Find decimal notation for $\frac{7}{20}$.

$$\frac{7}{20} = \frac{7}{20}\cdot\frac{5}{5} = \frac{35}{100} = 0.35$$ We use $\frac{5}{5}$ for 1 to get a denominator of 100. ◀

▶ **EXAMPLE 3** Find decimal notation for $\frac{9}{40}$.

$$\frac{9}{40} = \frac{9}{40}\cdot\frac{25}{25} = \frac{225}{1000} = 0.225$$ We use $\frac{25}{25}$ for 1 to get a denominator of 1000. ◀

▶ **EXAMPLE 4** Find decimal notation for $\frac{87}{25}$.

$$\frac{87}{25} = \frac{87}{25}\cdot\frac{4}{4} = \frac{348}{100} = 3.48$$ We use $\frac{4}{4}$ for 1 to get a denominator of 100. ◀

DO EXERCISES 1–4.

We can also divide to find decimal notation.

▶ **EXAMPLE 5** Find decimal notation for $\frac{3}{5}$.

$$\frac{3}{5} = 3 \div 5 \qquad \begin{array}{r} 0.6 \\ 5\overline{)3.0} \\ \underline{3\ 0} \\ 0 \end{array} \qquad \frac{3}{5} = 0.6$$ ◀

▶ **EXAMPLE 6** Find decimal notation for $\frac{7}{8}$.

$$\frac{7}{8} = 7 \div 8 \qquad \begin{array}{r} 0.8\ 7\ 5 \\ 8\overline{)7.0\ 0\ 0} \\ \underline{6\ 4} \\ 6\ 0 \\ \underline{5\ 6} \\ 4\ 0 \\ \underline{4\ 0} \\ 0 \end{array} \qquad \frac{7}{8} = 0.875$$ ◀

DO EXERCISES 5 AND 6.

OBJECTIVES

After finishing Section 5.3, you should be able to:

- a Convert from fractional notation to decimal notation.
- b Round numbers named by repeating decimals.
- c Calculate using fractional and decimal notation together.

FOR EXTRA HELP

Tape 7C

Tape 8A

MAC: 5
IBM: 5

Find decimal notation. Use multiplying by 1.

1. $\frac{4}{5}$ 0.8

2. $\frac{9}{20}$ 0.45

3. $\frac{11}{40}$ 0.275

4. $\frac{33}{25}$ 1.32

Find decimal notation.

5. $\frac{2}{5}$ 0.4

6. $\frac{3}{8}$ 0.375

ANSWERS ON PAGE A-3

Find decimal notation.

7. $\frac{1}{6}$ $0.1\overline{6}$

8. $\frac{2}{3}$ $0.\overline{6}$

ANSWERS ON PAGE A-3

In Examples 5 and 6, the division **terminated,** meaning that eventually we got a remainder of 0. A terminating decimal occurs when the denominator has only 2's or 5's, or both, as factors. This assumes that the fractional notation has been simplified.

Consider a different situation:

$$\frac{5}{6} \quad \text{or} \quad \frac{5}{2 \cdot 3}.$$

Since 6 has a 3 as a factor, the division will not terminate. We can still use division to get decimal notation, but answers will be repeating decimals, as follows.

▶ **EXAMPLE 7** Find decimal notation for $\frac{5}{6}$.

We have

$$\frac{5}{6} = 5 \div 6 \qquad \begin{array}{r} 0.8\;3\;3 \\ 6\,\overline{)\,5.0\;0\;0} \\ \underline{4.8} \\ 2\;0 \\ \underline{1\;8} \\ 2\;0 \\ \underline{1\;8} \\ 2 \end{array}$$

Since 2 keeps reappearing as a remainder, the digits repeat and will continue to do so; therefore,

$$\frac{5}{6} = 0.83333\ldots.$$

The dots indicate an endless sequence of digits in the quotient. When there is a repeating pattern, the dots are often replaced by a bar to indicate the repeating part—in this case, only the 3:

$$\frac{5}{6} = 0.8\overline{3}.$$ ◀

DO EXERCISES 7 AND 8.

▶ **EXAMPLE 8** Find decimal notation for $\frac{4}{11}$.

$$\frac{4}{11} = 4 \div 11 \qquad \begin{array}{r} 0.3\;6\;3\;6 \\ 11\,\overline{)\,4.0\;0\;0\;0} \\ \underline{3\;3} \\ 7\;0 \\ \underline{6\;6} \\ 4\;0 \\ \underline{3\;3} \\ 7\;0 \\ \underline{6\;6} \\ 4 \end{array}$$

Since 7 and 4 keep reappearing as remainders, the sequence of digits "36" repeats in the quotient, and

$$\frac{4}{11} = 0.363636\ldots, \quad \text{or} \quad 0.\overline{36}. \quad ◀$$

DO EXERCISES 9 AND 10.

▶ **EXAMPLE 9** Find decimal notation for $\frac{5}{7}$.

We have

$$\begin{array}{r} 0.714285 \\ 7\overline{)5.000000} \\ \underline{4\;9} \\ 1\;0 \\ \underline{7} \\ 3\;0 \\ \underline{2\;8} \\ 2\;0 \\ \underline{1\;4} \\ 6\;0 \\ \underline{5\;6} \\ 4\;0 \\ \underline{3\;5} \\ 5 \end{array}$$

Since 5 appears as a remainder, the sequence of digits "714285" repeats in the quotient, and

$$\frac{5}{7} = 0.714285714285\ldots, \quad \text{or} \quad 0.\overline{714285}. \quad ◀$$

The length of a repeating part can be very long—too long to find on a calculator. An example is $\frac{5}{97}$, which has a repeating part with 96 digits.

DO EXERCISE 11.

b Rounding in Problem Solving

In applied problems, repeating decimals are rounded to get approximate answers.

▶ **EXAMPLES** Round each to the nearest tenth, hundredth, and thousandth.

	Nearest tenth	*Nearest hundredth*	*Nearest thousandth*
10. $0.8\overline{3} = 0.83333\ldots$	0.8	0.83	0.833
11. $0.\overline{09} = 0.090909\ldots$	0.1	0.09	0.091
12. $0.\overline{714285} = 0.714285714285\ldots$	0.7	0.71	0.714 ◀

DO EXERCISES 12–14.

Find decimal notation.

9. $\frac{5}{11}$ $0.\overline{45}$

10. $\frac{12}{11}$ $1.\overline{09}$

11. Find decimal notation for $\frac{3}{7}$.

$0.\overline{428571}$

Round each to the nearest tenth, hundredth, and thousandth.

12. $0.\overline{6}$ 0.7; 0.67; 0.667

13. $0.\overline{808}$ 0.8; 0.81; 0.808

14. $6.\overline{245}$ 6.2; 6.25; 6.245

ANSWERS ON PAGE A-3

Calculate.

15. $\frac{5}{6} \times 0.864$ 0.72

16. $\frac{1}{3} \times 0.384 + \frac{5}{8} \times 0.6784$ 0.552

ANSWERS ON PAGE A-3

C Calculations with Fractional and Decimal Notation Together

In certain kinds of calculations, fractional and decimal notation might occur together. In such cases, there are at least three ways in which we might proceed.

▶ **EXAMPLE 13** Calculate: $\frac{2}{3} \times 0.576$.

Method 1. One way to do this calculation is to convert the decimal notation to fractional notation so that both numbers are in fractional notation. The answer can be left in fractional notation and simplified, or we can convert back to decimal notation and round, if appropriate.

$$\begin{aligned}\frac{2}{3} \times 0.576 &= \frac{2}{3} \cdot \frac{576}{1000} = \frac{2 \cdot 576}{3 \cdot 1000} \\ &= \frac{2 \cdot 2 \cdot 2 \cdot 2 \cdot 2 \cdot 2 \cdot 2 \cdot 3 \cdot 3}{2 \cdot 2 \cdot 2 \cdot 3 \cdot 5 \cdot 5 \cdot 5} \\ &= \frac{2 \cdot 2 \cdot 2 \cdot 3}{2 \cdot 2 \cdot 2 \cdot 3} \cdot \frac{2 \cdot 2 \cdot 2 \cdot 2 \cdot 3}{5 \cdot 5 \cdot 5} \\ &= 1 \cdot \frac{2 \cdot 2 \cdot 2 \cdot 2 \cdot 3}{5 \cdot 5 \cdot 5} \\ &= \frac{2 \cdot 2 \cdot 2 \cdot 2 \cdot 3}{5 \cdot 5 \cdot 5} = \frac{48}{125}, \text{ or } 0.384.\end{aligned}$$

Method 2. A second way to do this calculation is to convert the fractional notation to decimal notation so that both numbers are in decimal notation. Since $\frac{2}{3}$ converts to repeating decimal notation, it is first rounded to some chosen decimal place. We choose three decimal places. Then, using decimal notation, we multiply. Note that the answer is not as accurate as that found by method 1, due to the rounding.

$$\frac{2}{3} \times 0.576 = 0.\overline{6} \times 0.576 \approx 0.667 \times 0.576 = 0.384192$$

Method 3. A third way to do this calculation is to treat 0.576 as $\frac{0.576}{1}$. Then we multiply 0.576 by 2, and divide the result by 3.

$$\frac{2}{3} \times 0.576 = \frac{2}{3} \times \frac{0.576}{1} = \frac{2 \times 0.576}{3} = \frac{1.152}{3} = 0.384$$ ◀

DO EXERCISES 15 AND 16.

NAME SECTION DATE

EXERCISE SET 5.3

a Find decimal notation.

1. $\frac{3}{5}$ **2.** $\frac{13}{20}$ **3.** $\frac{13}{40}$ **4.** $\frac{1}{16}$

5. $\frac{1}{5}$ **6.** $\frac{3}{20}$ **7.** $\frac{17}{20}$ **8.** $\frac{3}{40}$

9. $\frac{19}{40}$ **10.** $\frac{51}{40}$ **11.** $\frac{39}{40}$ **12.** $\frac{21}{40}$

13. $\frac{13}{25}$ **14.** $\frac{21}{125}$ **15.** $\frac{2502}{125}$ **16.** $\frac{121}{200}$

17. $\frac{1}{4}$ **18.** $\frac{1}{2}$ **19.** $\frac{23}{40}$ **20.** $\frac{11}{20}$

21. $\frac{18}{25}$ **22.** $\frac{37}{25}$ **23.** $\frac{19}{16}$ **24.** $\frac{5}{8}$

25. $\frac{4}{15}$ **26.** $\frac{7}{9}$ **27.** $\frac{1}{3}$ **28.** $\frac{1}{9}$

29. $\frac{4}{3}$ **30.** $\frac{8}{9}$ **31.** $\frac{7}{6}$ **32.** $\frac{7}{11}$

33. $\frac{4}{7}$ **34.** $\frac{14}{11}$ **35.** $\frac{11}{12}$ **36.** $\frac{5}{12}$

ANSWERS

1. 0.6
2. 0.65
3. 0.325
4. 0.0625
5. 0.2
6. 0.15
7. 0.85
8. 0.075
9. 0.475
10. 1.275
11. 0.975
12. 0.525
13. 0.52
14. 0.168
15. 20.016
16. 0.605
17. 0.25
18. 0.5
19. 0.575
20. 0.55
21. 0.72
22. 1.48
23. 1.1875
24. 0.625
25. $0.2\overline{6}$
26. $0.\overline{7}$
27. $0.\overline{3}$
28. $0.\overline{1}$
29. $1.\overline{3}$
30. $0.\overline{8}$
31. $1.1\overline{6}$
32. $0.\overline{63}$
33. $0.\overline{571428}$
34. $1.\overline{27}$
35. $0.91\overline{6}$
36. $0.41\overline{6}$

ANSWERS

37. 0.3; 0.27; 0.267
38. 0.8; 0.78; 0.778
39. 0.3; 0.33; 0.333
40. 0.1; 0.11; 0.111
41. 1.3; 1.33; 1.333
42. 0.9; 0.89; 0.889
43. 1.2; 1.17; 1.167
44. 0.6; 0.64; 0.636
45. 0.6; 0.57; 0.571
46. 1.3; 1.27; 1.273
47. 0.9; 0.92; 0.917
48. 0.4; 0.42; 0.417
49. 11.06
50. 307.84
51. $417.51\overline{6}$
52. $1.7737\overline{36}$
53. 0.20425
54. 3771.75
55. 21
56. 10
57. $3\frac{2}{5}$
58. $30\frac{7}{10}$
59. 325
60. $2 \cdot 2 \cdot 2 \cdot 2 \cdot 2 \cdot 2 \cdot 2$
61. $0.\overline{142857}$
62. $0.\overline{285714}$
63. $0.\overline{428571}$
64. $0.\overline{571428}$
65. $0.\overline{714285}$
66. $0.\overline{857142}$
67. $0.\overline{1}$
68. $0.\overline{01}$
69. $0.\overline{001}$
70. $0.\overline{0001}$
71. $0.\overline{012345679}$
72. $0.\overline{123456790}$
73. $1.\overline{2345678901}$
74. $12.\overline{345679012}$
75. $\frac{2}{3}, \frac{5}{7}, \frac{15}{19}, \frac{11}{13}, \frac{17}{20}, \frac{13}{15}$

b

37.–47. Round each answer of the odd-numbered Exercises 25–35 to the nearest tenth, hundredth, and thousandth.

38.–48. Round each answer of the even-numbered Exercises 26–36 to the nearest tenth, hundredth, and thousandth.

c Calculate.

49. $\frac{7}{8} \times 12.64$

50. $\frac{4}{5} \times 384.8$

51. $\frac{47}{9} \times 79.95$

52. $\frac{7}{11} \times 2.7873$

53. $\frac{5}{6} \times 0.0765 + \frac{5}{4} \times 0.1124$

54. $\frac{3}{5} \times 6384.1 - \frac{3}{8} \times 156.56$

SKILL MAINTENANCE

55. Multiply: $9 \cdot 2\frac{1}{3}$.

56. Divide: $84 \div 8\frac{2}{5}$.

57. Subtract: $20 - 16\frac{3}{5}$.

58. Add: $14\frac{3}{5} + 16\frac{1}{10}$.

59. Find the LCM of 25 and 65.

60. Find the prime factorization of 128.

SYNTHESIS

■ Each of the following is a calculator exercise. In Exercises 61–74, find decimal notation.

61. $\frac{1}{7}$ **62.** $\frac{2}{7}$ **63.** $\frac{3}{7}$ **64.** $\frac{4}{7}$ **65.** $\frac{5}{7}$

66. From the pattern of Exercises 61–65, guess the decimal notation for $\frac{6}{7}$. Check on your calculator.

67. $\frac{1}{9}$ **68.** $\frac{1}{99}$ **69.** $\frac{1}{999}$

70. From the pattern of Exercises 67–71, guess the decimal notation for $\frac{1}{9999}$. Check on your calculator.

71. $\frac{1}{81}$ **72.** $\frac{10}{81}$ **73.** $\frac{100}{81}$ **74.** $\frac{1000}{81}$

75. Find decimal notation for each fraction. Use that notation to arrange the fractions in order from smallest to largest.

$$\frac{2}{3}, \frac{15}{19}, \frac{11}{13}, \frac{5}{7}, \frac{13}{15}, \frac{17}{20}$$

5.4 Estimating

a Estimating Sums, Differences, Products, and Quotients

Estimating has many uses. It can be done before a problem is even attempted to get an idea of the answer. It can be done afterward as a check, even when we are using a calculator. In many situations, an estimate is all we need. We usually estimate by rounding the numbers so that there are 1 or 2 nonzero digits. Consider the following advertisements for Examples 1–4.

SALE 89.95

Reg. 114.95. AM/FM stereo electronic clock radio with stereo cassette tape recorder/player. Clock has 24 hour set and forget alarm. Cassette with 2 LED indicators for stereo/record.

SALE 349.95

Reg. 399.95. 19" (meas. diag.) color TV features matrix picture tube. VHF/UHF tuning.

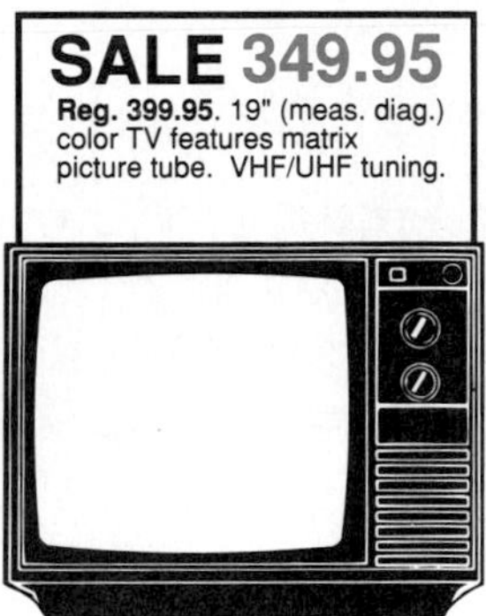

SALE 219.95

Reg. 299.95. 3.4 HP powerhead vac with 9 handy tools. Motorized beater bar has a headlight to aid cleaning.

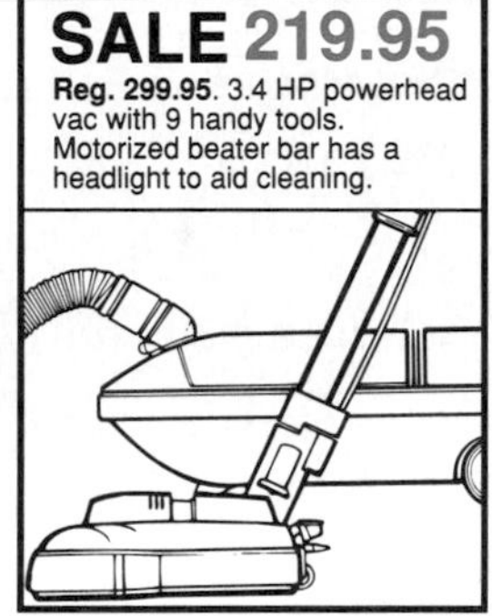

▶ **EXAMPLE 1** Estimate to the nearest ten the total cost of one clock radio and one TV.

We are estimating the sum

$$\$89.95 + \$349.95 = \text{Total cost.}$$

The estimate to the nearest ten is

$$\$90 + \$350. \quad \text{(Estimated total cost)}$$

We rounded 89.95 to the nearest ten and 349.95 to the nearest ten. The estimated sum is $440. ◀

DO EXERCISE 1.

▶ **EXAMPLE 2** About how much more does the TV cost than the clock radio? Estimate to the nearest ten.

We are estimating the difference

$$\$349.95 - \$89.95 = \text{Price difference.}$$

The estimate to the nearest ten is

$$\$350 - \$90 = \$260. \quad \text{(Estimated price difference)}$$ ◀

DO EXERCISE 2.

▶ **EXAMPLE 3** Estimate the total cost of 4 vacuum cleaners.

We are estimating the product

$$4 \times \$219.95 = \text{Total cost.}$$

The estimate is found by rounding 219.95 to the nearest ten: $4 \times \$220 = \$880$. ◀

DO EXERCISE 3.

OBJECTIVE

After finishing Section 5.4, you should be able to:

a Estimate sums, differences, products, and quotients.

FOR EXTRA HELP

Tape 7D Tape 8A MAC: 5 IBM: 5

1. Estimate to the nearest ten the total cost of one TV and one vacuum cleaner. Which of the following is an appropriate estimate?

a) $5700 **b)** $570
c) $790 **d)** $57

(b)

2. About how much more does the TV cost than the vacuum cleaner? Estimate to the nearest ten. Which of the following is an appropriate estimate?

a) $130 **b)** $1300
c) $580 **d)** $13

(a)

3. Estimate the total cost of 6 clock radios. Which of the following is an appropriate estimate?

a) $5400 **b)** $760
c) $54 **d)** $540

(d)

4. About how many vacuum cleaners can be bought for $1100? Which of the following is an appropriate estimate?

a) 8 **b)** 5
c) 11 **d)** 124

(b)

ANSWERS ON PAGE A-3

Estimate the product. Do not find the actual product. Which of the following is an appropriate estimate?

5. 2.4×8

a) 16 **b)** 34 **c)** 125 **d)** 5

(a)

6. 24×0.6

a) 200 **b)** 5 **c)** 110 **d)** 20

(d)

7. 0.86×0.432

a) 0.04 **b)** 0.4 **c)** 1.1 **d)** 4

(b)

8. 0.82×0.1

a) 800 **b)** 8 **c)** 0.08 **d)** 80

(c)

9. 0.12×18.248

a) 180 **b)** 1.8 **c)** 0.018 **d)** 18

(b)

10. 24.234×5.2

a) 200 **b)** 125 **c)** 12.5 **d)** 234

(b)

Estimate the quotient. Which of the following is an appropriate estimate?

11. $59.78 \div 29.1$

a) 200 **b)** 20 **c)** 2 **d)** 0.2

(c)

12. $82.08 \div 2.4$

a) 40 **b)** 4.0 **c)** 400 **d)** 0.4

(a)

13. $0.1768 \div 0.08$

a) 8 **b)** 10 **c)** 2 **d)** 20

(c)

Estimate. Which of the following is an appropriate estimate?

14. $0.0069 \div 0.15$

a) 0.5 **b)** 50 **c)** 0.05 **d)** 23.4

(c)

ANSWERS ON PAGE A-3

▶ **EXAMPLE 4** About how many clock radios can be bought for $350?

We estimate the quotient

$$\$350 \div \$89.95.$$

We want a whole-number estimate. We choose our rounding appropriately. Rounding $89.95 to the nearest one, we get $90. Since $350 is close to $360, which is a multiple of $90, we estimate

$$\$360 \div \$90,$$

so the answer is about 4. ◀

DO EXERCISE 4 ON THE PRECEDING PAGE.

▶ **EXAMPLE 5** Estimate: 4.8×52.

We have

$$5 \times 50 = 250. \qquad \text{(Estimated product)}$$

We rounded 4.8 to the nearest one and 52 to the nearest ten. ◀

Compare these estimates for the product 4.94×38:

$$5 \times 40 = 200, \qquad 5 \times 38 = 190, \qquad 4.9 \times 40 = 196.$$

The first estimate was the easiest. You could probably do it mentally. The others had more nonzero digits.

DO EXERCISES 5–10.

▶ **EXAMPLE 6** Estimate: $82.08 \div 24$.

This is about $80 \div 20$, so the answer is about 4. ◀

▶ **EXAMPLE 7** Estimate: $94.18 \div 3.2$.

This is about $90 \div 3$, so the answer is about 30. ◀

▶ **EXAMPLE 8** Estimate: $0.0156 \div 1.3$.

This is about $0.02 \div 1$, so the answer is about 0.02. ◀

DO EXERCISES 11–13.

In some cases, it is easier to estimate a quotient directly rather than by rounding the divisor and the dividend.

▶ **EXAMPLE 9** Estimate: $0.0074 \div 0.23$.

We estimate 3 for a quotient. We check by multiplying.

$$0.23 \times 3 = 0.69$$

We make the estimate smaller. We estimate 0.3. We check by multiplying.

$$0.23 \times 0.3 = 0.069$$

We make the estimate smaller. We estimate 0.03. We check by multiplying.

$$0.23 \times 0.03 = 0.0069$$

This is about 0.0074, so the quotient is about 0.03. ◀

DO EXERCISE 14.

NAME SECTION DATE

EXERCISE SET 5.4

a Consider the ads below for Exercises 1–8. Estimate the sums, differences, products, or quotients involved in these problems. Indicate which of the choices is an appropriate estimate.

1. Estimate the total cost of one audio-video cabinet and one stereo system.

a) $36
b) $72
c) $3.60
d) $360

2. Estimate the total cost of one audio-video cabinet and one TV.

a) $410
b) $820
c) $41
d) $4.10

3. About how much more does the TV cost than the stereo system?

a) $500
b) $80
c) $50
d) $5

4. About how much more does the TV cost than the audio-video cabinet?

a) $410
b) $190
c) $19
d) $9

5. Estimate the total cost of 9 TVs.

a) $2700
b) $27
c) $270
d) $540

6. Estimate the total cost of 16 stereo systems.

a) $5010
b) $4000
c) $40
d) $410

7. About how many TVs can be bought for $1700?

a) 600
b) 72
c) 6
d) 60

8. About how many stereo systems can be bought for $1300?

a) 10
b) 5
c) 50
d) 500

Estimate by rounding as directed.

9. 0.02 + 1.31 + 0.34; nearest tenth

10. 0.88 + 2.07 + 1.54; nearest one

11. 6.03 + 0.007 + 0.214; nearest one

ANSWERS

1. (d)
2. (a)
3. (c)
4. (b)
5. (a)
6. (b)
7. (c)
8. (b)
9. 1.6
10. 5
11. 6

ANSWERS

12. 110
13. 60
14. 14
15. 2.3
16. 12
17. 180
18. (d)
19. (a)
20. (b)
21. (c)
22. (a)
23. (b)
24. (c)
25. (b)
26. (c)
27. $2 \cdot 2 \cdot 3 \cdot 3 \cdot 3$
28. $2 \cdot 2 \cdot 2 \cdot 2 \cdot 5 \cdot 5$
29. $\frac{5}{16}$
30. $\frac{129}{251}$
31. Yes
32. No

12. $1.11 + 8.888 + 99.94$; nearest one

13. $52.367 + 1.307 + 7.324$; nearest one

14. $12.9882 + 1.0115$; nearest tenth

15. $2.678 - 0.445$; nearest tenth

16. $12.9882 - 1.0115$; nearest one

17. $198.67432 - 24.5007$; nearest ten

Estimate. Choose a rounding digit that gives 1 or 2 nonzero digits. Indicate which of the choices is an appropriate estimate.

18. $234.12321 - 200.3223$
a) 600
b) 60
c) 300
d) 30

19. 49×7.89
a) 400
b) 40
c) 4
d) 0.4

20. 7.4×8.9
a) 95
b) 63
c) 124
d) 6

21. 98.4×0.083
a) 80
b) 12
c) 8
d) 0.8

22. 78×5.3
a) 400
b) 800
c) 40
d) 8

23. $3.6 \div 4$
a) 10
b) 1
c) 0.1
d) 0.01

24. $0.0713 \div 1.94$
a) 4
b) 0.4
c) 0.04
d) 40

25. $74.68 \div 24.7$
a) 9
b) 3
c) 12
d) 120

26. $914 \div 0.921$
a) 9
b) 90
c) 900
d) 0.9

SKILL MAINTENANCE

Find the prime factorization.

27. 108

28. 400

Simplify.

29. $\frac{125}{400}$

30. $\frac{3225}{6275}$

SYNTHESIS

The following were done on a calculator. Estimate to see if the decimal point was placed correctly.

31. $178.9462 \times 61.78 = 11{,}055.29624$

32. $14{,}973.35 \div 298.75 = 501.2$

5.5 Solving Problems

a Solving problems using decimals is similar to solving problems with whole numbers. We translate to an equation that corresponds to the situation. Then we solve the equation.

▶ **EXAMPLE 1** The Internal Revenue Service allows a tax deduction of 26.5¢ per mile for mileage driven for business purposes. What deduction, in dollars, would be allowed for driving 127 miles?

1. *Familiarize.* We first draw a picture or at least visualize the situation. Repeated addition fits this situation. We let d = the deduction, in dollars, allowed for driving 127 miles.

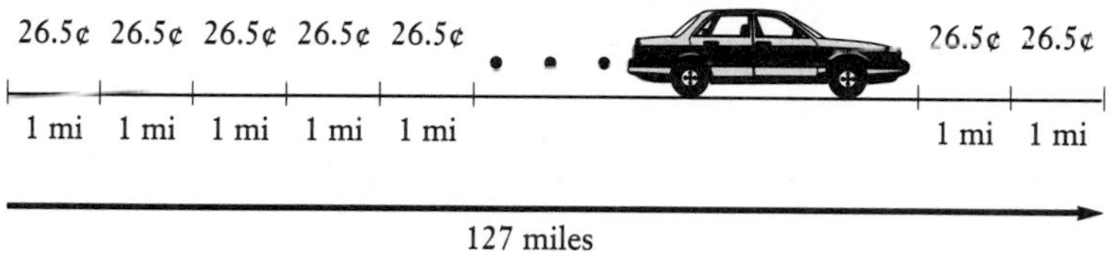

2. *Translate.* We translate as follows.

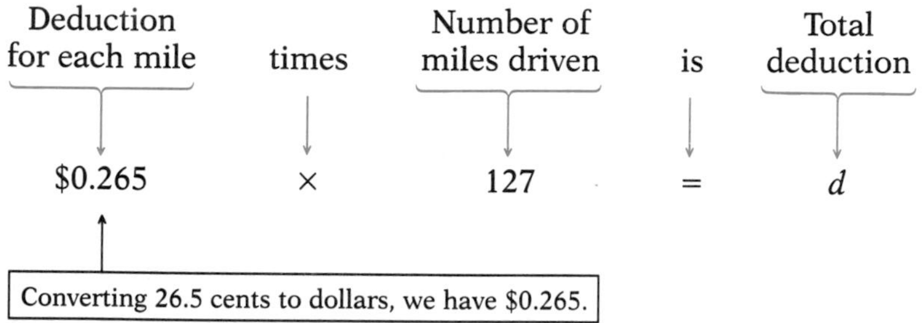

3. *Solve.* To solve the equation, we carry out the multiplication.

$$\begin{array}{r} 127 \\ \times\ 0.265 \\ \hline 33.655 \end{array}$$

Thus, $d = 33.655$.

4. *Check.* We can obtain a partial check by rounding and estimating:

$$127 \times 0.265 \approx 130 \times 0.3 = 39 \approx 33.655.$$

5. *State.* The total deduction would be \$33.66, rounded to the nearest cent. ◀

DO EXERCISE 1.

▶ **EXAMPLE 2** A loan of \$7382.52 is to be paid off in 36 monthly payments. How much is each payment?

1. *Familiarize.* We first draw a picture. We let n = the amount of each payment.

OBJECTIVE

After finishing Section 5.5, you should be able to:

a Solve problems involving multiplication and division with decimals.

FOR EXTRA HELP

Tape 7E | Tape 8B | MAC: 5 IBM: 5

1. At a printing company, the cost of copying is 8 cents per page. How much, in dollars, would it cost to make 466 copies? \$37.28

ANSWER ON PAGE A-3

2. A loan of $4425 is to be paid off in 12 monthly payments. How much is each payment? $368.75

ANSWER ON PAGE A-3

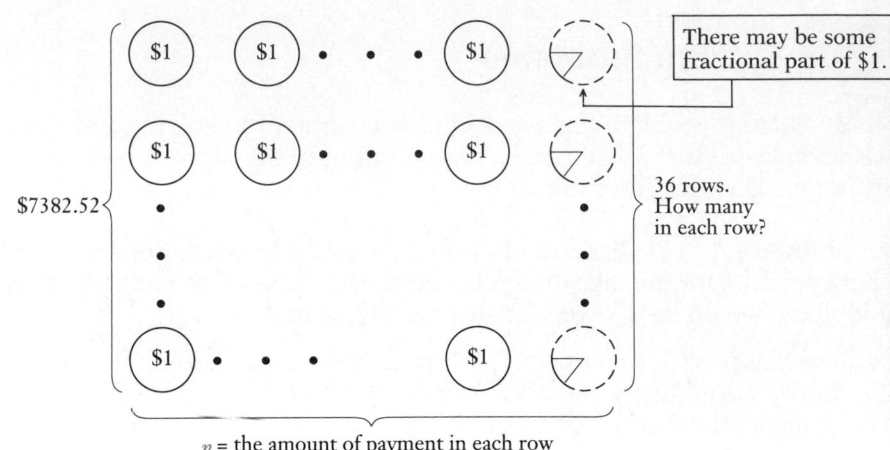

2. *Translate.* The problem can be translated to the following equation, thinking that

$$\text{(Total loan)} \div \text{(Number of payments)} = \text{Amount of each payment}$$
$$\$7382.52 \div 36 = n.$$

3. *Solve.* To solve the equation, we carry out the division.

$$\begin{array}{r} 205.07 \\ 36\overline{)7382.52} \\ 720000 \\ \hline 18252 \\ 18000 \\ \hline 252 \\ 252 \\ \hline 0 \end{array}$$

Thus, $n = 205.07$.

4. *Check.* A partial check can be obtained by estimating the quotient: $7382.56 ÷ 36 ≈ 8000 ÷ 40 = 200 ≈ 205.07. The estimate checks.
5. Each payment is $205.07. ◀

DO EXERCISE 2.

The area of a rectangular region is given by the formula *Area* = *Length* · *Width*, or $A = l \cdot w$. We can use this formula when numbers are named by decimals.

▶ **EXAMPLE 3** The rectangular page in a book measures 23.2 cm by 21.8 cm. Find the area.

1. *Familiarize.* We first draw a picture, letting A = the area.

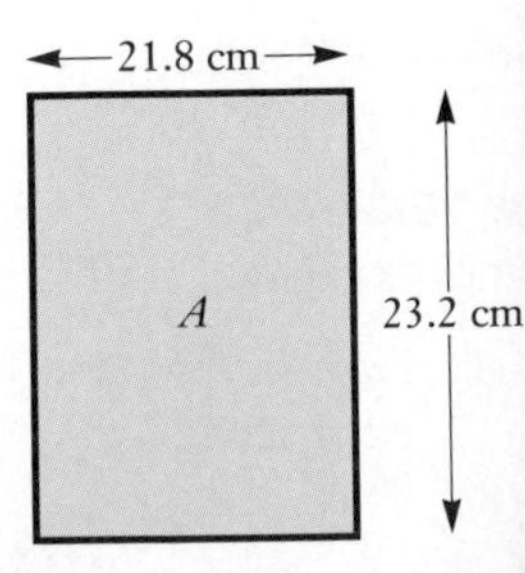

2. *Translate.* Then we use the formula $A = l \cdot w$ and translate.

$$A = 23.2 \times 21.8$$

3. *Solve.* We solve by carrying out the multiplication.

$$\begin{array}{r} 23.2 \\ \times\ 21.8 \\ \hline 1856 \\ 2320 \\ 46400 \\ \hline 505.76 \end{array}$$

Thus, $A = 505.76$.

4. *Check.* We obtain a partial check by estimating the product:

$$23.2 \times 21.8 \approx 20 \times 22 = 440 \approx 505.76.$$

Since this estimate is not too close, we might repeat our calculation or change our estimate to be more certain. We leave this to the student. We see that 505.76 checks.

5. *State.* The area is 505.76 sq cm. ◀

DO EXERCISE 3.

3. A standard-size file card measures 12.7 cm by 7.6 cm. Find its area. 96.52 sq cm

▶ **EXAMPLE 4** One pound of rump roast contains 3 servings. It costs \$2.89 a pound. What is the cost per serving? Round to the nearest cent.

1. *Familiarize.* We let c = the cost per serving.

2. *Translate.* We translate as follows.

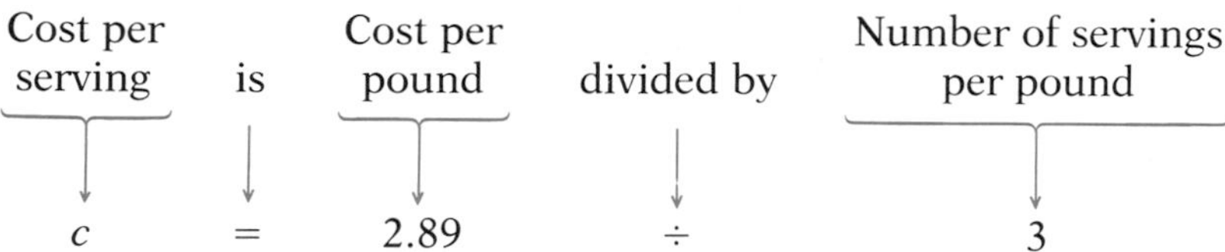

3. *Solve.* To solve, we carry out the division.

$$\begin{array}{r} 0.9633 \\ 3\,\overline{)\,2.8900} \\ 27 \\ \hline 19 \\ 18 \\ \hline 10 \\ 9 \\ \hline 10 \end{array} \qquad c = 0.96\overline{3}$$

Though we did not, a store will usually round up in a problem like this so that it does not lose money by selling in smaller quantities.

4. *Check.* We check by estimating the quotient:

$$2.89 \div 3 \approx 3 \div 3 = 1 \approx 0.96\overline{3}.$$

In this case, our check provides a good estimate.

5. *State.* We round $0.96\overline{3}$ and get the cost per serving to be about \$0.96. ◀

DO EXERCISE 4.

4. One pound of shankless ham contains 4.5 servings. It costs \$1.59. What is the cost per serving? Round to the nearest cent. \$0.35

ANSWERS ON PAGE A-3

5. A driver filled the gasoline tank and noted that the odometer read 38,320.8. After the next filling, the odometer read 38,735.5. It took 14.5 gal to fill the tank. How many miles per gallon did the driver get?

28.6 miles per gallon

Multistep Problems

▶ **EXAMPLE 5** *Gas mileage.* A driver filled the gasoline tank and noted that the odometer read 67,507.8. After the next filling, the odometer read 68,006.1. It took 16.5 gal to fill the tank. How many miles per gallon did the driver get?

1. *Familiarize.* We draw a picture.

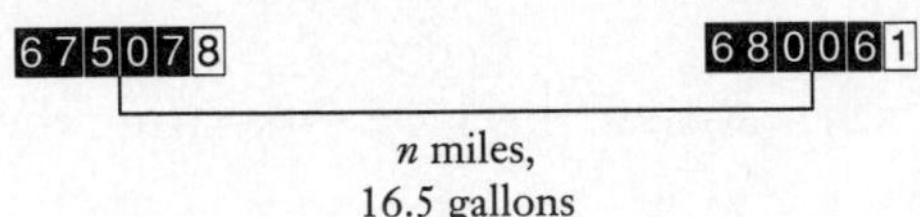

This is a two-step problem. First, we find the number of miles that have been driven between fillups. We let n = the number of miles driven.

2., 3. *Translate* and *Solve.* This is a "how-much-more" situation. We translate and solve as follows.

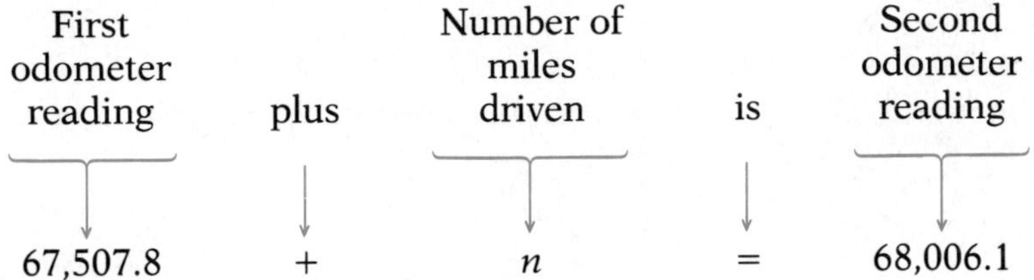

$$67{,}507.8 + n = 68{,}006.1$$

To solve the equation, we subtract 67,507.8 on both sides:

$n = 68{,}006.1 - 67{,}507.8$

$n = 498.3.$

$$\begin{array}{r} 68{,}006.1 \\ -\ 67{,}507.8 \\ \hline 498.3 \end{array}$$

Second, we divide the total number of miles driven by the number of gallons. This gives us m = the number of miles per gallon—that is, the mileage. The division that corresponds to the situation is

$$498.3 \div 16.5 = m.$$

To find the number m, we divide.

$$\begin{array}{r} 30.2 \\ 16.5\overline{)498.30} \\ 4950 \\ \hline 330 \\ 330 \\ \hline 0 \end{array}$$

Thus, $m = 30.2$.

4. *Check.* To check, we first multiply the number of miles per gallon times the number of gallons:

$$16.5 \times 30.2 = 498.3.$$

Then we add 498.3 to 67,507.8:

$$67{,}507.8 + 498.3 = 68{,}006.1.$$

The mileage 30.2 checks.

5. *State.* The driver gets 30.2 miles per gallon. ◀

DO EXERCISE 5.

ANSWER ON PAGE A-3

NAME SECTION DATE

EXERCISE SET 5.5

a Solve.

1. What is the cost of 7 blouses at $32.98 each?

2. What is the cost of 8 shirts at $26.99 each?

3. What is the cost, in dollars, of 17.7 gal of regular gasoline at 169.9 cents per gallon? (169.9 cents = $1.699) Round to the nearest cent.

4. What is the cost, in dollars, of 20.4 gal of lead-free gasoline at 179.9 cents per gallon? Round to the nearest cent.

5. A car went 250 mi in 4 hr. How far did it go in 1 hr?

6. A car went 325.8 mi in 6 hr. How far did it go in 1 hr?

7. A group of 8 students pays $47.60 for food for a picnic. What is each person's share?

8. A rectangular field measures 40.3 yd by 65.7 yd. Find its area.

9. A rectangular parking lot measures 800.4 ft by 312.6 ft. Find its area.

10. A group of 4 students pays $40.76 for a snack. What is each person's share of the cost?

11. What is the cost of 6 record albums at $8.88 each?

12. How much income does a state collect from the sale of 100,000 lottery tickets at $1.50 each?

ANSWERS

1. $230.86
2. $215.92
3. $30.07
4. $36.70
5. 62.5 mi
6. 54.3 mi
7. $5.95
8. 2647.71 sq yd
9. 250,205.04 sq ft
10. $10.19
11. $53.28
12. $150,000

ANSWERS

13. $139.36

14. 700 mg

15. $465.78

16. $389.75

17. 887.4 km

18. 2.117 billion lb

19. $57.35

20. $188.75

21. 20.2 mpg

22. 18.5 mpg

23. 11.9752 cu ft

24. 9

13. A family of five can save $2.68 per week by eating cereal to be cooked, such as oatmeal, rather than ready-to-eat cereal. How much would be saved in 1 year? Use 52 weeks for 1 year.

14. A medical assistant prepares 200 injections, each with 3.5 mg of penicillin. How much penicillin is used in all?

15. A loan of $11,178.72 is to be paid off in 24 monthly payments. How much is each payment?

16. A loan of $11,692.50 is to be paid off in 30 monthly payments. How much is each payment?

17. A plane flies 147.9 kilometers per hour for 6 hr. How far does it go?

18. Each day we consume 5.8 million lb of chocolate. How much chocolate do we consume in one year? Assume 365 days in one year.

19. It costs $24.95 a day plus 27 cents per mile to rent a compact car at Acme. How much, in dollars, would it cost to drive the car 120 mi in one day?

20. A student worked 53 hr during a week one summer. The student earned $3.50 per hour for the first 40 hr and $3.75 per hour for overtime. How much was earned during the week?

21. A driver filled the gasoline tank and noted that the odometer read 26,342.8. After the next filling, the odometer read 26,736.7. It took 19.5 gal to fill the tank. How many miles per gallon did the driver get?

22. A driver filled the gasoline tank and noted that the odometer read 18,943.2. After the next filling, the odometer read 19,305.8. It took 19.6 gal to fill the tank. How many miles per gallon did the driver get?

23. The water in a tank weighs 748.45 lb. One cubic foot of water weighs 62.5 lb. How many cubic feet of water does the tank hold?

24. A family bought $25 saving bonds for $18.75 each. They spent $168.75. How many bonds did they buy?

25. The average video game costs 25 cents and runs for 1.5 min. Assuming a player does not win any free games and plays continuously, how much money, in dollars, does it cost to play a video game for one hour?

26. A family owns a house with an assessed value of $94,500. For every $1000 of assessed value, they pay $7.68 in taxes. How much in taxes do they have to pay?

27. A person weighing 170 lb burns 8.6 calories per minute while mowing a lawn. How many calories would be burned in 2 hr of mowing?

28. Lot A measures 250.1 ft by 302.7 ft. Lot B measures 389.4 ft by 566.2 ft. What is the total area of the two lots?

29. A driver wants to estimate gas mileage per gallon. At 36,057.1 mi, the tank is filled with 10.7 gal. At 36,217.6 mi, the tank is filled with 11.1 gal. Find the mileage per gallon. Round to the nearest tenth.

30. A baseball player gets 1 hit in 3 "at bats." What part of the "at bats" were hits? Give decimal notation. Round to the nearest thousandth. (This is a player's "batting average.")

31. In a recent year, Jose Canseco got 187 hits in 610 at bats. What was his batting average? Round to the nearest thousandth.

32. A person earned $189.37 during a 40-hr week. What was the hourly wage? Round to the nearest cent.

33. A student earned $78.27 working part-time for 9 days. How much was earned each day? Round to the nearest cent. Assume that the student worked the same number of hours each day.

34. A golfer paid $10.75 for a dozen golf balls. What was the cost for each? Round to the nearest cent. (*Hint:* 1 dozen = 12.)

35. A 4-kg canned ham cost $16.58. What is the cost per kilogram? Round to the nearest cent.

36. A restaurant owner bought 20 dozen eggs for $13.80. Find the cost of each egg to the nearest tenth of a cent (thousandth of a dollar).

ANSWERS

25. $10

26. $725.76

27. 1032 cal

28. 296,183.55 sq ft

29. 14.5 mpg

30. 0.333

31. 0.307

32. $4.73

33. $8.70

34. $0.90

35. $4.15

36. $0.058

ANSWERS

37. $394.03

38. 1032 cal; 0.3 lb

39. 6020.48 sq m

40. $661.50

41. $196,987.20

42. $581.44

43. $13\frac{7}{12}$

44. 2

45. $34\frac{11}{12}$

46. $36\frac{1}{8}$

47. a) $13.38

b) $14.49

c) (a)

d) $13.78

e) (d)

37. A realtor paid $18,716.47 for 47.5 acres of land. What was the cost per acre? Round to the nearest cent.

38. A person weighing 170 lb burns 8.6 calories per minute while mowing a lawn. How many calories would be burned in 2 hr of mowing? One has to burn about 3500 calories in order to lose 1 lb. How many pounds would be lost by mowing for 2 hr? Round to the nearest tenth.

39. A lot measures 50.8 meters (m) by 120.2 meters. A swimming pool measuring 10.2 m by 8.4 m is built on the lot. How much area is left over?

40. A construction worker gets $13.50 per hour for the first 40 hours of work, and time and a half, or $20.25 per hour, for any overtime work exceeding 40 hours per week. One week she works 46 hours. How much was her pay?

41. You borrow $120,000 to buy a house. Financial institutions get back more money than they loan so that they make money on loans. You agree to make monthly payments of $880.52 for 30 years. How much more do you pay back than the amount of the loan?

42. A car rental agency charges $59.95 per day plus $0.39 per mile for a certain type of car. How much is the rental charge for a 4-day trip of 876 miles?

SKILL MAINTENANCE

43. Subtract: $24\frac{1}{4} - 10\frac{2}{3}$.

44. Divide: $8\frac{1}{2} \div 4\frac{1}{4}$.

45. Add: $24\frac{1}{4} + 10\frac{2}{3}$.

46. Multiply: $8\frac{1}{2} \cdot 4\frac{1}{4}$.

SYNTHESIS

47. A car wash charges $4.50 for a wash, but reduces its rates (as shown) in relation to how much gasoline you buy from them.

Cost to wash	Gas bought
$4.50 wash	No gas
$3.59	8–9.9 gal
$3.39	10–11.9 gal
$3.19	12–14.9 gal
$2.99	15–fill

a) Suppose you buy 10 gal of gas at 99.9 cents per gallon at the car wash and get your car washed. How much does it cost? (99.9 cents = $0.999)

b) Suppose you buy 10 gal of gas at 99.9 cents per gallon at a service station and then go get your car washed. What is the total cost?

c) Which method, (a) or (b), is cheaper?

d) Suppose you buy 10 gal of gas at 103.9 cents per gallon at the car wash and get your car washed. How much does it cost?

e) Which method, (b) or (d), is cheaper?

SUMMARY AND REVIEW EXERCISES: CHAPTER 5

The review sections and objectives to be tested in addition to the material in this chapter are [2.1d], [2.5b], [3.5a, b], and [3.6a, b].

Multiply.

1. $\begin{array}{r} 48 \\ \times\, 0.27 \\ \hline \end{array}$

2. $\begin{array}{r} 0.174 \\ \times\quad 0.83 \\ \hline \end{array}$

3. $\begin{array}{r} 0.043 \\ \times\quad 100 \\ \hline \end{array}$

4. $\begin{array}{r} 0.087 \\ \times\quad 3.2 \\ \hline \end{array}$

5. $\begin{array}{r} 3.7 \\ \times\, 0.29 \\ \hline \end{array}$

6. $\begin{array}{r} 2.08 \\ \times\, 0.105 \\ \hline \end{array}$

7. $\begin{array}{r} 24.68 \\ \times\, 0.001 \\ \hline \end{array}$

8. $\begin{array}{r} 24.68 \\ \times\, 1000 \\ \hline \end{array}$

Divide.

9. $8\overline{)60}$

10. $25\overline{)80}$

11. $52\overline{)23.4}$

12. $2.6\overline{)117.52}$

13. $7.2\overline{)11.52}$

14. $2.14\overline{)2.18708}$

15. $\dfrac{276.3}{100}$

16. $\dfrac{276.3}{1000}$

17. $\dfrac{0.1274}{0.1}$

18. $\dfrac{13.892}{0.01}$

Solve.

19. $3 \cdot x = 20.85$

20. $10 \cdot y = 425.4$

21. Estimate the product 7.82×34.487 by rounding to the nearest one.

22. Estimate the quotient $82.304 \div 17.287$ by rounding to the nearest ten.

23. Estimate the difference $219.875 - 4.478$ by rounding to the nearest one.

24. Estimate the sum \$45.78 + \$78.99 by rounding to the nearest one.

Find decimal notation. Use multiplying by 1.

25. $\dfrac{13}{5}$

26. $\dfrac{32}{25}$

27. $\dfrac{11}{4}$

Find decimal notation. Use division.

28. $\frac{13}{4}$

29. $\frac{7}{6}$

30. $\frac{17}{11}$

Round the answer to Exercise 30 to the nearest:

31. Tenth.

32. Hundredth.

33. Thousandth.

Solve.

34. Four dresses, each costing $59.95, were bought. What was the total spent?

35. A train traveled 496.02 km in 6 hr. How far did it travel in 1 hr?

36. A florist sold 13 potted palms for a total of $423.65. What was the cost for each palm? Round to the nearest cent.

37. The average person drinks 3.48 cups of tea per day. How many cups of tea are drunk in a week? in a month (30 days)?

38. A student buys 6 records for $53.88. How much was each record?

Convert from cents to dollars.

39. 8273 cents

40. 487 cents

Convert from dollars to cents.

41. $24.93

42. $9.86

Convert the number in the sentence to standard notation.

43. We breathe 3.4 billion cubic feet of oxygen each day.

44. Your blood travels 1.2 million miles in a week.

Calculate.

45. $(8 - 1.23) \div 4 + 5.6 \times 0.02$

46. $(1 + 0.07)^2 + 10^3 \div 10^2 + [4(10.1 - 5.6) + 8(11.3 - 7.8)]$

47. $\frac{3}{4} \times 20.85$

48. $\frac{1}{3} \times 123.7 + \frac{4}{9} \times 0.684$

SKILL MAINTENANCE

49. Multiply: $8\frac{1}{3} \cdot 5\frac{1}{4}$.

50. Divide: $20 \div 5\frac{1}{3}$.

51. Add: $12\frac{1}{2} + 7\frac{3}{10}$.

52. Subtract: $24 - 17\frac{2}{5}$.

53. Simplify: $\frac{28}{56}$.

54. Find the prime factorization of 192.

❖ THINKING IT THROUGH

1. Consider finding decimal notation for $\frac{44}{61}$. Discuss as many ways as you can for finding such notation and give the answer.
2. Discuss the role of estimating in calculating with decimal notation.
3. Explain how fractional notation can be used to justify multiplication with decimal notation.

NAME SECTION DATE

TEST: CHAPTER 5

Multiply.

1. $\begin{array}{r} 32 \\ \times 0.25 \end{array}$

2. $\begin{array}{r} 0.125 \\ \times \quad 0.24 \end{array}$

3. $\begin{array}{r} 0.037 \\ \times \quad 100 \end{array}$

4. $\begin{array}{r} 0.099 \\ \times \quad 2.1 \end{array}$

5. $\begin{array}{r} 3.4 \\ \times 0.32 \end{array}$

6. $\begin{array}{r} 3.06 \\ \times 0.104 \end{array}$

7. $\begin{array}{r} 213.45 \\ \times \quad 0.001 \end{array}$

8. $\begin{array}{r} 73.962 \\ \times \quad 10 \end{array}$

Divide.

9. $4\overline{)19}$

10. $25\overline{)11}$

11. $42\overline{)10.08}$

12. $3.3\overline{)100.32}$

13. $82\overline{)15.58}$

14. $\dfrac{346.89}{1000}$

15. $\dfrac{346.89}{0.01}$

16. $\dfrac{0.00123}{100}$

17. Solve: $4.8 \cdot y = 404.448$.

18. Estimate the product 8.91×22.457 by rounding to the nearest one.

19. Estimate the quotient $78.2209 \div 16.09$ by rounding to the nearest ten.

Find decimal notation. Use multiplying by 1.

20. $\dfrac{8}{5}$

21. $\dfrac{22}{25}$

22. $\dfrac{21}{4}$

Find decimal notation. Use division.

23. $\dfrac{3}{4}$

24. $\dfrac{11}{9}$

25. $\dfrac{15}{7}$

ANSWERS

1. [5.1a] 8
2. [5.1a] 0.03
3. [5.1a] 3.7
4. [5.1a] 0.2079
5. [5.1a] 1.088
6. [5.1a] 0.31824
7. [5.1a] 0.21345
8. [5.1a] 739.62
9. [5.2a] 4.75
10. [5.2a] 0.44
11. [5.2a] 0.24
12. [5.2a] 30.4
13. [5.2a] 0.19
14. [5.2a] 0.34689
15. [5.2a] 34,689
16. [5.2a] 0.0000123
17. [5.2b] 84.26
18. [5.4a] 198
19. [5.4a] 4
20. [5.3a] 1.6
21. [5.3a] 0.88
22. [5.3a] 5.25
23. [5.3a] 0.75
24. [5.3a] $1.\overline{2}$
25. [5.3a] $2.\overline{142857}$

Round the answer to Exercise 25 to the nearest:

26. Tenth.

27. Hundredth.

✓ **28.** Thousandth.

Solve.

29. A student bought 6 books at \$19.95 each. How much was spent? ✓

30. A person walked 11.85 km in 5 hr. How far did the person walk in 1 hr?

31. A consumer paid \$23,456.98 for 14 acres of land. How much did it cost for 1 acre? Round to the nearest cent.

32. Convert from dollars to cents: \$87.95.

33. Convert from cents to dollars: 949 cents. ✓

34. Convert to standard notation: 38.7 trillion.

Calculate.

35. $256 \div 3.2 \div 2 - 1.56 + 78.325 \times 0.02$ ✓

36. $(1 - 0.08)^2 + \{6[5(12.1 - 8.7) + 10(14.3 - 9.6)]\}$

37. $\frac{7}{8} \times 345.6$ ✓

38. $\frac{2}{3} \times 79.95 - \frac{7}{9} \times 1.235$

SKILL MAINTENANCE

39. Subtract: $28\frac{2}{3} - 2\frac{1}{6}$.

40. Add: $2\frac{3}{16} + \frac{1}{2}$.

41. Divide: $3\frac{3}{8} \div 3$.

42. Multiply: $2\frac{1}{10} \cdot 6\frac{2}{3}$.

43. Simplify: $\frac{33}{54}$.

44. Find the prime factorization of 360.

ANSWERS

26. [5.3b] 2.1

27. [5.3b] 2.14

28. [5.3b] 2.143

29. [5.5a] \$119.70

30. [5.5a] 2.37 km

31. [5.5a] \$1675.50

32. [5.1b] 8795¢

33. [5.1b] \$9.49

34. [5.1b] 38,700,000,000,000

35. [5.2c] 40.0065

36. [5.2c] 384.8464

37. [5.3c] 302.4

38. [5.3c] $52.339\overline{4}$

39. [3.5b] $26\frac{1}{2}$

40. [3.5a] $2\frac{11}{16}$

41. [3.6b] $1\frac{1}{8}$

42. [3.6b] 14

43. [2.5b] $\frac{11}{18}$

44. [2.1d] $2 \cdot 2 \cdot 2 \cdot 3 \cdot 3 \cdot 5$

CUMULATIVE REVIEW: CHAPTERS 1–5

Convert to fractional notation.

1. $2\frac{2}{9}$

2. 3.052

Find decimal notation.

3. $\frac{7}{5}$

4. $\frac{6}{11}$

5. Determine whether 43 is prime, composite, or neither.

6. Determine whether 2,053,752 is divisible by 4.

Calculate.

7. $48 + 12 \div 4 - 10 \times 2 + 6892 \div 4$

8. $4.7 - \{0.1[1.2(3.95 - 1.65) + 1.5 \div 2.5]\}$

Round to the nearest hundredth.

9. 584.903

10. $218.\overline{5}$

11. Estimate the product 16.392×9.715 by rounding to the nearest one.

12. Estimate by rounding to the nearest tenth:
$$2.714 + 4.562 - 3.31 - 0.0023.$$

13. Estimate the product 6418×1984 by rounding to the nearest hundred.

14. Estimate the quotient $717.832 \div 124.998$ by rounding to the nearest ten.

Add and simplify.

15. $2\frac{1}{4} + 3\frac{4}{5}$

16.
$$\begin{array}{r} 34{,}921 \\ 93{,}092 \\ +\ 11{,}103 \\ \hline \end{array}$$

17. $\frac{1}{6} + \frac{2}{3} + \frac{8}{9}$

18. $143.9 + 2.053$

Subtract and simplify.

19. $723{,}041 - 12{,}904$

20. $19 - 5.903$

21. $5\frac{1}{7} - 4\frac{3}{7}$

22. $\frac{10}{11} - \frac{9}{10}$

Multiply and simplify.

23. $\frac{3}{8} \cdot \frac{4}{9}$

24. $\begin{array}{r} 2532 \\ \times 2100 \\ \hline \end{array}$

25. $\begin{array}{r} 23.9 \\ \times \quad 0.2 \\ \hline \end{array}$

26. $\begin{array}{r} 27.9431 \\ \times \quad 0.001 \\ \hline \end{array}$

Divide and simplify.

27. $16.5 \overline{)35.013}$

28. $26 \overline{)47{,}918}$

29. $13.8621 \div 0.001$

30. $\frac{4}{9} \div \frac{8}{15}$

Solve.

31. $8.32 + x = 9.1$

32. $75 \cdot x = 2100$

33. $y \cdot 9.47 = 81.6314$

34. $1062 + y = 368{,}313$

35. $t + \frac{5}{6} = \frac{8}{9}$

36. $\frac{7}{8} \cdot t = \frac{7}{16}$

Solve.

37. In 1986 there were 1368 heart transplants, 8800 kidney transplants, 924 liver transplants, and 130 pancreas transplants. How many transplants of these four organs were there in 1986?

38. After paying a \$150 down payment on a sofa, $\frac{3}{10}$ of the total cost was paid. How much did the sofa cost?

39. There are 60 seconds in a minute and 60 minutes in an hour. How many seconds are in a day?

40. A student's tuition was \$3600. A loan was obtained for $\frac{2}{3}$ of the tuition. How much was the loan?

41. The balance in a checking account is \$314.79. After a check is written for \$56.02, what is the balance in the account?

42. A clerk in a deli sold $1\frac{1}{2}$ lb of ham, $2\frac{3}{4}$ lb of turkey, and $2\frac{1}{4}$ lb of roast beef. How many pounds of meat was sold?

43. A baker used $\frac{1}{2}$ lb of sugar for cookies, $\frac{2}{3}$ lb of sugar for pie, and $\frac{5}{6}$ lb of sugar for cake. How much sugar was used in all?

44. A rectangular room measures 19.8 ft by 23.6 ft. Find its area.

SYNTHESIS

45. A box of sauce mixes weighs $15\frac{3}{4}$ lb. Each package weighs $1\frac{3}{4}$ ounces. How many packages are in the box?

46. A customer in a grocery store used a manufacturer's coupon to buy juice. With the coupon, if 5 cartons of juice were purchased, the sixth carton was free. The price of each carton was \$1.09. What was the cost per carton with the coupon? Round to the nearest cent.

INTRODUCTION In the mathematics below, we have what is called a *proportion.* The expressions on either side are called *ratios.* In this chapter, ratios and proportions are used to solve problems such as this one. A topic of interest to consumers, *unit pricing,* will also be studied.

The review sections to be tested in addition to the material in this chapter are 2.5, 4.4, 5.1, and 5.2. ❖

Ratio and Proportion

AN APPLICATION

Estimating wildlife populations. To determine the number of fish in a lake, a conservationist catches 225 fish, tags them, and throws them back into the lake. Later, 108 fish are caught, and it is found that 15 of them are tagged. Estimate how many fish are in the lake.

THE MATHEMATICS

Let F = the number of fish in the lake. We can then translate the problem to this equation.

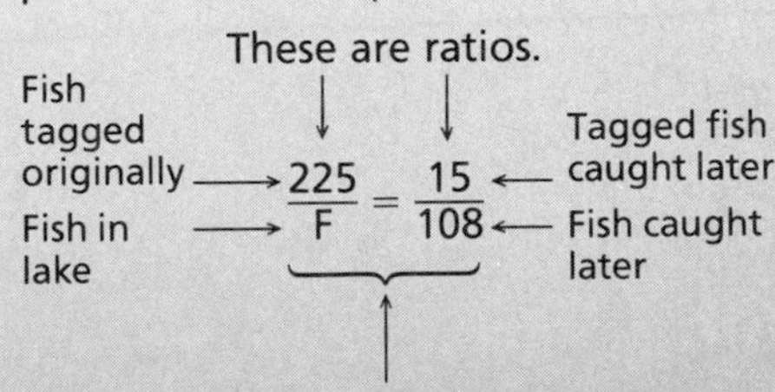

$$\frac{225}{F} = \frac{15}{108}$$

❖ POINTS TO REMEMBER: CHAPTER 6

Test for Equality of Fractions:	$\frac{a}{b} = \frac{c}{d}$ if $a \cdot d = b \cdot c$
Equation-Solving Skills:	Sections 2.1–2.3
Inequality-Solving Skills:	Section 2.7
Formula-Solving Skills:	Section 2.6

PRETEST: CHAPTER 6

Write fractional notation for the ratio.

1. 35 to 43

2. 0.079 to 1.043

Solve.

3. $\frac{5}{6} = \frac{x}{27}$

4. $\frac{y}{0.25} = \frac{0.3}{0.1}$

5. What is the rate in miles per gallon?

408 miles, 16 gallons

6. A student picked 3 bushels of apples in 40 minutes. What is the rate in bushels per minute?

7. A 24-oz loaf of bread costs $1.39. Find the unit price in cents per ounce. Round to the nearest hundredth of a cent.

8. Which has the lower unit price?

Orange juice
Brand A: 93¢ for 12 oz
Brand B: $1.07 for 16 oz

Solve.

9. A person traveled 216 km in 6 hr. At this rate, how far will the person go in 54 hr?

10. If 4 cans of peaches cost $1.98, how many cans of peaches can you buy for $10.89?

11. A watch loses 5 min in 10 hr. At this rate, how much will it lose in 24 hr?

12. On a map, 4 in. represents 225 actual miles. If two cities are 7 in. apart on the map, how far are they actually apart?

6.1 Introduction to Ratio and Proportion

a Ratio

A *ratio* is the quotient of two quantities.

For example, each day in this country about 5200 people die. Of these, 1070 die of cancer. The *ratio* of those who die of cancer to those who die is shown by the fractional notation

$$\frac{1070}{5200} \quad \text{or by the notation} \quad 1070{:}5200.$$

We read such notation as "the ratio of 1070 to 5200," listing the numerator first and the denominator second. Some other numbers whose ratio is $\frac{1070}{5200}$ are 107 and 520, 2140 and 10,400, and 10.7 and 52.

CAUTION! Remember that the ratio of a to b is given by $\frac{a}{b}$, where a is the numerator and b is the denominator.

DO EXERCISE 1.

Since

$$\frac{30}{20} = \frac{3}{2} \quad \text{and} \quad \frac{6}{4} = \frac{3}{2},$$

it follows that

$$\frac{30}{20} = \frac{6}{4}.$$

We say that the pair of numbers 30 and 20 has the same ratio as the pair of numbers 6 and 4.

DO EXERCISE 2.

▶ **EXAMPLE 1** In Michigan, the ratio of lawyers to people is 2.3 per 1000. Write fractional notation for the ratio of lawyers to people.

$$\frac{2.3}{1000}$$ ◀

▶ **EXAMPLE 2** Hank Aaron had 12,364 "at bats" in his career and 755 home runs. Write fractional notation for the ratio of at bats to home runs.

$$\frac{12{,}364}{755}$$ ◀

DO EXERCISES 3–6.

OBJECTIVES

After finishing Section 6.1, you should be able to:

a Write fractional notation for ratios.

b Determine whether two pairs of numbers are proportional.

c Solve proportions.

FOR EXTRA HELP

Tape 8A

Tape 8B

MAC: 6
IBM: 6

1. Find three pairs of numbers whose ratio is $\frac{2}{1}$.
   4 and 2, 10 and 5, 16 and 8; answers may vary.

2. Find three pairs of numbers whose ratio is $\frac{3}{2}$.
   60 and 40, 12 and 8, 24 and 16; answers may vary.

3. We drink 182.5 gal of liquid each year. Of this, 21.1 gal is milk. Write fractional notation for the ratio of milk drunk to total amount drunk. $\frac{21.1}{182.5}$

4. Of the 365 days in each year, it takes 107 days of work for the average person to pay his or her taxes. Write fractional notation for the ratio of days worked for taxes to total number of days worked. $\frac{107}{365}$

Write fractional notation for the ratio.

5. 7 to 11 $\frac{7}{11}$

6. 3.4 to 0.189 $\frac{3.4}{0.189}$

ANSWERS ON PAGE A-3

For a standard television screen, there are 4 units of length for every 3 units of width.

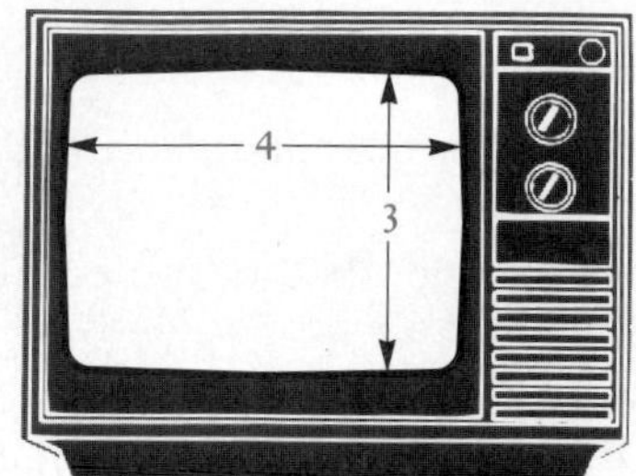

7. What is the ratio of the length to the width? $\frac{4}{3}$

8. What is the ratio of the width to the length? $\frac{3}{4}$

9. A family earning \$11,400 per year will spend about \$2964 for food. What is the ratio of food expenses to yearly income? $\frac{2964}{11,400}$

10. A pitcher gives up 4 earned runs in $7\frac{2}{3}$ innings of pitching. What is the ratio of earned runs to the number of innings pitched? $\frac{4}{7\frac{2}{3}}$

Determine whether the two pairs of numbers are proportional.

11. 3, 4 and 6, 8 Yes

12. 1, 4 and 10, 39 No

13. 1, 2 and 20, 39 No

ANSWERS ON PAGE A-3

▶ **EXAMPLE 3** In the triangle at the right:

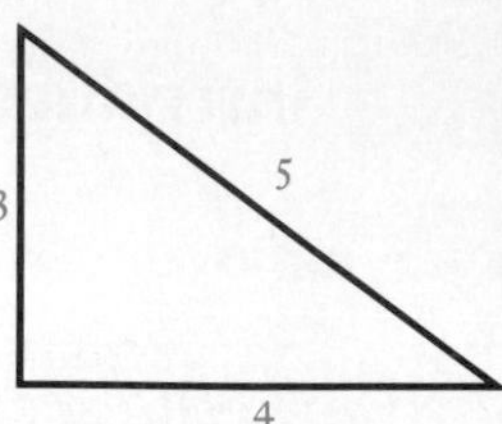

a) What is the ratio of the length of the longest side to the length of the shortest side?

$$\frac{5}{3}$$

b) What is ratio of the length of the shortest side to the longest side?

$$\frac{3}{5}$$ ◀

DO EXERCISES 7 AND 8.

▶ **EXAMPLE 4** A family earning \$21,400 per year will spend about \$3210 for car expenses. What is the ratio of car expenses to yearly income?

$$\frac{3210}{21,400}$$ ◀

DO EXERCISES 9 AND 10.

b Proportion

When two pairs of numbers (such as 3, 2 and 6, 4) have the same ratio, we say that they are **proportional.** The equation

$$\frac{3}{2} = \frac{6}{4}$$

states that 3, 2 and 6, 4 are proportional. Such an equation is called a **proportion.** We sometimes read

$$\frac{3}{2} = \frac{6}{4}$$

as "3 is to 2 as 6 is to 4."

Since ratios are represented by fractional notation, we can test whether two ratios are the same by using the test for equality discussed in Section 2.5. (It is also a skill to review for the chapter test.)

▶ **EXAMPLE 5** Determine whether 1, 2 and 3, 6 are proportional.

We can use cross-products:

$$1 \cdot 6 = 6 \qquad \frac{1}{2} \times \frac{3}{6} \qquad 2 \cdot 3 = 6.$$

Since the cross-products are the same, 6 = 6, we know that $\frac{1}{2} = \frac{3}{6}$, so the numbers are proportional. ◀

▶ **EXAMPLE 6** Determine whether 2, 5 and 4, 7 are proportional.

We can use cross-products:

$$2 \cdot 7 = 14 \qquad \frac{2}{5} \times \frac{4}{7} \qquad 5 \cdot 4 = 20$$

Since the cross-products are not the same, $14 \neq 20$, we know that $\frac{2}{5} \neq \frac{4}{7}$, so the numbers are not proportional. ◀

DO EXERCISES 11–13.

C Solving Proportions

Let us see how to solve proportions. Consider the proportion

$$\frac{x}{8} = \frac{3}{5}.$$

One way to solve a proportion is to use cross-products. Then we divide on both sides to get the variable alone.

$$5 \cdot x = 8 \cdot 3 \quad \text{Finding cross-products}$$

$$x = \frac{8 \cdot 3}{5} \quad \text{Dividing by 5 on both sides}$$

$$x = \frac{24}{5} \quad \text{Multiplying}$$

$$x = 4.8$$

We can check that 4.8 is the solution by replacing x by 4.8 and using cross-products:

$$4.8 \cdot 5 = 24 \qquad \frac{4.8}{8} \;\; \frac{3}{5} \qquad 8 \cdot 3 = 24$$

Since the cross-products are the same, it follows that $\frac{4.8}{8} = \frac{3}{5}$, so the numbers 4.8, 8 and 3, 5 are proportional, and 4.8 is the solution of the equation.

To solve $\frac{x}{a} = \frac{c}{d}$, find cross-products and divide on both sides to get x alone.

DO EXERCISE 14.

▶ **EXAMPLE 7** Solve: $\frac{x}{7} = \frac{5}{3}$. Write a mixed numeral for the answer.

We have

$$\frac{x}{7} = \frac{5}{3}$$

$$3 \cdot x = 7 \cdot 5 \quad \text{Finding cross-products}$$

$$x = \frac{7 \cdot 5}{3} \quad \text{Dividing by 3}$$

$$x = \frac{35}{3}.$$

The solution is $\frac{35}{3}$, or $11\frac{2}{3}$. ◀

DO EXERCISE 15.

14. Solve: $\frac{x}{63} = \frac{2}{9}$. 14

15. Solve: $\frac{x}{9} = \frac{5}{4}$. $11\frac{1}{4}$

ANSWERS ON PAGE A-3

16. Solve: $\frac{21}{5} = \frac{n}{2.5}$. 10.5

17. Solve: $\frac{2}{3} = \frac{6}{x}$. 9

18. Solve: $\frac{0.4}{0.9} = \frac{4.8}{t}$. 10.8

ANSWERS ON PAGE A-3

▶ **EXAMPLE 8** Solve: $\frac{7.7}{15.4} = \frac{y}{2.2}$. Write decimal notation for the answer.

$$\frac{7.7}{15.4} = \frac{y}{2.2}$$

$7.7 \times 2.2 = 15.4 \times y$ Finding cross-products

$\frac{7.7 \times 2.2}{15.4} = y$ Dividing by 15.4

$\frac{16.94}{15.4} = y$ Multiplying

$1.1 = y.$ Dividing: $15.4\overline{)16.94}$ → $154\overline{)169.4}$: quotient 1.1; 154; 15 4; 15 4; 0

The solution is 1.1. ◀

DO EXERCISE 16.

▶ **EXAMPLE 9** Solve: $\frac{3}{x} = \frac{6}{4}$.

$$\frac{3}{x} = \frac{6}{4}$$

$3 \cdot 4 = x \cdot 6$ Finding cross-products

$\frac{3 \cdot 4}{6} = x$ Dividing by 6

$\frac{12}{6} = x$ Multiplying

$2 = x.$ Simplifying

The solution is 2. ◀

DO EXERCISE 17.

▶ **EXAMPLE 10** Solve: $\frac{3.4}{4.93} = \frac{10}{n}$. Write decimal notation for the answer.

$$\frac{3.4}{4.93} = \frac{10}{n}$$

$n \times 3.4 = 4.93 \times 10$ Finding cross-products

$n = \frac{4.93 \times 10}{3.4}$ Dividing by 3.4

$n = \frac{49.3}{3.4}$ Multiplying

$n = 14.5.$ Dividing: $3.4\overline{)49.30}$ → $34\overline{)493.0}$: quotient 14.5; 340; 153 0; 136 0; 17 0; 17 0; 0

The solution is 14.5. ◀

DO EXERCISE 18.

NAME SECTION DATE

EXERCISE SET 6.1

a Write fractional notation for the ratio.

1. 4 to 5
2. 178 to 572
3. 0.4 to 12
4. 0.078 to 3.456
5. In a bread recipe, there are 2 cups of milk to 12 cups of flour. What is the ratio of cups of milk to cups of flour?
6. One person in four plays a musical instrument. What is the ratio of those of us who play an instrument to the total number of people? What is the ratio of those of us who do not play an instrument to the total number of people?
7. Foreign tourists spend $13.1 billion in this country annually. The most money, $2.7 billion, is spent in Florida. What is the ratio of the amount spent in Florida to the total amount spent? What is the ratio of the total amount spent to the amount spent in Florida?
8. In Washington, D.C., there are 36.1 lawyers for every 1000 people. What is the ratio of lawyers to people? What is the ratio of people to lawyers?

b Determine whether the two pairs of numbers are proportional.

9. 5, 6 and 7, 9
10. 7, 5 and 6, 4
11. 1, 2 and 10, 20
12. 7, 3 and 21, 9

c Solve.

13. $\frac{18}{4} = \frac{x}{10}$
14. $\frac{x}{45} = \frac{20}{25}$
15. $\frac{x}{8} = \frac{9}{6}$
16. $\frac{8}{10} = \frac{n}{5}$
17. $\frac{t}{12} = \frac{5}{6}$
18. $\frac{12}{4} = \frac{x}{3}$
19. $\frac{2}{5} = \frac{8}{n}$
20. $\frac{10}{6} = \frac{5}{x}$
21. $\frac{n}{15} = \frac{10}{30}$
22. $\frac{2}{24} = \frac{x}{36}$
23. $\frac{16}{12} = \frac{24}{x}$
24. $\frac{7}{11} = \frac{2}{x}$

ANSWERS

1. $\frac{4}{5}$
2. $\frac{178}{572}$
3. $\frac{0.4}{12}$
4. $\frac{0.078}{3.456}$
5. $\frac{2}{12}$
6. $\frac{1}{4}$; $\frac{3}{4}$
7. $\frac{2.7}{13.1}$; $\frac{13.1}{2.7}$
8. $\frac{36.1}{1000}$; $\frac{1000}{36.1}$
9. No
10. No
11. Yes
12. Yes
13. 45
14. 36
15. 12
16. 4
17. 10
18. 9
19. 20
20. 3
21. 5
22. 3
23. 18
24. $3\frac{1}{7}$

ANSWERS

25. 22

26. 36

27. 28

28. $12\frac{1}{2}$

29. $9\frac{1}{3}$

30. $21\frac{1}{3}$

31. $2\frac{8}{9}$

32. 2.7

33. 0.06

34. 39.05

35. 5

36. 25

37. 1

38. 1

39. 1

40. $\frac{9}{14}$

41. 14

42. 0

43. $2\frac{3}{16}$

44. $\frac{3}{4}$

45. =

46. ≠

47. 50

48. 9.5

49. 14.5

50. 152

51. $\frac{13{,}339{,}000}{145{,}304{,}000} \approx 0.09$; $\frac{145{,}304{,}000}{13{,}339{,}000} \approx 10.89$

25. $\frac{6}{11} = \frac{12}{x}$ **26.** $\frac{8}{9} = \frac{32}{n}$ **27.** $\frac{20}{7} = \frac{80}{x}$ **28.** $\frac{5}{x} = \frac{4}{10}$

29. $\frac{12}{9} = \frac{x}{7}$ **30.** $\frac{x}{20} = \frac{16}{15}$ **31.** $\frac{x}{13} = \frac{2}{9}$ **32.** $\frac{1.2}{4} = \frac{x}{9}$

33. $\frac{t}{0.16} = \frac{0.15}{0.40}$ **34.** $\frac{x}{11} = \frac{7.1}{2}$ **35.** $\frac{25}{100} = \frac{n}{20}$ **36.** $\frac{35}{125} = \frac{7}{m}$

37. $\frac{7}{\frac{1}{4}} = \frac{28}{x}$ **38.** $\frac{x}{6} = \frac{1}{6}$ **39.** $\frac{\frac{1}{4}}{\frac{1}{2}} = \frac{\frac{1}{2}}{x}$ **40.** $\frac{1}{7} = \frac{x}{4\frac{1}{2}}$

41. $\frac{1}{2} = \frac{7}{x}$ **42.** $\frac{x}{3} = \frac{0}{9}$ **43.** $\frac{\frac{2}{7}}{\frac{3}{4}} = \frac{\frac{5}{6}}{y}$ **44.** $\frac{\frac{5}{4}}{\frac{5}{8}} = \frac{\frac{3}{2}}{Q}$

SKILL MAINTENANCE

Use = or ≠ for ▒ to write a true sentence.

45. $\frac{12}{8}$ ▒ $\frac{6}{4}$ **46.** $\frac{4}{7}$ ▒ $\frac{5}{9}$

Divide. Write decimal notation for the answer.

47. 200 ÷ 4 **48.** 95 ÷ 10 **49.** 232 ÷ 16 **50.** 342 ÷ 2.25

SYNTHESIS

51. ▦ In Australia, there are 13,339,000 people and 145,304,000 sheep. What is the ratio of people to sheep? What is the ratio of sheep to people?

6.2 Rates

a When a ratio is used to compare two different kinds of measure, we call it a **rate.** Suppose that a car is driven 200 km in 4 hr. The ratio

$$\frac{200\text{ km}}{4\text{ hr}}, \quad \text{or } 50\,\frac{\text{km}}{\text{hr}}, \quad \text{or 50 kilometers per hour,} \quad \text{or 50 km/h}$$

Recall that "per" means "division," or "for each."

is the rate traveled in kilometers per hour, which is the division of the number of kilometers by the number of hours. A ratio of distance traveled to time is also called **speed.**

▶ **EXAMPLE 1** A student drives 145 km on 2.5 L of gas. What is the rate in kilometers per liter?

$$\frac{145\text{ km}}{2.5\text{ L}}, \quad \text{or} \quad 58\frac{\text{km}}{\text{L}}.$$ ◀

▶ **EXAMPLE 2** It takes 60 oz of grass seed to seed 3000 sq ft of lawn. What is the rate in ounces per square foot?

$$\frac{60\text{ oz}}{3000\text{ sq ft}}, \quad \text{or} \quad 0.02\frac{\text{oz}}{\text{sq ft}}.$$ ◀

▶ **EXAMPLE 3** A cook buys 10 lb of potatoes for \$1.69. What is the rate in cents per pound?

$$\frac{\$1.69}{10\text{ lb}} = \frac{169\text{ cents}}{10\text{ lb}}, \quad \text{or} \quad 16.9\frac{\text{cents}}{\text{lb}}.$$ ◀

▶ **EXAMPLE 4** A student earned \$3690 for working 3 months one summer. What was the rate of pay?

The rate of pay is the ratio of money earned per time worked, or

$$\frac{\$3690}{3\text{ mo}} = \$1230 \text{ per month,} \quad \text{or} \quad 1230\frac{\text{dollars}}{\text{month}}.$$ ◀

DO EXERCISES 1–8.

OBJECTIVES

After finishing Section 6.2, you should be able to:

a Give the ratio of two different kinds of measure as a rate.

b Find unit prices and use them to determine which of two possible purchases has the lower unit price.

FOR EXTRA HELP

Tape 8B | Tape 9A | MAC: 6 IBM: 6

What is the rate, or speed, in kilometers per hour?

1. 45 km, 9 hr 5 km/h

2. 120 km, 10 hr 12 km/h

3. 3 km, 10 hr 0.3 km/h

What is the rate, or speed, in meters per second?

4. 2200 m, 2 sec 1100 m/sec

5. 52 m, 13 sec 4 m/sec

6. 232 m, 16 sec 14.5 m/sec

7. A well-hit golf ball can travel 500 ft in 2 sec. What is the rate, or speed, of the golf ball in feet per second? 250 ft/sec

8. A leaky faucet can lose 14 gal of water in a week. What is the rate in gallons per day? 2 gal/day

ANSWERS ON PAGE A-3

9. A customer bought a 14-oz package of oat bran for \$2.89. What is the unit price in cents per ounce? Round to the nearest hundredth of a cent. 20.64¢/oz

10. Which has the lower unit price? [*Note:* 1 qt = 32 fluid ounces (fl oz).] Can A

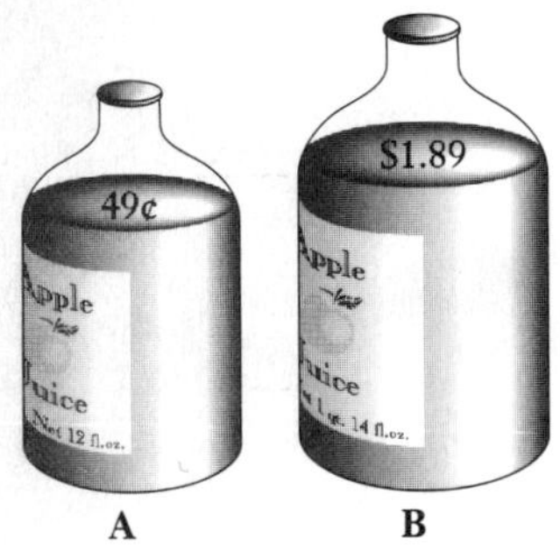

ANSWERS ON PAGE A-3

b Unit Pricing

A **unit price** is the ratio of price to the number of units. A unit price is applied so that the price of one unit can be determined.

▶ **EXAMPLE 5** A customer bought a 20-lb box of powdered detergent for \$9.47. What is the unit price in dollars per pound?

The unit price is the price in dollars for each pound.

$$\text{Unit price} = \frac{\text{Price}}{\text{Number of units}}$$
$$= \frac{\$9.47}{20 \text{ lb}} = \frac{9.47}{20} \cdot \frac{\$}{\text{lb}}$$
$$= 0.4735 \text{ dollars per pound}$$ ◀

DO EXERCISE 9.

For comparison shopping, it helps to find unit prices.

▶ **EXAMPLE 6** Which has the lower unit price?

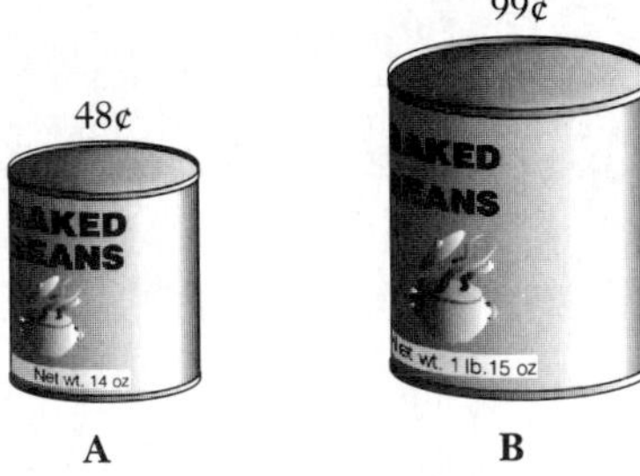

To find out, we compare the unit prices—in this case, the price per ounce.

For can A: $\frac{48 \text{ cents}}{14 \text{ oz}} \approx 3.429 \frac{\text{cents}}{\text{oz}}$

For can B: We need to find the total number of ounces:

$$1 \text{ lb}, 15 \text{ oz} = 16 \text{ oz} + 15 \text{ oz} = 31 \text{ oz}.$$

Then

$$\frac{99 \text{ cents}}{31 \text{ oz}} \approx 3.194 \frac{\text{cents}}{\text{oz}}.$$

Thus can B has the lower unit price. ◀

In many stores, unit prices are now listed on the items or the shelves.

DO EXERCISE 10.

NAME SECTION DATE

EXERCISE SET 6.2

a In Exercises 1–6, find the rate as a ratio of distance to time.

1. 120 km, 3 hr

2. 18 km, 9 hr

3. 440 m, 40 sec

4. 200 mi, 25 sec

5. 342 yd, 2.25 days

6. 492 m, 60 sec

7. A car is driven 500 km in 20 hr. What is the rate in kilometers per hour? in hours per kilometer?

8. A student eats 3 hamburgers in 15 min. What is the rate in hamburgers per minute? in minutes per hamburger?

9. To water a lawn adequately, it takes 623 gal of water for every 1000 sq ft. What is the rate in gallons per square foot?

10. An 8-lb shankless ham contains 36 servings of meat. What is the ratio in servings per pound?

11. A 12-lb boneless rib roast contains 30 servings of meat. What is the ratio in servings per pound?

12. A car is driven 200 km on 40 L of gasoline. What is the rate in kilometers per liter?

13. Light travels 186,000 mi in 1 sec. What is its rate, or speed, in miles per second?

14. Sound travels 1100 ft in 1 sec. What is its rate, or speed, in feet per second?

ANSWERS

1. 40 km/h
2. 2 km/h
3. 11 m/sec
4. 8 mi/sec
5. 152 yd/day
6. 8.2 m/sec
7. 25 km/hr; 0.04 hr/km
8. 0.2 hamburgers/min; 5 min/hamburger
9. 0.623 gal/sq ft
10. 4.5 servings/lb
11. 2.5 servings/lb
12. 5 km/L
13. 186,000 mi/sec
14. 1100 ft/sec

ANSWERS

15. 2.3 km/h

16. 124 km/h

17. 560 mi/h

18. 0.168 mi/h

19. \$9.50/yd

20. \$0.32375/oz, or 32.375¢/oz

21. 21.46¢/oz

22. 8.04¢/oz

15. A black racer snake can travel 4.6 km in 2 hr. What is its rate, or speed, in kilometers per hour?

16. Impulses in nerve fibers travel 310 km in 2.5 hr. What is the rate, or speed, in kilometers per hour?

17. A jet flew 2660 mi in 4.75 hr. What was its speed?

18. A turtle traveled 0.42 mi in 2.5 hr. What was its speed?

b

19. The fabric for a dress costs \$33.25 for 3.5 yd. Find the unit price.

20. An 8-oz bottle of shampoo costs \$2.59. Find the unit price.

21. A 13-oz can of coffee costs \$2.79. What is the unit price in cents per ounce? Round to the nearest hundredth of a cent.

22. A 6-can carton of 12-oz cans of soda costs \$5.79. What is the unit price in cents per ounce? Round to the nearest hundredth of a cent.

23. A $1\frac{1}{4}$-lb container of cottage cheese costs $1.35. Find the unit price in dollars per pound.

24. A $\frac{2}{3}$-lb package of Colby cheese costs $1.69. Find the unit price in dollars per pound. Round to the nearest hundredth of a dollar.

Which has the lower unit price?

25.

Chili sauce
Brand A: 18 oz for $1.79 Brand B: 16 oz for $1.65

26.

Napkins
Brand A: 140 napkins for 61 cents Brand B: 125 napkins for 44 cents

27.

Tomato juice
Brand A: $1.29 for 1 qt, 14 oz Brand B: 49 cents for 18 oz

28.

Evaporated milk
Brand A: 56 cents for 13 oz Brand B: $1.99 for 1 qt, 8 oz

29.

Soap
Brand A: $1.00 for 3 bars Brand B: $1.29 for 4 bars

30.

Chicken noodle soup
Brand A: 10.5 oz, 2 for 79 cents Brand B: 11 oz for 41 cents

ANSWERS

23. $1.08/lb

24. $2.54/lb

25. A

26. B

27. B

28. A

29. B

30. B

ANSWERS

31. B

32. A

33. Eight 16-oz bottles

34. 1.5 kg

35. 1.7 million

36. 109.608

37. 67,819

38. 0.67819

39. Approx. 11.54 m/sec; 0.08666 sec/m

31.

Fancy tuna
Brand A: $1.29 for 7 oz Brand B: $3.96 for 1 lb, 8 oz

32.

Flour
Brand A: $1.25 for 3 lb, 2 oz Brand B: $0.99 for 28 oz

33.

Soda
The same kind of soda comes in two types of cartons. Which type has the lower unit price? Six 16-oz bottles for $4.49, or Eight 16-oz bottles for $5.49

34.

Jelly
The same kind of jelly comes in two sizes. Which size has the lower unit price? $2.48 for 1 kg, or $3.69 for 1.5 kg

SKILL MAINTENANCE

35. There are 20.6 million people in this country who play the piano and 18.9 million who play the guitar. How many more play the piano than the guitar?

Multiply.

36. $\begin{array}{r} 45.67 \\ \times \quad 2.4 \\ \hline \end{array}$

37. $\begin{array}{r} 678.19 \\ \times \quad 100 \\ \hline \end{array}$

38. $\begin{array}{r} 678.19 \\ \times \quad 0.001 \\ \hline \end{array}$

SYNTHESIS

39. ▦ Anne Henning set an Olympic record in speed skating with a time of 43.33 sec in the 500-m race. What was her rate, or speed, in meters per second? in seconds per meter?

6.3 Proportion Problems

a Proportions have applications in many fields such as business, chemistry, biology, health sciences, and home economics, as well as to areas of daily life.

▶ **EXAMPLE 1** A car travels 800 km in 3 days. At this rate, how far will it travel in 15 days?

We let x = the distance traveled in 15 days. Then we translate to a proportion. We make each side the ratio of distance to time, with distance in the numerator and time in the denominator.

$$\text{Distance in 15 days} \longrightarrow \frac{x}{15} = \frac{800}{3} \longleftarrow \text{Distance in 3 days}$$
(Time → 15; 3 ← Time)

Each side of the equation represents the same ratio. That is the meaning of the equation. It may be helpful in setting up a proportion to read it, in the case above, as "the unknown distance x is to 15 days, as the known distance 800 kilometers is to 3 days."

Solve: $3 \cdot x = 15 \cdot 800$ Finding cross-products

$x = \frac{15 \cdot 800}{3}$ Dividing by 3 on both sides

$x = \frac{5 \cdot 3 \cdot 800}{3}$ Factoring

$x = 5 \cdot 800$ Simplifying

$x = 4000$

Thus the car travels 4000 km in 15 days. ◀

DO EXERCISE 1.

▶ **EXAMPLE 2** If 2 shirts can be bought for \$47, how many shirts can be bought for \$188?

We let x = the number of shirts that can be bought for \$188. Then we translate to a proportion. We make each side the ratio of the number of shirts to cost, with the number of shirts in the numerator and the cost, in dollars, in the denominator.

$$\text{Shirts} \longrightarrow \frac{2}{47} = \frac{x}{188} \longleftarrow \text{Shirts}$$
(Dollars → 47; 188 ← Dollars)

Solve: $2 \cdot 188 = 47 \cdot x$ Finding cross-products

$\frac{2 \cdot 188}{47} = x$ Dividing by 47 on both sides

$\frac{2 \cdot 47 \cdot 4}{47} = x$ Factoring

$2 \cdot 4 = x$ Simplifying

$8 = x$

Thus, 8 shirts can be bought for \$188. ◀

DO EXERCISE 2.

OBJECTIVE

After finishing Section 6.3, you should be able to:

a Solve problems involving proportions.

FOR EXTRA HELP

Tape 8C

Tape 9A

MAC: 6
IBM: 6

1. A car travels 700 km in 5 days. At this rate, how far will it travel in 24 days? 3360 km

2. If 7 tickets cost \$45.50, what is the cost of 17 tickets? \$110.50

ANSWERS ON PAGE A-3

3. Kirk McCaskill, a pitcher for the California Angels, gave up 69 earned runs in 212 innings of pitching. What was the pitcher's earned run average? Round to the nearest hundredth. 2.93

4. In the rectangles below, the ratio of length to width is the same. Find the width of the larger rectangle. 21 cm

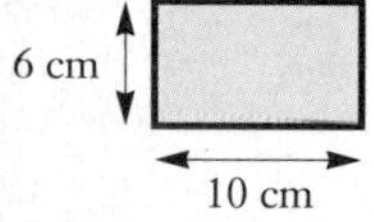

ANSWERS ON PAGE A-3

▶ **EXAMPLE 3** *Earned run average.* Dwight Gooden, a pitcher for the New York Mets, gave up 41 earned runs in 211 innings. At this rate, how many runs did he give up every 9 innings (there are 9 innings in a baseball game)?

We have the following proportion.

$$\begin{array}{l} \text{Earned runs each 9 innings} \longrightarrow \\ \text{Innings in one game} \longrightarrow \end{array} \frac{E}{9} = \frac{41}{211} \begin{array}{l} \longleftarrow \text{Earned runs} \\ \longleftarrow \text{Innings pitched} \end{array}$$

Solve: $211 \cdot E = 9 \cdot 41$ **Finding cross-products**

$E = \frac{9 \cdot 41}{211}$ **Dividing by 211 on both sides**

$E = \frac{369}{211}$ **Multiplying**

$E \approx 1.75$ **Dividing and rounding to the nearest hundredth**

We know that E = the **earned run average.** Then $E = 1.75$ means that, on the average, Dwight Gooden gave up 1.75 runs every 9 innings (every game) that he pitched. ◀

DO EXERCISE 3.

▶ **EXAMPLE 4** In the rectangles below, the ratio of length to width is the same. Find the width of the larger rectangle.

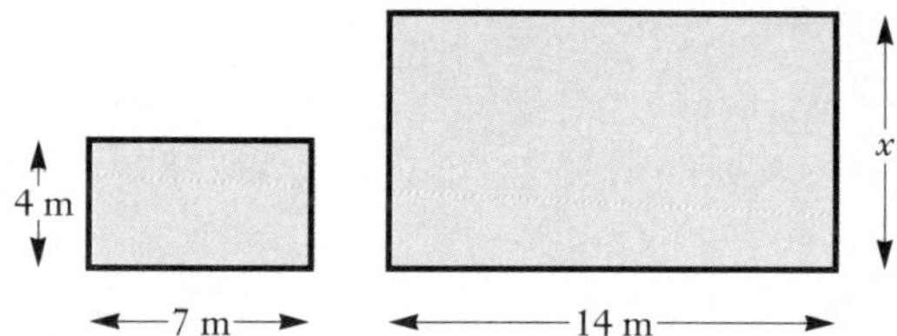

We let x = the width of the larger rectangle. Then we translate into a proportion.

$$\begin{array}{l} \text{Width} \longrightarrow \\ \text{Length} \longrightarrow \end{array} \frac{4}{7} = \frac{x}{14} \begin{array}{l} \longleftarrow \text{Width} \\ \longleftarrow \text{Length} \end{array}$$

Solve: $4 \cdot 14 = 7 \cdot x$ **Finding cross-products**

$\frac{4 \cdot 14}{7} = x$ **Dividing by 7 on both sides**

$\frac{4 \cdot 2 \cdot 7}{7} = x$ **Factoring**

$4 \cdot 2 = x$ **Simplifying**

$8 = x$ **Dividing**

Thus the width is 8 m. ◀

DO EXERCISE 4.

► **EXAMPLE 5** *Estimating wildlife populations.* To determine the number of deer in a forest, a conservationist catches 612 deer, tags them, and releases them. Later 244 deer are caught, and it is found that 72 of them are tagged. Estimate how many deer are in the forest.

We let D = the number of deer in the forest. Then we translate to a proportion.

$$\text{Deer tagged originally} \longrightarrow \frac{612}{D} = \frac{72}{244} \longleftarrow \text{Tagged deer caught later}$$
$$\text{Deer in forest} \longrightarrow D \qquad 244 \longleftarrow \text{Deer caught later}$$

Solve:

$$244 \cdot 612 = D \cdot 72 \quad \text{Finding cross-products}$$
$$\frac{244 \cdot 612}{72} = D \quad \text{Dividing by 72 on both sides}$$
$$\frac{2 \cdot 2 \cdot 2 \cdot 2 \cdot 3 \cdot 3 \cdot 17 \cdot 61}{2 \cdot 2 \cdot 2 \cdot 3 \cdot 3} = D \quad \text{Factoring}$$
$$2 \cdot 17 \cdot 61 = D \quad \text{Simplifying}$$
$$2074 = D.$$

Thus we estimate that there are 2074 deer in the forest. ◄

DO EXERCISE 5.

► **EXAMPLE 6** It takes 60 oz of grass seed to seed 3000 sq ft of lawn. At this rate, how much would be needed for 5000 sq ft of lawn?

We let g = the number of ounces of grass seed. Then we translate to a proportion.

$$\text{Grass seed needed} \longrightarrow \frac{g}{5000} = \frac{60}{3000} \longleftarrow \text{Grass seed needed}$$
$$\text{Amount of lawn} \longrightarrow 5000 \qquad 3000 \longleftarrow \text{Amount of lawn}$$

Solve:

$$3000 \cdot g = 5000 \cdot 60 \quad \text{Finding cross-products}$$
$$g = \frac{5000 \cdot 60}{3000} \quad \text{Dividing by 3000 on both sides}$$
$$g = \frac{1000 \cdot 5 \cdot 3 \cdot 20}{1000 \cdot 3} \quad \text{Factoring}$$
$$g = 5 \cdot 20 \quad \text{Simplifying}$$
$$g = 100$$

Thus, 100 oz are needed for 5000 sq ft of lawn. ◄

DO EXERCISE 6.

5. To determine the number of fish in a lake, a conservationist catches 225 fish, tags them, and throws them back into the lake. Later, 108 fish are caught, and it is found that 15 of them are tagged. Estimate how many fish are in the lake. 1620

6. In Example 6, how much seed would be needed for 7000 sq ft of lawn? 140 oz

ANSWERS ON PAGE A-3

❖ SIDELIGHTS

An Application of Ratio: State Lottery Profits

The chart below shows the profits of state lotteries in a recent year. Use the information to do the exercises.

State	Profit, in millions
Arizona	$ 22.0
Colorado	32.0
Connecticut	148.8
Delaware	15.0
Illinois	517.8
Maine	4.4
Maryland	263.7
Massachusetts	284.0
Michigan	320.0
New Hampshire	4.3
New Jersey	388.0
New York	615.0
Ohio	338.0
Pennsylvania	572.6
Rhode Island	18.6
Vermont	1.2
Washington	58.8

EXERCISES

1. Which state made the most from lotteries? New York
2. Which state made the least from lotteries? Vermont
3. How much more did the state with the most lottery income make than the state with the least income? $618.3 million
4. How much, in billions, did these states make together from lotteries? $3.6042 billion
5. What is the ratio of the lottery income of New York to the entire amount taken in by lotteries? Use your calculator. 0.1706
6. How much did New England take in from lotteries? $461.3 million
7. How much more did Ohio take in than Maryland? $74.3 million
8. The population of Washington is 4,300,000. At what rate, in dollars per person, did the people of Washington contribute to their lottery? $13.67 per person
9. The population of New York is 17,667,000. At what rate, in dollars per person, did the people of New York contribute to their lottery? $34.81 per person
10. The population of Illinois is 11,486,000. At what rate, in dollars per person, did the people of Illinois contribute to their lottery? $45.08
11. Which state, Washington, New York, or Illinois, has the highest ratio of lottery contributions per person? Illinois

NAME SECTION DATE

EXERCISE SET 6.3

ANSWERS

a Solve.

1. A car travels 234 km in 14 days. At this rate, how far would it travel in 42 days?

2. An automobile went 84 mi on 6.5 gal of gasoline. At this rate, how many gallons would be needed to go 126 mi?

3. If 2 sweatshirts cost $18.80, how much would 9 sweatshirts cost?

4. If 2 cans of beans cost $0.49, how many cans of beans can you buy for $6.37?

5. Tom Browning, a pitcher for the Cincinnati Reds, gave up 71 earned runs in 179 innings. What was his earned run average? Round to the nearest hundredth.

6. Bret Saberhagen, a pitcher for the Kansas City Royals, gave up 54 earned runs in 173 innings. What was his earned run average? Round to the nearest hundredth.

7. To determine the number of deer in a game preserve, a conservationist catches 318 deer, tags them, and releases them. Later, 168 deer are caught, and it is found that 56 of them are tagged. Estimate how many deer there are in the game preserve.

8. To determine the number of trout in a lake, a conservationist catches 112 trout, tags them, and throws them back into the lake. Later, 82 trout are caught, and it is found that 32 of them are tagged. Estimate how many trout there are in the lake.

9. An 8-lb shankless ham contains 36 servings of meat. How many pounds of ham would be needed for 54 servings?

10. A 12-lb boneless rib roast contains 30 servings of meat. How many servings would be provided by a 15-lb boneless rib roast?

11. Coffee beans from 14 trees are required to produce 17 lb of coffee, which each person in the United States drinks each year. How many trees are required to produce 391 lb of coffee?

12. A student bought a car. In the first 8 months, it was driven 10,000 km. At this rate, how many kilometers will the car be driven in 1 year? Use 12 months for 1 year.

Answers

1. 702 km
2. 9.75 gal
3. $84.60
4. 26
5. 3.57
6. 2.81
7. 954
8. 287
9. 12 lb
10. 37.5
11. 322
12. 15,000 km

ANSWERS

13. 120 lb

14. 1.25 m

15. 1980

16. 256 min

17. 58.1 mi

18. 650 mi

19. $7\frac{1}{3}$ in.

20. $45.44

21. No; the money will be gone in 24 weeks; $133.33 more

22. a)

b)

23. 2150

13. In a metal alloy, the ratio of zinc to copper is 3 to 13. If there are 520 lb of copper, how much zinc is there?

14. When a tree 8 m tall casts a shadow 5 m long, how long a shadow is cast by a person 2 m tall?

15. A quality-control inspector examined 200 light bulbs and found 18 defective. At this rate, how many defective bulbs would there be in a lot of 22,000?

16. A student has to read 32 essays in a literature class. The student reads 5 essays in 40 minutes. How long will it take the student to read all 32 essays?

17. On a map, 1 in. represents 16.6 mi. If two cities are 3.5 in. apart on the map, how far are they apart in reality?

18. On a map, $\frac{1}{4}$ in. represents 50 mi. If two cities are $3\frac{1}{4}$ in. apart on the map, how far are they apart in reality?

19. Under typical conditions, $1\frac{1}{2}$ ft of snow will melt to 2 in. of water. To how many inches of water will $5\frac{1}{2}$ ft of snow melt?

20. Tires are often priced according to the number of miles that they can be expected to be driven. Suppose a tire priced at $39.76 is expected to be driven 35,000 mi. How much would you expect to pay for a tire that is expected to be driven 40,000 mi?

21. A student attends a university whose academic year consists of two 16-week semesters. The student budgets $400 for incidental expenses for the academic year. After 3 weeks, the student has spent $50 for incidental expenses. Assuming the student continues to spend at the same rate, will the budget for incidental expenses be adequate? If not, when will the money be exhausted and how much more will be needed to complete the year?

22. A basic stereo system consists of a compact-disc player, a receiver–amplifier, and two speakers. A standard rule of thumb on the relative investment in these components is 1:3:2. That is, the receiver–amplifier should cost three times the amount of the compact-disc player and the speakers should cost twice as much as the amount of the compact-disc.

a) You have $1800 to spend. How should you allocate the funds if you use this rule of thumb?

b) How should you allocate $3000?

a) $300 for the CD player, $900 for the receiver/amplifier, $600 for the speakers; b) $500 for the CD player, $1500 for the receiver/amplifier, $1000 for the speakers

SYNTHESIS

23. Cy Young, one of the greatest pitchers of all time, had an earned run average of 2.63. He pitched more innings, 7356, than anyone in the history of baseball. How many earned runs did he give up?

SUMMARY AND REVIEW EXERCISES: CHAPTER 6

The review sections and objectives to be tested in addition to the material in this chapter are [2.5c], [5.1a], [5.2a], and [4.4a].

Write fractional notation for the ratio.

1. 47 to 84

2. 46 to 1.27

3. 83 to 100

4. 0.72 to 197

5. Each day in the United States, 5200 people die. Of these, 1070 die of cancer. Write fractional notation for the ratio of the number of people who die to the number of people who die of cancer.

Solve.

6. $\frac{8}{9} = \frac{x}{36}$

7. $\frac{120}{\frac{3}{7}} = \frac{7}{x}$

8. $\frac{6}{x} = \frac{48}{56}$

9. $\frac{4.5}{120} = \frac{0.9}{x}$

10. What is the rate in kilometers per hour?

117.7 kilometers, 5 hours

11. A lawn requires 319 gal of water for every 500 sq ft. What is the rate in gallons per square foot?

12. What is the rate in dollars per kilogram?

$355.04, 14 kilograms

13. A 25-lb turkey serves 18 people. What is the rate in servings per pound?

14. A 1 lb, 7 oz package of flour costs $1.30. Find the unit price in cents per ounce. Round to the nearest tenth of a cent.

15. It costs 79 cents for a $14\frac{1}{2}$-oz can of tomatoes. Find the unit price in cents per ounce. Round to the nearest hundredth of a cent.

Which has the lower unit price?

16.

White bread
Brand A: 16 oz for 89 cents Brand B: 12 oz for 65 cents

17.

Canned pineapple juice
Brand A: 12 oz for 99 cents Brand B: 18 oz for $1.26

Solve.

18. If 3 dozen eggs cost \$2.67, how much will 5 dozen eggs cost?

19. In a factory, it was discovered that 39 circuits out of a lot of 65 were defective. At this rate, how many defective circuits can be expected in a lot of 585 circuits?

20. A train travels 448 km in 7 hr. At this rate, how far would it go in 13 hr?

21. Fifteen acres are required to produce 54 bushels of tomatoes. At this rate, how many acres would be required to produce 97.2 bushels of tomatoes?

22. It is known that 5 people produce 13 kg of garbage in one day. San Diego, California, has 920,000 people. How many kilograms of garbage are produced in San Diego in one day?

23. Under typical conditions, $1\frac{1}{2}$ ft of snow will melt to 2 in. of water. To how many inches of water will $4\frac{1}{2}$ ft of snow melt?

24. In Michigan, there are 2.3 lawyers for every 1000 people. The population of Detroit is 1,140,000. How many lawyers are there in Detroit?

SKILL MAINTENANCE

25. A family has \$2347.89 in its checking account. It writes checks for \$678.95 and \$38.54. How much is left in the checking account?

Use $=$ or $\neq$ for ▢ to write a true sentence.

26. $\frac{5}{2}$ ▢ $\frac{10}{4}$

27. $\frac{4}{6}$ ▢ $\frac{8}{10}$

28. Multiply.

$$\begin{array}{r} 456.1 \\ \times\ 23.4 \\ \hline \end{array}$$

29. Divide. Write decimal notation for the answer.

$$5.6\overline{)254.8}$$

❖ THINKING IT THROUGH

1. Discuss the relationships among ratios, rates, and proportions.
2. In what way is a unit price a rate?

NAME SECTION DATE

TEST: CHAPTER 6

ANSWERS

Write fractional notation for the ratio.

1. 85 to 97

2. 0.34 to 124

Solve.

3. $\frac{9}{4} = \frac{27}{x}$

4. $\frac{150}{2.5} = \frac{x}{6}$

5. What is the rate in meters per second?

10 meters, 16 seconds

6. A 12-lb shankless ham contains 16 servings. What is the rate in servings per pound?

7. A 1 lb, 2 oz package of mahi mahi fish costs $3.49. Find the unit price in cents per ounce. Round to the nearest hundredth of a cent.

8. Which has the lower unit price?

Orange juice
Brand A: $1.19 for 12 oz Brand B: $1.33 for 16 oz

1. [6.1a] $\frac{85}{97}$

2. [6.1a] $\frac{0.34}{124}$

3. [6.1c] 12

4. [6.1c] 360

5. [6.2a] 0.625 m/sec

6. [6.2a] $1\frac{1}{3}$ servings/lb

7. [6.2b] 19.39¢/oz

8. [6.2b] B

ANSWERS

9. [6.3a] 1512 km

10. [6.3a] 44

11. [6.3a] 4.8 min

12. [6.3a] 525 mi

13. [4.4a] 25.8 million lb

14. [2.5c] $\neq$

15. [5.1a] 17,324.14

16. [5.2a] 0.9944

Solve.

9. A person traveled 432 km in 12 hr. At this rate, how far would the person go in 42 hr?

10. If 2 cans of apricots cost $1.19, how many cans of apricots can you buy for $26.18?

11. A watch loses 2 min in 10 hr. At this rate, how much would it lose in 24 hr?

12. On a map, 3 in. represents 225 mi. If two cities are 7 in. apart on the map, how far are they apart in reality?

SKILL MAINTENANCE

13. Kellogg's sells 146.2 million lb of Corn Flakes and 120.4 million lb of Frosted Flakes each year. How many more pounds of Corn Flakes do they sell than Frosted Flakes?

14. Use $=$ or $\neq$ for ▒ to write a true sentence.

$$\frac{6}{5} \ \square \ \frac{11}{9}$$

15. Multiply:

$$\begin{array}{r} 2\,3\,4.1\,1 \\ \times \quad\quad 7\,4 \\ \hline \end{array}$$

16. Divide: $\dfrac{99.44}{100}$.

CUMULATIVE REVIEW: CHAPTERS 1–6

Add and simplify.

1. $\begin{array}{r} 27.68 \\ 3.019 \\ +\ 483.297 \\ \hline \end{array}$

2. $\begin{array}{r} 2\frac{1}{3} \\ +\ 4\frac{5}{12} \\ \hline \end{array}$

3. $\frac{6}{35} + \frac{5}{28}$

Subtract and simplify.

4. $\begin{array}{r} 40.2 \\ -\ 9.709 \\ \hline \end{array}$

5. $73.82 - 0.908$

6. $\frac{4}{15} - \frac{3}{20}$

Multiply and simplify.

7. $\begin{array}{r} 37.64 \\ \times\ 5.9 \\ \hline \end{array}$

8. 5.678×100

9. $2\frac{1}{3} \cdot 1\frac{2}{7}$

Divide and simplify.

10. $2.3\overline{)98.9}$

11. $54\overline{)48,546}$

12. $\frac{7}{11} \div \frac{14}{33}$

13. Write expanded notation: 30,074.

14. Write a word name for 120.07.

Which number is larger?

15. 0.7, 0.698

16. 0.799, 0.8

17. Find the prime factorization of 144.

18. Find the LCM of 28 and 35.

19. What part is shaded?

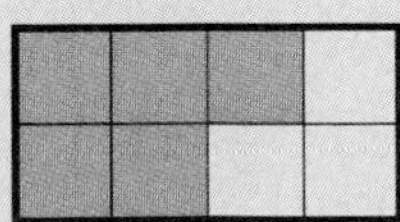

20. Simplify: $\frac{90}{144}$.

Calculate.

21. $\frac{3}{5} \times 9.53$

22. $\frac{1}{3} \times 0.645 - \frac{3}{4} \times 0.048$

23. Write fractional notation for the ratio 0.3 to 15.

24. Determine whether the pairs 3, 9 and 25, 75 are proportional.

25. Find the rate in meters per second:

660 m, 12 sec.

26. A 14-oz jar of applesauce costs $0.39. A 30-oz jar of applesauce costs $0.99. Which has the lower unit price?

Solve.

27. $\frac{14}{25} = \frac{x}{54}$

28. $423 = 16 \cdot t$

29. $\frac{2}{3} \cdot y = \frac{16}{27}$

30. $\frac{7}{16} = \frac{56}{x}$

31. $34.56 + n = 67.9$

32. $t + \frac{7}{25} = \frac{5}{7}$

Solve.

33. A car travels 337.62 km in 8 hr. How far does it travel in 1 hr?

34. A machine can stamp out 925 washers in 5 min. How much time would be needed to stamp out 1295 washers?

35. A salesperson drove 347.6 mi, 249.8 mi, and 379.5 mi on three separate trips. What was the total mileage?

36. In 1987, 1,098,323 people stayed overnight at Yosemite National Park. In the same year, 621,337 camped at Yellowstone National Park. How many more people stayed at Yosemite than at Yellowstone?

37. A person receives $52 for working a full day. How much is received for working $\frac{3}{4}$ of a day?

38. It takes a carpenter $\frac{2}{3}$ hr to hang a door. How many doors can the carpenter hang in 8 hr?

39. A 46-oz juice can contains $5\frac{3}{4}$ cups of juice. A recipe calls for $3\frac{1}{2}$ cups of juice. How many cups are left?

40. A recent space shuttle made 16 orbits a day during an 8.25-day mission. How many orbits were made during the entire mission?

SYNTHESIS

41. A car travels 88 ft in 1 sec. What is the rate in miles per hour?

42. A 12-oz bag of shredded mozzarella cheese costs $2.07. Blocks of mozzarella cheese are sold for $2.79 per pound. Which is the better buy?

INTRODUCTION This chapter introduces a new kind of notation for numbers: percent notation. It will be seen that $\frac{1}{2}$, 0.5, and 50% are all names for the same number. Finally, percent notation and equations will be used to solve problems.

The review sections to be tested in addition to the material in this chapter are 3.4, 5.2, 5.3, and 6.1. ❖

7 Percent Notation

AN APPLICATION

Have you ever wondered why you receive so much junk mail? One reason offered by the U.S. Postal Service is that we open and read about 78% of the advertising that we receive in the mail. Suppose a business sends out 9500 advertising brochures. How many of them can it expect to be opened and read?

THE MATHEMATICS

We let a = the number opened and read. The problem can be translated to an equation and solved as follows.

This is percent notation. (↓ pointing to 78%)

Restate: What number is 78% of 9500?

Translate: $a = 78\% \times 9500.$

(↑ pointing to =) This is the kind of equation we will consider.

❖ POINTS TO REMEMBER: CHAPTER 7

Skills at Working with Ratios and Proportions:	Chapter 6
Conversion between Fractional and Decimal Notation:	Sections 4.1 and 5.3
Equation Solving Skills:	Sections 1.7 and 5.2

PRETEST: CHAPTER 7

1. Find decimal notation for 87%.

2. Find percent notation for 0.537.

3. Find percent notation for $\frac{3}{4}$.

4. Find fractional notation for 37%.

5. Translate to an equation. Then solve.
What is 60% of 75?

6. Translate to a proportion. Then solve.
What percent of 50 is 35?

Solve.

7. The weight of muscles in a human body is 40% of total body weight. A person weighs 225 lb. What do the muscles weigh?

8. The population of a town increased from 3000 to 3600. Find the percent of increase in population.

9. The sales tax rate in Maryland is 5%. How much tax is charged on a purchase of $286? What is the total price?

10. A salesperson's commission rate is 28%. What is the commission from the sale of $18,400 worth of merchandise?

11. The marked price of a stereo is $450 and is on sale at Lowland Appliances for 25% off. What are the discount and the sale price?

12. What is the simple interest on $1200 principal at the interest rate of 8.3% for one year?

13. What is the simple interest on $500 at 8% for $\frac{1}{2}$ year?

14. Interest is compounded annually. Find the amount in an account if $6000 is invested at 9% for 2 years.

7.1 Percent Notation

a Understanding Percent Notation

Of the people in this country, 7% claim to have seen a UFO (Unidentified Flying Object). What does this mean? It means that, on the average, out of every 100 people, 7 of them claim to have seen a UFO. Thus, 7% is a ratio of 7 to 100, or $\frac{7}{100}$.

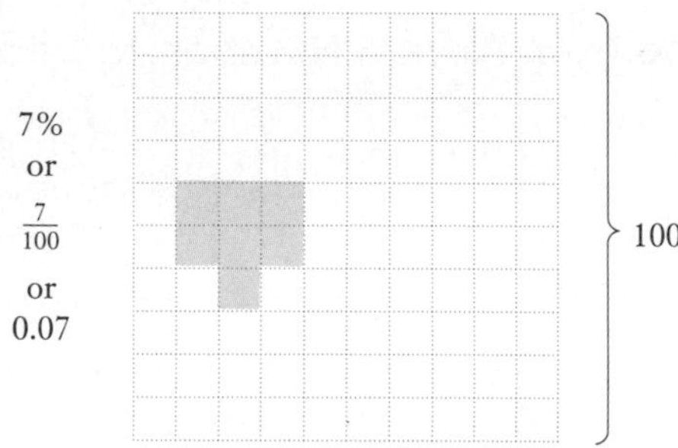

Percent notation is used extensively in our lives. Here are some examples:

51.6% of all new marriages will end in divorce;

95% of hair spray is alcohol.

38.7% of those accidents requiring medical attention occur in the home;

23% of us go to the movies once a month;

50% of us choose pepperoni as a pizza topping;

45.8% of us sleep between 7 and 8 hours per night;

88.6% of us prefer to live in a single-family home;

7.7% financing is sometimes available on new-car loans.

The notation $n\%$ arose historically meaning "n per hundred." This leads us to the following equivalent ways of defining percent.

Percent notation, $n\%$, is defined

using ratio as: $n\%$ = the ratio of n to 100 = $\frac{n}{100}$;

using fractional notation as: $n\% = n \times \frac{1}{100}$;

using decimal notation as: $n\% = n \times 0.01$.

▶ **EXAMPLE 1** Write three kinds of notation for 38%.

Using ratio: $38\% = \frac{38}{100}$ A ratio of 38 to 100

Using fractional notation: $38\% = 38 \times \frac{1}{100}$ Replacing % by $\times \frac{1}{100}$

Using decimal notation: $38\% = 38 \times 0.01$ Replacing % by $\times 0.01$ ◀

OBJECTIVES

After finishing Section 7.1, you should be able to:

a Write three kinds of notation for a percent.

b Convert from percent notation to decimal notation.

c Convert from decimal notation to percent notation.

FOR EXTRA HELP

Tape 8D

Tape 9B

MAC: 7
IBM: 7

Write three kinds of notation as in Examples 1 and 2.

1. 70% $\frac{70}{100}$; $70 \times \frac{1}{100}$; 70×0.01

2. 23.4% $\frac{23.4}{100}$; $23.4 \times \frac{1}{100}$; 23.4×0.01

3. 100% $\frac{100}{100}$; $100 \times \frac{1}{100}$; 100×0.01

Find decimal notation.

4. 34% 0.34

5. 78.9% 0.789

6. One year the rate of inflation was 12.08%. Find decimal notation for 12.08%. 0.1208

7. The present world population growth rate is 2.1% per year. Find decimal notation for 2.1%.

0.021

ANSWERS ON PAGE A-4

▶ **EXAMPLE 2** Write three kinds of notation for 67.8%.

Using ratio: $67.8\% = \frac{67.8}{100}$ A ratio of 67.8 to 100

Using fractional notation: $67.8\% = 67.8 \times \frac{1}{100}$ Replacing % by $\times \frac{1}{100}$

Using decimal notation: $67.8\% = 67.8 \times 0.01$ Replacing % by $\times 0.01$ ◀

DO EXERCISES 1–3.

b Converting from Percent Notation to Decimal Notation

Consider 78%.

$$78\% = \frac{78}{100} \quad \text{Using the definition of percent as a ratio}$$
$$= 0.78 \quad \text{Converting to decimal notation}$$

Dividing by 100 amounts to moving the decimal point two places to the left. Thus a quick way to convert from percent notation to decimal notation is to drop the percent symbol and move the decimal point two places to the left.

To convert from percent notation to decimal notation:	36.5%	
a) drop the percent symbol, and	36.5	
b) divide by 100, which means to move the decimal point two places to the left.	0.36.5	Move 2 places to the left.
	36.5% = 0.365	

▶ **EXAMPLE 3** Find decimal notation for 99.44%.

a) Drop the percent symbol. 99.44

b) Move the decimal point two places to the left. 0.99.44

Thus, 99.44% = 0.9944. ◀

▶ **EXAMPLE 4** The population growth rate of Europe is 1.1%. Find decimal notation for 1.1%.

a) Drop the percent symbol. 1.1

b) Move the decimal point two places to the left. 0.01.1

Thus, 1.1% = 0.011. ◀

DO EXERCISES 4–7.

C Converting from Decimal Notation to Percent Notation

Consider 0.38.

$$0.38 = \frac{38}{100} \quad \text{Converting to fractional notation}$$

$$= 38\% \quad \text{Using the definition of percent as a ratio}$$

We can convert from decimal notation to percent notation by moving the decimal point two places to the right and writing a percent symbol.

To convert from decimal notation to percent notation,	**0.675**	
a) move the decimal point two places to the right, and	**0.67.5**	**Move 2 places to the right.**
b) write a % symbol.	**67.5%**	
	0.675 = 67.5%	

▶ **EXAMPLE 5** Find percent notation for 1.27.

a) Move the decimal point two places to the right. 1.27.

b) Write a % symbol. 127%

Thus, 1.27 = 127%. ◀

▶ **EXAMPLE 6** Television sets are on 0.25 of the time. Find percent notation for 0.25.

a) Move the decimal point two places to the right. 0.25.

b) Write a % symbol. 25%

Thus, 0.25 = 25%. ◀

DO EXERCISES 8–12.

Find percent notation.

8. 0.24 24%

9. 3.47 347%

10. 1 100%

11. Muscles make up 0.4 of a person's body. Find percent notation for 0.4. 40%

12. Of those who buy music, 0.38 purchase prerecorded cassettes. Find percent notation for 0.38.
38%

It is thought that the Roman Emperor Augustus began percent notation by taxing goods sold at a rate of $\frac{1}{100}$. In time, the symbol "%" evolved by interchanging the parts of the symbol "100" to "0/0" and then to "%".

ANSWERS ON PAGE A-4

❖ SIDELIGHTS

Calculator Corner: Finding Whole-Number Remainders in Division

▶ **EXAMPLE** Find the quotient and the whole-number remainder:

$$567 \div 13.$$

We are using a calculator with a 10-digit readout.

a) Find decimal notation for the quotient using your calculator:

$$567 \div 13 \approx 43.61538462.$$

b) Subtract the whole-number part of the answer to (a):

$$43.61538462 - 43 = 0.61538462.$$

c) Multiply the answer to (b) by the divisor, 13:

$$0.61538462 \times 13 = 8.00000006.$$

Note the rounding error on the result. This will sometimes happen when approximating using a calculator.

d) The answer is

$$43 \text{ R } 8, \quad \text{or} \quad 43\tfrac{8}{13}.$$

EXERCISES

Find the quotient and the whole-number remainder.

1. $478 \div 17$ 28 R 2

2. $815 \div 7$ 116 R 3

3. $824 \div 11$ 74 R 10

4. $7888 \div 19$ 415 R 3

NAME SECTION DATE

EXERCISE SET 7.1

a Write three kinds of notation as in Examples 1 and 2 on pp. 279–280.

1. 90%

$\frac{90}{100}$; $90 \times \frac{1}{100}$; 90×0.01

2. 43.8%

$\frac{43.8}{100}$; $43.8 \times \frac{1}{100}$; 43.8×0.01

3. 12.5%

$\frac{12.5}{100}$; $12.5 \times \frac{1}{100}$; 12.5×0.01

4. 120%

$\frac{120}{100}$; $120 \times \frac{1}{100}$; 120×0.01

b Find decimal notation.

5. 67% **6.** 13% **7.** 45.6% **8.** 88.9%

9. 59.01% **10.** 20.08% **11.** 10% **12.** 20%

13. 1% **14.** 100% **15.** 200% **16.** 300%

17. 0.1% **18.** 0.4% **19.** 0.09% **20.** 0.12%

21. 0.18% **22.** 5.5% **23.** 23.19% **24.** 87.99%

25. Blood is 90% water. Find decimal notation for 90%.

26. Of all college football players, 2.6% play professional football. Find decimal notation for 2.6%.

27. Of those accidents requiring medical attention, 10.8% of them occur on roads. Find decimal notation for 10.8%.

28. Of all records that are purchased, 58.1% of them are pop/rock. Find decimal notation for 58.1%.

ANSWERS

1.

2.

3.

4.

5. 0.67

6. 0.13

7. 0.456

8. 0.889

9. 0.5901

10. 0.2008

11. 0.1

12. 0.2

13. 0.01

14. 1

15. 2

16. 3

17. 0.001

18. 0.004

19. 0.0009

20. 0.0012

21. 0.0018

22. 0.055

23. 0.2319

24. 0.8799

25. 0.9

26. 0.026

27. 0.108

28. 0.581

ANSWERS

29. 0.458
30. 0.23
31. 47%
32. 87%
33. 3%
34. 1%
35. 100%
36. 400%
37. 33.4%
38. 88.9%
39. 75%
40. 99%
41. 40%
42. 50%
43. 0.6%
44. 0.8%
45. 1.7%
46. 2.4%
47. 27.18%
48. 89.11%
49. 2.39%
50. 0.073%
51. 2.5%
52. 95%
53. 24%
54. 6%
55. $33\frac{1}{3}$
56. $37\frac{1}{2}$
57. $0.\overline{6}$
58. $0.\overline{3}$
59. Multiply by 100.
60. Divide by 100, or multiply by 0.01.

29. It is known that 45.8% percent of us sleep between 7 and 8 hours. Find decimal notation for 45.8%.

30. It is known that 23% of us go to the movies once a month. Find decimal notation for 23%.

c Find percent notation.

31. 0.47
32. 0.87
33. 0.03
34. 0.01
35. 1.00
36. 4.00
37. 0.334
38. 0.889
39. 0.75
40. 0.99
41. 0.4
42. 0.5
43. 0.006
44. 0.008
45. 0.017
46. 0.024
47. 0.2718
48. 0.8911
49. 0.0239
50. 0.00073

51. A person's brain is 0.025 of the body weight. Find percent notation for 0.025.

52. Of all school children, 0.95 have some tooth decay. Find percent notation for 0.95.

53. It is known that 0.24 of all children choose pizza as their favorite food. Find percent notation for 0.24.

54. It is known that 0.06 of all children choose hamburger as their favorite food. Find percent notation for 0.06.

SKILL MAINTENANCE

Convert to a mixed numeral.

55. $\frac{100}{3}$

56. $\frac{75}{2}$

Convert to decimal notation.

57. $\frac{2}{3}$

58. $\frac{1}{3}$

SYNTHESIS

59. ▦ What would you do to an entry on a calculator in order to get percent notation?

60. ▦ What would you do to percent notation on a calculator in order to get decimal notation?

7.2 Percent Notation and Fractional Notation

a Converting from Fractional Notation to Percent Notation

To convert from fractional notation to percent notation,	$\frac{3}{5}$ Fractional notation
a) find decimal notation by division, and	$5\overline{)3.0}$ = **0.6**; 3 0; 0
b) convert the decimal notation to percent notation.	**0.6 = 0.60 = 60%** Percent notation

▶ **EXAMPLE 1** Find percent notation for $\frac{3}{8}$.

a) Find decimal notation by division.

$$\begin{array}{r} 0.375 \\ 8\overline{)3.000} \\ \underline{2\,4} \\ 60 \\ \underline{56} \\ 40 \\ \underline{40} \\ 0 \end{array}$$

b) Convert the decimal notation to percent notation. Move the decimal point two places to the right, and write a % symbol.

0.37.5

$$\frac{3}{8} = 37.5\%, \text{ or } 37\frac{1}{2}\%$$

Don't forget the % symbol.

◀

DO EXERCISES 1 AND 2.

OBJECTIVES

After finishing Section 7.2, you should be able to:

a Convert from fractional notation to percent notation.

b Convert from percent notation to fractional notation.

FOR EXTRA HELP

Tape 8E

Tape 9B

MAC: 7
IBM: 7

Find percent notation.

1. $\frac{1}{4}$ 25%

2. $\frac{7}{8}$ 87.5%, or $87\frac{1}{2}\%$

ANSWERS ON PAGE A-4

3. The human body is $\frac{2}{3}$ water. Find percent notation for $\frac{2}{3}$.

66.$\overline{6}$%, or $66\frac{2}{3}$%

4. Find percent notation: $\frac{5}{6}$.

83.$\overline{3}$%, or $83\frac{1}{3}$%

Find percent notation.

5. $\frac{57}{100}$

57%

6. $\frac{19}{25}$

76%

ANSWERS ON PAGE A-4

▶ **EXAMPLE 2** Of all meals, $\frac{1}{3}$ are eaten outside the home. Find percent notation for $\frac{1}{3}$.

a) Find decimal notation by division.

$$\begin{array}{r} 0.3\,3\,3 \\ 3\,\overline{)\,1.0\,0\,0} \\ \underline{9} \\ 1\,0 \\ \underline{9} \\ 1\,0 \\ \underline{9} \\ 1 \end{array}$$

We get a repeating decimal: $0.33\overline{3}$.

b) Convert the answer to percent notation.

$$0.33.\overline{3}$$

$$\frac{1}{3} = 33.\overline{3}\%, \text{ or } 33\frac{1}{3}\%$$ ◀

DO EXERCISES 3 AND 4.

In some cases, division is not the easiest way to convert. The following are some optional ways this might be done.

▶ **EXAMPLE 3** Find percent notation for $\frac{69}{100}$.

We use the definition of percent as a ratio.

$$\frac{69}{100} = 69\%$$ ◀

▶ **EXAMPLE 4** Find percent notation for $\frac{17}{20}$.

We multiply by 1 to get 100 in the denominator. We think of what we have to multiply 20 by in order to get 100. That number is 5, so we multiply by 1 using $\frac{5}{5}$.

$$\frac{17}{20} \cdot \frac{5}{5} = \frac{85}{100} = 85\%$$ ◀

DO EXERCISES 5 AND 6.

b Converting from Percent Notation to Fractional Notation

To convert from percent notation to fractional notation,	**30%**	**Percent notation**
a) use the definition of percent as a ratio, and	$\frac{30}{100}$	
b) simplify, if possible.	$\frac{3}{10}$	**Fractional notation**

▶ **EXAMPLE 5** Find fractional notation for 75%.

$$75\% = \frac{75}{100} \quad \text{Using the definition of percent}$$

$$= \frac{3 \cdot 25}{4 \cdot 25} = \frac{3}{4} \cdot \frac{25}{25} = \frac{3}{4} \quad \text{Simplifying}$$

▶ **EXAMPLE 6** Find fractional notation for 62.5%.

$$62.5\% = \frac{62.5}{100} \quad \text{Using the definition of percent}$$

$$= \frac{62.5}{100} \times \frac{10}{10} \quad \text{Multiplying by 1 to eliminate the decimal point in the numerator}$$

$$= \frac{625}{1000}$$

$$= \frac{5 \cdot 125}{8 \cdot 125} = \frac{5}{8} \cdot \frac{125}{125} = \frac{5}{8} \quad \text{Simplifying}$$

◀

▶ **EXAMPLE 7** Find fractional notation for $16\frac{2}{3}\%$.

$$16\frac{2}{3}\% = \frac{50}{3}\% \quad \text{Converting from the mixed numeral to fractional notation}$$

$$= \frac{50}{3} \times \frac{1}{100} \quad \text{Using the definition of percent}$$

$$= \frac{50 \cdot 1}{3 \cdot 50 \cdot 2} = \frac{1}{6} \cdot \frac{50}{50} = \frac{1}{6} \quad \text{Simplifying}$$

◀

DO EXERCISES 7–9.

The table on the inside front cover contains decimal, fractional, and percent equivalents that are used so often that it would speed up your work if you learned them. For example, $\frac{1}{3} = 0.\overline{3}$, so we say that the **decimal equivalent** of $\frac{1}{3}$ is $0.\overline{3}$, or that $0.\overline{3}$ has the **fractional equivalent** $\frac{1}{3}$.

DO EXERCISE 10.

Find fractional notation.

7. 60% $\frac{3}{5}$

8. 3.25% $\frac{13}{400}$

9. $66\frac{2}{3}\%$ $\frac{2}{3}$

10. Complete this table.

Fractional notation	$\frac{1}{5}$	$\frac{5}{6}$	$\frac{3}{8}$
Decimal notation	0.2	$0.83\overline{3}$	0.375
Percent notation	20%	$83.\overline{3}\%$, or $83\frac{1}{3}\%$	$37\frac{1}{2}\%$

ANSWERS ON PAGE A-4

❖ SIDELIGHTS

Applications of Ratio and Percent: The Price-Earnings Ratio and Stock Yields

The Price-Earnings Ratio

If a company in one year has total earnings of \$5,000,000 and has issued 100,000 shares of stock, the earnings per share are \$50. The **price-earnings ratio, P/E,** is the price of the stock divided by the earnings per share. At one time the price per share of IBM was $\$263\frac{1}{8}$ and the earnings per share were \$17.60. For the IBM stock, the price-earnings ratio, P/E, is given by

$$\frac{P}{E} = \frac{\text{Price of stock}}{\text{Earnings per share}}$$

$$= \frac{263\frac{1}{8}}{17.60}$$

$$= \frac{263.125}{17.60} \qquad \text{Converting to decimal notation}$$

$$\approx 15.0. \qquad \text{Dividing, using a calculator, and rounding to the nearest tenth}$$

Stock Yields

The price per share of IBM stock was $\$263\frac{1}{8}$ and the company was paying a yearly dividend of \$10 per share. It is helpful to those interested in stocks to know what percent the dividend is of the price of the stock. The percent is called the **yield.** For the IBM stock the yield is given by

$$\text{Yield} = \frac{\text{Dividend}}{\text{Price per share}}$$

$$= \frac{10}{263\frac{1}{8}}$$

$$= \frac{10}{263.125} \qquad \text{Converting to decimal notation}$$

$$\approx 0.038 \qquad \text{Dividing and rounding to the nearest thousandth}$$

$$= 3.8\%. \qquad \text{Converting to percent notation}$$

EXERCISES

Compute the price-earnings ratio and the yield for the given stock.

	Stock	*Price per Share*	*Earnings*	*Dividend*	
1.	General Motors	$\$68\frac{5}{8}$	\$11.50	\$5.55	6.0, 8.1%
2.	K-Mart	31	2.60	0.56	11.9, 1.8%
3.	United Airlines	$18\frac{5}{8}$	4.00	0.60	4.7, 3.2%
4.	AT&T	62	6.60	4.20	9.4, 6.8%

NAME SECTION DATE

EXERCISE SET 7.2

a Find percent notation.

1. $\frac{41}{100}$ **2.** $\frac{36}{100}$ **3.** $\frac{1}{100}$ **4.** $\frac{5}{100}$ **5.** $\frac{2}{10}$ **6.** $\frac{7}{10}$

7. $\frac{3}{10}$ **8.** $\frac{9}{10}$ **9.** $\frac{1}{2}$ **10.** $\frac{3}{4}$ **11.** $\frac{5}{8}$ **12.** $\frac{1}{8}$

13. $\frac{2}{5}$ **14.** $\frac{4}{5}$ **15.** $\frac{2}{3}$ **16.** $\frac{1}{3}$ **17.** $\frac{1}{6}$ **18.** $\frac{5}{6}$

19. $\frac{4}{25}$ **20.** $\frac{17}{25}$ **21.** $\frac{1}{20}$ **22.** $\frac{31}{50}$ **23.** $\frac{17}{50}$ **24.** $\frac{3}{20}$

25. Bread is $\frac{9}{25}$ water. Find percent notation for $\frac{9}{25}$.

26. Milk is $\frac{7}{8}$ water. Find percent notation for $\frac{7}{8}$.

b Find fractional notation.

27. 80% **28.** 50% **29.** 62.5% **30.** 12.5%

31. $33\frac{1}{3}\%$ **32.** $83\frac{1}{3}\%$ **33.** $16.\overline{6}\%$ **34.** $66.\overline{6}\%$

ANSWERS

1. 41%
2. 36%
3. 1%
4. 5%
5. 20%
6. 70%
7. 30%
8. 90%
9. 50%
10. 75%
11. 62.5%
12. 12.5%
13. 40%
14. 80%
15. $66.\overline{6}\%$, or $66\frac{2}{3}\%$
16. $33.\overline{3}\%$, or $33\frac{1}{3}\%$
17. $16.\overline{6}\%$, or $16\frac{2}{3}\%$
18. $83.\overline{3}\%$, or $83\frac{1}{3}\%$
19. 16%
20. 68%
21. 5%
22. 62%
23. 34%
24. 15%
25. 36%
26. 87.5%
27. $\frac{4}{5}$
28. $\frac{1}{2}$
29. $\frac{5}{8}$
30. $\frac{1}{8}$
31. $\frac{1}{3}$
32. $\frac{5}{6}$
33. $\frac{1}{6}$
34. $\frac{2}{3}$

ANSWERS

35. $\frac{29}{400}$

36. $\frac{97}{2000}$

37. $\frac{1}{125}$

38. $\frac{1}{500}$

39. $\frac{7}{20}$

40. $\frac{3}{50}$

41. See table.

42. 72.5

43. 5

44. 400

45. 18.75

46. $11.\overline{1}\%$

47. $5.\overline{405}\%$

35. 7.25% **36.** 4.85%

37. 0.8% **38.** 0.2%

39. The United States uses 35% of the world's energy. Find fractional notation for 35%.

40. The United States has 6% of the world's population. Find fractional notation for 6%.

41. Complete the table.

Fractional notation	Decimal notation	Percent notation
$\frac{1}{8}$	0.125	$12\frac{1}{2}\%$, or 12.5%
$\frac{1}{6}$	$0.1\overline{6}$	$16\frac{2}{3}\%$, or $16.\overline{6}\%$
$\frac{1}{5}$	0.2	20%
$\frac{1}{4}$	0.25	25%
$\frac{1}{3}$	$0.\overline{3}$	$33\frac{1}{3}\%$, or $33.\overline{3}\%$
$\frac{3}{8}$	0.375	$37\frac{1}{2}\%$, or 37.5%
$\frac{2}{5}$	0.4	40%
$\frac{1}{2}$	0.5	50%
$\frac{3}{5}$	0.6	60%
$\frac{5}{8}$	0.625	$62\frac{1}{2}\%$, or 62.5%
$\frac{2}{3}$	$0.\overline{6}$	$66\frac{2}{3}\%$, or $66.\overline{6}\%$
$\frac{3}{4}$	0.75	75%
$\frac{4}{5}$	0.8	80%
$\frac{5}{6}$	$0.8\overline{3}$	$83\frac{1}{3}\%$, or $83.\overline{3}\%$
$\frac{7}{8}$	0.875	$87\frac{1}{2}\%$, or 87.5%
$\frac{1}{1}$	1	100%

SKILL MAINTENANCE

Solve.

42. $10 \cdot x = 725$ **43.** $15 \cdot y = 75$ **44.** $0.05 \times b = 20$ **45.** $3 = 0.16 \times b$

SYNTHESIS

Find percent notation.

46. 🖩 $\frac{41}{369}$

47. 🖩 $\frac{54}{999}$

7.3 Solving Percent Problems Using Equations

a Translating to Equations

To solve a problem involving percents, it is helpful to translate first to an equation.

EXAMPLE 1 Translate:

23% of 5 is what?
↓ ↓ ↓ ↓ ↓
$23\% \cdot 5 = a$ ◀

> **"Of" translates to "·", or "×".**
> **"What" translates to some letter.**
> **"Is" translates to "=".**
> **% translates to "$\times \frac{1}{100}$" or "× 0.01".**

EXAMPLE 2 Translate:

What is 11% of 49?
↓ ↓ ↓ ↓ ↓
$a = 11\% \cdot 49$ ◀

DO EXERCISES 1 AND 2.

EXAMPLE 3 Translate:

3 is 10% of what?
↓ ↓ ↓ ↓ ↓
$3 = 10\% \cdot b$ Any letter can be used. ◀

EXAMPLE 4 Translate:

45% of what is 23?
↓ ↓ ↓ ↓ ↓
$45\% \times b = 23$ ◀

DO EXERCISES 3 AND 4.

EXAMPLE 5 Translate:

10 is what percent of 20?
↓ ↓ ↓ ↓ ↓
$10 = n \times 20$ ◀

EXAMPLE 6 Translate:

What percent of 50 is 7?
↓ ↓ ↓ ↓ ↓
$n \cdot 50 = 7$ ◀

DO EXERCISES 5 AND 6.

OBJECTIVES

After finishing Section 7.3, you should be able to:

a Translate percent problems to equations.

b Solve basic percent problems.

FOR EXTRA HELP

Tape 8F

Tape 10A

MAC: 7
IBM: 7

Translate to an equation. Do not solve.

1. 12% of 50 is what?
$12\% \times 50 = a$

2. What is 40% of 60?
$a = 40\% \times 60$

Translate to an equation. Do not solve.

3. 45 is 20% of what?
$45 = 20\% \times t$

4. 120% of what is 60?
$120\% \times y = 60$

Translate to an equation. Do not solve.

5. 16 is what percent of 40?
$16 = n \times 40$

6. What percent of 84 is 10.5?
$b \times 84 = 10.5$

ANSWERS ON PAGE A-4

7. Solve:

What is 12% of 50? 6

ANSWER ON PAGE A-4

b Solving Percent Problems

In solving percent problems, we use the same strategy that we have used for solving problems throughout this text.

> **To solve percent problems,**
> **a) translate to an equation, and**
> **b) solve the equation.**

Percent problems are actually of three different types. Although the method we present does *not* require that you be able to identify which type we are studying, it is helpful to know them.

We know that

15 is 25% of 60, or

$15 = 25\% \times 60.$

We can think of this as:

> **Amount = Percent number × Base.**

Each of the three types of percent problems depends on which of the three pieces of information is missing.

1. Finding the amount

Example: What is 25% of 60?

Translation: $y = 25\% \cdot 60$

2. Finding the base

Example: 15 is 25% of what number?

Translation: $15 = 25\% \cdot y$

3. Finding the percent number

Example: 15 is what percent of 60?

Translation: $15 = y \cdot 60$

Finding the Amount

▶ **EXAMPLE 7** What is 11% of 49?

Translate: $a = 11\% \times 49.$

The letter is by itself. To solve the equation, we just convert 11% to decimal notation and multiply.

$$\begin{array}{r} 4\,9 \\ \times\, 0.1\,1 \\ \hline 4\,9 \\ 4\,9\,0 \\ \hline a = 5.3\,9 \end{array}$$

$11\% = 0.11$

> A way of checking answers is by estimating as follows:
> $11\% \times 49 \approx 10\% \times 50$
> $= 0.10 \times 50 = 5.$
> Since 5 is close to 5.39, our answer is reasonable.

Thus, 5.39 is 11% of 49. The answer is 5.39. ◀

DO EXERCISE 7.

▶ **EXAMPLE 8** 120% of $42 is what?

Translate: $120\% \times 42 = a$.

The letter is by itself. To solve the equation, we carry out the calculation.

$$\begin{array}{r} 4\,2 \\ \times\ 1.2 \\ \hline 8\,4 \\ 4\,2\,0 \\ \hline a = 5\,0.4 \end{array} \qquad 120\% = 1.20 = 1.2$$

Thus, 120% of $42 is $50.40. The answer is $50.40. ◀

DO EXERCISE 8.

8. Solve:
64% of $55 is what? $35.20

Finding the Base

▶ **EXAMPLE 9** 5% of what is 20?

Translate: $5\% \times b = 20$.

This time the letter is *not* by itself. To solve the equation, we divide on both sides by 5%:

$b = 20 \div 5\%$ Dividing on both sides by 5%

$b = 20 \div 0.05$ 5% = 0.05

$b = 400.$

$$0.05\overline{)20.00} = 400. \quad \begin{array}{r} 2000 \\ \hline 0 \end{array}$$

Thus, 5% of 400 is 20. The answer is 400. ◀

DO EXERCISE 9.

9. Solve:
20% of what is 45? 225

▶ **EXAMPLE 10** $3 is 16% of what?

Translate: $3 is 16% of what?

$$3 = 16\% \times b.$$

Again, the letter is not by itself. To solve the equation, we divide on both sides by 16%:

$3 \div 16\% = b$ Dividing on both sides by 16%

$3 \div 0.16 = b$ 16% = 0.16

$18.75 = b.$

$$0.16\overline{)3.0000} = 18.75 \quad \begin{array}{r} 16 \\ \hline 140 \\ 128 \\ \hline 120 \\ 112 \\ \hline 80 \\ 80 \\ \hline 0 \end{array}$$

Thus, $3 is 16% of $18.75. The answer is $18.75. ◀

DO EXERCISE 10.

10. Solve:
$60 is 120% of what? $50

ANSWERS ON PAGE A-4

11. Solve:

16 is what percent of 40?

40%

12. Solve:

What percent of $84 is $10.50?

12.5%

ANSWERS ON PAGE A-4

Finding the Percent Number

In solving these problems, you must remember to convert to percent notation after you have solved the equation.

▶ **EXAMPLE 11** 10 is what percent of 20?

Translate: 10 is what percent of 20?

$$10 = n \times 20.$$

To solve the equation, we divide on both sides by 20 and convert the result to percent notation:

$$n \cdot 20 = 10$$

$$\frac{n \cdot 20}{20} = \frac{10}{20} \qquad \text{Dividing on both sides by 20}$$

$$n = 0.50 = 50\%. \qquad \text{Converting to percent notation}$$

Thus, 10 is 50% of 20. The answer is 50%. ◀

DO EXERCISE 11.

▶ **EXAMPLE 12** What percent of $50 is $16?

Translate: What percent of $50 is $16?

$$n \times 50 = 16.$$

To solve the equation, we divide on both sides by 50 and convert the answer to percent notation:

$$n = 16 \div 50 \qquad \text{Dividing on both sides by 50}$$

$$n = \frac{16}{50}$$

$$n = \frac{16}{50} \cdot \frac{2}{2}$$

$$n = \frac{32}{100}$$

$$n = 32\%.$$

Thus, 32% of $50 is $16. The answer is 32%. ◀

DO EXERCISE 12.

NAME SECTION DATE

EXERCISE SET 7.3

a Translate to an equation. Do not solve.

1. What is 41% of 89?
2. 87% of 41 is what?
3. 89 is what percent of 99?
4. What percent of 25 is 8?
5. 13 is 25% of what?
6. 21.4% of what is 20?

b Solve.

7. What is 120% of 75?
8. What is 65% of 480?
9. 150% of 30 is what?
10. 100% of 13 is what?
11. What is 5% of \$300?
12. What is 3% of \$45?
13. 2.1% of 50 is what?
14. $33\frac{1}{3}$ % of 240 is what? (*Hint:* $33\frac{1}{3}\% = \frac{1}{3}$.)
15. \$12 is what percent of \$50?
16. \$15 is what percent of \$60?
17. 20 is what percent of 10?
18. 90 is what percent of 30?

ANSWERS

1. $y = 41\% \times 89$
2. $87\% \times 41 = p$
3. $89 = a \times 99$
4. $y \times 25 = 8$
5. $13 = 25\% \times y$
6. $21.4\% \times m = 20$
7. 90
8. 312
9. 45
10. 13
11. \$15
12. \$1.35
13. 1.05
14. 80
15. 24%
16. 25%
17. 200%
18. 300%

ANSWERS

19. 50%

20. 80%

21. 125%

22. 50%

23. 40

24. 225

25. $40

26. $89

27. 88

28. 78

29. 20

30. 50

31. 6.25

32. 423

33. $846.60

34. $515.20

35. $\frac{9}{100}$

36. $\frac{179}{100}$

37. 0.89

38. 0.07

39. $880 (can vary); $843.20

40. 62.5% (can vary); 64.5%

19. What percent of $300 is $150?

20. What percent of $50 is $40?

21. What percent of 80 is 100?

22. What percent of 30 is 15?

23. 20 is 50% of what?

24. 45 is 20% of what?

25. 40% of what is $16?

26. 100% of what is $89?

27. 56.32 is 64% of what?

28. 34.32 is 44% of what?

29. 70% of what is 14?

30. 70% of what is 35?

31. What is $62\frac{1}{2}\%$ of 10?

32. What is $35\frac{1}{4}\%$ of 1200?

33. What is 8.3% of $10,200?

34. What is 9.2% of $5600?

SKILL MAINTENANCE

Write fractional notation.

35. 0.09

36. 1.79

Write decimal notation.

37. $\frac{89}{100}$

38. $\frac{7}{100}$

SYNTHESIS

Solve.

39. ▦ What is 7.75% of $10,880?

Estimate ______________

Calculate ______________

40. ▦ 50,951.775 is what percent of 78,995?

Estimate ______________

Calculate ______________

7.4 Solving Percent Problems Using Proportions*

a Translating to Proportions

A percent is a ratio of some number to 100. For example, 75% is the ratio

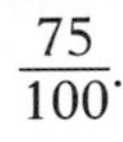

$$\frac{75}{100}.$$

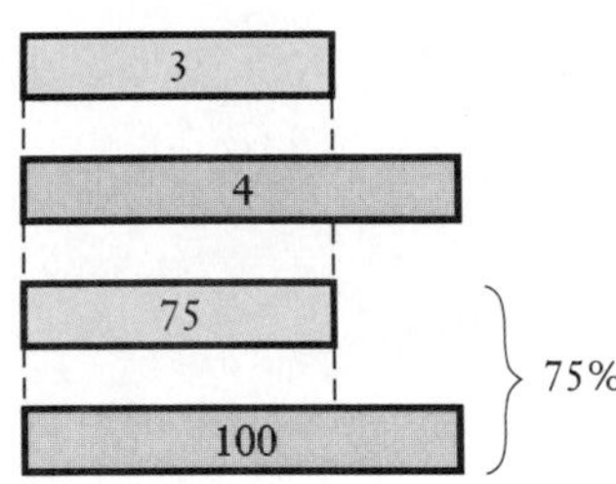

We also know that 3 and 4 have the same ratio as 75 and 100. Thus,

$$\frac{3}{4} = \frac{75}{100} = 75\%.$$

To solve a percent problem using a proportion, we translate as follows:

$$\text{Amount} \longrightarrow \frac{a}{b} = \frac{n}{100} \longleftarrow \text{Number}$$
$$\text{Base} \longrightarrow b \qquad 100 \longleftarrow 100$$

You might find it helpful to read this as "part is to whole as part is to whole."

For example,

75% of 48 is 36

translates to

$$\frac{36}{48} = \frac{75}{100}.$$

A clue in translating is that the base, b, corresponds to 100 and usually follows the wording "percent of." Also, $n\%$ always translates to $n/100$. Another aid in translating is to make a comparison drawing. We usually start with the percent side. We have 0% at the top and 100% at the bottom. Then we estimate where the 75% would be located. The numbers, or quantities, that correspond are then filled in. The base—in this case, 48—always corresponds to 100% and the amount—in this case, 36—corresponds to 75%.

Percents	Quantities
0%	0
100%	

Percents	Quantities
0%	0
75%	
100%	

Percents	Quantities
0%	0
75%	36
100%	48

The proportion can then be read easily from the drawing.

OBJECTIVES

After finishing Section 7.4, you should be able to:

a Translate percent problems to proportions.

b Solve basic percent problems.

FOR EXTRA HELP

Tape 8F — Tape 10A — MAC: 7 IBM: 7

*Note: This section presents an alternative method for solving basic percent problems. You can use either equations or proportions to solve percent problems, but you might prefer one method over the other, or your instructor may direct you to use one method over the other.

Translate to a proportion. Do not solve.

1. 12% of 50 is what? $\frac{12}{100}=\frac{a}{50}$

2. What is 40% of 60? $\frac{40}{100}=\frac{a}{60}$

▶ **EXAMPLE 1** Translate to a proportion.

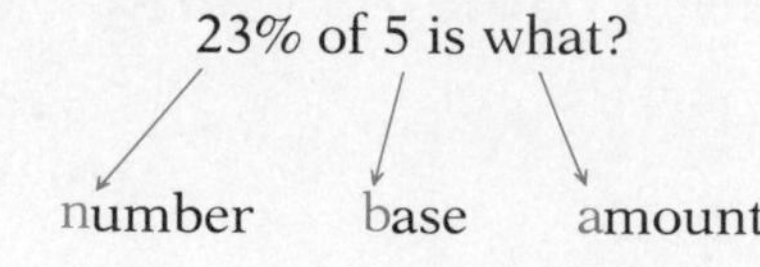

$$\frac{23}{100}=\frac{a}{5}$$

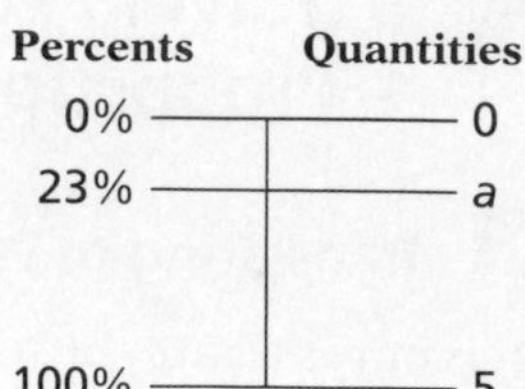

▶ **EXAMPLE 2** Translate to a proportion.

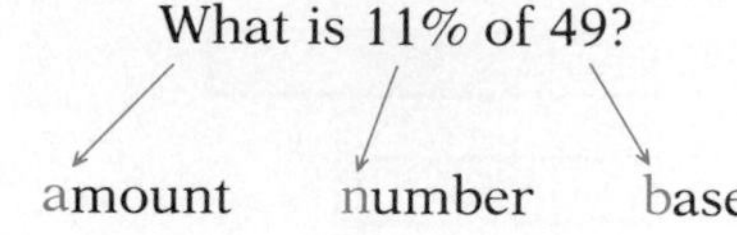

Percents Quantities
0% 0
11% a
100% 49

$$\frac{11}{100}=\frac{a}{49}$$

DO EXERCISES 1 AND 2.

Translate to a proportion. Do not solve.

3. 45 is 20% of what? $\frac{20}{100}=\frac{45}{b}$

4. 120% of what is 60? $\frac{120}{100}=\frac{60}{b}$

▶ **EXAMPLE 3** Translate to a proportion.

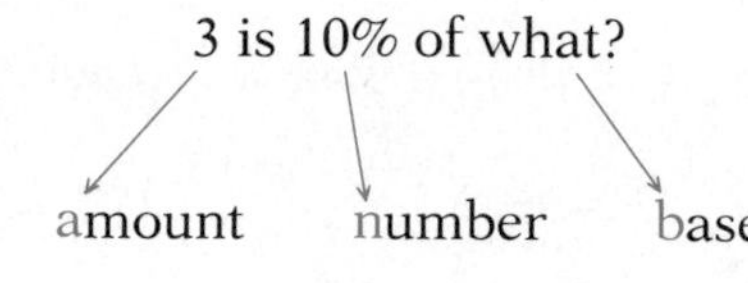

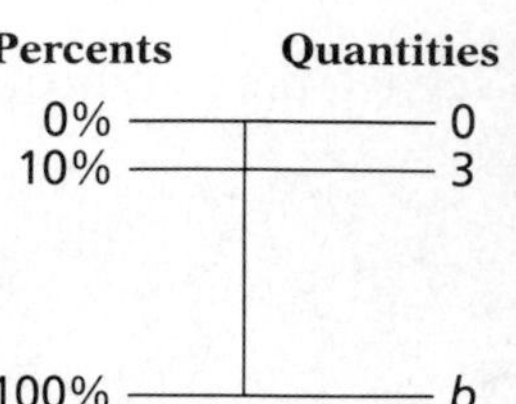

$$\frac{10}{100}=\frac{3}{b}$$

▶ **EXAMPLE 4** Translate to a proportion.

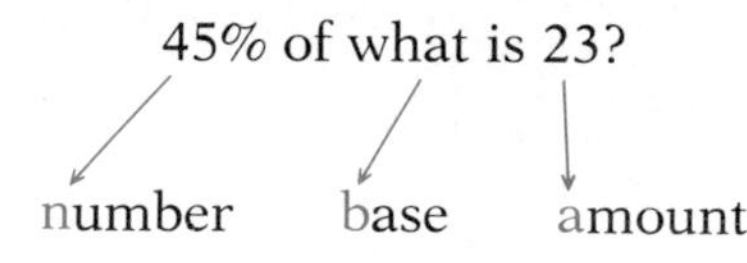

Percents Quantities
0% 0
45% 23
100% b

$$\frac{45}{100}=\frac{23}{b}$$

DO EXERCISES 3 AND 4.

▶ **EXAMPLE 5** Translate to a proportion.

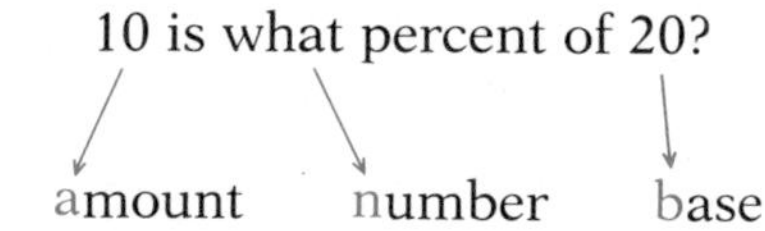

Percents Quantities
0% 0
n% 10
100% 20

$$\frac{n}{100}=\frac{10}{20}$$

ANSWERS ON PAGE A-4

7.4 Solving Percent Problems Using Proportions*

OBJECTIVES

After finishing Section 7.4, you should be able to:

a Translate percent problems to proportions.

b Solve basic percent problems.

FOR EXTRA HELP

Tape 8F | Tape 10A | MAC: 7 IBM: 7

a Translating to Proportions

A percent is a ratio of some number to 100. For example, 75% is the ratio

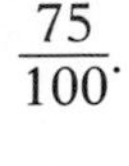

$\frac{75}{100}$.

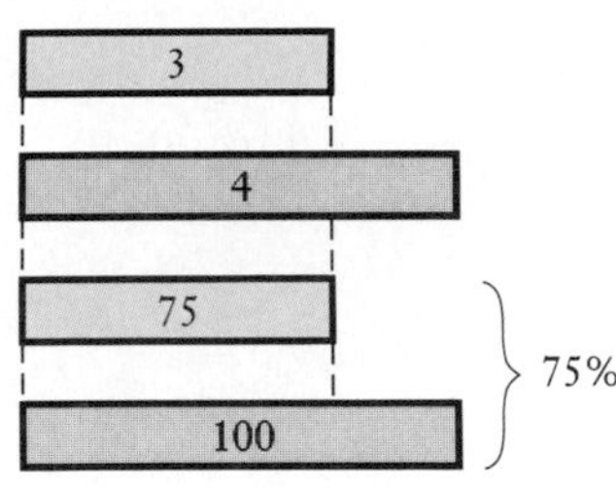

We also know that 3 and 4 have the same ratio as 75 and 100. Thus,

$$\frac{3}{4} = \frac{75}{100} = 75\%.$$

To solve a percent problem using a proportion, we translate as follows:

$$\text{Amount} \longrightarrow \frac{a}{b} = \frac{n}{100} \longleftarrow \text{Number}$$
$$\text{Base} \longrightarrow b \qquad 100 \longleftarrow 100$$

You might find it helpful to read this as "part is to whole as part is to whole."

For example,

75% of 48 is 36

translates to

$$\frac{36}{48} = \frac{75}{100}.$$

A clue in translating is that the base, b, corresponds to 100 and usually follows the wording "percent of." Also, $n\%$ always translates to $n/100$. Another aid in translating is to make a comparison drawing. We usually start with the percent side. We have 0% at the top and 100% at the bottom. Then we estimate where the 75% would be located. The numbers, or quantities, that correspond are then filled in. The base—in this case, 48—always corresponds to 100% and the amount—in this case, 36—corresponds to 75%.

Percents	Quantities
0%	0
100%	

Percents	Quantities
0%	0
75%	
100%	

Percents	Quantities
0%	0
75%	36
100%	48

The proportion can then be read easily from the drawing.

*Note: This section presents an alternative method for solving basic percent problems. You can use either equations or proportions to solve percent problems, but you might prefer one method over the other, or your instructor may direct you to use one method over the other.

Translate to a proportion. Do not solve.

1. 12% of 50 is what? $\frac{12}{100} = \frac{a}{50}$

2. What is 40% of 60? $\frac{40}{100} = \frac{a}{60}$

Translate to a proportion. Do not solve.

3. 45 is 20% of what? $\frac{20}{100} = \frac{45}{b}$

4. 120% of what is 60? $\frac{120}{100} = \frac{60}{b}$

▶ EXAMPLE 1 Translate to a proportion.

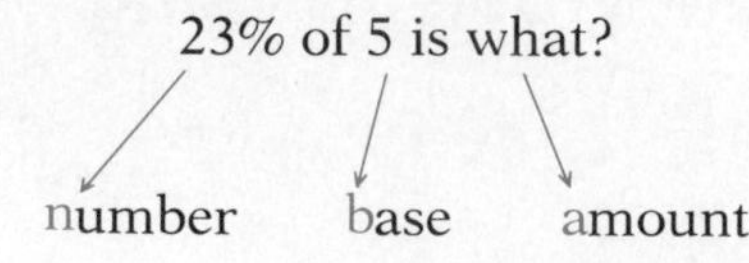

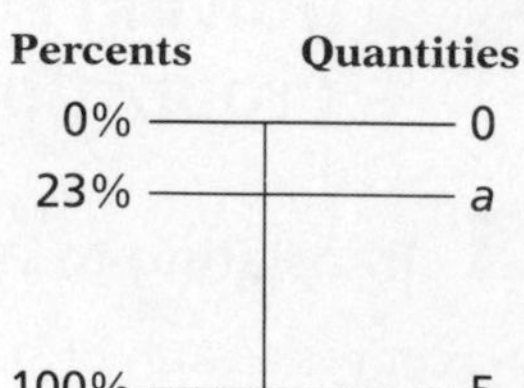

$$\frac{23}{100} = \frac{a}{5}$$

▶ EXAMPLE 2 Translate to a proportion.

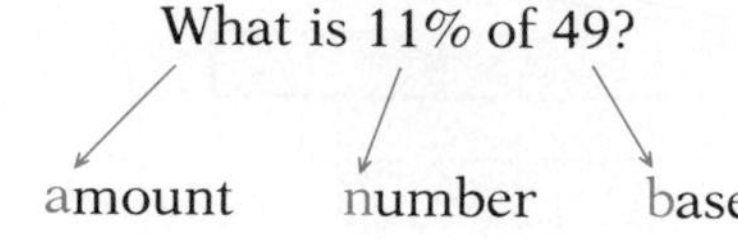

Percents	Quantities
0%	0
11%	a
100%	49 ◀

$$\frac{11}{100} = \frac{a}{49}$$

DO EXERCISES 1 AND 2.

▶ EXAMPLE 3 Translate to a proportion.

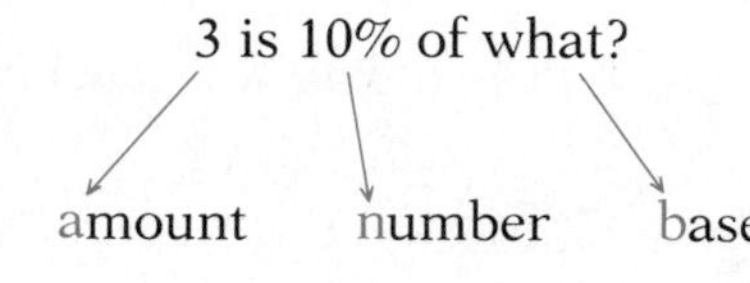

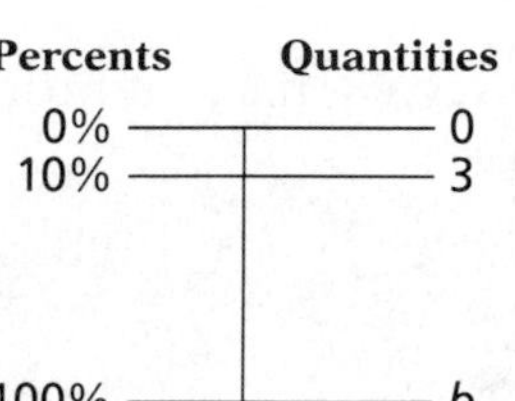

$$\frac{10}{100} = \frac{3}{b}$$

▶ EXAMPLE 4 Translate to a proportion.

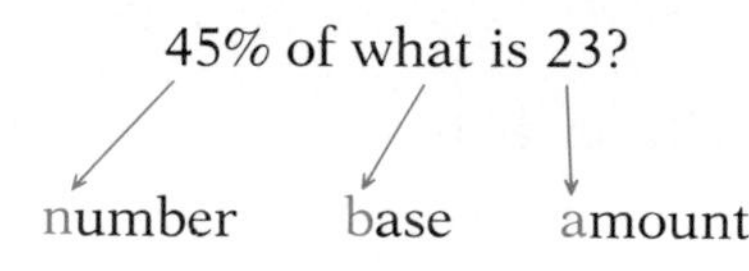

Percents	Quantities
0%	0
45%	23
100%	b ◀

$$\frac{45}{100} = \frac{23}{b}$$

DO EXERCISES 3 AND 4.

▶ EXAMPLE 5 Translate to a proportion.

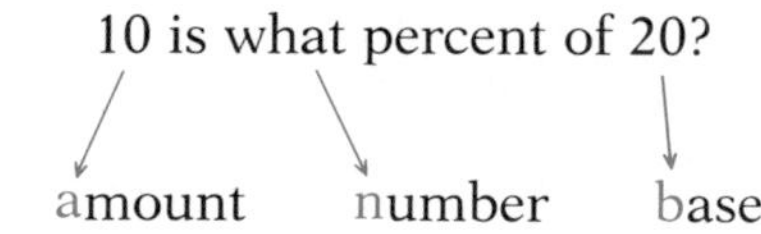

Percents	Quantities
0%	0
n%	10
100%	20 ◀

$$\frac{n}{100} = \frac{10}{20}$$

ANSWERS ON PAGE A-4

▶ **EXAMPLE 6** Translate to a proportion.

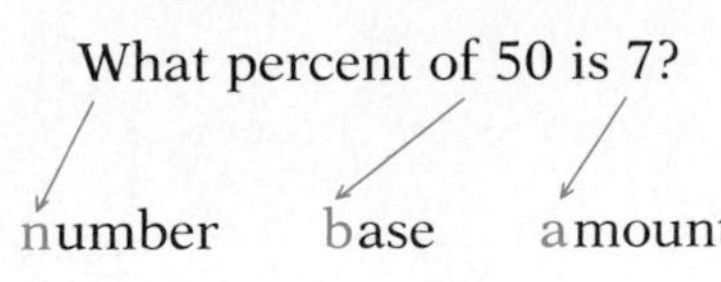

Percents	Quantities
0%	0
n%	7
100%	50

$$\frac{n}{100} = \frac{7}{50}$$ ◀

DO EXERCISES 5 AND 6.

Translate to a proportion. Do not solve.

5. 16 is what percent of 40?

$\frac{n}{100} = \frac{16}{40}$

6. What percent of 84 is 10.5?

$\frac{n}{100} = \frac{10.5}{84}$

b Solving Percent Problems

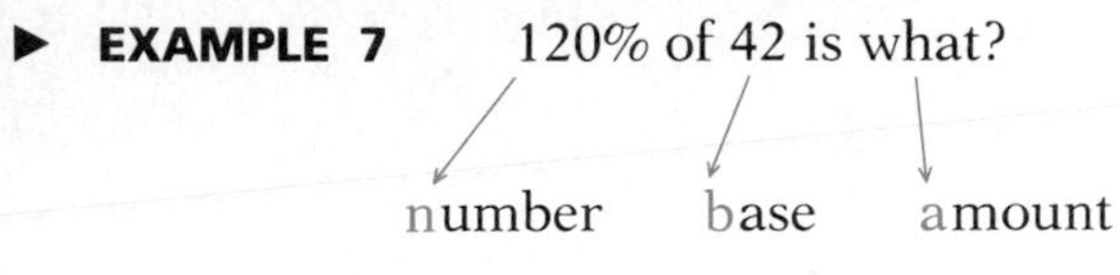

Percents	Quantities
0%	0
100%	42
120%	a

Translate: $\frac{120}{100} = \frac{a}{42}$

Solve: $120 \cdot 42 = 100 \cdot a$ **Finding cross-products**

$\frac{120 \cdot 42}{100} = a$ **Dividing by 100**

$\frac{5040}{100} = a$

$50.4 = a$ **Simplifying**

Thus, 120% of 42 is 50.4. The answer is 50.4. ◀

DO EXERCISES 7 AND 8.

Solve.

7. What is 12% of 50? 6

8. 64% of 55 is what? 35.2

▶ **EXAMPLE 8** 5% of what is \$20?

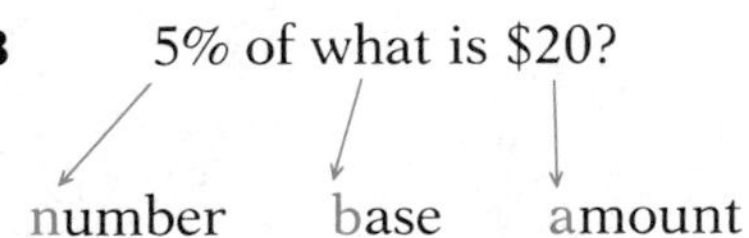

Percents	Quantities
0%	0
5%	20
100%	b

Translate: $\frac{5}{100} = \frac{20}{b}$

Solve: $5 \cdot b = 100 \cdot 20$ **Finding cross-products**

$b = \frac{100 \cdot 20}{5}$ **Dividing by 5**

$b = \frac{5 \cdot 20 \cdot 20}{5}$ **Factoring**

$b = 400$ **Simplifying**

Thus, 5% of \$400 is \$20. The answer is \$400. ◀

DO EXERCISE 9.

9. Solve:

20% of what is \$45? \$225

ANSWERS ON PAGE A-4

10. Solve:

60 is 120% of what? 50

▶ **EXAMPLE 9** 3 is 16% of what?

amount number base

Percents | Quantities
0% — 0
16% — 3
100% — b

Translate: $\dfrac{16}{100} = \dfrac{3}{b}$

Solve: $16 \cdot b = 100 \cdot 3$ **Finding cross-products**

$b = \dfrac{100 \cdot 3}{16}$ **Dividing by 16**

$b = \dfrac{300}{16}$

$b = 18.75$

Thus, 3 is 16% of 18.75. The answer is 18.75. ◀

DO EXERCISE 10.

11. Solve:

$16 is what percent of $40?

40%

▶ **EXAMPLE 10** $10 is what percent of $20?

amount number base

Percents | Quantities
0% — 0
n% — $10
100% — $20

Translate: $\dfrac{n}{100} = \dfrac{10}{20}$

Solve: $20 \cdot n = 100 \cdot 10$ **Finding cross-products**

$n = \dfrac{100 \cdot 10}{20}$ **Dividing by 20**

$n = \dfrac{20 \cdot 5 \cdot 10}{20}$

$n = 50$

Thus, $10 is 50% of $20. The answer is 50%. ◀

DO EXERCISE 11.

12. Solve:

What percent of 84 is 10.5?

12.5%

▶ **EXAMPLE 11** What percent of 50 is 16?

number base amount

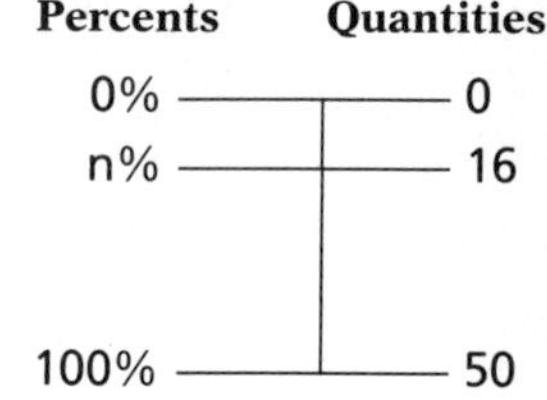

Translate: $\dfrac{n}{100} = \dfrac{16}{50}$

Solve: $50 \cdot n = 100 \cdot 16$ **Finding cross-products**

$n = \dfrac{100 \cdot 16}{50}$ **Dividing by 50**

$n = \dfrac{50 \cdot 2 \cdot 16}{50}$

$n = 32$

Thus, 32% of 50 is 16. The answer is 32%. ◀

DO EXERCISE 12.

ANSWERS ON PAGE A-4

NAME SECTION DATE

EXERCISE SET 7.4

a Translate to a proportion. Do not solve.

1. What is 82% of 74?
2. 58% of 65 is what?
3. 4.3 is what percent of 5.9?
4. What percent of 6.8 is 5.3?
5. 14 is 25% of what?
6. 22.3% of what is 40?

b Solve.

7. What is 84% of $50?
8. What is 78% of $90?
9. 80% of 550 is what?
10. 90% of 740 is what?
11. What is 8% of 1000?
12. What is 9% of 2000?
13. 4.8% of 60 is what?
14. 63.1% of 80 is what?
15. $24 is what percent of $96?
16. $14 is what percent of $70?
17. 102 is what percent of 100?
18. 103 is what percent of 100?
19. What percent of $480 is $120?
20. What percent of $80 is $60?

ANSWERS

1. $\frac{82}{100} = \frac{a}{74}$
2. $\frac{58}{100} = \frac{a}{65}$
3. $\frac{n}{100} = \frac{4.3}{5.9}$
4. $\frac{n}{100} = \frac{5.3}{6.8}$
5. $\frac{25}{100} = \frac{14}{b}$
6. $\frac{22.3}{100} = \frac{40}{b}$
7. $42
8. $70.20
9. 440
10. 666
11. 80
12. 180
13. 2.88
14. 50.48
15. 25%
16. 20%
17. 102%
18. 103%
19. 25%
20. 75%

ANSWERS

21. 93.75%

22. $33.\overline{3}\%$, or $33\frac{1}{3}\%$

23. $72

24. $375

25. 90

26. 120

27. 88

28. 98

29. 20

30. 12.5

31. 25

32. 1124.5

33. $780.2

34. $6612

35. $1134 (can vary); $1118.64

36. 12,500 (can vary); 12,448

21. What percent of 160 is 150?

22. What percent of 24 is 8?

23. $18 is 25% of what?

24. $75 is 20% of what?

25. 60% of what is 54?

26. 80% of what is 96?

27. 65.12 is 74% of what?

28. 63.7 is 65% of what?

29. 80% of what is 16?

30 80% of what is 10?

31. What is $62\frac{1}{2}\%$ of 40?

32. What is $43\frac{1}{4}\%$ of 2600?

33. What is 9.4% of $8300?

34. What is 8.7% of $76,000?

SYNTHESIS

Solve.

35. ▦ What is 8.85% of $12,640?

Estimate ______________

Calculate ______________

36. ▦ 78.8% of what is 9809.024?

Estimate ______________

Calculate ______________

7.5 Applications of Percent

a Percent Problems

Problems involving percent are not always stated in a manner easily translated to an equation. In such cases, it is helpful to restate the problem before translating. Sometimes it also helps to draw a picture.

▶ **EXAMPLE 1** The FBI annually receives 16,000 applications for the position of an agent. It accepts 600 of these applicants. What percent does it accept?

Method 1. Solve using an equation.

Restate: 600 is what percent of 16,000?

Translate: $600 = n \times 16{,}000$

To solve the equation, we divide on both sides by 16,000:

$$600 \div 16{,}000 = n$$
$$0.0375 = n$$
$$3.75\% = n.$$

The FBI accepts 3.75% of its applicants.

Method 2*. Solve using a proportion.

Restate: 600 is what percent of 16,000?

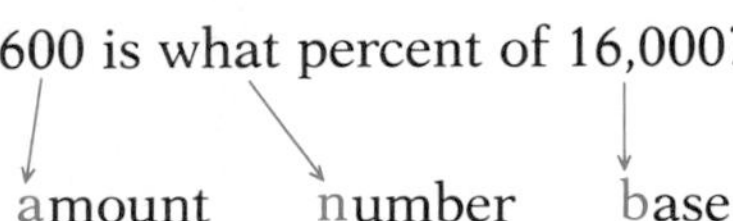

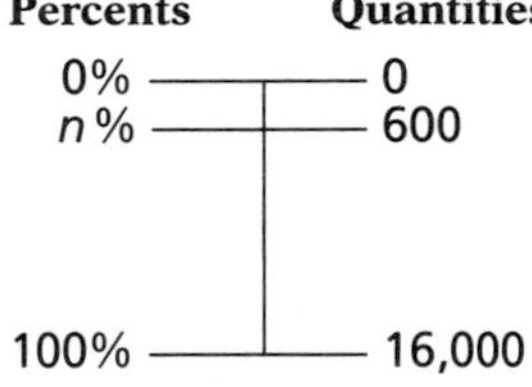

Translate: $\frac{n}{100} = \frac{600}{16{,}000}$

Solve: $16{,}000 \cdot n = 100 \cdot 600$ **Finding cross-products**

$n = \frac{100 \cdot 600}{16{,}000}$ **Dividing by 16,000**

$n = \frac{60{,}000}{16{,}000}$

$n = 3.75$

The FBI accepts 3.75% of its applicants. ◀

DO EXERCISE 1.

*Note: If you skipped Section 7.4, then you should ignore method 2.

OBJECTIVES

After finishing Section 7.5, you should be able to:

a Solve applied percent problems.

b Solve percent problems involving percent increase or decrease.

FOR EXTRA HELP

Tape 9A Tape 10B MAC: 7 IBM: 7

1. A college basketball team won 11 of its 25 games. What percent of its games did it win? 44%

ANSWER ON PAGE A-4

2. The weight of a human brain is 2.5% of total body weight. A person weighs 200 lb. What does the brain weigh? 5 lb

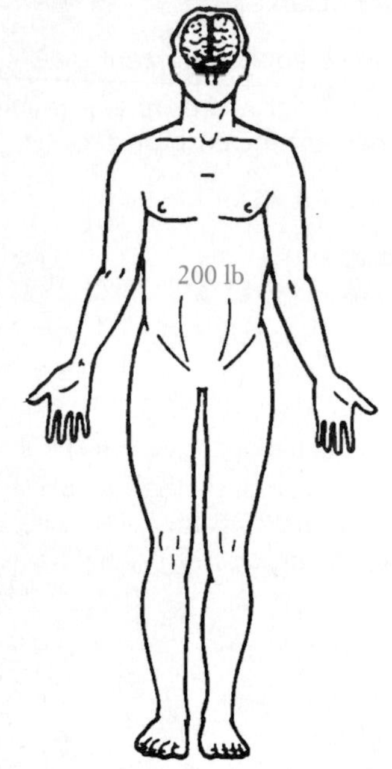

▶ **EXAMPLE 2** Have you ever wondered why you receive so much junk mail? One reason offered by the U.S. Postal Service is that we open and read 78% of the advertising we receive in the mail. Suppose that a business sends out 9500 advertising brochures. How many of them can it expect to be opened and read?

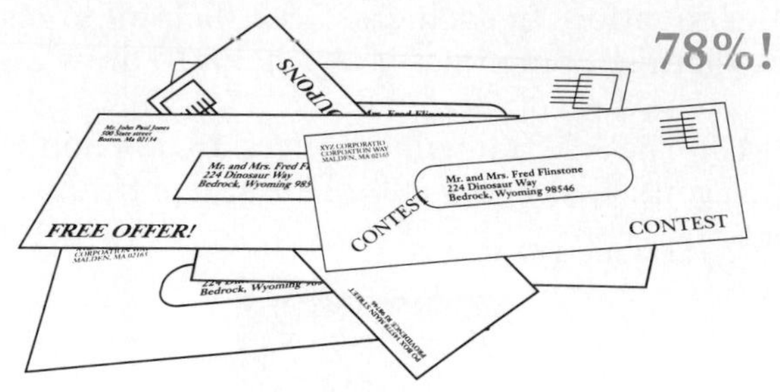

Solve using an equation.

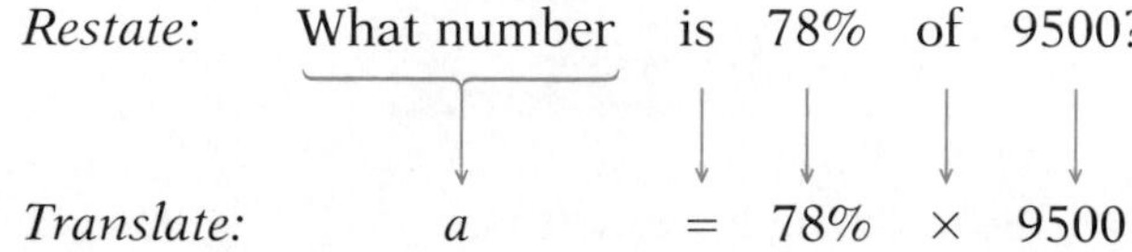

This tells us what to do. We convert 78% to decimal notation and multiply:

$$a = 78\% \times 9500 = 0.78 \times 9500 = 7410.$$

The business can expect 7410 of its brochures to be opened and read.

Method 2. Solve using a proportion.

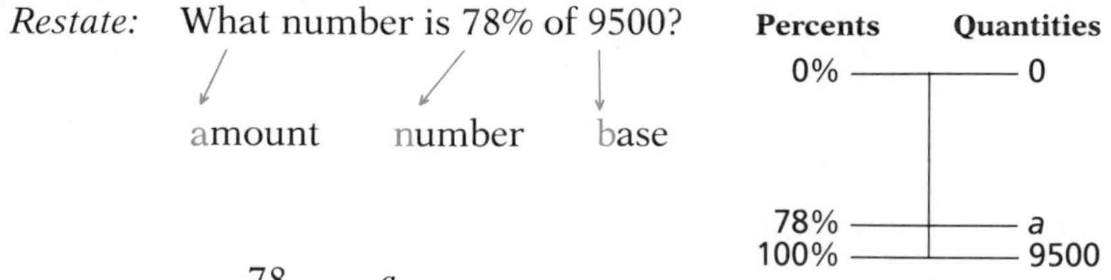

Translate: $\frac{78}{100} = \frac{a}{9500}$

Solve: $78 \cdot 9500 = 100 \cdot a$ **Finding cross-products**

$\frac{78 \cdot 9500}{100} = a$ **Dividing by 100**

$\frac{741{,}000}{100} = a$

$7410 = a$

The business can expect 7410 of its brochures to be opened and read. ◀

DO EXERCISE 2.

ANSWER ON PAGE A-4

b Percent Increase or Decrease

Percent is often used to state increases or decreases. Suppose the population of a town has *increased* 70%. This means that the increase was 70% of the former population. The population of a town is 2340 and it increases 70%. The increase is 70% of 2340, or 1638. The new population is 2340 + 1638, or 3978, which is shown below.

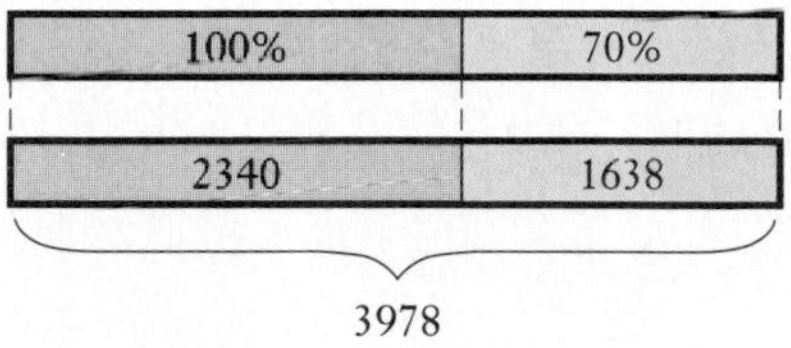

100%	70%
2340	1638

3978

What do we mean when we say that the price of Swiss cheese has decreased 8%? If the price was $1.00 a pound and it went down to $0.92 a pound, then the decrease is $0.08, which is 8% of the original price. We can see this in the following figure.

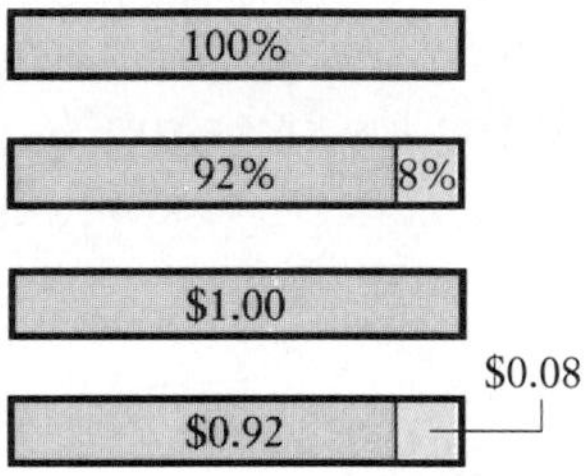

100%

92% | 8%

$1.00

$0.92 | $0.08

To find a percent of increase or decrease, find the amount of increase or decrease and then determine what percent this is of the *original* amount.

▶ **EXAMPLE 3** The price of milk increased from 40 cents per liter to 45 cents per liter. What was the percent of increase?

We make a drawing.

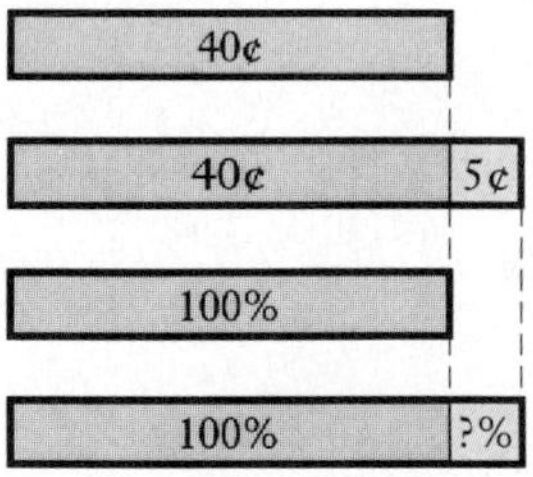

40¢

40¢ | 5¢

100%

100% | ?%

a) First, we find the increase by subtracting.

$$\begin{array}{rl} 45 & \text{New price} \\ -\,40 & \text{Original price} \\ \hline 5 & \text{Increase} \end{array}$$

The increase is 5 cents.

3. The price of an automobile increased from \$5800 to \$6322. What was the percent of increase? 9%

ANSWER ON PAGE A-4

b) Now we ask:

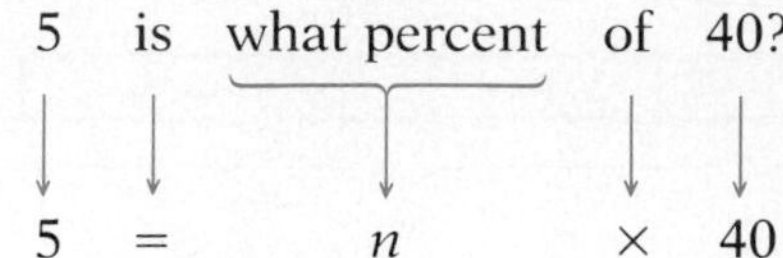

5 is what percent of 40 (the original price)?

A common error is to use 45 instead of 40, the *original* amount.

To find out, we use either of our two methods.

Method 1. Solve using an equation.

$$\underbrace{5}\ \ \text{is}\ \ \text{what percent}\ \ \text{of}\ \ 40?$$

$$5 = n \times 40$$

To solve the equation, we divide on both sides by 40:

$$5 \div 40 = n \qquad \textbf{Dividing by 40}$$
$$0.125 = n$$
$$12.5\% = n.$$

The percent of increase was 12.5%.

Method 2. Solve using a proportion.

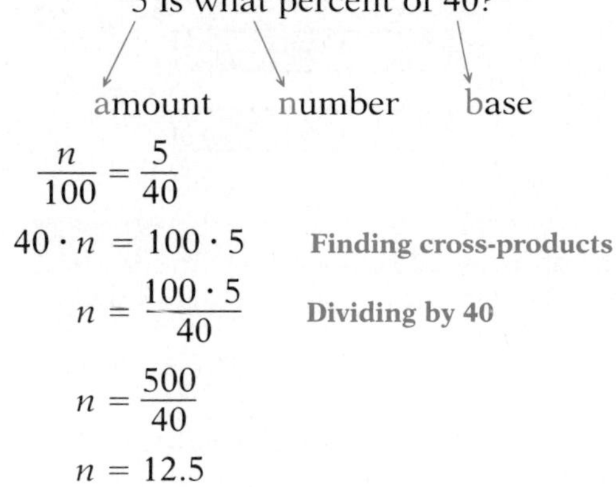

5 is what percent of 40?

amount number base

$$\frac{n}{100} = \frac{5}{40}$$
$$40 \cdot n = 100 \cdot 5 \qquad \textbf{Finding cross-products}$$
$$n = \frac{100 \cdot 5}{40} \qquad \textbf{Dividing by 40}$$
$$n = \frac{500}{40}$$
$$n = 12.5$$

The percent of increase was 12.5%. ◀

DO EXERCISE 3.

▶ **EXAMPLE 4** By proper furnace maintenance, a family that pays a monthly fuel bill of $78.00 can reduce their bill to $70.20. What is the percent of decrease?

We make a drawing.

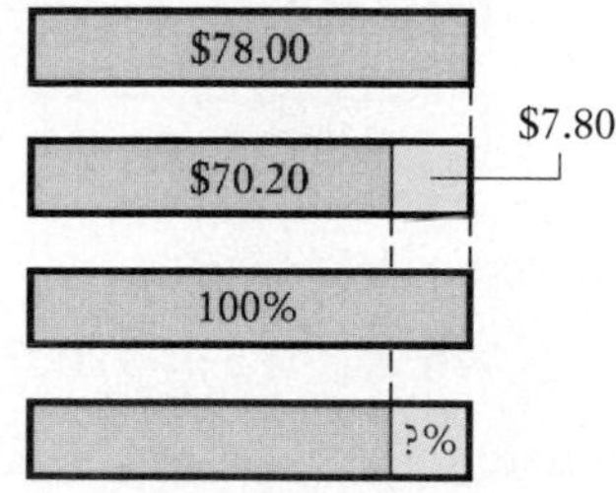

a) First, we find the decrease.

$$\begin{array}{r l} 78.00 & \text{Original bill} \\ -\ 70.20 & \text{New bill} \\ \hline 7.80 & \text{Decrease} \end{array}$$

The decrease is $7.80.

b) Now we ask:

7.80 is what percent of 78.00 (the original bill)?

A common error is to use $70.20 instead of $78.00, the *original* amount.

Method 1. Solve using an equation.

7.80 is what percent of 78.00?

$$7.80 = n \times 78.00$$

To solve the equation, we divide on both sides by 78:

$$7.8 \div 78 = n \quad \text{Dividing by 78}$$
$$0.1 = n$$
$$10\% = n.$$

The percent of decrease is 10%.

Method 2. Solve using a proportion.

7.80 is what percent of 78.00?

amount (7.80), number (what percent), base (78.00)

$$\frac{n}{100} = \frac{7.80}{78.00}$$
$$78.00 \times n = 100 \times 7.80 \quad \text{Finding cross-products}$$
$$n = \frac{100 \times 7.80}{78.00} \quad \text{Dividing by 78.00}$$
$$n = \frac{780}{78}$$
$$n = 10$$

The percent of decrease is 10%. ◀

DO EXERCISE 4.

4. By using only cold water in the washing machine, a family that pays a monthly fuel bill of $78.00 can reduce their bill to $74.88. What is the percent of decrease? 4%

ANSWER ON PAGE A-4

5. A consumer earns \$9800 one year and gets a 9% raise the next. What is the new salary?

\$10,682

▶ **EXAMPLE 5** A consumer earns \$9700 one year and gets a 6% raise the next. What is the new salary?

We make a drawing.

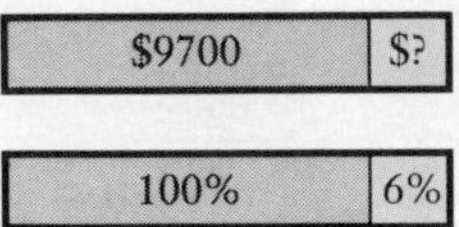

a) First, we find the increase. We ask:

What is 6% of 9700?

Method 1. Solve using an equation.

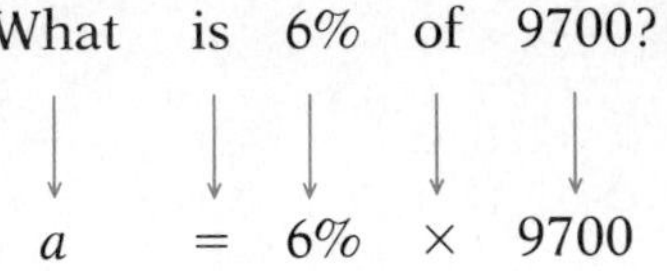

This tells us what to do. We convert 6% to decimal notation and multiply:

$$a = 0.06 \times 9700 = 582.$$

The increase is \$582.00.

Method 2. Solve using a proportion.

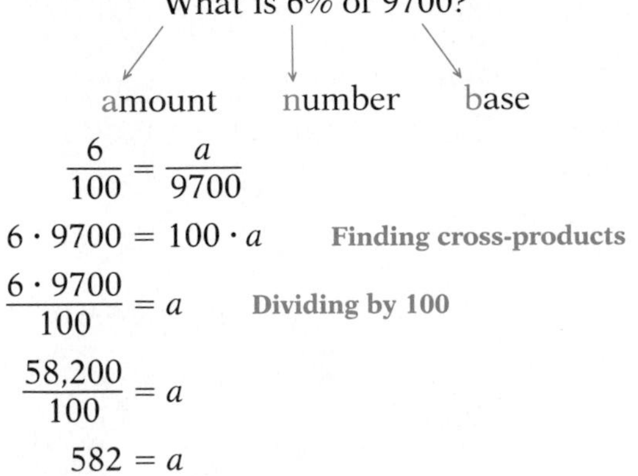

$$\frac{6}{100} = \frac{a}{9700}$$

$$6 \cdot 9700 = 100 \cdot a \qquad \text{Finding cross-products}$$

$$\frac{6 \cdot 9700}{100} = a \qquad \text{Dividing by 100}$$

$$\frac{58{,}200}{100} = a$$

$$582 = a$$

The increase is \$582.00.

b) The new salary is

$$\$9700 + \$582 = \$10{,}282.$$ ◀

DO EXERCISE 5.

ANSWER ON PAGE A-4

NAME SECTION DATE

EXERCISE SET 7.5

ANSWERS

a Solve.

1. It has been determined by sociologists that 17% of the population is left-handed. Each week 160 men enter a tournament conducted by the Professional Bowlers Association. How many would you expect to be left-handed? not left-handed? Round to the nearest one.

2. A guideline commonly used by businesses is to use 5% of their operating budget for advertising. A business has an operating budget of $8000 per week. How much should it spend each week for advertising? for other expenses?

3. Of all moviegoers, 67% are in the 12–29 age group. A theater contained 800 people for a showing of *Teenage Ninja Mathematics Professors*. How many were in the 12–29 age group? not in this age group?

4. Deming, New Mexico, claims to have the purest drinking water in the world. It is 99.9% pure. If you had 240 L of water from Deming, how much of it, in liters, would be pure? impure?

5. A baseball player gets 13 hits in 40 at bats. What percent are hits? not hits?

6. On a test of 80 items, a student had 76 correct. What percent were correct? incorrect?

7. A lab technician has 680 mL of a solution of water and acid; 3% is acid. How many milliliters are acid? water?

8. A lab technician has 540 mL of a solution of alcohol and water; 8% is alcohol. How many milliliters are alcohol? water?

Answers

1. 27; 133
2. $400; $7600
3. 536; 264
4. 239.76 L; 0.24 L
5. 32.5%; 67.5%
6. 95%; 5%
7. 20.4 mL; 659.6 mL
8. 43.2 mL; 496.8 mL

ANSWERS

9. 25%

10. 7%

11. 45%; $37\frac{1}{2}$%; $17\frac{1}{2}$%

12. a) $15,000; $25,000

b) 62.5%

13. 8%

14. 5%

9. Of the 8760 hours in a year, most television sets are on for 2190 hours. What percent is this?

10. In a medical study, it was determined that if 800 people kiss someone else who has a cold, only 56 will actually catch a cold. What percent is this?

11. A nut dealer has 1800 lb of peanuts, 1500 lb of cashews, and 700 lb of almonds. What percent are peanuts? cashews? almonds?

12. It costs an oil company $40,000 a day to operate two refineries. Refinery A takes 37.5% of the cost, and refinery B takes the rest of the cost.

a) What is the cost of operating refinery A? refinery B?

b) What percent of the cost does it take to run refinery B?

b Solve.

13. The amount in a savings account increased from $200 to $216. What was the percent of increase?

14. The population of a small town increased from 840 to 882. What was the percent of increase?

15. During a sale, a dress decreased in price from \$70 to \$56. What was the percent of decrease?

16. A person on a diet goes from a weight of 125 lb to a weight of 110 lb. What is the percent of decrease?

17. A person earns \$8600 one year and gets a 5% raise in salary. What is the new salary?

18. A person earns \$10,400 one year and gets an 8% raise in salary. What is the new salary?

19. The value of a car typically decreases by 30% in the first year. A car is bought for \$12,000. What is its value one year later?

20. One year the pilots of Pan American Airlines shocked the business world by taking an 11% pay cut. The former salary was \$55,000. What was the reduced salary?

21. World population is increasing by 1.6% each year. In 1990, it was 5.2 billion. How much will it be in 1991? 1992? 1993?

22. By increasing the thermostat from 72° to 78°, a family can reduce its cooling bill by 50%. If the cooling bill was \$106.00, what would the new bill be? By what percent has the temperature been increased?

ANSWERS

15. 20%

16. 12%

17. \$9030

18. \$11,232

19. \$8400

20. \$48,950

21. 5.2832 billion; about 5.3677 billion; about 5.4536 billion

22. \$53.00; $8.\overline{3}\%$, or $8\frac{1}{3}\%$

ANSWERS

23. $12,500

24. 34.375%, or $34\frac{3}{8}\%$

25. 166; 156; 146; 136; 122

26. $5779.20; $7525; $11,076.80

27. Neither; they are the same.

28. $83.\overline{3}\%$, or $83\frac{1}{3}\%$

29. About 5 ft, 6 in.

30. About 6 ft, 7 in.

23. A car normally depreciates 30% of its original value in the first year. A car is worth $8750 after the first year. What was its original cost?

24. A standard or nominal "two by four" actually measures $1\frac{1}{2}$ in. by $3\frac{1}{2}$ in. The rough board is 2 in. by 4 in., but is planed and dried to the finished size. What percent of the wood is removed in planing and drying?

25. *Treadmill test.* Treadmill tests are often administered to diagnose heart ailments. A guideline in such a test is to try to get you to reach what is called your **maximal heartbeat,** in beats per minute. The maximal heartbeat is found by subtracting a person's age from 220 and then multiplying by 85%. What is the maximal heartbeat of a person of age 25? 36? 48? 60? 76? Round to the nearest one.

26. *Car depreciation.* Given normal use, an American-made car will depreciate 30% of its original cost the first year and 14% of its remaining value in the second year. What is the value of a car at the end of the second year if its original cost was $9600? $12,500? $18,400?

SYNTHESIS

27. Which is higher, if either?

a) $1000 increased by 15%, then that amount decreased by 15%, or,

b) $1000 decreased by 15%, then that amount increased by 15%.

28. If p is 120% of q, q is what percent of p?

29. It has been determined that at the age of 10, a girl has reached 84.4% of her final adult growth. A girl is 4 ft, 8 in. at the age of 10. What will be her final adult height?

30. It has been determined that at the age of 15 a boy has reached 96.1% of his final adult height. A boy is 6 ft, 4 in. at the age of 15. What will be his final adult height?

7.6 Consumer Applications: Sales Tax

a Percent is used in sales tax computations. The sales tax rate in Arkansas is 3%. This means that the tax is 3% of the purchase price. Suppose the purchase price on a coat is $124.45. The sales tax is then

3% of $124.45 or 0.03×124.45,

or

3.7335, or about $3.73.

The total that you pay is the price plus the sales tax:

$124.45 + $3.73, or $128.18.

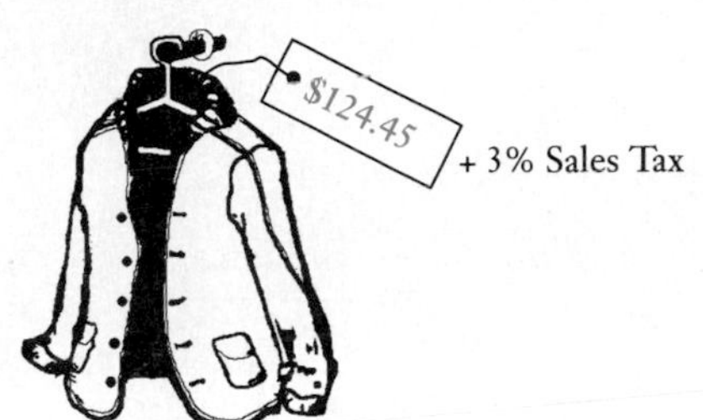

Bill:		
Purchase price	=	$124.45
Sales tax (3% of $124.45)	=	+ 3.73
Final price		$128.18

> **Sales tax = Sales tax rate × Purchase price**
> **Total price = Purchase price + Sales tax**

▶ **EXAMPLE 1** The sales tax rate in California is 6%. How much tax is charged on the purchase of a coat for $124.45? What is the total price?

a) We first find the sales tax. It is

6% of $124.45, or 0.06×124.45,

$$\begin{array}{r} 124.45 \\ \times \quad 0.06 \\ \hline 7.4670 \end{array}$$

which is

7.467, or about $7.47.

b) The total price is purchase price plus sales tax, or

$$\begin{array}{r} 124.45 \\ + \quad 7.47 \\ \hline 131.92 \end{array}$$

The total price is $131.92. ◀

DO EXERCISE 1.

OBJECTIVES

After finishing Section 7.6, you should be able to:

a Solve problems involving percent and sales tax.

FOR EXTRA HELP

Tape 9B — Tape 10B — MAC: 7, IBM: 7

1. The sales tax rate in California is 6%. How much tax is charged on the purchase of a refrigerator for $368.95? What is the total price? $22.14; $391.09

ANSWER ON PAGE A-4

2. The sales tax is $33 on the purchase of a washing machine for $550. What is the sales tax rate? 6%

3. The sales tax on a stereo is $25.20 and the sales tax rate is 6%. Find the purchase price (the price before taxes are added).

$420

ANSWERS ON PAGE A-4

▶ **EXAMPLE 2** The sales tax is $32 on the purchase of a sofa for $800. What is the sales tax rate?

Think: Sales tax is what percent of purchase price?

Translate: $32 = r \times 800$

To solve the equation, we divide on both sides by 800:

$$32 \div 800 = r$$
$$0.04 = r$$
$$4\% = r.$$

The sales tax rate is 4%. ◀

DO EXERCISE 2.

▶ **EXAMPLE 3** The sales tax on a chair is $12.69 and the sales tax rate is 2%. Find the purchase price (the price before taxes are added).

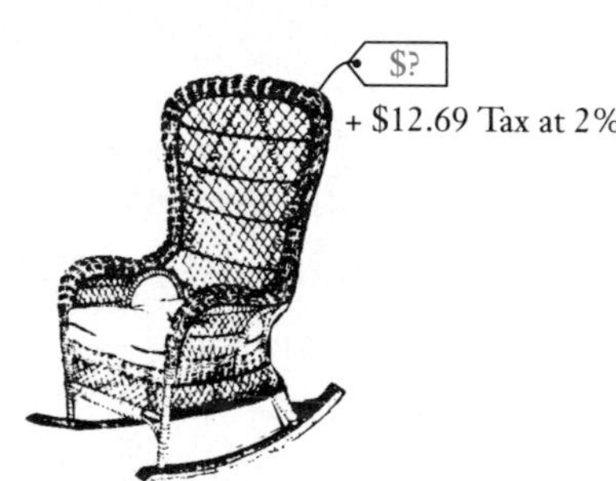

Think: Sales tax is 2% of what?

Translate: $12.69 = 2\% \times p$, or $12.69 = 0.02 \times p$.

To solve the equation, we divide on both sides by 0.02:

$$12.69 \div 0.02 = p$$
$$634.5 = p.$$

$$\begin{array}{r} 634.5 \\ 0.02\overline{)12.690} \\ \underline{1200} \\ 69 \\ \underline{60} \\ 9 \\ \underline{8} \\ 10 \\ \underline{10} \\ 0 \end{array}$$

The purchase price is $634.50. ◀

DO EXERCISE 3.

NAME SECTION DATE

EXERCISE SET 7.6

ANSWERS

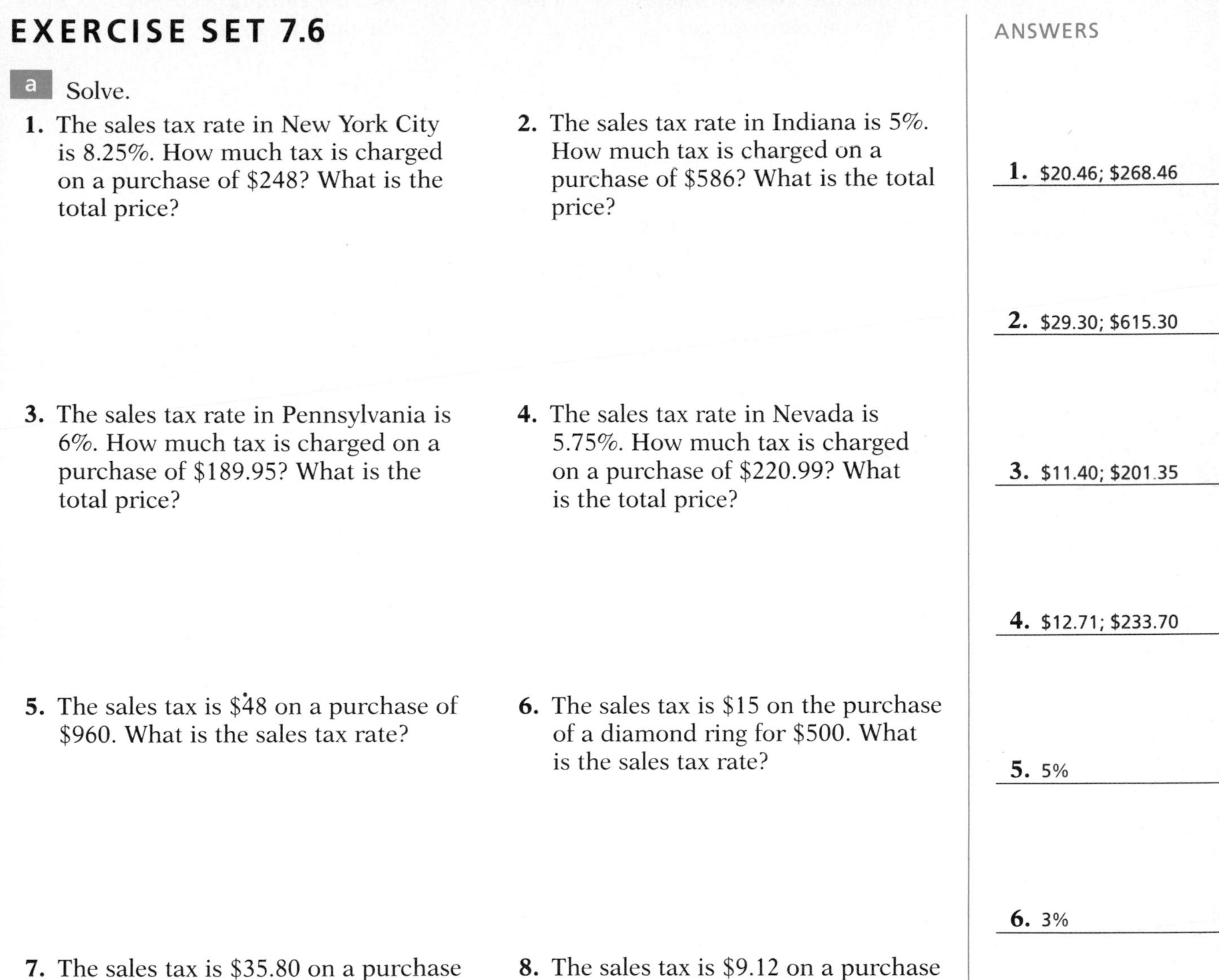

a Solve.

1. The sales tax rate in New York City is 8.25%. How much tax is charged on a purchase of $248? What is the total price?

2. The sales tax rate in Indiana is 5%. How much tax is charged on a purchase of $586? What is the total price?

3. The sales tax rate in Pennsylvania is 6%. How much tax is charged on a purchase of $189.95? What is the total price?

4. The sales tax rate in Nevada is 5.75%. How much tax is charged on a purchase of $220.99? What is the total price?

5. The sales tax is $48 on a purchase of $960. What is the sales tax rate?

6. The sales tax is $15 on the purchase of a diamond ring for $500. What is the sales tax rate?

7. The sales tax is $35.80 on a purchase of $895. What is the sales tax rate?

8. The sales tax is $9.12 on a purchase of $456. What is the sales tax rate?

9. The sales tax on a car is $168 and the sales tax rate is 6%. Find the purchase price (the price before taxes are added).

10. The sales tax on a purchase of $112 and the sales tax rate is 2%. Find the purchase price.

1. $20.46; $268.46

2. $29.30; $615.30

3. $11.40; $201.35

4. $12.71; $233.70

5. 5%

6. 3%

7. 4%

8. 2%

9. $2800

10. $5600

ANSWERS

11. \$800
12. \$1200
13. \$33.25
14. \$39
15. 5.6%
16. 6.2%
17. 37
18. $6\frac{1}{9}$
19. $1.\overline{18}$
20. $2\frac{7}{11}$
21. \$5214.72
22. 6.5%

11. The sales tax on a purchase is \$28 and the sales tax rate is 3.5%. Find the purchase price.

12. The sales tax on a purchase is \$66 and the sales tax rate is 5.5%. Find the purchase price.

13. The sales tax rate in Dallas is 1% for the city and 4% for the state. How much tax is charged on a purchase of \$665?

14. The sales tax rate in Omaha is 1.5% for the city and 3.5% for the state. How much tax is charged on a purchase of \$780?

15. The sales tax is \$1030.40 on an automobile purchase of \$18,400. What is the sales tax rate?

16. The sales tax is \$979.60 on an automobile purchase of \$15,800. What is the sales tax rate?

SKILL MAINTENANCE

17. $2.3 \times y = 85.1$

18. $\frac{5}{x} = \frac{9}{11}$

19. Convert to decimal notation: $\frac{13}{11}$.

20. Convert to a mixed numeral: $\frac{29}{11}$.

SYNTHESIS

21. 🖩 The sales tax rate on a purchase is 5.4%. How much tax is charged on a purchase of \$96,568.95?

22. 🖩 The sales tax is \$3811.88 on a purchase of \$58,644.24. What is the sales tax rate?

7.7 Consumer Applications: Commission and Discount

a Commission

When you work for a **salary,** you get the same amount of money each week or month. When you work for a **commission,** you get paid a percentage of the amount that you sell. To find commission, take a certain percentage of sales.

Commission = Commission rate × Sales

▶ **EXAMPLE 1** A salesperson's commission rate is 20%. What is the commission from the sale of $25,560 worth of vacuum cleaners?

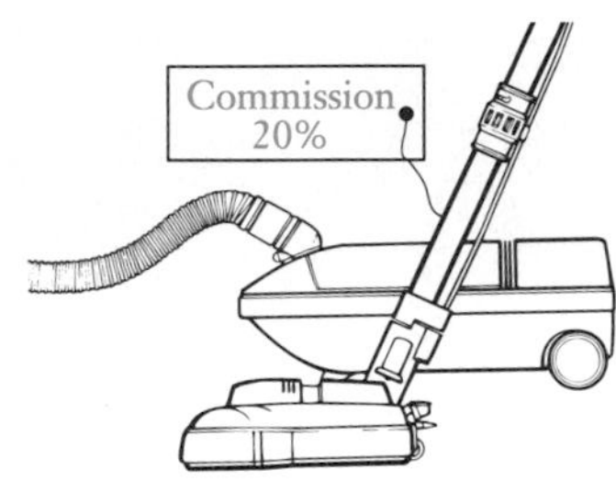

Commission	=	*Commission rate*	×	*Sales*
C	=	20%	×	25,560

This tells us what to do. We multiply.

$$\begin{array}{r} 25{,}560 \\ \times \quad 0.2 \\ \hline 5112.0 \end{array} \qquad 20\% = 0.20 = 0.2$$

The commission is $5112. ◀

DO EXERCISE 1.

▶ **EXAMPLE 2** A salesperson earns a commission of $3000 selling $60,000 worth of farm machinery. What is the commission rate?

Commission	=	*Commission rate*	×	*Sales*
3000	=	*r*	×	60,000

OBJECTIVES

After finishing Section 7.7, you should be able to:

a Solve problems involving commission and percent.

b Solve problems involving discount and percent.

FOR EXTRA HELP

 Tape 9C

 Tape 11A

 MAC: 7 IBM: 7

1. A salesperson's commission rate is 30%. What is the commission from the sale of $18,760 worth of air conditioners? **$5628**

ANSWER ON PAGE A-4

2. A salesperson earns a commission of $6000 selling $24,000 worth of refrigerators. What is the commission rate?

25%

To solve this equation, we divide on both sides by 60,000:

$$3000 \div 60{,}000 = r.$$

We can divide, but this time we simplify by removing a factor of 1:

$$r = \frac{3000}{60{,}000} = \frac{1}{20} \cdot \frac{3000}{3000} = \frac{1}{20} = 0.05 = 5\%.$$

The commission rate is 5%. ◀

DO EXERCISE 2.

▶ **EXAMPLE 3** A motorcycle salesperson's commission rate is 25%. A commission of $425 is received. How many dollars worth of motorcycles were sold?

$$\begin{array}{ccccc} \textit{Commission} & = & \textit{Commission rate} & \times & \textit{Sales} \\ 425 & = & 25\% & \times & S \end{array}$$

3. A clothing salesperson's commission is 16%. A commission of $268 is received. How many dollars worth of clothing were sold? $1675

To solve this equation, we divide on both sides by 25%:

$$425 \div 25\% = S$$
$$425 \div 0.25 = S$$
$$1700 = S.$$

$$\begin{array}{r} 1\,7\,0\,0. \\ 0.2\,5\,\overline{)4\,2\,5.0\,0_\wedge} \\ 2\,5\,0 \\ \hline 1\,7\,5 \\ 1\,7\,5 \\ \hline 0 \end{array}$$

There were $1700 worth of motorcycles sold. ◀

DO EXERCISE 3.

ANSWERS ON PAGE A-4

b Discount

The regular price of a rug is \$60. It is on sale at 25% off. Since 25% of \$60 is \$15, the sale price is \$60 − \$15, or \$45. We call \$60 the **marked price,** 25% the **rate of discount,** \$15 the **discount,** and \$45 the **sale price.** These are related as follows.

> **Discount = Rate of discount × Marked price**
> **Sale price = Marked price − Discount**

▶ **EXAMPLE 4** A rug is marked \$240 and is on sale at 25% off. What is the discount? the sale price?

a) *Discount* = *Rate of discount* × *Marked price*
$D = 25\% \times 240$

This tells us what to do. We convert 25% to decimal notation and multiply.

$$\begin{array}{r} 240 \\ \times\ 0.25 \\ \hline 1200 \\ 4800\ \\ \hline 60.00 \end{array} \qquad 25\% = 0.25$$

The discount is \$60.

b) *Sale price* = *Market price* − *Discount*
$S = 240 - 60$

This tells us what to do. We subtract.

$$\begin{array}{r} 240 \\ -\ 60 \\ \hline 180 \end{array}$$

The sale price is \$180. ◀

DO EXERCISE 4.

4. A suit is marked \$140 and is on sale at 24% off. What is the discount? the sale price?

\$33.60; \$106.40

ANSWER ON PAGE A-4

❖ SIDELIGHTS

An Application: Water Loss

The human body is $\frac{2}{3}$ water.

If you lose

1% of your body water, you will be thirsty;
8% of your body water, you will almost collapse;
10% of your body water, you will be unconscious; and
20% of your body water, you will die.

EXERCISES

You weigh 180 lb.

1. How much of your body weight, in pounds, is water? 120 lb
2. A loss of how many pounds of water would make you thirsty? 1.2 lb
3. A loss of how many pounds of water would make you almost collapse? 9.6 lb
4. A loss of how many pounds of water would make you lose consciousness? 12 lb
5. A loss of how many pounds of water would cause you to die? 24 lb

NAME SECTION DATE

EXERCISE SET 7.7

a Solve.

1. A salesperson's commission rate is 20%. What is the commission from the sale of \$18,450 worth of furnaces?

2. A salesperson's commission rate is 32%. What is the commission from the sale of \$12,500 worth of dictionaries?

3. A salesperson earns \$120 selling \$2400 worth of television sets. What is the commission rate?

4. A salesperson earns \$408 selling \$3400 worth of stereos. What is the commission rate?

5. A sweeper salesperson's commission rate is 40%. A commission of \$392 is received. How many dollars worth of sweepers were sold?

6. A real estate agent's commission rate is 7%. A commission of \$2800 is received on the sale of a home. How much did the home sell for?

7. A real estate commission is 7%. What is the commission on the sale of a \$98,000 home?

8. A real estate commission is 8%. What is the commission on the sale of a piece of land for \$68,000?

9. An encyclopedia salesperson earns a salary of \$500 a month, plus a 2% commission on sales. One month \$990 worth of encyclopedias were sold. What were the wages that month?

10. Some salespersons have their commission increased according to how much they sell. A salesperson gets a commission of 5% for the first \$2000 and 8% on the amount over \$2000. What is the total commission on sales of \$6000?

b Find what is missing.

11.

Marked price	*Rate of discount*	*Discount*	*Sale price*
\$300	10%		

12.

\$2000	40%		

ANSWERS

1. \$3690
2. \$4000
3. 5%
4. 12%
5. \$980
6. \$40,000
7. \$6860
8. \$5440
9. \$519.80
10. \$420
11. \$30; \$270
12. \$800; \$1200

ANSWERS

13. $3; $2

14. $5; $15

15. $12.50; $112.50

16. $438.00; $372.30

17. 40%; $360

18. 15%; $10,880

19. $387; $30\frac{6}{17}$%

20. The discount is actually $30\frac{30}{199}$%; a 30% discount from 9.95 is 6.97.

21. $0.\overline{5}$

22. $2.\overline{09}$

23. $0.91\overline{6}$

24. $1.\overline{857142}$

25. $5924.25; $73,065.75

26. $1.79

	Marked price	Rate of discount	Discount	Sale price
13.	$5.00	60%		
14.	$20.00	25%		
15.	$125.00	10%		
16.		15%	$65.70	
17.	$600		$240	
18.	$12,800		$1920	

19. Find the discount and the rate of discount for the ring in this ad.

20. What is the mathematical error in this ad?

SKILL MAINTENANCE

Find decimal notation.

21. $\frac{5}{9}$

22. $\frac{23}{11}$

23. $\frac{11}{12}$

24. $\frac{13}{7}$

SYNTHESIS

25. ▦ A real estate commission rate is 7.5%. A house sells for $78,990. What is the commission? How much does the seller get for the house after paying the commission?

26. In a recent subscription drive, *People* offered a subscription of 104 weekly issues for a price of $1.29 per issue. They advertised that this was a savings of 27.9% off the newsstand price. What was the newsstand price?

7.8 Consumer Applications: Interest

a Simple Interest

You put \$100 in a savings account for 1 year. The \$100 is called the **principal.** The **interest rate** is 8%. This means you get back 8% of the principal, which is

$$8\% \text{ of } \$100, \quad \text{or} \quad 0.08 \times 100, \quad \text{or} \quad \$8.00,$$

in addition to the principal. The \$8.00 is called the **interest.**

OBJECTIVES

After finishing Section 7.8, you should be able to:

a Solve problems involving simple interest and percent.

b Solve problems involving compound interest.

FOR EXTRA HELP

Tape 9D Tape 11A MAC: 7 IBM: 7

▶ **EXAMPLE 1** What is the interest on \$2500 principal at the interest rate of 6% for 1 year?

We take 6% of \$2500:

$$6\% \times 2500 = 0.06 \times \$2500 = 150$$

$$\begin{array}{r} 2\,5\,0\,0 \\ \times \quad 0.0\,6 \\ \hline 1\,5\,0.0\,0 \end{array}$$

The interest for 1 year is \$150. ◀

DO EXERCISE 1.

1. What is the interest on \$4300 principal at the interest rate of 14% for 1 year? \$602

To find interest for a fraction t of a year, we compute the interest for 1 year and multiply by t.

▶ **EXAMPLE 2** What is the interest on \$2500 principal at the interest rate of 6% for $\frac{1}{4}$ year?

a) We find the interest for 1 year. We take 6% of \$2500:

$$6\% \times 2500 = 0.06 \times 2500 = 150.$$

b) We multiply by $\frac{1}{4}$:

$$\frac{1}{4} \times 150 = \frac{150}{4} = 37.50.$$

$$\begin{array}{r} 3\,7.5 \\ 4\,)\overline{1\,5\,0.0} \\ 1\,2\,0 \\ \hline 3\,0 \\ 2\,8 \\ \hline 2\,0 \\ 2\,0 \\ \hline 0 \end{array}$$

The interest for $\frac{1}{4}$ year is \$37.50. ◀

DO EXERCISE 2.

2. What is the interest on \$4300 principal at the interest rate of 14% for $\frac{3}{4}$ year? \$451.50

ANSWERS ON PAGE A-4

3. What is the interest on \$4800 at 7% for 60 days? \$56

Money is often borrowed for 30, 60, or 90 days even though the interest rate is given **per year.** To simplify calculations, businesspeople consider there to be 360 days in a year. If a loan is for 30 days, it is for 30/360 of a year. The actual interest is found by finding interest for 1 year and taking 30/360 of it.

▶ **EXAMPLE 3** What is the interest on \$400 at 8% for 30 days?

We convert 30 days to a fractional part of one year.

$$\text{Interest} = (\text{Interest for 1 year}) \times \frac{30}{360}$$

$$= (8\% \times \$400) \times \frac{30}{360}$$

$$= 0.08 \times 400 \times \frac{1}{12} \qquad \frac{30}{360} = \frac{1}{12}$$

$$= 32 \times \frac{1}{12} \qquad \begin{array}{r} 400 \\ \times\ 0.08 \\ \hline 32.00 \end{array}$$

$$= \frac{32}{12}$$

$$= \frac{8}{3} \qquad \begin{array}{r} 2.666 \\ 3\,)\overline{8.000} \\ \underline{6} \\ 20 \\ \underline{18} \\ 20 \\ \underline{18} \\ 20 \\ \underline{18} \\ 2 \end{array}$$

$$= 2.66\overline{6}$$

$$\approx 2.67 \qquad \text{Rounding to the nearest hundredth}$$

The interest for 30 days is \$2.67. ◀

A general formula for interest is as follows.

> **Interest = Rate · Principal · Time (expressed in some part of a year), or**
>
> $I = (r \cdot P) \cdot t$, **or, more commonly,** $I = P \cdot r \cdot t$.

Interest computed in this way is called **simple interest.**

DO EXERCISE 3.

ANSWER ON PAGE A-4

b Compound Interest

When interest is paid *on interest,* we call it **compound interest.** This type of interest is usually paid on savings accounts. Suppose you have $100 in a savings account at 6%. In 1 year, you earn

6% of $100, or $6 interest.

Then you have $106. If you leave the interest in your account, the next year you earn interest on $106, which is

6% of $106, or 0.06×106, or $6.36.

You then have $106 + $6.36, or $112.36 in your account. When this happens, we say that interest is **compounded annually.** The interest of $6 the first year earned $0.36 the second year.

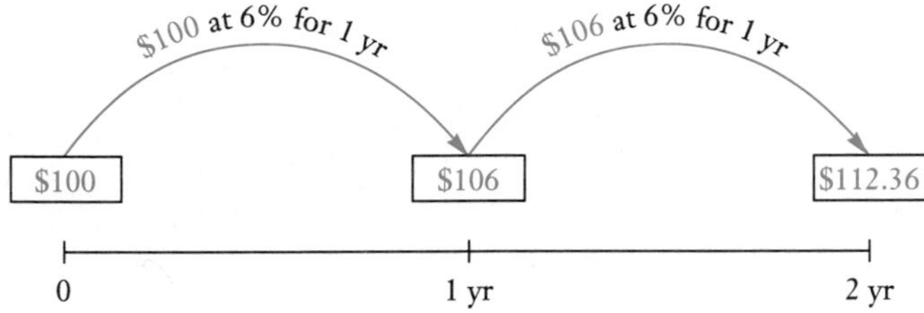

▶ **EXAMPLE 4** Interest is compounded annually. Find the amount in an account if $2000 is invested at 8% for 2 years.

a) We find the interest at the end of 1 year:

$$\begin{aligned} I &= 8\% \times \$2000 \\ &= 0.08 \times \$2000 \\ &= \$160. \end{aligned}$$

b) We then find the new principal after 1 year:

$$\$2000 + \$160 = \$2160.$$

c) Going into the second year, the principal is $2160. We now find the interest for 1 year after that:

$$\begin{aligned} I &= 8\% \times \$2160 \\ &= 0.08 \times \$2160 \\ &= \$172.80 \end{aligned} \qquad \begin{array}{r} 2\,1\,6\,0 \\ \times \quad 0.0\,8 \\ \hline 1\,7\,2.8\,0 \end{array}$$

d) Next we find the new principal after 2 years:

$$\$2160 + \$172.80 = \$2332.80.$$

The amount in the account after 2 years is $2332.80. ◀

DO EXERCISE 4.

4. Interest is compounded annually. Find the amount in an account if $2000 is invested at 11% for 2 years. $2464.20

ANSWER ON PAGE A-4

5. Interest is compounded semiannually. Find the amount in an account if \$2000 is invested at 5% for 1 year.

\$2101.25

Interest added to an account every half year is **compounded semiannually.** Suppose you have \$100 in a savings account at 6%. In $\frac{1}{2}$ year, you earn

$$6\% \times \$100 \times \frac{1}{2}, \quad \text{or} \quad \$3 \text{ interest.}$$

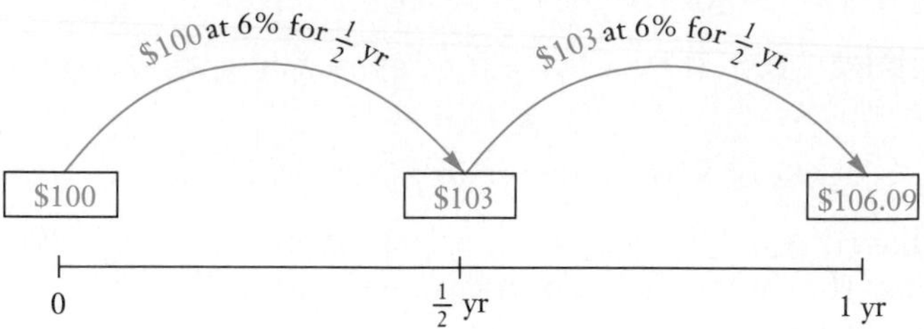

Then you have \$103 if you leave the interest in your account. The last half of the year, you earn interest on \$103, which is

$$6\% \times \$103 \times \frac{1}{2}, \quad \text{or} \quad \$3.09.$$

You then have \$103 + \$3.09, or \$106.09 in your account. If interest were compounded annually, you would have \$106, but with interest compounded semiannually, you have \$106.09. The more often interest is compounded, the more interest your money earns.

▶ **EXAMPLE 5** Interest is compounded semiannually. Find the amount in an account if \$2000 is invested at 8% for 1 year.

a) We find the interest at the end of $\frac{1}{2}$ year:

$$\begin{aligned} I &= 8\% \times \$2000 \times \frac{1}{2} \\ &= 0.08 \times \$2000 \times \frac{1}{2} \\ &= \$160 \times \frac{1}{2} \\ &= \$80. \end{aligned}$$

b) We then find the new principal after $\frac{1}{2}$ year:

$$\$2000 + \$80 = \$2080.$$

c) Going into the last half of the year, the principal is \$2080. Now we find the interest for $\frac{1}{2}$ year after that:

$$\begin{aligned} I &= 8\% \times \$2080 \times \frac{1}{2} \\ &= 0.08 \times \$2080 \times \frac{1}{2} \\ &= \$166.40 \times \frac{1}{2} \\ &= \$83.20. \end{aligned}$$

d) Next we find the new principal after 1 year:

$$\$2080 + \$83.20 = \$2163.20.$$

The amount in the account after 1 year is \$2163.20. ◀

DO EXERCISE 5.

ANSWER ON PAGE A-4

NAME SECTION DATE

EXERCISE SET 7.8

a Find the *simple* interest.

	Principal	*Rate of interest*	*Time*
1.	$200	13%	1 year
2.	$450	18%	1 year
3.	$2000	12.4%	1 year
4.	$200	7.7%	$\frac{1}{2}$ year
5.	$4300	14%	$\frac{1}{4}$ year
6.	$2000	15%	30 days
7.	$5000	14.5%	60 days
8.	$3300	12%	90 days
9.	$4400	9.4%	1 year
10.	$8800	8.8%	1 year

ANSWERS

1. $26

2. $81

3. $248

4. $7.70

5. $150.50

6. $25

7. $120.83

8. $99

9. $413.60

10. $774.40

ANSWERS

11. $484

12. $463.97

13. $236.75

14. $1322.50

15. $466.56

16. $1188.10

17. $2184.05

18. $5356.13

19. 4.5

20. $8\frac{3}{4}$

21. $33\frac{1}{3}$

22. $3\frac{13}{17}$

23. $1434.53

24. $\$1000 \times 8\% \times \frac{30}{360}$

b Interest is compounded annually. Find the amount in the account after the given time. Round to the nearest cent.

	Principal	*Rate of interest*	*Time*
11.	$400	10%	2 years
12.	$400	7.7%	2 years
13.	$200	8.8%	2 years
14.	$1000	15%	2 years

Interest is compounded semiannually. Find the amount in the account after the given time. Round to the nearest cent.

	Principal	*Rate of interest*	*Time*
15.	$400	16%	1 year
16.	$1000	18%	1 year
17.	$2000	9%	1 year
18.	$5000	7%	1 year

SKILL MAINTENANCE

Solve.

19. $\frac{9}{10} = \frac{x}{5}$

20. $\frac{7}{x} = \frac{4}{5}$

Convert to a mixed numeral.

21. $\frac{100}{3}$

22. $\frac{64}{17}$

SYNTHESIS

23. ▦ What is the simple interest on $24,680 at 7.75% for $\frac{3}{4}$ year?

24. ▦ Which gives the most interest, $\$1000 \times 8\% \times \frac{30}{360}$, or $\$1000 \times 8\% \times \frac{30}{365}$?

SUMMARY AND REVIEW EXERCISES: CHAPTER 7

The review sections and objectives to be tested in addition to the material in this chapter are [3.4b], [5.3a], [5.2b], and [6.1c].

Find percent notation.

1. 0.483

2. 0.36

3. $\frac{3}{8}$

4. $\frac{1}{3}$

Find decimal notation.

5. 73.5%

6. $6\frac{1}{2}\%$

Find fractional notation.

7. 24%

8. 6.3%

Translate to an equation. Then solve.

9. 30.6 is what percent of 90?

10. 63 is 84 percent of what?

11. What is $38\frac{1}{2}\%$ of 168?

Translate to a proportion. Then solve.

12. 24 percent of what is 16.8?

13. 22.2 is what percent of 30?

14. What is $38\frac{1}{2}\%$ of 168?

15. Food expenses take 26% of the average family's budget. A family makes $700 one month. How much do they spend for food?

16. The price of a color television set was reduced from $350 to $308. Find the percent of decrease in price.

17. A county has a population that is increasing 3% each year. This year the population is 80,000. What will it be next year?

18. The price of a box of cookies increased from 85 cents to $1.02. What was the percent of increase in the price?

19. A college has a student body with 960 students. Of these, 17.5% are seniors. How many students are seniors?

20. A state charges a sales tax of $4\frac{1}{2}\%$. What is the state tax charged on a purchase of $320?

21. In a certain state, a sales tax of \$378 is collected on the purchase of a car for \$7560. What is the sales tax rate?

22. A salesperson earns \$753.50 selling \$6850 worth of televisions. What is the commission rate?

23. An item has a marked price of \$350. It is placed on sale at 12% off. What are the discount and the sale price?

24. An item priced at \$305 is discounted at the rate of 14%. What are the discount and the sale price?

25. An insurance salesperson receives a 7% commission. If \$420 worth of insurance is sold, what is the commission?

26. What is the simple interest on \$180 at 16% for $\frac{1}{3}$ year?

27. What is the simple interest on \$180 principal at the interest rate of 12% for 1 year?

28. What is the simple interest on \$220 principal at the interest rate of 14.5% for 1 year?

29. What is the simple interest on \$250 at 12.2% for $\frac{1}{2}$ year?

30. Interest is compounded semiannually. Find the amount in an account if \$200 is invested at 12% for 1 year.

31. Interest is compounded annually. Find the amount in an account if \$150 is invested at 12% for 2 years.

SKILL MAINTENANCE

Solve.

32. $\frac{3}{8} = \frac{7}{x}$

33. $10.4 \times y = 665.6$

34. $100 \times x = 761.23$

35. $\frac{1}{6} = \frac{7}{x}$

Convert to decimal notation.

36. $\frac{11}{3}$

37. $\frac{11}{7}$

Convert to a mixed numeral.

38. $\frac{11}{3}$

39. $\frac{121}{7}$

❖ THINKING IT THROUGH

1. Discuss as many daily uses of percent as you can.

2. Describe each method used for solving basic percent problems.

3. Explain percent increase and percent decrease and a common error when doing such problems.

NAME SECTION DATE

TEST: CHAPTER 7

ANSWERS

1. Find decimal notation for 89%.

2. Find percent notation for 0.674.

3. Find percent notation for $\frac{7}{8}$.

4. Find fractional notation for 65%.

5. Translate to an equation. Then solve.

   What is 40% of 55?

6. Translate to a proportion. Then solve.

   What percent of 80 is 65?

Solve.

7. The weight of muscles in a human body is 40% of total body weight. A person weighs 125 lb. What do the muscles weigh?

8. The population of a town increased from 2000 to 2400. Find the percent of increase in population.

Answers

1. [7.1b] 0.89
2. [7.1c] 67.4%
3. [7.2a] 87.5%
4. [7.2b] $\frac{13}{20}$
5. [7.3a, b] $m = 40\% \times 55$; 22
6. [7.4a, b] $\frac{n}{100} = \frac{65}{80}$; 81.25%
7. [7.5a] 50 lb
8. [7.5b] 20%

ANSWERS

9. [7.6a] \$16.20; \$340.20

10. [7.7a] \$630

11. [7.7b] \$40; \$160

12. [7.8a] \$8.52

13. [7.8a] \$4.30

14. [7.8b] \$127.69

15. [5.2b] 222

16. [6.1c] 16

17. [5.3a] $1.41\overline{6}$

18. [3.4b] $3\frac{21}{44}$

9. The sales tax rate in Maryland is 5%. How much tax is charged on a purchase of \$324? What is the total price?

10. A salesperson's commission rate is 15%. What is the commission from the sale of \$4200 worth of merchandise?

11. The marked price of an item is \$200 and is on sale at 20% off. What are the discount and the sale price?

12. What is the simple interest rate on \$120 principal at the interest rate of 7.1% for 1 year?

13. What is the simple interest on \$100 at 8.6% for $\frac{1}{2}$ year?

14. Interest is compounded annually. Find the amount in an account if \$100 is invested at 13% for 2 years.

SKILL MAINTENANCE

Solve.

15. $8.4 \times y = 1864.8$

16. $\frac{5}{8} = \frac{10}{x}$

17. Convert to decimal notation: $\frac{17}{12}$.

18. Convert to a mixed numeral: $\frac{153}{44}$.

CUMULATIVE REVIEW: CHAPTERS 1–7

1. Find fractional notation for 0.091.

2. Find decimal notation for $\frac{13}{6}$.

3. Find decimal notation for 3%.

4. Find percent notation for $\frac{9}{8}$.

5. Write fractional notation for the ratio 5 to 0.5.

6. Find the rate in kilometers per hour: 350 km, 15 hr.

Use $<$, $>$, or $=$ for ■ to write a true sentence.

7. $\frac{5}{7}$ ■ $\frac{6}{8}$

8. $\frac{6}{14}$ ■ $\frac{15}{25}$

Estimate the sum or difference by rounding to the nearest hundred.

9. $263{,}961 + 32{,}090 + 127.89$

10. $73{,}510 - 23{,}450$

Calculate.

11. $46 - [4(6 + 4 \div 2) + 2 \times 3 - 5]$

12. $[0.8(1.5 - 9.8 \div 49) + (1 + 0.1)^2] \div 1.5$

Add and simplify.

13. $\frac{6}{5} + 1\frac{5}{6}$

14. $46.9 + 2.84$

15. $$\begin{array}{r} 487{,}094 \\ 6{,}936 \\ +\ 21{,}120 \\ \hline \end{array}$$

Subtract and simplify.

16. $35 - 34.98$

17. $3\frac{1}{3} - 2\frac{2}{3}$

18. $\frac{8}{9} - \frac{6}{7}$

Multiply and simplify.

19. $\frac{7}{9} \cdot \frac{3}{14}$

20. $$\begin{array}{r} 236{,}984 \\ \times\ 3{,}600 \\ \hline \end{array}$$

21. $$\begin{array}{r} 46.012 \\ \times\ 0.03 \\ \hline \end{array}$$

Divide and simplify.

22. $6\frac{3}{5} \div 4\frac{2}{5}$

23. $431.2 \div 35.2$

24. $15\overline{)1850}$

Solve.

25. $36 \cdot x = 3420$

26. $y + 142.87 = 151$

27. $\frac{2}{15} \cdot t = \frac{6}{5}$

28. $\frac{3}{4} + x = \frac{5}{6}$

29. $\frac{y}{25} = \frac{24}{15}$

30. $\frac{16}{n} = \frac{21}{11}$

Solve.

31. On a checking account of \$7428.63, a check was drawn for \$549.79. What was left in the account?

32. A total of \$57.50 was paid for 5 neckties. How much did each cost?

33. A 12-oz box of cereal costs 90¢. Find the unit price in cents per ounce.

34. A bus travels 456 km in 6 hr. At this rate, how far would the bus travel in 8 hr?

35. In 1984, Americans threw away 50 million pounds of paper. It is projected that this will increase to 65 million pounds in the year 2000. Find the percent of increase.

36. A state charges a sales tax of $6\frac{3}{4}\%$. What is the sales tax charged on a purchase of \$84?

37. In 1987, there were 7625 McDonald's restaurants and 5415 Pizza Hut restaurants in the United States. How many more McDonald's restaurants were there?

38. How many pieces of ribbon $1\frac{4}{5}$ yd long can be cut from a length of ribbon 9 yd long?

39. A student walked $\frac{7}{10}$ km to school and then $\frac{8}{10}$ km to the library. How far did the student walk?

40. On a map, 1 in. represents 80 mi. How much does $\frac{3}{4}$ in. represent?

SYNTHESIS

41. A student wishes to invest money in a bank. Bank A offers 10% simple interest. Bank B offers 9.75% interest compounded semiannually. Which bank offers the highest rate of return on the investment?

42. On a trip through the mountains, a car traveled 240 mi on $7\frac{1}{2}$ gal of gasoline. On a trip across the plains, the same car traveled 351 mi on $9\frac{3}{4}$ gal of gasoline. What was the percent of increase or decrease in miles per gallon?

INTRODUCTION There are many ways in which to analyze or describe data. One is to look at certain numbers or *statistics* related to the data. We will consider three kinds of statistics, the *average,* the *median,* and the *mode.* Another way is to make graphs. We will consider several types of graphs, pictographs, bar graphs, line graphs, and circle graphs.

The review sections to be tested in addition to the material in this chapter are 2.7, 6.3, 7.3, 7.4, and 7.5. ❖

Descriptive Statistics

AN APPLICATION

Hollis Stacy, shown in the photograph here, won a recent U.S. Women's Open with scores of 74, 72, 75, and 69. What was her average score?

THE MATHEMATICS

We add the scores and divide by the number of addends. The average is as follows:

$$\text{Average} = \frac{74 + 72 + 75 + 69}{4}.$$

❖ POINTS TO REMEMBER: CHAPTER 8

Addition using decimal notation: Section 4.3
Division using decimal notation: Section 5.2

PRETEST: CHAPTER 8

Find the (a) average, (b) median, and (c) mode.

1. 46, 50, 53, 55

2. 5, 4, 3, 2, 1

3. 4, 17, 4, 18, 4, 17, 18, 20

4. A car traveled 660 km in 12 hr. What was the average number of kilometers per hour?

5. To get a C in chemistry, a student must average 70 on four tests. Scores on the first three tests were 68, 71, and 65. What is the lowest score that the student can make on the last test and still get a C?

6. The following data show the percentage of women selecting a particular reason for exercising. Make a circle graph to show the data.

Health: 51%
Lose weight: 38%
Relieve stress: 11%

7. The following table shows the comparison of the cost of a $100,000 life insurance policy for female smokers and nonsmokers at certain ages.

a) How much does it cost a female smoker, age 32, for insurance?
b) How much does it cost a female nonsmoker, age 32, for insurance?
c) How much more does it cost a female smoker, age 35, than a nonsmoker at the same age?

8. Using the data in Exercise 7, draw a vertical bar graph showing the cost of insurance for a female smoker at various ages. Use age on the horizontal scale and cost on the vertical scale.

Life Insurance: Female

Age	Cost (Smoker)	Cost (Nonsmoker)
31	$294	$170
32	298	172
33	302	176
34	310	178
35	316	182

The line graph below shows the number of first-year students enrolled in law school for various years.

9. In what year was enrollment largest?

10. How many more were in law school in 1988 than in 1965?

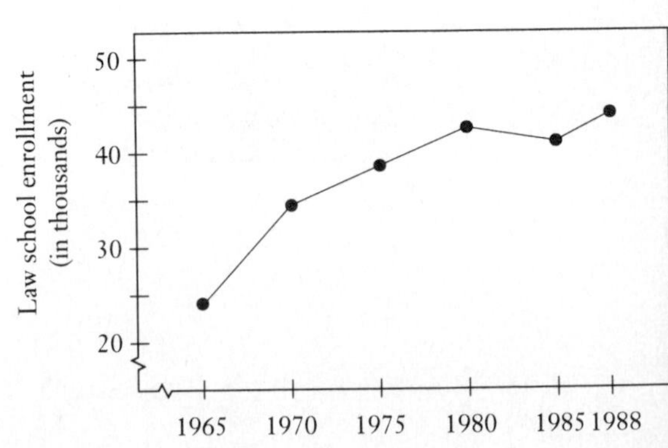

8.1 Averages, Medians, and Modes

a Averages

A **statistic** is a number that describes a set of data. One way to describe or examine data is to look for a number or *"center point"* that characterizes the data. The most common kind of center point is the *mean* or *average* of the set of numbers.

Suppose a student made the following scores on four tests.

Test 1: 78
Test 2: 81
Test 3: 82
Test 4: 79

What is the *average* of the scores? First we add the scores:

$$78 + 81 + 82 + 79 = 320.$$

Second, we divide by the number of addends:

$$\frac{320}{4} = 80.$$

Note that $78 + 81 + 82 + 79 = 320$, and that

$$80 + 80 + 80 + 80 = 320.$$

The number 80 is called the **average** of the set of test scores. It's also called the **arithmetic mean** or, simply, the **mean.**

> **To find the *average* of a set of numbers, add them. Then divide by the number of addends.**

▶ **EXAMPLE 1** On a four-day trip, a car was driven the following number of kilometers each day: 240, 302, 280, 320. What was the average number of kilometers per day?

$$\frac{240 + 302 + 280 + 320}{4} = \frac{1142}{4}, \text{ or } 285.5$$

The car was driven an average of 285.5 km per day. The average is such that if the car had been driven exactly 285.5 km each day, it would have completed the trip in 4 days. ◀

DO EXERCISES 1–4.

▶ **EXAMPLE 2** Hank Aaron holds the record for the most home runs in a career, with 755. He played for 22 seasons. What was the average number of home runs he hit per year? Round to the nearest tenth.

We already have the sum of the home runs. It is 755. We divide the total, 755, by the number of seasons, 22, and round:

$$\frac{755}{22} \approx 34.3.$$

Hank Aaron hit an average of 34.3 home runs per year. ◀

DO EXERCISE 5.

OBJECTIVES

After finishing Section 8.1, you should be able to:

- a Find the average of a set of numbers and solve problems involving averages.
- b Find the median of a set of numbers and solve problems involving medians.
- c Find the mode of a set of numbers and solve problems involving modes.

FOR EXTRA HELP

Tape 10A | Tape 11B | MAC: 8 IBM: 8

Find the average.

1. 14, 175, 36 75

2. 75, 36.8, 95.7, 12.1 54.9

3. A student made the following scores on five tests: 68, 85, 82, 74, 96. What was the average score? 81

4. In the first five games, a basketball player scored points as follows: 26, 21, 13, 14, 23. Find the average number of points scored per game. 19.4

5. O. J. Simpson set an NFL rushing record, gaining 2003 yd in a 14-game season. What was the average number of yards he gained per game? Round to the nearest tenth. 143.1 yd/game

ANSWERS ON PAGE A-4

6. According to EPA estimates in a recent year, a Honda Civic was expected to travel 700 mi (city) on 25 gal of gasoline. What was the average number of miles expected per gallon? 28 mpg

7. A student obtained the following grades one semester.

Grade	Number of credit hours in course
B	3
C	4
C	4
A	2

What was the student's grade point average? Assume that the grade point values are 4.00 for an A, 3.00 for a B, and so on.

2.54

ANSWERS ON PAGE A-4

▶ **EXAMPLE 3** According to EPA estimates in a recent year, a Chevrolet Corvette was expected to travel 375 mi (city) on 25 gal of gasoline. What was the average number of miles expected per gallon?

We divide the total number of miles, 375, by the number of gallons, 25:

$$\frac{375}{25} = 15.$$

The average was 15 miles per gallon. ◀

DO EXERCISE 6.

▶ **EXAMPLE 4** *Grade point average, GPA.* In most colleges students are assigned grade point values for grades obtained. The **grade point average,** or **GPA,** is the average of the grade point values for each hour taken. Suppose at a certain college grade point values are assigned as follows:

A: 4.00 D: 1.00
B: 3.00 F: 0.00
C: 2.00

A student obtained the following grades for one semester. What was the student's grade point average?

Course	Grade	Number of credit hours in course
Accounting	B	4
Calculus	A	5
English	A	5
French	C	3
Physical education	F	1

To find the GPA, we first add all the grade point values for each hour taken. We do this by first multiplying the grade point value (in color below) by the number of hours in the course and then adding as follows:

Accounting	$3.00 \cdot 4 = 12$
Calculus	$4.00 \cdot 5 = 20$
English	$4.00 \cdot 5 = 20$
French	$2.00 \cdot 3 = 6$
Physical education	$0.00 \cdot 1 = 0$
	58 (Total)

The total number of hours taken is $4 + 5 + 5 + 3 + 1$, or 18. We divide 58 by 18 and round to the nearest hundredth:

$$\frac{58}{18} \approx 3.22.$$

The student's grade point average was 3.22. ◀

DO EXERCISE 7.

▶ **EXAMPLE 5** To get a B in math, a student must score an average of 80 on the tests. On the first four tests, the scores were 79, 88, 64, and 78. What is the lowest score that the student can get on the last test and still get a B?

We can find the total of the five scores needed as follows:

$$80 + 80 + 80 + 80 + 80 = 5 \cdot 80, \quad \text{or} \quad 400.$$

The total of the scores on the first four tests is

$$79 + 88 + 64 + 78 = 309.$$

Thus the student needs to get at least

$$400 - 309, \quad \text{or} \quad 91$$

to get a B. We can check this as follows:

$$\frac{79 + 88 + 64 + 78 + 91}{5} = \frac{400}{5}, \quad \text{or} \quad 80.$$ ◀

DO EXERCISE 8.

8. To get an A in math, a student must score an average of 90 on the tests. On the first three tests, the scores were 80, 100, and 86. What is the lowest score that the student can get on the last test and still get an A? 94

b Medians

Another kind of center point is a *median*. Suppose a student made the following scores on five tests.

Test 1: 78
Test 2: 81
Test 3: 82
Test 4: 76
Test 5: 84

Let's first list the scores in order from smallest to largest:

76, 78, 81, 82, 84.

The middle score is called the **median.** Thus, 81 is the median of the scores.

▶ **EXAMPLE 6** What is the median of this set of numbers?

99, 870, 91, 98, 106, 90, 98

We first rearrange the numbers in order from smallest to largest. Then we locate the middle number, 98.

90, 91, 98, 98, 99, 106, 870

↑ Middle number

The median is 98. ◀

DO EXERCISES 9–11.

Find the median.

9. 17, 13, 18, 14, 19 17

10. 20, 14, 13, 19, 16, 18, 17 17

11. 78, 81, 83, 91, 103, 102, 122, 119, 88 91

The *median* of a set of data is the middle number if there is an odd number of numbers. If there is an even number of numbers, then there are two numbers in the middle and the *median* is the number that is halfway between the two middle numbers.

ANSWERS ON PAGE A-4

Find the median.

12. 13, 20, 19, 16, 18, 14 17

13. 68, 34, 67, 69, 34, 70 67.5

Find the mode.

14. 23, 45, 45, 45, 78 45

15. 34, 34, 67, 67, 68, 70 34, 67

16. 13, 24, 27, 28, 67, 89

13, 24, 27, 28, 67, 89

17. A student received the following tests scores:

74, 86, 96, 67, 82.

a) What is the median score? 82

b) What is the mean? 81

c) What is the mode?

74, 86, 96, 67, 82

ANSWERS ON PAGE A-4

▶ **EXAMPLE 7** What is the median of this set of numbers?

69, 80, 61, 63, 62, 65

We first rearrange the numbers in order from smallest to largest. There is an even number of numbers. We look for the middle two, which are 63 and 65. The median is halfway between 63 and 65. It is 64.

61, 62, 63, ↑ 65, 69, 80

Median 64

Note that the number halfway between two numbers is their average. In this example, the number halfway between 63 and 65 is found as follows:

$$\text{Median} = \frac{63 + 65}{2} = \frac{128}{2} = 64.$$ ◀

▶ **EXAMPLE 8** What is the median of this set of numbers?

25, 26, 24, 23

We first rearrange the numbers in order from smallest to largest. There is an even number of numbers. The two middle numbers are 24 and 25. The median is halfway between 24 and 25. We find it as follows:

23, 24, ↑ 25, 26

Median 24.5

$$\text{Median} = \frac{24 + 25}{2} = \frac{49}{2} = 24.5.$$ ◀

DO EXERCISES 12 AND 13.

C Modes

The final type of center point sometimes used to analyze data is the **mode.**

> **The *mode* of a set of data is the number or numbers that occur most often.**

▶ **EXAMPLE 9** Find the mode of this set of data.

13, 14, 17, 17, 18, 19

The number that occurs most often is 17. Thus the mode is 17. ◀

A set of data has just one mean and just one median, but it can have more than one mode. If no number repeats, then each number in the set is a mode.

▶ **EXAMPLE 10** Find the mode, or modes, of this set of data.

33, 34, 34, 34, 35, 36, 37, 37, 37, 38, 39, 40

There are two numbers that occur most often, 34 and 37. Thus the modes are 34 and 37. ◀

DO EXERCISES 14–17.

NAME SECTION DATE

EXERCISE SET 8.1

a, **b**, **c** For each set of numbers, find the average, the median, and the mode.

1. 8, 7, 15, 15, 15, 12

2. 72, 83, 85, 88, 92

3. 5, 10, 15, 20, 25, 30, 35

4. 13, 13, 25, 27, 32

5. 1.2, 4.3, 5.7, 7.4, 7.4

6. 13.4, 13.4, 12.6, 42.9

7. 234, 228, 234, 229, 234, 278

8. \$29.95, \$28.79, \$30.95, \$29.95

The following are the weights of the defensive linemen of the Dallas Cowboys. Use the data for Exercises 9 and 10.

Weight (lb)	*Weight (kg)*
250	113
255	116
260	118
260	118

9. What are the average, median, and mode of the weights in pounds?

10. What are the average, median, and mode of the weights in kilograms?

11. The following temperatures were recorded for seven days:

43°, 40°, 23°, 38°, 54°, 35°, 47°.

What was the average temperature? the median? the mode?

12. Hollis Stacy, a professional golfer, scored 74, 72, 75, and 69 to win the U.S. Women's Open in a recent year. What was the average score? the median? the mode?

13. According to EPA estimates in a recent year, a Triumph TR-7 was expected to get 522 mi (highway) on 18 gal of gasoline. What was the average number of miles per gallon?

14. According to EPA estimates in a recent year, a Ford Wagon was expected to get 432 mi (highway) on 24 gal of gasoline. What was the average number of miles per gallon?

In Exercises 15 and 16 are the grades of a student for one semester. In each case find the grade point average. Assume that the grade point values are 4.00 for an A, 3.00 for a B, and so on.

15.

Grades	Number of credit hours in course
B	4
B	5
B	3
C	4

16.

Grades	Number of credit hours in course
A	5
B	4
B	3
C	5

ANSWERS

1. Average: 12; median: 13.5; mode: 15

2. Average: 84; median: 85; mode: 72, 83, 85, 88, 92

3. Average: 20; median: 20; mode: 5, 10, 15, 20, 25, 30, 35

4. Average: 22; median: 25; mode: 13

5. Average: 5.2; median: 5.7; mode; 7.4

6. Average: 20.575; median: 13.4; mode: 13.4

7. Average: 239.5; median: 234; mode: 234

8. Average: \$29.91; median: \$29.95; mode: \$29.95

9. Average: 256.25 lb; median: 257.5 lb; mode: 260 lb

10. Average: 116.25 kg; median: 117 kg; mode: 118 kg

11. Average: 40°; median: 40°; mode: 43°, 40°, 23°, 38°, 54°, 35°, 47°

12. Average: 72.5; median: 73; mode: 74, 72, 75, 69

13. 29 mpg

14. 18 mpg

15. 2.75

16. 3.00

ANSWERS

17. Average: $9.75; median: $9.79; mode: $9.79

18. Average: $2.25; median: $2.29; mode: $1.99, $2.09, $2.29, $2.39, $2.49

19. 90

20. 96

21. $18,460

22. 196

23. $\frac{4}{9}$

24. 1.96

25. 1.999396

26. 182

27. 171

17. The following prices per pound of steak were found at five supermarkets:

$9.79, $9.59, $9.69, $9.79, $9.89.

What was the average price per pound of steak? the median price? the mode?

18. The following prices per pound of ground beef were found at five supermarkets:

$2.39, $2.29, $2.49, $2.09, $1.99.

What was the average price per pound of ground beef? the median price? the mode?

19. To get a B in math, a student must average 80 on five tests. Scores on the first four tests were 80, 74, 81, and 75. What is the lowest score that the student can get on the last test and still get a B?

20. To get an A in math, a student must average 90 on five tests. Scores on the first four tests were 90, 91, 81, and 92. What is the lowest score that the student can get on the last test and still get an A?

21. The following are the salaries of the employees at the Suitemup Clothing Store. What is the average salary?

Number	*Type*	*Salary*
1	Owner	$29,200
5	Salesperson	19,600
3	Secretary	14,800
1	Custodian	13,000

SKILL MAINTENANCE

Multiply.

22. $14 \cdot 14$

23. $\frac{2}{3} \cdot \frac{2}{3}$

24. 1.4×1.4

25. 1.414×1.414

SYNTHESIS

■ *Bowling averages.* Computing a bowling average involves a special kind of rounding. In effect, we never round up. For example, suppose a bowler gets a total of 599 for 3 games. To find the average, we divide 599 by 3 and drop the amount to the right of the decimal point:

$$\frac{599}{3} \approx 199.67. \qquad \text{The bowler's average is 199.}$$

In each case, find the bowling average.

26. 547 pins in 3 games

27. 4621 in 27 games

8.2 Tables, Charts, and Pictographs

a Tables and Charts

Another way to analyze data is to present it first in a **table** or **chart.**

▶ **EXAMPLE 1** The following chart is a sample rate schedule for dialing direct, long-distance, and interstate telephone calls.

		MON	TUE	WED	THU	FRI	SAT	SUN
8 A.M.	Weekday	Full rate						
5 P.M.	Evening	45% discount from full rate						
11 P.M. – 8 A.M.	Night and Weekend	60% discount from full rate						

a) Which would be less expensive: placing a call at 9 P.M. on Wednesday or 7 A.M. on Friday?

b) On which days of the week is an 11 A.M. call less expensive than on other days?

c) What discount from the full rate will you get if you call at 6 P.M. on Sunday?

Careful attention to the chart will give us the answers.

a) Placing a call at 9 P.M. puts us in the evening-rate portion of the chart. We then read across the chart to Wednesday and find that we will receive a 45% discount. Placing a call at 7 A.M. puts us in the night-and-weekend portion of the chart. Again, we read across the chart, this time to Friday, and we find a discount of 60%. It is less expensive to place the call on Friday.

b) Calling at 11 A.M. puts us in the weekday rate (full rate) on Monday through Friday. On Saturday and Sunday, however, an 11 A.M. call places us in the night-and-weekend rate (60% discount). Thus calling on Saturday and Sunday at 11 A.M. is less expensive than on other days at the same time.

c) A call at 6 P.M. puts us in the evening-rate schedule. When we read across the chart to Sunday (passing over the lightly shaded area on Saturday), we find that we are still using the evening rate between 5 P.M. and 11 P.M. Therefore, we would receive a 45% discount. ◀

DO EXERCISES 1–3.

OBJECTIVES

After finishing Section 8.2, you should be able to:

a Read and interpret data from tables and charts.

b Read and interpret data from pictographs.

c Draw simple pictographs.

FOR EXTRA HELP

Tape 10B | Tape 11B | MAC: 8 IBM: 8

Use the chart in Example 1 to answer each of the following.

1. Which of the following times has the lowest rate: 10 P.M. on Wednesday, 10 A.M. on Tuesday, or 2 P.M. on Sunday?

2 P.M. on Sunday

2. Suppose the full rate to call an out-of-state friend is \$0.50 for the first minute and \$0.34 for each additional minute. You decide to call your friend at midnight on Saturday and you talk for 7 min. What is the cost of the call? \$1.016

3. When, during the weekdays, is the rate the highest?

From 8 A.M. to 5 P.M.

ANSWERS ON PAGE A-4

Use the table in Example 2 to answer each of the following.

4. In which cities will you have to pay $50 or more, per week, for family day care?

   Dallas, Denver, and San Francisco

5. If you live in St. Louis, what will be the combined weekly cost range to place your 1-year-old and your 3-year-old in a day-care center? $140–$190

6. In which cities are you sure to spend under $150 per week for your child to be cared for in your own home? None

ANSWERS ON PAGE A-4

► **EXAMPLE 2** According to the U.S. Bureau of Labor Statistics, 56% of women with children under the age of 6, or about 9 million women, work outside the home. The following table shows the weekly cost ranges, per child, for several day-care programs in seven major U.S. cities.

	Family day care		**Day-care center**		**Caregiver comes to child's home**
	Age	**Cost**	**Age**	**Cost**	
Boston	0–2 2–5	$45–160 $40–160	0–2 2–5	$90–150 $75–110	$260–340
New York	0–2 2–5	$35–140 $40–160	0–2 2–5	$60–150 $75–110	$165–300
Atlanta	0–2 2–5	$30–60 $30–55	0–2 2–5	$35–70 $50–70	$165–230
St. Louis	0–2 2–5	$45–50 $40–160	0–2 2–5	$65–80 $75–110	$165 and up
Dallas	0–2 2–5	$50–70 $50–70	0–2 2–5	$60–90 $50–70	$165–200
Denver	0–2 2–5	$65–105 $55–105	0–2 2–5	$65–105 $55–105	$165–200
San Francisco	0–2 2–5	$55–90 $55–85	0–2 2–5	$90–120 $65–90	$165–200

a) In which city is it most expensive to have the caregiver come to your home?

b) How much will it cost you each week to have your 4-year-old child cared for in a day-care center in New York?

c) What is the maximum, per child, that you would expect to pay for a 4-year-old child cared for in family day care in Atlanta?

We look at the table to answer the questions.

a) We go to the last column, since the caregiver will come to your home, and read down the column, looking for the greatest entry. When we find it ($260–$340), we read back across, all the way to the left of that entry, and find that that is the rate in Boston.

b) This time we find New York in the first column. We then read across to the column under the heading "Day-Care Center." Since your child is 4 years old, we select the 2–5 age range, and the corresponding cost entry is $75–$110.

c) Locating Atlanta on the left, we read across to the column headed "Family Day Care." Since the range indicated is $30–$55, the maximum that you would expect to pay is $55. ◄

DO EXERCISES 4–6.

b Reading and Interpreting Pictographs

Pictographs (or picture graphs) are another way to show information. Instead of actually listing the amounts to be considered, a pictograph uses visually appropriate symbols to represent the amounts. In addition, a *key* is given telling what each symbol represents, so there is no need to read from a second scale.

▶ **EXAMPLE 3** This pictograph shows an approximation of the number of productive oil wells in eight Middle-Eastern countries. Just below the graph, there is a key that tells you that each symbol represents 50 oil wells.

Productive Oil Wells in the Middle East

Iran
Iraq
Israel
Kuwait
Oman
Saudi Arabia
Syria
Turkey

= 50 wells

a) How many productive oil wells are there in Saudi Arabia?

b) How many more productive oil wells does Oman have than Turkey?

c) What is the total number of productive oil wells in the top four oil-producing countries?

We can compute the answers by first reading the pictograph.

a) Saudi Arabia's wells are represented by 11 symbols. Since each symbol stands for 50 wells, we can multiply 11×50 and get 550 oil wells.

b) Oman has 9 symbols representing about 450 oil wells (9×50). Turkey has about 400 oil wells, as shown by 8 symbols (8×50). Therefore, Oman has approximately 50 more wells than Turkey does.

c) By counting the symbols, you can see that Iran (11 symbols), Kuwait (11 symbols), Saudi Arabia (11 symbols), and Syria (12 symbols) are the top four oil-producing countries. The total number of symbols for these countries is 45, which represents a total of 2250 oil wells (45×50). ◀

DO EXERCISES 7–9.

Use the pictograph in Example 3 to answer each of the following.

7. How many productive oil wells are there in Syria? 600

8. Turkey has more productive oil wells than which two countries combined? Iraq and Israel

9. What is the average number of productive oil wells for these 8 countries? 8.625×50, or 431.25

ANSWERS ON PAGE A-4

Use the graph in Example 4 to answer each of the following.

10. India has a greater population than the combined populations of which other countries?

Brazil, France, USSR, US

11. Which two countries are closest in population?

USSR and US

12. Which two countries have the greatest difference in population?

India and France

ANSWERS ON PAGE A-4

You should realize by now that, although pictographs seem to be very easy to read, they are difficult to draw accurately because whole symbols reflect loose approximations due to significant rounding. In pictographs, you also need to use some mathematics to find the actual amounts.

► **EXAMPLE 4** This pictograph shows fairly recent estimates of the population of five countries.

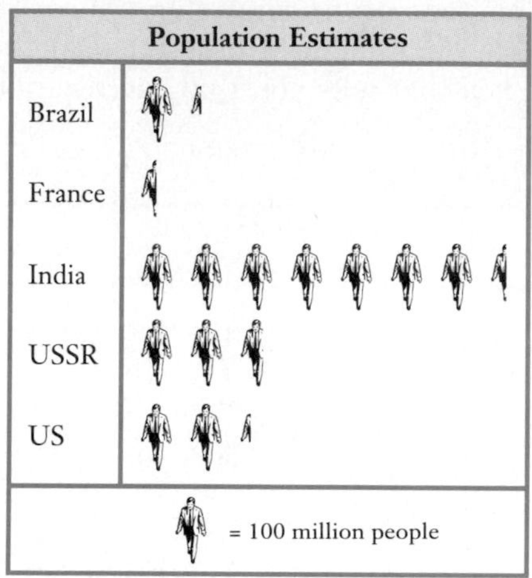

Give the approximate population of each country.

Brazil's population is represented by 1 whole symbol (100 million people) and about $\frac{1}{3}$ of another symbol (33 million people) for a total of 133 million people.

France's population is shown by about $\frac{1}{2}$ of a symbol, representing a total of 50 million people.

For India, we have 7 whole symbols (7×100 million people) and $\frac{1}{2}$ of another (50 million people), giving a total of 750 million people.

The population of the USSR is represented by 2 whole symbols (2×100 million people) and about $\frac{3}{4}$ of another (75 million people) for a total of 275 million people.

The population of the United States is shown by 2 whole symbols (2×100 million people) and about $\frac{1}{3}$ of another (33 million people), giving a total of 233 million people. ◄

One advantage of pictographs is that the appropriate choice of a symbol will tell you, at a glance, the kind of measurement being made. Another advantage is that the comparison of amounts represented in the graph can be expressed more easily by just counting symbols. For instance, in Example 3, the ratio of Iraq's oil wells to Saudi Arabia's oil wells is 6:11.

One disadvantage of pictographs is that, to make a pictograph easy to read, the amounts must be rounded significantly to the unit that a symbol represents. This makes it difficult to accurately represent an amount. Another problem is that it is difficult to determine very accurately how much a partial symbol represents. A third disadvantage is that you must use some mathematics to finally compute the amount represented, since there is usually no explicit statement of the amount.

DO EXERCISES 10–12.

C Drawing Pictographs

▶ **EXAMPLE 5** Draw a pictograph showing the U.S production of apples in five different years. Let the apple symbol represent 50,000,000 boxes of apples and use the following information.

a) 1965: 150,000,000 boxes
b) 1970: 200,000,000 boxes
c) 1975: 225,000,000 boxes
d) 1980: 250,000,000 boxes
e) 1985: 275,000,000 boxes

Some computation is necessary.

a) Since 3 × 50 million is 150 million, three apple symbols will show the boxes for 1965.

b) 4 × 50 million is 200 million, so four apple symbols are used to show the boxes for 1970.

c) Since 4 × 50 million is only 200 million, we need not only these four apple symbols, but also $\frac{1}{2}$ of another ($\frac{1}{2}$ of 50 million is 25 million) to show 225 million boxes for 1975.

d) Five apple symbols will be used to show the boxes for 1980, since 5 × 50 million is 250 million.

e) We will need five whole apple symbols to represent 250 million boxes plus $\frac{1}{2}$ of another to show an additional 25 million boxes. This will give a total of 275 million boxes for 1985.

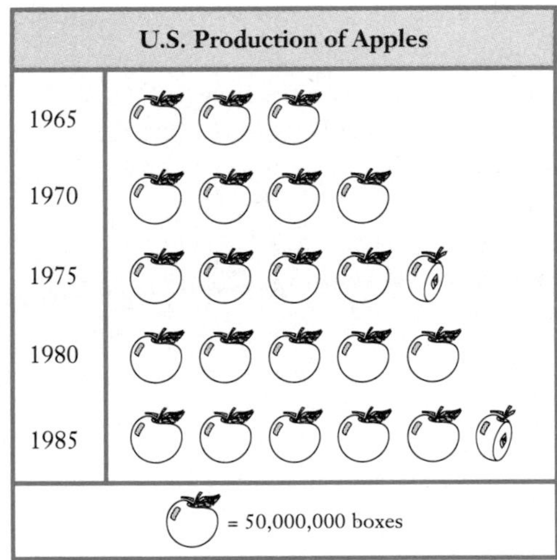

◀

DO EXERCISE 13.

13. Draw a pictograph to represent the following data regarding the number of television sets in various cities.

a) Honolulu, Hawaii: 200,000
b) Columbus, Ohio: 400,000
c) Houston, Texas: 550,000
d) Pittsburgh, Pennsylvania: 467,000

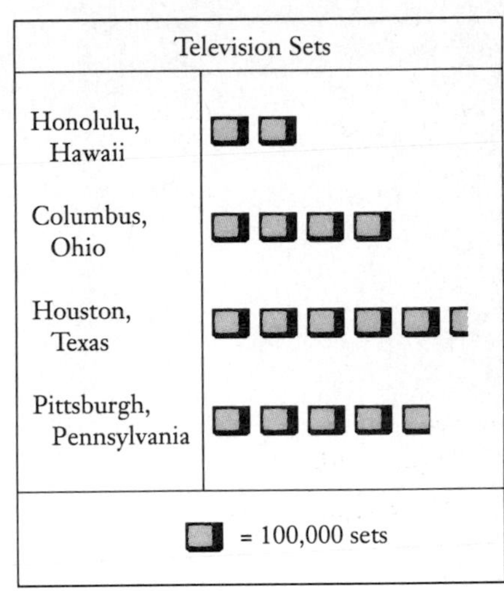

ANSWERS ON PAGE A-4

❖ SIDELIGHTS

A Problem-Solving Extra

Consider the following bar graph for the exercises below.

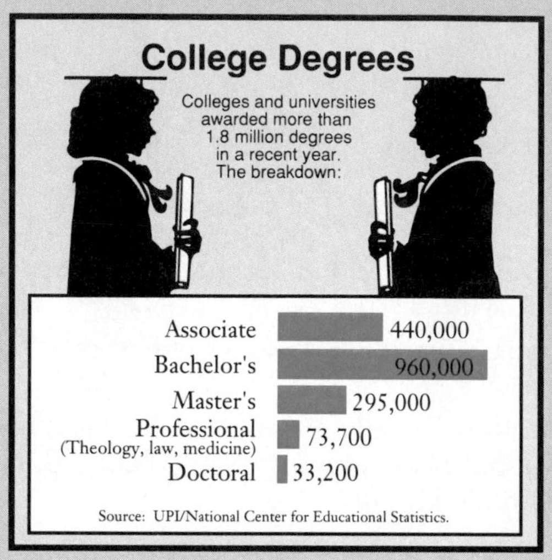

EXERCISES

1. How many more bachelor's degrees than associate degrees were awarded? 520,000
2. How many more master's degrees than doctoral degrees were awarded? 261,800
3. In all, how many graduate degrees were awarded; that is, how many master's, doctoral, and professional degrees were awarded? 401,900

NAME SECTION DATE

EXERCISE SET 8.2

a The following table gives information about various types of nails. Use the table for Exercises 1–10.

NAIL SIZES

		Approximate number per pound		
Penny number	**Length (inches)**	**Common nails**	**Box nails**	**Finishing nails**
4	$1\frac{1}{2}$	316	437	548
6	2	181	236	309
8	$2\frac{1}{2}$	106	145	189
10	3	69	94	121
12	$3\frac{1}{4}$	64	87	113
16	$3\frac{1}{2}$	49	71	90
20	4	31	52	62
30	$4\frac{1}{2}$	20		
40	5			
50	$5\frac{1}{2}$			

1. How long is a 20-penny nail?

2. What penny number is given to a nail that is $2\frac{1}{2}$ in. long?

3. How many 10-penny box nails are there in a pound?

4. What type of nail comes 309 to the pound?

5. How many more 16-penny finishing nails can you get in a pound than 10-penny common nails?

6. How many fewer 8-penny common nails will you get in a pound than 6-penny common nails?

7. How many 30-penny box nails will you get in a pound?

8. What type of nail comes 40 to the pound?

9. How many nails will you get if you buy 5 pounds of 4-penny finishing nails?

10. You need approximately 448 12-penny common nails. How many pounds should you buy?

ANSWERS

1. 4 in.

2. 8

3. 94

4. 6-penny finishing nail

5. 21

6. 75

7. Not given in table

8. None listed in table

9. 2740

10. 7 lb

ANSWERS

11. 294

12. 378

13. Calisthenics

14. Moderate walking

15. Aerobic dance

16. Moderate bicycling

17. 660

18. 60 min

19. Moderate walking

20. Calisthenics and moderate jogging

21. 121

22. 306

This table shows the number of calories burned in 30 min of exercise for various types of activities and several weight categories. Use the table for Exercises 11–22.

Activity	Calories burned in 30 min		
	110 lb	**132 lb**	**154 lb**
Aerobic dance	201	237	282
Calisthenics	216	261	351
Racquetball	213	252	294
Tennis	165	192	222
Moderate bicycling	138	171	198
Moderate jogging	321	378	453
Moderate walking	111	132	159

11. How many calories are burned by a 154-lb person after 30 min of racquetball?

12. How many calories are burned by a 132-lb person after 30 min of moderate jogging?

13. What activity burns 216 calories in 30 min for a 110-lb person?

14. What activity burns calories at the rate of 132 every 30 min for a 132-lb person?

15. Which burns more calories in 30 min for a 154-lb person: aerobic dance or tennis?

16. Which burns more calories in 30 min: moderate bicycling or moderate walking?

17. How many calories will have been burned by a 110-lb person after 2 hr of tennis?

18. How many minutes of moderate walking will it take for a 110-lb person to burn as many calories as a 154-lb person will burn playing 30 min of tennis?

19. Which activity burns the least number of calories for a 132-lb person?

20. A 110-lb person needs to burn at least 215 calories every 30 min. What activities will provide at least that rate of burn?

21. How many calories would you expect a 120-lb person to burn during 30 min of moderate walking?

22. How many calories would you expect a 143-lb person to burn during 30 min of calisthenics?

b This pictograph shows sales of shampoo for a soap company for six consecutive years. Use the pictograph for Exercises 23–30.

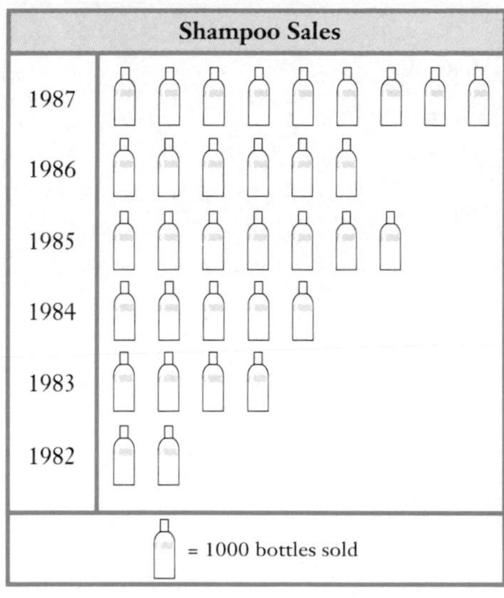

23. In which year was the greatest number of bottles sold?

24. Between what two consecutive years was there the greatest growth?

25. Between what two years was the amount of positive growth the least?

26. How many sales does one bottle represent?

27. How many bottles were sold in 1985?

28. How many more bottles were sold in 1987 than in 1983?

29. In which year was there actually a decline in the number of bottles sold?

30. The sales for 1987 were how many times the sales for 1982?

ANSWERS

23. 1987

24. 1986 and 1987

25. 1983 and 1984

26. 1000

27. 7000

28. 5000

29. 1986

30. $4\frac{1}{2}$

ANSWERS

31. 3

32. 5

33. 16

34. 4

35. 13

36. 33

37. Out

38. Singled

39. See chart.

40. 12

41. 27,859.5 sq mi

This pictograph shows a baseball player's "at-bats" in one month. Use the pictograph for Exercises 31–38.

At-Bat Record

Home Runs

Triples

Doubles

Singles

Walks

Outs

Each = 3 times at bat

31. How many times at bat does one baseball symbol represent?

32. How many of the player's hits were home runs?

33. How many hits were singles?

34. How many more doubles than home runs did the player hit?

35. How many fewer triples than singles did the player hit?

36. What was the total number of hits for the month?

37. What happened exactly 12 times to the batter?

38. What did the player do most during the month?

C **39.** Draw a pictograph representing the number of U.S. automobile registrations shown by the information below. Be sure to put in all of the appropriate labels. Use a license plate symbol to represent 10,000,000 cars.

1960:	62.5 million cars
1965:	75.0 million cars
1970:	90.0 million cars
1975:	110.0 million cars
1980:	135.0 million cars
1985:	157.5 million cars

U.S. Automobile Registrations

1960

1965

1970

1975

1980

1985

412CNG = 10,000,000 cars

SKILL MAINTENANCE

Solve.

40. A football team has won 3 out of its first 4 games. At this rate, how many games will it win in a 16-game season?

41. The state of Maine is 90% forest. The area of Maine is 30,955 sq mi. How many square miles of Maine are forest?

8.3 Bar Graphs and Line Graphs

A **bar graph** is convenient for showing comparisons because you can tell at a glance which amount represents the largest or smallest quantity. Of course, since bar graphs are a more abstract form of pictographs, this is true of pictographs as well. However, with bar graphs, a *second scale* is usually included so that a more accurate determination of the amount can be made.

a Reading and Interpreting Bar Graphs

▶ **EXAMPLE 1** A recent National Assessment of Educational Progress Survey showed these reasons given by students for dropping out of high school.

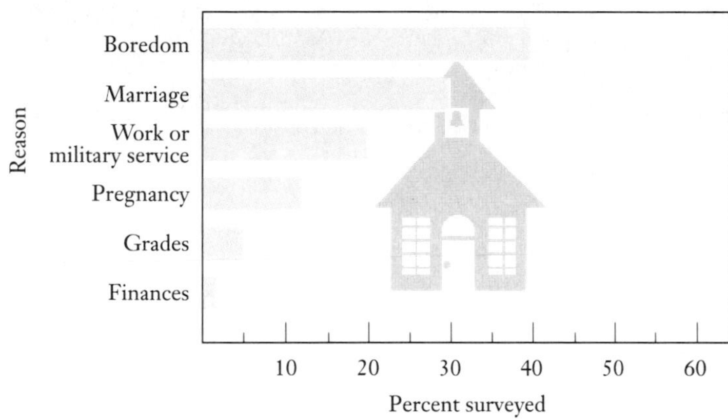

a) Approximately how many in the survey dropped out because of pregnancy?

b) What reason was given least often for dropping out?

c) What reason was given by about 30% for dropping out?

We look at the graph to answer the questions.

a) We go to the right end of the bar representing pregnancy and then go down to the percent scale. We can read, fairly accurately, that approximately 12% dropped out because of pregnancy.

b) We look for the shortest bar and find that it represents finances.

c) We go to the right on the percent scale to find the 30% mark and then up until we reach a bar that ends at approximately 30%. We then go across to the left and read the reason. The reason given by about 30% was marriage. ◀

DO EXERCISES 1–3.

OBJECTIVES

After finishing Section 8.3, you should be able to:

- a Read and interpret data from bar graphs.
- b Draw bar graphs.
- c Read and interpret data from line graphs.
- d Draw simple line graphs.

FOR EXTRA HELP

Tape 10C Tape 12A MAC: 8 IBM: 8

Use the bar graph in Example 1 to answer each of the following.

1. What reason was given most often for dropping out?

   Boredom

2. What reason was given by about 20% for dropping out?

   Work or military service

3. How many in the survey dropped out because of grades?

   About 5%

ANSWERS ON PAGE A-4

Use the bar graph in Example 2 to answer each of the following.

4. What is the average annual income for a person who has completed only high school?

Approximately $17,000

5. What is the greatest average annual income for a person who doesn't finish high school?

Approximately $14,000

6. How much more can you expect to earn annually if you complete high school than if you complete only the 8th grade?

Approximately $6000

ANSWERS ON PAGE A-4

Of course, the bars can be drawn vertically as well.

► **EXAMPLE 2** A recent survey of 2000 individuals produced the following information on average income based on years of schooling.

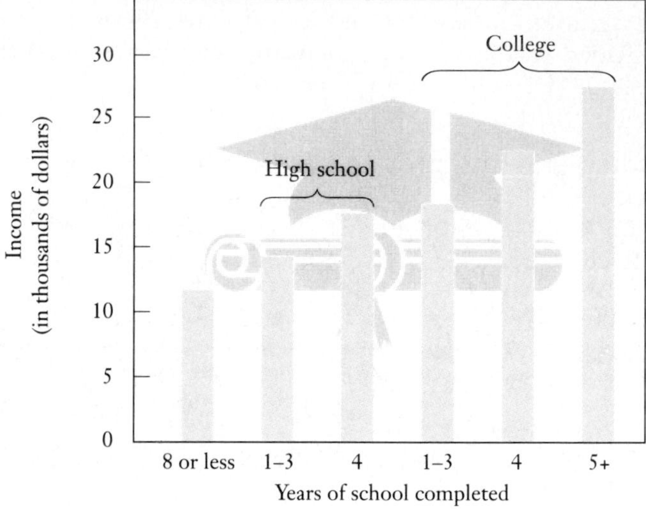

a) What is the average annual income for a person who has completed two years of high school?

b) How many years of schooling does it take to expect an average annual income of at least $20,000?

c) How much more income, on the average, can be expected after completing three years of college than after completing only one year of college?

Interpreting the graph carefully will give us the answers.

a) We go to the right, across the bottom, to the bar representing income for a person with 1–3 years of high school. We then go up to the top of the bar and, from there, back to the left to read approximately $14,000 on the income scale.

b) We go up the left-hand scale of the graph to the $20,000 mark and read to the right, until we come to a bar crossing our path. Moving down on that bar, we find that at least 4 years of college are needed.

c) There is only one bar representing 1–3 years of college. Therefore, this graph shows no difference in income between the two groups, though it may exist. ◄

DO EXERCISES 4–6.

b Drawing Bar Graphs

▶ **EXAMPLE 3** Make a vertical bar graph to show the following information about the number of cricket chirps per minute as that relates to the temperature.

56°F: 69 chirps per minute
59°F: 76 chirps per minute
62°F: 88 chirps per minute
65°F: 100 chirps per minute

First, we indicate on the base or horizontal scale the different degree markings. (See the figure at the left below.)

Then we label the marks on the vertical scale appropriately by 10's to represent the number of cricket chirps (see the graph at the left below). The jagged lines at the start of the horizontal and vertical scales indicate that we have left out a portion of the scales (to save space) since it was not necessary in providing information.

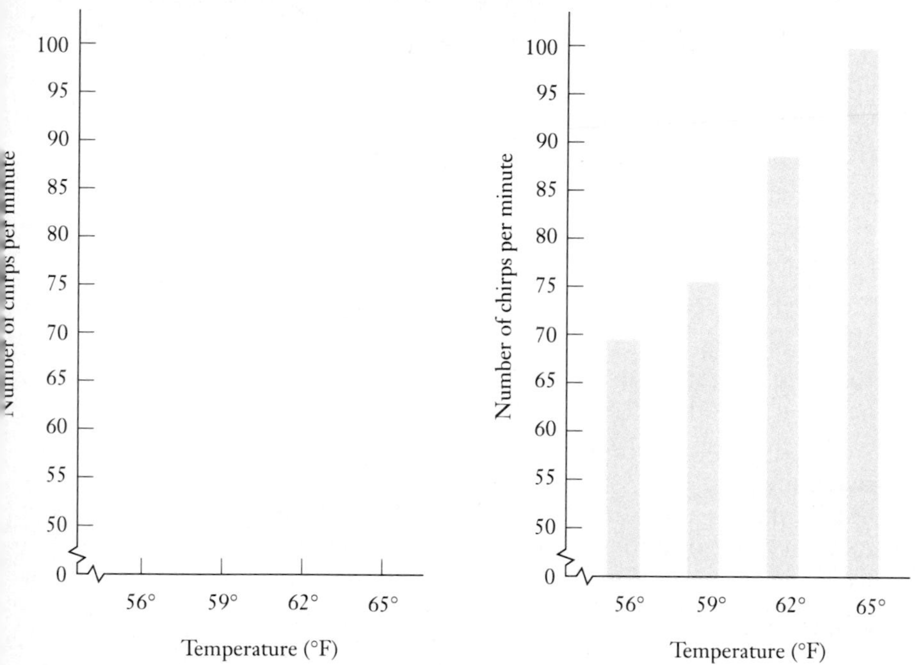

Finally, we draw vertical bars to show the number of chirps per minute for each temperature reading, as shown in the figure at the right above. ◀

DO EXERCISE 7.

7. Make a horizontal bar graph to show the loudness of various sounds listed below. (*Hint:* See Example 1.) A decibel is a measure of the loudness of sounds.

Sound	Loudness (in decibels)
Whisper	15
Tick of watch	30
Speaking aloud	60
Noisy factory	90
Moving car	80
Car horn	98
Subway	104

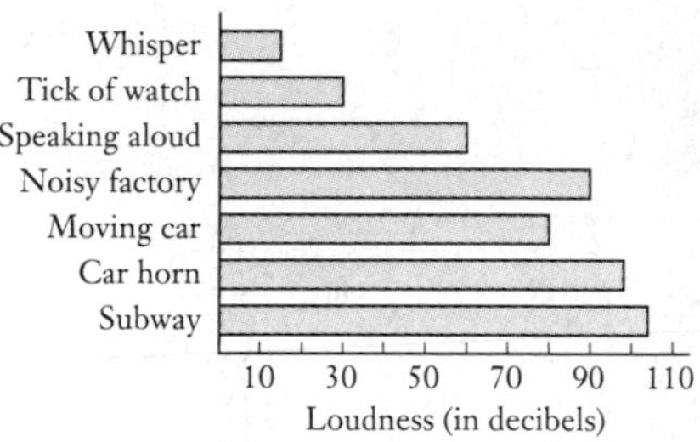

ANSWERS ON PAGE A-4

Use the line graph in Example 4 to answer each of the following.

8. For which week was the DJIA closing the highest?

Sixth week

9. For which week was the DJIA closing about 2100?

Second week

10. About how many points did the DJIA increase between weeks 1 and 6? About 225

ANSWERS ON PAGE A-4

C Reading and Interpreting Line Graphs

Line graphs are often used to show a change over time as well as to indicate patterns or trends.

▶ **EXAMPLE 4** This line graph shows the closing Dow Jones Industrial Average (DJIA) for each of six weeks. Note that, again, we have a jagged line at the base of the vertical scale indicating an unnecessary portion of the scale. Note, too, that the vertical scale differs from the horizontal scale so that numbers fit reasonably.

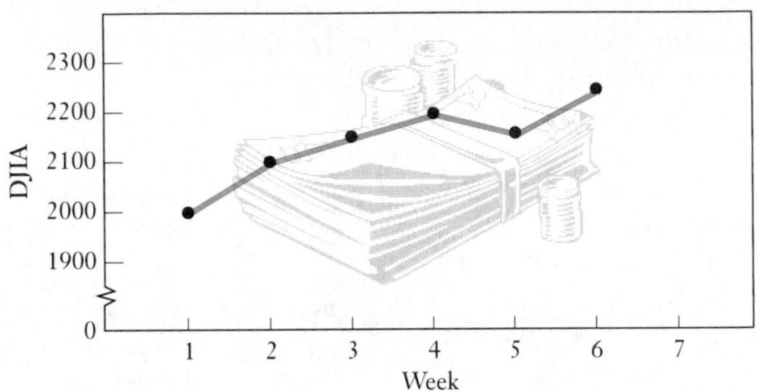

a) For which week was the DJIA closing the lowest?

b) Between which two weeks did the DJIA closing decrease?

c) For which week was the DJIA closing about 2200?

We look at the graph to find the answers.

a) For the first week, the line is at its lowest point, representing a close of about 2000.

b) Reading the graph from left to right, we see that the line went down between the 4th and 5th weeks.

c) We locate 2200 on the DJIA scale and then move to the right until we reach the point representing a closing that is closest to our position. At that point, we move down to the "Week" scale and see which week is indicated. We find that the DJIA closing was closest to 2200 for the 4th week. ◀

DO EXERCISES 8–10.

▶ **EXAMPLE 5** The line graph below gives information from a 14-year study about the average daily cost of a semiprivate hospital room. The averages were figured every two years.

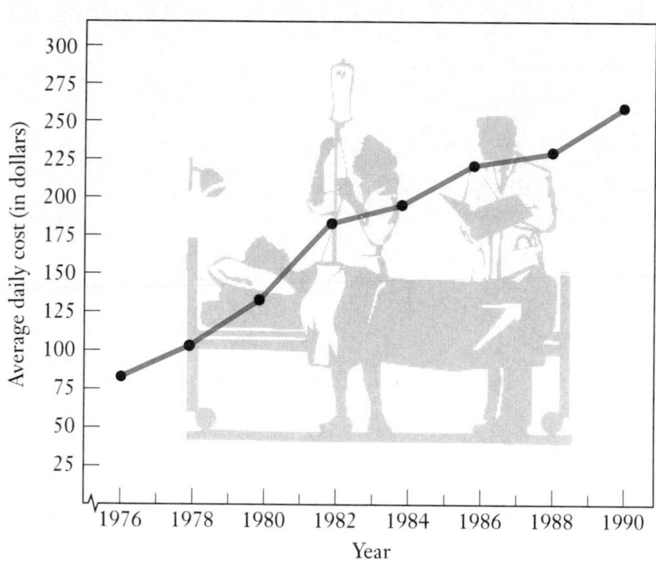

a) Give the average daily cost for a semiprivate room in 1978.

b) In which year was the average daily cost $181?

c) How much did the average daily cost increase from 1984 to 1986?

We read the graph to find the answers.

a) We find the year 1978 on the bottom scale and move up from that point to the line. We then go straight across to the left from the line and find that we are slightly above the $100 mark. A reasonable estimate would be $104.

b) Going up the left scale to a point slightly above $175, we move straight across to the right until we cross the line. At that point, we go down to the scale for "years" on the bottom and find that this occurred in 1982.

c) The graph shows an approximate average daily cost of $193 in 1984 and $218 in 1986. This gives an increase of $25 between these two years. ◀

DO EXERCISES 11–13.

Use the line graph in Example 5 to answer each of the following.

11. Between which two years did the average daily cost increase the most? 1980 and 1982

12. On the basis of the trend indicated in the graph, predict how much the average daily cost will increase from 1990 to 1994.

Approximately $50

13. On the basis of the indicated pattern, in approximately what year did the average daily cost go over $200? 1985

ANSWERS ON PAGE A-4

14. Draw a line graph to show how the average price per acre of farmland has changed in four years. Use the following information.

1986: $548
1987: $563
1988: $597
1989: $645

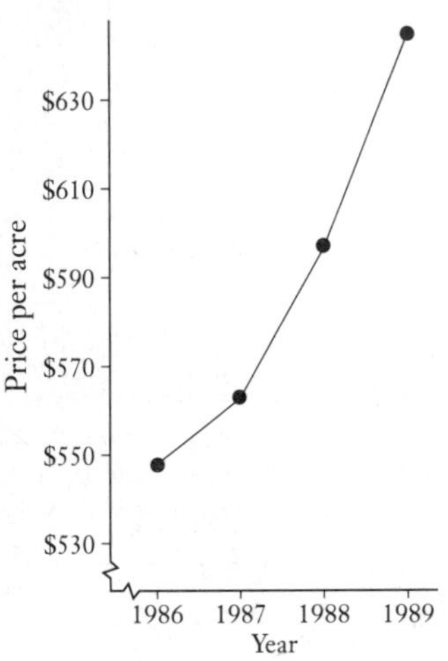

ANSWER ON PAGE A-5

d Drawing Line Graphs

▶ **EXAMPLE 6** Draw a line graph to show how the total number of inches of rainfall has changed in five years. Use the following information.

1982: 30 inches of rainfall
1983: 28 inches of rainfall
1984: 25 inches of rainfall
1985: 30 inches of rainfall
1986: 27 inches of rainfall

First, we indicate on the horizontal scale the different years and title it "Years." (See the following graph.)

Then we mark the vertical scale appropriately by 5's to show the number of inches of rainfall (see the graph below) and title it "Number of inches of rainfall."

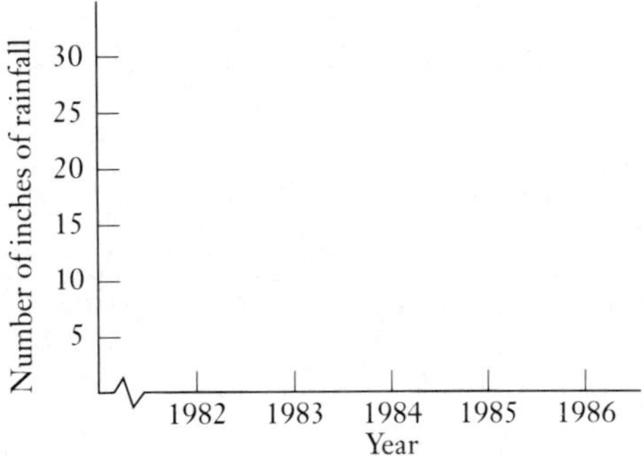

Now, we mark the points above each of the years at the appropriate level to indicate the number of inches of rainfall, and draw line segments connecting them to show the change.

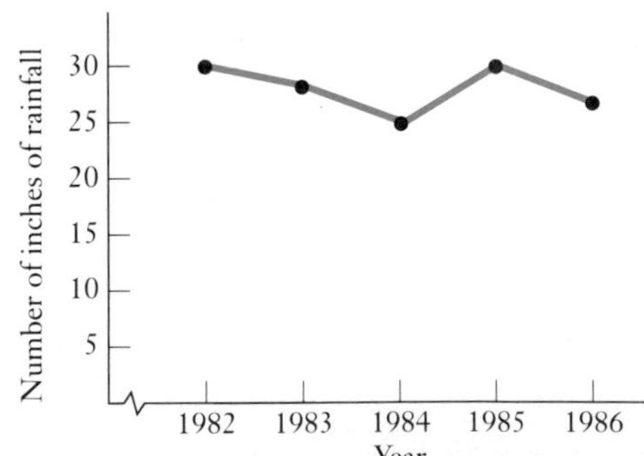

◀

DO EXERCISE 14.

NAME SECTION DATE

EXERCISE SET 8.3

ANSWERS

a This horizontal bar graph shows the average length, in weeks, in the growing season for 8 U.S. cities.

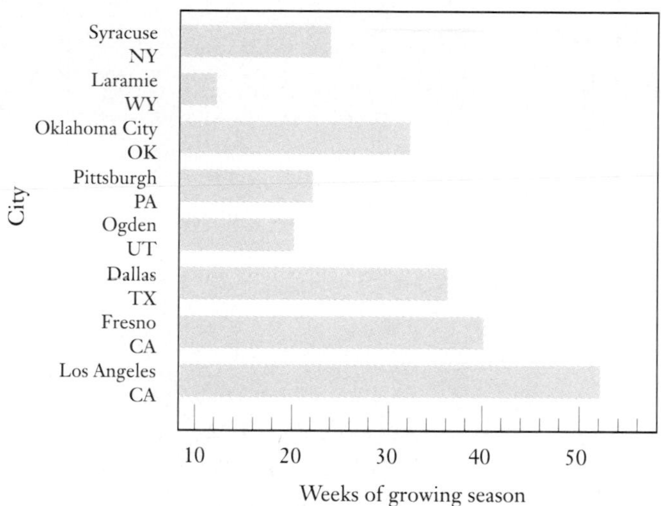

1. Which city has the longest growing season?

2. Which city has the shortest growing season?

3. How many weeks long is the growing season in Pittsburgh?

4. How many weeks long is the growing season in Dallas?

5. How many times longer is the growing season in Fresno than in Ogden?

6. How many times longer is the growing season in Dallas than in Laramie?

7. Which city most closely approximates one half of the growing season of Los Angeles?

8. Which city has approximately $2\frac{1}{2}$ times the growing season of Laramie?

1. Los Angeles

2. Laramie

3. 22

4. 36

5. 2

6. 3

7. Syracuse

8. Oklahoma City

ANSWERS

9. Approximately $270

10. Approximately $212

11. San Francisco

12. New York

13. Approximately $50

14. Approximately $28

15. See graph.

16. 580

This vertical bar graph shows the average daily expenses for lodging, food, and rental car for traveling executives.

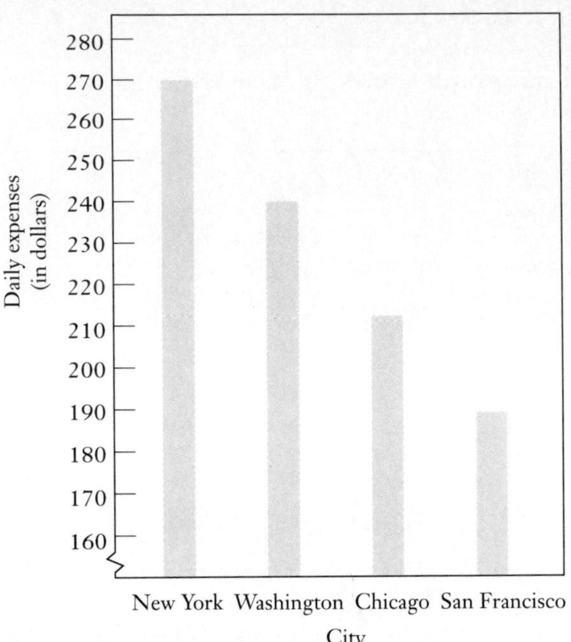

9. What are the average daily expenses in New York?

10. What are the average daily expenses in Chicago?

11. Which city is the least expensive of the four?

12. Which city is the most expensive of the four?

13. How much more are the average daily expenses in Washington than in San Francisco?

14. How much less are the average daily expenses in Chicago than in Washington?

b

15. Use the following information to make a bar graph showing the number of calories burned during each activity by a person weighing 152 lb. Use the blank graph below.

Tennis:	420 calories per hour
Jogging:	650 calories per hour
Hiking:	590 calories per hour
Office work:	180 calories per hour
Sleeping:	70 calories per hour

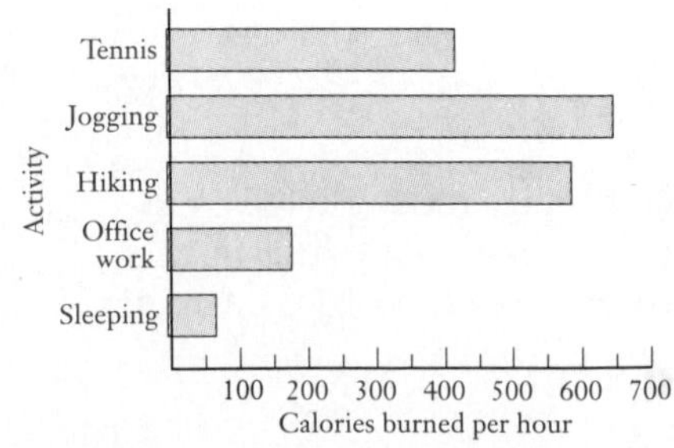

16. What is the difference in calories burned per hour between sleeping and jogging?

C The line graph below reflects the hourly readings of the outside temperature during one fall day.

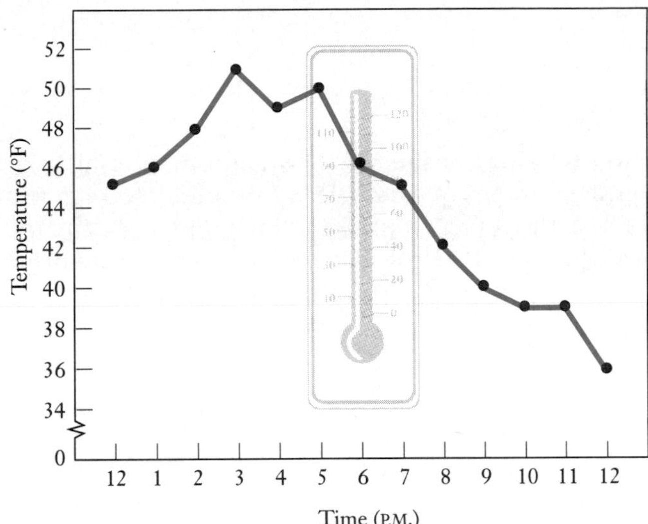

17. At what time was the temperature the highest?

18. At what time was the temperature the lowest?

19. What was the difference in temperature between the highest and lowest readings?

20. Between which two hours did the temperature increase the most?

21. Between which two hours did the temperature decrease the most?

22. How much colder was it at midnight than at noon?

The following line graph shows the estimated sales (in millions of dollars) for several years for a company. Again, note the jagged line at the base of the vertical scale.

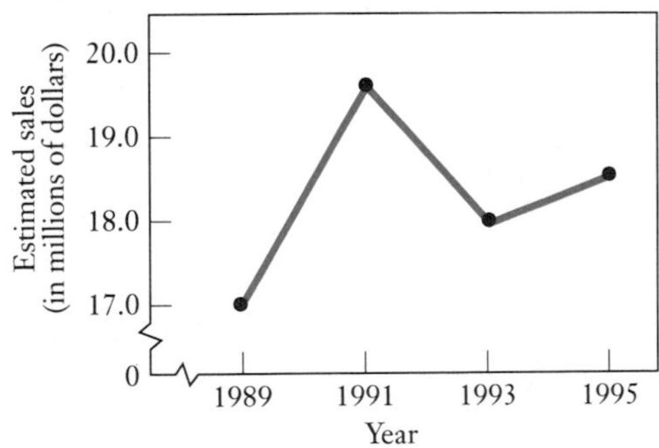

23. In what year are estimated sales the greatest?

24. In what year are estimated sales the least?

25. What are estimated sales in 1989?

26. What are estimated sales in 1995?

ANSWERS

17. 3:00 P.M.

18. 12:00 midnight

19. Approximately 15°

20. Between 2:00 P.M. and 3:00 P.M.

21. Between 5:00 P.M. and 6:00 P.M.

22. Approximately 9°

23. 1991

24. 1989

25. Approximately $17 million

26. Approximately $18.5 million

ANSWERS

27. Approximately $1.5 million

28. Approximately $1.5 million

29. See graph.

30. Between 4 P.M. and 5 P.M.

31. Between 8 P.M. and 9 P.M.

32. 72 cars

33. 18%

34. 18 min

35. $66.\overline{6}\%$, or $66\frac{2}{3}\%$

36. 82.5

27. How much greater are estimated sales in 1991 than in 1993?

28. How much less are estimated sales in 1989 than in 1995?

d

29. A rural intersection is being considered for an automatic traffic signal. A traffic survey, recording the number of cars passing through the intersection, gave the following results. Use these facts to make a line graph showing the number of cars counted for each hour. Be sure to label the two scales appropriately.

12 noon:	50 cars	5 P.M.	100 cars
1 P.M.:	40 cars	6 P.M.	112 cars
2 P.M.:	60 cars	7 P.M.	88 cars
3 P.M.:	65 cars	8 P.M.	70 cars
4 P.M.:	77 cars	9 P.M.	35 cars

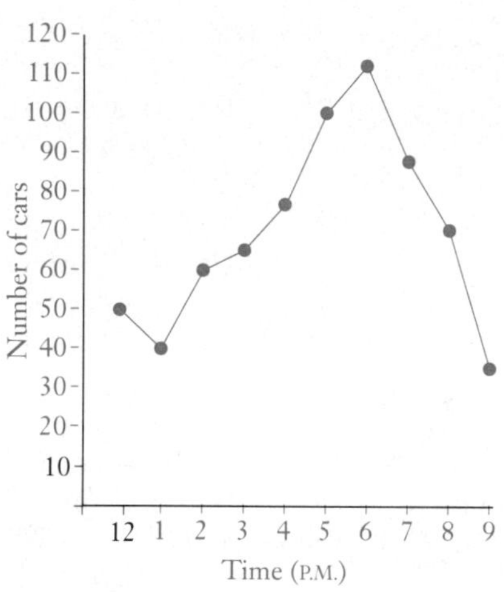

30. Between which two hours did traffic increase the most?

31. Between which two hours did traffic decrease the most?

32. How much did traffic increase from 1 P.M. to 6 P.M.?

SKILL MAINTENANCE

33. It is known to operators of pizza restaurants that if 50 pizzas are ordered in an evening, people will request extra cheese on 9 of them. What percent of the pizzas sold are ordered with extra cheese?

34. A clock loses 3 min every 12 hr. At this rate, how much time will the clock lose in 72 hr?

35. 34 is what percent of 51?

36. 110% of 75 is what?

8.4 Circle Graphs

We often use **circle graphs** to show the percent of a quantity used in different categories. Circle graphs can also be used very effectively to show visually the *ratio* of one category to another. In either case, it is quite often necessary to use mathematics to find the actual amounts represented for each specific category.

a Reading and Interpreting Circle Graphs

▶ **EXAMPLE 1** This circle graph shows expenses as a percent of income in a family of four, according to a recent study of the Bureau of Labor Statistics. (*Note:* Due to rounding, the sum of the percents is 101% instead of 100%.)

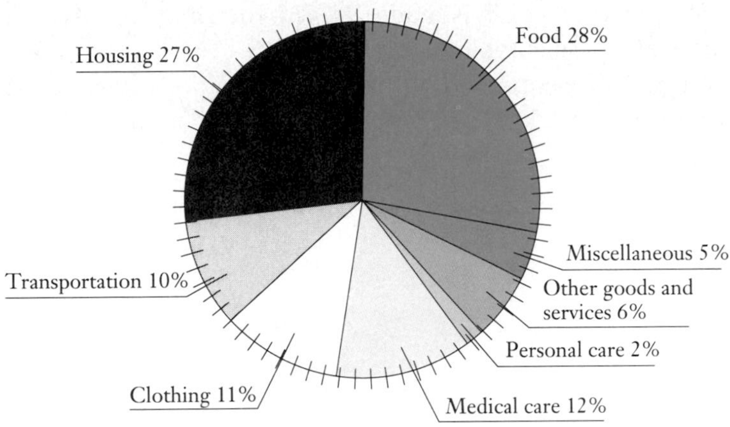

a) Which item accounts for the greatest expense?

b) For a family with a $2000 monthly income, how much is spent for transportation?

c) Some surveys combine medical care with personal care. What percent would be spent on those two items combined?

We look at the sections of the graph to find the answers.

a) It is immediately apparent that there are two sections that are larger than the rest. Of those two sections, the one representing food is the larger, at 28%.

b) The section of the circle representing transportation shows a 10% expense; 10% of $2000 is $200.

c) In a circle graph, we can add percents safely for problems of this type. Therefore, 12% (medical care) + 2% (personal care) = 14%. ◀

DO EXERCISES 1–3.

OBJECTIVES

After finishing Section 8.4, you should be able to:

a Read and interpret data from circle graphs.

b Draw circle graphs.

FOR EXTRA HELP

Tape 10D

Tape 12A

MAC: 8
IBM: 8

Consider a family with a $2000 monthly income and use the circle graph in Example 1 to answer each of the following.

1. How much would this family typically spend on housing each month? $540

2. What percent of the income is spent on housing and clothing combined? 38%

3. Compare the amount spent on medical care with the amount spent on personal care. What is the ratio? 6 to 1

ANSWERS ON PAGE A-5

b Drawing Circle Graphs

▶ **EXAMPLE 2** In a quick inventory, it was found that the types of books listed below made up the indicated percent of available books in a library. Use this information to draw a circle graph reflecting the different types of books available.

a) History books: 25%

b) Science books: 10%

c) Fiction: 45%

d) Reference books: 5%

e) Other: 15%

We will first draw each section in a separate working circle to illustrate more clearly how each is made. We will then combine them in a single circle to show the complete graph.

Our circles will be marked off in 5-degree sections. We are providing you with circles marked off, but you could also use a protractor. (Remember that there are 360 degrees in a circle.)

a) History books account for 25%, so they should be shown by 25% of the circle. Mathematically, this is 25% of 360° (0.25 × 360), or 90°.

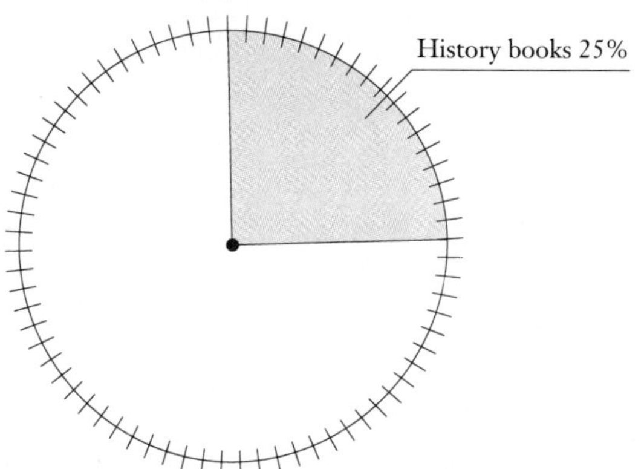

b) Science books account for 10% of the books; 10% of 360° (0.10 × 360) is 36° more.

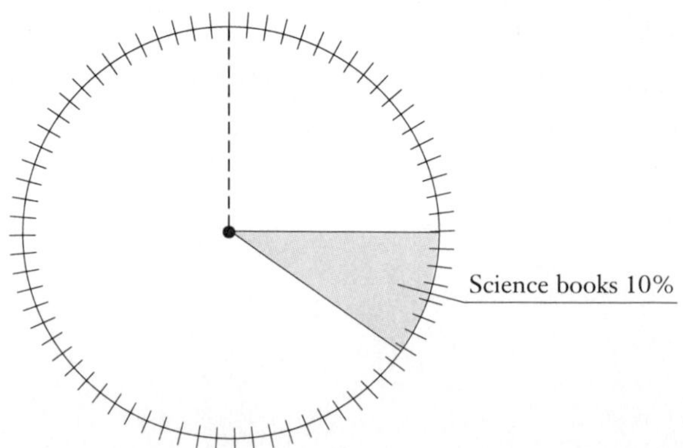

c) Fiction represents 45% of the books, and 45% of 360° (0.45 × 360) is another 162°.

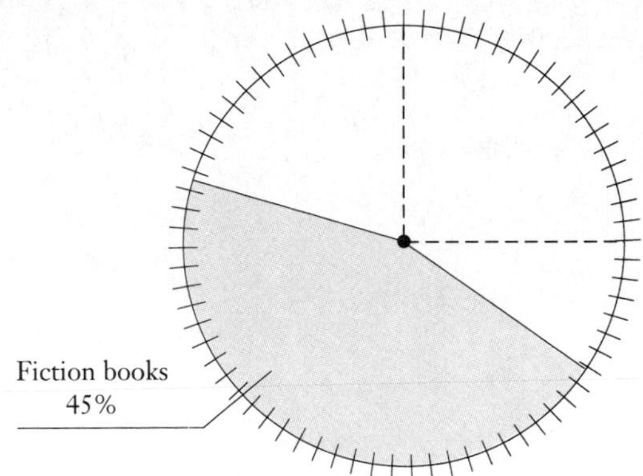

d) Reference books represent 5% of the books; 5% of 360° (0.05 × 360) is 18° more.

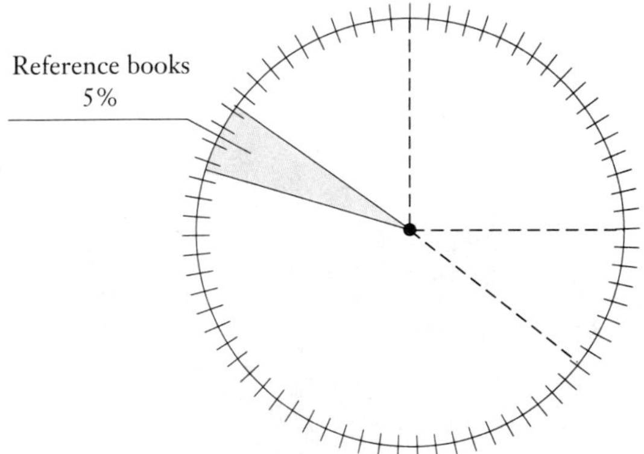

4. For each of the following uses of paper, find the number of degrees needed, to the nearest degree, to draw a circle graph. Then draw the graph.

a) Packaging: 48% 173°

b) Writing paper: 30% 108°

c) Tissues: 8% 29°

d) Other: 14% 50°

e) Draw the graph.

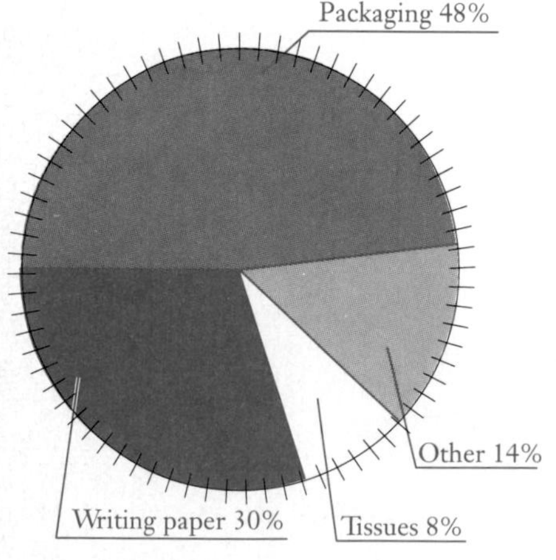

ANSWERS ON PAGE A-5

e) Other books make up the remaining 15%. We find that 15% of 360° (0.15×360) is 54°. Note that this last section accounts exactly for the remainder of the circle.

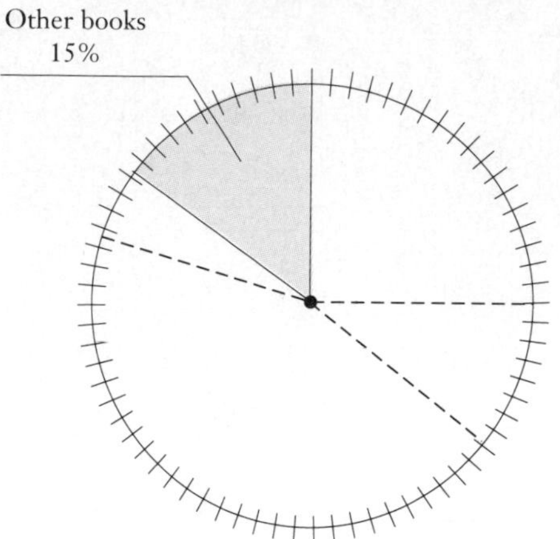

Now we can combine all of these sections in a single circle, which results in the circle graph below.

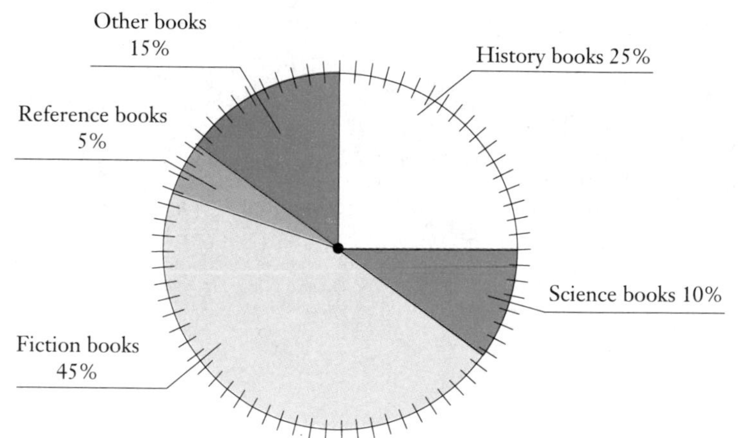

DO EXERCISE 4.

NAME SECTION DATE

EXERCISE SET 8.4

ANSWERS

a This circle graph, in the shape of a record, shows music preferences of customers on the basis of record store sales, according to the National Association of Recording Merchandisers.

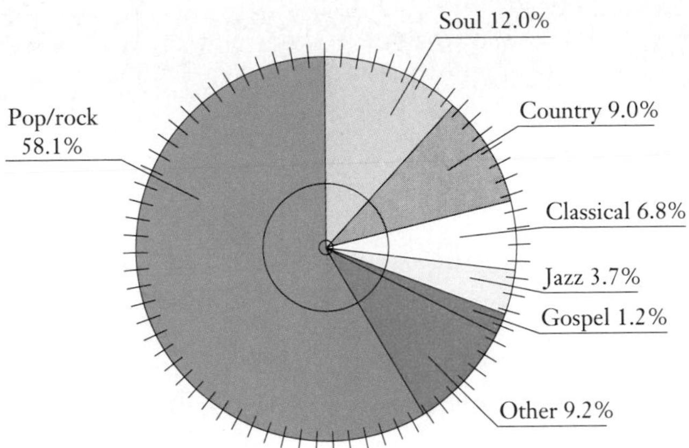

1. What percent of all records sold are jazz?

2. Together, what percent of all records sold are either soul or pop/rock?

3. A music store sells 3000 records a month. How many are country?

4. A music store sells 2500 records a month. How many are gospel?

5. What percent of all records sold are classical?

6. Together, what percent of all records sold are either classical or jazz?

1. 3.7%

2. 70.1%

3. 270

4. 30

5. 6.8%

6. 10.5%

ANSWERS

7. Gas purchased

8. Gas produced

9. 4¢

10. 41¢

11. 16¢

12. Interest and other operations

13. See graph.

14. 300.6

15. 25%

16. 115

17. $\frac{3}{4}$

This circle graph shows how each customer's dollar is spent by the Indiana Gas Company on an annual basis.

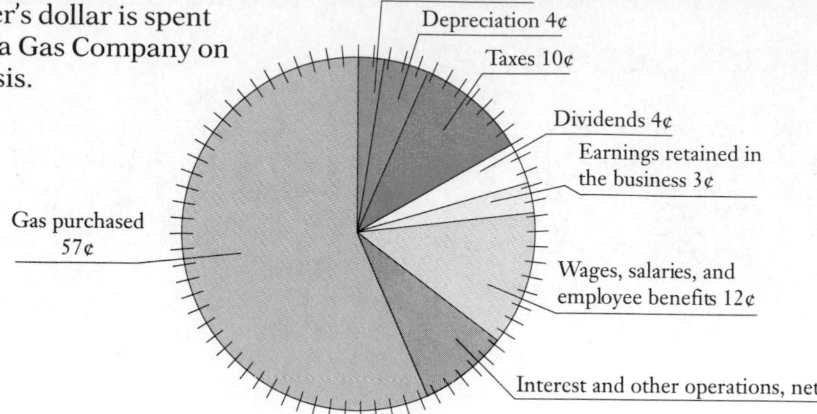

7. On which item is the most spent?

8. On which item is the least spent?

9. How much of each dollar is spent on dividends?

10. How much of each dollar is left after expenses for gas purchased and gas produced?

11. How much of each dollar is spent on dividends, wages, salaries, and employee benefits all together?

12. The total amount spent on depreciation and dividends is the same as the amount spent on what other item?

b

13. Use this information on vacation expenditures to find the number of degrees required to represent each type of expenditure, and then draw and label an appropriate circle graph.

Transportation: 15%
Meals: 20%
Lodging: 32%
Recreation: 18%
Other: 15%

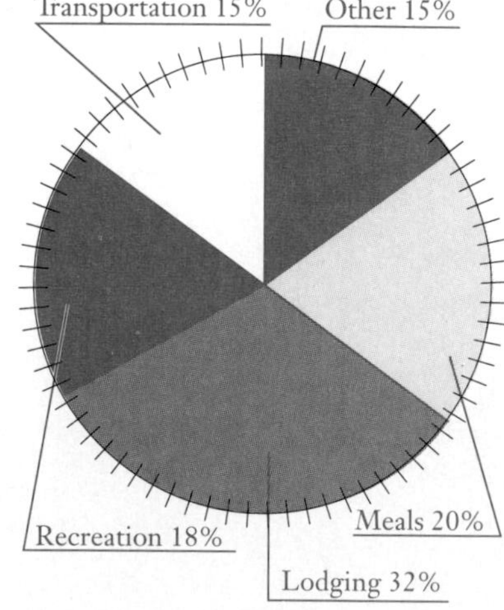

SKILL MAINTENANCE

Solve.

14. What is 45% of 668?

15. 16 is what percent of 64?

16. 23 is 20 percent of what?

17. Divide and simplify: $\frac{4}{7} \div \frac{16}{21}$.

SUMMARY AND REVIEW EXERCISES: CHAPTER 8

The review sections and objectives to be tested in addition to the material in this chapter are [2.7b], [6.3a], [7.3a], [7.4a], and [7.5a].

Find the average.

1. 26, 34, 43, 51

2. 7, 11, 14, 17, 18

3. 0.2, 1.7, 1.9, 2.4

4. 700, 900, 1900, 2700, 3000

5. \$2, \$14, \$17, \$17, \$21, \$29

6. 20, 190, 280, 470, 470, 500

Find the median.

7. 26, 34, 43, 51

8. 7, 11, 14, 17, 18

9. 0.2, 1.7, 1.9, 2.4

10. 700, 900, 1900, 2700, 3000

11. \$2, \$14, \$17, \$17, \$21, \$29

12. 20, 190, 280, 470, 470, 500

Find the mode.

13. 26, 34, 43, 26, 51

14. 7, 11, 11, 14, 17, 17, 18

15. 0.2, 0.2, 0.2, 1.7, 1.9, 2.4

16. 700, 700, 800, 2700, 800

17. \$2, \$14, \$17, \$17, \$21, \$29

18. 20, 20, 20, 20, 20, 500

19. One summer, a student earned the following amounts over a four-week period: \$102, \$112, \$130, and \$98. What was the average amount earned? the median?

20. The following temperatures were recorded every four hours on a certain day: 63°, 58°, 66°, 72°, 71°, 67°. What was the average temperature for that day?

21. To get an A in math, a student must average 90 on four tests. Scores on the first three tests were 94, 78, and 92. What is the lowest score that the student can make on the last test and still get an A?

This table illustrates various living expenses for urban areas in several U.S. cities. Use it for Exercises 22–27.

Urban area	1800-square-foot house	Rent for 2-bedroom apartment	Dry-clean suit	Woman's shampoo/trim	1 dozen eggs	1 game bowling
Washington, D.C.	$150,277	$ 848	$6.35	$27.20	$0.86	$2.14
Springfield, Missouri	$ 70,716	$ 292	$5.75	$19.10	$0.93	$1.50
Laurel, Mississippi	$ 75,400	$ 258	$4.17	$14.84	$0.68	$1.75
Highest	$290,000 (Boston MA)	$1600 (Boston MA)	$8.68 (Fairbanks AL)	$27.20 (Washington DC)	$1.42 (Bakersfield CA	$2.50 (Boston MA)
Lowest	$ 65,600 (Nevada MO)	$ 258 (Laurel MS)	$3.25 (Canton OH)	$6.50 (Sherman TX)	$0.50 (Kirksville MO)	$0.98 (Canton OH)

22. Where does a 2-bedroom apartment cost the most?

23. What is the lowest price that you will pay for one dozen eggs?

24. How much more will you pay to dry-clean a suit in Washington, D.C., than in Laurel, Mississippi?

25. How much does an 1800-square-foot home in Springfield, Missouri, cost?

26. Where will it cost $27.20 to get a woman's shampoo and trim?

27. Where will you pay the lowest price for one game of bowling?

This pictograph shows, for 1986, the number of space launches of several countries. Use it for Exercises 28–31.

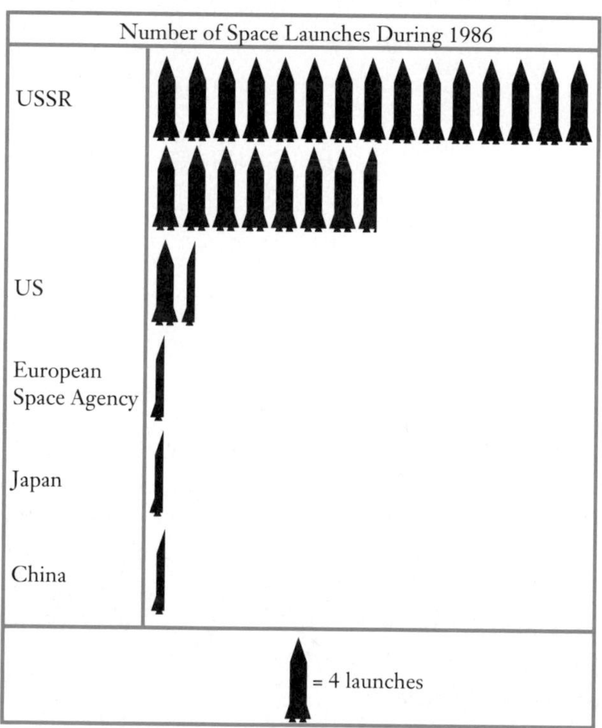

28. Which country had the most launches in 1986?

29. What was the greatest number of launches for one country?

30. What country had three times as many launches as China?

31. How many launches were made by all five countries combined?

This bar graph shows the Fast Food Price Index for several major cities. By definition for the purpose of this comparison, "fast food" consists of a quarter-pound cheeseburger, large order of fries, and medium-sized soft drink. Use it for Exercises 32–37.

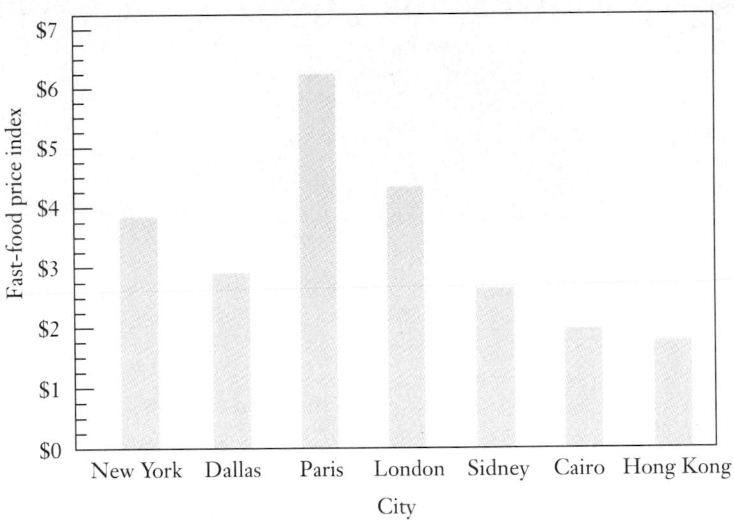

32. What is the most that you will pay for this meal?

33. Where will you pay the least for this meal?

34. In which of the given United States cities will you spend the most for this meal?

35. What is the least that you will spend for this meal considering the given U.S. cities?

36. How much more will you pay for this meal in Paris than in Hong Kong?

37. Where will you pay close to the same price for this meal that you would pay in Dallas?

This line graph shows the number of accidents per 100 drivers, by age. Use it for Exercises 38–43.

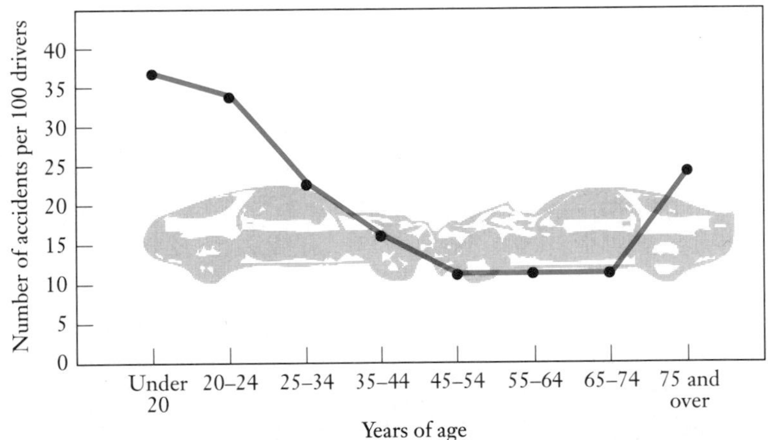

38. What age group has the most accidents per 100 drivers?

39. What is the fewest number of accidents per 100 in any age group?

40. How many more accidents do people over 75 years of age have than those in the age range of 65–74?

41. Between what ages does the number of accidents stay basically the same?

42. How many fewer accidents do persons 25–34 years of age have than those 20–24 years of age?

43. What age group has accidents more than three times as often as persons 55–64 years of age?

This circle graph shows the percent of homebuyers preferring various locations for their homes. Use it for Exercises 44–47.

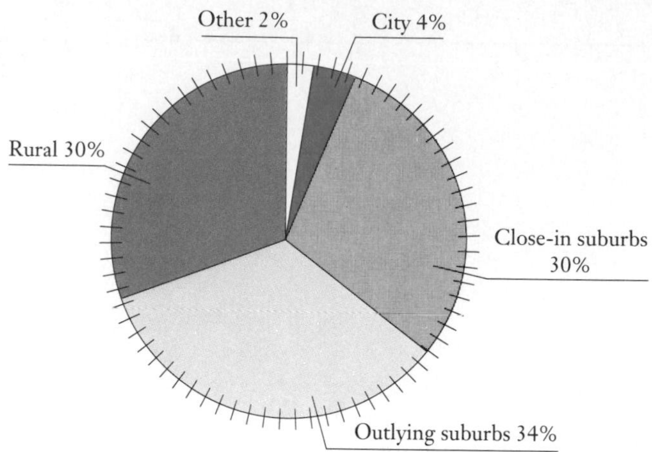

44. What percent of homebuyers prefer the rural areas?

45. What is the preference of 34% of the homebuyers?

46. What percent of homebuyers prefer to live somewhere in the suburbs?

47. Homebuyers prefer a rural area how many times more than the city?

SKILL MAINTENANCE

Solve.

48. A company car was driven 4200 miles in the first 4 months of a year. At this rate, how far will it be driven in 12 months?

49. 92% of the world population does not have a telephone. The population is about 234 million. How many do not have a telephone?

50. 789 is what percent of 355.05?

51. What percent of 98 is 49?

Divide and simplify.

52. $\frac{3}{4} \div \frac{5}{6}$

53. $\frac{5}{8} \div \frac{3}{2}$

❖ THINKING IT THROUGH

1. Compare averages, medians, and modes. Discuss why you might use one over the others to analyze a set of data.
2. Compare bar graphs and line graphs. Discuss why you might use one over the other to graph a set of data.
3. Compare bar graphs and circle graphs. Discuss why you might use one over the other to graph a set of data.

NAME SECTION DATE

TEST: CHAPTER 8

ANSWERS

Find the average.

1. 45, 49, 52, 54
2. 1, 2, 3, 4, 5
3. 3, 17, 17, 18, 18, 20

Find the median and the mode.

4. 45, 49, 52, 54
5. 1, 2, 3, 4, 5
6. 3, 17, 17, 18, 18, 20

7. A car went 754 km in 13 hr. What was the average number of kilometers per hour?

8. To get a C in chemistry, a student must average 70 on four tests. Scores on the first three tests were 68, 71, and 65. What is the lowest score that the student can make on the last test and still get a C?

Use the following table to answer the questions about the number of calories burned during various walking activities in Exercises 9–12.

	Calories burned in 30 min		
Walking activity	**110 lb**	**132 lb**	**154 lb**
Walking			
Fitness (5 mph)	183	213	246
Mildly energetic (3.5 mph)	111	132	159
Strolling (2 mph)	69	84	99
Hiking			
3 mph with 20-lb load	210	249	285
3 mph with 10-lb load	195	228	264
3 mph with no load	183	213	246

9. What activity provides the greatest benefit in burned calories for a person weighing 132 lb?

10. What is the least strenuous activity you must perform if you weigh 154 lb and you want to burn at least 250 calories every 30 min?

1. [8.1a] 50
2. [8.1a] 3
3. [8.1a] 15.5
4. [8.1b, c] Median: 50.5; mode: 45, 49, 52, 54
5. [8.1b, c] Median: 3; mode: 1, 2, 3, 4, 5
6. [8.1b, c] Median: 17.5; mode: 17, 18
7. [8.1a] 58 km/h
8. [8.1a] 76
9. [8.2a] Hiking with 20-lb load
10. [8.2a] Hiking with 10-lb load

ANSWERS

11. How is "mildly energetic walking" defined?

12. What type of walking can a person weighing 110 lb do that will give the same benefit as some type of hiking?

13. Draw a vertical bar graph using an appropriate set of scales, showing the percent of teachers holding a master's degree for the following years. Be sure to label the scales properly.

1961: 24%
1966: 23%
1971: 28%
1976: 38%
1981: 50%
1986: 51%

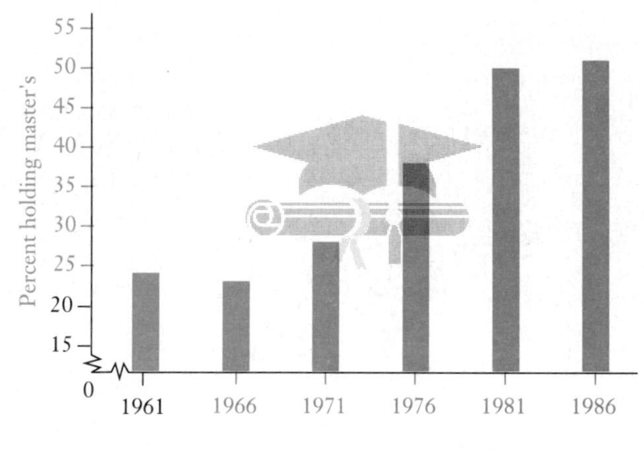

The following pictograph shows the number of hits in a season for several professional baseball players. Use it for Exercises 14–18.

Number of Hits in a Season for 5 Professional Players

Carew
Garr
Gross
McRae
Allen

= 25 hits

14. How many hits does each baseball symbol represent?

15. Who had the most hits?

[8.2a]
11. Walking at 3.5 mph

[8.2a]
12. Fitness walking

[8.3b]
13. See graph.

14. [8.2b] 25

15. [8.2b] Carew

16. How many hits did Allen get?

17. Who got 210 hits?

18. How many more hits did Gross get than McRae?

The following line graph shows the amount of money being spent by travelers to come to the United States. Use it for Exercises 19–22.

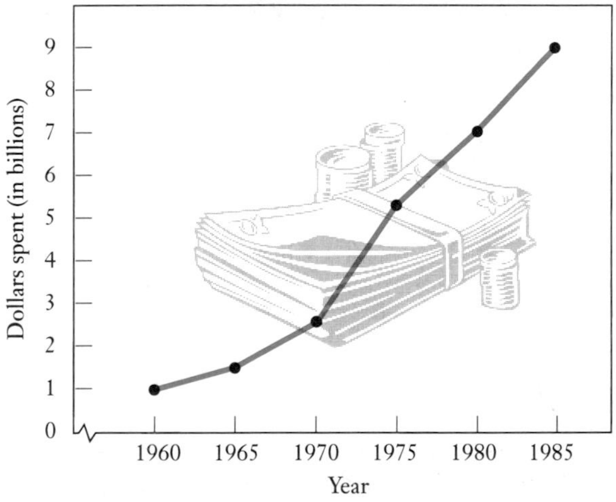

19. What trend, if any, is being shown in this graph?

20. In what five-year period was the increase the greatest?

21. How much was spent in 1960?

22. How much more was spent in 1985 than in 1980?

ANSWERS

16. [8.2b] 145

17. [8.2b] Garr

18. [8.2b] 10

19. [8.3c] Increasing

20. [8.3c] 1970 to 1975

21. [8.3c] $1 billion

22. [8.3c] $2 billion

ANSWERS

23. [8.4b] See graph.

24. [2.7b] $\frac{25}{4}$

25. [7.3a], [7.4a] 68

26. [7.5a] 15,600

27. [6.3a] 340

23. Use the following information to make a circle graph showing the percent of available money that people invest in certain ways. Be sure to label each section appropriately. (*Note:* The given circle is divided into 5-degree sections.)

Savings accounts:	48%
Stocks:	12%
Mutual funds:	16%
Retirement funds:	24%

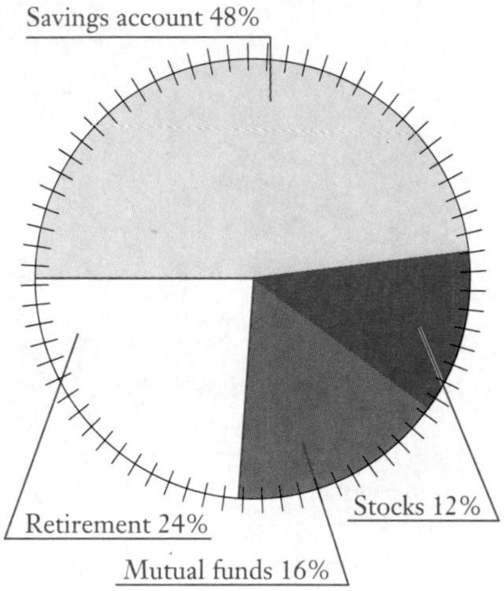

SKILL MAINTENANCE

24. Divide and simplify: $\frac{3}{5} \div \frac{12}{125}$.

25. 17 is 25% of what number?

26. 78% of the television sets that are on are tuned to one of the major networks. Suppose 20,000 TV sets in a town are being watched. How many are tuned to a major network?

27. A baseball player gets 7 hits in the first 20 times at bat. At this rate, how many times at bat will it take to get 119 hits?

CUMULATIVE REVIEW: CHAPTERS 1–8

1. In 402,513 what does the digit 5 mean?

2. Evaluate: $3 + 5^3$.

3. Find all the factors of 60.

4. Round 52.045 to the nearest tenth.

5. Convert to fractional notation: $3\frac{3}{10}$.

6. Convert from cents to dollars: 210¢.

7. Convert to standard notation: $3.25 billion.

8. Determine whether 11, 30 and 4, 12 are proportional.

Add and simplify.

9. $2\frac{2}{5} + 4\frac{3}{10}$

10. $41.063 + 3.5721$

Subtract and simplify.

11. $\frac{14}{15} - \frac{3}{5}$

12. $350 - 24.57$

Multiply and simplify.

13. $3\frac{3}{7} \cdot 4\frac{3}{8}$

14. $12{,}456 \times 220$

Divide and simplify.

15. $\frac{13}{15} \div \frac{26}{27}$

16. $104{,}676 \div 24$

Solve.

17. $\frac{5}{8} = \frac{6}{x}$

18. $\frac{2}{5} \cdot y = \frac{3}{10}$

19. $21.5 \cdot y = 146.2$

20. $x = 398{,}112 \div 26$

Solve.

21. Eighteen ounces of cheese costs 99¢. Find the unit price in cents per ounce.

22. A college has a student body of 6000 students. Of these, 55.4% own a car. How many students own a car?

23. In any given year, Americans eat an average of 2.7 lb of peanut butter, 1.5 lb of salted peanuts, 1.2 lb of peanut candy, 0.7 lb of in-shell peanuts, and 0.1 lb of peanuts in other forms. How many pounds of peanuts does each American eat, on the average, in one year?

24. A piece of fabric $1\frac{3}{4}$ yd long is cut into 7 equal strips. What is the width of each strip?

25. In 1987, American utility companies generated 1464 billion kilowatt-hours of electricity using coal, 455 billion kilowatt-hours using nuclear power, 273 billion using natural gas, 250 billion using hydroelectric plants, 118 billion using petroleum, and 12 billion using geothermal technology and other methods. How many kilowatt-hours of electricity were produced in 1987?

26. A recipe calls for $\frac{3}{4}$ of a cup of sugar. How much sugar should be used for an amount that is $\frac{1}{2}$ of the recipe?

27. A business is owned by four people. One owns $\frac{1}{3}$, the second owns $\frac{1}{4}$, and the third owns $\frac{1}{6}$. How much does the fourth person own?

28. In manufacturing valves for engines, a factory was discovered to make 4 defective valves out of a lot of 18 valves. At this rate, how many defective valves can be expected in a lot of 5049 valves?

29. A landscaper bought 22 evergreen trees for $210. What was the cost for each tree? Round to the nearest cent.

30. A salesperson earns $182 selling $2600 worth of electronic equipment. What is the commission rate?

This circle graph shows the percent of 18-year-olds surveyed who planned to vote in an upcoming presidential election. Use it for Exercises 31–33.

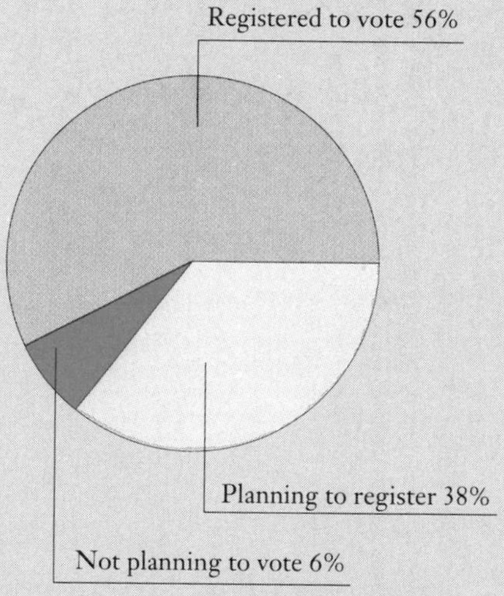

31. What percent of all 18-year-olds were registered to vote?

32. What percent of all 18-year-olds were planning to vote in the election?

33. In a class of 250 18-year-old freshmen, how many were not planning to vote?

34. Draw a vertical bar graph, using an appropriate set of scales, showing the percentage of people in a National Geographic Society survey who said it is absolutely necessary to know something about the following listed subjects. Be sure to label the scales properly.

Math:	83%
Computer skills:	64%
Science:	38%
Geography:	37%
History:	36%
Foreign languages:	20%

35. Find the mode of this set of numbers:

3, 5, 2, 5, 1, 3, 5, 2.

36. Find the median of this set of numbers:

61, 67, 60, 63.

SYNTHESIS

37. A photography club meets four times a month. In September, the attendance figures were 28, 23, 26, and 23. In October, the attendance figures were 26, 20, 14, and 28. What was the percent increase or decrease in average attendance from September to October?

INTRODUCTION This chapter introduces American and metric systems used to measure length, and presents conversion from one unit to another within as well as between each system. These concepts are then applied to finding areas of squares, rectangles, triangles, parallelograms, and circles. Right triangles are then studied using square roots and the Pythagorean theorem.

The review sections to be tested in addition to the material in this chapter are 7.1 and 7.2. ❖

Geometry and Measures: Length and Area

AN APPLICATION

A standard-sized softball diamond is a square whose sides are each 65 ft in length. What is the perimeter of a softball diamond?

THE MATHEMATICS

We find the perimeter by finding the distance around the square:

$$\underbrace{65 + 65 + 65 + 65.}_{\text{This is the perimeter.}}$$

❖ POINTS TO REMEMBER: CHAPTER 9

Exponential notation: Section 1.9
Computational skills with whole numbers, decimals, and fractions

PRETEST: CHAPTER 9

Complete.

1. 8 ft = ______ in.

2. 5 in. = ______ ft

3. 8.46 km = ______ m

4. 9.2 mm = ______ cm

5. Find the perimeter.

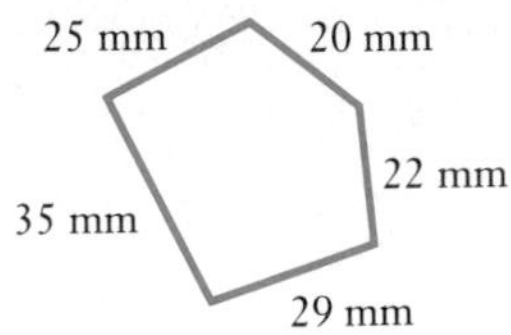

6. Find the area of a square whose sides have length 10 ft.

Find the area.

7.

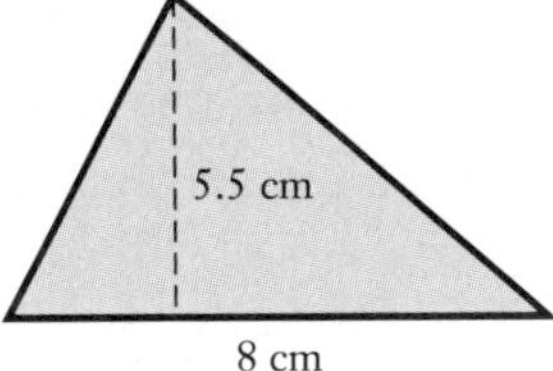

8.

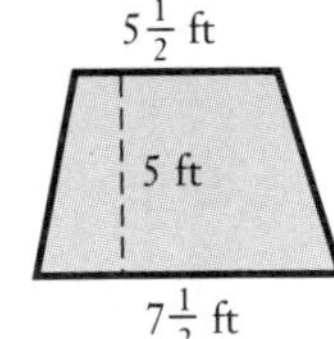

9.

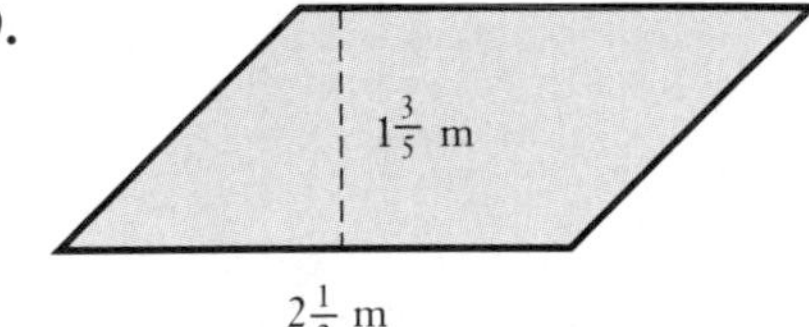

10. Find the length of a diameter of a circle with a radius of 4.8 m.

11. Find the circumference of the circle in Exercise 10. Use 3.14 for π.

12. Find the area of the circle in Exercise 10. Use 3.14 for π.

13. Find the area of the shaded region.

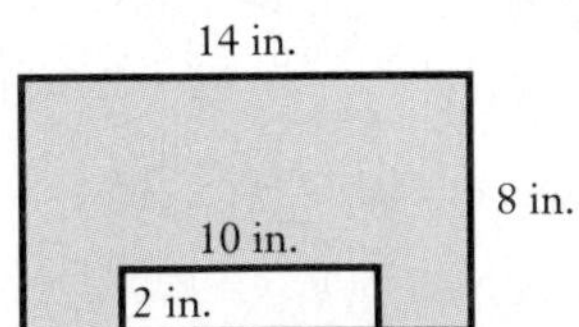

14. Simplify: $\sqrt{81}$.

15. Approximate to three decimal places: $\sqrt{97}$.

In a right triangle, find the length of the side not given. Find an exact answer and an approximation to three decimal places.

16. $a = 12, b = 16$

17. $a = 2, c = 7$

9.1 Linear Measures: American Units

Length, or distance, is one kind of measure. To find lengths, we *start* with some **unit segment** and assign to it a measure of 1. Suppose $\overline{AB}$ below is a unit segment.

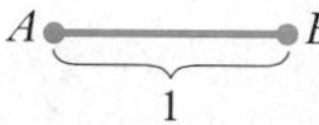

Let's measure segment $\overline{CD}$ below.

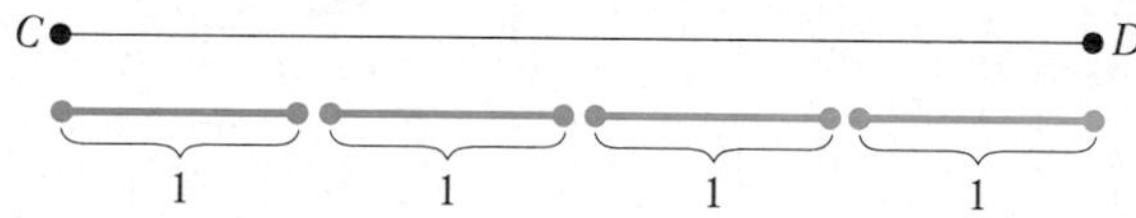

Since we can place 4 unit segments end to end along $\overline{CD}$, the measure of $\overline{CD}$ is 4.

Sometimes we have to use parts of units, called **subunits.** For example, the measure of the segment $\overline{MN}$ below is $1\frac{1}{2}$. We place one unit segment and one half-unit segment end to end.

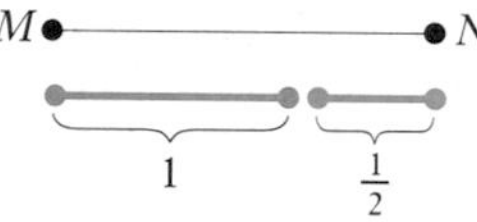

DO EXERCISES 1–4.

a American Measures

American units of length are related as follows:

American Units of Length

12 inches (in.) = 1 foot (ft);
3 feet = 1 yard (yd);
5280 feet = 1 mile (mi)

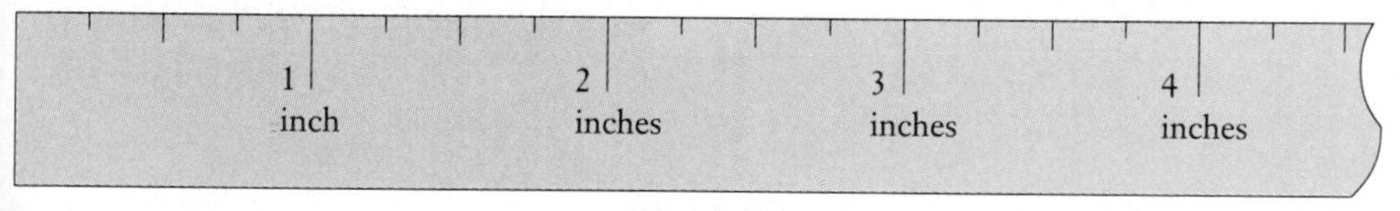

(Actual size)

These American units have also been called "English," or "British-American," because at one time they were used by both countries. Today, both Canada and England state that they have converted to the metric system. However, if you travel in England, you will still see units such as "miles" on road signs.

OBJECTIVE

After finishing Section 9.1, you should be able to:

a Convert from one American unit of length to another.

FOR EXTRA HELP

Tape 11A

Tape 12B

MAC: 9
IBM: 9

Use the unit below to measure the length of each segment or object.

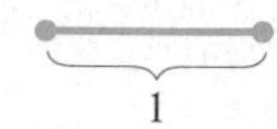

1. 

2. 

3. 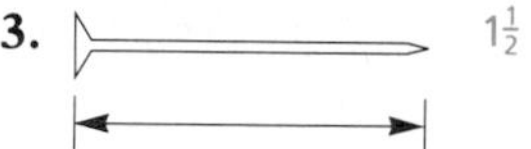

4. 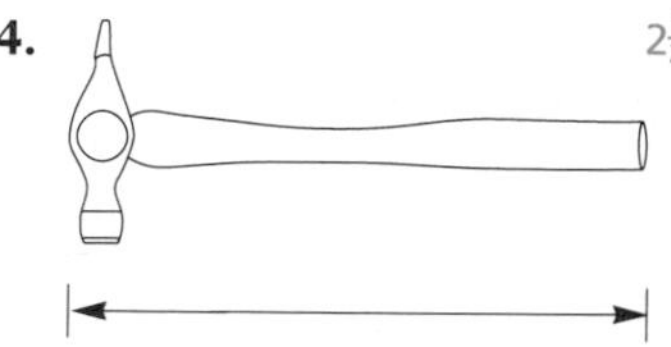

ANSWERS ON PAGE A-5

To change from certain American units to others, we make substitutions.

▶ **EXAMPLE 1** Complete: 1 yd = _______ in.

$$\begin{aligned} 1 \text{ yd} &= 3 \text{ ft} \\ &= 3 \times 1 \text{ ft} && \textbf{We think of 3 ft as } 3 \times \textbf{ft, or } 3 \times 1 \textbf{ ft.} \\ &= 3 \times 12 \text{ in.} && \textbf{Substituting 12 in. for 1 ft} \\ &= 36 \text{ in.} && \textbf{Multiplying} \end{aligned}$$ ◀

▶ **EXAMPLE 2** Complete: 7 yd = _______ in.

$$\begin{aligned} 7 \text{ yd} &= 7 \times 1 \text{ yd} \\ &= 7 \times 3 \text{ ft} && \textbf{Substituting 3 ft for 1 yd} \\ &= 7 \times 3 \times 1 \text{ ft} \\ &= 7 \times 3 \times 12 \text{ in.} && \textbf{Substituting 12 in. for 1 ft; } 7 \times 3 = 21;\ 21 \times 12 = 252 \\ &= 252 \text{ in.} \end{aligned}$$ ◀

DO EXERCISES 5 AND 6.

Complete.

5. 8 yd = ___288___ in.

6. 14.5 yd = ___43.5___ ft

Sometimes it helps to use multiplying by 1 in making conversions. For example, 12 in. = 1 ft, so

$$\frac{12 \text{ in.}}{1 \text{ ft}} = 1 \quad \text{and} \quad \frac{1 \text{ ft}}{12 \text{ in.}} = 1.$$

These symbols represent division. If we divide 12 in. by 1 ft or 1 ft by 12 in., we would expect to get 1 because the lengths are the same.

▶ **EXAMPLE 3** Complete: 48 in. = _______ ft.

We want to convert from "in." to "ft." We multiply by 1 using a symbol for 1 with "in." on the bottom and "ft" on the top to eliminate inches and to convert to feet.

$$\begin{aligned} 48 \text{ in.} &= \frac{48 \text{ in.}}{1} \times \frac{1 \text{ ft}}{12 \text{ in.}} && \textbf{Multiplying by 1 using } \frac{\textbf{1 ft}}{\textbf{12 in.}} \textbf{ to eliminate in.} \\ &= \frac{48 \text{ in}}{12 \text{ in.}} \times 1 \text{ ft} \\ &= \frac{48}{12} \times \frac{\text{in.}}{\text{in.}} \times 1 \text{ ft} \\ &= 4 \times 1 \text{ ft} && \textbf{The } \frac{\textbf{in.}}{\textbf{in.}} \textbf{ acts like 1, so we can omit it.} \\ &= 4 \text{ ft} \end{aligned}$$

We can also look at this conversion as "canceling" units:

$$48 \text{ in.} = \frac{48 \not{\text{in.}}}{1} \times \frac{1 \text{ ft}}{12 \not{\text{in.}}} = \frac{48}{12} \times 1 \text{ ft} = 4 \text{ ft.}$$ ◀

DO EXERCISES 7–9.

Complete.

7. 72 in. = ___6___ ft

8. 17 in. = ___$1\frac{5}{12}$___ ft

9. 24 ft = ___8___ yd

ANSWERS ON PAGE A-5

▶ **EXAMPLE 4** Complete: 25 ft = ______ yd.

Since we are converting from "ft" to "yd," we choose a symbol for 1 with "yd" on the top and "ft" on the bottom.

$$25 \text{ ft} = 25 \text{ ft} \times \frac{1 \text{ yd}}{3 \text{ ft}}$$ **3 ft = 1 yd, so $\frac{3 \text{ ft}}{1 \text{ yd}} = 1$, and $\frac{1 \text{ yd}}{3 \text{ ft}} = 1$. We use $\frac{1 \text{ yd}}{3 \text{ ft}}$ to eliminate ft.**

$$= \frac{25}{3} \times \frac{\text{ft}}{\text{ft}} \times 1 \text{ yd}$$

$$= 8\tfrac{1}{3} \times 1 \text{ yd}$$ **The $\frac{\text{ft}}{\text{ft}}$ acts like 1, so we can omit it.**

$$= 8\tfrac{1}{3} \text{ yd, or } 8.3\overline{3} \text{ yd}$$

Again, in this example, we can consider conversion from the point of view of canceling:

$$25 \text{ ft} = 25 \cancel{\text{ft}} \times \frac{1 \text{ yd}}{3 \cancel{\text{ft}}} = \frac{25}{3} \times 1 \text{ yd} = 8\tfrac{1}{3} \text{ yd}, \quad \text{or } 8.\overline{3} \text{ yd}.$$ ◀

DO EXERCISES 10 AND 11.

Complete.

10. 99 ft = __33__ yd

11. 35 ft = __$11\frac{2}{3}$, or $11.\overline{6}$__ yd

▶ **EXAMPLE 5** Complete: 23,760 ft = ______ mi.

We choose a symbol for 1 with "mi" on the top and "ft" on the bottom.

$$23{,}760 \text{ ft} = 23{,}760 \text{ ft} \times \frac{1 \text{ mi}}{5280 \text{ ft}}$$ **5280 ft = 1 mi, so $\frac{1 \text{ mi}}{5280 \text{ ft}} = 1$.**

$$= \frac{23{,}760}{5280} \times \frac{\text{ft}}{\text{ft}} \times 1 \text{ mi}$$

$$= 4.5 \times 1 \text{ mi}$$ **Dividing**

$$= 4.5 \text{ mi}$$

Let us also consider this example using canceling:

$$23{,}760 \text{ ft} = 23{,}760 \cancel{\text{ft}} \times \frac{1 \text{ mi}}{5280 \cancel{\text{ft}}} = \frac{23{,}760}{5280} \times 1 \text{ mi} = 4.5 \times 1 \text{ mi} = 4.5 \text{ mi}.$$ ◀

DO EXERCISES 12 AND 13.

Complete.

12. 26,400 ft = __5__ mi

13. 6 mi = __31,680__ ft

ANSWERS ON PAGE A-5

❖ SIDELIGHTS

Applications to Baseball

There are many applications of mathematics to baseball. We studied one when we considered *earned run average* in Example 3 of Section 6.3. Here we consider several more.

EXERCISES

1. The *batting average* of a player is the number of hits divided by the number of times at bat. The result is usually rounded to three decimal places.

a) Ty Cobb holds the record for the highest career batting average. He had 4191 hits in 11,429 at bats. What was his batting average?
b) Ted Williams won the batting title in 1941 with a batting average of 0.406. He was also the last major league player to win a batting title with a 0.400 average. He had 456 at bats. How many hits did he have? Round to the nearest one.
c) What is the highest batting average that a player can have?

2. The *slugging average* of a player is the total bases divided by the total number of at bats. The result is usually rounded to three decimal places.

a) Rogers Hornsby holds the single season record for slugging average. This happened in 1925 when he got 381 total bases in 504 at bats. What was his slugging average?
b) Willie Mays got 1960 singles, 523 doubles, 140 triples, and 660 home runs in 10,881 at bats. What was his career slugging average? (*Hint:* Total bases $= 1960 \cdot 1 + 523 \cdot 2 + 140 \cdot 3 + 660 \cdot 4$.)
c) What is the highest slugging average that a player can have?

3. Many excellent baseball players have had their career statistics lessened because of military service. Examples are such Hall of Fame players as Bob Feller, Ted Williams, and Joe DiMaggio. Use ratio and proportion to answer the following questions.

a) Bob Feller won 266 games in 16 actual seasons of major league pitching. He would have pitched 4 more seasons had he not served in World War II. Estimate how many career wins he would have had if he had played all 20 years. Round to the nearest one.
b) Feller had 3 no-hitters in his career. Estimate how many he would have had if he had played 4 more seasons. Round to the nearest one.
c) In Ted Williams' 19-year career in the major leagues, he had 1839 runs batted in (RBIs). He missed 3 years of playing due to his military service. Estimate how many career RBIs he would have had if he had played the additional 3 years. Round to the nearest one.

4. Hank Aaron holds the major league career record for home runs with 755. This was 41 more than Babe Ruth had in his career. Ruth hit 8.5 home runs for every 100 times at bat, but Aaron hit 6.5 home runs for every 100 times at bat. The difference is that Aaron took better care of his physical health and had 3965 more at bats than Ruth did.

a) Ruth had 8399 at bats. How many did Aaron have?
b) If Ruth had had the same number of at bats as Aaron, how many career home runs would he have hit? Round to the nearest one.
c) Would Ruth have had the major league home run record if he had had the same number of at bats as Aaron?

5. The *designated hitter rule,* used in the American League but not in the National League, has been a source of great controversy since its inception. A poll was recently conducted among 234,832 fans to see whether they liked the rule. Of these, 41% were for the rule and 59% were against the rule. How many were for the rule? How many against?

1.(a) 0.367; **(b)** 185; **(c)** 1.000 **2.(a)** 0.756; **(b)** 0.557; **(c)** 4.000 **3.(a)** 333; **(b)** 4; **(c)** 2129 **4.(a)** 12,364; **(b)** 1051; **(c)** yes **5.** About 96,281; about 138,551

NAME SECTION DATE

EXERCISE SET 9.1

a Complete.

1. 1 ft = ______ in. **2.** 1 yd = ______ ft **3.** 1 in. = ______ ft

4. 1 mi = ______ yd **5.** 1 mi = ______ ft **6.** 1 ft = ______ yd

7. 13 yd = ______ in. **8.** 10 yd = ______ ft **9.** 84 in. = ______ ft

10. 48 ft = ______ yd **11.** 18 in. = ______ ft **12.** 29 ft = ______ yd

13. 3 mi = ______ ft **14.** 3 mi = ______ yd **15.** 3 in. = ______ ft

16. 11,616 ft = ______ mi **17.** 10 ft = ______ yd

18. 4.6 yd = ______ ft **19.** 10 mi = ______ ft

20. 15,840 ft = ______ mi **21.** $4\frac{1}{2}$ ft = ______ yd

ANSWERS

1. 12
2. 3
3. $\frac{1}{12}$
4. 1760
5. 5280
6. $\frac{1}{3}$
7. 468
8. 30
9. 7
10. 16
11. $1\frac{1}{2}$
12. $9\frac{2}{3}$
13. 15,840
14. 5280
15. $\frac{1}{4}$
16. 2.2
17. $3\frac{1}{3}$
18. 13.8
19. 52,800
20. 3
21. $1\frac{1}{2}$

ANSWERS

22. 3
23. 1
24. 360
25. 110
26. 1
27. 2
28. 132,000
29. 300
30. 20
31. 30
32. 10
33. $\frac{1}{36}$
34. $1\frac{1}{12}$
35. 126,720
36. 1
37. 238.7
38. 2387
39. 23,870
40. 0.02387
41. 0.0041 in.

22. 36 in. = ______ ft **23.** 36 in. = ______ yd **24.** 10 yd = ______ in.

25. 330 ft = ______ yd **26.** 1760 yd = ______ mi **27.** 3520 yd = ______ mi

28. 25 mi = ______ ft **29.** 100 yd = ______ ft **30.** 240 in. = ______ ft

31. 360 in. = ______ ft **32.** 360 in. = ______ yd **33.** 1 in. = ______ yd

34. 13 in. = ______ ft **35.** 2 mi = ______ in. **36.** 63,360 in. = ______ mi

SKILL MAINTENANCE

37. 23.87 × 10 **38.** 23.87 × 100

39. 23.87 × 1000 **40.** 23.87 × 0.001

SYNTHESIS

41. ▦ Recently the national debt was $1.824 trillion. To get an idea of this amount, picture that if that many $1 bills were stacked on top of each other, they would reach halfway to the moon. Halfway to the moon is 119,433 mi. How thick, in inches, is a $1 bill?

9.2 Linear Measures: The Metric System

The **metric system** is used in most countries of the world, and the United States is now making greater use of it as well. The metric system does not use inches, feet, pounds, and so on, although units for time and electricity are the same as those you use now.

An advantage of the metric system is that it is easier to convert from one unit to another. That is because the metric system is based on the number 10.

The basic unit of length is the **meter.** It is just over a yard. In fact, 1 meter $\approx$ 1.1 yd.

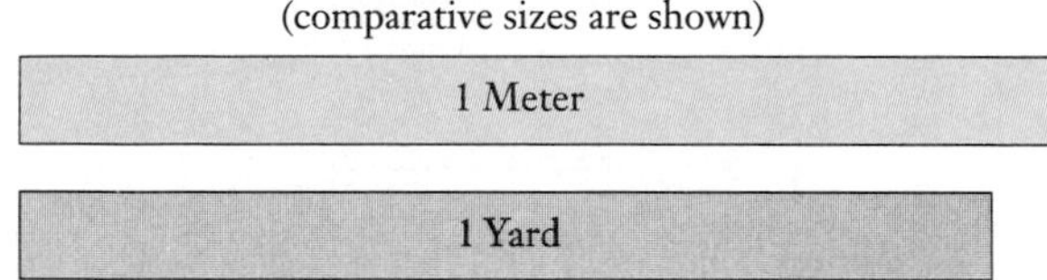

The other units of length are multiples of the length of a meter:

10 times a meter, 100 times a meter, 1000 times a meter, and so on,

or fractions of a meter:

$\frac{1}{10}$ of a meter, $\frac{1}{100}$ of a meter, $\frac{1}{1000}$ of a meter, and so on.

Metric Units of Length

1 *kilo*meter (km) = 1000 meters (m)
1 *hecto*meter (hm) = 100 meters (m)
1 *deka*meter (dam) = 10 meters (m)
1 meter (m)
1 *deci*meter (dm) = $\frac{1}{10}$ meter (m)
1 *centi*meter (cm) = $\frac{1}{100}$ meter (m)
1 *milli*meter (mm) = $\frac{1}{1000}$ meter (m)

dam and *dm* are not used much.

You should memorize these names and abbreviations. Think of *kilo-* for 1000, *hecto-* for 100, and so on. We will use these prefixes when considering units of area, capacity, and mass (weight).

Thinking Metric

To familiarize yourself with metric units, consider the following.

1 kilometer (1000 meters)	is slightly more than $\frac{1}{2}$ mile (0.6 mi).
1 meter	is just over a yard (1.1 yd).
1 centimeter (0.01 meter)	is a little more than the width of a paper-clip (about 0.4 inch).

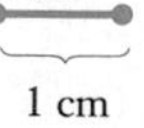
1 cm

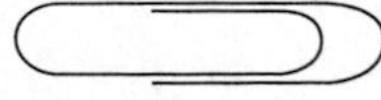

OBJECTIVES

After finishing Section 9.2, you should be able to:

a Convert from one metric unit to another.

b Convert between American and metric units of length.

FOR EXTRA HELP

Tape 11B

Tape 12B

MAC: 9
IBM: 9

1 inch is about 2.54 centimeters.

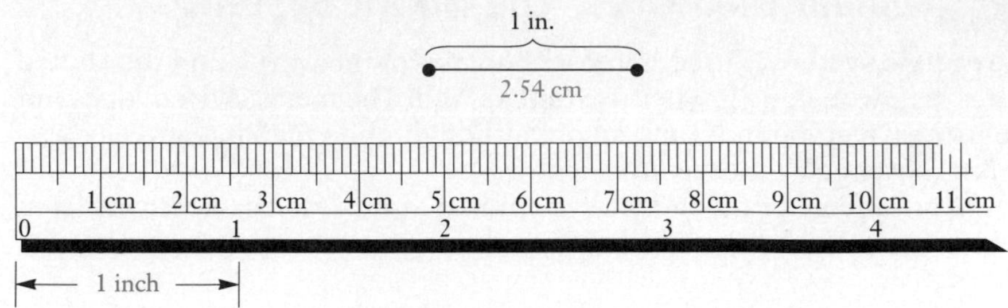

1 millimeter is about the diameter of a paperclip wire.

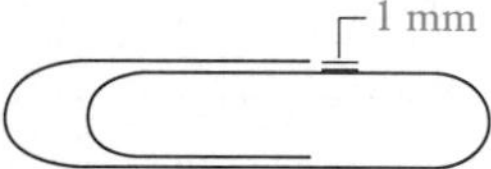

The millimeter (mm) is used to measure small distances, especially in industry.

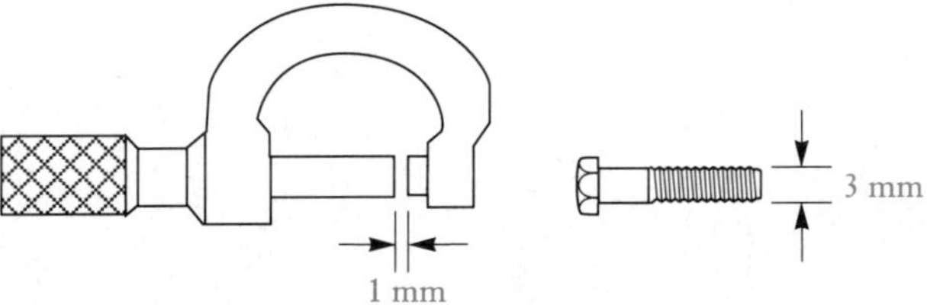

The centimeter (cm) is used for body dimensions and clothing sizes, mostly in places where inches are now being used.

The meter (m) is used for expressing dimensions of larger objects—say, the length of a building—and for shorter distances, such as the length of a rug.

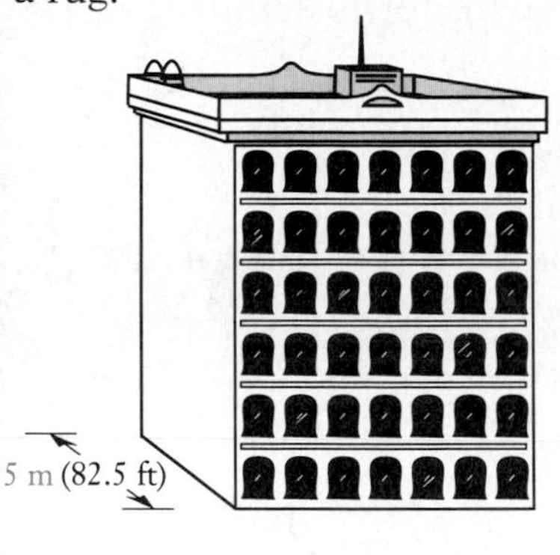

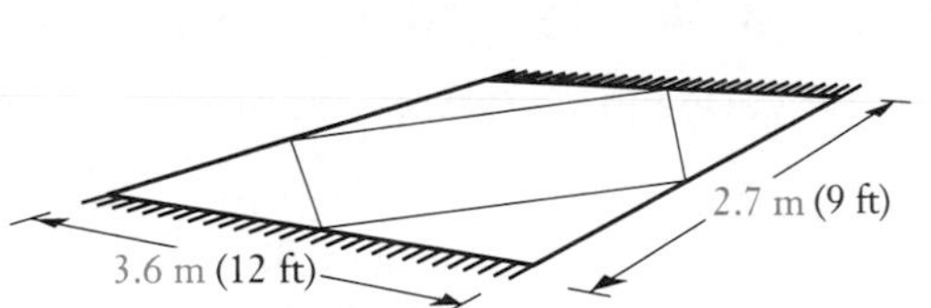

The kilometer (km) is used for longer distances, mostly in places where miles are now being used.

DO EXERCISES 1–6.

Complete with mm, cm, m, or km.

1. A stick of gum is 7 __cm__ long.
2. Minneapolis is 3213 __km__ from San Francisco.
3. A penny is 1 __mm__ thick.
4. The halfback ran 7 __m__.
5. The book is 3 __cm__ thick.
6. The desk is 2 __m__ long.

a Changing Metric Units

▶ **EXAMPLE 1** Complete: 4 km = ______ m.

$$\begin{aligned} 4\text{ km} &= 4 \times 1\text{ km} \\ &= 4 \times 1000\text{ m} \qquad \textbf{Substituting 1000 m for 1 km} \\ &= 4000\text{ m} \end{aligned}$$ ◀

DO EXERCISES 7 AND 8.

Complete.

7. 23 km = __23,000__ m
8. 4 hm = __400__ m

Since

$$\frac{1}{10}\text{ m} = 1\text{ dm}, \quad \frac{1}{100}\text{ m} = 1\text{ cm}, \quad \text{and} \quad \frac{1}{1000}\text{ m} = 1\text{ mm},$$

it follows that

1 m = 10 dm, 1 m = 100 cm, and 1 m = 1000 mm.

It will help to memorize these.

ANSWERS ON PAGE A-5

Complete.

9. 1.78 m = 178 cm

10. 9.04 m = 9040 mm

Complete.

11. 7814 m = 7.814 km

12. 7814 m = 781.4 dam

ANSWERS ON PAGE A-5

▶ **EXAMPLE 2** Complete: 93.4 m = ______ cm.

We want to convert from "m" to "cm." We multiply by 1 using a symbol for 1 with "m" on the bottom and "cm" on the top to eliminate meters and convert to centimeters.

$$93.4 \text{ m} = 93.4 \text{ m} \times \frac{100 \text{ cm}}{1 \text{ m}}$$ **Multiplying by 1 using $\frac{100 \text{ cm}}{1 \text{ m}}$**

$$= 93.4 \times 100 \times \frac{\text{m}}{\text{m}} \times 1 \text{ cm}$$ **The $\frac{\text{m}}{\text{m}}$ acts like 1, so we omit it.**

$$= 9340 \text{ cm}$$ **Multiplying by 100 moves the decimal point two places to the right.**

We can also work this example by canceling:

$$93.4 \text{ m} = 93.4 \cancel{\text{m}} \times \frac{100 \text{ cm}}{1 \cancel{\text{m}}} = 93.4 \times 100 \times 1 \text{ cm} = 9340 \text{ cm}.$$ ◀

▶ **EXAMPLE 3** Complete: 0.248 m = ______ mm.

We are converting from "m" to "mm" so we choose a symbol for 1 with "mm" on the top and "m" on the bottom.

$$0.248 \text{ m} = 0.248 \text{ m} \times \frac{1000 \text{ mm}}{1 \text{ m}}$$ **Multiplying by 1 using $\frac{1000 \text{ mm}}{1 \text{ m}}$**

$$= 0.248 \times 1000 \times \frac{\text{m}}{\text{m}} \times 1 \text{ mm}$$ **The $\frac{\text{m}}{\text{m}}$ acts like 1, so we omit it.**

$$= 248 \text{ mm}$$ **Multiplying by 1000 moves the decimal point three places to the right.**

Using canceling, we can work this example as follows:

$$0.248 \text{ m} = 0.248 \cancel{\text{m}} \times \frac{1000 \text{ mm}}{1 \cancel{\text{m}}} = 0.248 \times 1000 \times 1 \text{ mm} = 248 \text{ mm}.$$ ◀

DO EXERCISES 9 AND 10.

▶ **EXAMPLE 4** Complete: 2347 m = ______ km.

$$2347 \text{ m} = 2347 \text{ m} \times \frac{1 \text{ km}}{1000 \text{ m}}$$ **Multiplying by 1 using $\frac{1 \text{ km}}{1000 \text{ m}}$**

$$= \frac{2347}{1000} \times \frac{\text{m}}{\text{m}} \times 1 \text{ km}$$ **The $\frac{\text{m}}{\text{m}}$ acts like 1, so we omit it.**

$$= 2.347 \text{ km}$$ **Dividing by 1000 moves the decimal point three places to the left.**

Using canceling, we can work this example as follows:

$$2347 \text{ m} = 2347 \cancel{\text{m}} \times \frac{1 \text{ km}}{1000 \cancel{\text{m}}} = \frac{2347}{1000} \times 1 \text{ km} = 2.347 \text{ km}.$$ ◀

DO EXERCISES 11 AND 12.

Sometimes we multiply by 1 more than once.

▶ **EXAMPLE 5** Complete: 8.42 mm = ______ cm.

$$8.42 \text{ mm} = 8.42 \text{ mm} \times \frac{1 \text{ m}}{1000 \text{ mm}} \times \frac{100 \text{ cm}}{1 \text{ m}}$$

Multiplying by 1 using $\frac{1 \text{ m}}{1000 \text{ mm}}$ and $\frac{100 \text{ cm}}{1 \text{ m}}$

$$= \frac{8.42 \times 100}{1000} \times \frac{\text{mm}}{\text{mm}} \times \frac{\text{m}}{\text{m}} \times 1 \text{ cm}$$

$$= \frac{842}{1000} \text{ cm}$$

$$= 0.842 \text{ cm}$$

Using canceling, we can work this example as follows:

$$8.42 \text{ mm} = 8.42 \cancel{\text{mm}} \times \frac{1 \cancel{\text{m}}}{1000 \cancel{\text{mm}}} \times \frac{100 \text{ cm}}{1 \cancel{\text{m}}}$$

$$= \frac{8.42 \times 100}{1000} \times 1 \text{ cm} = 0.842 \text{ cm}.$$ ◀

DO EXERCISES 13 AND 14.

Complete.

13. 9.67 mm = 0.967 cm

14. 89 km = 8,900,000 cm

Mental Conversion

Look back over the examples and exercises done thus far and you will see that changing from one unit to another in the metric system amounts to only the movement of a decimal point. That is because the metric system is based on 10. Let's find a faster way to convert. Look at the following table.

1000 km	100 hm	10 dam	1 m	0.1 dm	0.01 cm	0.001 mm

Each place in the table has a value $\frac{1}{10}$ that to the left or 10 times that to the right. Thus moving one place in the table corresponds to one decimal place. Let us convert mentally.

▶ **EXAMPLE 6** Complete: 8.42 mm = ______ cm.

Think: To go from mm to cm in the table is a move of one place to the left. Thus we move the decimal point one place to the left.

8.42 0.8.42 8.42 mm = 0.842 cm ◀

▶ **EXAMPLE 7** Complete: 1.886 km = ______ cm.

Think: To go from km to cm is a move of five places to the right. Thus we move the decimal point five places to the right.

1.886 1.88600. 1.886 km = 188,600 cm ◀

ANSWERS ON PAGE A-5

Complete. Try to do this mentally using the table.

15. 6780 m = 6.78 km

16. 9.74 cm = 97.4 mm

17. 1 mm = 0.1 cm

18. 845.1 mm = 8.451 dm

Complete.

19. 100 yd = 90.909 m
(The length of a football field)

20. 500 mi = 804.5 km
(The Indianapolis 500-mile race)

21. 3213 km = 1995.273 mi
(The distance from Minneapolis to San Francisco)

ANSWERS ON PAGE A-5

▶ **EXAMPLE 8** Complete: 1 m = ______ cm.

Think: To go from m to cm in the table is a move of two places to the right. Thus we move the decimal point two places to the right.

1 1.00. 1 m = 100 cm ◀

Make metric conversions mentally as much as possible.

The fact that conversions can be done so easily is an important advantage of the metric system.

The most commonly used metric units of length are km, m, cm, and mm. We have purposely used these more often than the others in the exercises.

DO EXERCISES 15–18.

b Converting Between American and Metric Units

We can make conversions between American and metric units by using the following table. Again, we either make a substitution or multiply by 1 appropriately.

Metric	American
1 m	39.37 in.
1 m	3.3 ft
2.54 cm	1 in.
1 km	0.621 mi
1.609 km	1 mi

▶ **EXAMPLE 9** Complete: 26.2 mi = ______ km. (This is the length of the Olympic marathon.)

$$26.2 \text{ mi} = 26.2 \times 1 \text{ mi} \approx 26.2 \times 1.609 \text{ km} = 42.1558 \text{ km}$$ ◀

▶ **EXAMPLE 10** Complete: 100 m = ______ yd. (This is the length of a dash in track.)

$$100 \text{ m} = 100 \times 1 \text{ m} \approx 100 \times 3.3 \text{ ft} = 330 \text{ ft}$$

$$= 330 \text{ ft} \times \frac{1 \text{ yd}}{3 \text{ ft}} = \frac{330}{3} \text{ yd} = 110 \text{ yd}$$ ◀

DO EXERCISES 19–21.

NAME SECTION DATE

EXERCISE SET 9.2

a Complete. Do as much as possible mentally.

1. a) 1 km = ______ m **2. a)** 1 hm = ______ m **3. a)** 1 dam = ______ m

b) 1 m = ______ km **b)** 1 m = ______ hm **b)** 1 m = ______ dam

4. a) 1 dm = ______ m **5. a)** 1 cm = ______ m **6. a)** 1 mm = ______ m

b) 1 m = ______ dm **b)** 1 m = ______ cm **b)** 1 m = ______ mm

7. 6.7 km = ______ m **8.** 9 km = ______ m **9.** 98 cm = ______ m

10. 0.233 cm = ______ m **11.** 8921 m = ______ km **12.** 6770 m = ______ km

13. 56.66 m = ______ km **14.** 5.666 m = ______ km **15.** 5666 m = ______ cm

16. 435 m = ______ cm **17.** 477 cm = ______ m **18.** 3.45 mm = ______ m

19. 6.88 m = ______ cm **20.** 6.88 m = ______ dm **21.** 1 mm = ______ cm

22. 1 cm = ______ km **23.** 1 km = ______ cm **24.** 2 km = ______ cm

ANSWERS

1. a) 1000
b) 0.001
2. a) 100
b) 0.01
3. a) 10
b) 0.1
4. a) 0.1
b) 10
5. a) 0.01
b) 100
6. a) 0.001
b) 1000
7. 6700
8. 9000
9. 0.98
10. 0.00233
11. 8.921
12. 6.77
13. 0.05666
14. 0.005666
15. 566,600
16. 43,500
17. 4.77
18. 0.00345
19. 688
20. 68.8
21. 0.1
22. 0.00001
23. 100,000
24. 200,000

ANSWERS

25. 142
26. 138
27. 0.82
28. 0.73
29. 450
30. 6000
31. 0.000024
32. 8
33. 0.688
34. 0.0688
35. 230
36. 70
37. 3.92
38. 0.00013
39. 100
40. 30.48
41. 6.6
42. 64.947285
43. 88.495
44. 62.1
45. 1.75
46. 2.34
47. 0.234
48. 0.0234
49. 13.85
50. $80\frac{1}{2}$
51. 4

25. 14.2 cm = ______ mm

26. 13.8 cm = ______ mm

27. 8.2 mm = ______ cm

28. 7.3 mm = ______ cm

29. 4500 mm = ______ cm

30. 6,000,000 m = ______ km

31. 0.024 mm = ______ m

32. 80,000 mm = ______ dam

33. 6.88 m = ______ dam

34. 6.88 m = ______ hm

35. 2.3 dam = ______ dm

36. 7 km = ______ hm

37. 392 dam = ______ km

38. 0.013 mm = ______ dm

b Complete.

39. 330 ft = ______ m
(The length of most baseball foul lines)

40. 12 in. = ______ cm

41. 2 m = ______ ft
(The length of a desk)

42. 104.585 km = ______ mi

43. 55 mph = ______ km/h
(A common speed limit in the U.S.)

44. 100 km/h = ______ mph
(A common speed limit in Canada)

SKILL MAINTENANCE

Divide. Find decimal notation for the answer.

45. $21 \div 12$

46. $23.4 \div 10$

47. $23.4 \div 100$

48. $23.4 \div 1000$

49. Multiply 3.14×4.41. Round to the nearest hundredth.

50. Multiply: $4 \times 20\frac{1}{8}$.

51. Multiply: $48 \times \frac{1}{12}$.

9.3 Perimeter

a Finding Perimeters

A *polygon* is a geometric figure with three or more sides. The *perimeter* of a polygon is the distance around it, or the sum of the lengths of its sides.

▶ **EXAMPLE 1** Find the perimeter of this polygon.

We add the lengths of the sides. Since all the units are the same, we add the numbers, keeping meters (m) as the unit.

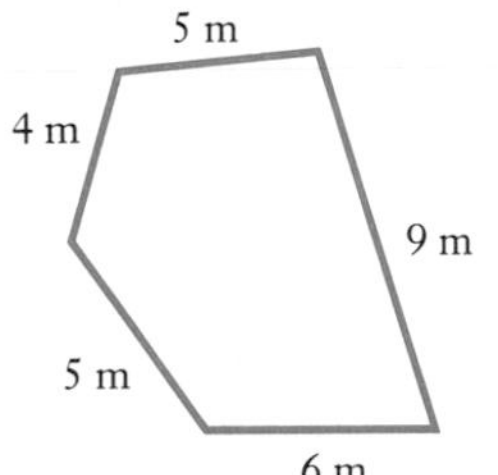

Perimeter = 6 m + 5 m + 4 m + 5 m + 9 m
= (6 + 5 + 4 + 5 + 9) m
= 29 m ◀

DO EXERCISES 1 AND 2.

▶ **EXAMPLE 2** Find the perimeter of a rectangle that is 3 cm by 4 cm.

Perimeter = 3 cm + 3 cm + 4 cm + 4 cm
= (3 + 3 + 4 + 4) cm = 14 cm.

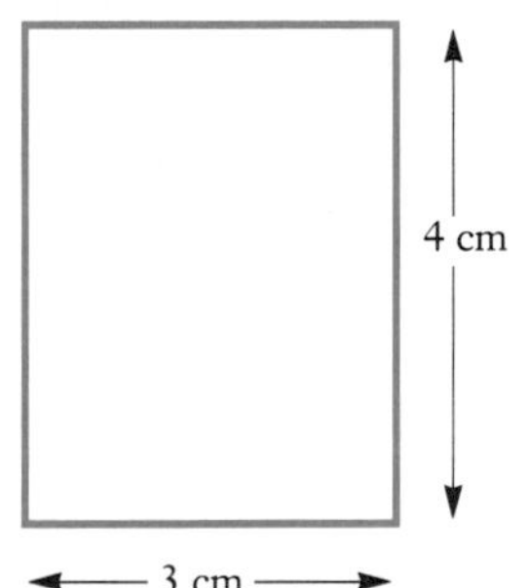

◀

DO EXERCISE 3.

The *perimeter* of a rectangle is twice the sum of the length and the width, or 2 times the length plus 2 times the width:

$$P = 2 \cdot (l + w), \text{ or } P = 2 \cdot l + 2 \cdot w.$$

▶ **EXAMPLE 3** Find the perimeter of a rectangle that is 4.3 ft by 7.8 ft.

$P = 2 \cdot (l + w) = 2 \cdot (4.3 \text{ ft} + 7.8 \text{ ft})$

$P = 2 \cdot (12.1 \text{ ft}) = 24.2 \text{ ft}$ ◀

DO EXERCISES 4 AND 5.

A **square** is a rectangle all of whose sides have the same length.

▶ **EXAMPLE 4** Find the perimeter of a square whose sides are 9 mm long.

$P = 9 \text{ mm} + 9 \text{ mm} + 9 \text{ mm} + 9 \text{ mm}$

$P = (9 + 9 + 9 + 9) \text{ mm} = 36 \text{ mm}$

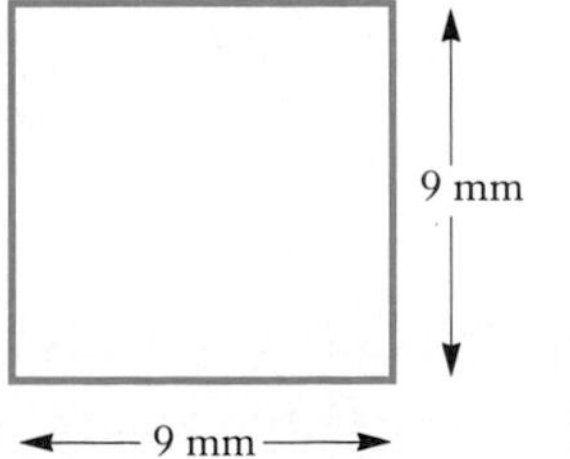

◀

DO EXERCISE 6 ON THE FOLLOWING PAGE.

OBJECTIVES

After finishing Section 9.3, you should be able to:

a Find the perimeter of a polygon.

b Solve problems involving perimeter.

FOR EXTRA HELP

 Tape 11c

Tape 13A

 MAC: 9 IBM: 9

Find the perimeter of the polygon.

1.

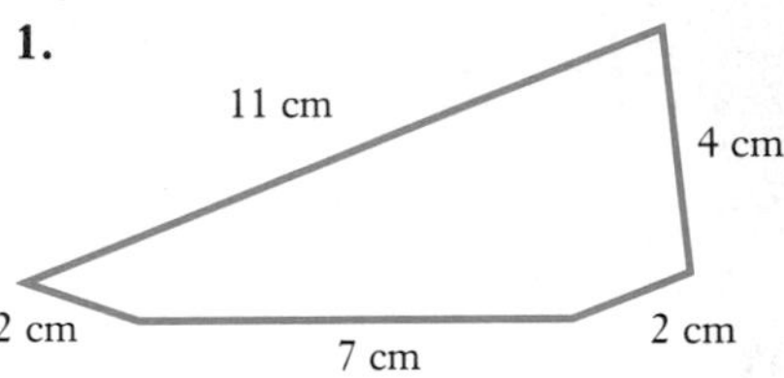

26 cm

2.

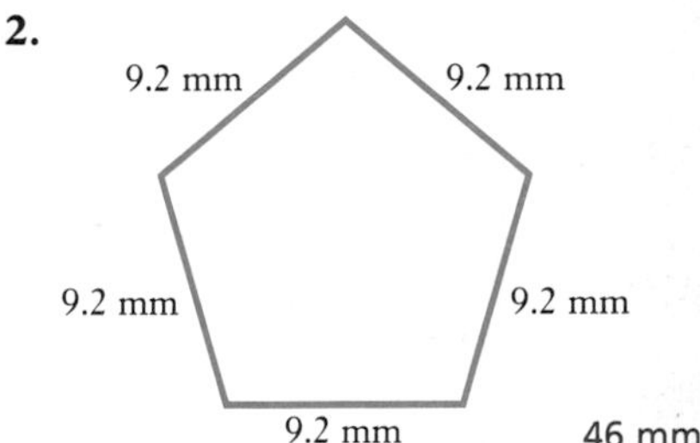

46 mm

3. Find the perimeter of a rectangle that is 2 cm by 4 cm.

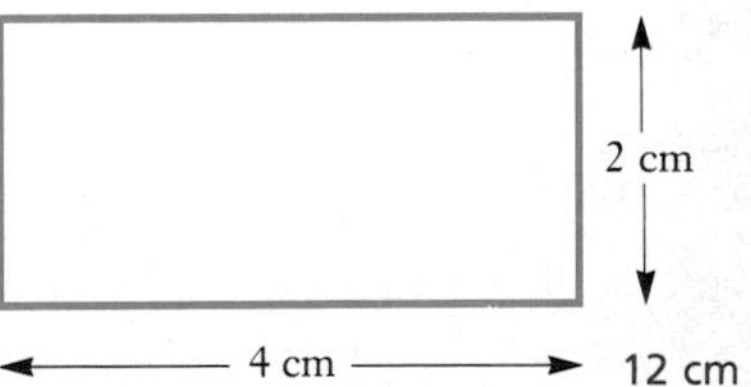

12 cm

4. Find the perimeter of a rectangle that is 5.25 yd by 3.5 yd. 17.5 yd

5. Find the perimeter of a rectangle that is 8 km by 8 km. 32 km

ANSWERS ON PAGE A-5

6. Find the perimeter of a square whose sides have length 10 km.

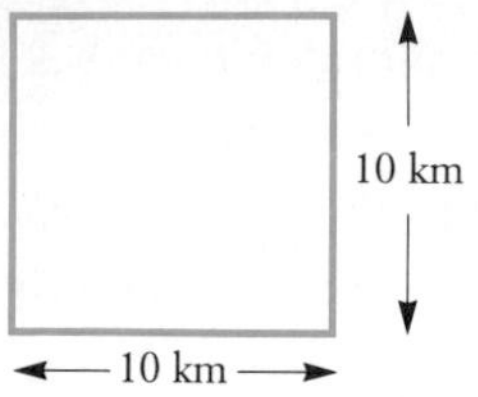

40 km

7. Find the perimeter of a square whose sides have length $5\frac{1}{4}$ yd.

21 yd

8. Find the perimeter of a square whose sides have length 7.8 km. 31.2 km

9. A garden is 25 ft by 10 ft. A fence is to be built around the garden. How many feet of fencing will be needed? Fencing costs \$4.95 a foot. What will be the cost of the fencing? 70 ft; \$346.50

ANSWERS ON PAGE A-5

The perimeter of a *square* is four times the length of a side:

$$P = 4 \cdot s.$$

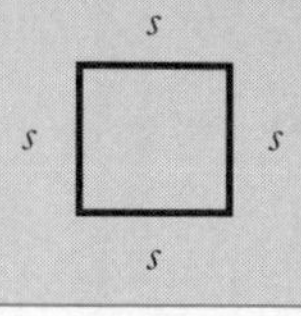

► **EXAMPLE 5** Find the perimeter of a square whose sides are $20\frac{1}{8}$ in. long.

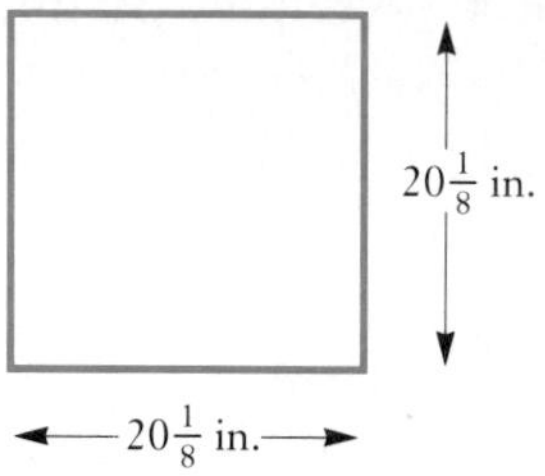

$$P = 4 \cdot s$$

$$P = 4 \cdot 20\frac{1}{8} \text{ in.}$$

$$P = 4 \cdot \frac{161}{8} \text{ in.}$$

$$P = \frac{4 \cdot 161}{4 \cdot 2} \text{ in.}$$

$$P = \frac{161}{2} \cdot \frac{4}{4} \text{ in.}$$

$$P = 80\frac{1}{2} \text{ in.}$$

◄

DO EXERCISES 7 AND 8.

b Solving Problems

► **EXAMPLE 6** A garden is 15 ft by 20 ft. A fence is to be built around the garden. How many feet of fence will be needed? Fencing sells for \$2.95 a foot. What will the fencing cost?

1. *Familiarize.* We make a drawing. We let P = the perimeter.

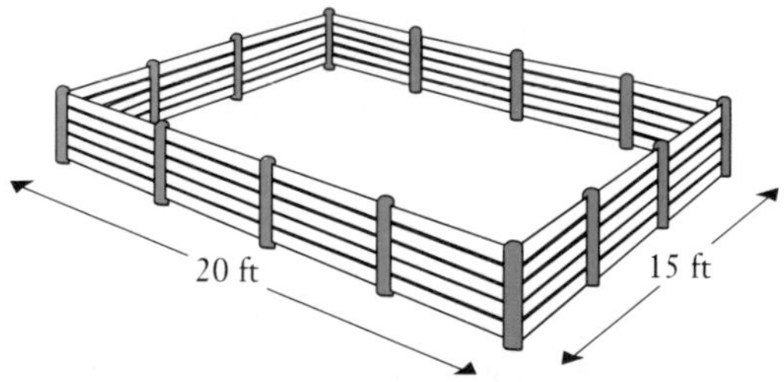

2. *Translate.* The perimeter of the garden is given by

$$P = 2 \cdot (l + w) = 2 \cdot (15 \text{ ft} + 20 \text{ ft}).$$

3. *Solve.* We calculate the perimeter as follows:

$$P = 2 \cdot (15 \text{ ft} + 20 \text{ ft}) = 2 \cdot (35 \text{ ft}) = 70 \text{ ft}.$$

Then we multiply by \$2.95 to find the cost of the fencing:

$$\text{Cost} = \$2.95 \times \text{Perimeter} = \$2.95 \times 70 \text{ ft} = \$206.50.$$

4. *Check.* The check is left to the student.
5. *State.* The fencing will cost \$206.50.

◄

DO EXERCISE 9.

NAME SECTION DATE

EXERCISE SET 9.3

a Find the perimeter of the polygon.

1.

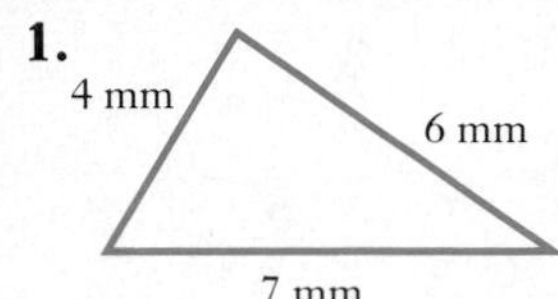

2.

3 m
1.2 m
1.2 m
3 m

3.

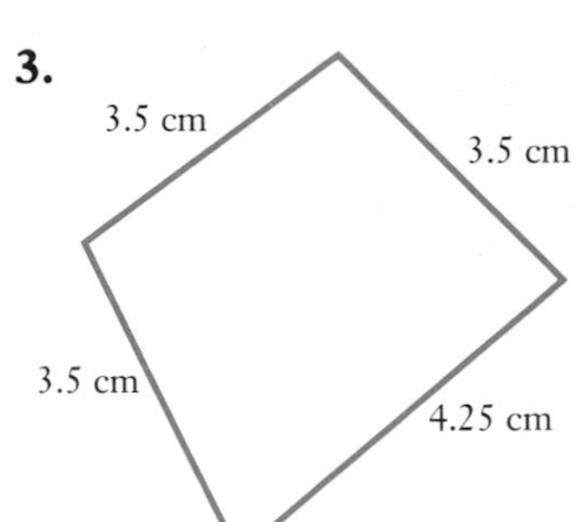

4.

3.4 km
5.6 km

5.

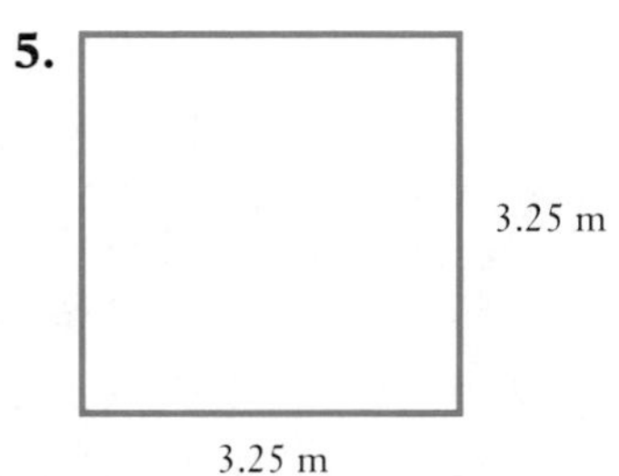

6.

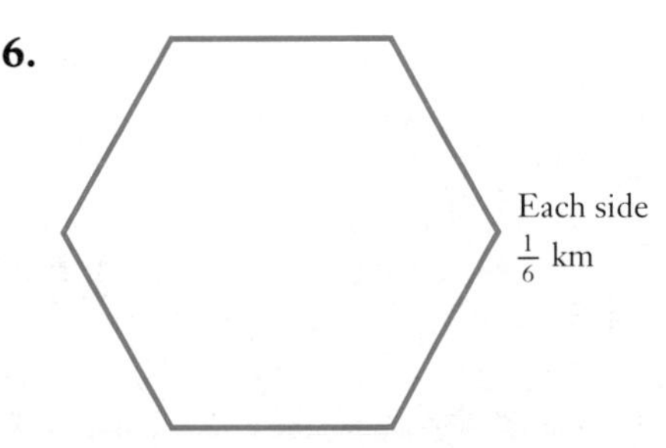

Find the perimeter of the rectangle.

7. 5 ft by 10 ft

8. 2.5 m by 100 m

9. 34.67 cm by 4.9 cm

10. $3\frac{1}{2}$ yd by $4\frac{1}{2}$ yd

Find the perimeter of the square.

11. 22 ft on a side

12. 56.9 km on a side

13. 45.5 mm on a side

14. $3\frac{1}{8}$ yd on a side

ANSWERS

1. 17 mm

2. 8.4 m

3. 15.25 cm

4. 18 km

5. 13 m

6. 1 km

7. 30 ft

8. 205 m

9. 79.14 cm

10. 16 yd

11. 88 ft

12. 227.6 km

13. 182 mm

14. $12\frac{1}{2}$ yd

ANSWERS

15. 826 m; $1197.70

16. 260 ft

17. 99 cm

18. 122 cm

19. a) 14

b) $33.60

c) 39 m

d) $33.15

e) $76.70

20. a) 228 ft

b) $1046.52

21. 0.561

22. 67.34%

23. 961

24. 112.5%, or $112\frac{1}{2}\%$

b Solve.

15. A fence is to be built around a 173-m–by–240-m field. What is the perimeter of the field? Fence wire costs $1.45 per meter. What will wire for the fence cost?

16. A standard-sized softball diamond is a square whose sides have length 65 ft. What is the perimeter of a softball diamond?

17. A standard sheet of typewriter paper is 21.6 cm by 27.9 cm. What is the perimeter of the paper?

18. A piece of flooring tile is a square 30.5 cm on a side. What is its perimeter?

19. A carpenter is to build a fence around a 9-m–by–12-m garden.

a) The posts are 3 m apart. How many posts will be needed?

b) The posts cost $2.40 each. How much will the posts cost?

c) The fence will surround all but 3 m of the garden, which will be a gate. How long will the fence be?

d) The fence costs $0.85 per meter. What will the cost of the fence be?

e) The gate costs $9.95. What is the total cost of the materials?

20. A rain gutter is to be installed around the house shown in the figure.

a) Find the perimeter of the house.

b) The gutter costs $4.59 per foot. Find the total cost of the gutter.

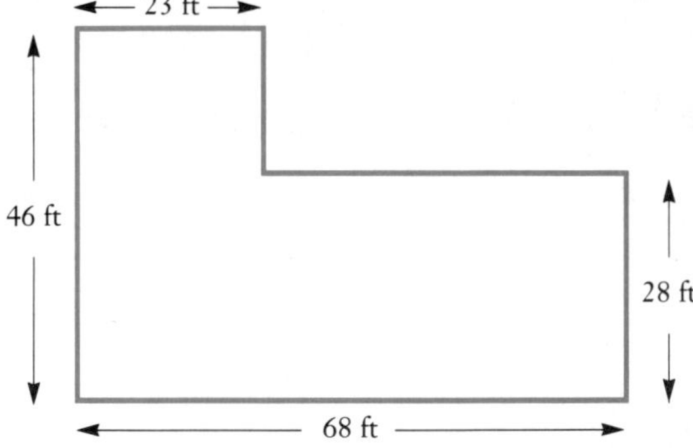

SKILL MAINTENANCE

21. Convert to decimal notation: 56.1%.

22. Convert to percent notation: 0.6734.

23. Evaluate: 31^2.

24. Convert to percent notation: $\frac{9}{8}$.

9.4 Area

OBJECTIVES

After finishing Section 9.4, you should be able to:

a Find the area of a rectangle or square.

b Solve problems involving areas of rectangles or squares.

FOR EXTRA HELP

Tape 11D

Tape 13A

MAC: 9
IBM: 9

a Rectangles

A polygon and its interior form a plane region. We can find the area of a *rectangular region* by filling it with square units. Two such units, a *square inch* and a *square centimeter,* are shown below.

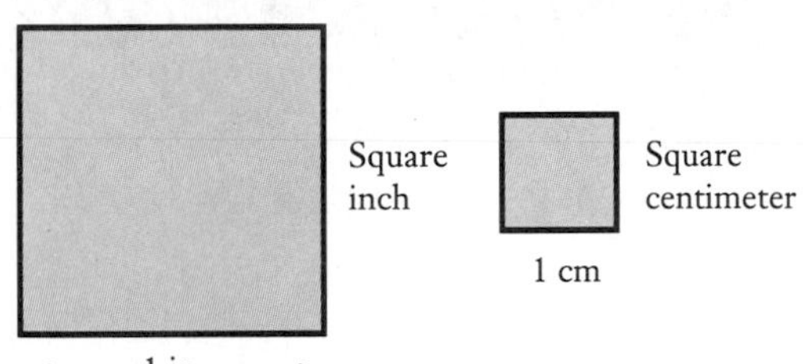

▶ **EXAMPLE 1** What is the area of this region?

We have a rectangular array. Since the region is filled with 12 square centimeters, its area is 12 square centimeters (sq cm), or 12 cm^2. The number of units is 3×4.

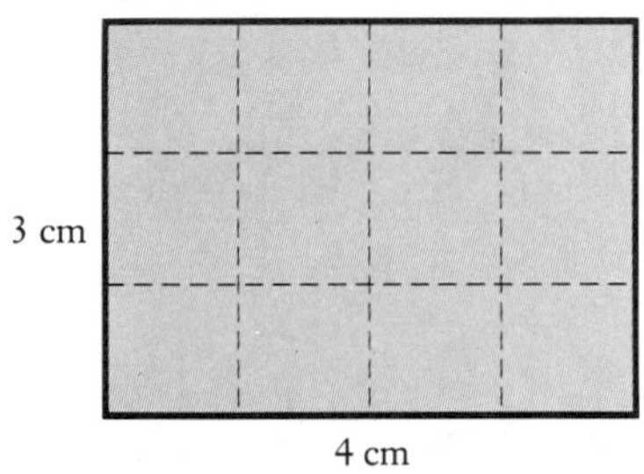

◀

DO EXERCISE 1.

1. What is the area of this region? Count the square centimeters.

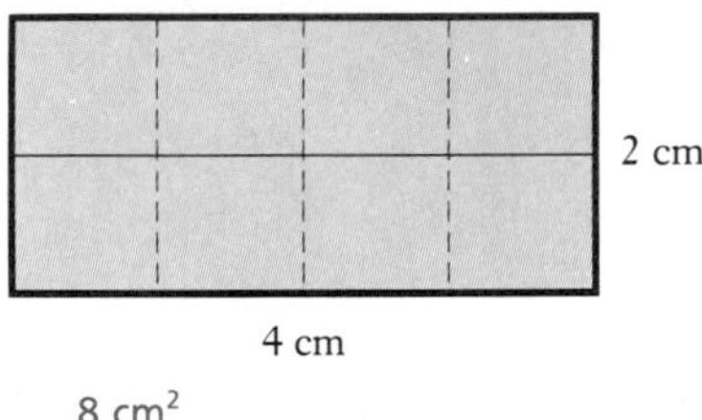

$8\ \text{cm}^2$

> **The area of a rectangular region is the product of the length l and the width w:**
>
> $$A = l \cdot w.$$
>
> 
>

▶ **EXAMPLE 2** Find the area of a rectangle that is 4 yd by 7 yd.

$$A = l \cdot w = 4\ \text{yd} \cdot 7\ \text{yd} = 4 \cdot 7 \cdot \text{yd} \cdot \text{yd} = 28\ \text{yd}^2$$

We think of $\text{yd} \cdot \text{yd}$ as $(\text{yd})^2$ and denote it yd^2. Thus we read "28 yd^2" as "28 square yards." ◀

DO EXERCISES 2 AND 3.

2. Find the area of a rectangle that is 7 km by 8 km. $56\ \text{km}^2$

3. Find the area of a rectangle that is $5\frac{1}{4}$ yd by $3\frac{1}{2}$ yd. $18\frac{3}{8}\ \text{yd}^2$

▶ **EXAMPLE 3** Find the area of a square whose sides are each 9 mm long.

$$A = (9\ \text{mm}) \cdot (9\ \text{mm})$$
$$A = 9 \cdot 9 \cdot \text{mm} \cdot \text{mm}$$
$$A = 81\ \text{mm}^2$$

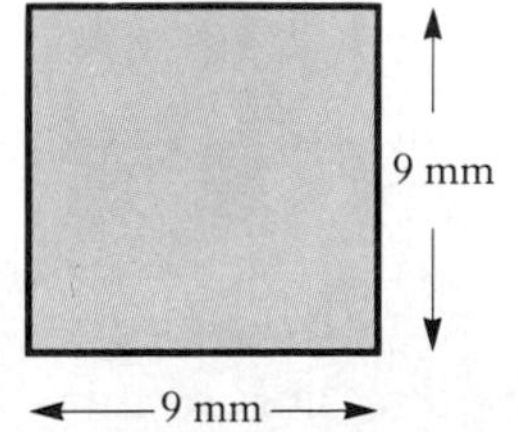

◀

DO EXERCISE 4.

4. Find the area of a square whose sides have length 12 km.

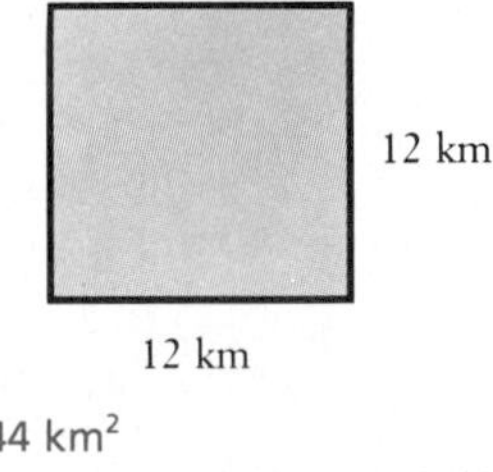

$144\ \text{km}^2$

ANSWERS ON PAGE A-5

The area of a square region is the square of the length of a side:

$$A = s \cdot s, \quad \text{or} \quad A = s^2.$$

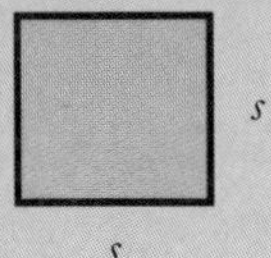

▶ **EXAMPLE 4** Find the area of a square whose sides have length 20.3 m.

$$A = s \cdot s$$
$$A = 20.3 \text{ m} \times 20.3 \text{ m}$$
$$A = 20.3 \times 20.3 \times \text{m} \times \text{m} = 412.09 \text{ m}^2$$ ◀

DO EXERCISES 5 AND 6.

b Solving Problems

▶ **EXAMPLE 5** A square sandbox 1.5 m on a side is placed on a 20-m–by–31.2-m lawn. It costs $0.04 per square meter to have the lawn mowed. What is the total cost of mowing?

1. *Familiarize.* We first draw a picture.

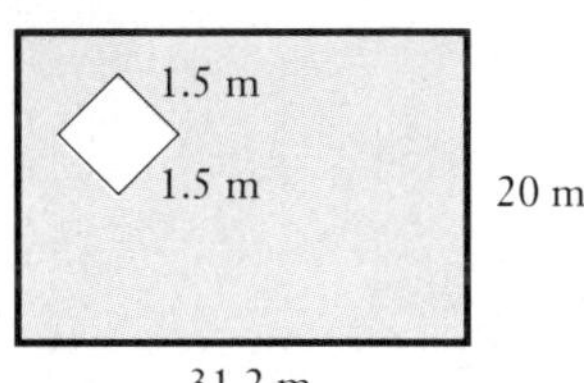

2. *Translate.* This is a two-step problem. We first find the area left over after the area of the sandbox is subtracted. Then we multiply by the cost per square meter. We let A = the area left over.

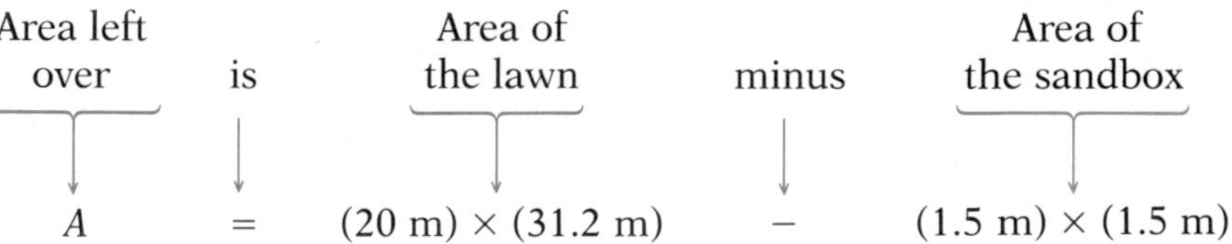

Area left over	is	Area of the lawn	minus	Area of the sandbox
A	$=$	$(20 \text{ m}) \times (31.2 \text{ m})$	$-$	$(1.5 \text{ m}) \times (1.5 \text{ m})$

3. *Solve.* The area of the lawn is

$$(20 \text{ m}) \times (31.2 \text{ m}) = 20 \times 31.2 \times \text{m} \times \text{m} = 624 \text{ m}^2.$$

The area of the sandbox is

$$(1.5 \text{ m}) \times (1.5 \text{ m}) = 1.5 \times 1.5 \times \text{m} \times \text{m} = 2.25 \text{ m}^2.$$

The area left over is

$$A = 624 \text{ m}^2 - 2.25 \text{ m}^2 = 621.75 \text{ m}^2.$$

Then we multiply by $0.04:

$$\$0.04 \times 621.75 = \$24.87.$$

4. *Check.* The check is left to the student.
5. *State.* The total cost of mowing the lawn is $24.87. ◀

DO EXERCISE 7.

5. Find the area of a square whose sides have length 10.9 m.

118.81 m^2

6. Find the area of a square whose sides have length $3\frac{1}{2}$ yd. $12\frac{1}{4}$ yd^2

7. A square flower bed 3.5 m on a side is dug on a 30-m by 22.4-m lawn. How much area is left over? Draw a picture first.

659.75 m^2

ANSWERS ON PAGE A-5

NAME SECTION DATE

EXERCISE SET 9.4

a Find the area.

1.

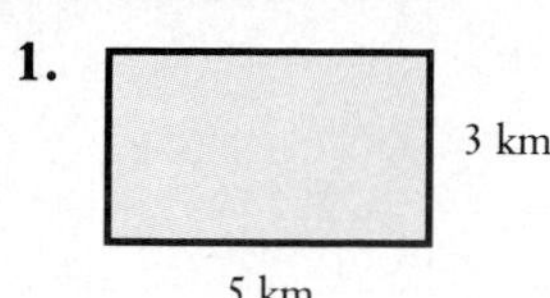

2. 1.5 m, 1.5 m

3. 2 cm, 0.7 cm

4.

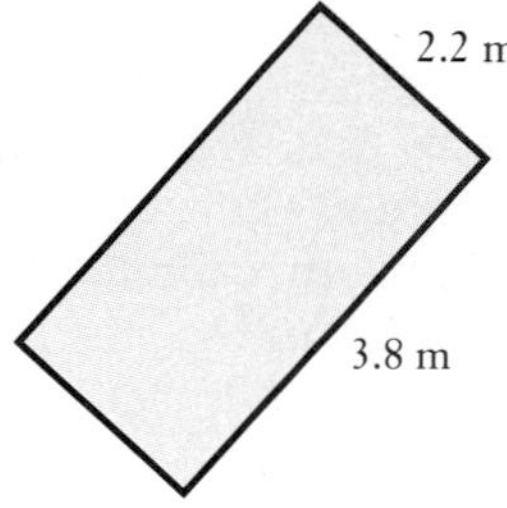

5. 2.5 mm, 2.5 mm

6.

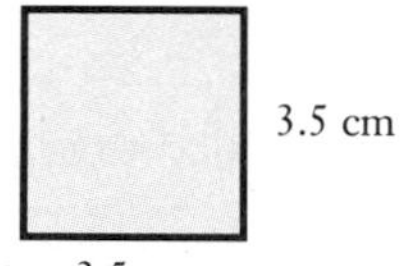

7.

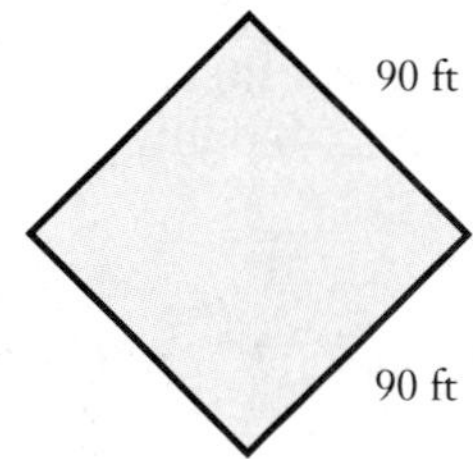

8.

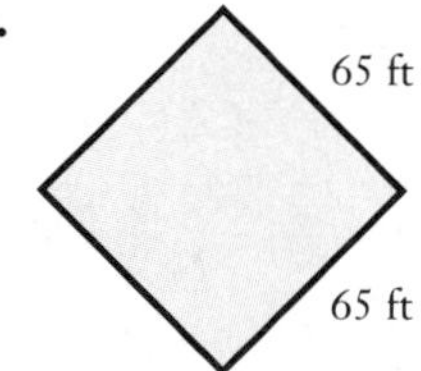

Find the area of the rectangle.

9. 5 ft by 10 ft

10. 14 yd by 8 yd

11. 34.67 cm by 4.9 cm

12. 2.45 km by 100 km

Find the area of the square.

13. 22 ft on a side

14. 18 yd on a side

15. 56.9 km on a side

16. 45.5 m on a side

b Solve.

17. A lot is 40 m by 36 m. A house 27 m by 9 m is built on the lot. How much area is left over?

18. A field is 240.8 m by 450.2 m. Part of the field, 160.4 m by 90.6 m, is paved for a parking lot. How much area is left over?

ANSWERS

1. 15 km^2
2. 2.25 m^2
3. 1.4 cm^2
4. 8.36 m^2
5. 6.25 mm^2
6. 12.25 cm^2
7. 8100 ft^2
8. 4225 ft^2
9. 50 ft^2
10. 112 yd^2
11. 169.883 cm^2
12. 245 km^2
13. 484 ft^2
14. 324 yd^2
15. 3237.61 km^2
16. 2070.25 m^2
17. 1197 m^2
18. 93,875.92 m^2

ANSWERS

19. 630.36 m^2

20. 26 in^2

21. a) 24.75 m^2

b) $207.90

22. a) 80.8 m^2

b) 4 L

c) $9.92

23. 107.5 mm^2

24. 80 cm^2

25. 45.2%

26. $33\frac{1}{3}\%$, or $33.\overline{3}\%$

27. 55%

28. 88%

19. A sidewalk is built around two sides of a building, as shown in the figure. What is the area of the sidewalk?

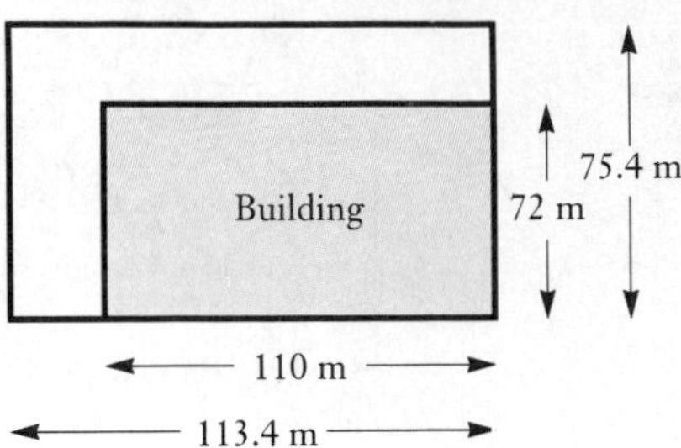

20. A standard sheet of typewriter paper is $8\frac{1}{2}$ in. by 11 in. We usually type on a $7\frac{1}{2}$ in.–by–9-in. area of the paper. What would be the area of the margin?

21. A family wants to carpet a 4.5-m–by–5.5-m room.

a) How many square meters of carpeting will they need?

b) The carpeting they want is $8.40 per square meter. How much will it cost?

22. A room is 4 m by 6 m. The ceiling is 3 m above the floor. There are two windows in the room, each 0.8 m by 1 m. The door is 0.8 m by 2 m.

a) What is the area of the walls and the ceiling?

b) A liter of paint will cover 20.2 sq m. How many liters will be needed for the room?

c) Paint costs $2.48 a liter. How much will it cost to paint the room?

Find the area of the shaded region.

23.

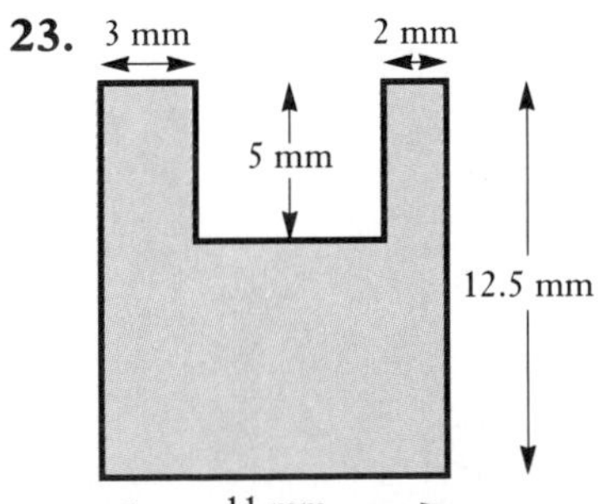

24.

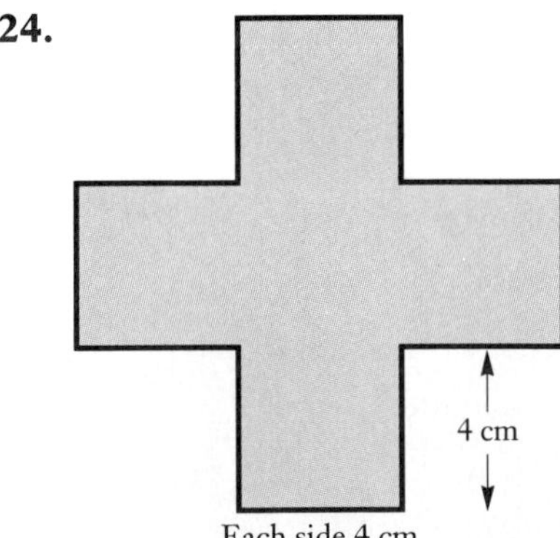

SKILL MAINTENANCE

Convert to percent notation.

25. 0.452

26. $\frac{1}{3}$

27. $\frac{11}{20}$

28. $\frac{22}{25}$

9.5 Areas of Parallelograms, Triangles, and Trapezoids

Parallelograms

A **parallelogram** is a four-sided figure with two pairs of parallel sides, as shown below.

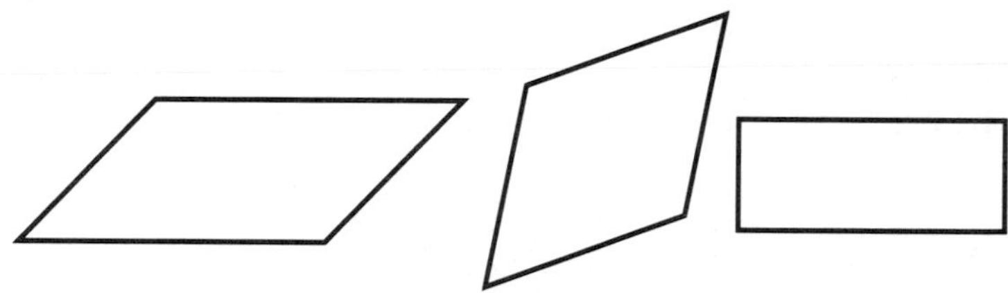

To find the area of a parallelogram, consider the one below.

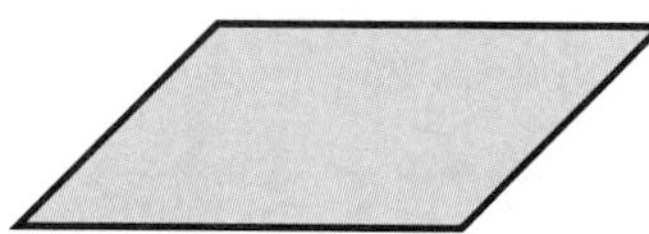

If we cut off a piece and move it to the other end, we get a rectangle.

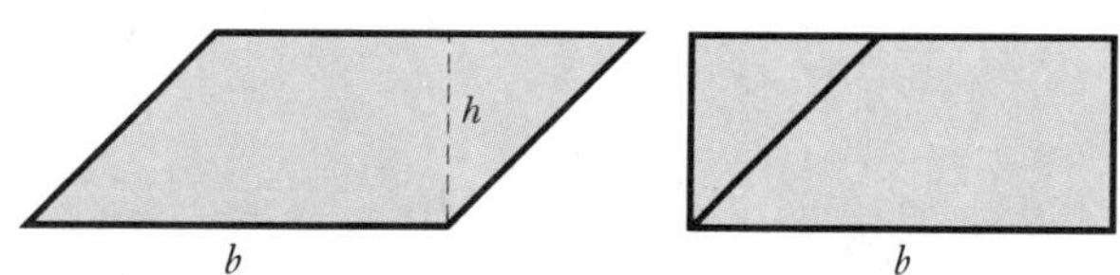

We can find the area by multiplying the length b, called a **base,** by h, called the **height.**

> **The area of a parallelogram is the product of the length of a base b and the height h:**
>
> $$A = b \cdot h.$$

▶ **EXAMPLE 1** Find the area of this parallelogram.

$A = b \cdot h$

$A = 7\text{ km} \cdot 5\text{ km}$

$A = 35\text{ km}^2$

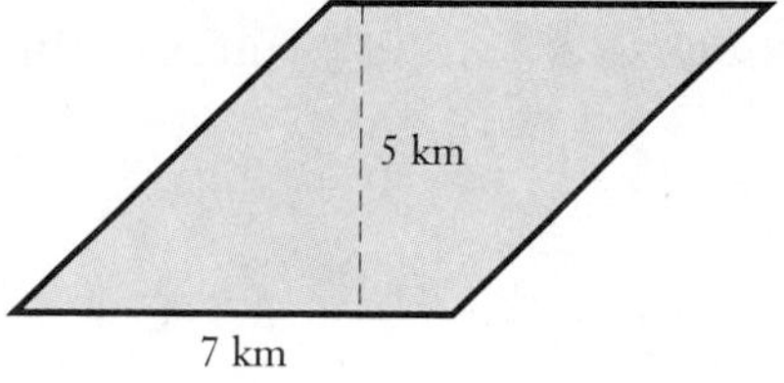

◀

OBJECTIVES

After finishing Section 9.5, you should be able to:

a Find areas of parallelograms, triangles, and trapezoids.

b Solve problems involving areas of parallelograms, triangles, and trapezoids

FOR EXTRA HELP

Tape 11E

Tape 13B

MAC: 9
IBM: 9

Find the area.

1.

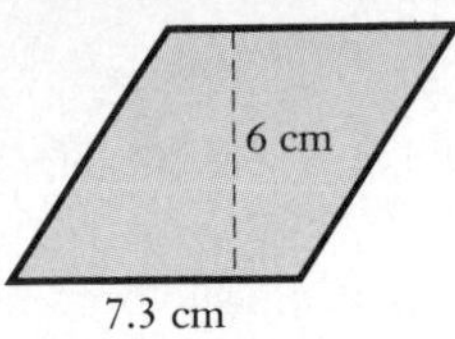

43.8 cm^2

2.

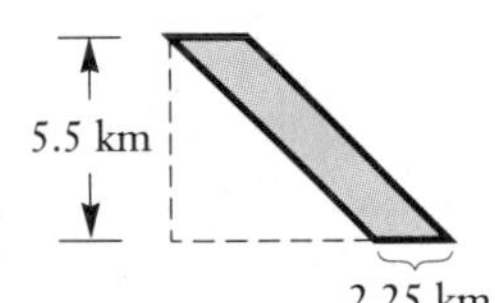

12.375 km^2

ANSWERS ON PAGE A-5

► **EXAMPLE 2** Find the area of this parallelogram.

$A = b \cdot h$

$A = (1.2\text{ m}) \times (6\text{ m})$

$A = 7.2\text{ m}^2$

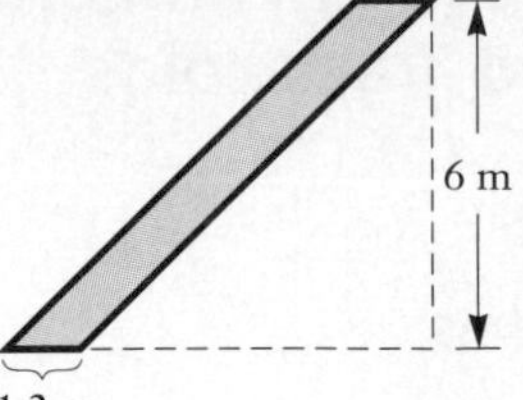

DO EXERCISES 1 AND 2.

Triangles

To find the area of a triangle, think of cutting out another just like it.

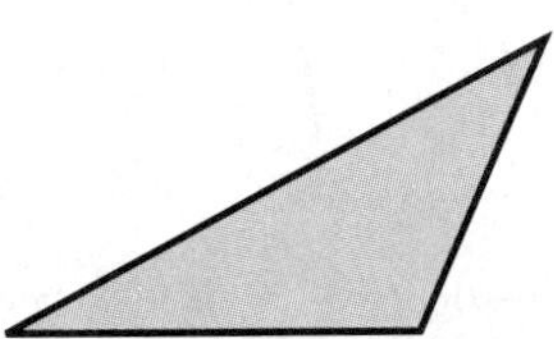

Then place the second one like this.

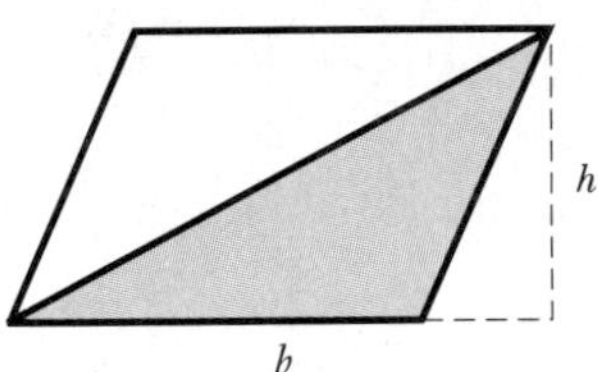

The resulting figure is a parallelogram whose area is

$$b \cdot h.$$

The triangle we started with has half the area of the parallelogram, or

$$\frac{1}{2} \cdot b \cdot h.$$

The area of a triangle is half the length of the base times the height:

$$A = \frac{1}{2} \cdot b \cdot h.$$

► **EXAMPLE 3** Find the area of this triangle.

$A = \frac{1}{2} \cdot b \cdot h$

$A = \frac{1}{2} \cdot 9\text{ m} \cdot 6\text{ m}$

$A = \frac{9 \cdot 6}{2}\text{ m}^2$

$A = 27\text{ m}^2$

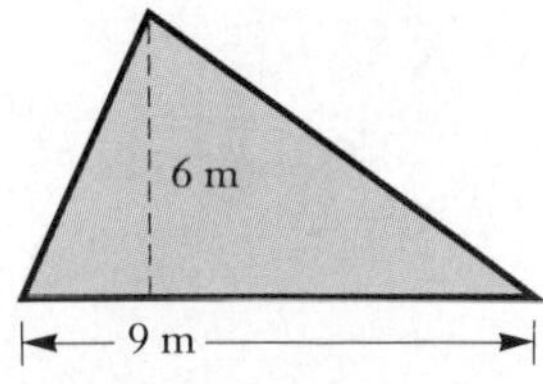

▶ **EXAMPLE 4** Find the area of this triangle.

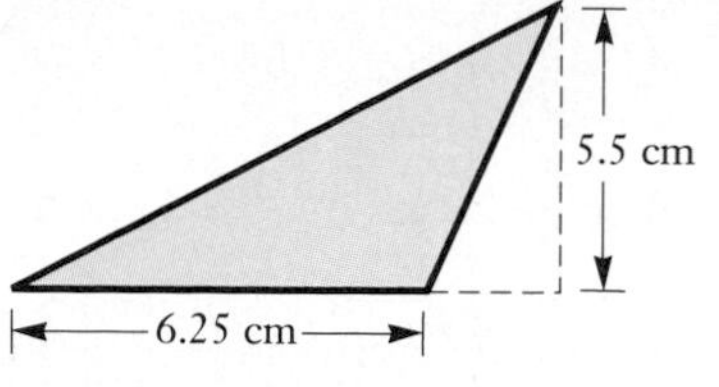

$$A = \frac{1}{2} \cdot b \cdot h$$

$$A = \frac{1}{2} \times 6.25 \text{ cm} \times 5.5 \text{ cm}$$

$$A = 0.5 \times 6.25 \times 5.5 \text{ cm}^2$$

$$A = 17.1875 \text{ cm}^2$$ ◀

DO EXERCISES 3 AND 4.

Trapezoids

A **trapezoid** is a four-sided figure with at least one pair of parallel sides, as shown below.

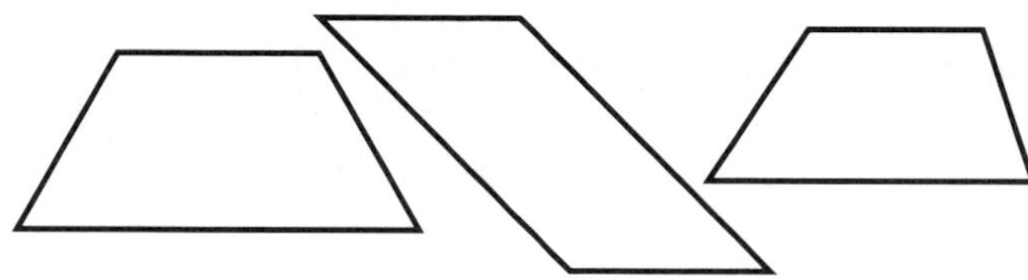

To find the area of a trapezoid, think of cutting out another just like it.

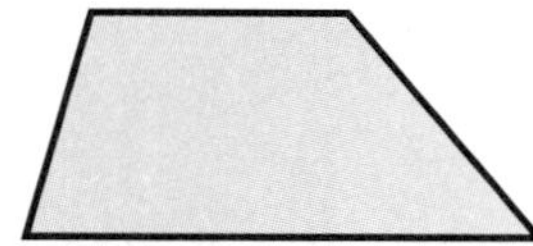

Then place the second one like this.

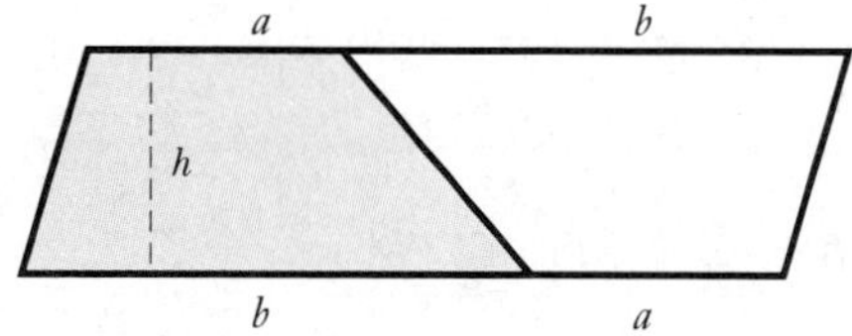

The resulting figure is a parallelogram whose area is

$$h \cdot (a + b).$$ **The base is $a + b$.**

The trapezoid we started with has half the area of the parallelogram, or

$$\frac{1}{2} \cdot h \cdot (a + b).$$

The area of a trapezoid is half the product of the height and the sum of the lengths of the parallel sides, or the product of the height and the average length of the bases:

$$A = \frac{1}{2} \cdot h \cdot (a + b) = h \cdot \frac{a + b}{2}.$$

Find the area.

3.

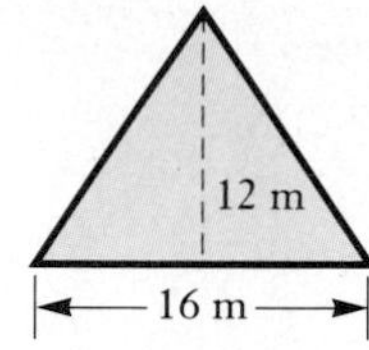

96 m^2

4.

3.4 cm

11 cm

18.7 cm^2

ANSWERS ON PAGE A-5

Find the area.

5.

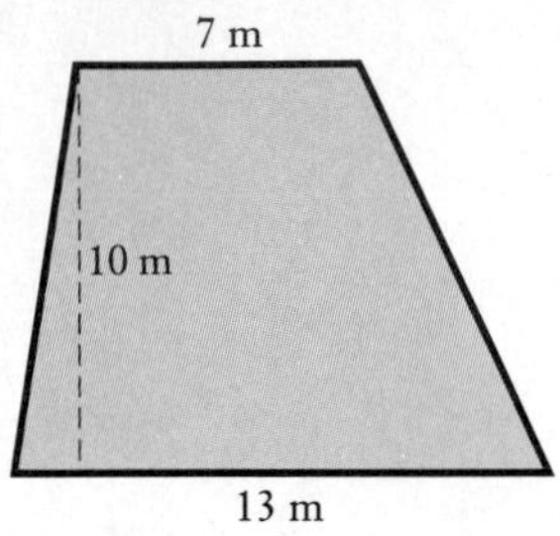

100 m^2

6.

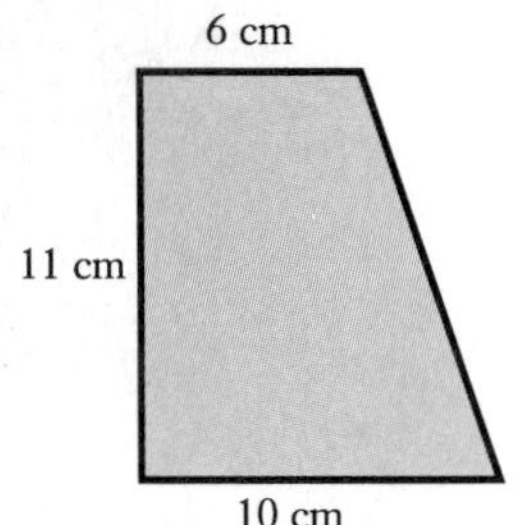

88 cm^2

7. Find the area of the shaded region.

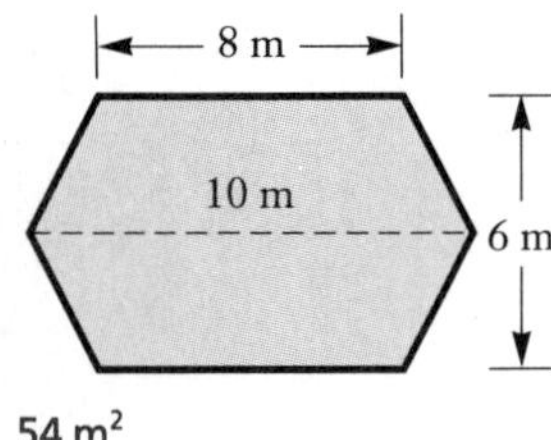

54 m^2

ANSWERS ON PAGE A-5

▶ **EXAMPLE 5** Find the area of this trapezoid.

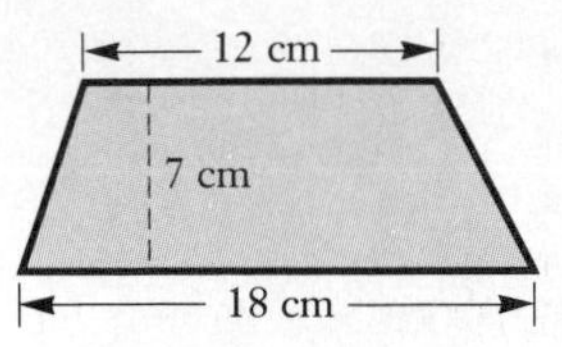

$$A = \frac{1}{2} \cdot h \cdot (a + b)$$

$$A = \frac{1}{2} \cdot 7 \text{ cm} \cdot (12 + 18) \text{ cm}$$

$$A = \frac{7 \cdot 30}{2} \cdot \text{cm}^2 = \frac{7 \cdot 15 \cdot 2}{1 \cdot 2} \text{ cm}^2$$

$$A = \frac{7 \cdot 15}{1} \cdot \frac{2}{2} \text{ cm}^2$$

$$A = 105 \text{ cm}^2$$ ◀

DO EXERCISES 5 AND 6.

b Solving Problems

▶ **EXAMPLE 6** Find the area of this kite.

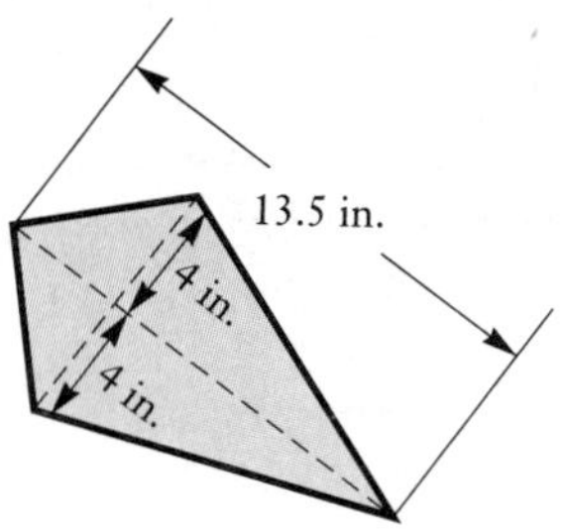

1. *Familiarize.* We look for the kinds of figures whose areas we can calculate using area formulas that we already know.
2. *Translate.* The shaded region consists of two triangles, each with a base of 13.5 in. and a height of 4 in. We can apply the formula $A = \frac{1}{2} \cdot b \cdot h$ for the area of a triangle and then multiply by 2.
3. *Solve.*

$$A = \frac{1}{2} \cdot (13.5 \text{ in.}) \cdot (4 \text{ in.}) = 27 \text{ in}^2$$

Then we multiply by 2:

$$2 \cdot 27 \text{ in}^2 = 54 \text{ in}^2.$$

4. *Check.* We can check by repeating the calculations.
5. *State.* The area of the shaded region is 54 in^2. ◀

DO EXERCISE 7.

NAME SECTION DATE

EXERCISE SET 9.5

a Find the area.

1.

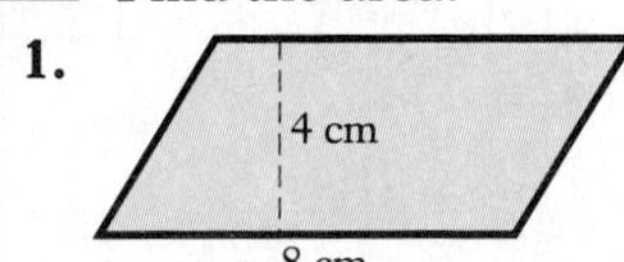

2.

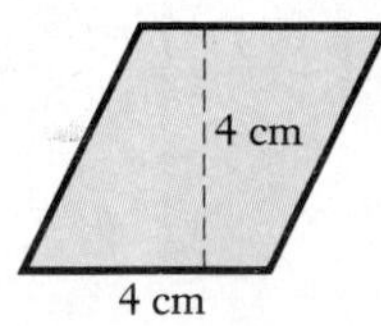

3.

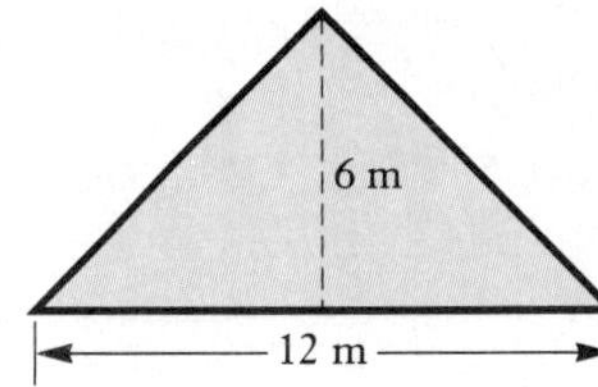

4.

18 km

18 km

5.

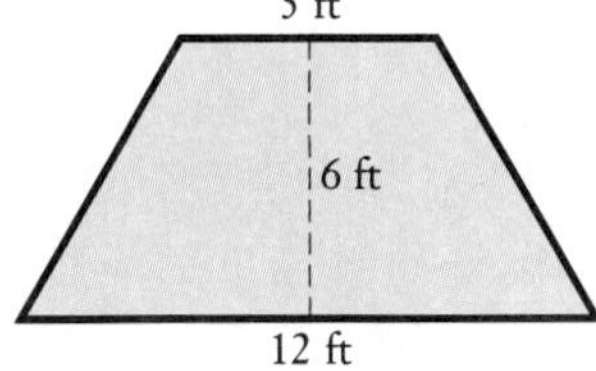

6.

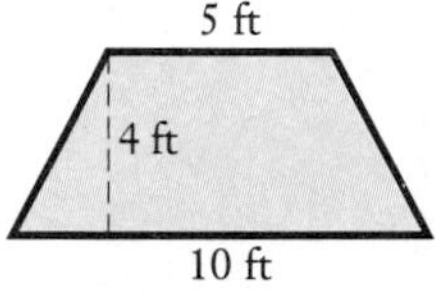

7.

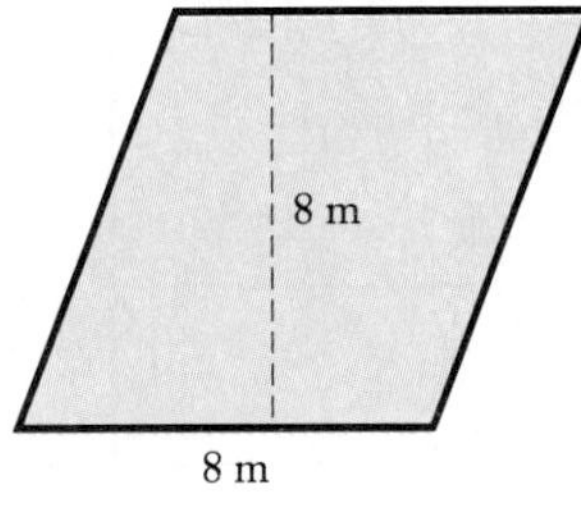

8.

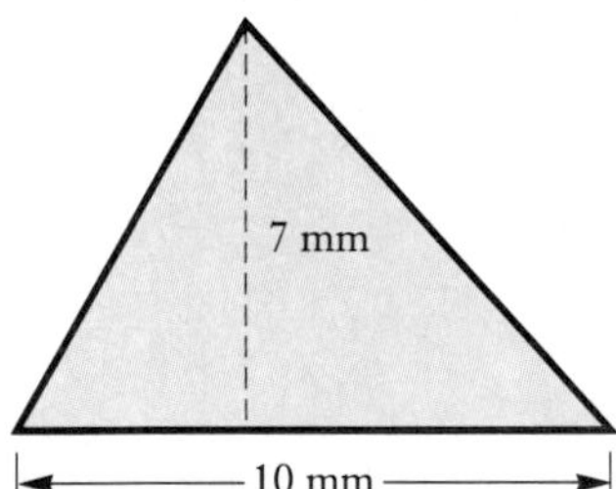

9.

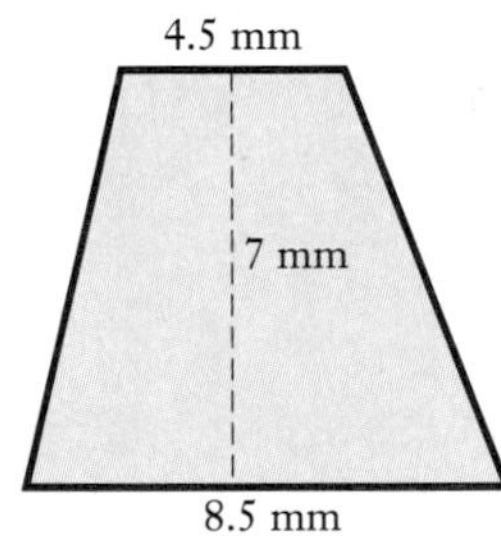

10.

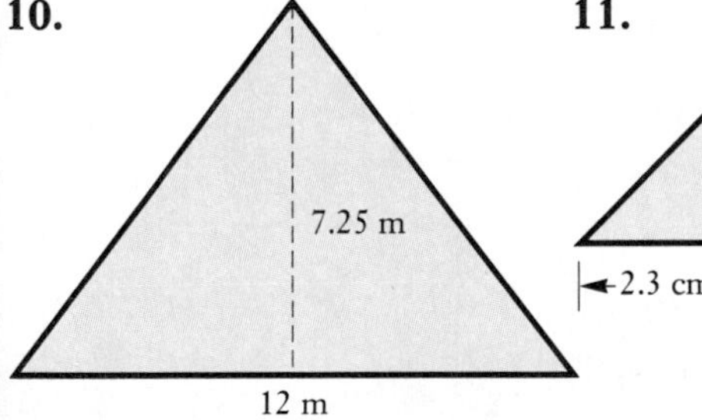

11.

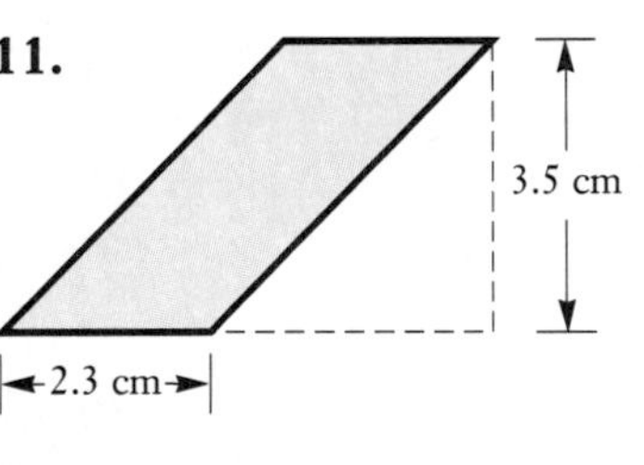

12.

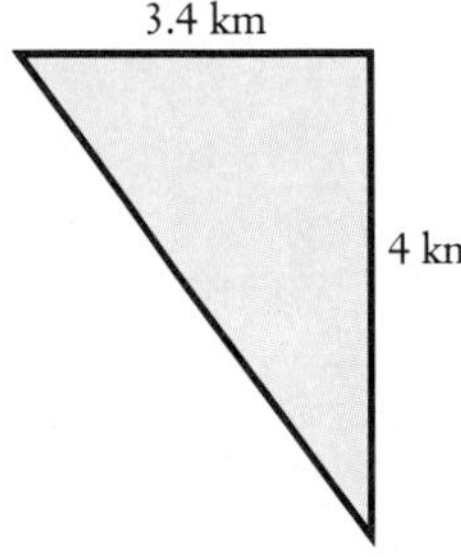

13.

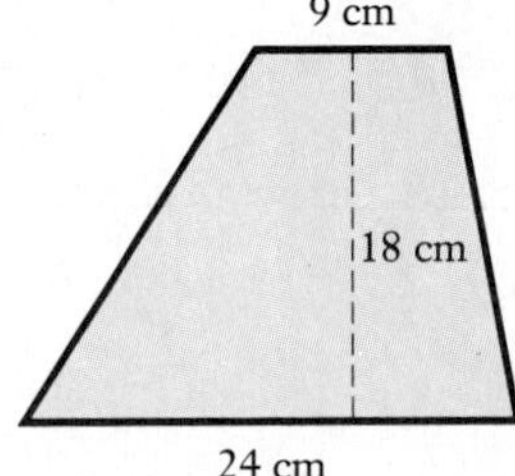

14.

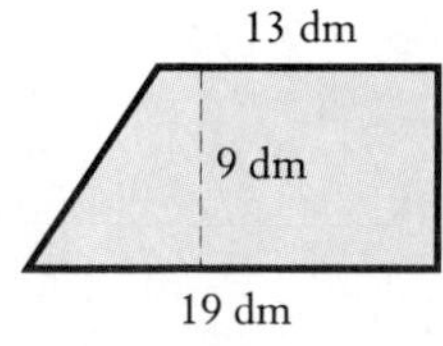

15.

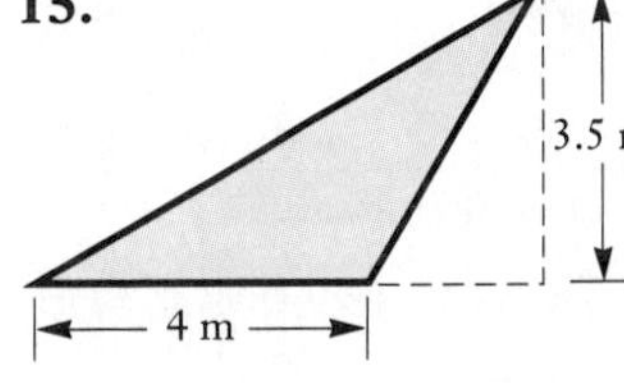

ANSWERS

1. 32 cm^2
2. 16 cm^2
3. 36 m^2
4. 162 km^2
5. 51 ft^2
6. 30 ft^2
7. 64 m^2
8. 35 mm^2
9. 45.5 mm^2
10. 43.5 m^2
11. 8.05 cm^2
12. 6.8 km^2
13. 297 cm^2
14. 144 dm^2
15. 7 m^2

ANSWERS

16. $9\frac{1}{24}$ yd²

17. $55\frac{1}{8}$ ft²

18. 35.04 mm²

19. 675 cm²

20. 61 in²

21. 10,816 in²

22. 133 m²

23. 852.04 m²

24. 6800 ft²

25. $\frac{37}{400}$

26. $\frac{7}{8}$

27. 137.5%

28. $66\frac{2}{3}\%$, or $66.\overline{6}\%$

16.

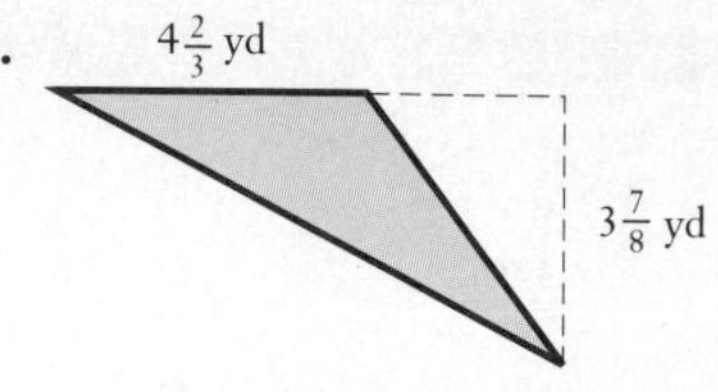

17.

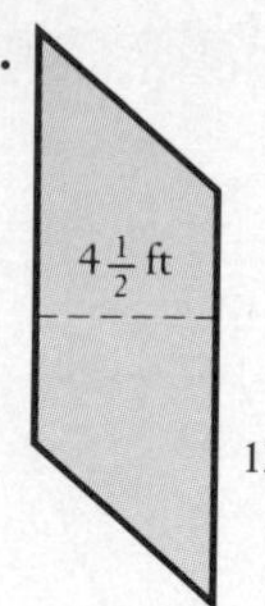

18.

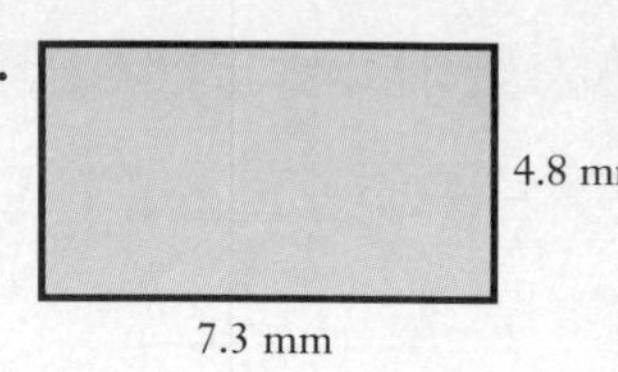

b Find the area of the shaded region.

19.

15 cm

30 cm

30 cm

20.

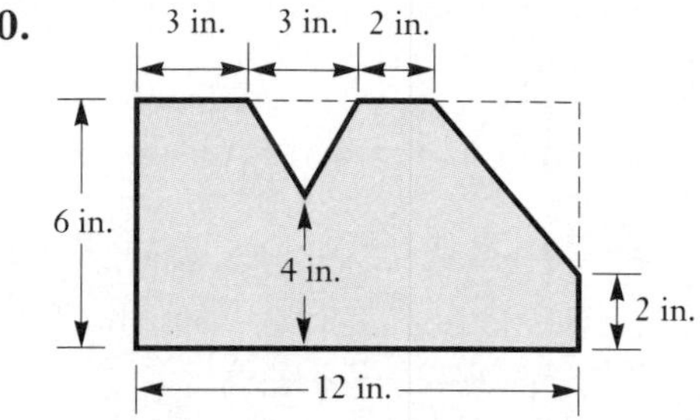

21.

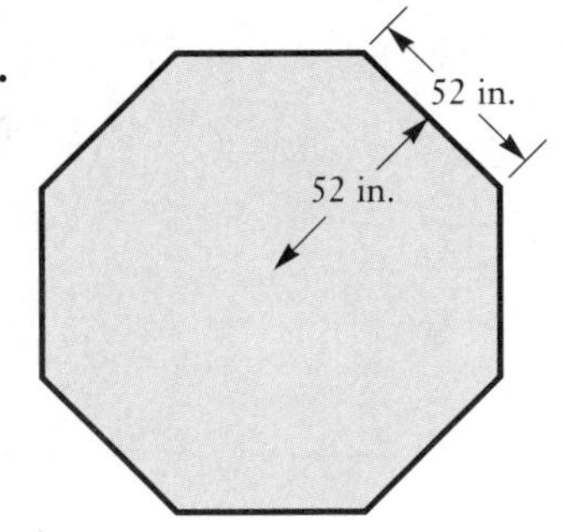

22.

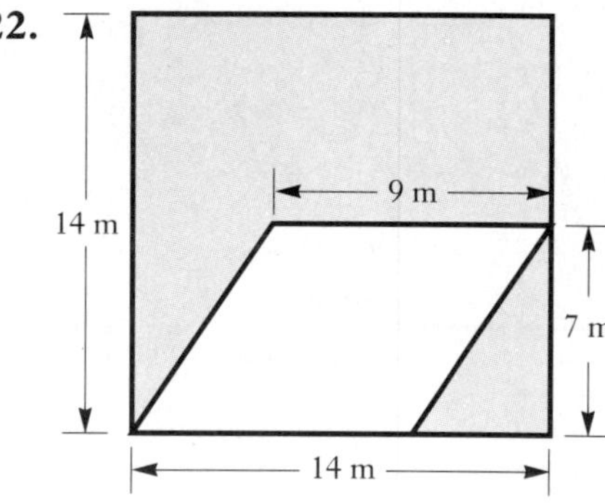

23. A lot is 36 m by 24 m. A triangular swimming pool with a height of 4.6 m and a base of 5.2 m is constructed on the lot. How much area is left over?

24. Find the total area of the sides and ends of the building.

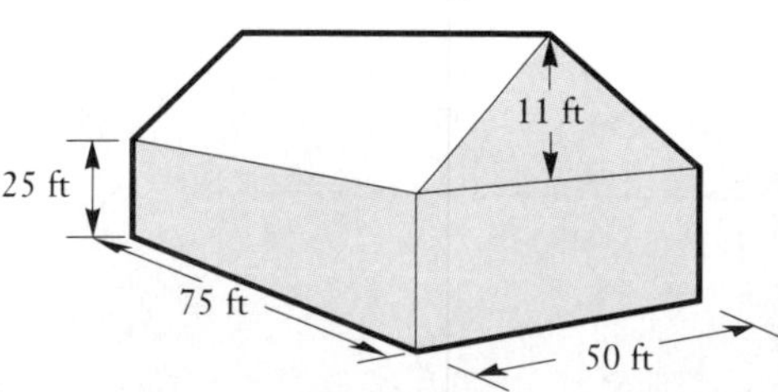

SKILL MAINTENANCE

Convert to fractional notation.

25. 9.25%

26. $87\frac{1}{2}\%$

Convert to percent notation.

27. $\frac{11}{8}$

28. $\frac{2}{3}$

9.6 Circles

a Radius and Diameter

At the right is a circle with center O. Segment $\overline{AC}$ is a *diameter.* A **diameter** is a segment that passes through the center of the circle and has endpoints on the circle. Segment $\overline{OB}$ is called a *radius.* A **radius** is a segment with one endpoint on the center and the other endpoint on the circle.

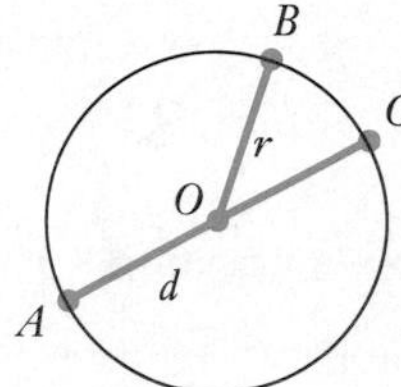

Suppose that d is the diameter of a circle and r is the radius. Then

$$d = 2 \cdot r \quad \text{and} \quad r = \frac{d}{2}.$$

▶ **EXAMPLE 1** Find the length of a radius of this circle.

$$r = \frac{d}{2}$$

$$r = \frac{12\text{ m}}{2}$$

$$r = 6\text{ m}$$

12 m ◀

▶ **EXAMPLE 2** Find the length of a diameter of this circle.

$$d = 2 \cdot r$$

$$d = 2 \cdot \frac{1}{4}\text{ ft}$$

$$d = \frac{1}{2}\text{ ft}$$

$\frac{1}{4}$ ft ◀

DO EXERCISES 1 AND 2.

b Circumference

The **circumference** of a circle is the distance around it and calculating it is similar to finding the perimeter of a polygon.

Take a 12-oz soft drink can and measure the circumference C of the lid with a tape measure. Then measure the diameter d. Then find the ratio C/d.

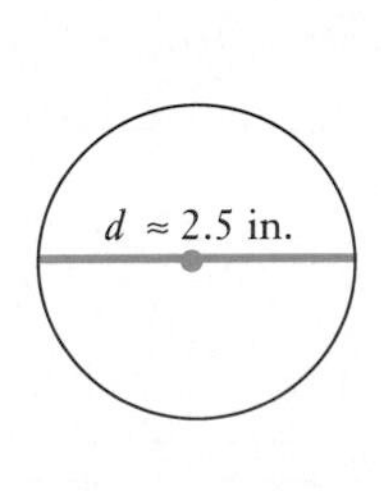

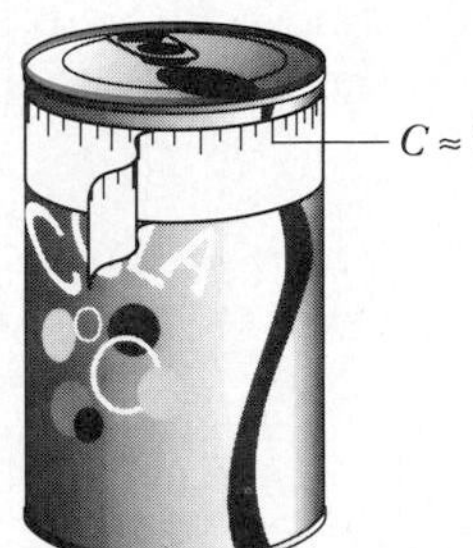

$$\frac{C}{d} \approx \frac{7.8\text{ in.}}{2.5\text{ in.}} \approx 3.1$$

OBJECTIVES

After finishing Section 9.6, you should be able to:

- a Find the length of a radius of a circle given the length of a diameter, and find the length of a diameter given the length of a radius.
- b Find the circumference of a circle given the length of a diameter or radius.
- c Find the area of a circle given the length of a radius.
- d Solve problems involving circles.

FOR EXTRA HELP

Tape 12A | Tape 13B | MAC: 9 IBM: 9

1. Find the length of a radius.

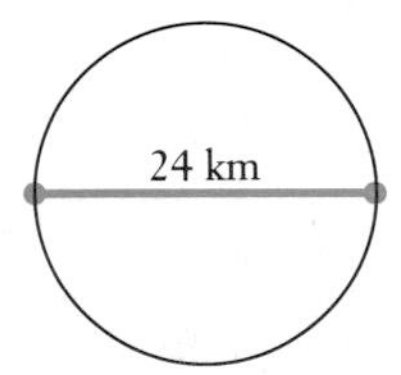

12 km

2. Find the length of a diameter.

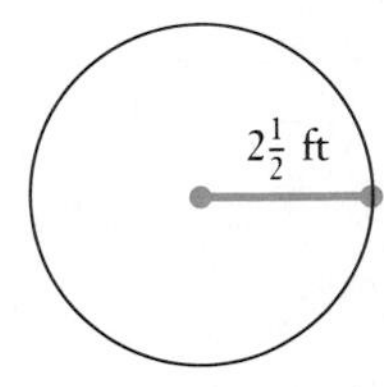

5 ft

ANSWERS ON PAGE A-5

3. Find the circumference of this circle. Use 3.14 for π. 62.8

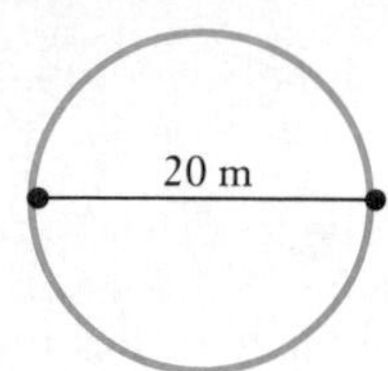

4. Find the circumference of this circle. Use $\frac{22}{7}$ for π. 88 m

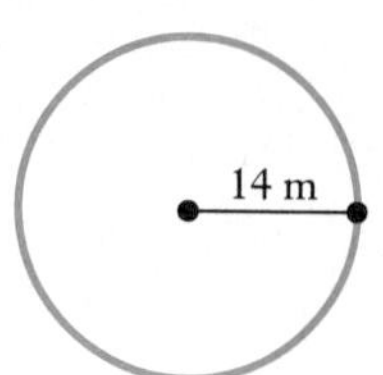

5. Find the circumference of this circle. Use 3.14 for π. 20.096 cm

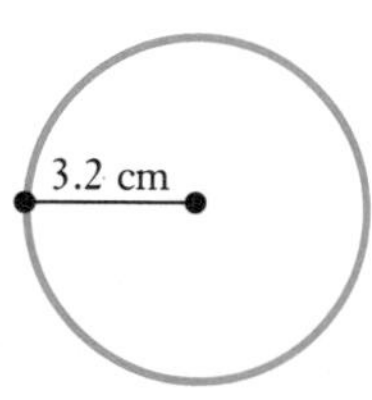

ANSWERS ON PAGE A-5

Suppose we did this with cans and circles of several sizes. We would get a number close to 3.1. For any circle, if we divide the circumference C by the diameter d, we get the same number. We call this number π (pi).

$\frac{C}{d} = \pi$ or $C = \pi \cdot d$. The number π is about 3.14, or about $\frac{22}{7}$.

▶ **EXAMPLE 3** Find the circumference of this circle. Use 3.14 for π.

$$C = \pi \cdot d$$
$$C \approx 3.14 \times 6 \text{ cm}$$
$$C = 18.84 \text{ cm}$$

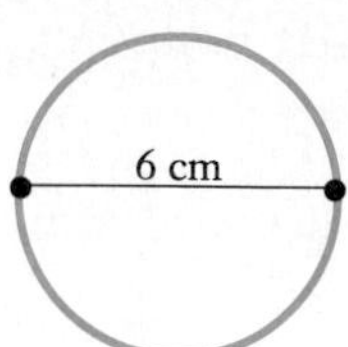

The circumference is about 18.84 cm. ◀

DO EXERCISE 3.

Since

$$d = 2 \cdot r,$$

where r is the length of a radius, it follows that

$$C = \pi \cdot d = \pi \cdot (2 \cdot r).$$

$$\mathbf{C = 2 \cdot \pi \cdot r}$$

▶ **EXAMPLE 4** Find the circumference of this circle. Use $\frac{22}{7}$ for π.

$$C = 2 \cdot \pi \cdot r$$
$$C \approx 2 \cdot \frac{22}{7} \cdot 70 \text{ m}$$
$$C = 2 \cdot 22 \cdot \frac{70}{7} \text{ m}$$
$$C = 44 \cdot 10 \text{ m}$$
$$C = 440 \text{ m}$$

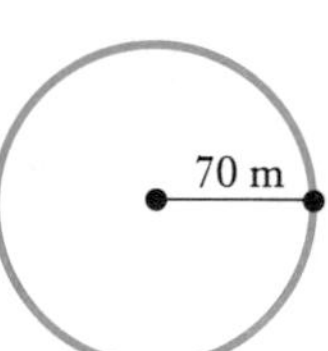

The circumference is about 440 m. ◀

▶ **EXAMPLE 5** Find the perimeter of this figure. Use 3.14 for π.

We let P = the perimeter. We have half a circle attached to a square. We add half the circumference to the lengths of the three line segments.

$$P = 3 \times 9.4 \text{ km} + \frac{1}{2} \times 2 \times \pi \times 4.7 \text{ km}$$
$$\approx 28.2 \text{ km} + 3.14 \times 4.7 \text{ km}$$
$$= 28.2 \text{ km} + 14.758 \text{ km}$$
$$= 42.958 \text{ km}$$

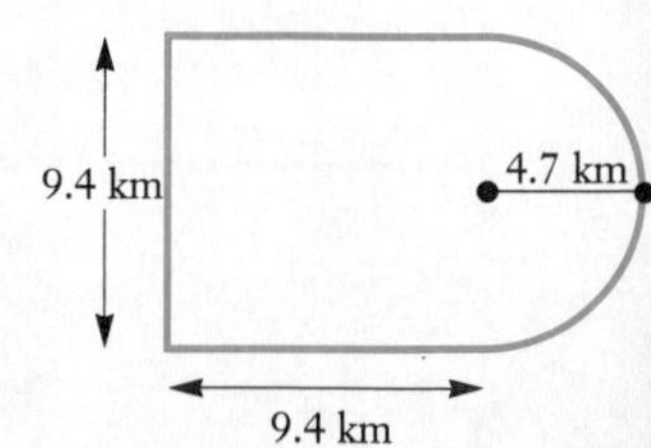

The perimeter is about 57.716 km. ◀

DO EXERCISES 4 AND 5.

C Area

Below is a circle of radius r.

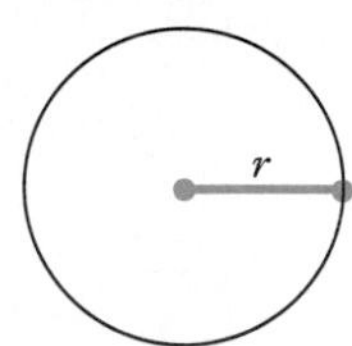

Think of cutting half the circular region into small pieces and arranging them as shown below.

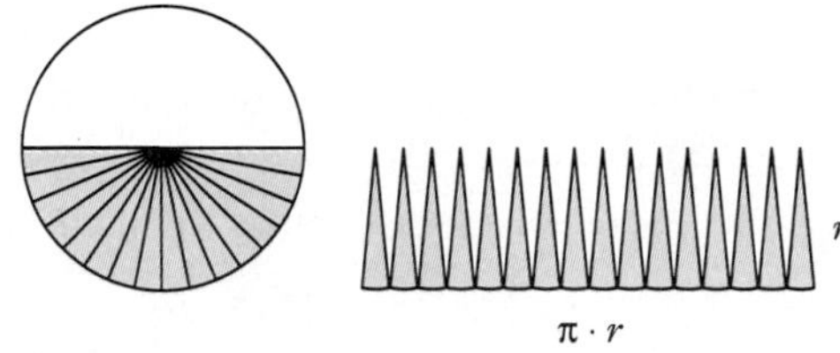

Then imagine cutting the other half of the circular region and arranging the pieces in with the others as shown below.

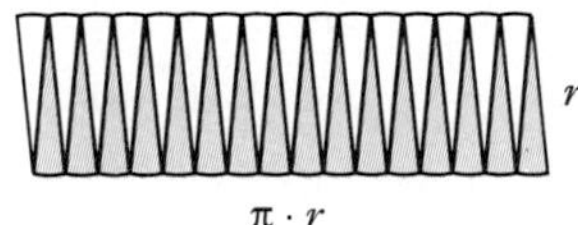

This is almost a parallelogram. The base has length $\frac{1}{2} \cdot 2 \cdot \pi \cdot r$, or $\pi \cdot r$ (half the circumference) and the height is r. Thus the area is about

$$(\pi \cdot r) \cdot r.$$

This is the area of a circle.

> **The area of a circle with radius of length r is given by**
> $$A = \pi \cdot r \cdot r, \quad \text{or} \quad A = \pi \cdot r^2.$$

▶ **EXAMPLE 6** Find the area of this circle. Use $\frac{22}{7}$ for π.

$$A = \pi \cdot r \cdot r$$

$$A \approx \frac{22}{7} \cdot 14 \text{ cm} \cdot 14 \text{ cm}$$

$$A = \frac{22}{7} \cdot 196 \text{ cm}^2$$

$$A = 616 \text{ cm}^2$$

14 cm

The area is about 616 cm^2. ◀

DO EXERCISE 6.

6. Find the area of this circle. Use $\frac{22}{7}$ for π. $78\frac{4}{7} \text{ km}^2$

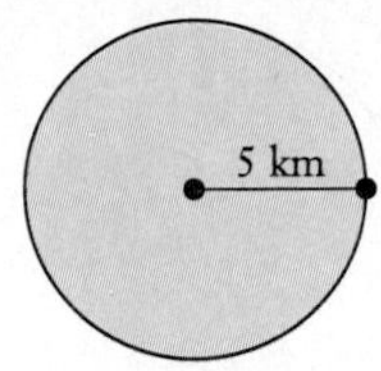

ANSWER ON PAGE A-5

7. Find the area of this circle. Use 3.14 for π. 339.62 cm^2

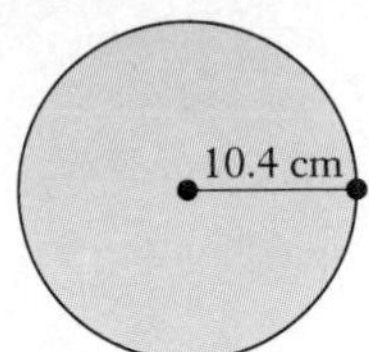

8. Which is larger and by how much: a 10-inch square cookie sheet or a 12-inch diameter pizza pan?

The pizza pan by 13.04 in^2

ANSWERS ON PAGE A-5

▶ **EXAMPLE 7** Find the area of this circle. Use 3.14 for π. Round to the nearest hundredth.

$$A = \pi \cdot r \cdot r$$
$$A \approx 3.14 \times 2.1 \text{ m} \times 2.1 \text{ m}$$
$$A = 3.14 \times 4.41 \text{ m}^2$$
$$A = 13.8474 \text{ m}^2$$
$$A \approx 13.85 \text{ m}^2$$

2.1 m

The area is about 13.85 m^2. ◀

DO EXERCISE 7.

d Solving Problems

▶ **EXAMPLE 8** Which is larger, and by how much: an 8-inch square cookie sheet or an 8-inch diameter circular pizza pan?

First, we draw a picture of each.

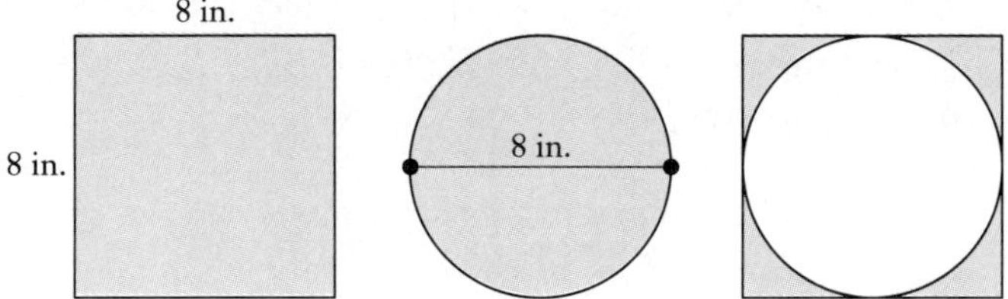

Then we compute areas.

The area of the square is

$$A = s \cdot s$$
$$= 8 \text{ in.} \times 8 \text{ in.}$$
$$= 64 \text{ in}^2.$$

The diameter of the circle is 8 in., so the radius is 8 in./2, or 4 in. The area of the circle is

$$A = \pi \cdot r \cdot r$$
$$\approx 3.14 \times 4 \text{ in.} \times 4 \text{ in.}$$
$$= 50.24 \text{ in}^2.$$

We see that the cookie sheet is larger by about

$$64 \text{ in}^2 - 50.24 \text{ in}^2, \quad \text{or} \quad 13.76 \text{ in}^2.$$

13.76 in^2 is actually the area of the shaded region shown to the right above. ◀

DO EXERCISE 8.

NAME SECTION DATE

EXERCISE SET 9.6

a Find the length of a diameter of the circle.

1.

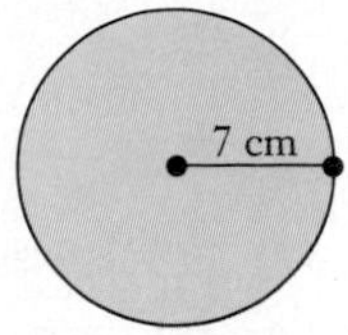

2.

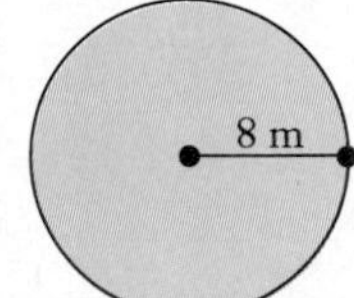

3.

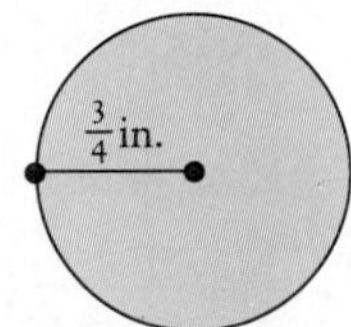

4.

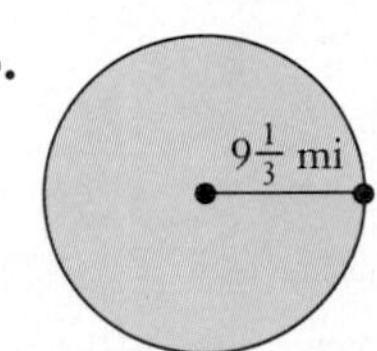

Find the length of a radius of the circle.

5.

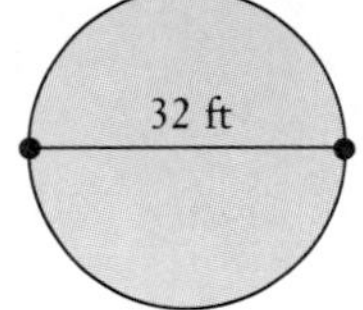

6.

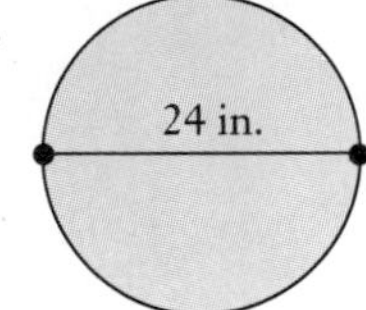

7.

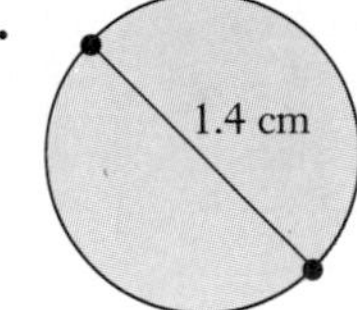

8.

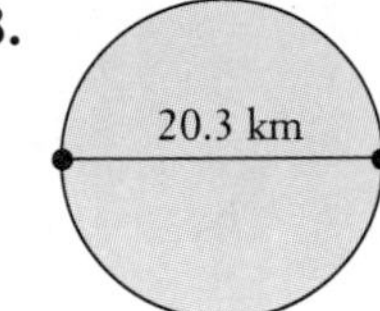

b Find the circumference of each circle in Exercises 1–4. Use $\frac{22}{7}$ for π.

9. Exercise 1 **10.** Exercise 2 **11.** Exercise 3 **12.** Exercise 4

Find the circumference of each circle in Exercises 5–8. Use 3.14 for π.

13. Exercise 5 **14.** Exercise 6 **15.** Exercise 7 **16.** Exercise 8

c Find the area of each circle in Exercises 1–4. Use $\frac{22}{7}$ for π.

17. Exercise 1 **18.** Exercise 2 **19.** Exercise 3 **20.** Exercise 4

Find the area of each circle in Exercises 5–8. Use 3.14 for π.

21. Exercise 5 **22.** Exercise 6 **23.** Exercise 7 **24.** Exercise 8

ANSWERS

1. 14 cm

2. 16 m

3. $1\frac{1}{2}$ in.

4. $18\frac{2}{3}$ mi

5. 16 ft

6. 12 in.

7. 0.7 cm

8. 10.15 km

9. 44 cm

10. $50\frac{2}{7}$ m

11. $4\frac{5}{7}$ in.

12. $58\frac{2}{3}$ mi

13. 100.48 ft

14. 75.36 in.

15. 4.396 cm

16. 63.742 km

17. 154 cm^2

18. $201\frac{1}{7}$ m^2

19. $1\frac{43}{56}$ in^2

20. $273\frac{7}{9}$ mi^2

21. 803.84 ft^2

22. 452.16 in^2

23. 1.5386 cm^2

24. 323.49065 km^2

ANSWERS

25. 3 cm; 18.84 cm; 28.26 cm^2

26. 2 cm; 6.28 cm; 3.14 cm^2

27. 151,976 km^2

28. The square pizza by 30.96 in^2

29. 3.454 m

30. 31.4 m

31. 2.5 cm; 1.25 cm; 4.90625 cm^2

32. 1.8 cm; 0.9 cm; 2.5434 cm^2

33. 65.94 m^2

34. 433.86 m^2; \$4555.53

d Solve. Use 3.14 for π.

25. The lid of a cola can has a 6-cm diameter. What is its radius? circumference? area?

26. A penny has a 1-cm radius. What is its diameter? circumference? area?

27. A radio station is allowed by the FCC to broadcast over an area with a radius of 220 km. How much area is this?

28. Which is larger and by how much: a 12-inch circular pizza or a 12-inch square pizza?

29. The trunk of an elm tree has a 1.1-m diameter. What is its circumference?

30. A silo has a 10-m diameter. What is its circumference?

31. The circumference of a quarter is 7.85 cm. What is the diameter? radius? area?

32. The circumference of a dime is 5.652 cm. What is the diameter? radius? area?

33. A circular swimming pool is surrounded by a walk that is 1 m wide. The diameter of the pool is 20 m. What is the area of the walk?

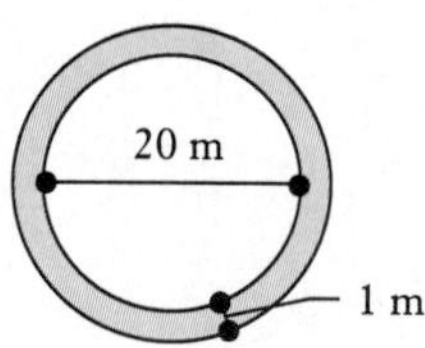

34. A roller rink is shown below. What is its area? Hardwood flooring costs \$10.50 per square meter. How much would flooring cost?

Find the perimeter. Use 3.14 for π.

35.

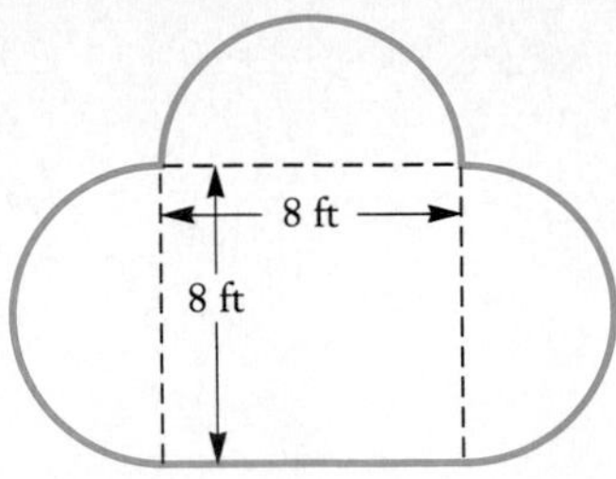

36.

2 cm
2 cm
2 cm

37.

4 yd
= radius

38.

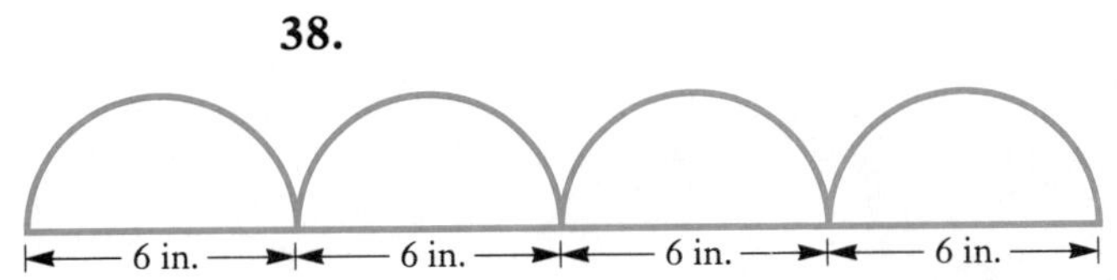

39.

10 yd
10 yd

40.

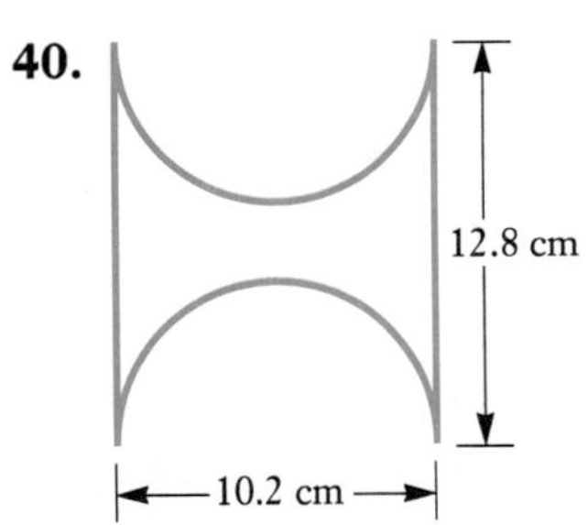

Find the area of the shaded region. Use 3.14 for π.

41.

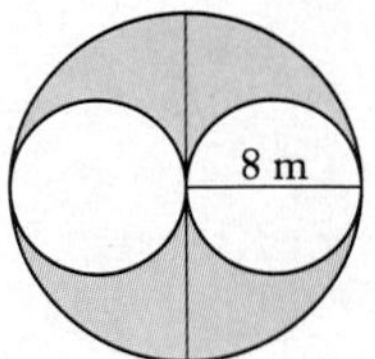

42.

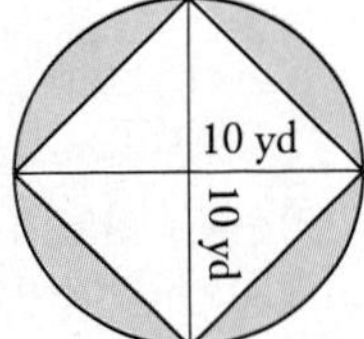

ANSWERS

35. 45.68 ft

36. 7.14 cm

37. 26.84 yd

38. 61.68 in.

39. 45.7 yd

40. 57.628 cm

41. 100.48 m^2

42. 114 yd^2

ANSWERS

43. 6.9972 cm^2

44. 150.72 ft^2

45. 48.8886 cm^2

46. 657.92 ft^2

47. 87.5%

48. 58%

49. $66.\overline{6}\%$

50. 43.61%

51. 3.142

52. 360,000

53. $3d$; πd; circumference of one ball, since $\pi > 3$

43.

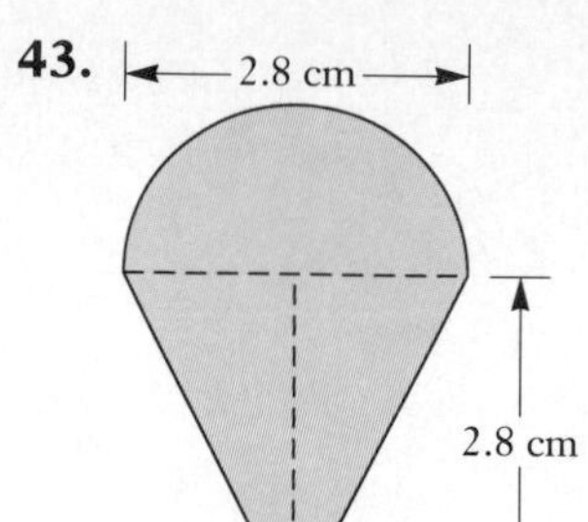

44.

8 ft

45.

12.8 cm

10.2 cm

46.

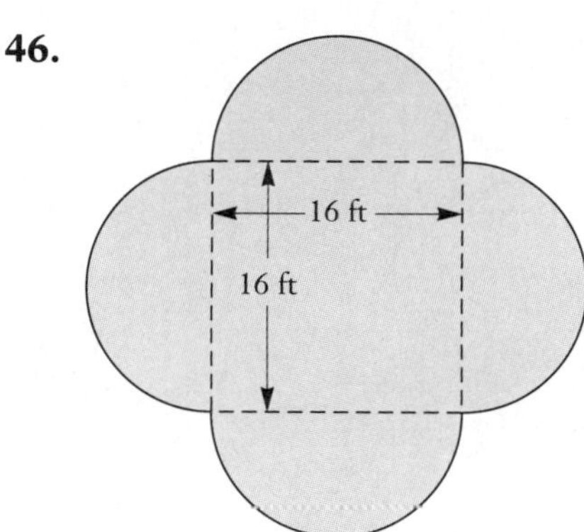

SKILL MAINTENANCE

Convert to percent notation.

47. 0.875

48. 0.58

49. $0.\overline{6}$

50. 0.4361

SYNTHESIS

51. ▦ $\pi \approx \frac{3927}{1250}$ is another approximation for π. Find decimal notation using your calculator. Round to the nearest thousandth.

52. ▦ The distance from Kansas City to Indianapolis is 500 mi. A car was driven this distance using tires with a radius of 14 in. How many revolutions of each tire occurred on the trip? Use $\frac{22}{7}$ for π.

53. Tennis balls are usually packed vertically three in a can, one on top of another. Suppose the diameter of a tennis ball is d. Find the height of the stack of balls. Find the circumference of one ball. Which is greater? Explain.

9.7 Square Roots and the Pythagorean Theorem

a Square Roots

If a number is a product of two identical factors, then either factor is called a *square root* of the number. (If $a = c^2$, then c is a square root of a.) The symbol $\sqrt{}$ (called a *radical* sign) is used in naming square roots.

For example, $\sqrt{36}$ is the square root of 36. It follows that

$$\sqrt{36} = \sqrt{6 \cdot 6} = 6 \qquad \text{The square root of 36 is 6.}$$

because $6^2 = 36$.

▶ **EXAMPLE 1** Simplify: $\sqrt{25}$.

$$\sqrt{25} = \sqrt{5 \cdot 5} = 5 \qquad \text{The square root of 25 is 5 because } 5^2 = 25.$$ ◀

▶ **EXAMPLE 2** Simplify: $\sqrt{144}$.

$$\sqrt{144} = \sqrt{12 \cdot 12} = 12 \qquad \text{The square root of 144 is 12 because } 12^2 = 144.$$ ◀

CAUTION! It is common to confuse squares and square roots. A number squared is that number times itself. A square root of a number is a number that when multiplied by itself gives the original number.

▶ **EXAMPLES** Simplify.

3. $\sqrt{4} = 2$
4. $\sqrt{256} = 16$
5. $\sqrt{361} = 19$ ◀

DO EXERCISES 1–24.

b Approximating Square Roots

Square roots of some numbers are not whole numbers or ordinary fractions. For example,

$$\sqrt{2}, \quad \sqrt{3}, \quad \sqrt{39}, \quad \text{and} \quad \sqrt{70}$$

are not whole numbers or ordinary fractions. We can approximate these square roots. For example, consider the following decimal approximations for $\sqrt{2}$. Each gives a closer approximation.

$$\sqrt{2} \approx 1.4 \quad \text{because} \quad (1.4)^2 = 1.96,$$
$$\sqrt{2} \approx 1.41 \quad \text{because} \quad (1.41)^2 = 1.9881$$
$$\sqrt{2} \approx 1.414 \quad \text{because} \quad (1.414)^2 = 1.999396$$
$$\sqrt{2} \approx 1.4142 \quad \text{because} \quad (1.4142)^2 = 1.99996164.$$

How do we find such approximations? The most common way is to use a calculator, but we can also use a table such as Table 1 at the back of the book.

OBJECTIVES

After finishing Section 9.7, you should be able to:

a Simplify square roots of squares such as

$$\sqrt{25}.$$

b Approximate square roots.

c Given the lengths of any two sides of a right triangle, find the length of the third side.

FOR EXTRA HELP

Tape 12B Tape 14A MAC: 9 IBM: 9

Find the square.

1. 9^2 81
2. 10^2 100
3. 11^2 121
4. 12^2 144

It would be helpful to memorize the squares of numbers from 1 to 25.

5. 13^2 169
6. 14^2 196
7. 15^2 225
8. 16^2 256
9. 17^2 289
10. 18^2 324
11. 25^2 625
12. 20^2 400

Simplify. Use the results of Exercises 1–12 above.

13. $\sqrt{9}$ 3
14. $\sqrt{16}$ 4
15. $\sqrt{121}$ 11
16. $\sqrt{100}$ 10
17. $\sqrt{81}$ 9
18. $\sqrt{64}$ 8
19. $\sqrt{324}$ 18
20. $\sqrt{400}$ 20
21. $\sqrt{225}$ 15
22. $\sqrt{169}$ 13
23. $\sqrt{1}$ 1
24. $\sqrt{0}$ 0

ANSWERS ON PAGE A-5

Approximate to three decimal places.

25. $\sqrt{5}$ 2.236

26. $\sqrt{78}$ 8.832

27. $\sqrt{168}$ 12.961

ANSWERS ON PAGE A-5

▶ **EXAMPLE 6** Approximate $\sqrt{3}$, $\sqrt{27}$, and $\sqrt{180}$ to three decimal places. Use a calculator or Table 1.

We use a calculator to find each square root. If more decimal places are given than we ask for, we round back to three places.

$$\sqrt{3} \approx 1.732,$$
$$\sqrt{27} \approx 5.196,$$
$$\sqrt{180} \approx 13.416$$ ◀

DO EXERCISES 25–27.

C The Pythagorean Theorem

A **right triangle** is a triangle with a 90° angle, as shown in the figure below. The small square in the corner indicates a 90° angle.

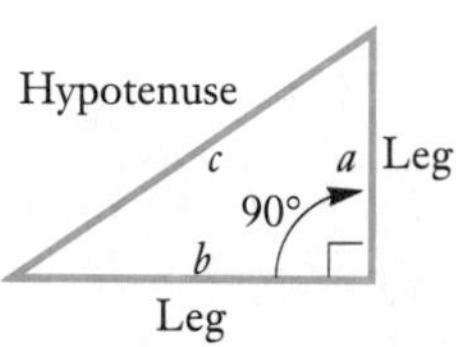

In a right triangle, the longest side is called the **hypotenuse.** It is also the side opposite the right angle. The other two sides are called **legs.** We generally use the letters a and b for the lengths of the legs and c for the length of the hypotenuse. They are related as follows.

> **The Pythagorean Theorem**
>
> **In any right triangle, if a and b are the lengths of the legs and c is the length of the hypotenuse, then**
>
> $$a^2 + b^2 = c^2, \quad \text{or}$$
> $$(\text{Leg})^2 + (\text{Other leg})^2 = (\text{Hypotenuse})^2.$$
>
> **The equation $a^2 + b^2 = c^2$ is called the *Pythagorean equation.***

The Pythagorean theorem is named after the ancient Greek mathematician Pythagoras (569?–500? B.C.). It is uncertain who actually proved this result the first time. We can think of this relationship as adding areas, as illustrated below:

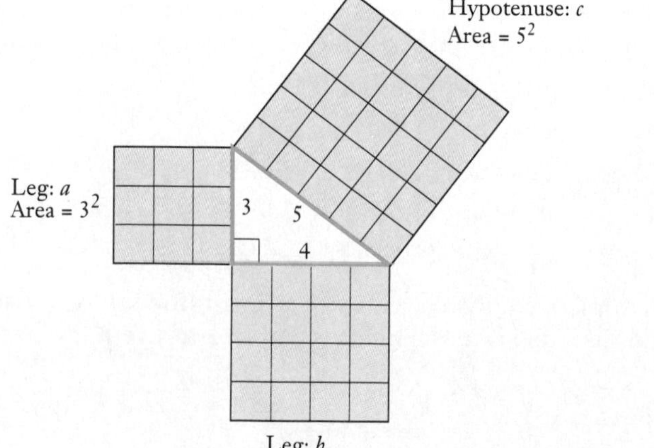

$$a^2 + b^2 = c^2$$
$$3^2 + 4^2 = 5^2$$
$$9 + 16 = 25.$$

If we know the lengths of any two sides of a right triangle, we can find the length of the third side.

▶ **EXAMPLE 7** Find the length of the hypotenuse of this right triangle. Give an exact answer and an approximation to three decimal places.

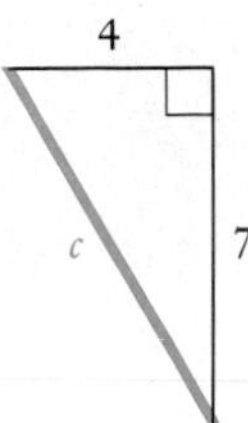

We substitute in the Pythagorean equation:

$$a^2 + b^2 = c^2$$
$$4^2 + 7^2 = c^2$$
$$16 + 49 = c^2$$
$$65 = c^2.$$

The solution of this equation is the square root of c. We approximate the square root using a calculator or Table 1.

Exact answer: $c = \sqrt{65}$

Approximate answer: $c \approx 8.062$ ◀

DO EXERCISE 28.

▶ **EXAMPLE 8** Find the length of the leg of this right triangle. Give an exact answer and an approximation to three decimal places.

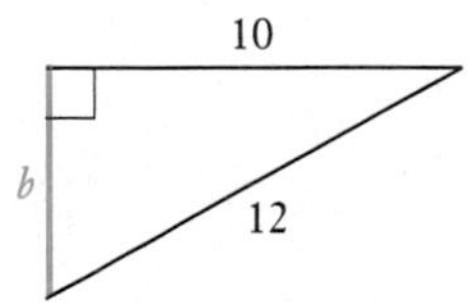

We substitute in the Pythagorean equation. Next, we solve for b^2 and then b, as follows:

$$a^2 + b^2 = c^2$$
$$10^2 + b^2 = 12^2$$
$$100 + b^2 = 144$$
$$b^2 = 144 - 100 = 44$$

Exact answer: $b = \sqrt{44}$

Approximation: $b \approx 6.633$. Using a calculator or Table 1 ◀

DO EXERCISES 29–31.

28. Find the length of the hypotenuse of this right triangle. Give an exact answer and an approximation to three decimal places. $c = \sqrt{41}$; $c \approx 6.403$

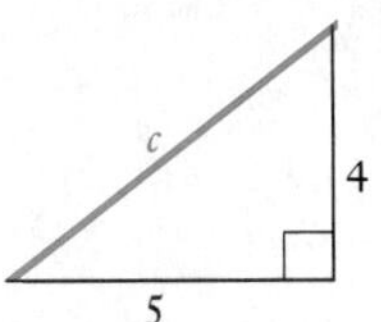

Find the length of the leg of the right triangle. Give an exact answer and an approximation to three decimal places.

29.

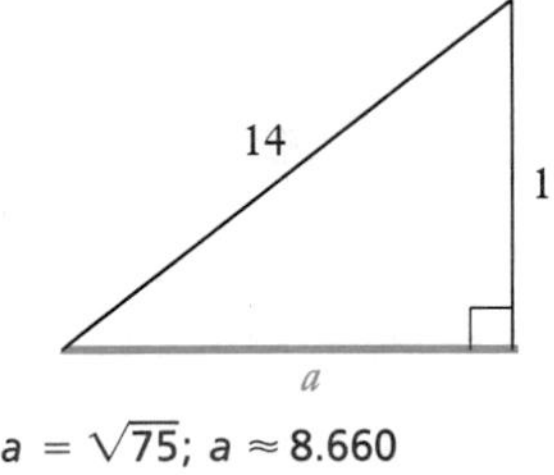

$a = \sqrt{75}$; $a \approx 8.660$

30.

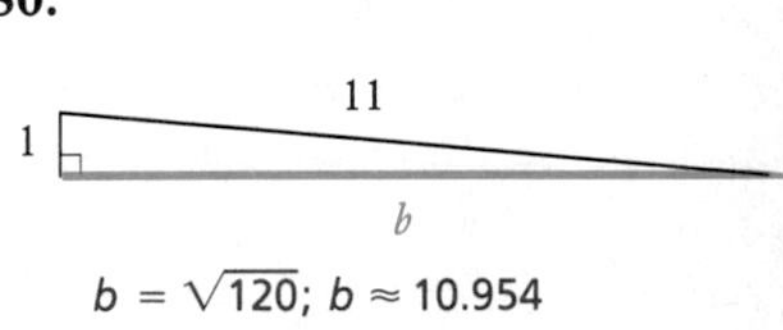

$b = \sqrt{120}$; $b \approx 10.954$

31.

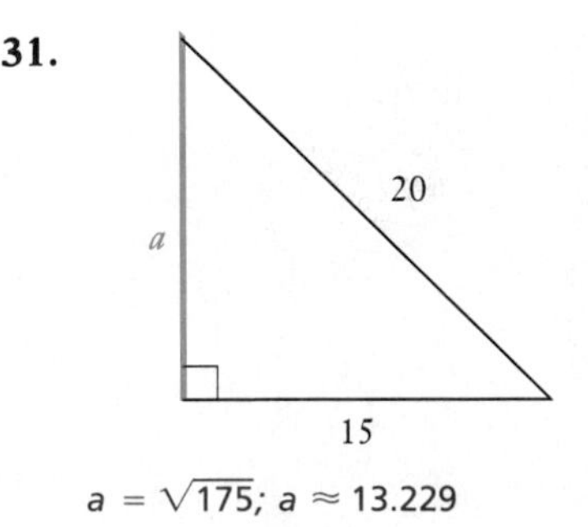

$a = \sqrt{175}$; $a \approx 13.229$

ANSWERS ON PAGE A-5

32. How long is a guy wire reaching from the top of an 18-ft pole to a point on the ground 10 ft from the pole? Give an exact answer and an approximation to three decimal places.

$\sqrt{424}$ ft $\approx$ 20.591 ft

ANSWER ON PAGE A-5

► **EXAMPLE 9** A 12-ft ladder leans against a building. The bottom of the ladder is 7 ft from the building. How high is the top of the ladder? Give an exact answer and an approximation to three decimal places.

1. *Familiarize.* We first make a drawing. In it we see a right triangle. We let h = the unknown height.

2. *Translate.* We substitute 7 for a, h for b, and 12 for c in the Pythagorean equation:

$$a^2 + b^2 = c^2 \quad \text{Pythagorean equation}$$
$$7^2 + h^2 = 12^2.$$

3. *Solve.* We solve for h^2 and then h:

$$49 + h^2 = 144$$
$$h^2 = 144 - 49$$
$$h^2 = 95$$
$$\textit{Exact answer:} \quad h = \sqrt{95}$$
$$\textit{Approximation:} \quad h \approx 9.747 \text{ ft}.$$

4. *Check:* $7^2 + (\sqrt{95})^2 = 49 + 95 = 144 = 12^2$.

5. *State.* The top of the ladder is $\sqrt{95}$, or about 9.747 ft from the ground. ◄

DO EXERCISE 32.

NAME SECTION DATE

EXERCISE SET 9.7

a Simplify.

1. $\sqrt{100}$ **2.** $\sqrt{25}$ **3.** $\sqrt{225}$ **4.** $\sqrt{441}$

5. $\sqrt{625}$ **6.** $\sqrt{576}$ **7.** $\sqrt{484}$ **8.** $\sqrt{361}$

9. $\sqrt{529}$ **10.** $\sqrt{169}$ **11.** $\sqrt{10,000}$ **12.** $\sqrt{1,000,000}$

b Approximate to three decimal places.

13. $\sqrt{48}$ **14.** $\sqrt{17}$ **15.** $\sqrt{8}$ **16.** $\sqrt{7}$

17. $\sqrt{18}$ **18.** $\sqrt{3}$ **19.** $\sqrt{6}$ **20.** $\sqrt{12}$

ANSWERS

1. 10
2. 5
3. 15
4. 21
5. 25
6. 24
7. 22
8. 19
9. 23
10. 13
11. 100
12. 1000
13. 6.928
14. 4.123
15. 2.828
16. 2.646
17. 4.243
18. 1.732
19. 2.449
20. 3.464

ANSWERS

21. 3.162

22. 4.359

23. 8.660

24. 10.488

25. 14

26. 11.269

27. 13.528

28. 12.247

29. $c = \sqrt{34}$; $c \approx 5.831$

30. $c = 17$

31. $c = \sqrt{98}$; $c \approx 9.899$

32. $c = \sqrt{32}$; $c \approx 5.657$

33. $a = 5$

34. $b = 12$

21. $\sqrt{10}$ **22.** $\sqrt{19}$ **23.** $\sqrt{75}$ **24.** $\sqrt{110}$

25. $\sqrt{196}$ **26.** $\sqrt{127}$ **27.** $\sqrt{183}$ **28.** $\sqrt{150}$

C Find the length of the third side of the right triangle. Give an exact answer and an approximation to three decimal places.

29.

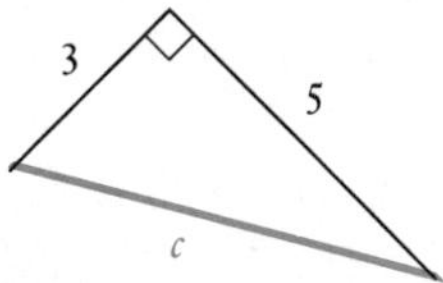

30.

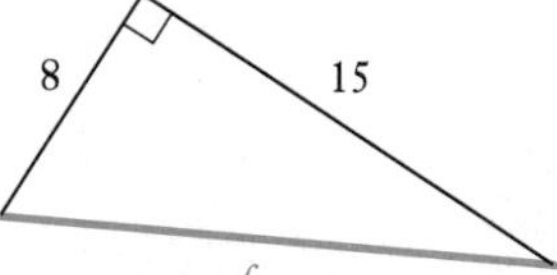

31.

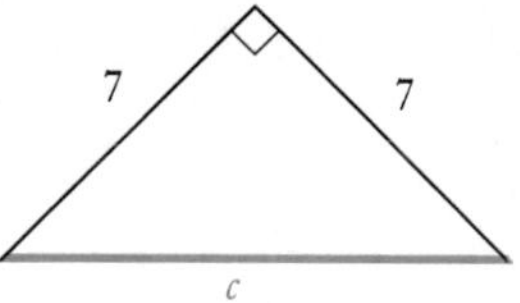

32.

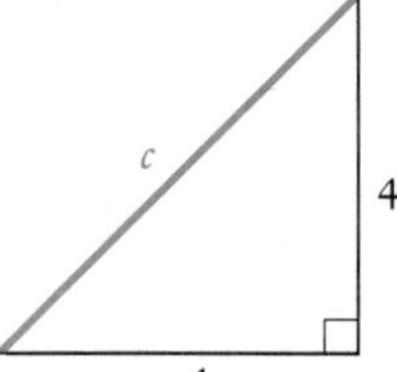

33.

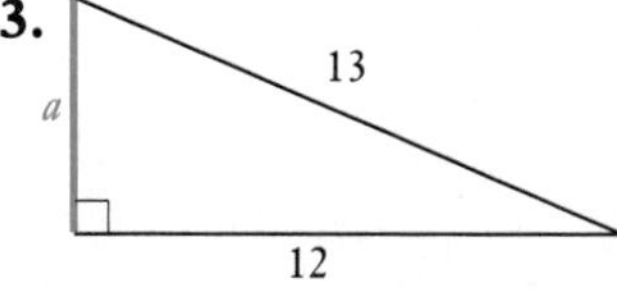

34.

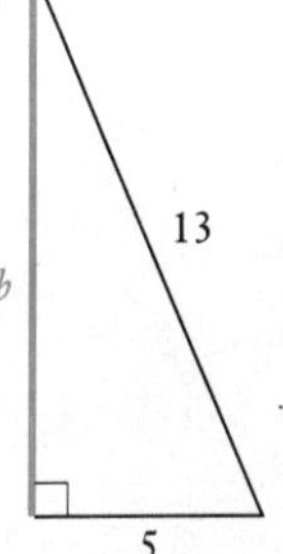

35.

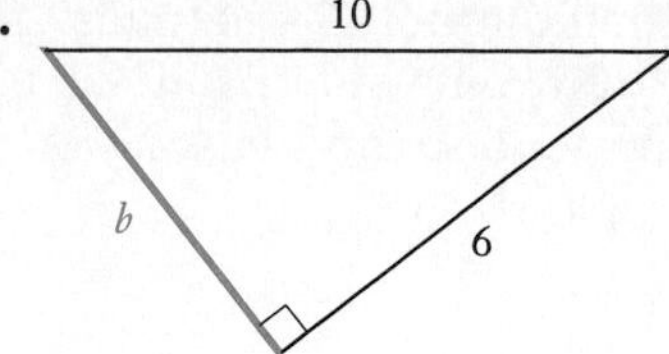

36.

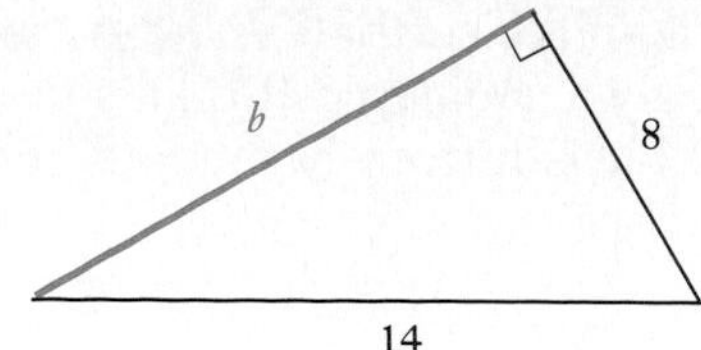

In a right triangle, find the length of the side not given. Give an exact answer and an approximation to three decimal places.

37. $a = 5, b = 12$

38. $a = 10, b = 24$

39. $a = 18, c = 30$

40. $a = 9, c = 15$

41. $b = 1, c = 20$

42. $a = 1, c = 32$

43. $a = 1, c = 15$

44. $a = 3, b = 4$

45. $a = 12, b = 5$

46. $a = 21, c = 43$

47. How long must a wire be to reach from the top of a 13-m telephone pole to a point on the ground 9 m from the base of the pole?

48. How long is a wire reaching from the top of a 12-ft pole to a point 8 ft from the pole?

ANSWERS

35. $b = 8$

36. $b = \sqrt{132}$; $b \approx 11.489$

37. $c = 13$

38. $c = 26$

39. $b = 24$

40. $b = 12$

41. $a = \sqrt{399}$; $a \approx 19.975$

42. $b = \sqrt{1023}$; $b \approx 31.984$

43. $b = \sqrt{224}$; $b \approx 14.967$

44. $c = 5$

45. $c = 13$

46. $b = \sqrt{1408}$; $b \approx 37.523$

47. $\sqrt{250}$ m ≈ 15.811 m

48. $\sqrt{208}$ ft ≈ 14.422 ft

ANSWERS

49. $\sqrt{8450}$ ft ≈ 91.924 ft

50. $\sqrt{16{,}200}$ ft ≈ 127.279 ft

51. $h = \sqrt{500}$ ft; $h \approx 22.361$ ft

52. $a = \sqrt{39}$ ft; $a \approx 6.245$ ft

53. 0.456

54. 0.1634

55. 1.23

56. 0.99

57. The areas are the same.

58. Length: 15.2 in.; width: 11.4 in.

49. A slow-pitch softball diamond is actually a square 65 ft on a side. How far is it from home to second base?

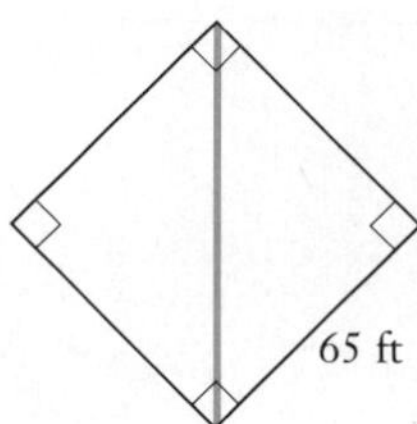

50. A baseball diamond is actually a square 90 ft on a side. How far is it from home to second base?

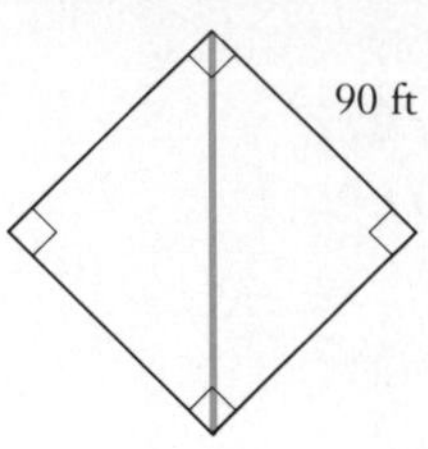

51. How high is this tree?

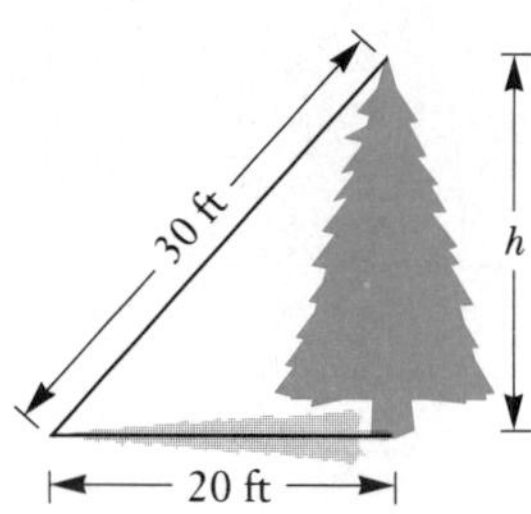

52. How far is the base of the fence post from point *A*?

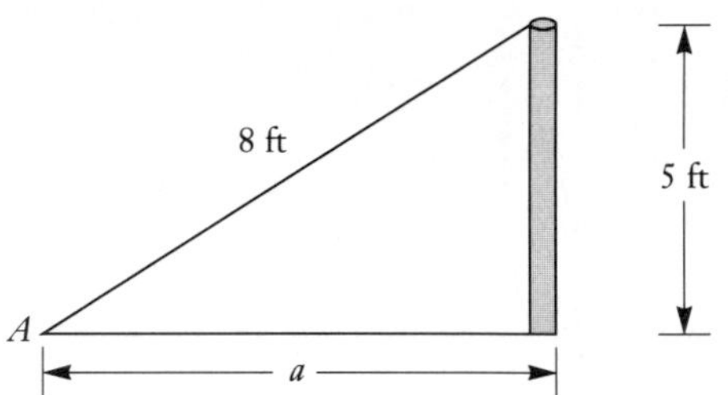

SKILL MAINTENANCE

Convert to decimal notation.

53. 45.6%

54. 16.34%

55. 123%

56. 99%

SYNTHESIS

57. Which of the triangles below has the larger area?

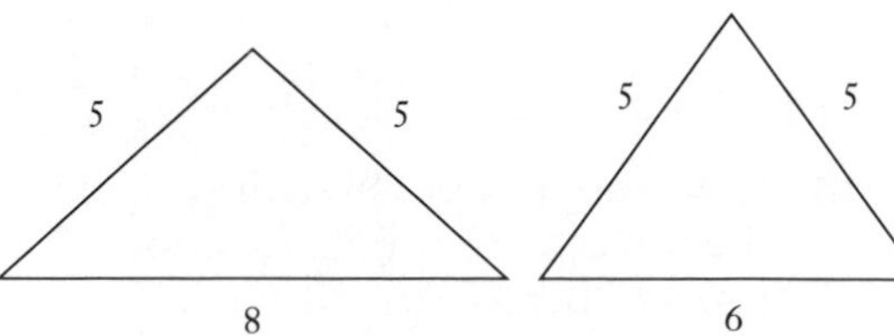

58. A 19-in. television set has a rectangular screen whose diagonal is 19 in. The ratio of length to width in any television set is 4 to 3. Find the length and the width of a 19-in. set.

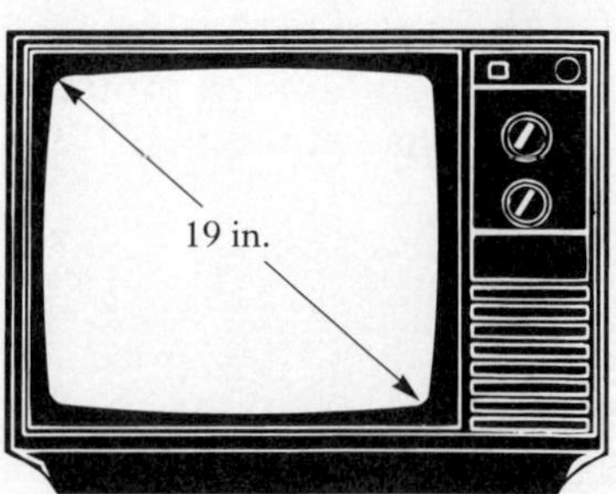

SUMMARY AND REVIEW: CHAPTER 9

IMPORTANT PROPERTIES AND FORMULAS

American Units of Length:	12 in. = 1 ft; 3 ft = 1 yd; 5280 ft = 1 mi
Metric Units of Length:	1 km = 1000 m; 1 hm = 100 m; 1 dam = 10 m
	1 dm = 0.1 m; 1 cm = 0.01 m; 1 mm = 0.001 m
American–Metric Conversion:	1 m = 39.37 in.; 1 m = 3.3 ft; 2.54 cm = 1 in.
	1 km = 0.621 mi; 1.609 km = 1 mi
Perimeter of a Rectangle:	$P = 2 \cdot (l + w)$, or $P = 2 \cdot l + 2 \cdot w$
Perimeter of a Square:	$P = 4 \cdot s$
Area of a Rectangle:	$A = l \cdot w$
Area of a Square:	$A = s \cdot s$, or $A = s^2$
Area of a Parallelogram:	$A = b \cdot h$
Area of a Triangle:	$A = \frac{1}{2} b \cdot h$
Area of a Trapezoid:	$A = \frac{1}{2} h \cdot (a + b)$
Radius and Diameter of a Circle:	$d = 2 \cdot r$
Circumference of a Circle:	$C = 2 \cdot \pi \cdot r$
Area of a Circle:	$A = \pi \cdot r \cdot r$, or $A = \pi \cdot r^2$
Pythagorean Equation:	$a^2 + b^2 = c^2$

REVIEW EXERCISES

The review sections and objectives to be tested in addition to the material in this chapter are [7.1a, b] and [7.2a, b].

Complete.

1. 8 ft = ______ yd

2. $\frac{5}{6}$ yd = ______ in.

3. 0.3 mm = ______ cm

4. 4 m = ______ km

5. 2 yd = ______ in.

6. 4 km = ______ cm

7. 14 in. = ______ ft

8. 15 cm = ______ m

Find the perimeter.

9.

5 m, 7 m, 4 m, 4 m, 3 m

10.

0.5 m, 1.9 m, 1.2 m, 0.8 m

11. The dimensions of a standard-sized tennis court are 78 ft by 36 ft. Find the perimeter of the tennis court.

Find the area.

12.

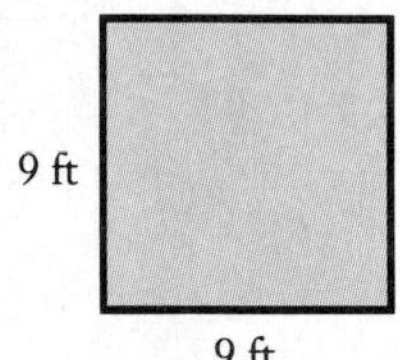

13.

14.

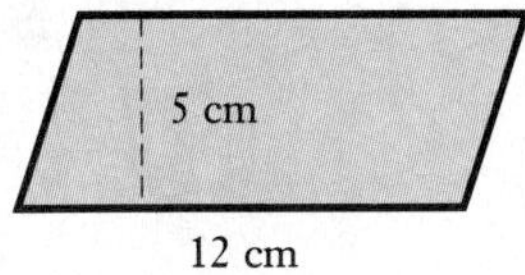

15.

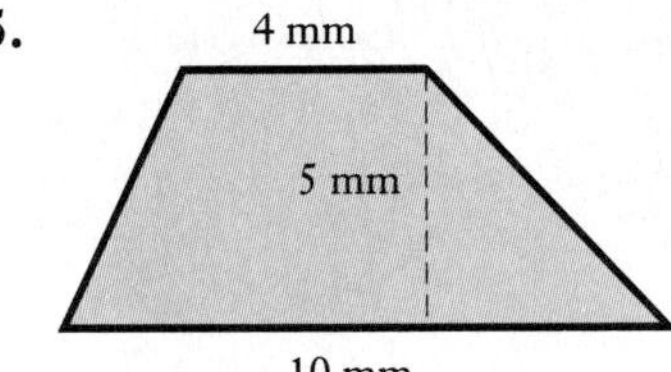

16.

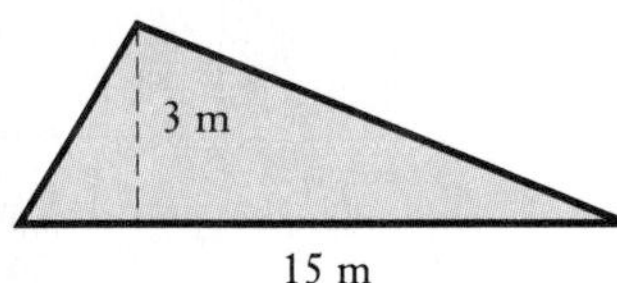

17.

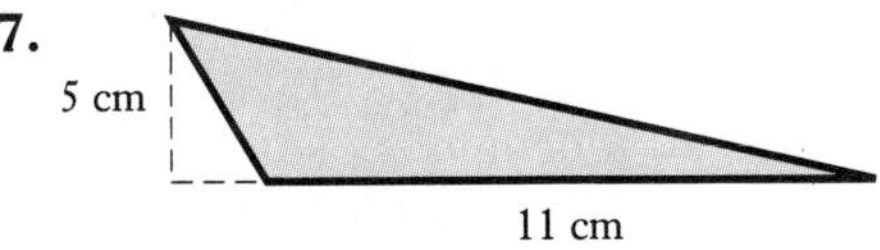

18.

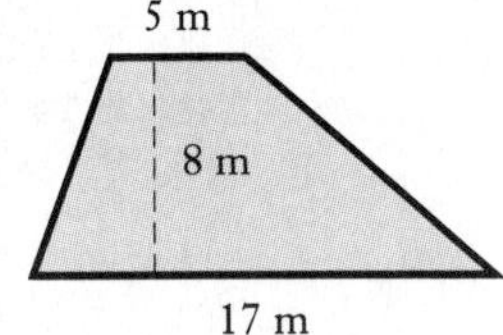

19.

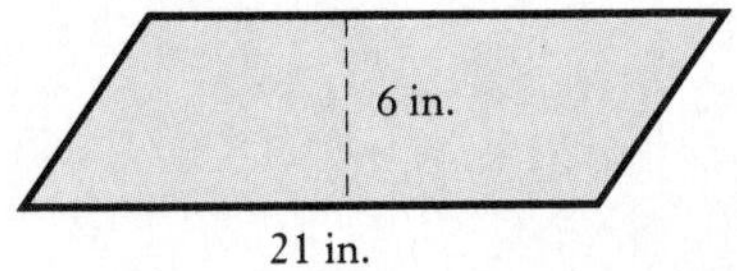

20. A sidewalk is built around three sides of a building and has equal width on the three sides, as shown at right. What is the area of the sidewalk?

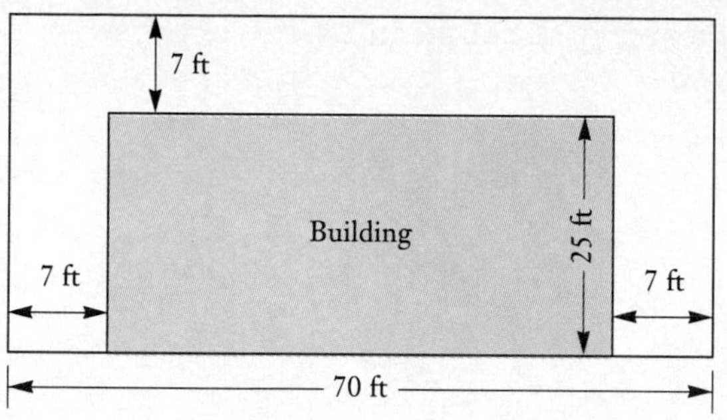

Find the length of a radius of the circle.

21.

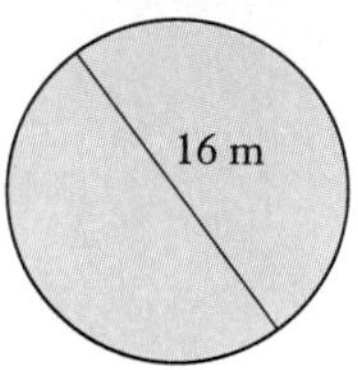

22.

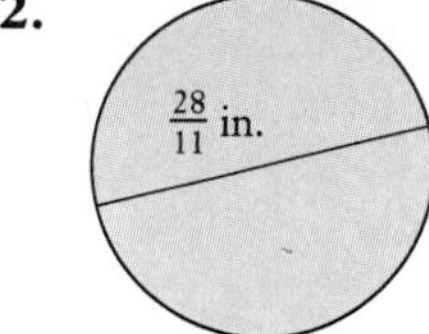

Find the length of a diameter of the circle.

23.

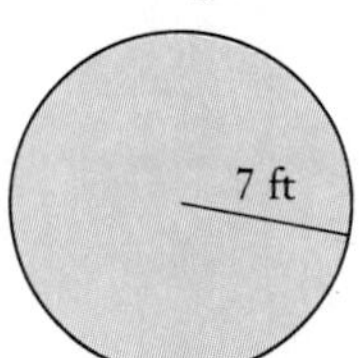

24.

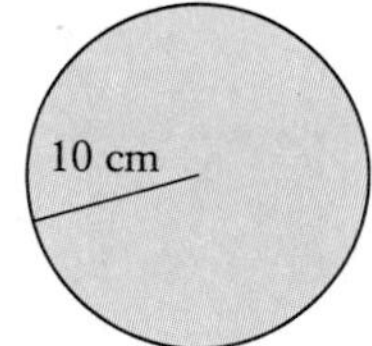

25. Find the circumference of the circle in Exercise 21. Use 3.14 for π.

26. Find the circumference of the circle in Exercise 22. Use $\frac{22}{7}$ for π.

27. Find the area of the circle in Exercise 21. Use 3.14 for π.

28. Find the area of the circle in Exercise 22. Use $\frac{22}{7}$ for π.

29. Find the area of the shaded region. Use 3.14 for π.

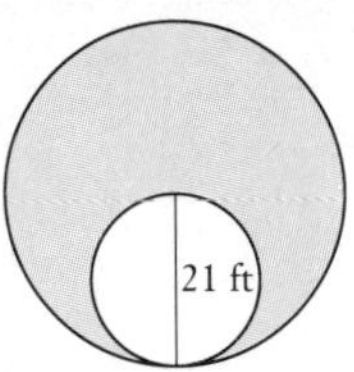

30. Simplify: $\sqrt{64}$.

In a right triangle, find the length of the side not given. Find an exact answer and an approximation to three decimal places.

31. $a = 15, b = 25$

32. $a = 7, c = 10$

33.

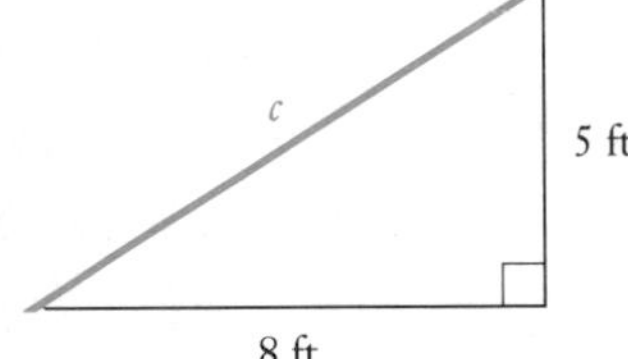

34.

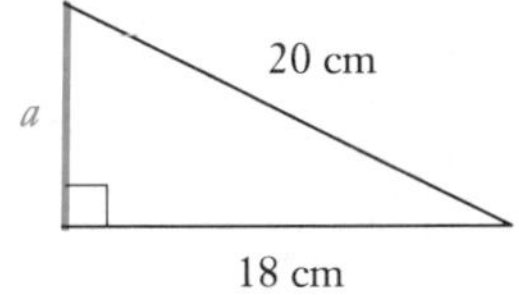

SKILL MAINTENANCE

35. Convert to percent notation: 0.47.

36. Convert to percent notation: $\frac{23}{25}$.

37. Convert to decimal notation: 56.7%.

38. Convert to fractional notation: 73%.

❖ THINKING IT THROUGH

1. List as many reasons as you can for using the metric system exclusively.
2. List as many reasons as you can for continuing our use of the American system.
3. Napoleon is credited with influencing the use of the metric system. Research this possibility and make a report.

NAME SECTION DATE

TEST: CHAPTER 9

Complete.

1. 4 ft = ______ in.

2. 4 in. = ______ ft

3. 6 km = ______ m

4. 8.7 mm = ______ cm

5. Find the perimeter.

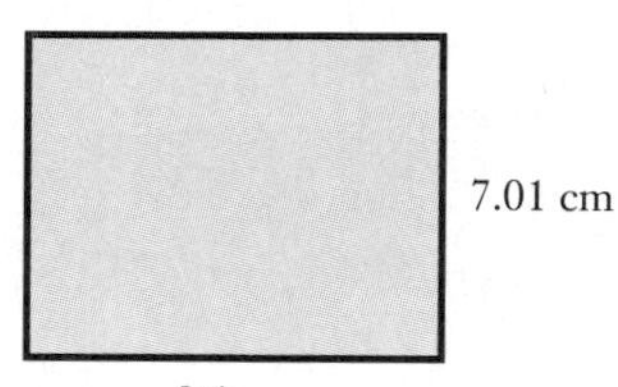

6. Find the area.

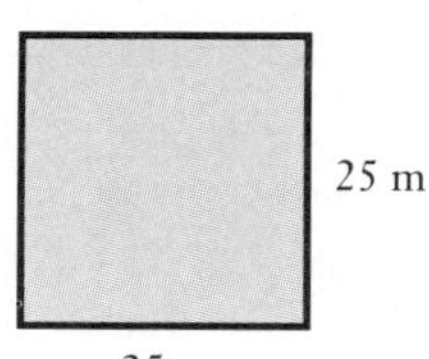

Find the area.

7.

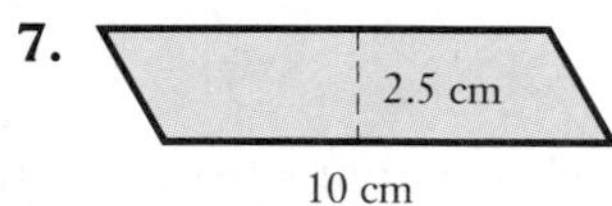

8.

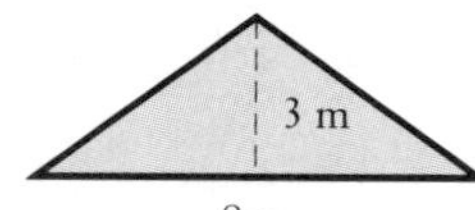

9.

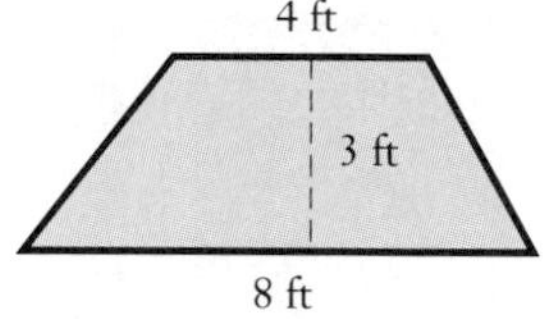

10. Find the length of a diameter of this circle.

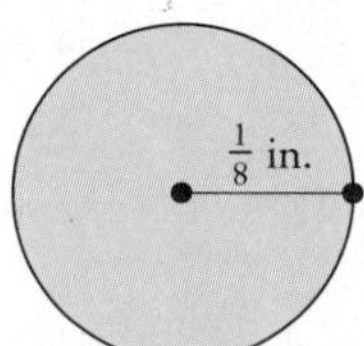

11. Find the length of a radius of this circle.

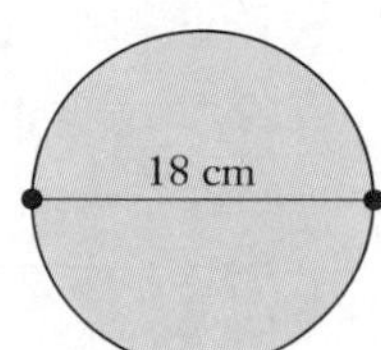

ANSWERS

1. [9.1a] 48

2. [9.1a] $\frac{1}{3}$

3. [9.2a] 6000

4. [9.2a] 0.87

5. [9.3a] 32.82 cm

6. [9.4a] 625 m^2

7. [9.5a] 25 cm^2

8. [9.5a] 12 m^2

9. [9.5a] 18 ft^2

10. [9.6a] $\frac{1}{4}$ in.

11. [9.6a] 9 cm

ANSWERS

12. [9.6b] $\frac{11}{14}$ in.

13. [9.6c] 254.34 cm^2

14. [9.6d] 103.815 km^2

15. [9.7a] 15

16. [9.7b] 9.327

17. [9.7c] $c = 40$

18. [9.7c] $b = \sqrt{60}$; $b \approx 7.746$

19. [9.7c] $c = \sqrt{2}$; $c \approx 1.414$

20. [9.7c] $b = \sqrt{51}$; $b \approx 7.141$

21. [7.1c] 93%

22. [7.2a] 81.25%

23. [7.1b] 0.932

24. [7.2b] $\frac{1}{3}$

12. Find the circumference of the circle in Exercise 10. Use $\frac{22}{7}$ for π.

13. Find the area of the circle in Exercise 11. Use 3.14 for π.

14. Find the area of the shaded region.

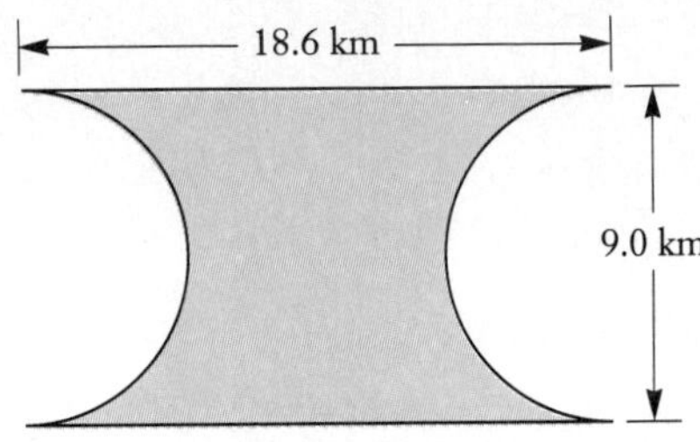

15. Simplify: $\sqrt{225}$.

16. Approximate to three decimal places: $\sqrt{87}$.

In a right triangle, find the length of the side not given. Find an exact answer and an approximation to three decimal places.

17. $a = 24$, $b = 32$

18. $a = 2$, $c = 8$

19.

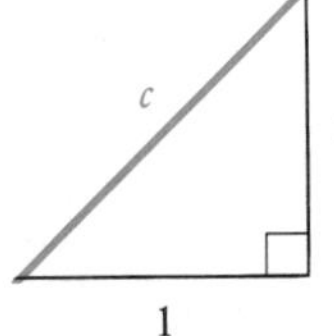

20.

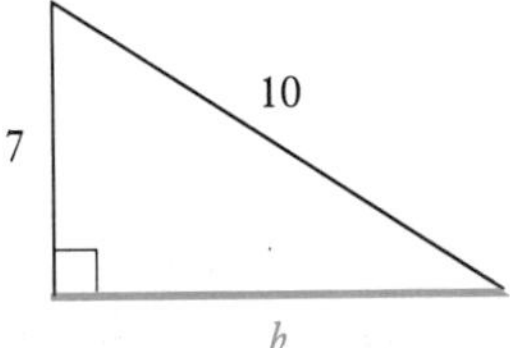

SKILL MAINTENANCE

21. Convert to percent notation: 0.93.

22. Convert to percent notation: $\frac{13}{16}$.

23. Convert to decimal notation: 93.2%.

24. Convert to fractional notation: $33\frac{1}{3}\%$.

CUMULATIVE REVIEW: CHAPTERS 1–9

Perform the indicated operation and simplify.

1. $46{,}231 \times 1100$

2. $\frac{1}{10} \cdot \frac{5}{6}$

3. $14.5 + \frac{4}{5} - 0.1$

4. $2\frac{3}{5} \div 3\frac{9}{10}$

5. $0.1\overline{)3.56}$

6. $3\frac{1}{2} - 2\frac{2}{3}$

7. Determine whether 1,298,032 is divisible by 8.

8. Determine whether 5,024,120 is divisible by 3.

9. Find the prime factorization of 99.

10. Find the LCM of 35 and 49.

11. Round $35.\overline{7}$ to the nearest tenth.

12. Write a word name for 103.064.

13. Find the average and median of this set of numbers:

9, 13, 17, 18, 21, 29.

Find percent notation.

14. 0.08

15. $\frac{3}{5}$

16. Simplify: $\sqrt{121}$.

17. Approximate to two decimal places: $\sqrt{29}$.

18. Complete: 2 yd = ______ ft.

19. Find the perimeter.

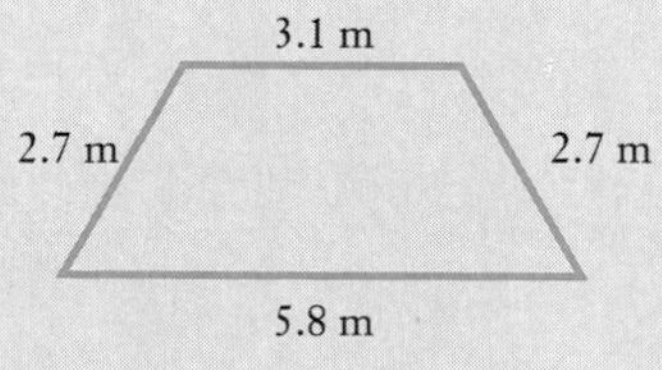

20. Find the area.

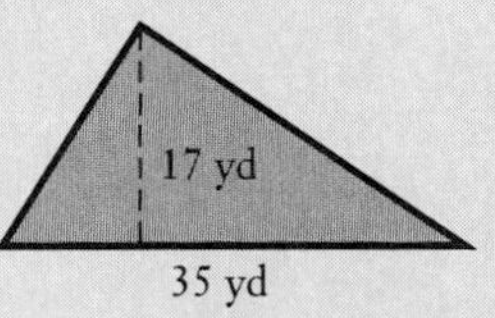

Solve.

21. $0.07 \cdot x = 10.535$

22. $x + 12{,}843 = 32{,}091$

23. $\frac{2}{3} \cdot y = 5$

24. $\frac{4}{5} + y = \frac{6}{7}$

This table shows typical sleep requirements in childhood.

Age	Hours of daytime sleep	Hours of nighttime sleep
1 week	8.0	8.5
1 month	7.0	8.5
3 months	5.5	9.5
6 months	3.3	11.0
9 months	2.5	11.5
12 months	2.3	11.5
18 months	2.0	11.5
2 years	1.5	11.5

25. How many hours of daytime sleep does a 3-month-old child need?

26. How many total hours of sleep does a 1-year-old child need?

27. How many more hours will a 6-month-old child sleep at night than a 1-week-old child?

28. How many hours would you expect a 2-month-old child to sleep at night?

Solve.

29. An item marked $220 was discounted to a sale price of $194. What was the rate of discount?

30. There are 11 million milk cows in America, each producing on the average 15,000 lb of milk per year. How many pounds of milk are produced each year in America?

31. A family has a rectangular kitchen table measuring 52 in. by 30 in. They replace it with a circular table with a 48-in. diameter. How much bigger is the new table? Use 3.14 for π.

32. A person on a diet loses $3\frac{1}{2}$ lb in 2 weeks. At this rate, how many pounds will be lost in 5 weeks?

33. The U.S. Department of Agriculture requires that 80% of the seeds that a company produces must sprout. To find out about the quality of the seeds it has produced, a company takes 500 seeds and plants them. It finds that 417 of the seeds sprout. Did the seeds pass government standards?

34. A mechanic spent $\frac{1}{3}$ hr changing a car's oil, $\frac{1}{2}$ hr rotating the tires, $\frac{1}{10}$ hr changing the air filter, $\frac{1}{4}$ hr adjusting the idle speed, and $\frac{1}{15}$ hr checking the brake and transmission fluids. How many hours did the mechanic spend working on the car?

35. A driver bought gasoline when the odometer read 86,897.2. At the next gasoline purchase, the odometer read 87,153.0. How many miles were driven?

36. A family has an annual income of $16,400. Of this, $\frac{1}{4}$ is spent for food. How much does the family spend for food?

SYNTHESIS

37. A homeowner is having a one-story addition built to an existing house. The addition measures 25 ft by 32 ft. The existing house measures 30 ft by 32 ft and has two stories. What is the percent increase in living area provided by the addition?

INTRODUCTION Several other measures considered in this chapter are volume, capacity, weight, mass, time, temperature, and units of area. In terms of the remainder of the text, any of these topics can be omitted without loss of continuity to the last two chapters.

The review sections to be tested in addition to the material in this chapter are 1.9, 7.8, 9.1, and 9.2. ❖

More on Measures

AN APPLICATION

How "big" is one million dollars? This photo shows one million one-dollar bills assembled by the Bureau of Engraving. The width of a dollar bill is 2.3125 in., the length is 6.0625 in., and the thickness is 0.0041 in. Find the volume occupied by one million one-dollar bills.

THE MATHEMATICS

To find the volume of a single one-dollar bill, we multiply the length times the width and then by the height of thickness:

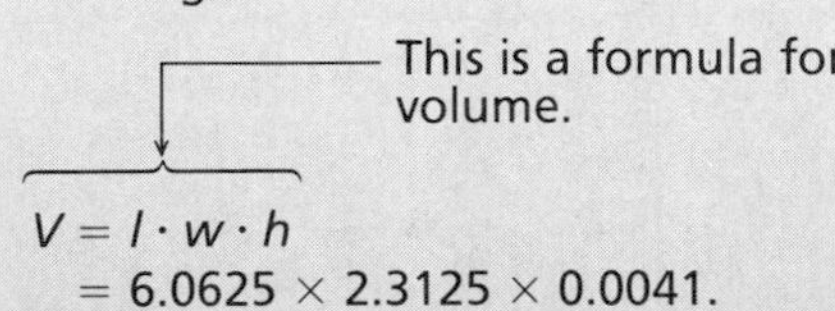

$$V = l \cdot w \cdot h$$
$$= 6.0625 \times 2.3125 \times 0.0041.$$

Then we multiply the result by one million.

❖ POINTS TO REMEMBER: CHAPTER 10

Unit Conversion Skills: Chapter 9

PRETEST: CHAPTER 10

Complete.

1. 2304 mL = ______ L

2. 2.4 L = ______ mL

3. 5 lb = ______ oz

4. 4.4 T = ______ lb

5. 4.8 kg = ______ g

6. 6.2 mg = ______ cg

7. 3400 mg = ______ g

8. 7 hr = ______ min

9. 16 day = ______ hr

10. 128 pt = ______ qt

11. 20 gal = ______ oz

12. 3 cups = ______ oz

13. Convert 77°F to Celsius.

14. Convert 37°C to Fahrenheit.

Complete.

15. 1 ft^2 = ______ in^2

16. 2 km^2 = ______ m^2

17. A bundle of test booklets weighs 1 kg. How many grams does 1 bundle weigh?

Find the volume. Use 3.14 for π.

18.

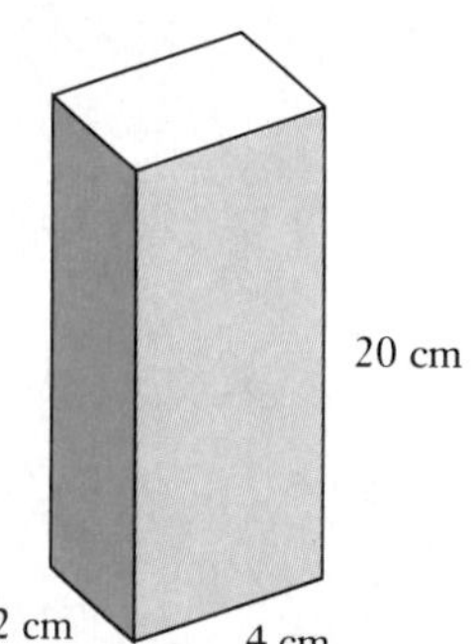

19.

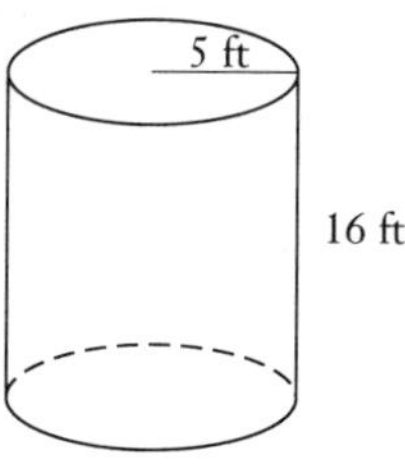

20.

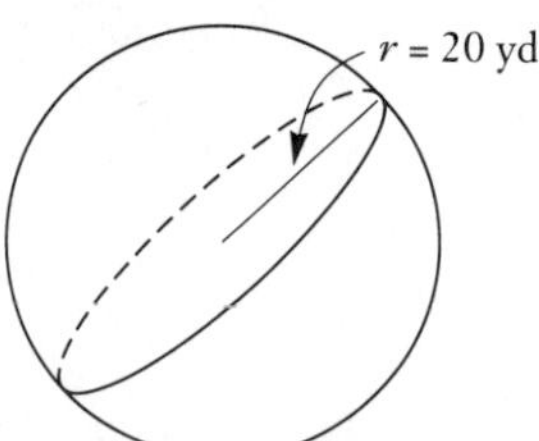

21.

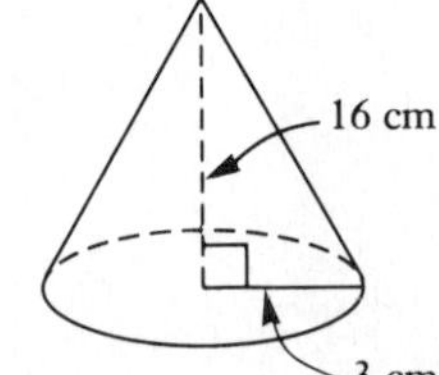

10.1 Volume and Capacity

a Volume

The **volume** of a **rectangular solid** is the number of unit cubes needed to fill it.

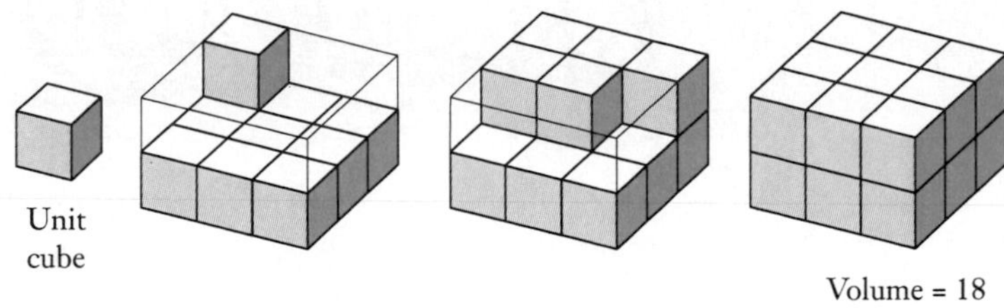

Two other units are shown below.

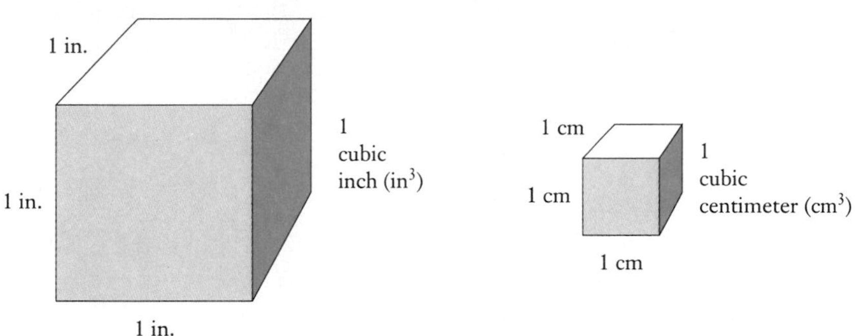

▶ **EXAMPLE 1** Find the volume.

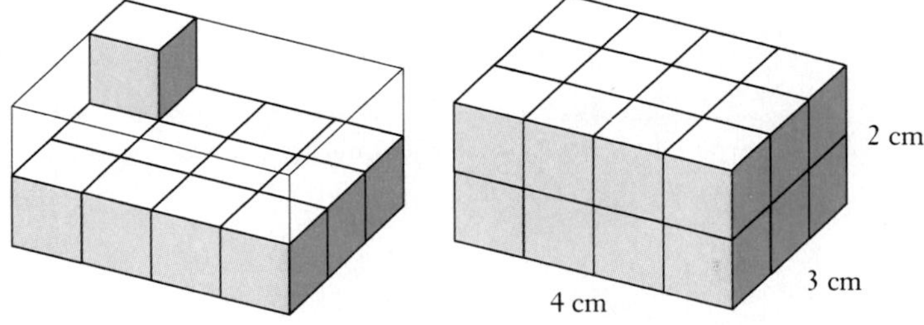

The figure is made up of 2 layers of 12 cubes each, so its volume is 24 cubic centimeters (cm^3). ◀

DO EXERCISE 1.

> **The volume of a rectangular solid is found by multiplying length by width by height:**
>
> $$V = l \cdot w \cdot h.$$

OBJECTIVES

After finishing Section 10.1, you should be able to:

a Find the volume of a rectangular solid using the

b Convert from one unit of capacity to another.

c Solve problems involving capacity.

FOR EXTRA HELP

 Tape 12C

 Tape 14A

 MAC: 10 IBM: 10

1. Find the volume. 12 cm^3

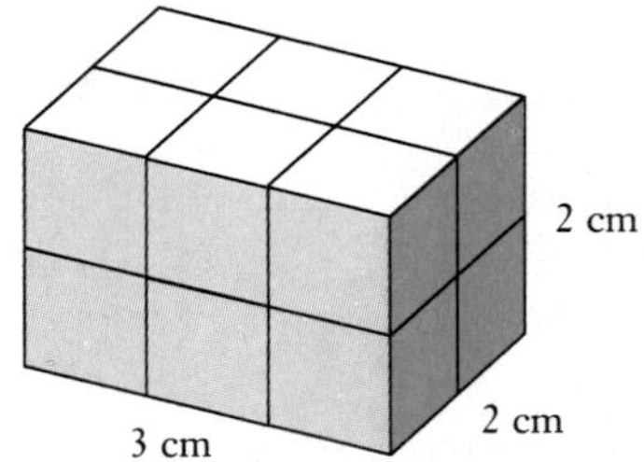

ANSWER ON PAGE A-5

2. Find the volume. 38.4 m^3

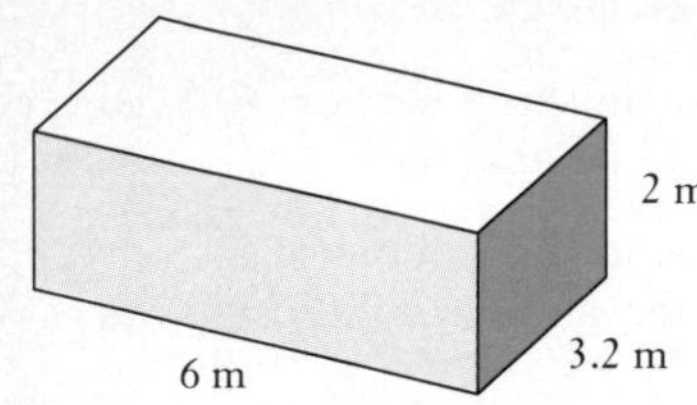

3. A cord of wood is 4 ft by 4 ft by 8 ft. What is the volume of a cord of wood? 128 ft^3

Complete.

4. 5 gal = __40__ pt

5. 80 qt = __20__ gal

ANSWERS ON PAGE A-5

▶ **EXAMPLE 2** The largest sized piece of luggage that you can carry on an airplane is 21 in. by 8 in. by 16 in. Find the volume of this solid.

$$\begin{aligned} V &= l \cdot w \cdot h \\ &= 21 \text{ in.} \cdot 8 \text{ in.} \cdot 16 \text{ in.} \\ &= 168 \cdot 16 \text{ in}^3 \\ &= 2688 \text{ in}^3 \end{aligned}$$

16 in. 21 in. 8 in. ◀

DO EXERCISES 2 AND 3.

b Capacity

To answer a question like "How much soda is in the bottle?" we need measures of **capacity.** American units of capacity are cups, pints, quarts, and gallons. These units are related as follows.

> American Units of Capacity
>
> **1 gallon (gal) = 4 quarts (qt)**
> **1 qt = 2 pints (pt)**
> **1 pt = 2 cups = 16 ounces (oz)**
> **1 cup = 8 ounces**

▶ **EXAMPLE 3** Complete: 9 gal = ______ oz.

We convert as follows:

$$\begin{aligned} 9 \text{ gal} &= 9 \cdot 1 \text{ gal} \\ &= 9 \cdot 4 \text{ qt} && \text{Substituting 4 qt for 1 gal} \\ &= 9 \cdot 4 \cdot 1 \text{ qt} \\ &= 9 \cdot 4 \cdot 2 \text{ pt} && \text{Substituting 2 pt for 1 qt} \\ &= 9 \cdot 4 \cdot 2 \cdot 1 \text{ pt} \\ &= 9 \cdot 4 \cdot 2 \cdot 16 \text{ oz} && \text{Substituting 16 oz for 1 pt} \\ &= 1152 \text{ oz.} \end{aligned}$$

◀

▶ **EXAMPLE 4** Complete: 24 qt = ______ gal.

In this case, we multiply by 1 using 1 gal in the numerator, since we are converting to gallons, and 4 qt in the denominator, since we are converting from quarts.

$$24 \text{ qt} = 24 \text{ qt} \cdot \frac{1 \text{ gal}}{4 \text{ qt}} = \frac{24}{4} \cdot 1 \text{ gal} = 6 \text{ gal}$$ ◀

DO EXERCISES 4 AND 5.

The metric system has a unit of capacity called a **liter.** A liter is just a bit more than a quart. It is defined as follows.

Metric Units of Capacity

1 liter (L) = 1000 cubic centimeters (1000 cm^3)
= 1 cubic decimeter (dm^3)
The script letter ℓ is also used for "liter."

The metric prefixes are also used with liters. The most common is **milli.** The milliliter (mL) is, then, $\frac{1}{1000}$ liter. Thus,

1 L = 1000 mL = 1000 cm^3;
0.001 L = 1 mL = 1 cm^3.

A preferred unit for drug dosage is the milliliter (mL) or the cubic centimeter (cm^3). The notation "cc" is also used for cubic centimeter, especially in medicine. The milliliter and the cubic centimeter are the same size.

1 mL = 1 cm^3 = 1 cc

A milliliter is about $\frac{1}{5}$ of a teaspoon.

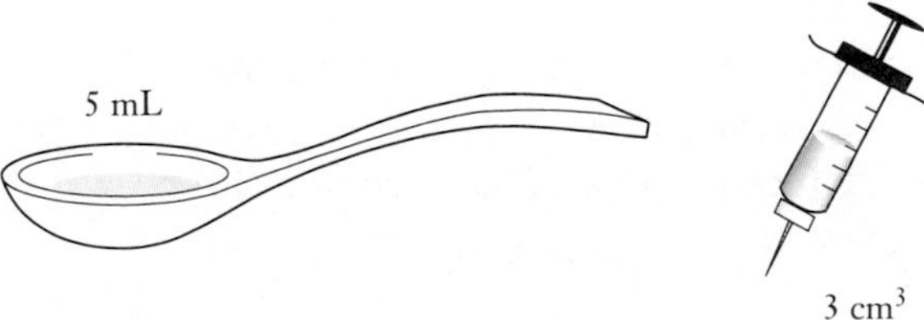

Volumes for which quarts and gallons are used are expressed in liters. Large volumes in business and industry use cubic meters (m^3).

DO EXERCISES 6–9.

Complete with mL or L.

6. The patient got an injection of 2 ___mL___ of penicillin.

7. There are 250 ___mL___ in a coffee cup.

8. The gas tank holds 80 ___L___.

9. Bring home 8 ___L___ of milk.

ANSWERS ON PAGE A-5

Complete.

10. 0.97 L = ___970___ mL

11. 8990 mL = ___8.99___ L

12. A physician ordered 4800 mL of 0.9% saline solution. How many liters were ordered? 4.8

13. A prescription calls for 4 oz of ephedrine.

a) How many milliliters is the prescription? 118.28 mL

b) How many liters is the prescription? 0.11828 L

14. At the same station, the price of lead-free gasoline is 39.9 cents a liter. Estimate the price of 1 gallon in dollars. $1.60

ANSWERS ON PAGE A-5

▶ **EXAMPLE 5** Complete: 4.5 L = ______ mL.

$$\begin{aligned} 4.5\text{ L} &= 4.5 \times (1\text{ L}) \\ &= 4.5 \times (1000\text{ mL}) \quad \text{Substituting 1000 mL for 1 L} \\ &= 4500\text{ mL} \end{aligned}$$

◀

▶ **EXAMPLE 6** Complete: 280 mL = ______ L.

$$\begin{aligned} 280\text{ mL} &= 280 \times (1\text{ mL}) \\ &= 280 \times (0.001\text{ L}) \quad \text{Substituting 0.001 L for 1 mL} \\ &= 0.28\text{ L} \end{aligned}$$

◀

DO EXERCISES 10 AND 11.

C Solving Problems

The metric system has extensive usage in medicine.

▶ **EXAMPLE 7** A physician ordered 3.5 L of 5% dextrose in water. How many milliliters were ordered?

We convert 3.5 L to milliliters:

$$\begin{aligned} 3.5\text{ L} &= 3.5 \times (1\text{ L}) \\ &= 3.5 \times (1000\text{ mL}) \\ &= 3500\text{ mL.} \end{aligned}$$

◀

DO EXERCISE 12.

▶ **EXAMPLE 8** In pharmaceutical work, liquids at the drugstore are given in liters or milliliters, but a physician's prescription is given in ounces. For conversion, a druggist knows that 1 oz = 29.57 mL. A prescription calls for 3 oz of ephedrine. How many milliliters is the prescription?

We convert as follows:

$$\begin{aligned} 3\text{ oz} &= 3 \times (1\text{ oz}) \\ &= 3 \times (29.57\text{ mL}) \quad \text{Substituting 29.57 mL for 1 oz} \\ &= 88.71\text{ mL.} \end{aligned}$$

◀

DO EXERCISE 13.

▶ **EXAMPLE 9** At a certain gasoline station, regular gasoline sold for 27.3¢ a liter. Estimate the cost of 1 gallon in dollars.

Since 1 liter is about 1 quart and there are 4 quarts in a gallon, the price of a gallon is about 4 times the price of a liter:

$$4 \times 27.3¢ = 109.2¢ = \$1.092.$$

Thus regular gasoline is about $1.09 a gallon. ◀

DO EXERCISE 14.

NAME SECTION DATE

EXERCISE SET 10.1

a Find the volume.

1.

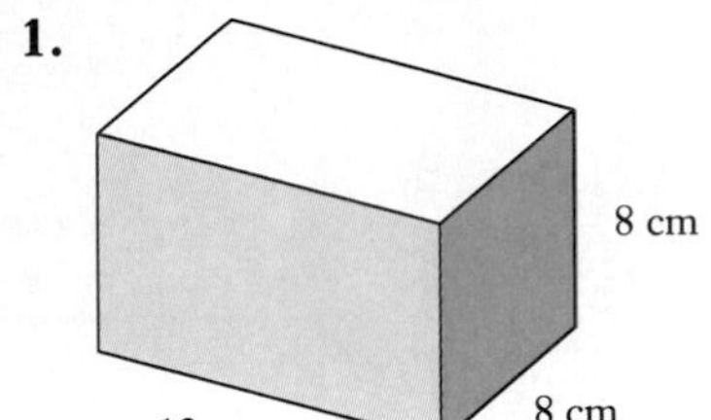

2.

3.

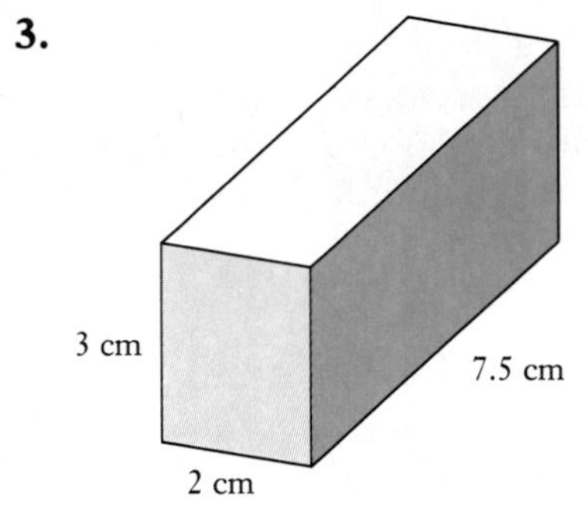

4.

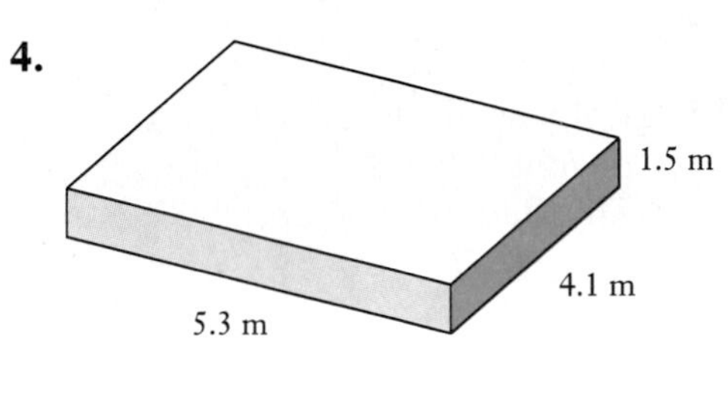

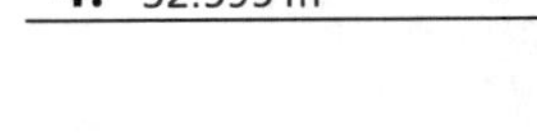

5.

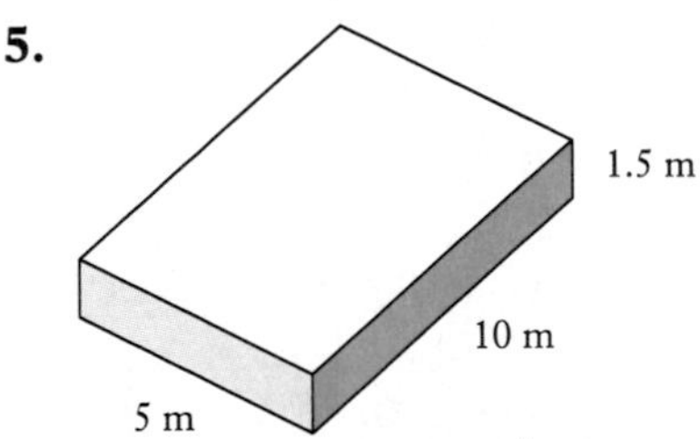

6.

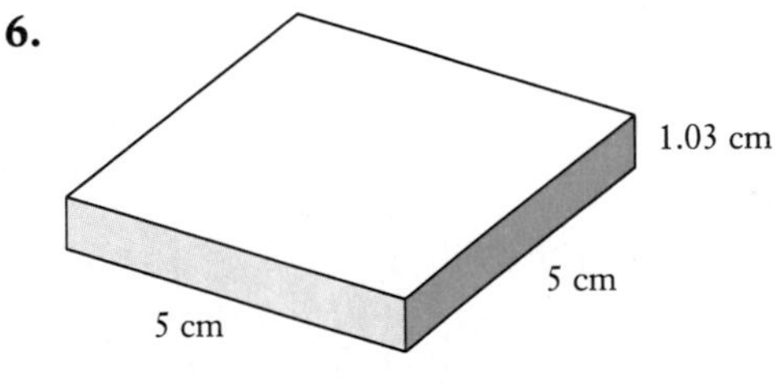

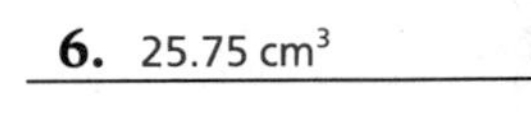

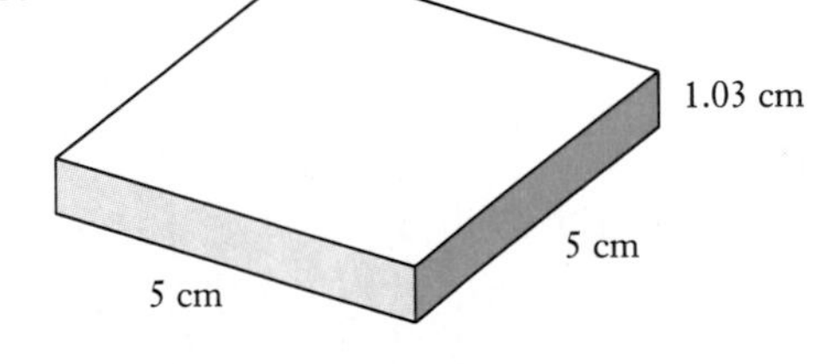

7.

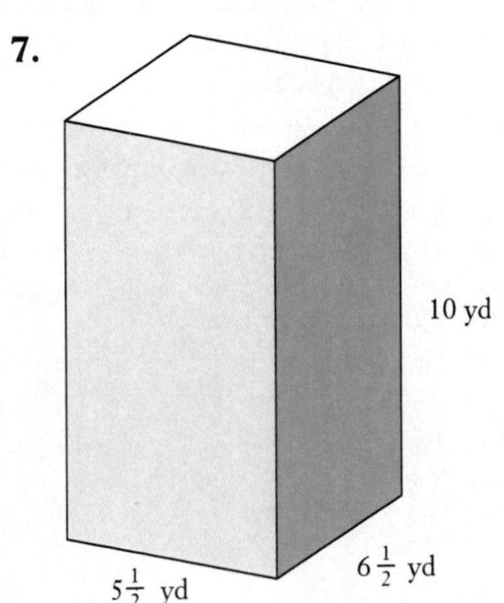

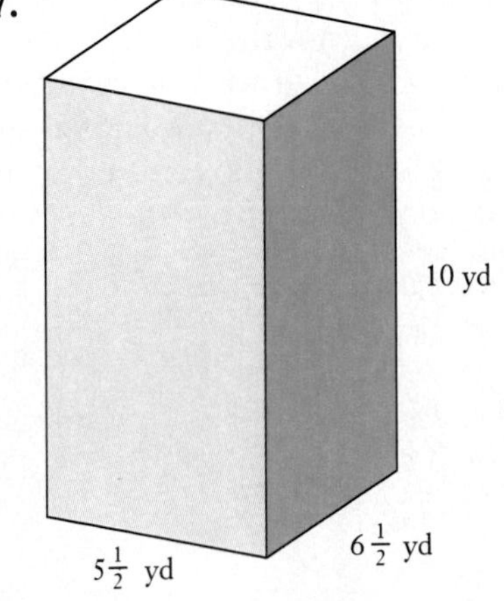

8.

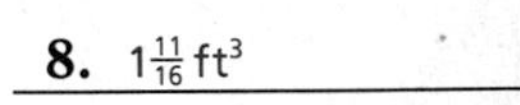

b Complete.

9. 1 L = ______ mL = ______ cm^3

10. ______ L = 1 mL = ______ cm^3

11. 87 L = ______ mL

12. 901 L = ______ mL

ANSWERS

1. 768 cm^3

2. 0.512 m^3

3. 45 cm^3

4. 32.595 m^3

5. 75 m^3

6. 25.75 cm^3

7. $357\frac{1}{2}$ yd^3

8. $1\frac{11}{16}$ ft^3

9. 1000; 1000

10. 0.001; 1

11. 87,000

12. 901,000

ANSWERS

13. 0.049
14. 0.017
15. 0.000401
16. 0.000013
17. 78,100
18. 49,200
19. 320
20. 2
21. 128
22. 10
23. 32
24. 16
25. 500 mL
26. 2.016 L
27. 125 mL
28. 16 oz
29. 5832 yd^3
30.
31. $39
32. $96
33. 1000
34. 225
35. 57,480 in^3; 33.3 ft^3

13. 49 mL = ______ L

14. 17 mL = ______ L

15. 0.401 mL = ______ L

16. 0.013 mL = ______ L

17. 78.1 L = ______ cm^3

18. 49.2 L = ______ cm^3

19. 10 qt = ______ oz

20. 32 oz = ______ pt

21. 1 gal = ______ oz

22. 20 cups = ______ pt

23. 8 gal = ______ qt

24. 1 gal = ______ cups

c Solve.

25. A physician ordered 0.5 L of normal saline solution. How many milliliters were ordered?

26. A patient receives 84 mL per hour of normal saline solution. How many liters did the patient receive in a 24-hr period?

27. A doctor wants a patient to receive 3 L of a normal saline solution in a 24-hr period. How many milliliters per hour must the nurse administer?

28. A doctor tells a patient to purchase 0.5 L of hydrogen peroxide. Commercially, hydrogen peroxide is found on the shelf in bottles that hold 4 oz, 8 oz, and 16 oz. Which bottle comes closest to filling the prescription? (1 qt = 32 oz)

29. If all the gold in the world could be gathered together, it would form a cube 18 yd on a side. Find the volume of the world's gold.

30. Many people leave the water running while brushing their teeth. Suppose that 32 oz of water is wasted in such a way each day. How much water, in gallons, is wasted in a week? a month? a year? Assuming each of the 244 million people in this country wastes water this way, estimate how much water is wasted in a year.
1.75 gal/week; 7.5 gal/month; 91.25 gal/year; 22,265,000,000 gal/year

SKILL MAINTENANCE

31. Find the simple interest on $600 at 13% for $\frac{1}{2}$ yr.

32. Find the simple interest on $600 at 8% for 2 yr.

Evaluate.

33. 10^3

34. 15^2

SYNTHESIS

35. Solve the problem regarding the volume of one million one-dollar bills as given on the chapter opening page. Give the answer in both cubic inches and cubic feet.

10.2 Volume of Cylinders, Spheres, and Cones

a Cylinders

A rectangular solid with a shaded base is shown below. Note that we can think of the volume as the product of the area of the base times the height:

$$V = l \cdot w \cdot h$$
$$= (l \cdot w) \cdot h$$
$$= (\text{Area of the base}) \cdot h$$
$$= B \cdot h,$$

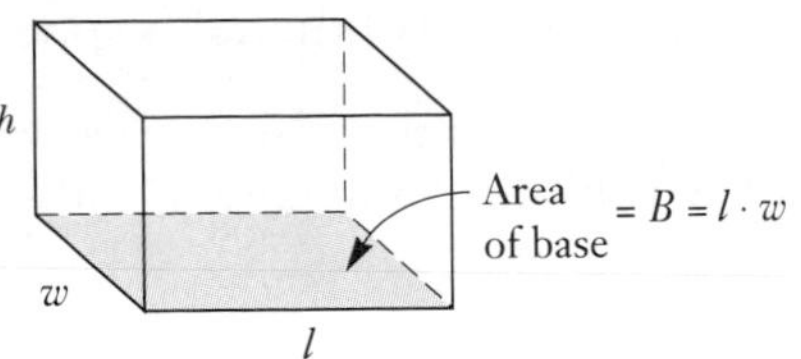

where B represents the area of the base.

Like rectangular solids, **circular cylinders** have bases of equal area that lie in parallel planes. The bases of circular cylinders are circular regions.

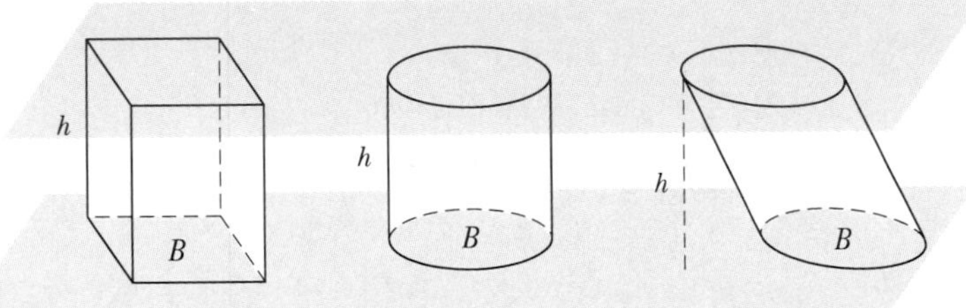

The volume of a circular cylinder is found in a manner similar to finding the volume of a rectangular solid. The volume is the product of the area of the base times the height. The height is always measured perpendicular to the base.

The volume of a circular cylinder is the product of the area of the base B and the height h:

$$V = B \cdot h, \quad \text{or} \quad V = \pi \cdot r^2 \cdot h.$$

▶ **EXAMPLE 1** Find the volume of this circular cylinder. Use 3.14 for π.

$$V = Bh = \pi \cdot r^2 \cdot h$$
$$\approx 3.14 \times 4\text{ cm} \times 4\text{ cm} \times 12\text{ cm}$$
$$= 602.88\text{ cm}^3$$

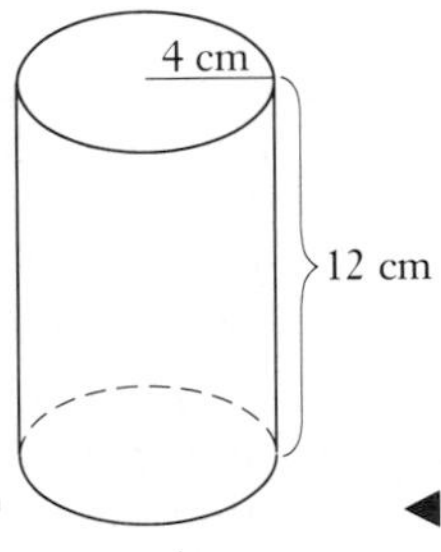

◀

DO EXERCISES 1 AND 2.

b Spheres

A **sphere** is the three-dimensional counterpart of a circle. It is the set of all points in space that are a given distance (the radius) from a given point (the center).

OBJECTIVES

After finishing Section 10.2, you should be able to:

a Given the radius and the height, find the volume of a circular cylinder.

b Given the radius, find the volume of a sphere.

c Given the radius, find the volume of a circular cone.

FOR EXTRA HELP

Tape 13A

Tape 14B

MAC: 10
IBM: 10

1. Find the volume of the cylinder. Use 3.14 for π. 785 ft^3

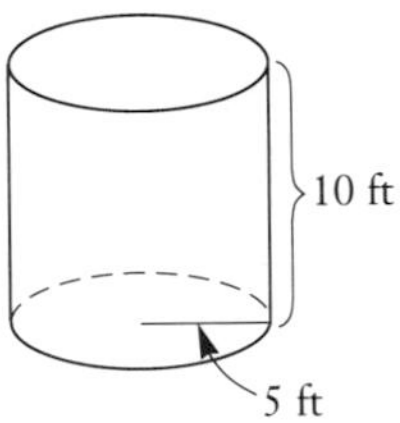

2. Find the volume of the cylinder. Use $\frac{22}{7}$ for π. 67,914 m^3

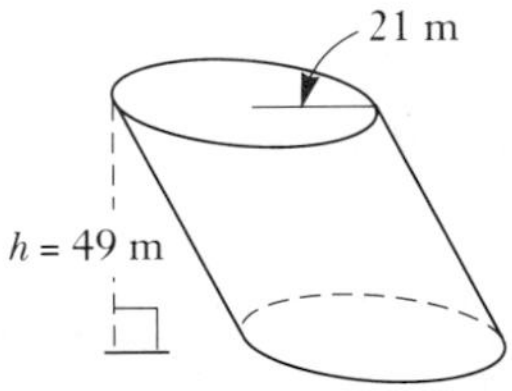

ANSWERS ON PAGE A-5

3. Find the volume of the sphere. Use $\frac{22}{7}$ for π. $91{,}989\frac{1}{3}$ ft^3

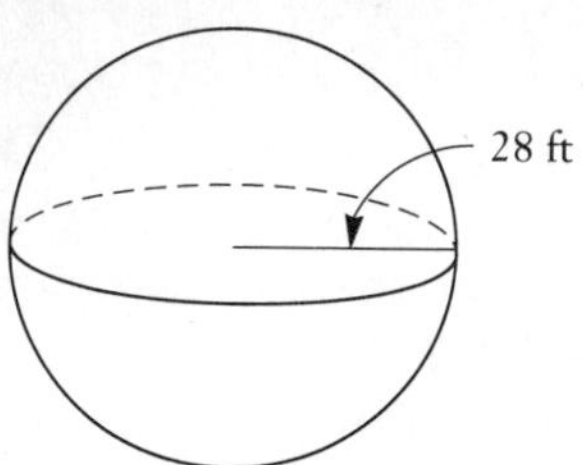

4. The radius of a standard-sized golf ball is 2.1 cm. Find its volume. Use 3.14 for π. 38.77272 cm^3

5. Find the volume of this cone. Use 3.14 for π. 1695.6 m^3

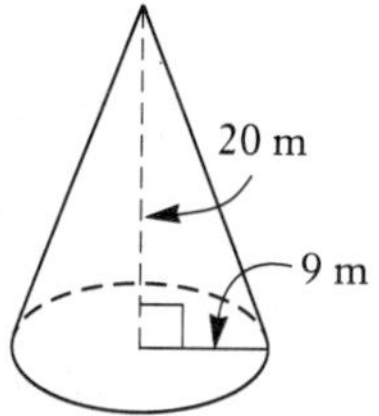

6. Find the volume of this cone. Use $\frac{22}{7}$ for π. 528 in^3

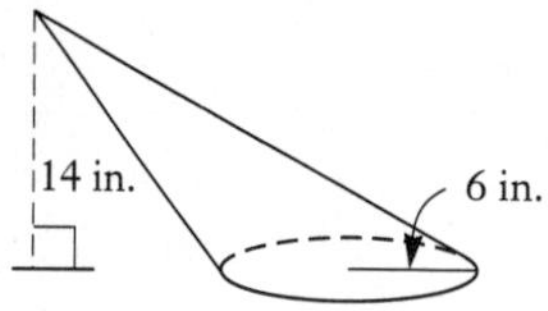

ANSWERS ON PAGE A-5

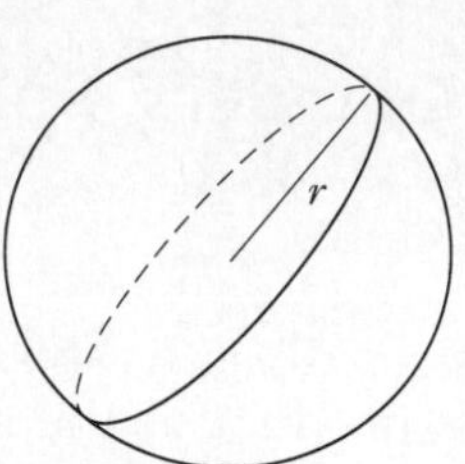

We find the volume of a sphere as follows.

The volume of a sphere of radius r is given by

$$V = \frac{4}{3} \cdot \pi \cdot r^3.$$

► **EXAMPLE 2** The radius of a standard-sized bowling ball is 4.2915 in. Find the volume of a bowling ball. Round to the nearest hundredth of a cubic inch. Use 3.14 for π.

$$V = \frac{4}{3} \cdot \pi \cdot r^3 \approx \frac{4}{3} \times 3.14 \times (4.2915 \text{ in.})^3$$
$$\approx 1.33 \times 3.14 \times 79.0364 \text{ in}^3 \approx 330.07 \text{ in}^3$$ ◄

DO EXERCISES 3 AND 4.

C Cones

Consider a circle in a plane and choose any point not in the plane. The circular region, together with the set of all segments connecting P to a point on the circle, is called a **circular cone.**

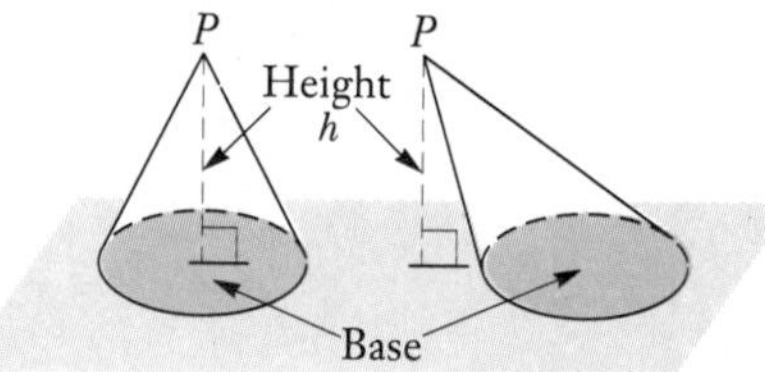

We find the volume of a cone as follows.

The volume of a circular cone with base radius r is one-third the product of the base area and the height:

$$V = \frac{1}{3} \cdot B \cdot h = \frac{1}{3}\pi \cdot r^2 \cdot h.$$

► **EXAMPLE 3** Find the volume of this cone. Use $\frac{22}{7}$ for π.

$$V = \frac{1}{3}\pi \cdot r^2 \cdot h$$
$$\approx \frac{1}{3} \times \frac{22}{7} \times 3 \text{ cm} \times 3 \text{ cm} \times 7 \text{ cm}$$
$$= 66 \text{ cm}^3$$

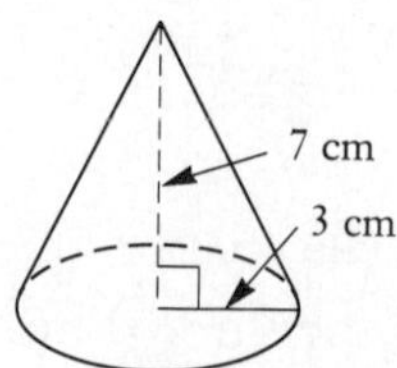

◄

DO EXERCISES 5 AND 6.

NAME SECTION DATE

EXERCISE SET 10.2

a Find the volume of the circular cylinder. Use 3.14 for π in Exercises 1–4. Use $\frac{22}{7}$ for π in Exercises 5 and 6.

1.
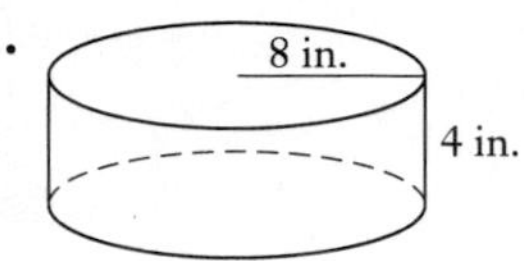

2.
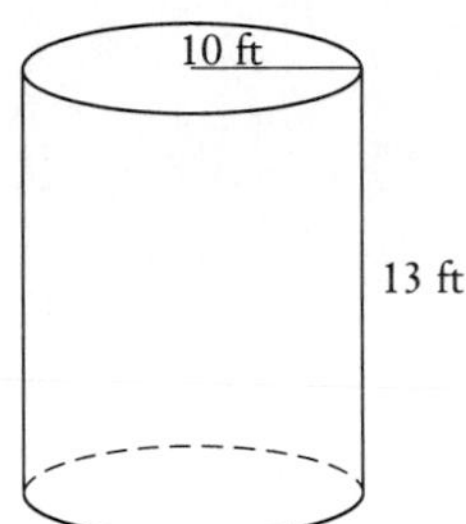

3.
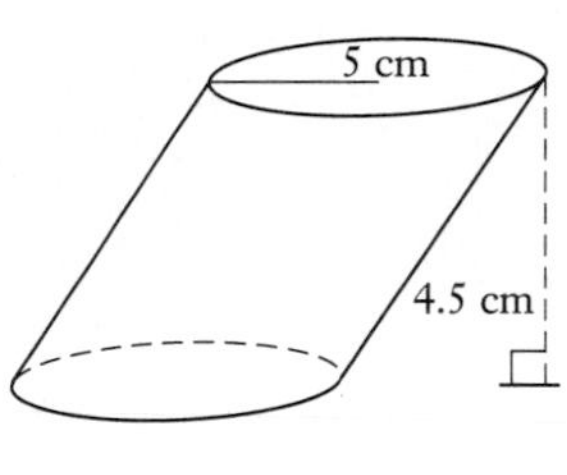

4.
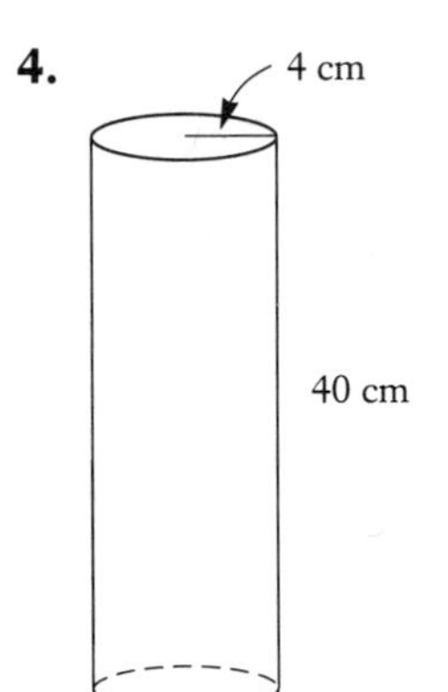

5.
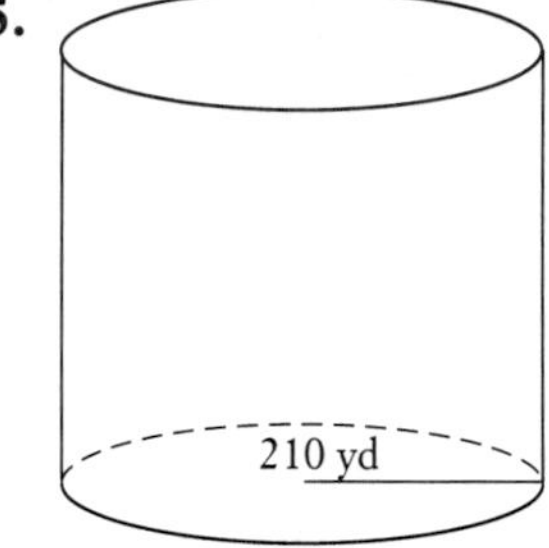

6.
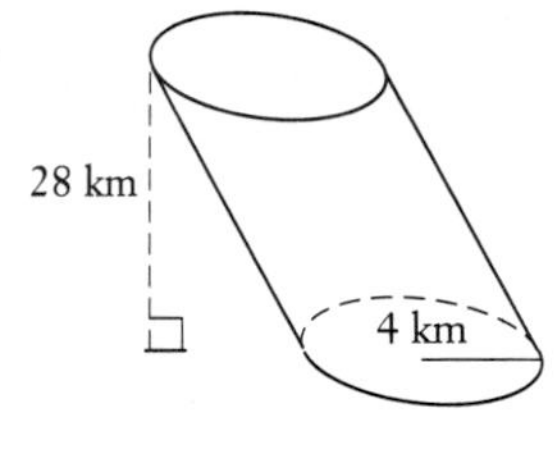

b Find the volume of the sphere. Use 3.14 for π in Exercises 7–10. Use $\frac{22}{7}$ for π in Exercises 11 and 12.

7.
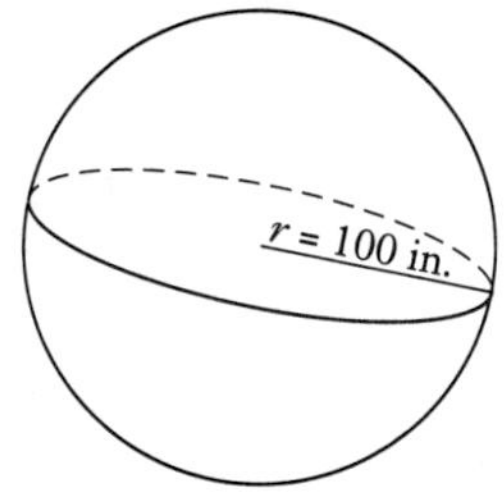

8.
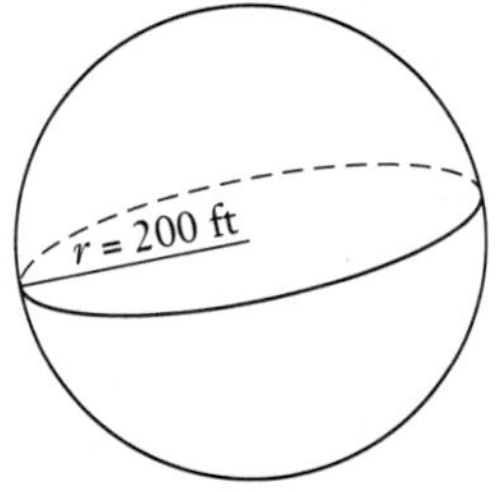

9.
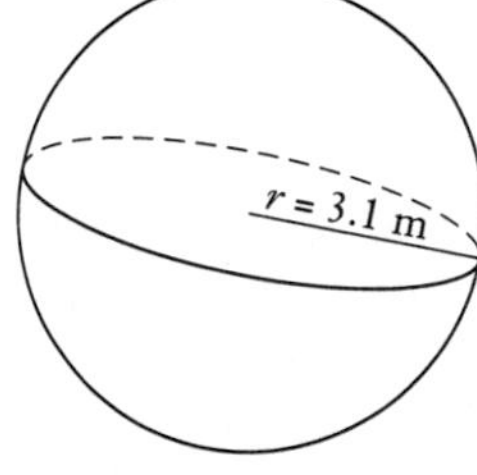

10.
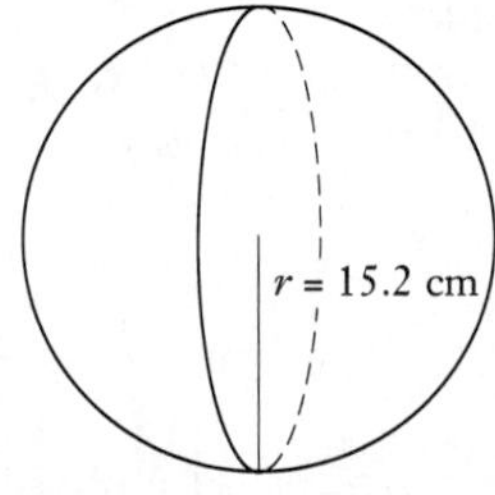

11.
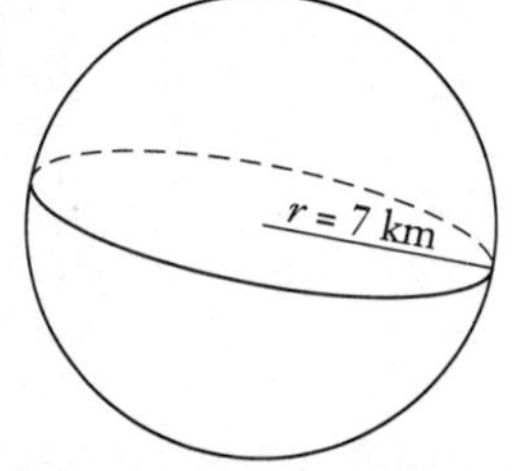

12.
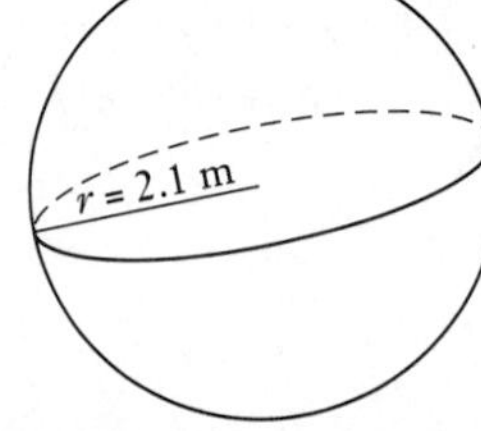

ANSWERS

1. 803.84 in^3
2. 4082 ft^3
3. 353.25 cm^3
4. 2009.6 cm^3
5. 41,580,000 yd^3
6. 1408 km^3
7. $4{,}186{,}666\frac{2}{3}$ in^3
8. $33{,}493{,}333\frac{1}{3}$ ft^3
9. 124.72 m^3
10. 14,702.77 cm^3
11. $1437\frac{1}{3}$ km^3
12. 38.808 m^3

ANSWERS

13. $113{,}982 \text{ ft}^3$

14. 94.2 m^3

15. 24.64 cm^3

16. $38{,}500 \text{ mm}^3$

17. $33{,}880 \text{ m}^3$

18. 50.24 in^3

19. 367.38 m^3

20. 4747.68 cm^3

21. 113.0 m^3

22. 143.72 cm^3

23.

24. 6 cm by 6 cm by 6 cm

25. $6.9\overline{7} \text{ ft}^3$; $52\frac{1}{3}$ gal

c Find the volume of the cone. Use 3.14 for π in Exercises 13 and 14. Use $\frac{22}{7}$ for π in Exercises 15 and 16.

13.

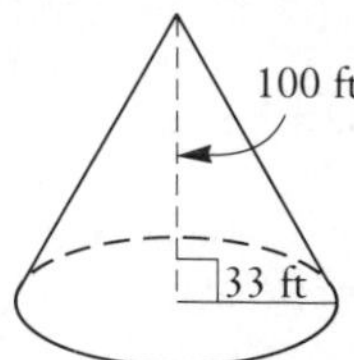

14.

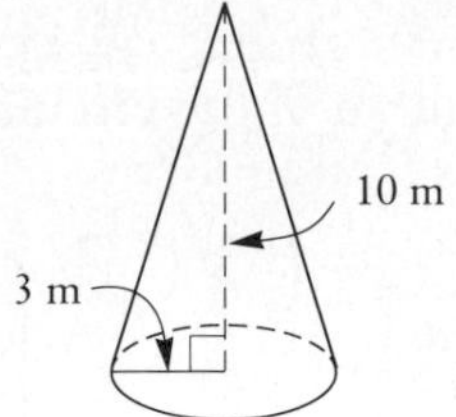

15.

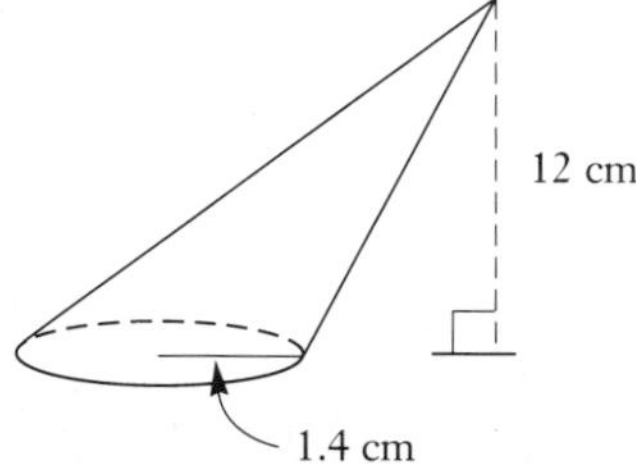

16.

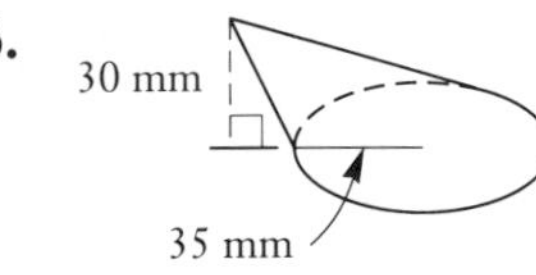

a, b Solve.

17. The diameter of the base of a circular cylinder is 14 m. The height is 220 m. Find the volume. Use $\frac{22}{7}$ for π.

18. A rung of a ladder is 2 in. in diameter and 16 in. long. Find the volume. Use 3.14 for π.

19. A barn silo, excluding the top, is a circular cylinder. The silo is 6 m in diameter and the height is 13 m. Find the volume. Use 3.14 for π.

20. A log of wood has a diameter of 12 cm and a height of 42 cm. Find the volume. Use 3.14 for π.

21. The diameter of a spherical gas tank is 6 m. Find the volume to the nearest tenth of a cubic meter. Use 3.14 for π.

22. The diameter of a tennis ball is 6.5 cm. Find the volume. Use 3.14 for π.

23. The diameter of the earth is 6400 km. Find the volume of the earth. Use 3.14 for π.

$137{,}188{,}693{,}333.33 \text{ km}^3$

24. The volume of a ball is $36\pi \text{ cm}^3$. Find the dimensions of a rectangular box that is just large enough to hold the ball.

SYNTHESIS

25. A hot water tank is a right circular cylinder that has a base with a diameter of 16 in. and a height of 5 ft. Find the volume of the tank in cubic feet. Use 3.14 for π. One cubic foot of water is about 7.5 gal. About how many gallons will the tank hold?

10.3 Weight, Mass, and Time

a Weight: The American System

The American units of weight are as follows.

> American Units of Weight
>
> **1 ton (T) = 2000 pounds (lb)**
> **1 lb = 16 ounces (oz)**

▶ **EXAMPLE 1** A well-known hamburger is called a "quarter-pounder." Find its name in ounces: a "______ ouncer."

$$\frac{1}{4}\text{ lb} = \frac{1}{4}\cdot 1\text{ lb}$$
$$= \frac{1}{4}\cdot 16\text{ oz} \quad \text{Substituting 16 oz for 1 lb}$$
$$= 4\text{ oz}$$

A "quarter-pounder" can also be called a "four-ouncer." ◀

▶ **EXAMPLE 2** Complete: 7.68 T = ______ lb.

$$7.68\text{ T} = 7.68 \times 1\text{ T}$$
$$= 7.68 \times 2000\text{ lb} \quad \text{Substituting 2000 lb for 1 T}$$
$$= 15{,}360\text{ lb}$$ ◀

DO EXERCISES 1–3.

b Mass: The Metric System

There is a difference between **mass** and **weight,** but the terms are often used interchangeably. People sometimes use the word "weight" instead of "mass." Weight is related to the force of gravity. The farther you are from the center of the earth, the less you weigh. Your mass stays the same no matter where you are.

The basic unit of mass is the **gram** (g), which is the mass of 1 cubic centimeter (1 cm^3 or 1 mL) of water. Since a cubic centimeter is small, a gram is a small unit of mass.

$$1\text{ g} = 1\text{ gram} = \text{the mass of } 1\text{ cm}^3\text{ (1 mL) of water}$$

OBJECTIVES

After finishing Section 10.3, you should be able to:

a Convert from one American unit of weight to another.

b Convert from one metric unit of mass to another.

c Convert from one unit of time to another.

FOR EXTRA HELP

Tape 13B | Tape 14B | MAC: 10 IBM: 10

1. 5 lb = 80 oz

2. 4.32 T = 8640 lb

3. 1 T = 32,000 oz

ANSWERS ON PAGE A-5

The following table shows the metric units of mass. The prefixes are the same as those for length.

Metric Units of Mass

1 metric ton (t) = 1000 kilograms (kg)
1 *kilo*gram (kg) = 1000 grams (g)
1 *hecto*gram (hg) = 100 grams (g)
1 *deka*gram (dag) = 10 grams (g)
1 gram (g)
1 *deci*gram (dg) = $\frac{1}{10}$ gram (g)
1 *centi*gram (cg) = $\frac{1}{100}$ gram (g)
1 *milli*gram (mg) = $\frac{1}{1000}$ gram (g)

Thinking Metric

One gram is about the mass of 1 raisin or 1 paperclip. Since 1 metric ton is 1000 kg and 1 kg is about 2.2 lb, it follows that 1 metric ton (t) is about 2200 lb, which is just a little more than 1 American ton (T).

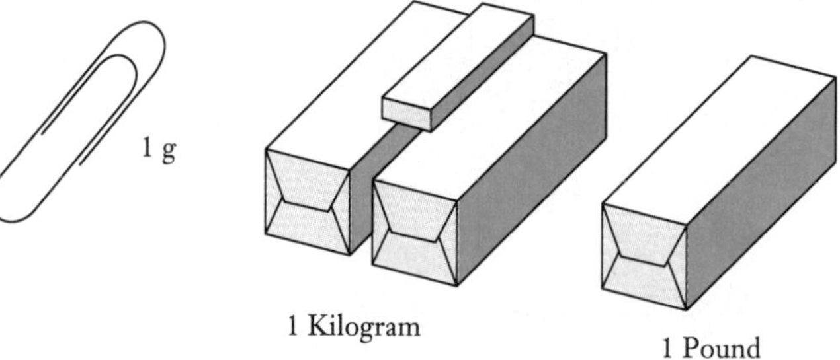

Small masses, such as dosages of medicine and vitamins, may be measured in milligrams (mg).

Each 2.5 mg

Grams (g) are used for objects normally given in ounces, such as the mass of a letter, a piece of candy, a coin, or a small package of food.

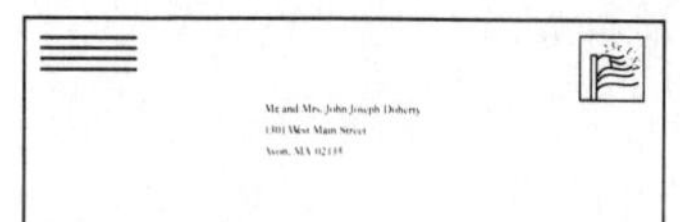

15 g

2 g

Kilograms (kg) are used for larger food packages, such as meat, or for masses of people.

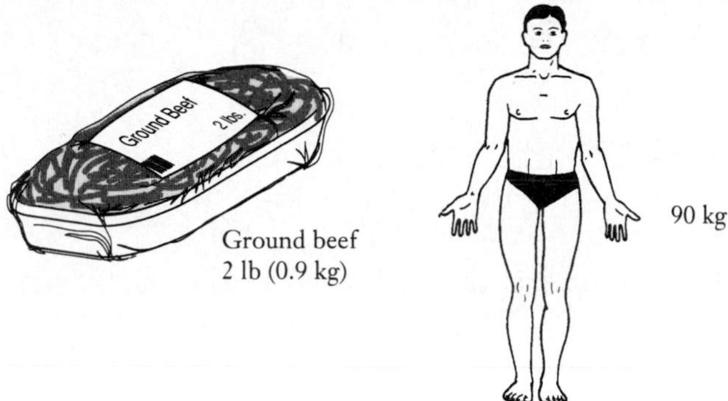

Ground beef
2 lb (0.9 kg)

90 kg

The metric ton (t) is used for very large masses, such as the mass of an automobile, a truckload of gravel, or an airliner.

DO EXERCISES 4–8.

Changing Units Mentally

As before, changing from one metric unit to another amounts to only the movement of a decimal point. We use this table.

1000	100	10	1	0.1	0.01	0.001
kg	**hg**	**dag**	**g**	**dg**	**cg**	**mg**

▶ **EXAMPLE 3** Complete: 8 kg = ______ g.

Think: To go from kg to g in the table is a move of three places to the right. Thus we move the decimal point three places to the right.

8.0   8.000.   8 kg = 8000 g ◀

▶ **EXAMPLE 4** Complete: 4235 g = ______ kg.

Think: To go from g to kg in the table is a move of three places to the left. Thus we move the decimal point three places to the left.

4235.0   4.235.0   4235 g = 4.235 kg ◀

DO EXERCISES 9 AND 10.

▶ **EXAMPLE 5** Complete: 6.98 cg = ______ mg.

Think: To go from cg to mg is a move of one place to the right. Thus we move the decimal point one place to the right.

6.98   6.9.8   6.98 cg = 69.8 mg ◀

The most commonly used metric units of mass are kg, g, cg, and mg. We have purposely used those more often than the others in the exercises.

Complete with mg, g, kg, or t.

4. A typewriter has a mass of 6 kg.

5. That person has a mass of 85.4 kg.

6. This is a 3 mg vitamin.

7. A pencil has a mass of 12 g.

8. A stationwagon has a mass of 3 t.

Complete.

9. 6.2 kg = 6200 g

10. 304.8 g = 0.3048 kg

ANSWERS ON PAGE A-5

▶ **EXAMPLE 6** Complete: 89.21 mg = ______ g.

Think: To go from mg to g is a move of three places to the left. Thus we move the decimal point three places to the left.

$$89.21 \qquad 0.089.21 \qquad 89.21 \text{ mg} = 0.08921 \text{ g}$$ ◀

DO EXERCISES 11–13.

C Time

A table of units of time is shown below. The metric system sometimes uses "h" for hour and "s" for second, but we will use the more familiar "hr" and "sec."

1 day = 24 hours (hr)	**1 year (yr) = $365\frac{1}{4}$ days**
1 hr = 60 minutes (min)	**1 week (wk) = 7 days**
1 min = 60 seconds (sec)	

Since we cannot have $\frac{1}{4}$ day on the calendar, we give each year 365 days and every fourth year 366 days (a leap year), unless it is a year at the beginning of a century not divisible by 400.

▶ **EXAMPLE 7** Complete: 1 hr = ______ sec.

$$\begin{aligned} 1 \text{ hr} &= 60 \text{ min} \\ &= 60 \cdot 1 \text{ min} \\ &= 60 \cdot 60 \text{ sec} \qquad \text{Substituting 60 sec for 1 min} \\ &= 3600 \text{ sec} \end{aligned}$$ ◀

▶ **EXAMPLE 8** Complete: 5 yr = ______ days.

$$\begin{aligned} 5 \text{ yr} &= 5 \cdot 1 \text{ yr} \\ &= 5 \cdot 365\tfrac{1}{4} \text{ days} \qquad \text{Substituting } 365\tfrac{1}{4} \text{ days for 1 yr} \\ &= 1826\tfrac{1}{4} \text{ days} \end{aligned}$$ ◀

▶ **EXAMPLE 9** Complete: 3 days = ______ min.

$$\begin{aligned} 3 \text{ days} &= 3 \cdot 1 \text{ day} \\ &= 3 \cdot 24 \text{ hr} \qquad \text{Substituting 24 hr for 1 day} \\ &= 3 \cdot 24 \cdot 1 \text{ hr} \\ &= 3 \cdot 24 \cdot 60 \text{ min} \qquad \text{Substituting 60 min for 1 hr} \\ &= 4320 \text{ min} \end{aligned}$$ ◀

DO EXERCISES 14–17.

Complete.

11. 7.7 cg = <u>77</u> mg

12. 2344 mg = <u>234.4</u> cg

13. 67 dg = <u>6700</u> mg

Complete.

14. 2 hr = <u>7200</u> sec

15. 4 yr = <u>1461</u> days

16. 1 day = <u>1440</u> min

17. 1 wk = <u>168</u> hr

ANSWERS ON PAGE A-5

NAME SECTION DATE

EXERCISE SET 10.3

a Complete.

1. 1 T = ______ lb
2. 1 lb = ______ oz
3. 6000 lb = ______ T
4. 5 T = ______ lb
5. 4 lb = ______ oz
6. 20 lb = ______ oz
7. 3.5 T = ______ lb
8. 3.01 T = ______ lb
9. 3200 oz = ______ T
10. 6400 oz = ______ T
11. 96 oz = ______ lb
12. 960 oz = ______ lb

b Complete.

13. 1 kg = ______ g
14. 1 hg = ______ g
15. 1 dag = ______ g
16. 1 dg = ______ g
17. 1 cg = ______ g
18. 1 mg = ______ g
19. 1 g = ______ mg
20. 1 g = ______ cg
21. 1 g = ______ dg
22. 25 kg = ______ g
23. 234 kg = ______ g
24. 678 g = ______ kg
25. 5200 g = ______ kg
26. 0.809 kg = ______ g
27. 67 hg = ______ kg

ANSWERS

1. 2000
2. 16
3. 3
4. 10,000
5. 64
6. 320
7. 7000
8. 6020
9. 0.1
10. 0.2
11. 6
12. 60
13. 1000
14. 100
15. 10
16. $\frac{1}{10}$, or 0.1
17. $\frac{1}{100}$, or 0.01
18. $\frac{1}{1000}$, or 0.001
19. 1000
20. 100
21. 10
22. 25,000
23. 234,000
24. 0.678
25. 5.2
26. 809
27. 6.7

ANSWERS

28. 0.45
29. 0.0502
30. 0.025
31. 6.78
32. 5.677
33. 6.9
34. 7.61
35. 800,000
36. 20,000
37. 1000
38. 2000
39. 0.0034
40. 430
41. 24
42. 60
43. 60
44. 7
45. $365\frac{1}{4}$
46. $730\frac{1}{2}$
47. 336
48. 14,400
49. 8.2
50. 18,000

28. 45 cg = ______ g

29. 0.502 dg = ______ g

30. 0.0025 cg = ______ mg

31. 6780 g = ______ kg

32. 5677 g = ______ kg

33. 69 mg = ______ cg

34. 76.1 mg = ______ cg

35. 8 kg = ______ cg

36. 0.02 kg = ______ mg

37. 1 t = ______ kg

38. 2 t = ______ kg

39. 3.4 cg = ______ dag

40. 4.3 dg = ______ mg

C Complete.

41. 1 day = ______ hr

42. 1 hr = ______ min

43. 1 min = ______ sec

44. 1 wk = ______ days

45. 1 yr = ______ days

46. 2 yr = ______ days

47. 2 wk = ______ hr

48. 4 hr = ______ sec

49. 492 sec = ______ min
(the amount of time it takes for the rays of the sun to reach the earth)

50. 5 hr = ______ sec

SKILL MAINTENANCE

Evaluate.

51. 2^4

52. 17^2

53. 5^3

54. 8^2

SYNTHESIS

Complete. Use 1 kg = 2.205 lb and 453.5 g = 1 lb. Round to four decimal places.

55. ▦ 1 lb = ______ kg

56. ▦ 1 g = ______ lb

Another metric unit used in medicine is the microgram (μg). It is defined as follows.

1 microgram = 1μg = $\frac{1}{1,000,000}$ **g;   1,000,000 μg = 1 g**

Thus a microgram is one millionth of a gram, and one million micrograms is one gram.

Complete.

57. 1 mg = ______ μg

58. 1μg = ______ mg

59. A physician orders 125 μg of digoxin. How many milligrams is the prescription?

60. A physician orders 0.25 mg of reserpine. How many micrograms is the prescription?

ANSWERS

51. 16

52. 289

53. 125

54. 64

55. 0.4535

56. 0.0022

57. 1000

58. 0.001

59. 0.125 mg

60. 250 μg

ANSWERS

61. 4

62. 2.5 mL

63. 8 mL

64. a) 200

b) 3

65. $\frac{2}{3}$ oz

66. About 0.03 yr

67. About 31.7 yr

68. About 31,700 yr

61. A medicine called sulfisoxazole usually comes in tablets that are 500 mg each. A standard dosage is 2 g. How many tablets would have to be taken in order to achieve this dosage?

62. Quinidine is a liquid mixture, part medicine and part water. There is 80 mg of Quinidine for every milliliter of liquid. A standard dosage is 200 mg. How much of the liquid mixture would be required in order to achieve the dosage?

63. A medicine called cephalexin is obtainable in a liquid mixture, part medicine and part water. There is 250 mg of cephalexin in 5 mL of liquid. A standard dosage is 400 mg. How much of the liquid would be required in order to achieve the dosage?

64. A medicine called Albuterol is used for the treatment of asthma. It typically comes in an inhaler that contains 18 g. One actuation, or spray, is 90 mg.

a) How many actuations are in one inhaler?

b) A student is going away for 4 months of college and wants to take enough Albuterol to last for that time. Assuming that the student will need 4 actuations per day, estimate about how many inhalers the student will need for the 4-month period.

65. At $0.75 a dozen, the cost of eggs is $1.50 per pound. How much does an egg weigh?

66. Estimate the number of years in one million seconds.

67. Estimate the number of years in one billion seconds.

68. Estimate the number of years in one trillion seconds.

10.4 Temperature

a Estimated Conversions

Below are two temperature scales: **Fahrenheit** for American measure, and **Celsius** for metric measure.

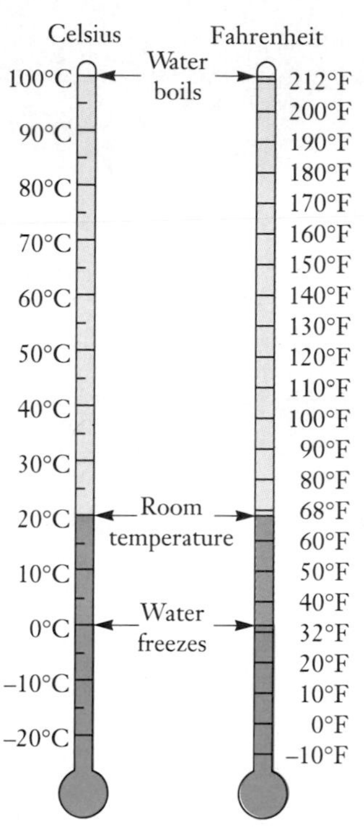

25°F

25°C

By laying a ruler or a piece of paper horizontally between the scales, we can make an approximate conversion from one measure of temperature to another.

▶ **EXAMPLES** Convert to Celsius. Approximate to the nearest ten degrees.

1. 212°F (Boiling point of water) 100°C This is exact.
2. 32°F (Freezing point of water) 0°C This is exact.

DO EXERCISES 1–3. ◀

▶ **EXAMPLES** Make an approximate conversion.

3. 44°C (Hot bath) 110°C This is approximate.
4. 20°C (Room temperature) 68°F This is approximate. ◀

DO EXERCISES 4–6.

OBJECTIVES

After finishing Section 10.4, you should be able to:

a. Make an approximate conversion from Celsius temperature to Fahrenheit, and from Fahrenheit temperature to Celsius.

b. Convert from Celsius temperature to Fahrenheit and from Fahrenheit to Celsius using the formulas $F = \frac{9}{5} \cdot C + 32$ and $C = \frac{5}{9} \cdot (F - 32)$.

FOR EXTRA HELP

Tape 13C

Tape 15A

MAC: 10
IBM: 10

Convert to Celsius. Approximate to the nearest ten degrees.

1. 180°F (Brewing coffee) 80°C

2. 25°F (Cold day) 0°C

3. – 10°F (Miserably cold day)
   – 20°C

Convert to Fahrenheit. Approximate to the nearest ten degrees.

4. 25°C (Warm day at the beach)
   80°F

5. 40°C (Temperature of a patient with a high fever) 100°F

6. 10°C (A cold bath) 50°F

ANSWERS ON PAGE A-5

Convert to Fahrenheit.

7. 80°C 176°F

8. 35°C 95°F

Convert to Celsius.

9. 95°F 35°C

10. 113°F 45°C

ANSWERS ON PAGE A-5

b Exact Conversions

The following formula allows exact conversion from Celsius to Fahrenheit.

$$F = \frac{9}{5} \cdot C + 32, \quad \text{or} \quad F = 1.8 \cdot C + 32$$

(Multiply by $\frac{9}{5}$, or 1.8, and add 32.)

▶ **EXAMPLES** Convert to Fahrenheit.

5. 0°C $F = \frac{9}{5} \cdot 0 + 32 = 0 + 32 = 32°$

Thus, 0°C = 32°F.

6. 37°C $F = 1.8 \cdot 37 + 32 = 66.6 + 32 = 98.6°$

Thus, 37°C = 98.6°F. This is normal body temperature. ◀

DO EXERCISES 7 AND 8.

The following formula allows exact conversion from Fahrenheit to Celsius.

$$C = \frac{5}{9} \cdot (F - 32)$$

(Subtract 32 and multiply by $\frac{5}{9}$.)

▶ **EXAMPLES** Convert to Celsius.

7. 212°F $C = \frac{5}{9} \cdot (212 - 32) = \frac{5}{9} \cdot 180 = 100°$

Thus, 212°F = 100°C.

8. 77°F $C = \frac{5}{9} \cdot (77 - 32) = \frac{5}{9} \cdot 45 = 25°$

Thus, 77°F = 25°C. ◀

DO EXERCISES 9 AND 10.

NAME SECTION DATE

EXERCISE SET 10.4

a Convert to Celsius. Approximate to the nearest ten degrees. Use the scales on p. 455.

1. 178°F **2.** 195°F **3.** 140°F **4.** 107°F

5. 88°F **6.** 50°F **7.** 10°F **8.** 120°F

Convert to Fahrenheit. Approximate to the nearest ten degrees. Use the scales on p. 455.

9. 86°C **10.** 93°C **11.** 58°C **12.** 35°C

13. −10°C **14.** − 5°C **15.** 5°C **16.** 15°C

ANSWERS

1. 80°C

2. 90°C

3. 60°C

4. 40°C

5. 30°C

6. 10°C

7. −10°C

8. 50°C

9. 190°F

10. 200°F

11. 140°F

12. 100°F

13. 10°F

14. 20°F

15. 40°F

16. 60°F

ANSWERS

17. 77°F
18. 185°F
19. 104°F
20. 194°F
21. 5432°F
22. 1832°F
23. 30°C
24. 15°C
25. 55°C
26. 60°C
27. 37°C
28. 40°C
29. 234
30. 230
31. 336
32. 24
33. 260.6°F

b Convert to Fahrenheit. Use the formula $F = \frac{9}{5} \cdot C + 32$.

17. 25°C **18.** 85°C **19.** 40°C **20.** 90°C

21. 3000°C, the melting point of iron **22.** 1000°C, the melting point of gold

Convert to Celsius. Use the formula $C = \frac{5}{9} \cdot (F - 32)$.

23. 86°F **24.** 59°F **25.** 131°F **26.** 140°F

27. 98.6°F, normal body temperature **28.** 104°F, high-fevered body temperature

SKILL MAINTENANCE

Complete.

29. 23.4 cm = ______ mm **30.** 0.23 km = ______ m

31. 28 ft = ______ in. **32.** 72 ft = ______ yd

SYNTHESIS

33. Another temperature scale often used is called **Kelvin.** Conversions from Celsius to Kelvin can be carried out using the formula

$$K = C + 273.$$

A chemistry textbook describes an experiment in which a reaction takes place at a temperature of 400°Kelvin. A student wishes to perform the experiment, but has only a Fahrenheit thermometer. At what Fahrenheit temperature will the reaction take place?

10.5 Converting Units of Area

a American Units

Let's do some conversions from one American unit of area to another.

▶ **EXAMPLE 1** Complete: $1 \text{ ft}^2 =$ _______ in^2.

$$1 \text{ ft}^2 = 1 \cdot (12 \text{ in.})^2 \quad \text{Substituting 12 in. for 1 ft}$$
$$= (12 \text{ in.}) \cdot (12 \text{ in.})$$
$$= 144 \text{ in}^2$$ ◀

▶ **EXAMPLE 2** Complete: $8 \text{ yd}^2 =$ _______ ft^2.

$$8 \text{ yd}^2 = 8 \cdot (3 \text{ ft})^2 \quad \text{Substituting 3 ft for 1 yd}$$
$$= 8 \cdot (3 \text{ ft}) \cdot (3 \text{ ft})$$
$$= 8 \cdot 3 \cdot 3 \cdot \text{ft} \cdot \text{ft}$$
$$= 72 \text{ ft}^2$$ ◀

DO EXERCISES 1–3.

American units are related as follows.

1 square yard (yd^2) = 9 square feet (ft^2)
1 square foot (ft^2) = 144 square inches (in^2)
1 square mile (mi^2) = 640 acres
1 acre = 43,560 ft^2

▶ **EXAMPLE 3** Complete: $36 \text{ ft}^2 =$ _______ yd^2.

We are converting from "ft^2" to "yd^2". We choose a symbol for 1 with yd^2 on top and ft^2 on the bottom.

$$36 \text{ ft}^2 = 36 \text{ ft}^2 \times \frac{1 \text{ yd}^2}{9 \text{ ft}^2} \quad \text{Multiplying by 1 using } \frac{1 \text{ yd}^2}{9 \text{ ft}^2}$$
$$= \frac{36}{9} \times \frac{\text{ft}^2}{\text{ft}^2} \times 1 \text{ yd}^2$$
$$= 4 \text{ yd}^2$$ ◀

▶ **EXAMPLE 4** Complete: $7 \text{ mi}^2 =$ _______ acres.

$$7 \text{ mi}^2 = 7 \cdot 1 \text{ mi}^2$$
$$= 7 \cdot 640 \text{ acres} \quad \text{Substituting 640 acres for 1 mi}^2$$
$$= 4480 \text{ acres}$$ ◀

DO EXERCISES 4 AND 5.

OBJECTIVES

After finishing Section 10.5, you should be able to:

a Convert from one American unit of area to another.

b Convert from one metric unit of area to another.

FOR EXTRA HELP

Tape 13D

Tape 15A

MAC: 10
IBM: 10

Complete.

1. $1 \text{ yd}^2 =$ ___9___ ft^2

2. $5 \text{ yd}^2 =$ ___45___ ft^2

3. $20 \text{ ft}^2 =$ ___2880___ in^2

Complete.

4. $360 \text{ in}^2 =$ ___2.5___ ft^2

5. $5 \text{ mi}^2 =$ ___3200___ acres

ANSWERS ON PAGE A-5

Complete.

6. $1 \text{ m}^2 = \underline{1{,}000{,}000} \text{ mm}^2$

7. $1 \text{ cm}^2 = \underline{\quad 100 \quad} \text{ mm}^2$

Complete.

8. $2.88 \text{ m}^2 = \underline{28{,}800} \text{ cm}^2$

9. $4.3 \text{ mm}^2 = \underline{0.043} \text{ cm}^2$

10. $678{,}000 \text{ m}^2 = \underline{0.678} \text{ km}^2$

ANSWERS ON PAGE A-5

b Metric Units

Let's convert from one metric unit of area to another.

▶ **EXAMPLE 5** Complete: $1 \text{ km}^2 = \underline{\qquad} \text{ m}^2$.

$$\begin{aligned} 1 \text{ km}^2 &= 1 \cdot (1000 \text{ m})^2 && \text{Substituting 1000 m for 1 km} \\ &= (1000 \text{ m}) \cdot (1000 \text{ m}) \\ &= 1{,}000{,}000 \text{ m}^2 \end{aligned}$$ ◀

▶ **EXAMPLE 6** Complete: $1 \text{ m}^2 = \underline{\qquad} \text{ cm}^2$.

$$\begin{aligned} 1 \text{ m}^2 &= 1 \cdot (100 \text{ cm})^2 && \text{Substituting 100 cm for 1 m} \\ &= (100 \text{ cm}) \cdot (100 \text{ cm}) \\ &= 10{,}000 \text{ cm}^2 \end{aligned}$$ ◀

DO EXERCISES 6 AND 7.

Mental Conversion

To convert mentally, we use the table as before and multiply the number of moves by 2 to determine the number of moves of the decimal point.

1000 km	100 hm	10 dam	1 m	0.1 dm	0.01 cm	0.001 mm

▶ **EXAMPLE 7** Complete: $3.48 \text{ km}^2 = \underline{\qquad} \text{ m}^2$.

Think: To go from km to m in the table is a move of 3 places to the right. So we move the decimal point $2 \cdot 3$, or 6 places to the right.

3.48   3.480000. $3.48 \text{ km}^2 = 3{,}480{,}000 \text{ m}^2$ ◀

▶ **EXAMPLE 8** Complete: $586.78 \text{ cm}^2 = \underline{\qquad} \text{ m}^2$.

Think: To go from cm to m in the table is a move of 2 places to the left. So we move the decimal point $2 \cdot 2$, or 4 places to the left.

586.78   0.0586.78 $586.78 \text{ cm}^2 = 0.058678 \text{ m}^2$ ◀

DO EXERCISES 8–10.

NAME SECTION DATE

EXERCISE SET 10.5

a Complete.

1. 1 ft^2 = _______ in^2

2. 1 yd^2 = _______ ft^2

3. 1 mi^2 = _______ acres

4. 1 acre = _______ ft^2

5. 1 in^2 = _______ ft^2

6. 1 ft^2 = _______ yd^2

7. 22 yd^2 = _______ ft^2

8. 40 ft^2 = _______ in^2

9. 44 yd^2 = _______ ft^2

10. 72 ft^2 = _______ yd^2

11. 20 mi^2 = _______ acres

12. 288 in^2 = _______ ft^2

13. 1 mi^2 = _______ ft^2

14. 1 mi^2 = _______ yd^2

15. 720 in^2 = _______ ft^2

16. 9 ft^2 = _______ yd^2

17. 144 in^2 = _______ ft^2

18. 36 in^2 = _______ ft^2

19. A factory is tooling a large sheet of metal for a machine. It is to be a 12-ft–by–18-ft rectangle with 20 squares cut out, each of which is 3 in. on a side. Find the area of the resulting sheet, in square feet.

20. A 30-ft–by–60-ft dance floor is to be turned into a discotheque by placing an 18-ft–by–42-ft dance floor in the middle and carpeting the rest of the room. The new dance floor is laid in pieces that are squares 8 in. by 8 in. How many such tiles are needed? What percent of the floor is the dance area?

ANSWERS

1. 144

2. 9

3. 640

4. 43,560

5. $\frac{1}{144}$

6. $\frac{1}{9}$

7. 198

8. 5760

9. 396

10. 8

11. 12,800

12. 2

13. 27,878,400

14. 3,097,600

15. 5

16. 1

17. 1

18. $\frac{1}{4}$

19. 214.75 ft^2

20. 1701; 42%

ANSWERS

21. 20,000,000

22. 65,000,000

23. 140

24. 28,000

25. 23.456

26. 838

27. 0.12

28. 0.000125

29. 2500

30. 2 ft^2

31. 1.875 ft^2

32. $4\frac{1}{3}$ ft^2

33. $13\frac{1}{3}$ ft^2

34. 7.83998704 m^2

35. 42.05915 cm^2

36. $240

37. 4.2

38. 0.022176

39. $\frac{1}{640}$, or 0.0015625

b Complete.

21. 20 km^2 = ______ m^2

22. 65 km^2 = ______ m^2

23. 0.014 m^2 = ______ cm^2

24. 0.028 m^2 = ______ mm^2

25. 2345.6 mm^2 = ______ cm^2

26. 8.38 cm^2 = ______ mm^2

27. 1200 cm^2 = ______ m^2

28. 125 mm^2 = ______ m^2

29. 250,000 mm^2 = ______ cm^2

Find the area of the shaded region. Give the answer in square feet.

30.

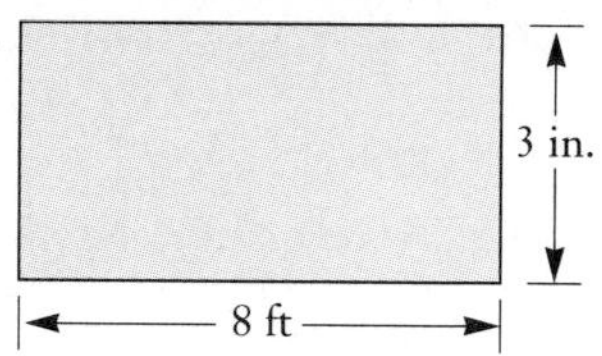

31.

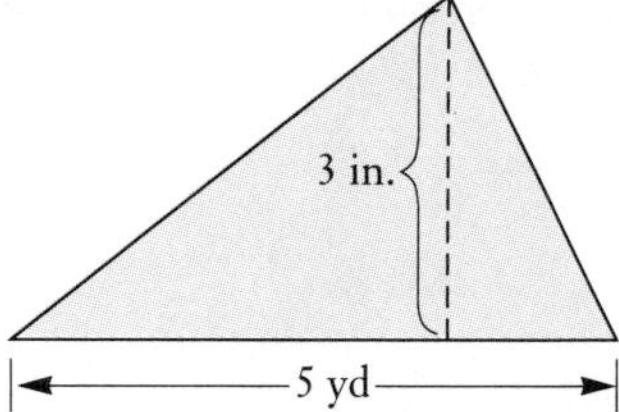

32.

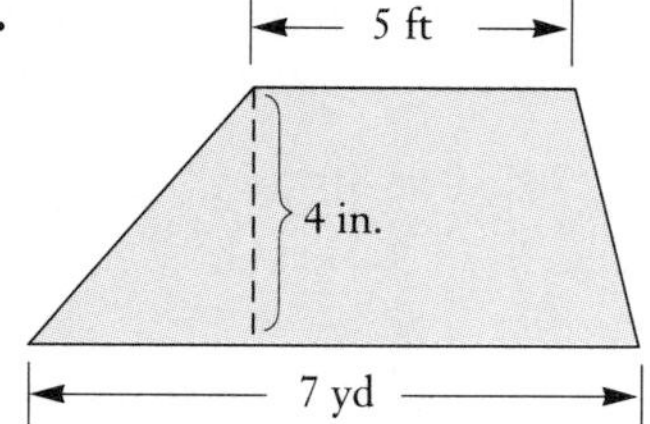

33.

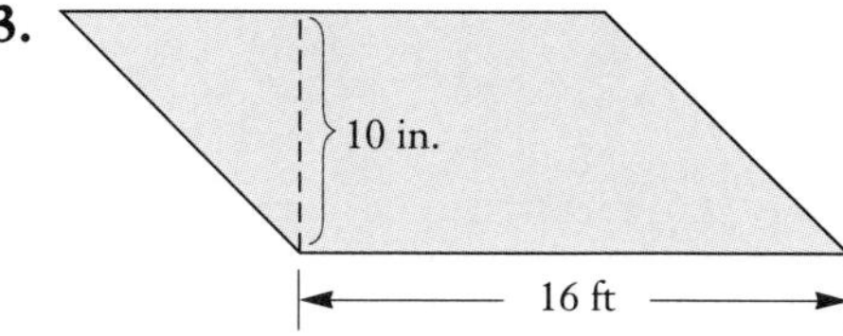

Find the area of the shaded region.

34.

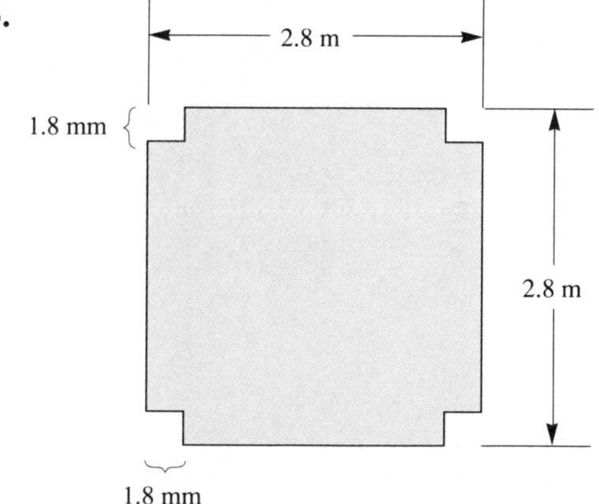

35.

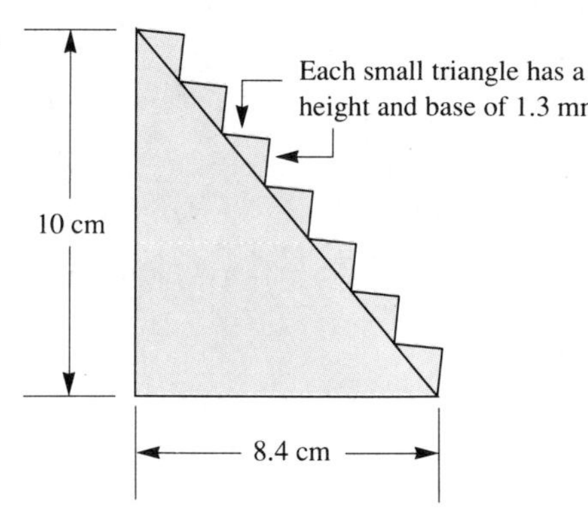

SKILL MAINTENANCE

36. Find the simple interest on $2000 at an interest rate of 8% for 1.5 years.

Complete.

37. 22,176 ft = ______ mi

38. 22,176 mm = ______ km

SYNTHESIS

Complete.

39. 1 acre = ______ mi^2

SUMMARY AND REVIEW: CHAPTER 10

IMPORTANT PROPERTIES AND FORMULAS

Volume of a Rectangular Solid:	$V = l \cdot w \cdot h$
American Units of Capacity:	1 gal = 4 qt; 1 qt = 2 pt; 1 pt = 16 oz; 1 pt = 2 cups; 1 cup = 8 oz
Metric Units of Capacity:	1 L = 1000 mL = 1000 cm^3
Volume of a Circular Cylinder:	$V = \pi \cdot r^2 \cdot h$
Volume of a Sphere:	$V = \frac{4}{3} \cdot \pi \cdot r^3$
Volume of a Cone:	$V = \frac{1}{3} \cdot \pi \cdot r^2 \cdot h$
American System of Weights:	1 T = 2000 lb; 1 lb = 16 oz
Metric System of Weights:	1 t = 1000 kg; 1 kg = 1000 g; 1 hg = 100 g; 1 dag = 10 g; 1 dg = 0.1 g; 1 cg = 0.01 g; 1 mg = 0.001 g
Units of Time:	1 min = 60 sec; 1 wk = 7 days; 1 hr = 60 min; 1 yr = 365.25 days; 1 day = 24 hr
Temperature Conversion:	$F = \frac{9}{5} \cdot C + 32$; $C = \frac{5}{9} \cdot (F - 32)$

REVIEW EXERCISES

The review sections and objectives to be tested in addition to the material in this chapter are [7.8a], [1.9b], [9.1a], and [9.2a].

Find the volume.

1.

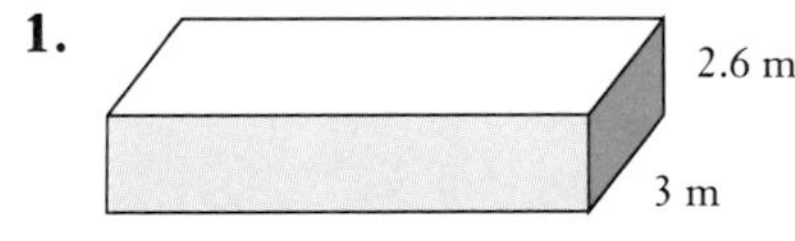

2.

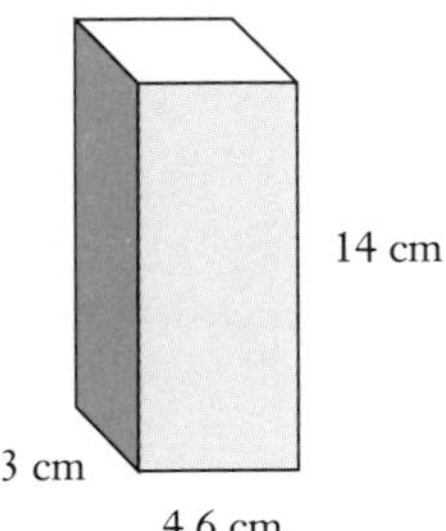

Complete.

3. 7 lb = ______ oz

4. 4 g = ______ kg

5. 16 min = ______ hr

6. 464 mL = ______ L

7. 3 min = ______ sec

8. 4.7 kg = ______ g

9. 8.07 T = ______ lb

10. 0.83 L = ______ mL

11. 6 hr = ______ days

12. 4 cg = ______ g

13. 0.2 g = ______ mg

14. 0.0003 kg = ______ cg

15. 60 mL = ______ L

16. 0.8 T = ______ lb

17. 0.4 L = ______ mL

18. 20 oz = ______ lb

19. $\frac{5}{6}$ min = ______ sec

20. 20 gal = ______ pt

21. 960 oz = ______ gal

22. 54 qt = ______ gal

23. Convert 27°C to Fahrenheit.

24. Convert 68°F to Celsius.

Find the volume. Use 3.14 for π.

25.

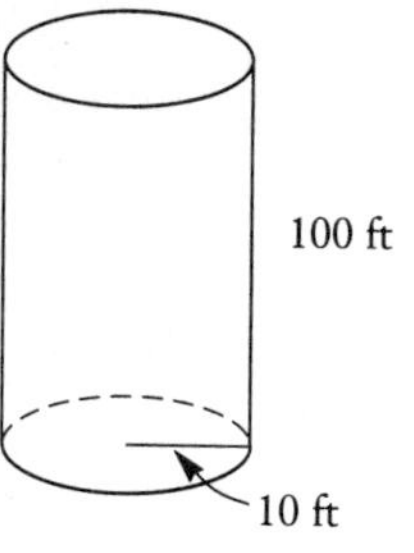

26.

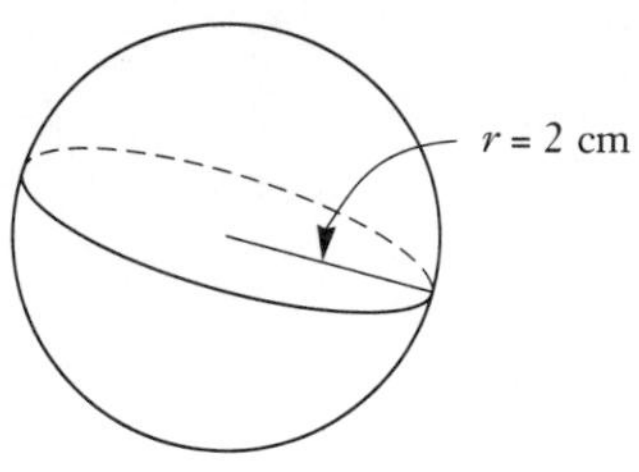

27.

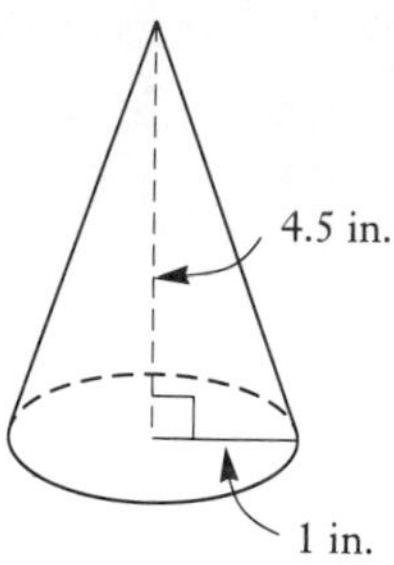

Complete.

28. $4\text{ yd}^2 = \_\_\_\_ \text{ ft}^2$

29. $0.3\text{ km}^2 = \_\_\_\_ \text{ m}^2$

30. $2070\text{ in}^2 = \_\_\_\_ \text{ ft}^2$

31. $600\text{ cm}^2 = \_\_\_\_ \text{ m}^2$

Solve.

32. A physician prescribed 780 mL per hour of a certain intravenous fluid for a patient. How many liters of fluid did this patient receive in one day?

SKILL MAINTENANCE

33. Find the simple interest on \$5000 at 9.5% for 30 days.

Evaluate.

34. 3^3

35. $(4.7)^2$

36. 4.7^3

37. $\left(\frac{1}{2}\right)^4$

Complete.

38. 2.5 mi = ____ ft

39. 144 in. = ____ yd

40. 4568 cm = ____ m

41. 4568 cm = ____ mm

❖ THINKING IT THROUGH

1. Do a report on where the names *Fahrenheit* and *Celsius* originated.
2. List and describe all the volume formulas that you have learned in this chapter.

NAME SECTION DATE

TEST: CHAPTER 10

Find the volume.

1.

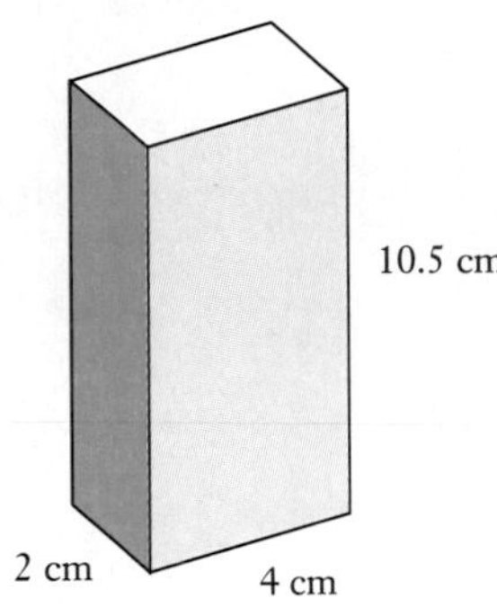

Complete.

2. 3080 mL = ______ L

3. 0.24 L = ______ mL

4. 4 lb = ______ oz

5. 4.11 T = ______ lb

6. 3.8 kg = ______ g

7. 4.325 mg = ______ cg

8. 2200 mg = ______ g

9. 5 hr = ______ min

10. 15 day = ______ hr

11. 64 pt = ______ qt

12. 10 gal = ______ oz

13. 5 cups = ______ oz

14. Convert 95°F to Celsius.

15. Convert 59°C to Fahrenheit.

ANSWERS

1. [10.1a] 84 cm^3

2. [10.1b] 3.08

3. [10.1b] 240

4. [10.3a] 64

5. [10.3a] 8220

6. [10.3b] 3800

7. [10.3b] 0.4325

8. [10.3b] 2.2

9. [10.3c] 300

10. [10.3c] 360

11. [10.1b] 32

12. [10.1b] 1280

13. [10.1b] 40

14. [10.4a] 35°C

15. [10.4b] 138.2°F

ANSWERS

16. [10.5a] 1728

17. [10.5b] 0.0003

18. [10.1a] 420 in^3

19. [10.2a] 1177.5 ft^3

20. [10.2b] $4186.\overline{6}$ yd^3

21. [10.2c] 113.04 cm^3

22. [7.8a] \$880

23. [1.9b] 1000

24. [1.9b] $\frac{1}{16}$

25. [1.9b] 9.8596

26. [1.9b] 0.00001

27. [9.1a] 42

28. [9.1a] 250

29. [9.2a] 2300

30. [9.2a] 3400

Complete.

16. $12 \text{ ft}^2 =$ ______ in^2

17. $3 \text{ cm}^2 =$ ______ m^2

18. A twelve-can carton of 12-oz soft drinks comes in a rectangular package $10\frac{1}{2}$ in. by 8 in. by 5 in. What is its volume?

Find the volume. Use 3.14 for π.

19.

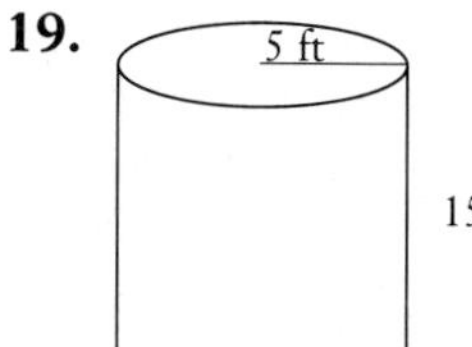

20.

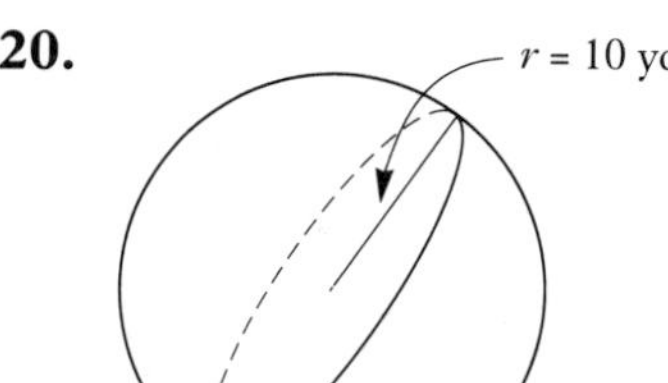

21.

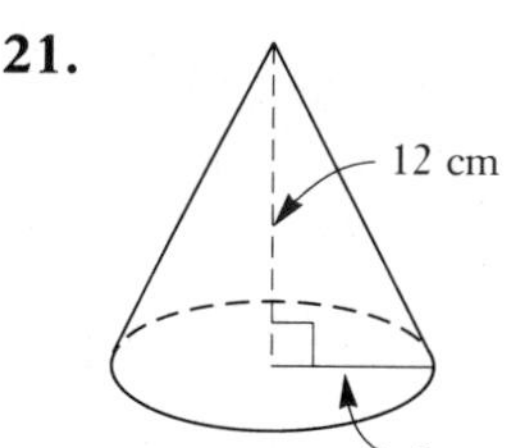

SKILL MAINTENANCE

22. Find the simple interest on \$10,000 at 8.8% for 1 year.

Evaluate.

23. 10^3

24. $\left(\frac{1}{4}\right)^2$

25. $(3.14)^2$

26. $(0.1)^5$

Complete.

27. 14 yd = ______ ft

28. 3000 in. = ______ ft

29. 2.3 km = ______ m

30. 34,000 mm = ______ cm

CUMULATIVE REVIEW: CHAPTERS 1–10

1. $1\frac{1}{2} + 2\frac{2}{3}$

2. $\left(\frac{1}{4}\right)^2 \div \left(\frac{1}{2}\right)^3 \times 2^4 + (10.3)(4)$

3. $120.5 - 32.98$

4. $22\overline{)27,148}$

5. $14 \div [33 \div 11 + 8 \times 2 - (15 - 3)]$

6. $8^3 + 45 \cdot 24 - 9^2 \div 3$

Find fractional notation.

7. 1.209

8. 17%

Use $<$, $>$, or $=$ for ■ to write a true sentence.

9. $\frac{5}{6}$ ■ $\frac{7}{8}$

10. $\frac{15}{18}$ ■ $\frac{10}{12}$

Complete.

11. 6 oz = ______ lb

12. 15°C = ______ °F

13. 0.087 L = ______ mL

14. 9 sec = ______ min

15. 3 yd^2 = ______ ft^2

16. 17 cm = ______ m

Find the perimeter and the area.

17.

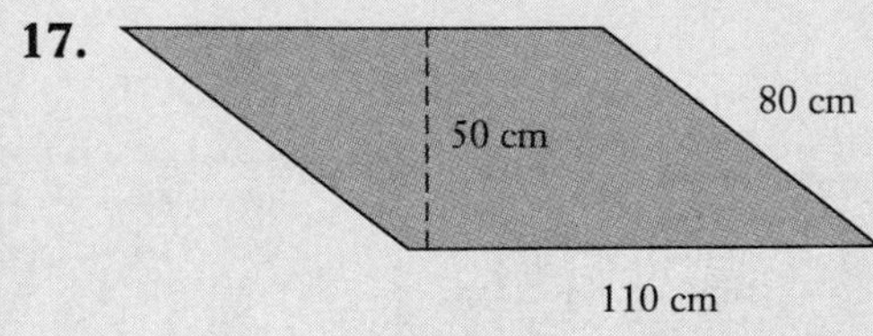

18.

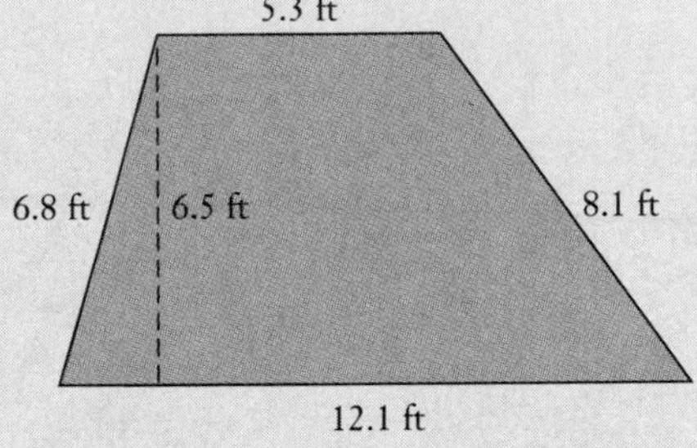

This line graph shows the average number of pounds of apples eaten per person in the United States.

19. What was the average number of pounds of apples that each person ate in 1987?

20. In what year did apple consumption decrease?

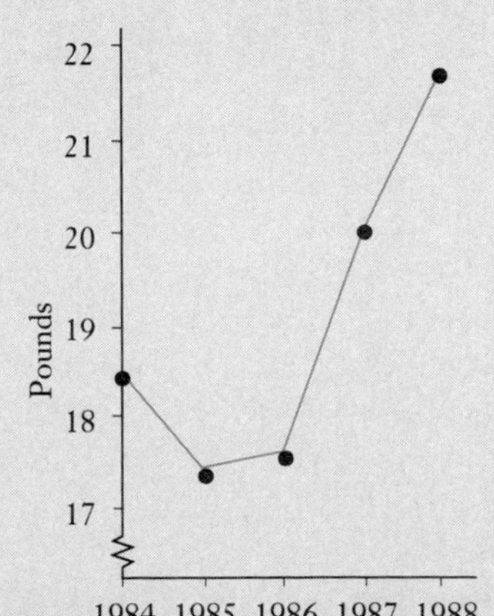

Solve.

21. $\frac{12}{15} = \frac{x}{18}$

22. $\frac{3}{x} = \frac{7}{10}$

23. $25 \cdot x = 2835$

24. $x + \frac{3}{4} = \frac{7}{8}$

25. To get an A in math, a student must score an average of 90 on five tests. On the first four tests, the scores were 85, 92, 79, and 95. What is the lowest score that the student can get on the last test and still get an A?

26. Americans own 52 million dogs, 56 million cats, 45 million birds, 250 million fish, and 125 million other creatures as house pets. How many pets do Americans own?

27. The diameter of a basketball is 20 cm. What is its volume? Use 3.14 for π.

28. What is the simple interest on \$800 at 12% for $\frac{1}{4}$ year?

29. How long must a rope be in order to reach from the top of an 8-m tree to a point on the ground 15 m from the bottom of the tree?

30. The sales tax on a purchase of \$5.50 is \$0.33. What is the sales tax rate?

31. A bolt of fabric in a fabric store has $10\frac{3}{4}$ yd on it. A customer purchases $8\frac{5}{8}$ yd. How many yards remain on the bolt?

32. What is the cost, in dollars, of 15.6 gal of gasoline at 139.9¢ per gallon? Round to the nearest cent.

33. A box of dry milk that makes 20 qt costs \$4.99. A box that makes 8 qt costs \$1.99. Which size has the lower unit price?

34. It is $\frac{7}{10}$ km from a student's house to the library. She starts to walk there, changes her mind after going $\frac{1}{4}$ of the distance, and returns home. How far did the student walk?

SYNTHESIS

35. A house sits on a lot measuring 75 ft by 200 ft. The lot is at the intersection of two streets, so there are sidewalks on two sides of the lot. In the winter, you have to shovel the snow off the sidewalks. If the sidewalks are 3 ft wide and the snow is 4 in. deep, what volume of snow must you shovel?

INTRODUCTION This chapter is the first of two that form an introduction to algebra. We expand on the numbers of arithmetic to new numbers called *real numbers.* We consider the rational numbers and the irrational numbers, which make up the real numbers. We will learn to add, subtract, multiply, and divide real numbers.

The review sections to be tested in addition to the material in this chapter are 2.1, 3.1, 7.3, and 9.4. ❖

The Real-Number System

AN APPLICATION

Death Valley is 280 ft below sea level. Express this fact using an integer.

THE MATHEMATICS

The integer that corresponds to this situation is -280. The elevation of Death Valley is

$$-280 \text{ ft.}$$

This is an integer. It is also a rational number and a real number.

❖ POINTS TO REMEMBER: CHAPTER 11

Skills at calculating with whole numbers
Skills at calculating with fractional and decimal notation

PRETEST: CHAPTER 11

Use either $<$ or $>$ for ▒ to write a true sentence.

1. 0 ▒ -5

2. 10 ▒ -5

3. -35 ▒ -45

4. $-\frac{2}{3}$ ▒ $\frac{4}{5}$

Find decimal notation.

5. $-\frac{5}{8}$

6. $-\frac{2}{3}$

7. $-\frac{10}{11}$

Find the absolute value.

8. $|-12|$

9. $|2.3|$

10. $|0|$

Find the opposite, or additive inverse.

11. 5.4

12. $-\frac{2}{3}$

Compute and simplify.

13. $-9 + (-8)$

14. $20.2 - (-18.4)$

15. $-\frac{5}{6} - \frac{3}{10}$

16. $-11.5 + 6.5$

17. $-9(-7)$

18. $\frac{5}{8}\left(-\frac{2}{3}\right)$

19. $-19.6 \div 0.2$

20. $-56 \div (-7)$

21. $12 - (-6) + 14 - 8$

22. $20 - 10 \div 5 + 2^3$

11.1 The Real Numbers

In this section, the *real numbers* are introduced. We begin with numbers called *integers* and build up to the real numbers.

We create the integers by starting with the whole numbers, 0, 1, 2, 3, and so on. For each natural number 1, 2, 3, and so on, we obtain a new number to the left of 0 on the number line:

For the number 1, there will be an *opposite* number -1 (negative 1).

For the number 2, there will be an *opposite* number -2 (negative 2).

For the number 3, there will be an *opposite* number -3 (negative 3), and so on.

The **integers** consist of the whole numbers and these new numbers. We picture them on a number line as follows.

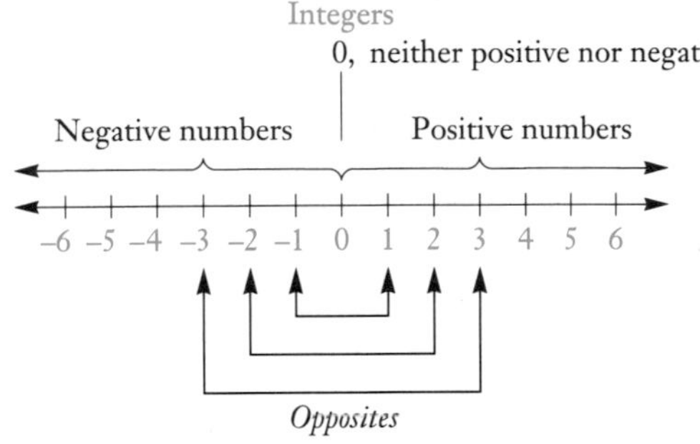

We call the newly obtained numbers **negative integers.** The natural numbers are called **positive integers.** Zero is neither positive nor negative. We call -1 and 1 opposites of each other. Similarly, -2 and 2 are opposites, -3 and 3 are opposites, -100 and 100 are opposites, and 0 is its own opposite. This gives us the integers, which extend infinitely to the left and right of 0.

The integers: . . . , $-5, -4, -3, -2, -1, 0, 1, 2, 3, 4, 5, \ldots$

a Integers and the Real World

Integers can be associated with many real-world problems and situations. The following examples will help you get ready to translate problem situations to mathematical language.

▶ **EXAMPLE 1** Tell which integer corresponds to this situation: The temperature is 3 degrees below zero.

3° below zero is $-3°$

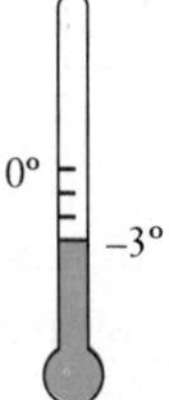

◀

OBJECTIVES

After finishing Section 11.1, you should be able to:

- a Tell which integers correspond to a real-world situation.
- b Graph rational numbers on a number line.
- c Convert from fractional notation for a rational number to decimal notation.
- d Determine which of two real numbers is greater and indicate which, using < or >.
- e Find the absolute value of a real number.

FOR EXTRA HELP

Tape 14A Tape 15B MAC: 11 IBM: 11

Tell which integers correspond to the situation.

1. The halfback gained 8 yd on the first down. The quarterback was sacked for a 5-yd loss on the second down. 8, −5

2. The highest temperature ever recorded in the United States was 134° in Death Valley on July 10, 1913. The coldest temperature ever recorded in the United States was 76° below zero in Tanana, Alaska, in January of 1886. 134, −76

3. At 10 sec before liftoff, ignition occurs. At 148 sec after liftoff, the first stage is detached from the rocket. −10, 148

4. A student owes $137 to the bookstore. The student has $289 in a savings account.

   −137, 289

ANSWERS ON PAGE A-6

▶ **EXAMPLE 2** Tell which integer corresponds to this situation: Death Valley is 280 feet below sea level.

The integer −280 corresponds to the situation. The elevation is −280 ft.

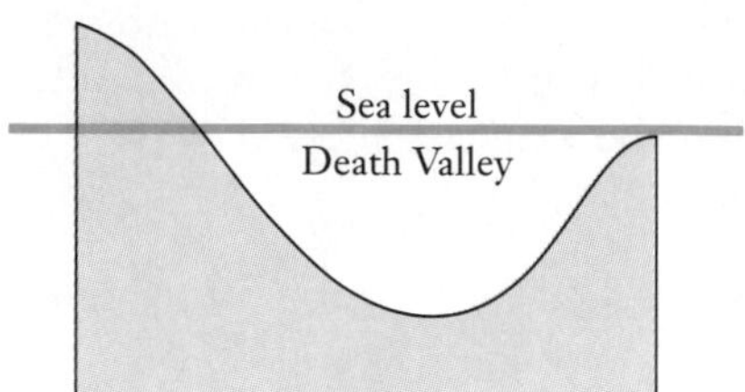

◀

▶ **EXAMPLE 3** Tell which integers correspond to this situation: A salesperson made $78 on Monday, but lost $57 on Tuesday.

The integers 78 and −57 correspond to the situation. The integer 78 corresponds to the profit on Monday and −57 corresponds to the loss on Tuesday. ◀

DO EXERCISES 1–4.

b The Rational Numbers

We created the integers by obtaining a negative number for each natural number. To create a larger number system, called the **rational numbers,** we consider quotients of integers with nonzero divisors. The following are rational numbers:

$$\frac{2}{3}, \quad -\frac{2}{3}, \quad \frac{7}{1}, \quad 4, \quad -3, \quad 0, \quad \frac{23}{-8}, \quad 2.4, \quad -0.17.$$

The number $-\frac{2}{3}$ (read "negative two-thirds") can also be named $\frac{2}{-3}$ or $\frac{-2}{3}$. The number 2.4 can be named $\frac{24}{10}$, or $\frac{12}{5}$, and −0.17 can be named $-\frac{17}{100}$.

Note that the rational numbers contain the whole numbers, the integers, and the numbers of arithmetic (also called the nonnegative rational numbers).

> **The *rational numbers* consist of all numbers that can be named in the form *a*/*b*, where *a* and *b* are integers and *b* is not 0.**

We picture the rational numbers on a number line, as follows. There is a point on the line for every rational number.

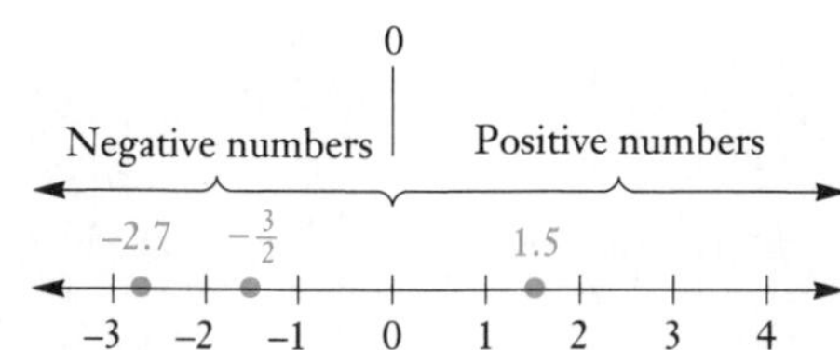

To **graph** a number means to find and mark its point on the line. Some numbers are graphed in the preceding figure.

► **EXAMPLE 4** Graph: $\frac{5}{2}$.

The number $\frac{5}{2}$ can be named $2\frac{1}{2}$ or 2.5. Its graph is halfway between 2 and 3.

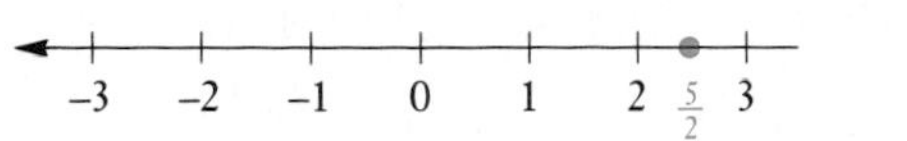

◄

► **EXAMPLE 5** Graph: −3.2.

The graph of −3.2 is $\frac{2}{10}$ of the way from −3 to −4.

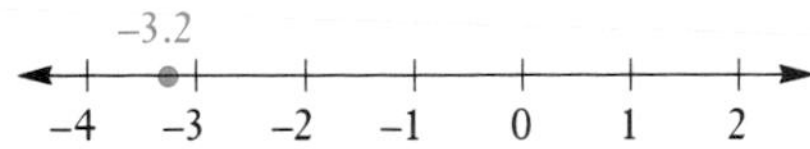

◄

► **EXAMPLE 6** Graph: $\frac{13}{8}$.

The number $\frac{13}{8}$ can be named $1\frac{5}{8}$ or 1.625. The graph is about $\frac{6}{10}$ of the way from 1 to 2.

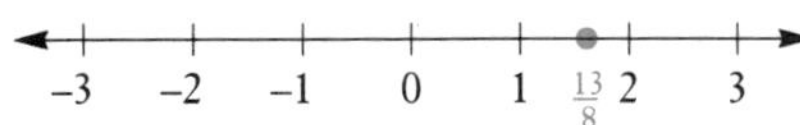

◄

DO EXERCISES 5–7.

C Notation for Rational Numbers

The rational numbers can be named using fractional or decimal notation.

► **EXAMPLE 7** Find decimal notation for $-\frac{5}{8}$.

We first find decimal notation for $\frac{5}{8}$. Since $\frac{5}{8}$ means 5 ÷ 8, we divide.

```
   0.6 2 5
8)5.0 0 0
  4 8
    2 0
    1 6
      4 0
      4 0
        0
```

Thus, $\frac{5}{8} = 0.625$, so $-\frac{5}{8} = -0.625$. ◄

Graph on a number line.

5. $-\frac{7}{2}$

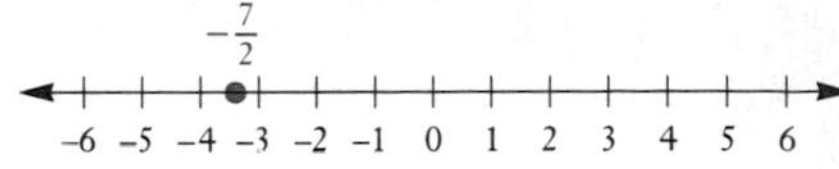

6. 1.4

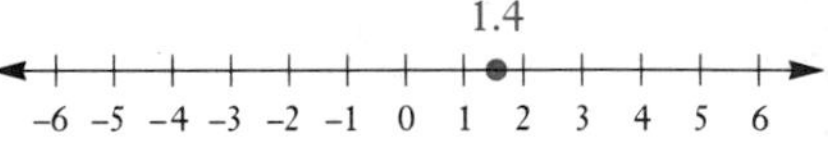

7. $-\frac{11}{4}$

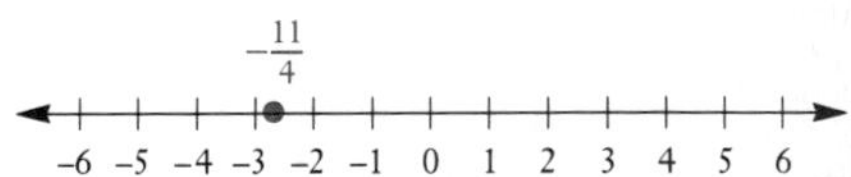

ANSWERS ON PAGE A-6

Find decimal notation.

8. $-\frac{3}{8}$ -0.375

9. $-\frac{6}{11}$ $-0.\overline{54}$

10. $\frac{4}{3}$ $1.\overline{3}$

ANSWERS ON PAGE A-6

Decimal notation for $-\frac{5}{8}$ is -0.625. We consider -0.625 to be a **terminating decimal.** Decimal notation for some numbers repeats.

▶ **EXAMPLE 8** Find decimal notation for $-\frac{7}{11}$.

We divide to find decimal notation for $\frac{7}{11}$.

$$\begin{array}{r} 0.6363\ldots \\ 11\overline{)7.0000} \\ \underline{66} \\ 40 \\ \underline{33} \\ 70 \\ \underline{66} \\ 40 \\ \underline{33} \\ 7 \end{array}$$

Thus, $\frac{7}{11} = 0.6363\ldots$, so $-\frac{7}{11} = -0.6363\ldots$. Repeating decimal notation can be abbreviated by writing a bar over the repeating part; in this case, we write $-0.\overline{63}$. ◀

DO EXERCISES 8–10.

d The Real Numbers and Order

The number line has a point for every rational number. However, there are some points on the line for which there are no rational numbers. These points correspond to what are called **irrational numbers.** Some examples of irrational numbers are π and $\sqrt{2}$.

Decimal notation for rational numbers *either* terminates *or* repeats. Decimal notation for irrational numbers *neither* terminates *nor* repeats. Some other examples of irrational numbers are $\sqrt{3}$, $-\sqrt{8}$, $\sqrt{11}$, and $0.121221222122221\ldots$. Whenever we take the square root of a number that is not a perfect square, we will get an irrational number.

> **The rational numbers and the irrational numbers together correspond to all the points on a number line and make up what is called the *real-number system.***

Real numbers are named in order on the number line, with larger numbers named further to the right. For any two numbers on the line, the one to the left is less than the one to the right.

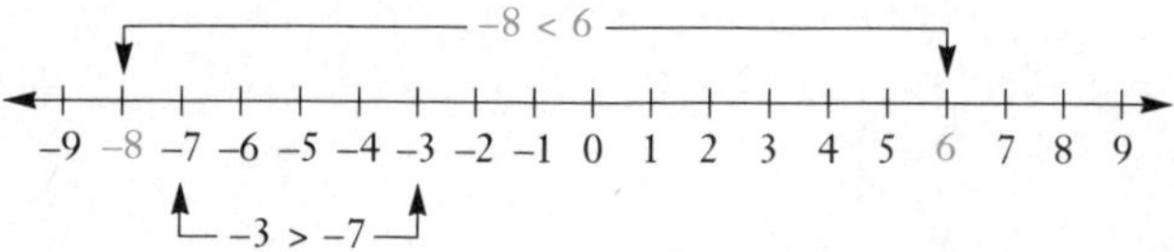

We use the symbol **<** to mean **"is less than."** The sentence $-8 < 6$ means "-8 is less than 6." The symbol **>** means **"is greater than."** The sentence $-3 > -7$ means "-3 is greater than -7."

▶ **EXAMPLES** Use either < or > for ▢ to write a true sentence.

9. -7 ▢ 3 Since -7 is to the left of 3, we have $-7 < 3$.

10. 6 ▢ -12 Since 6 is to the right of -12, then $6 > -12$.

11. -18 ▢ -5 Since -18 is to the left of -5, we have $-18 < -5$.

12. -2.7 ▢ $-\frac{3}{2}$ The answer is $-2.7 < -\frac{3}{2}$.

13. 1.5 ▢ -2.7 The answer is $1.5 > -2.7$.

14. -3.45 ▢ 1.32 The answer is $-3.45 < 1.32$.

15. $\frac{5}{8}$ ▢ $\frac{7}{11}$ We convert to decimal notation: $\frac{5}{8} = 0.625$, and $\frac{7}{11} = 0.6363\ldots$. Thus, $\frac{5}{8} < \frac{7}{11}$. ◀

DO EXERCISES 11–18.

e Absolute Value

From the number line, we see that numbers like 4 and -4 are the same distance from zero. Distance is always a nonnegative number. We call the distance from zero the **absolute value** of the number.

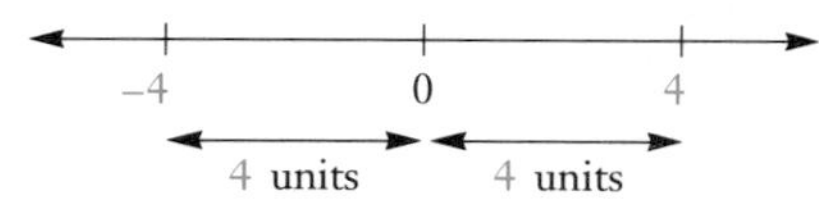

The *absolute value* of a number is its distance from zero on a number line. We use the symbol $|x|$ to represent the absolute value of a number x.

To find absolute value:

1. **If a number is negative, make it positive.**
2. **If a number is positive or zero, leave it alone.**

▶ **EXAMPLES** Find the absolute value.

16. $|-7|$ The distance of -7 from 0 is 7, so $|-7|$ is 7.

17. $|12|$ The distance of 12 from 0 is 12, so $|12|$ is 12.

18. $|0|$ The distance of 0 from 0 is 0, so $|0|$ is 0.

19. $\left|\frac{3}{2}\right| = \frac{3}{2}$

20. $|-2.73| = 2.73$ ◀

DO EXERCISES 19–23.

Use either < or > for ▢ to write a true sentence.

11. -3 ▢ 7 <

12. -8 ▢ -5 <

13. 7 ▢ -10 >

14. 3.1 ▢ -9.5 >

15. $-\frac{2}{3}$ ▢ -1 >

16. $-\frac{11}{8}$ ▢ $\frac{23}{15}$ <

17. $-\frac{2}{3}$ ▢ $-\frac{5}{9}$ <

18. -4.78 ▢ -5.01 >

Find the absolute value.

19. $|8|$ 8

20. $|0|$ 0

21. $|-9|$ 9

22. $\left|-\frac{2}{3}\right|$ $\frac{2}{3}$

23. $|5.6|$ 5.6

ANSWERS ON PAGE A-6

❖ SIDELIGHTS

An Application: Leap Years

Years divisible by 4 are leap years, except those indicating the beginning of a century (last two digits 00), in which case they must be divisible by 400 in order to be leap years. For example, the following are all leap years:

1980 (80 is divisible by 4);
1984 (84 is divisible by 4);
1988 (88 is divisible by 4);
2000 (2000 is divisible by 400).

But the following are not leap years:

1981 (81 is *not* divisible by 4);
2100 (2100 is *not* divisible by 400).

EXERCISES

Determine whether each year is a leap year.

1. 1992 Yes
2. 1990 No
3. 1492 Yes
4. 1941 No
5. 1950 No
6. 2200 No
7. 2001 No
8. 1996 Yes

NAME SECTION DATE

EXERCISE SET 11.1

a Tell which real numbers correspond to the situation.

1. In a game a player won 5 points. In the next game a player lost 12 points.

2. The temperature on Wednesday was 18° above zero. On Thursday, it was 2° below zero.

3. The Dead Sea, between Jordan and Israel, is 1286 ft below sea level, whereas Mt. Everest is 29,028 ft above sea level.

4. In bowling, team A is 34 pins behind team B after the first game. Describe the situation in two ways.

5. A student deposited \$750 in a savings account. Two weeks later, the student withdrew \$125.

6. During a certain time period, the United States had a deficit of \$3 million in foreign trade.

7. During a video game, a player intercepted a missile worth 20 points, lost a starship worth 150 points, and captured a base worth 300 points.

8. 3 seconds before liftoff of a rocket occurs. 128 seconds after liftoff of a rocket occurs.

b Graph the number on the number line.

9. $\frac{10}{3}$

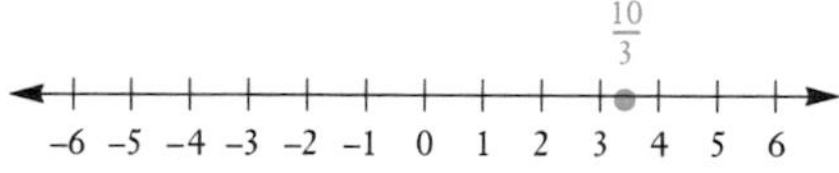

10. $-\frac{17}{5}$

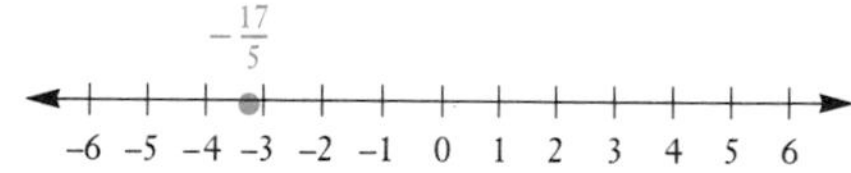

11. -4.3

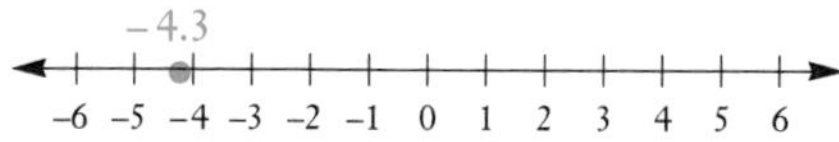

12. 3.87

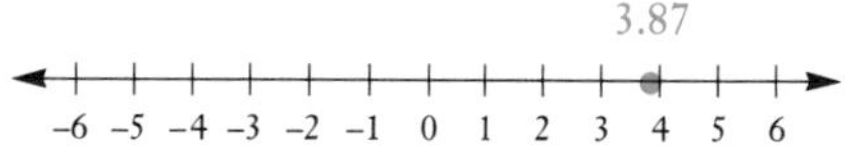

c Find decimal notation.

13. $-\frac{5}{8}$ 14. $-\frac{1}{8}$ 15. $-\frac{5}{3}$ 16. $-\frac{5}{6}$ 17. $-\frac{7}{6}$

18. $-\frac{5}{12}$ 19. $-\frac{7}{8}$ 20. $-\frac{1}{10}$ 21. $-\frac{7}{20}$

d Use either $<$ or $>$ for ▒ to write a true sentence.

22. 5 ▒ 0 23. 9 ▒ 0 24. -9 ▒ 5 25. 8 ▒ -8

ANSWERS

1. 5, −12
2. 18, −2
3. −1286, 29,028
4. A = B − 34; B = A + 34
5. 750, −125
6. −3,000,000
7. 20, −150, 300
8. −3, 128
9. See graph.
10. See graph.
11. See graph.
12. See graph.
13. −0.625
14. −0.125
15. $-1.\overline{6}$
16. $-0.8\overline{3}$
17. $-1.1\overline{6}$
18. $-0.41\overline{6}$
19. −0.875
20. −0.1
21. −0.35
22. $>$
23. $>$
24. $<$
25. $>$

ANSWERS

26. $<$
27. $>$
28. $<$
29. $<$
30. $>$
31. $>$
32. $<$
33. $>$
34. $>$
35. $<$
36. $<$
37. $>$
38. $<$
39. $>$
40. $<$
41. $<$
42. 3
43. 7
44. 10
45. 11
46. 0
47. 4
48. 24
49. 325
50. $\frac{2}{3}$
51. $\frac{10}{7}$
52. 0
53. 14.8
54. $2 \cdot 3 \cdot 3 \cdot 3$
55. $2 \cdot 2 \cdot 2 \cdot 2 \cdot 2 \cdot 2 \cdot 3$
56. 18
57. 96
58. $>$
59. $<$
60. $=$
61. $-\frac{5}{6}, -\frac{3}{4}, -\frac{2}{3}, \frac{1}{6}, \frac{3}{8}, \frac{1}{2}$
62.

26. $-6 \;\square\; 6$ **27.** $0 \;\square\; -7$ **28.** $-8 \;\square\; -5$ **29.** $-4 \;\square\; -3$

30. $-5 \;\square\; -11$ **31.** $-3 \;\square\; -4$ **32.** $-6 \;\square\; -5$ **33.** $-10 \;\square\; -14$

34. $2.14 \;\square\; 1.24$ **35.** $-3.3 \;\square\; -2.2$ **36.** $-14.5 \;\square\; 0.011$

37. $17.2 \;\square\; -1.67$ **38.** $-12.88 \;\square\; -6.45$ **39.** $-14.34 \;\square\; -17.88$

40. $\frac{5}{12} \;\square\; \frac{11}{25}$ **41.** $-\frac{14}{17} \;\square\; -\frac{27}{35}$

e Find the absolute value.

42. $|-3|$ **43.** $|-7|$ **44.** $|10|$ **45.** $|11|$

46. $|0|$ **47.** $|-4|$ **48.** $|-24|$ **49.** $|325|$

50. $\left|-\frac{2}{3}\right|$ **51.** $\left|-\frac{10}{7}\right|$ **52.** $\left|\frac{0}{4}\right|$ **53.** $|14.8|$

SKILL MAINTENANCE

Find the prime factorization.

54. 54 **55.** 192

Find the LCM.

56. 6, 18 **57.** 6, 24, 32

SYNTHESIS

Use either $<$, $>$, or $=$ for $\square$ to write a true sentence.

58. $|-5| \;\square\; |-2|$ **59.** $|4| \;\square\; |-7|$ **60.** $|-8| \;\square\; |8|$

List in order from the least to the greatest.

61. $-\frac{2}{3}, \frac{1}{2}, -\frac{3}{4}, -\frac{5}{6}, \frac{3}{8}, \frac{1}{6}$

62. $7^1, -5, |-6|, 4, |3|, -100, 0, 1^7, \frac{14}{4}$

$-100, -5, 0, 1^7, |3|, \frac{14}{4}, 4, |-6|, 7^1$

11.2 Addition of Real Numbers

In this section, we consider addition of real numbers. First, to gain an understanding, we add using a number line. Then we consider rules for addition.

OBJECTIVES

After finishing Section 11.2, you should be able to:

a Add real numbers without using a number line.

b Find the additive inverse, or opposite, of a real number.

FOR EXTRA HELP

Tape 14B Tape 15B MAC: 11 IBM: 11

Addition on a Number Line

Addition of numbers can be illustrated on a number line. To do the addition $a + b$, we start at a, and then move according to b.

a) If b is positive, we move to the right.

b) If b is negative, we move to the left.

c) If b is 0, we stay at a.

▶ **EXAMPLE 1** Add: $3 + (-5)$.

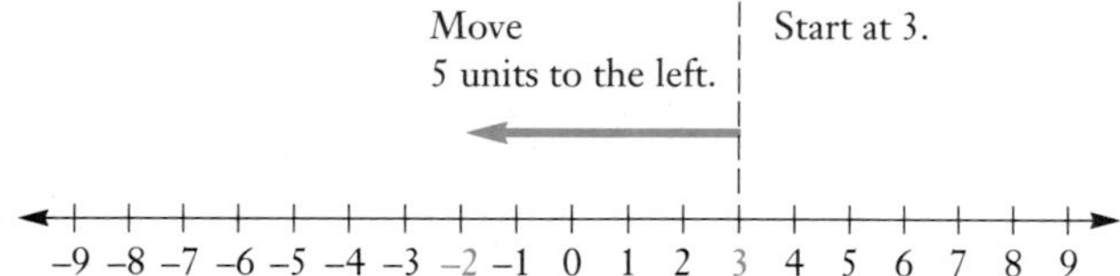

$3 + (-5) = -2$ ◀

▶ **EXAMPLE 2** Add: $-4 + (-3)$.

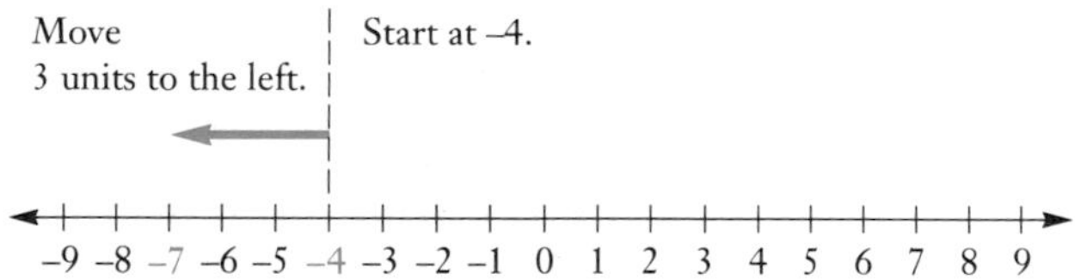

$-4 + (-3) = -7$ ◀

▶ **EXAMPLE 3** Add: $-4 + 9$.

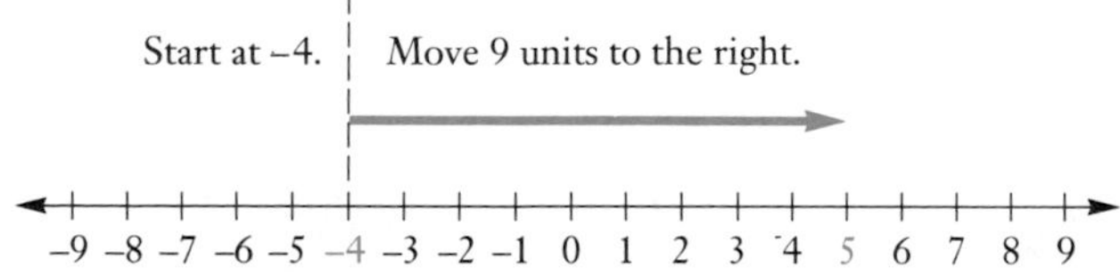

$-4 + 9 = 5$ ◀

▶ **EXAMPLE 4** Add: $-5.2 + 0$.

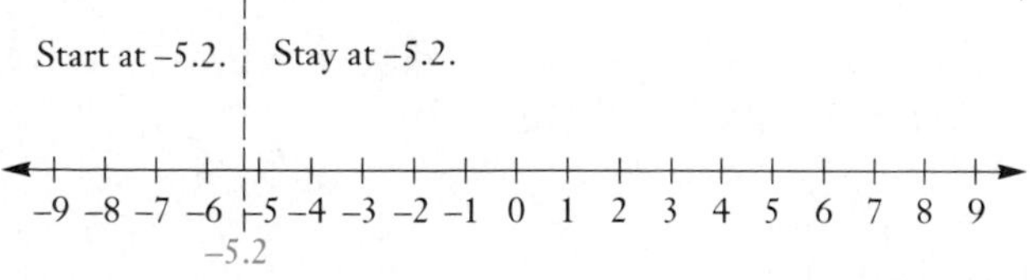

$-5.2 + 0 = -5.2$ ◀

DO EXERCISES 1–6.

Add using a number line.

1. $0 + (-8)$ -8

2. $1 + (-4)$ -3

3. $-3 + (-5)$ -8

4. $-3 + 7$ 4

5. $-5.4 + 5.4$ 0

6. $-\frac{5}{2} + \frac{1}{2}$ -2

ANSWERS ON PAGE A-6

Add without using a number line.

7. $-5 + (-6)$ -11

8. $-9 + (-3)$ -12

9. $-4 + 6$ 2

10. $-7 + 3$ -4

11. $5 + (-7)$ -2

12. $-20 + 20$ 0

13. $-11 + (-11)$ -22

14. $10 + (-7)$ 3

15. $-0.17 + 0.7$ 0.53

16. $-6.4 + 8.7$ 2.3

17. $-4.5 + (-3.2)$ -7.7

18. $-8.6 + 2.4$ -6.2

19. $\frac{5}{9} + \left(-\frac{7}{9}\right)$ $-\frac{2}{9}$

20. $-\frac{1}{5} + \left(-\frac{3}{4}\right)$ $-\frac{19}{20}$

ANSWERS ON PAGE A-6

a Addition Without a Number Line

You may have noticed some patterns in the previous examples. These lead us to rules for adding without using a number line that are more efficient for adding larger or more complicated numbers.

Rules for Addition of Real Numbers

1. ***Positive numbers:*** **Add the same as numbers of arithmetic. The answer is positive.**
2. ***Negative numbers:*** **Add absolute values. The answer is negative.**
3. ***A positive and a negative number:*** **Subtract absolute values. Then:**
   a) **If the positive number has the greater absolute value, the answer is positive.**
   b) **If the negative number has the greater absolute value, the answer is negative.**
   c) **If the numbers have the same absolute value, the answer is 0.**
4. ***One number is zero:*** **The sum is the other number.**

Rule 4 is known as the **Identity Property of Zero.** It says that for any real number a, $a + 0 = a$.

▶ **EXAMPLES** Add without using a number line.

5. $-12 + (-7) = -19$ Two negatives. Think: Add the absolute values, getting 19. Make the answer *negative*, -19.

6. $-1.4 + 8.5 = 7.1$ The absolute values are 1.4 and 8.5. The difference is 7.1. The positive number has the larger absolute value, so the answer is *positive*, 7.1.

7. $-36 + 21 = -15$ The absolute values are 36 and 21. The difference is 15. The negative number has the larger absolute value, so the answer is *negative*, -15.

8. $1.5 + (-1.5) = 0$ The numbers have the same absolute value. The sum is 0.

9. $-\frac{7}{8} + 0 = -\frac{7}{8}$ One number is zero. The sum is $-\frac{7}{8}$.

10. $-9.2 + 3.1 = -6.1$

11. $-\frac{3}{2} + \frac{9}{2} = \frac{6}{2} = 3$

12. $-\frac{2}{3} + \frac{5}{8} = -\frac{16}{24} + \frac{15}{24} = -\frac{1}{24}$ ◀

DO EXERCISES 7–20.

Suppose we wish to add several numbers, some positive and some negative, as follows. How can we proceed?

$$15 + (-2) + 7 + 14 + (-5) + (-12)$$

We can change grouping and order as we please when adding. For instance, we can group the positive numbers together and the negative numbers together and add them separately. Then we add the two results.

EXAMPLE 13 Add: $15 + (-2) + 7 + 14 + (-5) + (-12)$.

a) $15 + 7 + 14 = 36$ Adding the positive numbers

b) $-2 + (-5) + (-12) = -19$ Adding the negative numbers

c) $36 + (-19) = 17$ Adding the results

We can also add the numbers in any other order we wish, say, from left to right as follows:

$$\begin{aligned}15 + (-2) + 7 + 14 + (-5) + (-12) &= 13 + 7 + 14 + (-5) + (-12)\\ &= 20 + 14 + (-5) + (-12)\\ &= 34 + (-5) + (-12)\\ &= 29 + (-12)\\ &= 17\end{aligned}$$

DO EXERCISES 21–24.

b Opposites and Additive Inverses

Suppose we add two numbers that are opposites, such as 6 and -6. The result is 0. When opposites are added, the result is always 0. Such numbers are also called **opposites,** or **additive inverses.** Every real number has an opposite.

> **Two numbers whose sum is 0 are called *opposites,* or *additive inverses,* of each other.**

EXAMPLES Find the opposite.

14. 34 The opposite of 34 is -34 because $34 + (-34) = 0$.

15. -8 The opposite of -8 is 8 because $-8 + 8 = 0$.

16. 0 The opposite of 0 is 0 because $0 + 0 = 0$.

17. $-\frac{7}{8}$ The opposite of $-\frac{7}{8}$ is $\frac{7}{8}$ because $-\frac{7}{8} + \frac{7}{8} = 0$.

DO EXERCISES 25–30.

To name the opposite, we use the symbol $-$, as follows.

> **The opposite, or additive inverse, of a number *a* can be named $-a$ (read "the opposite of *a*," or "the additive inverse of *a*").**

Note that if we take a number, say, 8, and find its opposite, -8, and then find the opposite of the result, we will have the original number, 8, again.

> **The opposite of the opposite of a number is the number itself. The additive inverse of the additive inverse of a number is the number itself. That is, for any number *a*,**
>
> $$-(-a) = a.$$

Add.

21. $(-15) + (-37) + 25 + 42 + (-59) + (-14)$ -58

22. $42 + (-81) + (-28) + 24 + 18 + (-31)$ -56

23. $-2.5 + (-10) + 6 + (-7.5)$ -14

24. -35, 17, 14, -27, 31, -12 -12

Find the opposite.

25. -4 4

26. 8.7 -8.7

27. -7.74 7.74

28. $-\frac{8}{9}$ $\frac{8}{9}$

29. 0 0

30. 12 -12

ANSWERS ON PAGE A-6

Find $-x$ and $-(-x)$ when x is each of the following.

31. 14 −14, 14

32. 1 −1, 1

33. −19 19, −19

34. −1.6 1.6, −1.6

35. $\frac{2}{3}$ $-\frac{2}{3}, \frac{2}{3}$

36. $-\frac{9}{8}$ $\frac{9}{8}, -\frac{9}{8}$

Change the sign. (Find the opposite.)

37. −4 4

38. −13.4 13.4

39. 0 0

40. $\frac{1}{4}$ $-\frac{1}{4}$

ANSWERS ON PAGE A-6

▶ **EXAMPLE 18** Find $-x$ and $-(-x)$ when $x = 16$.

We replace x in each case by 16.

a) If $x = 16$, then $-x = -16$. **The opposite of 16 is −16.**

b) If $x = 16$, then $-(-x) = -(-16) = 16$. **The opposite of the opposite of 16 is 16.** ◀

▶ **EXAMPLE 19** Find $-x$ and $-(-x)$ when $x = -3$.

We replace x in each case by −3.

a) If $x = -3$, then $-x = -(-3) = 3$.

b) If $x = -3$, then $-(-x) = -(-(-3)) = -3$. ◀

Note that in Example 19 we used an extra set of parentheses to show that we are substituting the negative number −3 for x. Symbolism like $--x$ is not considered meaningful.

DO EXERCISES 31–36.

A symbol such as −8 is usually read "negative 8." It could be read "the additive inverse of 8," because the additive inverse of 8 is negative 8. It could also be read "the opposite of 8," because the opposite of 8 is −8. Thus a symbol like −8 can be read in more than one way. A symbol like $-x$, which has a variable, should be read "the opposite of x" or "the additive inverse of x" and *not* "negative x," because we do not know whether it represents a positive number, a negative number, or 0. You can verify this by referring to the preceding examples.

We can use the symbolism $-a$ to restate the definition of opposite, or additive inverse.

> **For any real number a, the opposite, or additive inverse, of a, $-a$, is such that**
>
> $$a + (-a) = (-a) + a = 0.$$

Signs of Numbers

A negative number is sometimes said to have a "negative sign." A positive number is said to have a "positive sign." When we replace a number by its opposite, we can say that we have "changed its sign."

▶ **EXAMPLES** Change the sign. (Find the opposite.)

20. −3 $-(-3) = 3$ **The opposite of −3 is 3.**

21. −10 $-(-10) = 10$

22. 0 $-(0) = 0$

23. 14 $-(14) = -14$ ◀

DO EXERCISES 37–40.

NAME SECTION DATE

EXERCISE SET 11.2

a Add. Do not use a number line except as a check.

1. $-9+2$ **2.** $2+(-5)$ **3.** $-10+6$ **4.** $8+(-3)$

5. $-8+8$ **6.** $6+(-6)$ **7.** $-3+(-5)$ **8.** $-4+(-6)$

9. $-7+0$ **10.** $-13+0$ **11.** $0+(-27)$ **12.** $0+(-35)$

13. $17+(-17)$ **14.** $-15+15$ **15.** $-17+(-25)$ **16.** $-24+(-17)$

17. $18+(-18)$ **18.** $-13+13$ **19.** $-18+18$ **20.** $11+(-11)$

21. $8+(-5)$ **22.** $-7+8$ **23.** $-4+(-5)$ **24.** $10+(-12)$

25. $13+(-6)$ **26.** $-3+14$ **27.** $-25+25$ **28.** $40+(-40)$

29. $63+(-18)$ **30.** $85+(-65)$ **31.** $-6.5+4.7$ **32.** $-3.6+1.9$

33. $-2.8+(-5.3)$ **34.** $-7.9+(-6.5)$ **35.** $-\frac{3}{5}+\frac{2}{5}$ **36.** $-\frac{4}{3}+\frac{2}{3}$

37. $-\frac{3}{7}+\left(-\frac{5}{7}\right)$ **38.** $-\frac{4}{9}+\left(-\frac{6}{9}\right)$ **39.** $-\frac{5}{8}+\frac{1}{4}$ **40.** $-\frac{5}{6}+\frac{2}{3}$

41. $-\frac{3}{7}+\left(-\frac{2}{5}\right)$ **42.** $-\frac{5}{8}+\left(-\frac{1}{3}\right)$ **43.** $-\frac{3}{5}+\left(-\frac{2}{15}\right)$ **44.** $-\frac{5}{9}+\left(-\frac{1}{18}\right)$

ANSWERS

1. -7
2. -3
3. -4
4. 5
5. 0
6. 0
7. -8
8. -10
9. -7
10. -13
11. -27
12. -35
13. 0
14. 0
15. -42
16. -41
17. 0
18. 0
19. 0
20. 0
21. 3
22. 1
23. -9
24. -2
25. 7
26. 11
27. 0
28. 0
29. 45
30. 20
31. -1.8
32. -1.7
33. -8.1
34. -14.4
35. $-\frac{1}{5}$
36. $-\frac{2}{3}$
37. $-\frac{8}{7}$
38. $-\frac{10}{9}$
39. $-\frac{3}{8}$
40. $-\frac{1}{6}$
41. $-\frac{29}{35}$
42. $-\frac{23}{24}$
43. $-\frac{11}{15}$
44. $-\frac{11}{18}$

ANSWERS

45. −6.3
46. −14.7
47. $\frac{7}{16}$
48. $\frac{1}{84}$
49. 39
50. −62
51. 50
52. 37.9
53. −1093
54. −1021
55. −24
56. 64
57. 26.9
58. −48.2
59. −9
60. 26
61. $\frac{14}{3}$
62. $-\frac{1}{328}$
63. −65
64. 29
65. $\frac{5}{3}$
66. −9.1
67. 14
68. 22.4
69. −10
70. $\frac{7}{8}$
71. All positive
72. All negative
73. Positive
74. Positive
75. Negative
76. Positive

45. $-5.7 + (-7.2) + 6.6$

46. $-10.3 + (-7.5) + 3.1$

47. $-\frac{7}{16} + \frac{7}{8}$

48. $-\frac{3}{28} + \frac{5}{42}$

49. $75 + (-14) + (-17) + (-5)$

50. $28 + (-44) + 17 + 31 + (-94)$

51. $-44 + \left(-\frac{3}{8}\right) + 95 + \left(-\frac{5}{8}\right)$

52. $24 + 3.1 + (-44) + (-8.2) + 63$

53. $98 + (-54) + 113 + (-998) + 44 + (-612) + (-18) + 334$

54. $-455 + (-123) + 1026 + (-919) + 213 + 111 + (-874)$

b Find the opposite, or additive inverse.

55. 24 **56.** −64 **57.** −26.9 **58.** 48.2

Find $-x$ when x is each of the following.

59. 9 **60.** −26 **61.** $-\frac{14}{3}$ **62.** $\frac{1}{328}$

Find $-(-x)$ when x is each of the following.

63. −65 **64.** 29 **65.** $\frac{5}{3}$ **66.** −9.1

Change the sign. (Find the opposite.)

67. −14 **68.** −22.4 **69.** 10 **70.** $-\frac{7}{8}$

SYNTHESIS

71. For what numbers x is $-x$ negative?

72. For what numbers x is $-x$ positive?

Tell whether the sum is positive, negative, or zero.

73. If n and m are positive, then $n + m$ is ________.

74. If n is positive and m is negative, then $n + (-m)$ is ________.

75. If n is positive and m is negative, then $-n + m$ is ________.

76. If $n = m$ and n and m are negative, then $-n + (-m)$ is ________.

11.3 Subtraction of Real Numbers

a We now consider subtraction of real numbers. Subtraction is defined as follows.

> **The difference $a - b$ is the number that when added to b gives a.**

For example, $45 - 17 = 28$ because $28 + 17 = 45$. Let us consider an example whose answer is a negative number.

▶ **EXAMPLE 1** Subtract: $5 - 8$.

Think: $5 - 8$ is the number that when added to 8 gives 5. What number can we add to 8 to get 5? The number must be negative. The number is -3:

$$5 - 8 = -3.$$

That is, $5 - 8 = -3$ because $5 = -3 + 8$. ◀

DO EXERCISES 1–3.

The definition above does *not* provide the most efficient way to do subtraction. From that definition, however, a faster way can be developed. Look for a pattern in the following examples.

Subtractions	*Adding an Opposite*
$5 - 8 = -3$	$5 + (-8) = -3$
$-6 - 4 = -10$	$-6 + (-4) = -10$
$-7 - (-10) = 3$	$-7 + 10 = 3$
$-7 - (-2) = -5$	$-7 + 2 = -5$

DO EXERCISES 4–7.

Perhaps you have noticed that we can subtract by adding the opposite of the number being subtracted. This can always be done.

> **For any real numbers a and b,**
>
> $$a - b = a + (-b).$$
>
> **(To subtract, add the opposite, or additive inverse, of the number being subtracted.)**

This is the method normally used for quick subtraction of real numbers.

▶ **EXAMPLES** Subtract. Check by addition.

2. $2 - 6 = 2 + (-6) = -4$ — The opposite of 6 is -6. We change the subtraction to addition and add the opposite. *Check:* $-4 + 6 = 2$.

3. $4 - (-9) = 4 + 9 = 13$ — The opposite of -9 is 9. We change the subtraction to addition and add the opposite. *Check:* $13 + (-9) = 4$.

OBJECTIVE

After finishing Section 11.3, you should be able to:

a Subtract real numbers and simplify combinations of additions and subtractions.

FOR EXTRA HELP

Tape 14C Tape 16A MAC: 11 IBM: 11

Subtract.

1. $-6 - 4$

Think: $-6 - 4$ is the number that when added to 4 gives -6. What number can be added to 4 to get -6? -10

2. $-7 - (-10)$

Think: $-7 - (-10)$ is the number that when added to -10 gives -7. What number can be added to -10 to get -7? 3

3. $-7 - (-2)$

Think: $-7 - (-2)$ is the number that when added to -2 gives -7. What number can be added to -2 to get -7? -5

Complete the addition and compare with the subtraction.

4. $4 - 6 = -2$;
$4 + (-6) =$ ______ -2

5. $-3 - 8 = -11$;
$-3 + (-8) =$ ______ -11

6. $-5 - (-9) = 4$;
$-5 + 9 =$ ______ 4

7. $-5 - (-3) = -2$;
$-5 + 3 =$ ______ -2

ANSWERS ON PAGE A-6

Subtract. Check by addition.

8. $2 - 8$ -6

9. $-6 - 10$ -16

10. $12.4 - 5.3$ 7.1

11. $-8 - (-11)$ 3

12. $-8 - (-8)$ 0

13. $\frac{2}{3} - \left(-\frac{5}{6}\right)$ $\frac{3}{2}$

Read each of the following. Then subtract by adding the opposite of the number being subtracted.

14. $3 - 11$ -8

15. $12 - 5$ 7

16. $-12 - (-9)$ -3

17. $-12.4 - 10.9$ -23.3

18. $-\frac{4}{5} - \left(-\frac{4}{5}\right)$ 0

Simplify.

19. $-6 - (-2) - (-4) - 12 + 3$

-9

20. $9 - (-6) + 7 - 11 - 14 - (-20)$

17

21. $-9.6 + 7.4 - (-3.9) - (-11)$

12.7

ANSWERS ON PAGE A-6

4. $-4.2 - (-3.6) = 4.2 + 3.6 = -0.6$ **Adding the opposite.** ***Check:*** $-0.6 + (-3.6) = -4.2$.

5. $-\frac{1}{2} - \left(-\frac{3}{4}\right) = -\frac{1}{2} + \frac{3}{4} = \frac{1}{4}$ **Adding the opposite.** ***Check:*** $\frac{1}{4} + \left(-\frac{3}{4}\right) = -\frac{1}{2}$. ◀

DO EXERCISES 8–13.

▶ **EXAMPLES** Read each of the following. Then subtract by adding the opposite of the number being subtracted.

6. $3 - 5$; Read "three minus 5"

$3 - 5 = 3 + (-5) = -2$ **Adding the opposite**

7. $\frac{1}{8} - \frac{7}{8}$; Read "one-eighth minus seven-eighths"

$\frac{1}{8} - \frac{7}{8} = \frac{1}{8} + \left(-\frac{7}{8}\right) = -\frac{6}{8}$, or $-\frac{3}{4}$

8. $-4.6 - (-9.8)$; Read "negative four point six minus negative nine point eight"

$-4.6 - (-9.8) = -4.6 + 9.8 = 5.2$

9. $-\frac{3}{4} - \frac{7}{5}$; Read "negative three-fourths minus seven-fifths"

$-\frac{3}{4} - \frac{7}{5} = -\frac{15}{20} + \left(-\frac{28}{20}\right) = -\frac{43}{20}$ ◀

DO EXERCISES 14–18.

When several additions and subtractions occur together, we can make them all additions.

▶ **EXAMPLES** Simplify.

10. $8 - (-4) - 2 - (-4) + 2 = 8 + 4 + (-2) + 4 + 2$
$= 16$

11. $8.2 - (-6.1) + 2.3 - (-4) = 8.2 + 6.1 + 2.3 + 4$
$= 20.6$ ◀

DO EXERCISES 19–21.

NAME SECTION DATE

EXERCISE SET 11.3

a Subtract.

1. $3 - 7$ **2.** $4 - 9$ **3.** $0 - 7$ **4.** $0 - 10$

5. $-8 - (-2)$ **6.** $-6 - (-8)$ **7.** $-10 - (-10)$ **8.** $-6 - (-6)$

9. $12 - 16$ **10.** $14 - 19$ **11.** $20 - 27$ **12.** $30 - 4$

13. $-9 - (-3)$ **14.** $-7 - (-9)$ **15.** $-11 - (-11)$ **16.** $-9 - (-9)$

17. $8 - (-3)$ **18.** $-7 - 4$ **19.** $-6 - 8$ **20.** $6 - (-10)$

21. $-4 - (-9)$ **22.** $-14 - 2$ **23.** $2 - 9$ **24.** $2 - 8$

25. $0 - 5$ **26.** $0 - 6$ **27.** $-5 - (-2)$ **28.** $-3 - (-1)$

29. $2 - 25$ **30.** $18 - 63$ **31.** $-42 - 26$ **32.** $-18 - 63$

33. $-71 - 2$ **34.** $-49 - 3$ **35.** $24 - (-92)$ **36.** $48 - (-73)$

37. $-2.8 - 0$ **38.** $6.04 - 1.1$ **39.** $\frac{3}{8} - \frac{5}{8}$ **40.** $\frac{3}{9} - \frac{9}{9}$

41. $\frac{3}{4} - \frac{2}{3}$ **42.** $\frac{5}{8} - \frac{3}{4}$ **43.** $-\frac{3}{4} - \frac{2}{3}$ **44.** $-\frac{5}{8} - \frac{3}{4}$

ANSWERS

1. -4
2. -5
3. -7
4. -10
5. -6
6. 2
7. 0
8. 0
9. -4
10. -5
11. -7
12. 26
13. -6
14. 2
15. 0
16. 0
17. 11
18. -11
19. -14
20. 16
21. 5
22. -16
23. -7
24. -6
25. -5
26. -6
27. -3
28. -2
29. -23
30. -45
31. -68
32. -81
33. -73
34. -52
35. 116
36. 121
37. -2.8
38. 4.94
39. $-\frac{1}{4}$
40. $-\frac{2}{3}$
41. $\frac{1}{12}$
42. $-\frac{1}{8}$
43. $-\frac{17}{12}$
44. $-\frac{11}{8}$

ANSWERS

45. $\frac{1}{8}$
46. $-\frac{1}{12}$
47. 19.9
48. 5
49. -9
50. -8.6
51. -0.01
52. -0.13
53. -2.7
54. 2.08
55. -3.53
56. 17.3
57. $-\frac{1}{2}$
58. $\frac{1}{8}$
59. $\frac{6}{7}$
60. 0
61. $-\frac{41}{30}$
62. 0
63. $-\frac{1}{156}$
64. $\frac{1}{42}$
65. 37
66. -22
67. -62
68. 22
69. 6
70. 4
71. 107
72. 116
73. 219
74. 190
75. 96.6 cm^2
76. $2\cdot2\cdot2\cdot2\cdot2\cdot3\cdot3\cdot3$
77. 116 m
78. 1767 m

45. $-\frac{5}{8}-\left(-\frac{3}{4}\right)$

46. $-\frac{3}{4}-\left(-\frac{2}{3}\right)$

47. $6.1-(-13.8)$

48. $1.5-(-3.5)$

49. $-3.2-5.8$

50. $-2.7-5.9$

51. $0.99-1$

52. $0.87-1$

53. $3-5.7$

54. $5.1-3.02$

55. $7-10.53$

56. $8-(-9.3)$

57. $\frac{1}{6}-\frac{2}{3}$

58. $-\frac{3}{8}-\left(-\frac{1}{2}\right)$

59. $-\frac{4}{7}-\left(-\frac{10}{7}\right)$

60. $\frac{12}{5}-\frac{12}{5}$

61. $-\frac{7}{10}-\frac{10}{15}$

62. $-\frac{4}{18}-\left(-\frac{2}{9}\right)$

63. $\frac{1}{13}-\frac{1}{12}$

64. $-\frac{1}{7}-\left(-\frac{1}{6}\right)$

Simplify.

65. $18-(-15)-3-(-5)+2$

66. $22-(-18)+7+(-42)-27$

67. $-31+(-28)-(-14)-17$

68. $-43-(-19)-(-21)+25$

69. $-93-(-84)-41-(-56)$

70. $84+(-99)+44-(-18)-43$

71. $-5-(-30)+30+40-(-12)$

72. $14-(-50)+20-(-32)$

73. $132-(-21)+45-(-21)$

74. $81-(-20)-14-(-50)+53$

SKILL MAINTENANCE

75. Find the area of a rectangle that is 8.4 cm by 11.5 cm.

76. Find the prime factorization of 864.

SYNTHESIS

77. The lowest point in Africa is Lake Assal, which is 156 m below sea level. The lowest point in South America is the Valdes Peninsula, which is 40 m below sea level. How much lower is Lake Assal than the Valdes Peninsula?

78. The deepest point in the Pacific Ocean is the Marianas Trench with a depth of 10,415 m. The deepest point in the Atlantic Ocean is the Puerto Rico Trench with a depth of 8648 m. How much higher is the Puerto Rico Trench than the Marianas Trench?

11.4 Multiplication of Real Numbers

a Multiplication of real numbers is very much like multiplication of numbers of arithmetic. The only difference is that we must determine whether the answer is positive or negative.

Multiplication of a Positive Number and a Negative Number

To see how to multiply a positive number and a negative number, consider the pattern of the following.

This number decreases by 1 each time. →

$$\begin{aligned} 4\cdot 5 &= 20 \\ 3\cdot 5 &= 15 \\ 2\cdot 5 &= 10 \\ 1\cdot 5 &= 5 \\ 0\cdot 5 &= 0 \\ -1\cdot 5 &= -5 \\ -2\cdot 5 &= -10 \\ -3\cdot 5 &= -15 \end{aligned}$$

← This number decreases by 5 each time.

DO EXERCISE 1.

According to this pattern, it looks as though the product of a negative number and a positive number is negative. That is the case, and we have the first part of the rule for multiplying numbers.

> **To multiply a positive number and a negative number, multiply their absolute values. The answer is negative.**

▶ **EXAMPLES** Multiply.

1. $8(-5) = -40$
2. $-\frac{1}{3}\cdot\frac{5}{7} = -\frac{5}{21}$
3. $(-7.2)5 = -36$ ◀

DO EXERCISES 2–7.

Multiplication of Two Negative Numbers

How do we multiply two negative numbers? Again we look for a pattern.

This number decreases by 1 each time. →

$$\begin{aligned} 4\cdot(-5) &= -20 \\ 3\cdot(-5) &= -15 \\ 2\cdot(-5) &= -10 \\ 1\cdot(-5) &= -5 \\ 0\cdot(-5) &= 0 \\ -1\cdot(-5) &= 5 \\ -2\cdot(-5) &= 10 \\ -3\cdot(-5) &= 15 \end{aligned}$$

← This number increases by 5 each time.

DO EXERCISE 8.

OBJECTIVES

After finishing Section 11.4, you should be able to:

a Multiply real numbers.

FOR EXTRA HELP

Tape 14D | Tape 16A | MAC: 11 IBM: 11

1. Complete, as in the example.

$$\begin{aligned} 4\cdot 10 &= 40 \\ 3\cdot 10 &= 30 \\ 2\cdot 10 &= 20 \\ 1\cdot 10 &= 10 \\ 0\cdot 10 &= 0 \\ -1\cdot 10 &= -10 \\ -2\cdot 10 &= -20 \\ -3\cdot 10 &= -30 \end{aligned}$$

Multiply.

2. $-3\cdot 6$ -18

3. $20\cdot(-5)$ -100

4. $4\cdot(-20)$ -80

5. $-\frac{2}{3}\cdot\frac{5}{6}$ $-\frac{5}{9}$

6. $-4.23(7.1)$ -30.033

7. $\frac{7}{8}\left(-\frac{4}{5}\right)$ $-\frac{7}{10}$

8. Complete, as in the example.

$$\begin{aligned} 3\cdot(-10) &= -30 \\ 2\cdot(-10) &= -20 \\ 1\cdot(-10) &= -10 \\ 0\cdot(-10) &= 0 \\ -1\cdot(-10) &= 10 \\ -2\cdot(-10) &= 20 \\ -3\cdot(-10) &= 30 \end{aligned}$$

ANSWERS ON PAGE A-6

Multiply.

9. $-3\cdot(-4)$ 12

10. $-16\cdot(-2)$ 32

11. $-7\cdot(-5)$ 35

12. $-\frac{4}{7}\left(-\frac{5}{9}\right)$ $\frac{20}{63}$

13. $-\frac{3}{2}\left(-\frac{4}{9}\right)$ $\frac{2}{3}$

14. $-3.25(-4.14)$ 13.455

Multiply.

15. $5(-6)$ -30

16. $(-5)(-6)$ 30

17. $(-3.2)\cdot 10$ -32

18. $\left(-\frac{4}{5}\right)\left(\frac{10}{3}\right)$ $-\frac{8}{3}$

Multiply.

19. $5\cdot(-3)\cdot 2$ -30

20. $-3\times(-4.1)\times(-2.5)$

-30.75

21. $-\frac{1}{2}\cdot\left(-\frac{4}{3}\right)\cdot\left(-\frac{5}{2}\right)$ $-\frac{5}{3}$

22. $-2\cdot(-5)\cdot(-4)\cdot(-3)$ 120

23. $(-4)(-5)(-2)(-3)(-1)$

-120

24. $(-1)(-1)(-2)(-3)(-1)(-1)$

6

ANSWERS ON PAGE A-6

According to the pattern, it looks as if the product of two negative numbers is positive. That is actually so, and we have the second part of the rule for multiplying real numbers.

To multiply two negative numbers, multiply their absolute values. The answer is positive.

DO EXERCISES 9–14.

The following is an alternative way to consider the rules we have for multiplication.

To multiply two real numbers:

1. **Multiply the absolute values.**
2. **If the signs are the same, the answer is positive.**
3. **If the signs are different, the answer is negative.**

▶ **EXAMPLES** Multiply.

4. $(-3)(-4) = 12$

5. $-1.6(2) = -3.2$

6. $9\cdot(-15) = -135$

7. $\left(-\frac{5}{6}\right)\left(-\frac{1}{9}\right) = \frac{5}{54}$ ◀

DO EXERCISES 15–18.

Multiplying More than Two Numbers

When multiplying more than two real numbers, we can choose order and grouping as we please.

▶ **EXAMPLES** Multiply.

8. $-8\cdot 2(-3) = -16(-3)$ Multiplying the first two numbers

$= 48$ Multiplying the results

9. $-8\cdot 2(-3) = 24\cdot 2$ Multiplying the negatives. Every pair of negative numbers gives a positive product.

$= 48$

10. $-3(-2)(-5)(4) = 6(-5)(4)$ Multiplying the first two numbers

$= (-30)4 = -120$

11. $(-\frac{1}{2})(8)(-\frac{2}{3})(-6) = (-4)4$ Multiplying the first two numbers and the last two numbers

$= -16$

12. $-5\cdot(-2)\cdot(-3)\cdot(-6) = 10\cdot 18 = 180$

13. $(-3)(-5)(-2)(-3)(-6) = (-30)(18) = -540$ ◀

We can see the following pattern in the results of Examples 12 and 13.

The product of an even number of negative numbers is positive. The product of an odd number of negative numbers is negative.

DO EXERCISES 19–24.

NAME SECTION DATE

EXERCISE SET 11.4

a Multiply.

1. $-8 \cdot 2$

2. $-2 \cdot 5$

3. $8 \cdot (-3)$

4. $9 \cdot (-5)$

5. $-9 \cdot 8$

6. $-10 \cdot 3$

7. $-8 \cdot (-2)$

8. $-2 \cdot (-5)$

9. $-7 \cdot (-6)$

10. $-9 \cdot (-2)$

11. $15 \cdot (-8)$

12. $-12 \cdot (-10)$

13. $-14 \cdot 17$

14. $-13 \cdot (-15)$

15. $-25 \cdot (-48)$

16. $39 \cdot (-43)$

17. $-3.5 \cdot (-28)$

18. $97 \cdot (-2.1)$

19. $4 \cdot (-3.1)$

20. $3 \cdot (-2.2)$

21. $-6 \cdot (-4)$

22. $-5 \cdot (-6)$

23. $-7 \cdot (-3.1)$

24. $-4 \cdot (-3.2)$

25. $\frac{2}{3} \cdot \left(-\frac{3}{5}\right)$

26. $\frac{5}{7} \cdot \left(-\frac{2}{3}\right)$

27. $-\frac{3}{8} \cdot \left(-\frac{2}{9}\right)$

28. $-\frac{5}{8} \cdot \left(-\frac{2}{5}\right)$

29. -6.3×2.7

30. -4.1×9.5

31. $-\frac{5}{9} \cdot \frac{3}{4}$

32. $-\frac{8}{3} \cdot \frac{9}{4}$

33. $7 \cdot (-4) \cdot (-3) \cdot 5$

ANSWERS

1. -16
2. -10
3. -24
4. -45
5. -72
6. -30
7. 16
8. 10
9. 42
10. 18
11. -120
12. 120
13. -238
14. 195
15. 1200
16. -1677
17. 98
18. -203.7
19. -12.4
20. -6.6
21. 24
22. 30
23. 21.7
24. 12.8
25. $-\frac{2}{5}$
26. $-\frac{10}{21}$
27. $\frac{1}{12}$
28. $\frac{1}{4}$
29. -17.01
30. -38.95
31. $-\frac{5}{12}$
32. -6
33. 420

34. $9 \cdot (-2) \cdot (-6) \cdot 7$

35. $-\frac{2}{3} \cdot \frac{1}{2} \cdot \left(-\frac{6}{7}\right)$

36. $-\frac{1}{8} \cdot \left(-\frac{1}{4}\right) \cdot \left(-\frac{3}{5}\right)$

37. $-3 \cdot (-4) \cdot (-5)$

38. $-2 \cdot (-5) \cdot (-7)$

39. $-2 \cdot (-5) \cdot (-3) \cdot (-5)$

40. $-3 \cdot (-5) \cdot (-2) \cdot (-1)$

41. $\frac{1}{5}\left(-\frac{2}{9}\right)$

42. $-\frac{3}{5}\left(-\frac{2}{7}\right)$

43. $-7 \cdot (-21) \cdot 13$

44. $-14 \cdot (34) \cdot 12$

45. $-4 \cdot (-1.8) \cdot 7$

46. $-8 \cdot (-1.3) \cdot (-5)$

47. $-\frac{1}{9}\left(-\frac{2}{3}\right)\left(\frac{5}{7}\right)$

48. $-\frac{7}{2}\left(-\frac{5}{7}\right)\left(-\frac{2}{5}\right)$

49. $4 \cdot (-4) \cdot (-5) \cdot (-12)$

50. $-2 \cdot (-3) \cdot (-4) \cdot (-5)$

51. $0.07 \cdot (-7) \cdot 6 \cdot (-6)$

52. $80 \cdot (-0.8) \cdot (-90) \cdot (-0.09)$

53. $\left(-\frac{5}{6}\right)\left(\frac{1}{8}\right)\left(-\frac{3}{7}\right)\left(-\frac{1}{7}\right)$

54. $\left(\frac{4}{5}\right)\left(-\frac{2}{3}\right)\left(-\frac{15}{7}\right)\left(\frac{1}{2}\right)$

55. $(-14) \cdot (-27) \cdot (-2)$

56. $7 \cdot (-6) \cdot 5 \cdot (-4) \cdot 3 \cdot (-2) \cdot 1 \cdot (-1)$

57. $(-8)(-9)(-10)$

58. $(-7)(-8)(-9) \cdot (-10)$

59. $(-6)(-7)(-8) \cdot (-9)(-10)$

60. $(-5)(-6)(-7)(-8) \cdot (-9)(-10)$

SKILL MAINTENANCE

61. Find the prime factorization of 4608. $2 \cdot 2 \cdot 2 \cdot 2 \cdot 2 \cdot 2 \cdot 2 \cdot 2 \cdot 2 \cdot 3 \cdot 3$

62. Find the LCM of 36 and 54.

63. 23 is what percent of 69?

64. What is 36% of 729?

SYNTHESIS

Simplify. Keep in mind the rules for order of operations given in Section 1.9.

65. $-6[(-5) + (-7)]$

66. $-3[(-8) + (-6)]\left(-\frac{1}{7}\right)$

67. $-(3^5) \cdot [-(2^3)]$

68. $4(2^4) \cdot [-(3^3)] \cdot 6$

69. $|(-2)^3 + 4^2| - (2 - 7)^2$

70. $|-11(-3)^2 - 5^3 - 6^2 - (-4)^2|$

71. What must be true of m and n if $-mn$ is to be (a) positive? (b) zero? (c) negative? (a) One must be negative and one must be positive. (b) Either or both must be zero. (c) Both must be negative or both must be positive.

ANSWERS

34. 756

35. $\frac{2}{7}$

36. $-\frac{3}{160}$

37. -60

38. -70

39. 150

40. 30

41. $-\frac{2}{45}$

42. $\frac{6}{35}$

43. 1911

44. -5712

45. 50.4

46. -52

47. $\frac{10}{189}$

48. -1

49. -960

50. 120

51. 17.64

52. -518.4

53. $-\frac{5}{784}$

54. $\frac{4}{7}$

55. -756

56. 5040

57. -720

58. 5040

59. $-30,240$

60. 151,200

61.

62. 108

63. $33\frac{1}{3}\%$, or $33.\overline{3}\%$

64. 262.44

65. 72

66. -6

67. 1944

68. $-10,368$

69. -17

70. 276

71. a)

b)

c)

11.5 Division and Order of Operations

We now consider division of real numbers. The definition of division results in rules for division very much like those for multiplication.

a Division of Integers

The quotient $\frac{a}{b}$ (or $a \div b$) is the number, if there is one, that when multiplied by b gives a.

Let us use the definition to divide integers.

▶ **EXAMPLES** Divide, if possible. Check your answer.

1. $14 \div (-7) = -2$ — We look for a number that when multiplied by -7 gives 14. That number is -2. *Check:* $(-2)(-7) = 14$.

2. $\frac{-32}{-4} = 8$ — We look for a number that when multiplied by -4 gives -32. The number is 8. *Check:* $8(-4) = -32$.

3. $\frac{-10}{7} = -\frac{10}{7}$ — We look for a number that when multiplied by 7 gives -10. That number is $-\frac{10}{7}$. *Check:* $-\frac{10}{7} \cdot 7 = -10$.

4. $\frac{-17}{0}$ **is undefined.** — We look for a number that when multiplied by 0 gives -17. There is no such number because the product of 0 and *any* number is 0. ◀

The rules for division are the same as those for multiplication. We state them together.

To multiply or divide two real numbers:

1. **Multiply or divide the absolute values.**
2. **If the signs are the same, the answer is positive.**
3. **If the signs are different, the answer is negative.**

DO EXERCISES 1–8.

Division by Zero

Example 4 shows why we cannot divide -17 by 0. We can use the same argument to show why we cannot divide any nonzero number b by 0. Consider $b \div 0$. We look for a number that when multiplied by 0 gives b. There is no such number because the product of 0 and any number is 0. Thus we cannot divide a nonzero number b by 0.

On the other hand, if we divide 0 by 0, we look for a number r such that $0 \cdot r = 0$. But $0 \cdot r = 0$ for any number r. Thus it appears that $0 \div 0$ could be any number we choose. Getting any answer we want when we divide 0 by 0 would be very confusing. Thus we agree that division by zero is undefined.

Division by zero is undefined. That is, $a \div 0$ is undefined for all real numbers a. But $0 \div a = 0$, when a is nonzero. That is, 0 divided by a nonzero number is 0.

OBJECTIVES

After finishing Section 11.5, you should be able to:

- a Divide integers.
- b Find the reciprocal of a real number.
- c Divide real numbers.
- d Simplify expressions using rules for order of operations.

FOR EXTRA HELP

Tape 14E

Tape 16B

MAC: 11
IBM: 11

Divide.

1. $6 \div (-3)$ -2

2. $\frac{-15}{-3}$ 5

3. $-24 \div 8$ -3

4. $\frac{-72}{-8}$ 9

5. $\frac{30}{-5}$ -6

6. $\frac{30}{-7}$ $-\frac{30}{7}$

7. $\frac{-5}{0}$ Undefined

8. $\frac{0}{-3}$ 0

ANSWERS ON PAGE A-6

Find the reciprocal.

9. $\frac{2}{3}$ $\frac{3}{2}$

10. $-\frac{5}{4}$ $-\frac{4}{5}$

11. -3 $-\frac{1}{3}$

12. $-\frac{1}{5}$ -5

13. 5.78 $\frac{1}{5.78}$

14. $-\frac{2}{3}$ $-\frac{3}{2}$

15. Complete the following table.

Number	Opposite	Reciprocal
$\frac{2}{3}$	$-\frac{2}{3}$	$\frac{3}{2}$
$-\frac{5}{4}$	$\frac{5}{4}$	$-\frac{4}{5}$
0	0	Undefined
1	-1	1
-4.5	4.5	$-\frac{1}{4.5}$

ANSWERS ON PAGE A-6

b Reciprocals

When two numbers such as $\frac{7}{8}$ and $\frac{8}{7}$ are multiplied, the result is 1. Such numbers are called **reciprocals** of each other. Every nonzero real number has a reciprocal, also called a **multiplicative inverse.**

Two numbers whose product is 1 are called *reciprocals* of each other.

▶ **EXAMPLES** Find the reciprocal.

5. -5 The reciprocal of -5 is $-\frac{1}{5}$ because $-5\left(-\frac{1}{5}\right) = 1$.

6. $-\frac{1}{2}$ The reciprocal of $-\frac{1}{2}$ is -2 because $\left(-\frac{1}{2}\right)(-2) = 1$.

7. $-\frac{2}{3}$ The reciprocal of $-\frac{2}{3}$ is $-\frac{3}{2}$ because $\left(-\frac{2}{3}\right)\left(-\frac{3}{2}\right) = 1$. ◀

For $a \neq 0$, the reciprocal of a can be named $\frac{1}{a}$ and the reciprocal of $\frac{1}{a}$ is a.

The reciprocal of a nonzero number $\frac{a}{b}$ can be named $\frac{b}{a}$.

The number 0 has no reciprocal.

DO EXERCISES 9–14.

The reciprocal of a positive number is also a positive number, because their product must be the positive number 1. The reciprocal of a negative number is also a negative number, because their product must be the positive number 1.

The reciprocal of a number has the same sign as the number itself.

It is important *not* to confuse *opposite* with *reciprocal.* Keep in mind that the opposite, or additive inverse, of a number is what we add to the number to get 0, whereas a reciprocal is what we multiply the number by to get 1. Compare the following.

Number	Opposite (Change the sign.)	Reciprocal (Invert but do not change the sign.)
$-\frac{3}{8}$	$\frac{3}{8}$	$-\frac{8}{3}$
19	-19	$\frac{1}{19}$
$\frac{18}{7}$	$-\frac{18}{7}$	$\frac{7}{18}$
-7.9	7.9	$-\frac{1}{7.9}$ or $-\frac{10}{79}$
0	0	Undefined

DO EXERCISE 15.

C Division of Real Numbers

We know that we can subtract by adding an opposite. Similarly, we can divide by multiplying by a reciprocal.

For any real numbers a and b, $b \neq 0$,

$$\frac{a}{b} = a \cdot \frac{1}{b}.$$

(To divide, we can multiply by the reciprocal of the divisor.)

▶ **EXAMPLES** Rewrite the division as a multiplication.

8. $-4 \div 3$ $\quad -4 \div 3$ is the same as $-4 \cdot \frac{1}{3}$

9. $\frac{6}{-7}$ $\quad \frac{6}{-7} = 6\left(-\frac{1}{7}\right)$

10. $\frac{3}{5} \div \left(-\frac{9}{7}\right)$ $\quad \frac{3}{5} \div \left(-\frac{9}{7}\right) = \frac{3}{5}\left(-\frac{7}{9}\right)$ ◀

DO EXERCISES 16–20.

When actually doing division calculations, we sometimes multiply by a reciprocal and we sometimes divide directly. With fractional notation, it is usually better to multiply by a reciprocal. With decimal notation, it is usually better to divide directly.

▶ **EXAMPLES** Divide by multiplying by the reciprocal of the divisor.

11. $\frac{2}{3} \div \left(-\frac{5}{4}\right) = \frac{2}{3} \cdot \left(-\frac{4}{5}\right) = -\frac{8}{15}$

12. $-\frac{5}{6} \div \left(-\frac{3}{4}\right) = -\frac{5}{6} \cdot \left(-\frac{4}{3}\right) = \frac{20}{18} = \frac{10 \cdot 2}{9 \cdot 2} = \frac{10}{9} \cdot \frac{2}{2} = \frac{10}{9}$

Be careful not to change the sign when taking a reciprocal!

13. $-\frac{3}{4} \div \frac{3}{10} = -\frac{3}{4} \cdot \left(\frac{10}{3}\right) = -\frac{30}{12} = -\frac{5}{2} \cdot \frac{6}{6} = -\frac{5}{2}$ ◀

With decimal notation, it is easier to carry out long division than to multiply by the reciprocal.

▶ **EXAMPLES** Divide.

14. $-27.9 \div (-3) = \frac{-27.9}{-3} = 9.3$ $\quad$ Do the long division $3\overline{)27.9}$ (quotient 9.3)
The answer is positive.

15. $-6.3 \div 2.1 = -3$ $\quad$ Do the long division $2.1\overline{)6.3\ 0}$ (quotient 3.0)
The answer is negative. ◀

DO EXERCISES 21–24.

Rewrite the division as a multiplication.

16. $\frac{4}{7} \div \left(-\frac{3}{5}\right)$ $\quad \frac{4}{7} \cdot \left(-\frac{5}{3}\right)$

17. $\frac{5}{-8}$ $\quad 5 \cdot \left(-\frac{1}{8}\right)$

18. $\frac{-10}{7}$ $\quad -10 \cdot \left(\frac{1}{7}\right)$

19. $-\frac{2}{3} \div \frac{4}{7}$ $\quad -\frac{2}{3} \cdot \frac{7}{4}$

20. $-5 \div 7$ $\quad -5 \cdot \left(\frac{1}{7}\right)$

Divide by multiplying by the reciprocal of the divisor.

21. $\frac{4}{7} \div \left(-\frac{3}{5}\right)$ $\quad -\frac{20}{21}$

22. $-\frac{8}{5} \div \frac{2}{3}$ $\quad -\frac{12}{5}$

23. $-\frac{12}{7} \div \left(-\frac{3}{4}\right)$ $\quad \frac{16}{7}$

24. $21.7 \div (-3.1)$ $\quad -7$

ANSWERS ON PAGE A-6

Simplify.

25. $23 - 42 \cdot 30$ -1237

26. $52 \cdot 5 + 5^3 - (4^2 - 48 \div 4)$ 381

ANSWERS ON PAGE A-6

d Order of Operations

When several operations are to be done in a calculation or a problem, we apply the same rules that we did in Section 1.9. We repeat them for review. If you did not study that section before, you should do so before continuing.

Rules for Order of Operations

1. **Do all calculations within parentheses before operations outside.**
2. **Evaluate all exponential expressions.**
3. **Do all multiplications and divisions in order from left to right.**
4. **Do all additions and subtractions in order from left to right.**

These rules are consistent with the way most computers do calculations.

▶ **EXAMPLE 16** Simplify: $-34 \cdot 56 - 17$.

There are no parentheses or powers so we start with the third step.

$-34 \cdot 56 - 17 = -1904 - 17$ Carrying out all multiplications and divisions in order from left to right

$= -1921$ Carrying out all additions and subtractions in order from left to right ◀

▶ **EXAMPLE 17** Simplify: $2^4 + 51 \cdot 4 - (37 + 23 \cdot 2)$.

$2^4 + 51 \cdot 4 - (37 + 23 \cdot 2)$

$= 2^4 + 51 \cdot 4 - (37 + 46)$ Carrying out all operations inside parentheses first, multiplying 23 by 2, following the rules for order of operations within the parentheses

$= 2^4 + 51 \cdot 4 - 83$ Completing the addition inside parentheses

$= 16 + 51 \cdot 4 - 83$ Evaluating exponential expressions

$= 16 + 204 - 83$ Doing all multiplications

$= 220 - 83$ Doing all additions and subtractions in order from left to right

$= 137$ ◀

DO EXERCISES 25 AND 26.

NAME SECTION DATE

EXERCISE SET 11.5

a Divide, if possible. Check each answer.

1. $36 \div (-6)$ **2.** $\frac{28}{-7}$ **3.** $\frac{26}{-2}$ **4.** $26 \div (-13)$

5. $\frac{-16}{8}$ **6.** $-22 \div (-2)$ **7.** $\frac{-48}{-12}$ **8.** $-63 \div (-9)$

9. $\frac{-72}{9}$ **10.** $\frac{-50}{25}$ **11.** $-100 \div (-50)$ **12.** $\frac{-200}{8}$

13. $-108 \div 9$ **14.** $\frac{-64}{-7}$ **15.** $\frac{200}{-25}$ **16.** $-300 \div (-13)$

17. $\frac{75}{0}$ **18.** $\frac{0}{-5}$ **19.** $\frac{81}{-9}$ **20.** $\frac{-145}{-5}$

b Find the reciprocal.

21. $-\frac{15}{7}$ **22.** $-\frac{3}{8}$ **23.** 13 **24.** -10

c Divide.

25. $\frac{3}{4} \div \left(-\frac{2}{3}\right)$ **26.** $\frac{7}{8} \div \left(-\frac{1}{2}\right)$ **27.** $-\frac{5}{4} \div \left(-\frac{3}{4}\right)$ **28.** $-\frac{5}{9} \div \left(-\frac{5}{6}\right)$

29. $-\frac{2}{7} \div \left(-\frac{4}{9}\right)$ **30.** $-\frac{3}{5} \div \left(-\frac{5}{8}\right)$ **31.** $-\frac{3}{8} \div \left(-\frac{8}{3}\right)$ **32.** $-\frac{5}{8} \div \left(-\frac{6}{5}\right)$

33. $-6.6 \div 3.3$ **34.** $-44.1 \div (-6.3)$ **35.** $\frac{-11}{-13}$ **36.** $\frac{-1.9}{20}$

37. $\frac{48.6}{-3}$ **38.** $\frac{-17.8}{3.2}$ **39.** $\frac{-9}{17-17}$ **40.** $\frac{-8}{-5+5}$

ANSWERS

1. -6
2. -4
3. -13
4. -2
5. -2
6. 11
7. 4
8. 7
9. -8
10. -2
11. 2
12. -25
13. -12
14. $\frac{64}{7}$
15. -8
16. $\frac{300}{13}$
17. Undefined
18. 0
19. -9
20. 29
21. $-\frac{7}{15}$
22. $-\frac{8}{3}$
23. $\frac{1}{13}$
24. $-\frac{1}{10}$
25. $-\frac{9}{8}$
26. $-\frac{7}{4}$
27. $\frac{5}{3}$
28. $\frac{2}{3}$
29. $\frac{9}{14}$
30. $\frac{24}{25}$
31. $\frac{9}{64}$
32. $\frac{25}{48}$
33. -2
34. 7
35. $\frac{11}{13}$
36. -0.095
37. -16.2
38. -5.5625
39. Undefined
40. Undefined

ANSWERS

41. -7
42. 11
43. -7
44. -36
45. -334
46. -160
47. 14
48. 23
49. 1880
50. 305
51. 12
52. 8
53. 8
54. 76
55. -86
56. -51
57. 37
58. 33
59. -1
60. -32
61. -10
62. 9
63. 25
64. -6
65. -7988
66. 12
67. -3000
68. -549
69. 60
70. -144
71. 1
72. 2
73. 10
74. -8
75. $-\frac{13}{45}$
76. $-\frac{21}{38}$
77. $-\frac{4}{3}$
78. $\frac{5}{9}$

d Simplify.

41. $8 - 2 \cdot 3 - 9$

42. $8 - (2 \cdot 3 - 9)$

43. $(8 - 2 \cdot 3) - 9$

44. $(8 - 2)(3 - 9)$

45. $16 \cdot (-24) + 50$

46. $10 \cdot 20 - 15 \cdot 24$

47. $2^4 + 2^3 - 10$

48. $40 - 3^2 - 2^3$

49. $5^3 + 26 \cdot 71 - (16 + 25 \cdot 3)$

50. $4^3 + 10 \cdot 20 + 8^2 - 23$

51. $4 \cdot 5 - 2 \cdot 6 + 4$

52. $4 \cdot (6 + 8)/(4 + 3)$

53. $4^3/8$

54. $5^3 - 7^2$

55. $8(-7) + 6(-5)$

56. $10(-5) + 1(-1)$

57. $19 - 5(-3) + 3$

58. $14 - 2(-6) + 7$

59. $9 \div (-3) + 16 \div 8$

60. $-32 - 8 \div 4 - (-2)$

61. $6 - 4^2$

62. $(2 - 5)^2$

63. $(3 - 8)^2$

64. $3 - 3^2$

65. $12 - 20^3$

66. $20 + 4^3 \div (-8)$

67. $2 \times 10^3 - 5000$

68. $-7(3^4) + 18$

69. $6[9 - (3 - 4)]$

70. $8[(6 - 13) - 11]$

71. $-1000 \div (-100) \div 10$

72. $256 \div (-32) \div (-4)$

73. $8 - (7 - 9)$

74. $(8 - 7) - 9$

75. $\dfrac{10 - 6^2}{9^2 + 3^2}$

76. $\dfrac{5^2 - 4^3 - 3}{9^2 - 2^2 - 1^5}$

77. $\dfrac{20(8 - 3) - 4(10 - 3)}{10(2 - 6) - 2(5 + 2)}$

78. $\dfrac{(3 - 5)^2 - (7 - 13)}{(12 - 9)^2 + (11 - 14)^2}$

SUMMARY AND REVIEW EXERCISES: CHAPTER 11

The review sections and objectives to be tested in addition to the material in this chapter are [2.1d], [3.1a], [7.3b], and [9.4a].

Find the absolute value.

1. $|-38|$

2. $|7.3|$

3. $\left|\frac{5}{2}\right|$

4. $-|-0.2|$

Find decimal notation.

5. $-\frac{5}{4}$

6. $-\frac{5}{6}$

7. $-\frac{5}{12}$

8. $-\frac{3}{11}$

Graph the number on a number line.

9. -2.5

10. $\frac{8}{9}$

Use either $<$ or $>$ for $\square$ to write a true sentence.

11. $-3 \ \square \ 10$

12. $-1 \ \square \ -6$

13. $0.126 \ \square \ -12.6$

14. $-\frac{2}{3} \ \square \ -\frac{1}{10}$

Find the opposite, or additive inverse, of the number.

15. 3.8

16. $-\frac{3}{4}$

17. Find $-x$ when x is -34.

18. Find $-(-x)$ when x is 5.

Find the reciprocal.

19. $\frac{3}{8}$

20. -7

21. $-\frac{1}{10}$

Compute and simplify.

22. $4 + (-7)$

23. $-\frac{2}{3} + \frac{1}{12}$

24. $6 + (-9) + (-8) + 7$

25. $-3.8 + 5.1 + (-12) + (-4.3) + 10$

26. $-3 - (-7)$

27. $-\frac{9}{10} - \frac{1}{2}$

28. $-3.8 - 4.1$

29. $-9 \cdot (-6)$

30. $-2.7(3.4)$

31. $\frac{2}{3} \cdot \left(-\frac{3}{7}\right)$

32. $3 \cdot (-7) \cdot (-2) \cdot (-5)$

33. $35 \div (-5)$

34. $-5.1 \div 1.7$

35. $-\frac{3}{5} \div \left(-\frac{4}{5}\right)$

36. $(-3.4 - 12.2) - 8(-7)$

37. $[-12(-3) - 2^3] - (-9)(-10)$

38. $625 \div (-25) \div 5$

SKILL MAINTENANCE

39. Find the area of a rectangle when the length is 10.5 cm and the width is 20 cm.

40. Find the LCM of 15, 27, and 30.

41. Find the prime factorization of 648.

42. 2016 is what percent of 5600?

SYNTHESIS

43. Simplify: $-\left|\frac{7}{8} - \left(-\frac{1}{2}\right) - \frac{3}{4}\right|$.

44. Simplify: $(|2.7 - 3| + 3^2 - |-3|) \div (-3)$.

❖ THINKING IT THROUGH

1. List three examples of rational numbers that are not integers.
2. What should happen if you enter a number on a calculator and press the reciprocal key twice? Why?

NAME SECTION DATE

TEST: CHAPTER 11

Use either < or > for ▒ to write a true sentence.

1. -4 ▒ 0
2. -3 ▒ -8
3. -0.78 ▒ -0.87
4. $-\frac{1}{8}$ ▒ $\frac{1}{2}$

Find decimal notation.

5. $-\frac{1}{8}$
6. $-\frac{4}{9}$
7. $-\frac{2}{11}$

Find the absolute value.

8. $|-7|$
9. $\left|\frac{9}{4}\right|$
10. $-|-2.7|$

Find the opposite, or additive inverse.

11. $\frac{2}{3}$
12. -1.4

13. Find $-x$ when x is -8.

Find the reciprocal.

14. -2
15. $\frac{4}{7}$

ANSWERS

1. [11.1d] <
2. [11.1d] >
3. [11.1d] >
4. [11.1d] <
5. [11.1c] -0.125
6. [11.1c] $-0.\overline{4}$
7. [11.1c] $-0.\overline{18}$
8. [11.1e] 7
9. [11.1e] $\frac{9}{4}$
10. [11.1e] -2.7
11. [11.2b] $-\frac{2}{3}$
12. [11.2b] 1.4
13. [11.2b] 8
14. [11.5b] $-\frac{1}{2}$
15. [11.5b] $\frac{7}{4}$

ANSWERS

16. [11.3a] 7.8

17. [11.2a] -8

18. [11.2a] $\frac{7}{40}$

19. [11.3a] 10

20. [11.3a] -2.5

21. [11.3a] $\frac{7}{8}$

22. [11.4a] -48

23. [11.4a] $\frac{3}{16}$

24. [11.5a] -9

25. [11.5c] $\frac{3}{4}$

26. [11.5c] -9.728

27. [11.5d] 109

28. [9.4a] 55.8 ft^2

29. [7.3b] 48%

30. [2.1d] $2 \cdot 2 \cdot 2 \cdot 5 \cdot 7$

31. [3.1a] 240

32. [11.1e], [11.3a] 15

Compute and simplify.

16. $3.1-(-4.7)$

17. $-8+4+(-7)+3$

18. $-\frac{1}{5}+\frac{3}{8}$

19. $2-(-8)$

20. $3.2-5.7$

21. $\frac{1}{8}-\left(-\frac{3}{4}\right)$

22. $4\cdot(-12)$

23. $-\frac{1}{2}\cdot\left(-\frac{3}{8}\right)$

24. $-45 \div 5$

25. $-\frac{3}{5}\div\left(-\frac{4}{5}\right)$

26. $4.864 \div (-0.5)$

27. $-2(16)-(2(-8)-5^3)$

SKILL MAINTENANCE

28. Find the area of a rectangle of length 12.4 ft and width 4.5 ft.

29. 24 is what percent of 50?

30. Find the prime factorization of 280.

31. Find the LCM of 16, 20, 30.

SYNTHESIS

32. Simplify: $|-27-3(4)| - |-36| + |-12|$.

CUMULATIVE REVIEW: CHAPTERS 1–11

Find decimal notation.

1. 26.3%

2. $-\frac{5}{11}$

Complete.

3. 83.4 cg = ______ mg

4. 2.75 mm^2 = ______ cm^2

5. Find the absolute value: $|-4.5|$.

6. Subtract: $2 - 13$.

7. What is the rate in meters per second?
150 meters, 12 seconds

8. Find the radius, the circumference, and the area of this circle. Use $\frac{22}{7}$ for π.

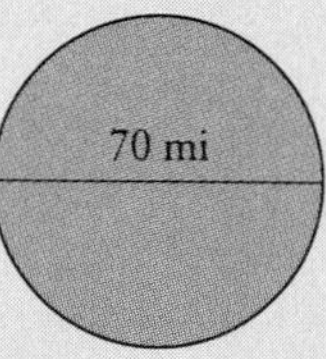

9. Simplify: $\sqrt{225}$.

10. Approximate to two decimal places: $\sqrt{69}$.

11. Multiply: $(-2)(5)$.

12. Divide: $\frac{-48}{-16}$.

13. Add: $-2 + 10$.

14. Draw a pictograph representing the number of hours that each type of farmer works each week using the information given below. Use a clock symbol to represent 10 hours. Be sure to put in all of the appropriate labels.

Dairy	70
Cash grain	40
Tobacco/cotton	35
Beef/hog/sheep	30

Compute and simplify.

15. 14.85×0.001

16. $36 - (-3) + (-42)$

17. $\frac{5}{22} - \frac{4}{11}$

18. $\frac{2}{27} \cdot \left(-\frac{9}{16}\right)$

19. $4\frac{2}{9} - 2\frac{7}{18}$

20. $-\frac{3}{14} \div \frac{6}{7}$

21. $3(-4.5) + (2^2 - 3 \cdot 4^2)$

22. $12{,}854 \cdot 750{,}000$

23. $35.1 + (-2.61)$

24. $32 \div [(-2)(-8) - (15 - (-1))]$

Solve.

25. 7 is what percent of 8?

26. 4 is $12\frac{1}{2}$% of what number?

27. Kerry had \$324.98 in a checking account. He wrote a check for \$12.76, deposited \$35.95, and wrote another check for \$213.09. The bank paid \$0.97 in interest and took out a service charge of \$3.00. How much was now in his checking account?

28. A can of fruit has a diameter of 7 cm and a height of 8 cm. Find the volume. Use 3.14 for π.

29. The following temperatures were recorded every four hours on a certain day: 42°, 40°, 45°, 52°, 50°, 40°. What was the average temperature for the day?

30. Thirteen percent of a student body of 600 received all A's on their grade reports. How many students received all A's?

31. A lot is 125.5 m by 75 m. A house 60 m by 40.5 m and a rectangular swimming pool 10 m by 8 m are built on the lot. How much area is left?

32. A recipe for a pie crust calls for $1\frac{1}{4}$ cups of flour, and a recipe for a cake calls for $1\frac{2}{3}$ cups of flour. How many cups of flour are needed to make both recipes?

33. The four top television game show winners in a recent year won \$74,834, \$58,253, \$57,200, and \$49,154. How much did these four win in all?

34. A jogger ran around a block 6.5 times. If the distance around the block is 0.7 km, how far did the jogger run?

INTRODUCTION In this chapter, we continue our introduction to algebra. We consider the manipulation of algebraic expressions. Then we use the manipulations to solve equations and problems.

The review sections to be tested in addition to the material in this chapter are 9.6, 10.1, 11.2, and 11.4. ❖

Algebra: Solving Equations and Problems

AN APPLICATION

The state of Colorado is a rectangle whose perimeter is 1300 mi. The length is 110 mi more than the width. Find the dimensions.

THE MATHEMATICS

Let w = the width of the state of Colorado. The problem can be translated to the following *equation:*

$$2(w + 110) + 2w = 1300.$$

❖ POINTS TO REMEMBER: CHAPTER 12

Identity Property of 1: $a \cdot 1 = a$
Simple Interest Formula: $I = Prt$
Perimeter of a Rectangle: $P = 2l + 2w$

PRETEST: CHAPTER 12

1. Evaluate $\frac{x}{2y}$ when $x = 5$ and $y = 8$.

2. Write an algebraic expression:
seventy-eight percent of some number.

Multiply.

3. $9(z - 2)$

4. $-2(2a + b - 5c)$

Factor.

5. $4x - 12$

6. $6y - 9z - 18$

Collect like terms.

7. $5x - 8x$

8. $6x - 9y - 4x + 11y + 18$

Solve.

9. $-7x = 49$

10. $4y + 9 = 2y + 7$

11. $6a - 2 = 10$

12. $x + (x + 1) + (x + 2) = 12$

Solve.

13. The perimeter of a rectangular field is 146 m. The width is 5 m less than the length. Find the dimensions.

14. Money is invested in a savings account at 9% simple interest. After one year, there is \$708.50 in the account. How much was originally invested?

12.1 Introduction to Algebra and Expressions

Many kinds of problems require the use of equations in order to be solved effectively. The study of algebra involves the use of equations to solve problems. Equations are constructed from algebraic expressions.

a Algebraic Expressions

In arithmetic you have worked with expressions such as

$$37 + 86, \quad 7 \times 8, \quad 19 - 7, \quad \text{and} \quad \frac{3}{8}.$$

In algebra we use certain letters for numbers and work with *algebraic expressions* such as

$$x + 86, \quad 7 \times t, \quad 19 - y, \quad \text{and} \quad \frac{a}{b}.$$

Expressions like these should be familiar from the equation and problem solving that we have already done.

Sometimes a letter can stand for various numbers. In that case, we call the letter a **variable.** Sometimes a letter can stand for just one number. In that case, we call the letter a **constant.** Let b = your date of birth. Then b is a constant. Let a = your age. Then a is a variable since a changes from year to year.

An **algebraic expression** consists of variables, numerals, and operation signs. When we replace a variable by a number, we say that we are **substituting** for the variable. This process is called **evaluating the expression.**

▶ **EXAMPLE 1** Evaluate $x + y$ for $x = 37$ and $y = 29$.

We substitute 37 for x and 29 for y and carry out the addition:

$$x + y = 37 + 29 = 66.$$

The number 66 is called the **value** of the expression. ◀

Algebraic expressions involving multiplication can be written in several ways. For example, "8 times a" can be written as $8 \times a$, $8 \cdot a$, $8(a)$, or simply $8a$. Two letters written together without a symbol, such as ab, also indicates a multiplication.

▶ **EXAMPLE 2** Evaluate $3y$ for $y = 14$.

$$3y = 3(14) = 42$$ ◀

DO EXERCISES 1–3.

OBJECTIVES

After finishing Section 12.1, you should be able to:

- a Evaluate an algebraic expression by substitution.
- b Use the distributive laws to multiply expressions like 8 and $x - y$.
- c Use the distributive laws to factor expressions like $4x - 12$.
- d Collect like terms.

FOR EXTRA HELP

Tape NC | Tape 16B | MAC: 12 IBM: 12

1. Evaluate $a + b$ for $a = 38$ and $b = 26$. 64

2. Evaluate $x - y$ for $x = 57$ and $y = 29$. 28

3. Evaluate $4t$ for $t = 15$. 60

ANSWERS ON PAGE A-6

4. Evaluate $\frac{a}{b}$ for $a = 200$ and $b = 8$. 25

5. Evaluate $\frac{10p}{q}$ when $p = 40$ and $q = 25$. 16

Complete the following tables by evaluating each expression for the given values.

6.

	$1 \cdot x$	x
$x = 3$	3	3
$x = -6$	−6	−6
$x = 4.8$	4.8	4.8

7.

	$2x$	$5x$
$x = 2$	4	10
$x = -6$	−12	−30
$x = 4.8$	9.6	24

ANSWERS ON PAGE A-6

Algebraic expressions involving division can also be written in several ways. For example, "8 divided by t" can be written as $8 \div t$, or $8/t$, or $\frac{8}{t}$, where the fraction bar is a division symbol.

▶ **EXAMPLE 3** Evaluate $\frac{a}{b}$ for $a = 63$ and $b = 9$.

We substitute 63 for a and 9 for b and carry out the division:

$$\frac{a}{b} = \frac{63}{9} = 7.$$ ◀

▶ **EXAMPLE 4** Evaluate $\frac{12m}{n}$ for $m = 8$ and $n = 16$.

$$\frac{12m}{n} = \frac{12 \cdot 8}{16} = \frac{96}{16} = 6$$ ◀

DO EXERCISES 4 AND 5.

b Equivalent Expressions and the Distributive Laws

In solving equations and doing other kinds of work in algebra, we manipulate expressions in various ways. To see how to do this, we consider some examples in which we evaluate expressions.

▶ **EXAMPLE 5** Evaluate $1 \cdot x$ for $x = 5$ and $x = -8$ and compare the results to x.

We substitute 5 for x:

$$1 \cdot x = 1 \cdot 5 = 5.$$

Then we substitute -8 for x:

$$1 \cdot x = 1 \cdot (-8) = -8.$$

We see that $1 \cdot x$ and x represent the same number. ◀

DO EXERCISES 6 AND 7.

We see in Example 5 and Margin Exercise 6 that the expressions represent the same number for any replacement of x that is meaningful. In that sense, the expressions $1 \cdot x$ and x are **equivalent.**

> **Two expressions that have the same value for all meaningful replacements are called *equivalent*.**

We see in Margin Exercise 7 that the expressions $2x$ and $5x$ are *not* equivalent.

The fact that $1 \cdot x$ and x are equivalent is a law of real numbers. It is called the **identity property of 1.** We often refer to the use of the identity property of 1 as "multiplying by 1." We have used multiplying by 1 for understanding many times in this text.

We now consider two other laws of real numbers called the **distributive laws.** They are the basis of many procedures in both arithmetic and algebra and are probably the most important laws that we use to manipulate algebraic expressions. The first distributive law involves two operations: addition and multiplication.

Let us begin by considering a multiplication problem from arithmetic:

```
    4 5
  ×   7
  -----
    3 5 ←— This is 7 · 5.
  2 8 0 ←— This is 7 · 40.
  -----
  3 1 5 ←— This is the sum 7 · 40 + 7 · 5.
```

To carry out the multiplication, we actually added two products. That is,

$$7 \cdot 45 = 7(40 + 5) = 7 \cdot 40 + 7 \cdot 5.$$

Let us examine this further. If we wish to multiply a sum of several numbers by a factor, we can either add and then multiply, or multiply and then add.

▶ **EXAMPLE 6** Evaluate $5(x + y)$ and $5x + 5y$ for $x = 2$ and $y = 8$ and compare the results.

We substitute 2 for x and 8 for y in each expression. Then we use our rules for order of operations to calculate.

a) $5(x + y) = 5(2 + 8)$
$= 5(10)$ **Adding within parentheses first, and then multiplying**
$= 50$

b) $5x + 5y = 5 \cdot 2 + 5 \cdot 8$
$= 10 + 40$ **Multiplying first and then adding**
$= 50$

We see that the expressions $5(x + y)$ and $5x + 5y$ are equivalent. ◀

DO EXERCISES 8–10.

> The Distributive Law of Multiplication over Addition
>
> **For any numbers *a*, *b*, and *c*,**
>
> $$a(b + c) = ab + ac.$$

In the statement of the distributive law, we know that in an expression such as $ab + ac$ the multiplications are to be done first according to our rules for order of operations. So, instead of writing $(4 \cdot 5) + (4 \cdot 7)$, we can write $4 \cdot 5 + 4 \cdot 7$. However, in $a(b + c)$, we cannot omit the parentheses. If we did we would have $ab + c$, which means $(ab) + c$. For example, $3(4 + 2) = 18$, but $3 \cdot 4 + 2 = 14$.

8. Evaluate $3(x + y)$ and $3x + 3y$ when $x = 5$ and $y = 7$. 36; 36

9. Evaluate $6x + 6y$ and $6(x + y)$ when $x = 10$ and $y = 5$. 90; 90

10. Evaluate $4(x + y)$ and $4x + 4y$ when $x = 11$ and $y = 5$. 64; 64

ANSWERS ON PAGE A-6

11. Evaluate $7(x - y)$ and $7x - 7y$ when $x = 9$ and $y = 7$. 14; 14

12. Evaluate $6x - 6y$ and $6(x - y)$ when $x = 10$ and $y = 5$. 30; 30

13. Evaluate $2(x - y)$ and $2x - 2y$ when $x = 11$ and $y = 5$. 12; 12

What are the terms of the expression?

14. $5x - 4y + 3$ $5x; -4y; 3$

15. $-4y - 2x + 3z$ $-4y; -2x; 3z$

Multiply.

16. $3(x - 5)$ $3x - 15$

17. $5(x - y + 4)$ $5x - 5y + 20$

18. $-2(x - 3)$ $-2x + 6$

19. $-5(x - 2y + 4z)$

$-5x + 10y - 20z$

ANSWERS ON PAGE A-6

The second distributive law relates multiplication and subtraction. This law says that to multiply by a difference, we can either subtract and then multiply or multiply and then subtract.

> The Distributive Law of Multiplication over Subtraction
>
> **For any numbers *a*, *b*, and *c*,**
>
> $$a(b - c) = ab - ac.$$

We often refer to "*the* distributive law" when we mean *either* of these laws.

DO EXERCISES 11–13.

What do we mean by the *terms* of an expression? **Terms** are separated by addition signs. If there are subtraction signs, we can find an equivalent expression that uses addition signs.

▶ **EXAMPLE 7** What are the terms of $3x - 4y + 2z$?

$$3x - 4y + 2z = 3x + (-4y) + 2z \qquad \text{Separating parts with + signs}$$

The terms are $3x$, $-4y$, and $2z$. ◀

DO EXERCISES 14 AND 15.

The distributive laws are the basis for a procedure in algebra called **multiplying**. In an expression such as $8(a + 2b - 7)$, we multiply each term inside the parentheses by 8:

$$8(a + 2b - 7) = 8 \cdot a + 8 \cdot 2b - 8 \cdot 7 = 8a + 16b - 56.$$

▶ **EXAMPLES** Multiply.

8. $9(x - 5) = 9x - 9(5)$ Using the distributive law of multiplication over subtraction

$= 9x - 45$

9. $-4(x - 2y + 3z) = -4 \cdot x - (-4)(2y) + (-4)(3z)$ Using both distributive laws

$= -4x - (-8y) + (-12z)$

$= -4x + 8y - 12z$

We can also do this problem by first finding an equivalent expression with all plus signs and then multiplying:

$$-4(x - 2y + 3z) = -4[x + (-2y) + 3z]$$
$$= -4 \cdot x + (-4)(-2y) + (-4)(3z) = -4x + 8y - 12z.$$

◀

DO EXERCISES 16–19.

C Factoring

Factoring is the reverse of multiplying. To factor, we can use the distributive laws in reverse:

$$ab + ac = a(b + c) \quad \text{and} \quad ab - ac = a(b - c).$$

To *factor* an expression is to find an equivalent expression that is a product.

Look at Example 8. To *factor* $9x - 45$, we find an equivalent expression that is a product, $9(x - 5)$. When all the terms of an expression have a factor in common, we can "factor it out" using the distributive laws. Note the following.

$9x$ has the factors $9, -9, 3, -3, 1, -1, x, -x, 3x, -3x, 9x, -9x$;
-45 has the factors $1, -1, 3, -3, 5, -5, 9, -9, 15, -15, 45, -45$.

We usually remove the largest common factor. That factor is 9.

Remember that an expression is factored when we find an equivalent expression that is a product.

▶ **EXAMPLES** Factor.

10. $5x - 10 = 5 \cdot x - 5 \cdot 2$ ← **Try to think of this step mentally.**
$= 5(x - 2)$ **You can check by multiplying.**

11. $9x + 27y - 9 = 9 \cdot x + 9 \cdot 3y - 9 \cdot 1 = 9(x + 3y - 1)$ ◀

CAUTION! Note that $3(3x + 9y - 3)$ is also equivalent to $9x + 27y - 9$, but it is *not* the desired form. However, we may complete the process by factoring out another factor of 3:

$$9x + 27y - 9 = 3(3x + 9y - 3) = 3 \cdot 3(x + 3y - 1) = 9(x + 3y - 1).$$

Remember to factor out the largest common factor.

▶ **EXAMPLES** Factor. Try to write just the answer, if you can.

12. $5x - 5y = 5(x - y)$

13. $-3x + 6y - 9z = -3(x - 2y + 3z)$

We might also factor the expression in Example 13 as follows:

$$-3x + 6y - 9z = 3(-x + 2y - 3z).$$

We usually factor out a negative when the first term is negative. The way we factor can depend on the situation in which we are working.

14. $18z - 12x - 24 = 6(3z - 2x - 4)$ ◀

Remember: An expression is factored when it is written as a product.

DO EXERCISES 20–23.

Factor.

20. $6z - 12$ $6(z - 2)$

21. $3x - 6y + 9$ $3(x - 2y + 3)$

22. $16a - 36b + 42$
$2(8a - 18b + 21)$

23. $-12x + 32y - 16z$
$-4(3x - 8y + 4z)$

ANSWERS ON PAGE A-6

d Collecting Like Terms

Terms such as $5x$ and $-4x$, whose variable factors are exactly the same, are called **like terms.** Similarly, $3y^2$ and $9y^2$ are like terms because the variables are raised to the same power. Terms such as $4y$ and $5y^2$ are not like terms, and $7x$ and $2y$ are not like terms.

The process of **collecting like terms** is also based on the distributive laws. We can also apply the distributive law "on the right."

▶ **EXAMPLES** Collect like terms. Try to write just the answer, if you can.

15. $4x + 2x = (4 + 2)x = 6x$ **Factoring out the *x* using a distributive law**

16. $2x + 3y - 5x - 2y = 2x - 5x + 3y - 2y$

$= (2 - 5)x + (3 - 2)y = -3x + y$

17. $3x - x = (3 - 1)x = 2x$

18. $x - 0.24x = 1 \cdot x - 0.24x = (1 - 0.24)x = 0.76x$

19. $x - 6x = 1 \cdot x - 6 \cdot x = (1 - 6)x = -5x$

20. $4x - 7y + 9x - 5 + 3y - 8 = 13x - 4y - 13$ ◀

DO EXERCISES 24–29.

Collect like terms.

24. $6x - 3x$ $3x$

25. $7x - x$ $6x$

26. $x - 9x$ $-8x$

27. $x - 0.41x$ $0.59x$

28. $5x + 4y - 2x - y$ $3x + 3y$

29. $3x - 7x - 11 + 8y + 4 - 13y$

$-4x - 5y - 7$

ANSWERS ON PAGE A-6

NAME SECTION DATE

EXERCISE SET 12.1

a Evaluate.

1. $6x$ for $x = 7$
2. $7y$ for $y = 7$
3. $\frac{x}{y}$ for $x = 9$ and $y = 3$
4. $\frac{m}{n}$ for $m = 14$ and $n = 2$
5. $\frac{3p}{q}$ for $p = 2$ and $q = 6$
6. $\frac{5y}{z}$ for $y = 15$ and $z = 25$
7. $\frac{x+y}{5}$ for $x = 10$ and $y = 20$
8. $\frac{p-q}{2}$ for $p = 16$ and $q = 2$

b Evaluate.

9. $10(x + y)$ and $10x + 10y$ for $x = 20$ and $y = 4$
10. $7(a + b)$ and $7a + 7b$ for $a = 16$ and $b = 6$
11. $10(x - y)$ and $10x - 10y$ for $x = 20$ and $y = 4$
12. $7(a - b)$ and $7a - 7b$ for $a = 16$ and $b = 6$

Multiply.

13. $2(b + 5)$
14. $4(x + 3)$
15. $7(1 - t)$
16. $4(1 - y)$
17. $6(5x + 2)$
18. $9(6m + 7)$
19. $7(x + 4 + 6y)$
20. $4(5x + 8 + 3p)$
21. $-7(y - 2)$
22. $-9(y - 7)$
23. $-9(-5x - 6y + 8)$
24. $-7(-2x - 5y + 9)$
25. $-4(x - 3y - 2z)$
26. $8(2x - 5y - 8z)$
27. $3.1(-1.2x + 3.2y - 1.1)$
28. $-2.1(-4.2x - 4.3y - 2.2)$

$-3.72x + 9.92y - 3.41$

c Factor. Check by multiplying.

29. $2x + 4$
30. $5y + 20$
31. $30 + 5y$
32. $7x + 28$

ANSWERS

1. 42
2. 49
3. 3
4. 7
5. 1
6. 3
7. 6
8. 7
9. 240; 240
10. 154; 154
11. 160; 160
12. 70; 70
13. $2b + 10$
14. $4x + 12$
15. $7 - 7t$
16. $4 - 4y$
17. $30x + 12$
18. $54m + 63$
19. $7x + 28 + 42y$
20. $20x + 32 + 12p$
21. $-7y + 14$
22. $-9y + 63$
23. $45x + 54y - 72$
24. $14x + 35y - 63$
25. $-4x + 12y + 8z$
26. $16x - 40y - 64z$
27.
28. $8.82x + 9.03y + 4.62$
29. $2(x + 2)$
30. $5(y + 4)$
31. $5(6 + y)$
32. $7(x + 4)$

ANSWERS

33. $7(2x + 3y)$
34. $6(3a + 4b)$
35. $5(x + 2 + 3y)$
36. $9(a + 3b + 9)$
37. $8(x - 3)$
38. $10(x - 5)$
39. $4(8 - y)$
40. $6(4 - m)$
41. $2(4x + 5y - 11)$
42. $3(3a + 2b - 5)$
43. $6(3x - 2y + 1)$
44.
45. $19a$
46. $14x$
47. $9a$
48. $-15x$
49. $8x + 9z$
50. $10a - 5b$
51. $-19a + 88$
52. $38x - 4$
53. $4t + 6y - 4$
54. $-92d - 39$
55. $8x$
56. $-8t$
57. $5n$
58. $9t$
59. $-16y$
60. $-6m + 4$
61. $17a - 12b - 1$
62. $3x + y + 2$
63. $4x + 2y$
64. $12y - 3z$
65. $7x + y$
66. $11a + 5b$
67. $0.8x + 0.5y$
68. $2.6a + 1.4b$

33. $14x + 21y$ **34.** $18a + 24b$ **35.** $5x + 10 + 15y$

36. $9a + 27b + 81$ **37.** $8x - 24$ **38.** $10x - 50$

39. $32 - 4y$ **40.** $24 - 6m$ **41.** $8x + 10y - 22$

42. $9a + 6b - 15$ **43.** $18x - 12y + 6$ **44.** $-14x + 21y + 7$

$7(-2x + 3y + 1)$, or $-7(2x - 3y - 1)$

d Collect like terms.

45. $9a + 10a$ **46.** $12x + 2x$ **47.** $10a - a$

48. $-16x + x$ **49.** $2x + 9z + 6x$ **50.** $3a - 5b + 7a$

51. $41a + 90 - 60a - 2$ **52.** $42x - 6 - 4x + 2$

53. $23 + 5t + 7y - t - y - 27$ **54.** $45 - 90d - 87 - 9d + 3 + 7d$

55. $11x - 3x$ **56.** $9t - 17t$

57. $6n - n$ **58.** $10t - t$

59. $y - 17y$ **60.** $3m - 9m + 4$

61. $-8 + 11a - 5b + 6a - 7b + 7$ **62.** $8x - 5x + 6 + 3y - 2y - 4$

63. $9x + 2y - 5x$ **64.** $8y - 3z + 4y$

65. $11x + 2y - 4x - y$ **66.** $13a + 9b - 2a - 4b$

67. $2.7x + 2.3y - 1.9x - 1.8y$ **68.** $6.7a + 4.3b - 4.1a - 2.9b$

12.2 The Addition Principle

a Solving Equations Using the Addition Principle

Consider the equation

$$x = 7.$$

We can easily "see" that the solution of this equation is 7. If we replace x by 7, we get

$$7 = 7, \quad \text{which is true.}$$

Now consider the equation

$$x + 6 = 13.$$

The solution of this equation is also 7, but the fact that 7 is the solution is not as obvious. We now begin to consider principles that allow us to start with an equation and end up with an equation like $x = 7$, in which the variable is alone on one side and for which the solution is easy to find. The equations $x + 6 = 13$ and $x = 7$ are **equivalent.**

Equations with the same solutions are called *equivalent equations.*

One principle that we use to solve equations concerns the addition principle, which we have used throughout this text.

The Addition Principle

If an equation $a = b$ is true, then

$$a + c = b + c$$

is true for any number c.

When we use the addition principle, we sometimes say that we "add the same number on both sides of an equation." Now we can add negative as well as positive numbers.

We can also subtract the same number on both sides. This is true since we can express every subtraction as an addition. That is, since

$$a - c = b - c \quad \text{means} \quad a + (-c) = b + (-c),$$

the addition principle tells us that we can "subtract the same number on both sides of an equation."

OBJECTIVE

After finishing Section 12.2, you should be able to:

a Solve equations using the addition principle.

FOR EXTRA HELP

Tape 15A

Tape 17A

MAC: 12
IBM: 12

1. Solve using the addition principle:

$x + 7 = 2.$ -5

Solve.

2. $8.7 = n - 4.5$ 13.2

3. $y + 17.4 = 10.9$ -6.5

ANSWERS ON PAGE A-6

► **EXAMPLE 1** Solve: $x + 5 = -7$.

We have

$$x + 5 = -7$$

$$x + 5 - 5 = -7 - 5 \quad \text{Using the addition principle, adding } -5 \text{ on both sides or subtracting 5 on both sides}$$

$$x + 0 = -12 \quad \text{Simplifying}$$

$$x = -12.$$

We can see that the solution of $x = -12$ is the number -12. To check the answer, we substitute -12 in the original equation.

Check:

$x + 5$	$= -7$	
$-12 + 5$	-7	
-7		TRUE

The solution of the original equation is -12. ◄

In Example 1, to get x alone, we used the addition principle and subtracted 5 on both sides. This eliminated the 5 on the left. We started with $x + 5 = -7$, and using the addition principle we found a simpler equation $x = -12$, for which it was easy to *"see"* the solution. The equations $x + 5 = -7$ and $x = -12$ are equivalent.

DO EXERCISE 1.

Now we solve an equation with a subtraction using the addition principle.

► **EXAMPLE 2** Solve: $-6.5 = y - 8.4$.

We have

$$-6.5 = y - 8.4$$

$$-6.5 + 8.4 = y - 8.4 + 8.4 \quad \text{Using the addition principle, adding 8.4 to eliminate } -8.4 \text{ on the right}$$

$$1.9 = y.$$

Check:

-6.5	$= y - 8.4$	
-6.5	$1.9 - 8.4$	
	-6.5	TRUE

The solution is 1.9. ◄

Note that equations are reversible. That is, if $a = b$ is true, then $b = a$ is true. Thus, when we solve $-6.5 = y - 8.4$, we can reverse it and solve $y - 8.4 = -6.5$ if we wish.

DO EXERCISES 2 AND 3.

NAME SECTION DATE

EXERCISE SET 12.2

a Solve using the addition principle. Don't forget to check!

1. $x + 2 = 6$

2. $x + 5 = 8$

3. $x + 15 = -5$

4. $y + 9 = 43$

5. $x + 6 = -8$

6. $t + 9 = -12$

7. $x + 16 = -2$

8. $y + 25 = -6$

9. $x - 9 = 6$

10. $x - 8 = 5$

11. $x - 7 = -21$

12. $x - 3 = -14$

13. $5 + t = 7$

14. $8 + y = 12$

15. $-7 + y = 13$

16. $-9 + z = 15$

17. $-3 + t = -9$

18. $-6 + y = -21$

19. $r + \frac{1}{3} = \frac{8}{3}$

20. $t + \frac{3}{8} = \frac{5}{8}$

21. $m + \frac{5}{6} = -\frac{11}{12}$

22. $x + \frac{2}{3} = -\frac{5}{6}$

23. $x - \frac{5}{6} = \frac{7}{8}$

24. $y - \frac{3}{4} = \frac{5}{6}$

ANSWERS

1. 4

2. 3

3. -20

4. 34

5. -14

6. -21

7. -18

8. -31

9. 15

10. 13

11. -14

12. -11

13. 2

14. 4

15. 20

16. 24

17. -6

18. -15

19. $\frac{7}{3}$

20. $\frac{1}{4}$

21. $-\frac{7}{4}$

22. $-\frac{3}{2}$

23. $\frac{41}{24}$

24. $\frac{19}{12}$

ANSWERS

25. $-\frac{1}{20}$
26. $-\frac{5}{8}$
27. 5.1
28. 4.7
29. 12.4
30. 17.8
31. -5
32. -10.6
33. $1\frac{5}{6}$
34. $\frac{7}{12}$
35. $-\frac{10}{21}$
36. $123\frac{1}{8}$
37. -11
38. 5
39. $-\frac{5}{12}$
40. $\frac{1}{3}$
41. 342.246
42. $\frac{13}{20}$
43. $-\frac{26}{15}$
44. -4
45. -10
46. 0
47. All real numbers
48. No solution
49. $-\frac{5}{17}$
50. 5, -5
51. 13, -13

25. $-\frac{1}{5} + z = -\frac{1}{4}$

26. $-\frac{1}{8} + y = -\frac{3}{4}$

27. $7.4 = x + 2.3$

28. $9.3 = 4.6 + x$

29. $7.6 = x - 4.8$

30. $9.5 = y - 8.3$

31. $-9.7 = -4.7 + y$

32. $-7.8 = 2.8 + x$

33. $5\frac{1}{6} + x = 7$

34. $5\frac{1}{4} = 4\frac{2}{3} + x$

35. $q + \frac{1}{3} = -\frac{1}{7}$

36. $47\frac{1}{8} = -76 + z$

SKILL MAINTENANCE

37. Add: $-3 + (-8)$.

38. Subtract: $-3 - (-8)$.

39. Multiply: $-\frac{2}{3} \cdot \frac{5}{8}$.

40. Divide: $-\frac{3}{7} \div \left(-\frac{9}{7}\right)$.

SYNTHESIS

Solve.

41. ▦ $-356.788 = -699.034 + t$

42. $-\frac{4}{5} + \frac{7}{10} = x - \frac{3}{4}$

43. $x + \frac{4}{5} = -\frac{2}{3} - \frac{4}{15}$

44. $8 - 25 = 8 + x - 21$

45. $16 + x - 22 = -16$

46. $x + x = x$

47. $x + 3 = 3 + x$

48. $x + 4 = 5 + x$

49. $-\frac{3}{2} + x = -\frac{5}{17} - \frac{3}{2}$

50. $|x| = 5$

51. $|x| + 6 = 19$

12.3 The Multiplication Principle

a Solving Equations Using the Multiplication Principle

Suppose that $a = b$ is true and we multiply a by some number c. We get the same answer if we multiply b by c, because a and b are the same number.

> The Multiplication Principle
>
> **If an equation $a = b$ is true, then**
>
> $$a \cdot c = b \cdot c$$
>
> **is true for any number c.**

When using the multiplication principle, we sometimes say that we "multiply on both sides by the same number."

▶ **EXAMPLE 1** Solve: $\frac{3}{8} = -\frac{5}{4}x$.

$$\frac{3}{8} = -\frac{5}{4}x$$

The reciprocal of $-\frac{5}{4}$ is $-\frac{4}{5}$. There is no sign change.

$$-\frac{4}{5} \cdot \frac{3}{8} = -\frac{4}{5} \cdot \left(-\frac{5}{4}x\right)$$ Multiplying by $-\frac{4}{5}$ to get $1 \cdot x$ and eliminate $-\frac{5}{4}$ on the right

$$-\frac{3}{10} = 1 \cdot x$$ Simplifying

$$-\frac{3}{10} = x$$ Identity property of 1: $1 \cdot x = x$

Check:

$\frac{3}{8} = -\frac{5}{4}x$	
$\frac{3}{8}$	$-\frac{5}{4}\left(-\frac{3}{10}\right)$
	$\frac{3}{8}$ TRUE

The solution is $-\frac{3}{10}$. ◀

Note that equations are reversible. That is, if $a = b$ is true, then $b = a$ is true. Thus, when we solve $\frac{3}{8} = -\frac{5}{4}x$, we can reverse it and solve $-\frac{5}{4}x = \frac{3}{8}$.

DO EXERCISES 1 AND 2.

In Example 1, to get x alone, we multiplied by the *multiplicative inverse,* or the *reciprocal,* of $-\frac{5}{4}$. When we multiplied, we got the *multiplicative identity,* 1 times x, or $1 \cdot x$, which simplified to x. This enabled us to eliminate the $-\frac{5}{4}$ on the right.

OBJECTIVE

After finishing Section 12.3, you should be able to:

a Solve equations using the multiplication principle.

FOR EXTRA HELP

Tape 15B Tape 17A MAC: 12 IBM: 12

Solve.

1. $\frac{2}{3} = -\frac{5}{6}y$ $-\frac{4}{5}$

2. $4x = -7$ $-\frac{7}{4}$

ANSWERS ON PAGE A-6

3. Solve: $5x = 40$. 8

4. Solve: $-6x = 108$. -18

ANSWERS ON PAGE A-6

The multiplication principle also tells us that we can "divide on both sides by a nonzero number." This is because division is the same as multiplying by a reciprocal. That is,

$$\frac{a}{c} = \frac{b}{c} \quad \text{means} \quad a \cdot \frac{1}{c} = b \cdot \frac{1}{c}, \quad \text{when } c \neq 0.$$

In an expression like $3x$, the number 3 is called the **coefficient.** In practice it is usually more convenient to "divide" on both sides of the equation if the coefficient of the variable is in decimal notation or is an integer. When the coefficient is in fractional notation, it is more convenient to "multiply" by a reciprocal.

► **EXAMPLE 2** Solve: $3x = 9$.

We have

$3x = 9$

$\frac{3x}{3} = \frac{9}{3}$ Using the multiplication principle, multiplying by $\frac{1}{3}$ on both sides or dividing by 3 on both sides

$1 \cdot x = 3$ Simplifying

$x = 3$. Identity property of 1

It is easy to see that the solution of $x = 3$ is 3.

Check:

$3x = 9$	
$3 \cdot 3$	9
9	TRUE

The solution of the original equation is 3. ◄

DO EXERCISE 3.

► **EXAMPLE 3** Solve: $-4x = 92$.

$-4x = 92$

$\frac{-4x}{-4} = \frac{92}{-4}$ Using the multiplication principle. Dividing on both sides by -4 is the same as multiplying by $-\frac{1}{4}$.

$1 \cdot x = -23$
$x = -23$ } Simplifying

Check:

$-4x = 92$	
$-4(-23)$	92
92	TRUE

The solution is -23. ◄

DO EXERCISE 4.

NAME SECTION DATE

EXERCISE SET 12.3

a Solve using the multiplication principle. Don't forget to check!

1. $6x = 36$
2. $3x = 39$
3. $5x = 45$
4. $9x = 72$
5. $84 = 7x$
6. $56 = 8x$
7. $-x = 40$
8. $100 = -x$
9. $-2x = -10$
10. $-68 = -34r$
11. $7x = -49$
12. $9x = -36$
13. $-12x = 72$
14. $-15x = 105$
15. $-21x = -126$
16. $-13x = -104$
17. $\frac{1}{7}t = -9$
18. $-\frac{1}{8}y = 11$
19. $\frac{3}{4}x = 27$
20. $\frac{4}{5}x = 16$
21. $-\frac{1}{3}t = 7$
22. $-\frac{1}{6}x = 9$
23. $-\frac{1}{3}m = \frac{1}{5}$
24. $\frac{1}{9} = -\frac{1}{7}z$

ANSWERS

1. 6
2. 13
3. 9
4. 8
5. 12
6. 7
7. -40
8. -100
9. 5
10. 2
11. -7
12. -4
13. -6
14. -7
15. 6
16. 8
17. -63
18. -88
19. 36
20. 20
21. -21
22. -54
23. $-\frac{3}{5}$
24. $-\frac{7}{9}$

ANSWERS

25. $-\frac{3}{2}$

26. $-\frac{2}{3}$

27. $\frac{9}{2}$

28. 1

29. 7

30. 20

31. −7

32. 2

33. 8

34. 8

35. 15.9

36. −9.38

37. 62.8 ft; 20 ft; 314 ft^2

38. 75.36 cm; 12 cm; 452.16 cm^2

39. 8000 ft^3

40. 31.2 cm^3

41. −8655

42. All real numbers

43. No solution

44. 12, −12

45. No solution

46. 250

25. $-\frac{3}{5}r = \frac{9}{10}$

26. $\frac{2}{5}y = -\frac{4}{15}$

27. $-\frac{3}{2}r = -\frac{27}{4}$

28. $-\frac{5}{7}x = -\frac{10}{14}$

29. $6.3x = 44.1$

30. $2.7y = 54$

31. $-3.1y = 21.7$

32. $-3.3y = 6.6$

33. $38.7m = 309.6$

34. $29.4m = 235.2$

35. $-\frac{2}{3}y = -10.6$

36. $-\frac{9}{7}y = 12.06$

SKILL MAINTENANCE

37. Find the circumference, the diameter, and the area of a circle whose radius is 10 ft. Use 3.14 for π.

38. Find the circumference, the radius, and the area of a circle whose diameter is 24 cm. Use 3.14 for π.

39. Find the volume of a rectangular solid of length 25 ft, width 10 ft, and height 32 ft.

40. Find the volume of a rectangular solid of length 1.3 cm, width 10 cm, and height 2.4 cm.

SYNTHESIS

Solve.

41. 🖩 $-0.2344m = 2028.732$

42. $0 \cdot x = 0$

43. $0 \cdot x = 9$

44. $4|x| = 48$

45. $2|x| = -12$

46. A student makes a calculation and gets an answer of 22.5. On the last step the student multiplies by 0.3 when a division by 0.3 should have been done. What should the correct answer be?

12.4 Using the Principles Together

a Applying Both Principles

Consider the equation $3x + 4 = 13$. It is more complicated than those in the preceding two sections. In order to solve such an equation, we first isolate the x-term, $3x$, using the addition principle. Then we apply the multiplication principle to get x by itself.

► **EXAMPLE 1** Solve: $3x + 4 = 13$.

$$3x + 4 = 13$$

$$3x + 4 - 4 = 13 - 4 \quad \text{Using the addition principle, subtracting 4 on both sides}$$

$$3x = 9 \quad \text{Simplifying}$$

$$\frac{3x}{3} = \frac{9}{3} \quad \text{Using the multiplication principle, dividing on both sides by 3}$$

$$x = 3 \quad \text{Simplifying}$$

Check:

$3x + 4 = 13$	
$3 \cdot 3 + 4$	13
$9 + 4$	
13	TRUE

The solution is 3. ◄

DO EXERCISE 1.

► **EXAMPLE 2** Solve: $-5x - 6 = 16$.

$$-5x - 6 = 16$$

$$-5x - 6 + 6 = 16 + 6 \quad \text{Adding 6 on both sides}$$

$$-5x = 22$$

$$\frac{-5x}{-5} = \frac{22}{-5} \quad \text{Dividing on both sides by } -5$$

$$x = -\frac{22}{5}, \text{ or } -4\frac{2}{5} \quad \text{Simplifying}$$

Check:

$-5x - 6 = 16$	
$-5(-\frac{22}{5}) - 6$	16
$22 - 6$	
16	TRUE

The solution is $-\frac{22}{5}$. ◄

DO EXERCISES 2 AND 3.

OBJECTIVES

After finishing Section 12.4, you should be able to:

a Solve equations using both the addition and multiplication principles.

b Solve equations in which like terms may need to be collected.

FOR EXTRA HELP

Tape 15C

Tape 17B

MAC: 12
IBM: 12

1. Solve: $9x + 6 = 51$. 5

Solve.

2. $8x - 4 = 28$ 4

3. $-\frac{1}{2}x + 3 = 1$ 4

ANSWERS ON PAGE A-6

4. Solve: $-18 - x = -57$. 39

Solve.

5. $-4 - 8x = 8$ $-\frac{3}{2}$

6. $41.68 = 4.7 - 8.6y$ -4.3

ANSWERS ON PAGE A-6

▶ **EXAMPLE 3** Solve: $45 - x = 13$.

$$45 - x = 13$$

$$-45 + 45 - x = -45 + 13 \quad \text{Adding } -45 \text{ on both sides}$$

$$-x = -32$$

$$-1 \cdot x = -32 \quad \text{Using the property of } -1\text{: } -x = -1 \cdot x$$

$$\frac{-1 \cdot x}{-1} = \frac{-32}{-1}$$

Dividing on both sides by -1. (You could have multiplied on both sides by -1 instead. That would also change the sign on both sides.)

$$x = 32$$

Check:

$45 - x = 13$	
$45 - 32$	13
13	TRUE

The solution is 32. ◀

DO EXERCISE 4.

As we improve our equation-solving skills, we begin to shorten some of our writing. Thus we may not always write a number being added, subtracted, multiplied, or divided on both sides. We simply write it on the opposite side.

▶ **EXAMPLE 4** Solve: $16.3 - 7.2y = -8.18$.

$$16.3 - 7.2y = -8.18$$

$$-7.2y = -16.3 + (-8.18)$$

Adding -16.3 on both sides. We just write the addition of -16.3 on the right side.

$$-7.2y = -24.48$$

$$y = \frac{-24.48}{-7.2}$$

Dividing by -7.2 on both sides. We just write the division by -7.2 on the right side.

$$y = 3.4$$

Check:

$16.3 - 7.2y = -8.18$	
$16.3 - 7.2(3.4)$	-8.18
$16.3 - 24.48$	
8.18	TRUE

The solution is 3.4. ◀

DO EXERCISES 5 AND 6.

b Collecting Like Terms

If there are like terms on one side of the equation, we collect them before using the addition or multiplication principles.

▶ **EXAMPLE 5** Solve: $3x + 4x = -14$.

$$3x + 4x = -14$$

$$7x = -14 \quad \text{Collecting like terms}$$

$$x = \frac{-14}{7} \quad \text{Dividing by 7 on both sides}$$

$$x = -2.$$

The number -2 checks, so the solution is -2. ◀

DO EXERCISES 7 AND 8.

If there are like terms on opposite sides of the equation, we get them on the same side by using the addition principle. Then we collect them. In other words, we get all terms with a variable on one side and all numbers on the other.

▶ **EXAMPLE 6** Solve: $2x - 2 = -3x + 3$.

$$\begin{aligned} 2x - 2 &= -3x + 3 \\ 2x - 2 + 2 &= -3x + 3 + 2 && \text{Adding 2} \\ 2x &= -3x + 5 && \text{Collecting like terms} \\ 2x + 3x &= -3x + 3x + 5 && \text{Adding } 3x \\ 5x &= 5 && \text{Simplifying} \\ \frac{5x}{5} &= \frac{5}{5} && \text{Dividing by 5} \\ x &= 1 && \text{Simplifying} \end{aligned}$$

Check:

$2x - 2$	$= -3x + 3$	
$2 \cdot 1 - 2$	$-3 \cdot 1 + 3$	
$2 - 2$	$-3 + 3$	
0	0	TRUE

The solution is 1. ◀

DO EXERCISE 9.

In Example 6, we used the addition principle to get all terms with a variable on one side and all numbers on the other side. Then we collected like terms and proceeded as before. If there are like terms on one side at the outset, they should be collected first.

▶ **EXAMPLE 7** Solve: $6x + 5 - 7x = 10 - 4x + 3$.

$$\begin{aligned} 6x + 5 - 7x &= 10 - 4x + 3 \\ -x + 5 &= 13 - 4x && \text{Collecting like terms} \\ 4x - x + 5 &= 13 - 4x + 4x && \text{Adding } 4x \\ 3x + 5 &= 13 && \text{Simplifying} \\ 3x + 5 - 5 &= 13 - 5 && \text{Subtracting 5} \\ 3x &= 8 && \text{Simplifying} \\ \frac{3x}{3} &= \frac{8}{3} && \text{Dividing by 3} \\ x &= \frac{8}{3}. && \text{Simplifying} \end{aligned}$$

The number $\frac{8}{3}$ checks, so it is the solution. ◀

DO EXERCISES 10–12.

Solve.

7. $4x + 3x = -21$ -3

8. $x - 0.09x = 728$ 800

9. Solve: $7y + 5 = 2y + 10$. 1

Solve.

10. $5 - 2y = 3y - 5$ 2

11. $7x - 17 + 2x = 2 - 8x + 15$ 2

12. $3x - 15 = 5x + 2 - 4x$ $\frac{17}{2}$

ANSWERS ON PAGE A-6

13. Solve: $\frac{7}{8}x - \frac{1}{4} + \frac{1}{2}x = \frac{3}{4} + x.$ $\frac{8}{3}$

ANSWER ON PAGE A-6

Clearing of Fractions and Decimals

We have stated that we generally use the addition principle first. There are, however, some situations in which it is to our advantage to use the multiplication principle first. Consider, for example,

$$\frac{1}{2}x = \frac{3}{4}.$$

If we multiply by 4 on both sides, we get $2x = 3$, which has no fractions. We have "cleared of fractions." Consider

$$2.3x = 5.$$

If we multiply by 10 on both sides, we get $23x = 50$, which has no decimal points. We have "cleared of decimals." The equations are then easier to solve. It is your choice whether to clear of fractions or decimals, but doing so often eases computations.

In what follows, we use the multiplication principle first to "clear of," or "eliminate," fractions or decimals. For fractions, the number by which we multiply is the **least common multiple of all the denominators.**

▶ **EXAMPLE 8** Solve:

$$\frac{2}{3}x - \frac{1}{6} + \frac{1}{2}x = \frac{7}{6} + 2x.$$

The number 6 is the least common multiple of all the denominators. We multiply on both sides by 6:

$$6\left(\frac{2}{3}x - \frac{1}{6} + \frac{1}{2}x\right) = 6\left(\frac{7}{6} + 2x\right) \quad \text{Multiplying by 6 on both sides}$$

$$6 \cdot \frac{2}{3}x - 6 \cdot \frac{1}{6} + 6 \cdot \frac{1}{2}x = 6 \cdot \frac{7}{6} + 6 \cdot 2x$$

Using the distributive laws. (*Caution*! Be sure to multiply all the terms by 6.)

$$4x - 1 + 3x = 7 + 12x \quad \text{Simplifying. Note that the fractions are cleared.}$$

$$7x - 1 = 7 + 12x \quad \text{Collecting like terms}$$

$$7x - 12x = 7 + 1 \quad \text{Subtracting } 12x \text{ and adding 1 to get all terms with variables on one side and all constant terms on the other side}$$

$$-5x = 8 \quad \text{Collecting like terms}$$

$$x = -\frac{8}{5}. \quad \text{Multiplying by } -\frac{1}{5} \text{ or dividing by } -5$$

The number $-\frac{8}{5}$ checks and is the solution. ◀

DO EXERCISE 13.

Here is a procedure for solving the equations of this section.

An Equation-Solving Procedure

1. **Multiply on both sides to clear the equation of fractions or decimals. (This is optional, but it can ease computations.)**
2. **Collect like terms on each side, if necessary.**
3. **Get all terms with variables on one side and all the constant terms on the other side, using the addition principle.**
4. **Collect like terms again, if necessary.**
5. **Multiply or divide to solve for the variable, using the multiplication principle.**
6. **Check all possible solutions in the original equation.**

We illustrate this by repeating Example 4, but we clear the equation of decimals first.

▶ **EXAMPLE 9** Solve: $16.3 - 7.2y = -8.18$.

The greatest number of decimal places in any one number is *two*. Multiplying by 100, which has *two* 0's, will clear of the decimals.

$100(16.3 - 7.2y) = 100(-8.18)$ **Multiplying by 100 on both sides**

$100(16.3) - 100(7.2y) = 100(-8.18)$ **Using a distributive law**

$1630 - 720y = -818$ **Simplifying**

$-720y = -818 - 1630$ **Subtracting 1630 on both sides**

$-720y = -2448$ **Collecting like terms**

$y = \frac{-2448}{-720} = 3.4$ **Dividing by −720 on both sides**

The number 3.4 checks and is the solution. ◀

DO EXERCISE 14.

14. Solve: $41.68 = 4.7 - 8.6y$.

−4.3

ANSWER ON PAGE A-6

❖ SIDELIGHTS

Careers Involving Mathematics

You are about to finish this course in basic mathematics. If you have done well, you might be considering a career in mathematics or one that involves mathematics. If either is the case, the following information may be valuable to you.

CAREERS INVOLVING MATHEMATICS. The following is the result of a survey conducted by *The Jobs Related Almanac,* published by the American References Inc., of Chicago. It used the criteria of salary, stress, work environment, outlook, security, and physical demands to rate the desirability of 250 jobs. The top 10 of the 250 jobs listed were:

1. Actuary
2. Computer programmer
3. Computer systems analyst
4. Mathematician
5. Statistician
6. Hospital administrator
7. Industrial engineer
8. Physicist
9. Astrologer
10. Paralegal.

(Items 1–5: The top 5 involve mathematics.)

Two things are interesting to note. First, the top five rated professions involve a heavy use of mathematics. The top, actuary, involves the application of mathematics to insurance. The second point of interest is that choices like doctor, lawyer, and astronaut are *not* in the top ten.

Perhaps you might be interested in a career in teaching mathematics. This profession will be expanding increasingly in the next ten years. The field of mathematics will need well-qualified mathematics teachers in all areas from elementary to junior high to secondary to two-year college to college instruction. Some questions you might ask yourself in making a decision about a career in mathematics teaching are the following.

1. Do you find yourself carefully observing the strengths and weaknesses of your teachers?
2. Are you deeply interested in mathematics?
3. Are you interested in the ways of learning? If a student is struggling with a topic, would it be challenging to you to discover two or three other ways to present the material so that the student might understand?
4. Are you able to put yourself in the place of the students in order to help them be successful in learning mathematics?

If you are interested in a career involving mathematics, the next courses you would take are *introductory algebra, intermediate algebra, precalculus algebra and trigonometry* and *calculus.* You might want to seek out a counselor in the mathematics department at your college for further assistance.

WHAT KIND OF SALARIES ARE THERE IN VARIOUS FIELDS? The College Placement Council published the following comparisons of the average salaries of graduating students with bachelors degrees who were taking the following jobs:

Subject area	Annual salary
All engineering	\$27,800
Computer science	\$26,400
Mathematics	\$25,900
Sciences other than math and computer science	\$22,200
Humanities and social science	\$21,800
Accounting	\$21,700
All business	\$21,300

Many people choose to go on to earn a masters degree. Here are salaries in the same fields for students just graduating with a masters degree:

Subject area	Annual salary
All engineering	\$34,000
Computer science	\$33,800
Mathematics	\$27,900
Sciences other than math and computer science	\$27,400
Humanities and social science	\$22,300
Accounting	\$26,000
Business administration	\$30,700

NAME SECTION DATE

EXERCISE SET 12.4

a Solve. Don't forget to check!

1. $5x + 6 = 31$

2. $3x + 6 = 30$

3. $8x + 4 = 68$

4. $7z + 9 = 72$

5. $4x - 6 = 34$

6. $6x - 3 = 15$

7. $3x - 9 = 33$

8. $5x - 7 = 48$

9. $7x + 2 = -54$

10. $5x + 4 = -41$

11. $-45 = 3 + 6y$

12. $-91 = 9t + 8$

13. $-4x + 7 = 35$

14. $-5x - 7 = 108$

15. $-7x - 24 = -129$

16. $-6z - 18 = -132$

b Solve.

17. $5x + 7x = 72$

18. $4x + 5x = 45$

19. $8x + 7x = 60$

20. $3x + 9x = 96$

21. $4x + 3x = 42$

22. $6x + 19x = 100$

23. $-6y - 3y = 27$

24. $-4y - 8y = 48$

25. $-7y - 8y = -15$

26. $-10y - 3y = -39$

27. $10.2y - 7.3y = -58$

28. $6.8y - 2.4y = -88$

29. $x + \frac{1}{3}x = 8$

30. $x + \frac{1}{4}x = 10$

31. $8y - 35 = 3y$

ANSWERS

1. 5
2. 8
3. 8
4. 9
5. 10
6. 3
7. 14
8. 11
9. −8
10. −9
11. −8
12. −11
13. −7
14. −23
15. 15
16. 19
17. 6
18. 5
19. 4
20. 8
21. 6
22. 4
23. −3
24. −4
25. 1
26. 3
27. −20
28. −20
29. 6
30. 8
31. 7

ANSWERS

32. −3
33. 2
34. 5
35. 5
36. 2
37. 2
38. 4
39. 10
40. 10
41. 4
42. 0
43. 0
44. 7
45. −1
46. $\frac{1}{2}$
47. $-\frac{4}{3}$
48. $-\frac{2}{3}$
49. $\frac{2}{5}$
50. −3
51. −2
52. −3
53. −4
54. $\frac{5}{3}$
55. $\frac{4}{5}$
56. 1
57. −6.5
58. $7(x - 3 - 2y)$
59. <
60. −14
61. 2
62. 0
63. −2
64. −2

32. $4x - 6 = 6x$

33. $8x - 1 = 23 - 4x$

34. $5y - 2 = 28 - y$

35. $2x - 1 = 4 + x$

36. $5x - 2 = 6 + x$

37. $6x + 3 = 2x + 11$

38. $5y + 3 = 2y + 15$

39. $5 - 2x = 3x - 7x + 25$

40. $10 - 3x = 2x - 8x + 40$

41. $4 + 3x - 6 = 3x + 2 - x$

42. $5 + 4x - 7 = 4x - 2 - x$

43. $4y - 4 + y + 24 = 6y + 20 - 4y$

44. $5y - 7 + y = 7y + 21 - 5y$

Solve. Clear of fractions or decimals first.

45. $\frac{7}{2}x + \frac{1}{2}x = 3x + \frac{3}{2} + \frac{5}{2}x$

46. $\frac{7}{8}x - \frac{1}{4} + \frac{3}{4}x = \frac{1}{16} + x$

47. $\frac{2}{3} + \frac{1}{4}t = \frac{1}{3}$

48. $-\frac{3}{2} + x = -\frac{5}{6} - \frac{4}{3}$

49. $\frac{2}{3} + 3y = 5y - \frac{2}{15}$

50. $\frac{1}{2} + 4m = 3m - \frac{5}{2}$

51. $\frac{5}{3} + \frac{2}{3}x = \frac{25}{12} + \frac{5}{4}x + \frac{3}{4}$

52. $1 - \frac{2}{3}y = \frac{9}{5} - \frac{y}{5} + \frac{3}{5}$

53. $2.1x + 45.2 = 3.2 - 8.4x$

54. $0.96y - 0.79 = 0.21y + 0.46$

55. $1.03 - 0.62x = 0.71 - 0.22x$

56. $1.7t + 8 - 1.62t = 0.4t - 0.32 + 8$

SKILL MAINTENANCE

57. Divide: $-22.1 \div 3.4$.

58. Factor: $7x - 21 - 14y$.

59. Use < or > for ▒ to write a true sentence: -15 ▒ -13.

60. Find $-(-x)$ when $x = -14$.

SYNTHESIS

Solve.

61. $\frac{y - 2}{3} = \frac{2 - y}{5}$

62. $3x = 4x$

63. $\frac{5 + 2y}{3} = \frac{25}{12} + \frac{5y + 3}{4}$

64. $0.05y - 1.82 = 0.708y - 0.504$

12.5 Solving Problems

a Translating to Algebraic Expressions

In algebra we translate problems to equations. The different parts of an equation are translations of word phrases to algebraic expressions. To translate it helps to learn what words translate to certain operation symbols.

KEY WORDS			
Addition (+)	**Subtraction (−)**	**Multiplication (·)**	**Division (÷)**
add sum plus more than increased by	subtract difference minus less than decreased by take from	multiply product times twice of	divide quotient divided by

▶ **EXAMPLE 1** Translate to an algebraic expression:

Twice (or two times) some number.

Think of some number, say, 8. What number is twice 8? It is 16. How did you get 16? You multiplied by 2. Do the same thing using a variable. We can use any variable we wish, such as x, y, m, or n. Let's use y to stand for some number. If we multiply by 2, we get an expression

$$y \times 2, \quad 2 \times y, \quad 2 \cdot y, \quad \text{or} \quad 2y.$$ ◀

▶ **EXAMPLE 2** Translate to an algebraic expression:

Seven less than some number.

We let

$$x = \text{the number.}$$

Now if the number were 23, then the translation would be $23 - 7$. If we knew the number to be 345, then the translation would be $345 - 7$. The translation is found as follows:

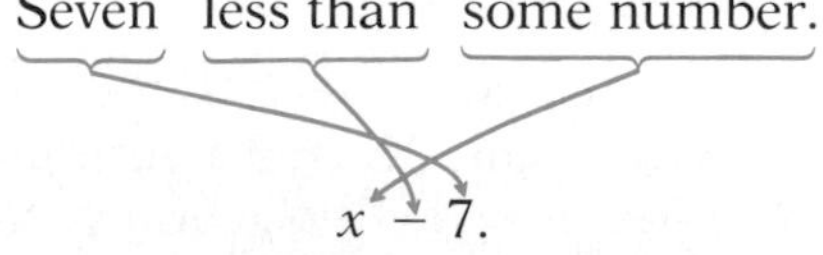

◀

Note that $7 - x$ is *not* a correct translation of the expression in Example 8. The expression $7 - x$ would be a translation of "seven minus some number" or "some number less than seven."

▶ **EXAMPLE 3** Translate to an algebraic expression:

Eighteen more than a number.

We let

$$t = \text{the number.}$$

OBJECTIVES

After finishing Section 12.5, you should be able to:

a Translate phrases to algebraic expressions.

b Solve problems by translating to equations.

FOR EXTRA HELP

Tape 15D Tape 17B MAC: 12 IBM: 12

Translate to an algebraic expression.

1. Twelve less than some number

$x - 12$

2. Twelve more than some number

$y + 12$, or $12 + y$

3. Four less than some number

$m - 4$

4. Half of some number $\frac{1}{2} \cdot p$

5. Six more than eight times some number $6 + 8x$, or $8x + 6$

6. The difference of two numbers

$a - b$

7. Fifty-nine percent of some number $59\%x$, or $0.59x$

8. Two hundred less than the product of two numbers

$xy - 200$

9. The sum of two numbers $p + q$

ANSWERS ON PAGE A-7

Now if the number were 26, then the translation would be 18 + 26. If we knew the number to be 174, then the translation would be 18 + 174. The translation is

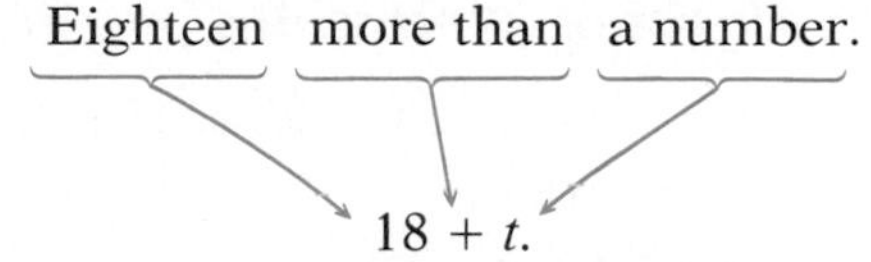

▶ **EXAMPLE 4** Translate to an algebraic expression:

A number divided by 5.

We let

m = the number.

Now if the number were 76, then the translation would be $76 \div 5$, or 76/5, or $\frac{76}{5}$. If the number were 213, then the translation would be $213 \div 5$, or 213/5, or $\frac{213}{5}$. The translation is found as follows:

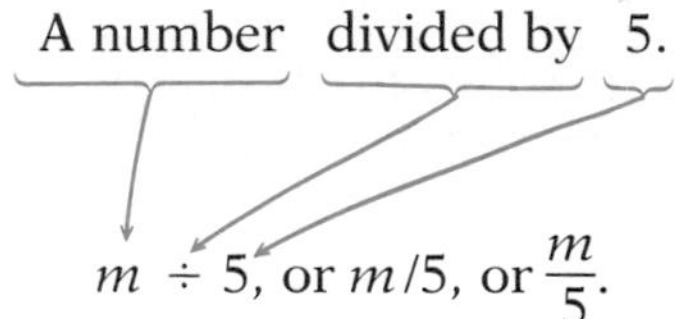

▶ **EXAMPLE 5** Translate to an algebraic expression.

Phrase	Algebraic expression
Five more than some number	$5 + n$, or $n + 5$
Half of a number	$\frac{1}{2}t$, or $\frac{t}{2}$
Five more than three times some number	$5 + 3p$, or $3p + 5$
The difference of two numbers	$x - y$
Six less than the product of two numbers	$mn - 6$
Seventy-six percent of some number	$76\%z$, or $0.76z$

DO EXERCISES 1–9.

b Five Steps for Solving Problems

We have studied many new equation-solving tools in this chapter. We now apply them to problem solving. We have purposely used the following strategy in order to introduce you to algebra. We will continue to use it now.

> Five Steps for Problem Solving in Algebra
>
> 1. ***Familiarize*** **yourself with the problem situation.**
> 2. ***Translate*** **to an equation.**
> 3. ***Solve*** **the equation.**
> 4. ***Check*** **your possible answer in the original problem.**
> 5. ***State*** **your answer clearly.**

Of the five steps, probably the most important is the first one: becoming familiar with the problem situation. Here are some hints for familiarization.

To familiarize yourself with the problem situation:

1. **If a problem is given in words, read it carefully.**
2. **Reread the problem, perhaps aloud. Try to verbalize the problem to yourself.**
3. **List the information given and the questions to be answered. Choose a variable (or variables) to represent the unknown and clearly state what the variable represents. Be descriptive! For example, let L = length, d = distance, and so on.**
4. **Find further information. Look up a formula in the back of the book or in a reference book. Talk to a reference librarian or an expert in the field.**
5. **Make a table of the given information and the information you have collected. Look for patterns that may help in the translation to an equation.**
6. **Make a drawing and label it with known information. Also, indicate unknown information, using specific units if given.**
7. **Guess or estimate the answer.**

▶ **EXAMPLE 6** A 72-in. board is cut into two pieces, and one piece is twice as long as the other. How long are the pieces?

1. *Familiarize.* We first draw a picture. We let

x = the length of the shorter piece.

Then $2x$ = the length of the longer piece.

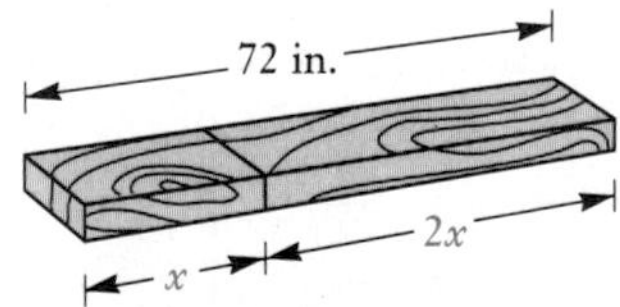

We can further familiarize ourselves with the problem by making some guesses. Let's suppose that $x = 31$ in. Then $2x = 62$ in., and $x + 2x = 93$ in. This is not correct but does help us to become familiar with the problem.

2. *Translate.* From the figure we can see that the lengths of the two pieces add up to 72 in. That gives us our translation.

Length of one piece	plus	Length of other	is	72
x	$+$	$2x$	$=$	72

3. *Solve.* We solve the equation:

$x + 2x = 72$

$3x = 72$ **Collecting like terms**

$x = 24.$ **Dividing by 3**

10. A 58-in. board is cut into two pieces. One piece is 2 in. longer than the other. How long are the pieces? 28 in., 30 in.

11. If 5 is subtracted from three times a certain number, the result is 10. What is the number? 5

ANSWERS ON PAGE A-7

4. *Check.* Do we have an answer to the *problem*? If one piece is 24 in. long, the other, to be twice as long, must be 48 in. long. The lengths of the pieces add up to 72 in. This checks.
5. *State.* One piece is 24 in. long, and the other is 48 in. long. ◀

DO EXERCISE 10.

▶ **EXAMPLE 7** Five plus three more than a number is nineteen. What is the number?

1. *Familiarize.* Let x = the number. Then "three more than the number" translates to $x + 3$ and "5 more than $x + 3$" translates to $5 + (x + 3)$.
2. *Translate.* The familiarization leads us to the following translation:

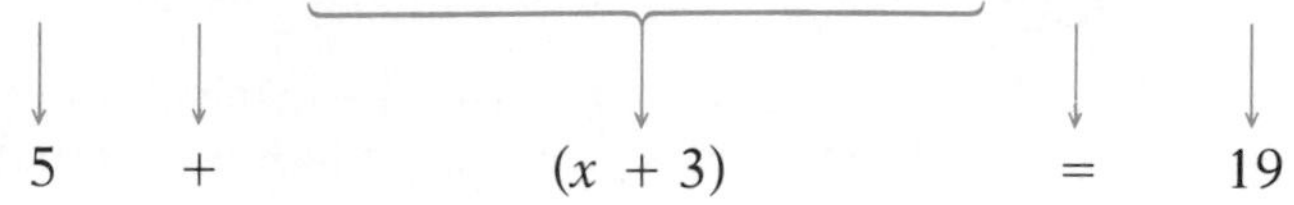

$$5 + (x + 3) = 19$$

3. *Solve.* We solve the equation:

$$5 + (x + 3) = 19$$
$$x + 8 = 19 \quad \text{Collecting like terms}$$
$$x = 11. \quad \text{Subtracting 8}$$

4. *Check.* Three more than 11 is 14. Adding 5 to 14, we get 19. This checks.
5. *State.* The number is 11. ◀

DO EXERCISE 11.

▶ **EXAMPLE 8** Acme Rent-A-Car rents an intermediate-sized car (such as a Chevrolet, Ford, or Plymouth) at a daily rate of \$44.95 plus 29 cents a mile. A salesperson can spend \$100 per day on car rental. How many miles can the person drive on the \$100?

1. *Familiarize.* Suppose the businessperson drives 75 miles. Then the cost is

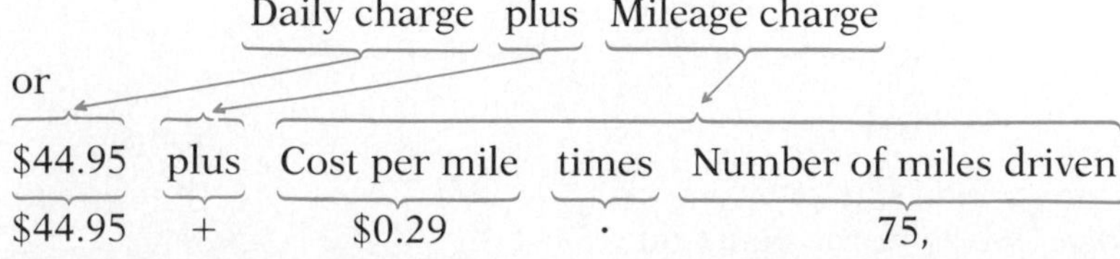

$$\$44.95 + \$0.29 \cdot 75,$$

which is \$44.95 + \$21.75, or \$66.70. This familiarizes us with the way in which a calculation is made. Note that we convert 29 cents to \$0.29 so that we have the same units, dollars. Otherwise, we will not get a correct answer.

Let m = the number of miles that can be driven on \$100.

2. *Translate.* We reword the problem and translate as follows.

Daily rate	plus	Cost per mile	times	Number of miles driven	is	Cost
$44.95	+	$0.29	·	m	=	$100

3. *Solve.* We solve the equation:

$$\begin{aligned} 44.95 + 0.29m &= 100 \\ 100(44.95 + 0.29m) &= 100(100) && \text{Multiplying by 100 on both sides to clear of the decimals} \\ 100(44.95) + 100(0.29m) &= 10{,}000 && \text{Using a distributive law} \\ 4495 + 29m &= 10{,}000 \\ 29m &= 5505 && \text{Subtracting 4495} \\ m &= \frac{5505}{29} && \text{Dividing by 29} \\ m &\approx 189.8. && \text{Rounding to the nearest tenth. "}\approx\text{" means "approximately equal to."} \end{aligned}$$

4. *Check.* We check in the original problem. We multiply 189.8 by $0.29, obtaining $55.042. Then we add $55.042 to $44.95 and get $99.992, which is just about the $100 allotted.

5. *State.* The person can drive about 189.8 miles on the car rental allotment of $100. ◀

DO EXERCISE 12.

12. Acme also rents compact cars at a rate of $34.95 plus 27 cents per mile. What mileage will allow the businessperson to stay within a budget of $100?

240.9 mi

▶ **EXAMPLE 9** The state of Colorado is a rectangle whose perimeter is 1300 mi. The length is 110 mi more than the width. Find the dimensions.

1. *Familiarize.* We first draw a picture. We let

$$w = \text{the width of the rectangle.}$$

Then $w + 110 = $ the length.

(We can also let l = the length and $l - 110$ = the width.)

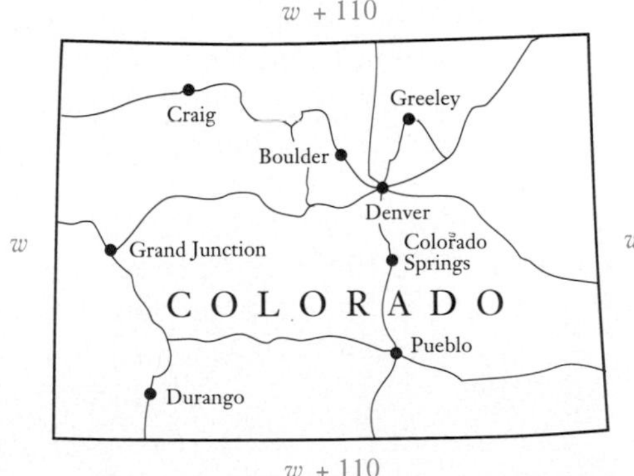

The perimeter P of a rectangle is the distance around it and is given by the formula $2l + 2w = P$, where l = the length and w = the width.

2. *Translate.* To translate the problem, we substitute $w + 110$ for l and 1300 for P, as follows:

$$2l + 2w = P$$
$$2(w + 110) + 2w = 1300.$$

ANSWER ON PAGE A-7

13. A standard-sized rug has a perimeter of 42 ft. The length is 3 ft more than the width. Find the dimensions of the rug.

Width: 9 ft; length: 12 ft

ANSWER ON PAGE A-7

3. *Solve.* We solve the equation:

$$2(w + 110) + 2w = 1300$$
$$2w + 220 + 2w = 1300 \quad \text{Multiplying using a distributive law}$$
$$4w + 220 = 1300$$
$$4w = 1080$$
$$w = 270.$$

Possible dimensions are $w = 270$ mi and $w + 110 = 380$ mi.

4. *Check.* If the width is 270 mi and the length is 110 mi + 270 mi, or 380 mi, the perimeter is 2(380 mi) + 2(270 mi), or 1300 mi. This checks.

5. *State.* The width is 270 mi, and the length is 380 mi. ◀

DO EXERCISE 13.

❖ SIDELIGHTS

Study Tips: Extra Tips on Problem Solving

We will often present some tips and guidelines to enhance your learning abilities. The following tips are focused on problem solving. They summarize some points already considered and propose some new tips.

- *The following are the five steps for problem solving:*

1. ***Familiarize* yourself with the problem situation.**
2. ***Translate* the problem to an equation.** As you study more mathematics, you will find that the translation may be to some other kind of mathematical language, such as an inequality.
3. ***Solve* the equation.** If the translation is to some other kind of mathematical language, you would carry out some kind of mathematical manipulation.
4. ***Check* the answer in the original equation.** This does not mean to check in the translated equation. It means to go back to the original worded problem.
5. ***State* the answer to the problem clearly.**

For Step 4 on checking, some further comment is appropriate. *You may be able to translate to an equation and to solve the equation, but none of the solutions of the equation are solutions of the original problem.* To see how this can happen, consider the following problem.

EXAMPLE The sum of two consecutive even integers is 537. Find the integers.

1. *Familiarize.* Suppose we let x = the first number. Then $x + 2$ = the second number.
2. *Translate.* The problem can be translated to the following equation: $x + (x + 2) = 537$.
3. *Solve.* We solve the equation as follows:

$$2x + 2 = 537$$
$$2x = 535$$
$$x = \tfrac{535}{2}, \quad \text{or } 267.5.$$

4. *Check.* Then $x + 2 = 269.5$. However, the numbers are not only not even, but they are not integers.
5. *State.* The problem has no solution. ◀

The following are some other tips.

- *To be good at problem solving, do lots of problems.* The situation is similar to what happens to some people learning a skill such as playing golf. At first they are not successful, but the more they practice and work at improving their skills the more successful they become. For problem solving, do more than just two or three odd-numbered problems assigned. Do them all and if you have time, do the even-numbered problems. Then find another book on the same subject and do problems in that book.
- *Look for patterns when solving problems.* You will eventually see patterns in similar kinds of problems. For example, there is a pattern in the way that you solve problems involving consecutive integers.
- *When translating to an equation, or some other mathematical language, consider the dimensions of the variables and the constants in the equation.* The variables that represent length should all be in the same unit, those that represent money should be all in dollars or all in cents, and so on.

NAME SECTION DATE

EXERCISE SET 12.5

a Translate to an algebraic expression.

1. 6 more than b

2. 8 more than t

3. 9 less than c

4. 4 less than d

5. 6 increased by q

6. 11 increased by z

7. b more than a

8. c more than d

9. x less than y

10. c less than h

11. x added to w

12. s added to t

13. The sum of r and s

14. The sum of d and f

15. Twice x

16. Three times p

17. 5 multiplied by t

18. The product of 3 and b

19. The product of 97% and some number

20. 43% of some number

b Solve.

21. What number added to 60 is 112?

22. Seven times what number is 2233?

23. A game board has 64 squares. If you win 35 squares and your opponent wins the rest, how many does your opponent win?

24. A consultant charges $80 an hour. How many hours did the consultant work to make $53,400?

25. In a recent year the cost of four 12-oz boxes of Post® Oat Flakes was $7.96. How much did one box cost?

26. The total amount spent on women's blouses in a recent year was $6.5 billion. This was $0.2 billion more than was spent on women's dresses. How much was spent on women's dresses?

27. When 18 is subtracted from six times a certain number, the result is 96. What is the number?

28. When 28 is subtracted from five times a certain number, the result is 232. What is the number?

ANSWERS

1. $b + 6$, or $6 + b$

2. $8 + t$, or $t + 8$

3. $c - 9$

4. $d - 4$

5. $6 + q$, or $q + 6$

6. $11 + z$, or $z + 11$

7. $b + a$, or $a + b$

8. $c + d$, or $d + c$

9. $y - x$

10. $h - c$

11. $x + w$, or $w + x$

12. $s + t$, or $t + s$

13. $r + s$, or $s + r$

14. $d + f$, or $f + d$

15. $2x$

16. $3p$

17. $5t$

18. $3b$

19. $97\%x$, or $0.97x$

20. $43\%m$, or $0.43m$

21. 52

22. 319

23. 29

24. 667.5

25. $1.99

26. $6.3 billion

27. 19

28. 52

ANSWERS

29. −10

30. −68

31. 20 m, 40 m, 120 m

32. 30 m, 90 m, 360 m

33. Width: 100 ft; length: 160 ft; area: 16,000 ft^2

34. Width: 165 ft; length: 265 ft; area: 43,725 ft^2

35. Length: 27.9 cm; width: 21.6 cm

36. Width: 275 mi; length: 365 mi

37. 450.5 mi

38. 460.5 mi

39. 20

40. 19

41. Length: 12 cm; width: 9 cm

42. 120

43. 5 half dollars, 10 quarters, 20 dimes, 60 nickels

44. 76

29. If you double a number and then add 16, you get $\frac{2}{5}$ of the original number. What is the original number?

30. If you double a number and then add 85, you get $\frac{3}{4}$ of the original number. What is the original number?

31. A 180-m rope is cut into three pieces. The second piece is twice as long as the first. The third piece is three times as long as the second. How long is each piece of rope?

32. A 480-m wire is cut into three pieces. The second piece is three times as long as the first. The third piece is four times as long as the second. How long is each piece?

33. The top of the John Hancock Building in Chicago is a rectangle whose length is 60 ft more than the width. The perimeter is 520 ft. Find the width and the length of the rectangle. Find the area of the rectangle.

34. The ground floor of the John Hancock Building is a rectangle whose length is 100 ft more than the width. The perimeter is 860 ft. Find the width and the length of the rectangle. Find the area of the rectangle.

35. The perimeter of a standard-sized piece of typewriter paper is 99 cm. The width is 6.3 cm less than the length. Find the length and the width.

36. The perimeter of the state of Wyoming is 1280 mi. The width is 90 mi less than the length. Find the width and the length.

37. Badger Rent-A-Car rents an intermediate-sized car at a daily rate of $34.95 plus 10 cents per mile. A businessperson is allotted $80 for car rental. How many miles can the businessperson travel on the $80?

38. Badger also rents compact cars at $43.95 plus 10 cents per mile. A businessperson has a car rental allotment of $90. How many miles can the businessperson travel on the $90?

SYNTHESIS

39. Abraham Lincoln's 1863 Gettysburg Address refers to the year 1776 as "Four *score* and seven years ago." Write an equation and find what a score is.

40. A student scored 78 on a test that had 4 seven-point fill-ins and 24 three-point multiple-choice questions. The student had one fill-in wrong. How many multiple-choice questions did the student get right?

41. The width of a rectangle is $\frac{3}{4}$ the length. The perimeter of the rectangle becomes 50 cm when the length and the width are each increased by 2 cm. Find the length and the width.

42. Apples are collected in a basket for six people. One third, one fourth, one eighth, and one fifth are given to four people, respectively. The fifth person gets ten apples with one apple remaining for the sixth person. Find the original number of apples in the basket.

43. A storekeeper goes to the bank to get $10 worth of change. The storekeeper requests twice as many quarters as half dollars, twice as many dimes as quarters, three times as many nickels as dimes, and no pennies or dollars. How many of each coin did the shopkeeper get?

44. A student has an average score of 82 on three tests. The student's average score on the first two tests is 85. What was the score on the third test?

SUMMARY AND REVIEW: CHAPTER 12

IMPORTANT PROPERTIES AND FORMULAS

The Addition Principle: If $a = b$ is true, then $a + c = b + c$ is true for any real number c.
The Multiplication Principle: If $a = b$ is true, then $ac = bc$ is true for any real number c.
The Distributive Laws: $a(b + c) = ab + ac$, $a(b - c) = ab - ac$

REVIEW EXERCISES

The review sections and objectives to be tested in addition to the material in this chapter are [9.6a, b, c], [10.1a], [11.2a], and [11.4a].

Solve.

1. $x + 5 = -17$

2. $-8x = -56$

3. $-\frac{x}{4} = 48$

4. $n - 7 = -6$

5. $15x = -35$

6. $x - 11 = 14$

7. $-\frac{2}{3} + x = -\frac{1}{6}$

8. $\frac{4}{5}y = -\frac{3}{16}$

9. $y - 0.9 = 9.09$

10. $5 - x = 13$

11. $5t + 9 = 3t - 1$

12. $7x - 6 = 25x$

13. $\frac{1}{4}x - \frac{5}{8} = \frac{3}{8}$

14. $14y = 23y - 17 - 10$

15. $0.22y - 0.6 = 0.12y + 3 - 0.8y$

16. $\frac{1}{4}x - \frac{1}{8}x = 3 - \frac{1}{16}x$

17. Translate to an algebraic expression: Nineteen percent of some number.

Solve.

18. A color television sold for \$629 in May. This was \$38 more than the January cost. Find the January cost.

19. Selma gets a \$4 commission for each appliance that she sells. One week she received \$108 in commissions. How many appliances did she sell?

20. A 8-m board is cut into two pieces. One piece is 2 m longer than the other. How long are the pieces?

21. If 14 is added to three times a certain number, the result is 41. Find the number.

22. The perimeter of a rectangle is 56 cm. The width is 6 cm less than the length. Find the width and the length.

23. A car rental agency rents compact cars at \$41.95 plus 12 cents per mile. A businessperson has a car rental allotment of \$258. How many miles can the businessperson travel on the \$258?

SKILL MAINTENANCE

24. Find the diameter, the circumference, and the area of a circle when $r = 20$ ft. Use 3.14 for π.

25. Find the volume of a rectangular solid when the length is 20 cm, the width is 18.5 cm, and the height is 4.6 cm.

26. Add: $-12 + 10 + (-19) + (-24)$.

27. Multiply: $(-2) \cdot (-3) \cdot (-5) \cdot (-2) \cdot (-1)$.

SYNTHESIS

28. The total length of the Nile and Amazon Rivers is 13,108 km. If the Amazon were 234 km longer, it would be as long as the Nile. Find the length of each river.

Solve.

29. $2|n| + 4 = 50$

30. $|3n| = 60$

❖ THINKING IT THROUGH

Explain all possible errors in each of the following.

1. Solve: $4 - 3x = 5$
$$3x = 9$$
$$x = 3.$$

2. Solve: $2x - 5 = 7$
$$2x = 2$$
$$x = 1.$$

NAME SECTION DATE

TEST: CHAPTER 12

Solve.

1. $x + 7 = 15$

2. $t - 9 = 17$

3. $3x = -18$

4. $-\frac{4}{7}x = -28$

5. $3t + 7 = 2t - 5$

6. $\frac{1}{2}x - \frac{3}{5} = \frac{2}{5}$

7. $8 - y = 16$

8. $-\frac{2}{5} + x = -\frac{3}{4}$

9. $0.4p + 0.2 = 4.2p - 7.8 - 0.6p$

ANSWERS

1. [12.2a] 8
2. [12.2a] 26
3. [12.3a] -6
4. [12.3a] 49
5. [12.4b] -12
6. [12.4a] 2
7. [12.4a] -8
8. [12.2a] $-\frac{7}{20}$
9. [12.4b] 2.5

ANSWERS

10. [12.5b] Width: 7 cm; length: 11 cm

11. [12.5b] 6

12. [12.5a] $x - 9$

13. [11.4a] 180

14. [11.2a] $-\frac{2}{9}$

15. [9.6a, b, c] 140 yd; 439.6 yd; 15,386 yd^2

16. [10.1a] 1320 ft^3

17. [11.1e], [12.4a] 15, −15

18. [12.5b] 60

Solve.

10. The perimeter of a rectangle is 36 cm. The length is 4 cm greater than the width. Find the width and the length.

11. If you triple a number and then subtract 14, you get $\frac{2}{3}$ of the original number. What is the original number?

12. Translate to an algebraic expression: Nine less than some number.

SKILL MAINTENANCE

13. Multiply:

$$(-9)\cdot(-2)\cdot(-2)\cdot(-5).$$

14. Add:

$$\frac{2}{3} + \left(-\frac{8}{9}\right).$$

15. Find the diameter, the circumference, and the area of a circle when the radius is 70 yd. Use 3.14 for π.

16. Find the volume of a rectangular solid when the length is 22 ft, the width is 10 ft, and the height is 6 ft.

SYNTHESIS

17. Solve: $3|w| - 8 = 37$.

18. A movie theater had a certain number of tickets to give away. Five people got the tickets. The first got $\frac{1}{3}$ of the tickets, the second got $\frac{1}{4}$ of the tickets, and the third got $\frac{1}{5}$ of the tickets. The fourth person got eight tickets, and there were five tickets left for the fifth person. Find the total number of tickets given away.

CUMULATIVE REVIEW: CHAPTERS 1–12

This is a review of the entire textbook. A question that may arise at this point is what notation to use for a particular problem or exercise. While there is no hard-and-fast rule, especially as you use mathematics outside the classroom, here is the guideline that we follow: Use the notation given in the problem. That is, if the problem is given using mixed numerals, give the answer in mixed numerals. If the problem is given in decimal notation, give the answer in decimal notation.

1. In 47201, what digit tells the number of thousands?

2. Write expanded notation for 7405.

Add and simplify, if appropriate.

3. $\begin{array}{r} 741 \\ +271 \\ \hline \end{array}$

4. $\begin{array}{r} 4903 \\ 5278 \\ 6391 \\ +4513 \\ \hline \end{array}$

5. $\frac{2}{13} + \frac{1}{26}$

6. $\begin{array}{r} 2\frac{4}{9} \\ +3\frac{1}{3} \\ \hline \end{array}$

7. $\begin{array}{r} 2.048 \\ 63.914 \\ +428.009 \\ \hline \end{array}$

8. $34.56 + 2.783 + 0.433 + 765.1$

Subtract and simplify, if possible.

9. $\begin{array}{r} 674 \\ -522 \\ \hline \end{array}$

10. $\begin{array}{r} 9465 \\ -8791 \\ \hline \end{array}$

11. $\frac{7}{8} - \frac{2}{3}$

12. $\begin{array}{r} 4\frac{1}{3} \\ -1\frac{5}{8} \\ \hline \end{array}$

13. $\begin{array}{r} 20.0 \\ -\;\;0.0027 \\ \hline \end{array}$

14. $40.03 - 5.789$

Multiply and simplify, if possible.

15. $\begin{array}{r} 297 \\ \times\;\;16 \\ \hline \end{array}$

16. $\begin{array}{r} 349 \\ \times 763 \\ \hline \end{array}$

17. $1\frac{3}{4} \cdot 2\frac{1}{3}$

18. $\frac{9}{7} \cdot \frac{14}{15}$

19. $12 \cdot \frac{5}{6}$

20. $\begin{array}{r} 34.09 \\ \times\;\;7.6 \\ \hline \end{array}$

Divide and simplify. State the answer using a whole-number quotient and remainder.

21. $6\overline{)3438}$

22. $34\overline{)1914}$

23. Give a mixed numeral for the quotient in Exercise 22.

24. $\frac{4}{5} \div \frac{8}{15}$

25. $2\frac{1}{3} \div 30$

26. $2.7\overline{)105.3}$

27. Round 68,489 to the nearest thousand.

28. Round 0.4275 to the nearest thousandth.

29. Round $21.\overline{83}$ to the nearest hundredth.

30. Determine whether 1368 is divisible by 8.

31. Find all the factors of 15.

32. Find the LCM of 16, 25, and 32.

Simplify.

33. $\frac{21}{30}$

34. $\frac{275}{5}$

35. Convert to a mixed numeral: $\frac{18}{5}$.

36. Use $=$ or $\neq$ for ■ to write a true sentence:

$\frac{4}{7}$ ■ $\frac{3}{5}$.

37. Use $<$ or $>$ for ■ to write a true sentence:

$\frac{4}{7}$ ■ $\frac{3}{5}$.

38. Which number is greater, 1.001 or 0.9976?

39. Use $<$ or $>$ for ■ to write a true sentence:

987 ■ 879.

40. What part is shaded?

Convert to decimal notation.

41. $\frac{37}{1000}$

42. $\frac{13}{25}$

43. $\frac{8}{9}$

44. 7%

Convert to fractional notation.

45. 4.63

46. $7\frac{1}{4}$

47. 40%

Convert to percent notation.

48. $\frac{17}{20}$

49. 1.5

Solve.

50. $234 + y = 789$

51. $3.9 \times y = 249.6$

52. $\frac{2}{3} \cdot t = \frac{5}{6}$

53. $\frac{8}{17} = \frac{36}{x}$

Solve.

54. A person receives gifts of $627 and $48. What was the total gift?

55. A machine wraps 134 candy bars per minute. How long does it take this machine to wrap 8710 bars?

56. A share of stock bought for $\$29\frac{5}{8}$ dropped $\$3\frac{7}{8}$ before it was resold. What was the price when it was resold?

57. At the start of a trip, a car's odometer read 27,428.6 miles and at the end of the trip the reading was 27,914.5 miles. How long was the trip?

58. From an income of $12,000, amounts of $2300 and $1600 are paid for federal and state taxes. How much remains after these taxes are paid?

59. A worker gets $47 a day for 9 days. How much was received?

60. A person walks $\frac{3}{5}$ km per hour. At this rate, how far would the person walk in $\frac{1}{2}$ hour?

61. Eight identical dresses cost a total of $679.68. What is the cost of each dress?

62. Eight gallons of paint covers 200 square feet. How much paint is needed to cover 325 square feet?

63. Eighteen ounces of a fruit drink cost $3.06. Find the unit price in cents per ounce.

64. What is the simple interest on $4000 principal at 16% for $\frac{3}{4}$ year?

65. A real estate agent received $5880 commission on the sale of an $84,000 home. What was the rate of commission?

66. The population of a city is 29,000 this year and is increasing at 4% per year. What will the population be next year?

67. Find the average, the median, and the mode of this set of numbers:

18, 21, 26, 31, 32, 18, 50.

Evaluate.

68. 18^2

69. 20^2

Simplify.

70. $\sqrt{9}$

71. $\sqrt{121}$

72. Approximate to three decimal places: $\sqrt{20}$.

73. Find the length of the third side of this right triangle. Give an exact answer and an approximation to three decimal places.

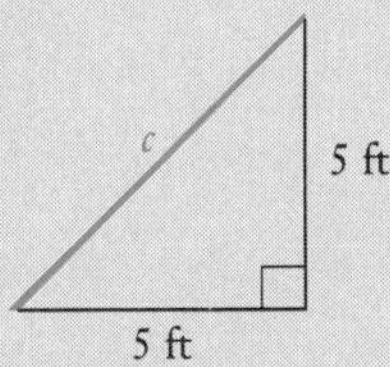

Complete.

74. $\frac{1}{3}$ yd = ______ in.

75. 4280 mm = ______ cm

76. 3 days = ______ hr

77. 20,000 g = ______ kg

78. 5 lb = ______ oz

79. 0.008 cg = ______ mg

80. 8190 mL = ______ L

81. 20 qt = ______ gal

82. Find the perimeter and the area.

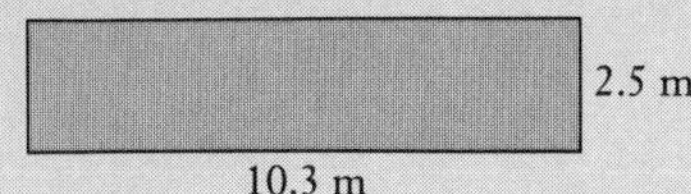

Find the area.

83.

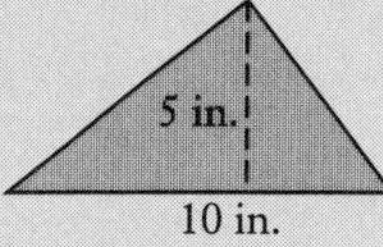

84.

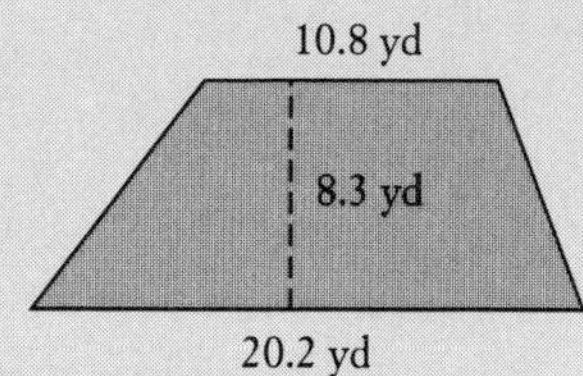

85.

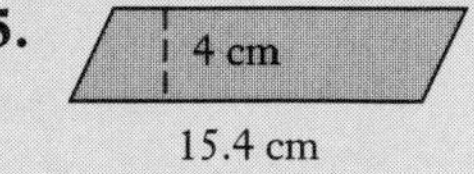

86. Find the diameter, the circumference, and the area of this circle. Use 3.14 for π.

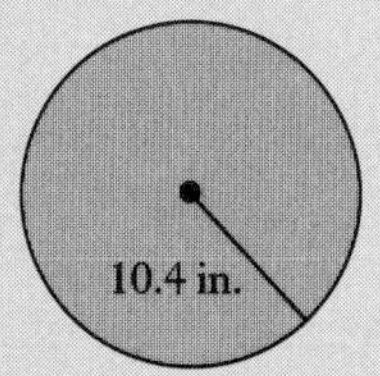

87. Find the volume.

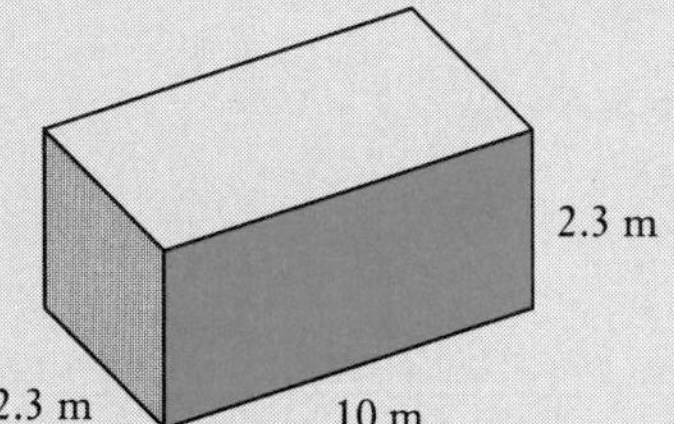

Find the volume. Use 3.14 for π.

88.

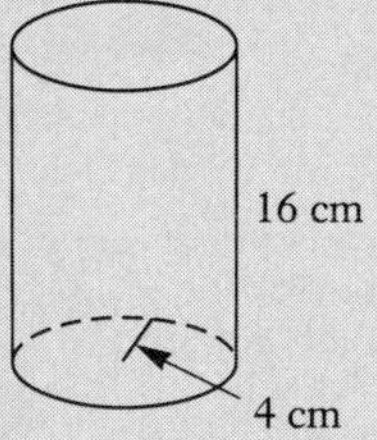

89.

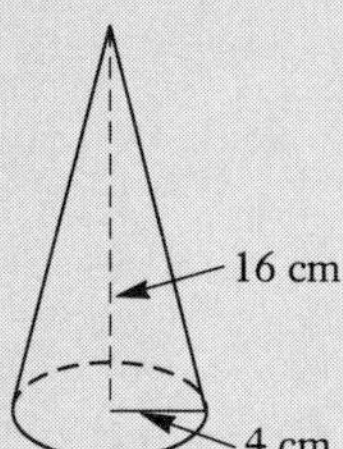

90.

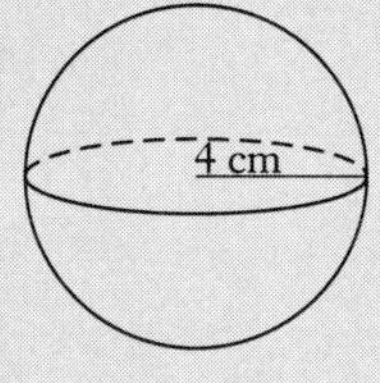

Simplify.

91. $12 \times 20 - 10 \div 5$

92. $4^3 - 5^2 + (16 \cdot 4 + 23 \cdot 3)$

93. $|(-1) \cdot 3|$

94. Add: $17 + (-3)$.

95. Subtract: $\left(-\frac{1}{3}\right) - \left(-\frac{2}{3}\right)$.

96. Multiply: $(-6) \cdot (-5)$.

97. Multiply: $-\frac{5}{7} \cdot \frac{14}{35}$.

98. Divide: $\frac{48}{-6}$.

Solve.

99. $7 - x = 12$

100. $-4.3x = -17.2$

101. $5x + 7 = 3x - 9$

Translate to an algebraic expression.

102. 17 more than some number

103. 38 percent of some number

104. A game board has 64 squares. If you win 25 squares and your opponent wins the rest, how many does your opponent get?

105. If you add one third of a number to the number itself, you get 48. What is the number?

NAME SECTION DATE

FINAL EXAMINATION

1. Write expanded notation for 8345.
[1.1a] 8 thousands + 3 hundreds + 4 tens + 5 ones

2. In 3784, what digit tells the number of hundreds?

Add and simplify, if possible.

3. $41.38 + 2.013 + 172.2247$

4. $3\frac{1}{4} + 5\frac{1}{2}$

5. $\frac{7}{5} + \frac{4}{15}$

6. $432 + 327$

7. $6209 + 2134 + 9187 + 4032$

8. $0.456 + 34.5 + 0.94 + 122.9877$

Subtract and simplify, if possible.

9. $8987 - 3426$

10. $9006 - 3069$

11. $31.2 - 0.808$

12. $3\frac{1}{2} - 2\frac{7}{8}$

13. $\frac{3}{4} - \frac{2}{3}$

14. $123.04 - 23.88$

Multiply and simplify, if possible.

15. $3\frac{1}{4} \cdot 7\frac{1}{2}$

16. $\frac{8}{9} \cdot \frac{3}{4}$

17. $\frac{2}{5} \cdot 15$

18. 342×17

19. 987×238

20. 25.43×8.9

Divide and simplify, if possible.

21. $8\overline{)4137}$
State the answer using a whole-number quotient and remainder.

22. Give a mixed numeral for the quotient in Exercise 21.

23. $21\overline{)4137}$

24. $\frac{3}{5} \div \frac{9}{10}$

25. $5\frac{2}{3} \div 4\frac{2}{5}$

26. $1.6\overline{)76.8}$

27. Round 42,574 to the nearest thousand.

28. Round 6.7892 to the nearest hundredth.

29. Round $7.\overline{38}$ to the nearest thousandth.

ANSWERS

1.
2. [1.1e] 7
3. [4.3a] 215.6177
4. [3.5a] $8\frac{3}{4}$
5. [3.2b] $\frac{5}{3}$
6. [1.2b] 759
7. [1.2b] 21,562
8. [4.3a] 158.8837
9. [1.3d] 5561
10. [1.3d] 5937
11. [4.3b] 30.392
12. [3.5b] $\frac{5}{8}$
13. [3.3a] $\frac{1}{12}$
14. [4.3b] 99.16
15. [3.6a] $24\frac{3}{8}$
16. [2.6a] $\frac{2}{3}$
17. [2.6a] 6
18. [1.5b] 5814
19. [1.5b] 234,906
20. [5.1a] 226.327
21. [1.6c] 517 R 1
22. [3.4b] $517\frac{1}{8}$
23. [1.6c] 197
24. [2.7b] $\frac{2}{3}$
25. [3.6b] $1\frac{19}{66}$
26. [5.2a] 48
27. [1.4a] 43,000
28. [4.2b] 6.79
29. [5.3b] 7.384

ANSWERS

30. [2.2a] Yes

31. [2.1a] 1, 2, 4, 8

32. [3.1a] 230

33. [2.5b] $\frac{3}{2}$

34. [2.5b] 10

35. [3.4b] $7\frac{2}{3}$

36. [2.5c] =

37. [3.3b] <

38. [4.2a] 0.9

39. [1.4c] <

40. [2.3b] $\frac{1}{3}$

41. [7.1b] 0.499

42. [5.3a] 0.24

43. [5.3a] $0.\overline{27}$

44. [4.1c] 7.86

45. [3.4a] $\frac{23}{4}$

46. [7.2b] $\frac{37}{100}$

47. [4.1b] $\frac{897}{1000}$

48. [7.1c] 77%

49. [7.2a] 96%

50. [6.1c] 3.84, or $3\frac{21}{25}$

51. [3.3c] $\frac{1}{25}$

52. [1.7b] 25

53. [4.3c] 245.7

54. [7.5b] 5%

55. [4.4a] $242.60

56. [2.7d] 80

30. Determine whether 3312 is divisible by 9.

31. Find all the factors of 8.

32. Find the LCM of 23, 46, and 10.

Simplify.

33. $\frac{63}{42}$

34. $\frac{100}{10}$

35. Convert to a mixed numeral: $\frac{23}{3}$.

36. Use = or ≠ for ■ to write a true sentence:

$$\frac{3}{5} \;■\; \frac{6}{10}.$$

37. Use < or > for ■ to write a true sentence:

$$\frac{6}{11} \;■\; \frac{5}{9}.$$

38. Which is greater, 0.089 or 0.9?

39. Use < or > for ■ to write a true sentence:

456 ■ 546.

40. What part is shaded?

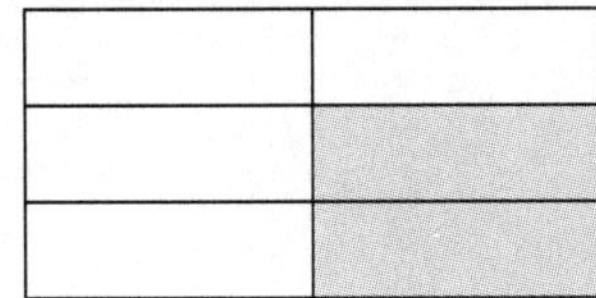

Convert to decimal notation.

41. 49.9%

42. $\frac{6}{25}$

43. $\frac{3}{11}$

44. $\frac{786}{100}$

Convert to fractional notation.

45. $5\frac{3}{4}$

46. 37%

47. 0.897

Convert to percent notation.

48. 0.77

49. $\frac{24}{25}$

Solve.

50. $\frac{25}{12} = \frac{8}{x}$

51. $y + \frac{2}{5} = \frac{11}{25}$

52. $78 \cdot t = 1950$

53. $3.9 + y = 249.6$

Solve.

54. The enrollment in a college increased from 3000 to 3150. Find the percent of increase.

55. A consumer spent $23 for groceries, $204.89 for clothes, and $14.71 for gasoline. How much was spent in all?

56. How many $\frac{1}{4}$-lb boxes of chocolate can be filled with 20 lb of chocolates?

57. A worker gets \$58 a day for 6 days. How much was received?

58. A $5\frac{1}{2}$-m pole was set $1\frac{3}{4}$ m into the ground. How much was above the ground?

59. A person got checks of \$324 and \$987. What was the total?

60. A student has \$75 in a checking account. Checks of \$17 and \$19 are written. How much is left in the account?

61. A driver travels 325 mi on 25 gal of gasoline. How many miles per gallon did the driver obtain?

62. A consumer paid \$101.94 for 6 identical blouses. How much did each blouse cost?

63. A driver traveled 216 km in 6 hr. At this rate, how far would the driver travel in 15 hr?

64. A 3-lb package of meat costs \$11.95. Find the unit cost in dollars per pound.

65. A student got 78% of the questions correct on a test. There were 50 questions. How many of the questions were correct?

66. What is the simple interest on \$2000 principal at 6% for $\frac{1}{2}$ year?

67. Find the average, the median, and the mode of this set of numbers:

\$11, \$12, \$12, \$12, \$19, \$25.

68. The circle graph shows color preference for a new car.

a) Which is the favorite color?
b) The survey considered 5000 people. How many preferred red?

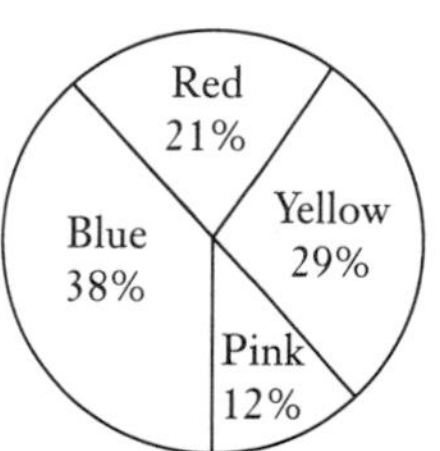

Evaluate.

69. 25^2

70. 16^2

Simplify.

71. $\sqrt{49}$

72. $\sqrt{625}$

73. Approximate to three decimal places: $\sqrt{24}$.

74. Find the length of the third side of this right triangle. Give an exact answer and an approximation to three decimal places.

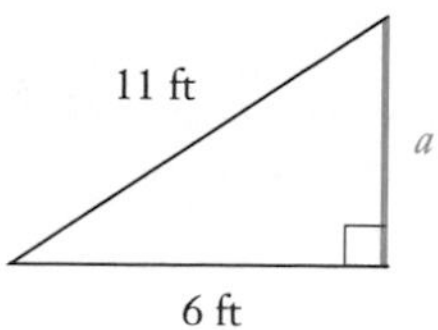

Complete.

75. 15 ft = ______ yd

76. 2371 m = ______ km

77. 5 L = ______ mL

78. 7 T = ______ lb

79. 24 hr = ______ min

80. 5.34 kg = ______ g

81. 75.4 mg = ______ cg

82. 80 oz = ______ pt

ANSWERS

57. [1.8a] \$348

58. [3.5c] $3\frac{3}{4}$ m

59. [1.8a] \$1311

60. [1.8a] \$39

61. [1.8a] 13 mpg

62. [5.5a] \$16.99

63. [6.3a] 540 km

64. [6.2b] \$3.98

65. [7.5a] 39

66. [7.8a] \$60

67. [8.1a, b, c] \$15.16; \$12; \$12

68. a) [8.4a] Blue

b) 1050

69. [1.9b] 625

70. [1.9b] 256

71. [9.7a] 7

72. [9.7a] 25

73. [9.7b] 4.899

74. [9.7c] $a = \sqrt{85}$ ft; $a \approx 9.220$ ft

75. [9.1a] 5

76. [9.2a] 2.371

77. [10.1b] 5000

78. [10.3a] 14,000

79. [10.3c] 1440

80. [10.3b] 5340

81. [10.3b] 7.54

82. [10.1b] 5

ANSWERS

83. [9.3a], [9.4a] 24.8 m; 26.88 m^2

84. [9.5a] 38.775 ft^2

85. [9.5a] 153 m^2

86. [9.5a] 216 dm^2

87. 27.004 yd; 58.0586 yd^2

88. [10.1a] 68.921 ft^3

89. [10.2a] 314,000 m^3

90. [10.2b] 4186.$\overline{6}$ m^3

91. [10.2c] 104,666.$\overline{6}$ m^3

92. [1.9c] 383

93. [1.9c] 99

94. [11.1e] 32

95. [11.3a] −22

96. [11.2a] −22

97. [11.4a] 30

98. [11.4a] $-\frac{5}{9}$

99. [11.5a] −6

100. [12.2a] −76.4

101. [12.3a] 27

102. [12.4b] −15

103. [12.5b] 532.5

104. [12.5b] 40

83. Find the perimeter and the area.

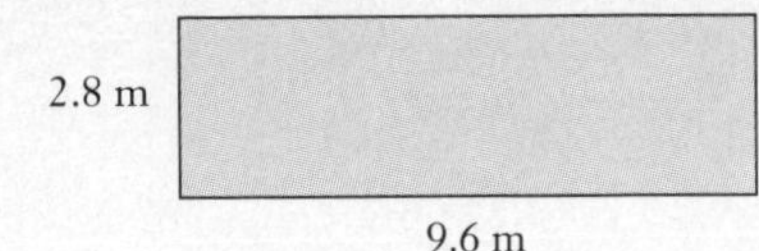

Find the area.

84.

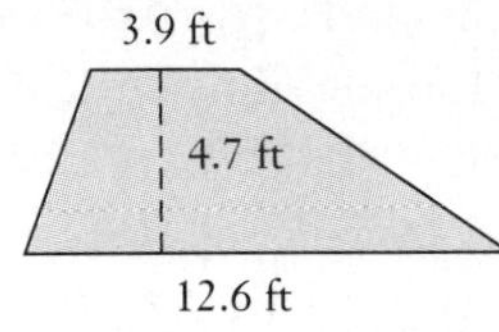

85.

86.

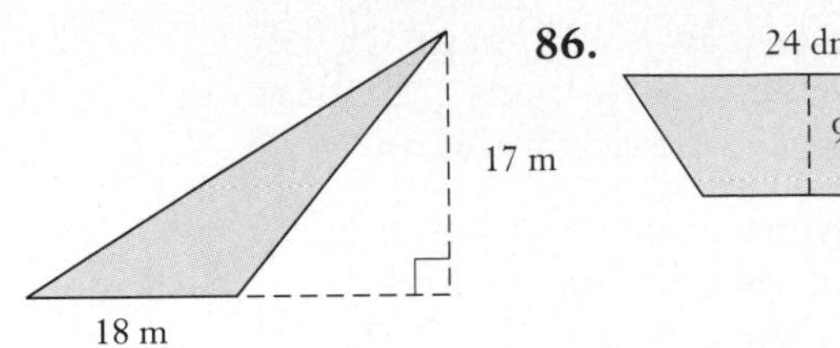

87. Find the radius, the circumference, and the area of this circle. Use 3.14 for π.

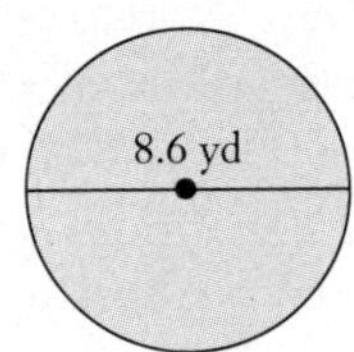

88. Find the volume.

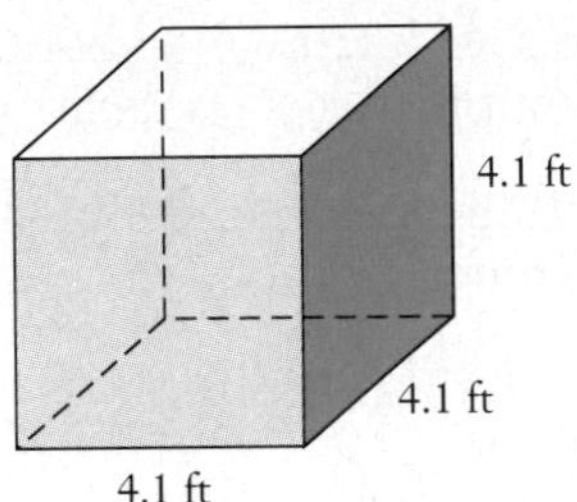

Find the volume. Use 3.14 for π.

89.

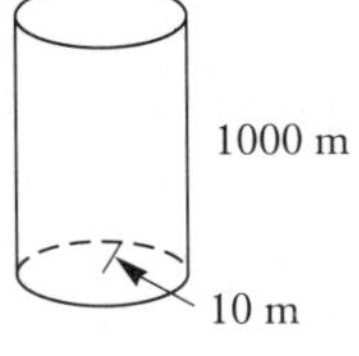

90.

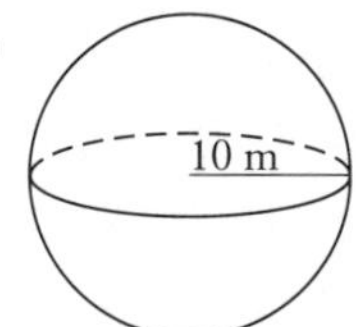

91.

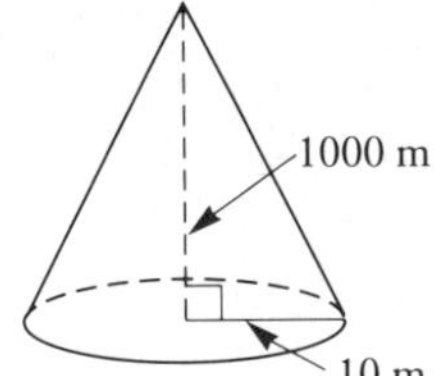

Simplify.

92. $200 \div 25 + 125 \cdot 3$

93. $(2 + 3)^3 - 4^3 + 19 \cdot 2$

94. $|-32|$

95. Subtract: $-7 - 15$.

96. Add: $-7 + (-15)$.

97. Multiply: $-5 \cdot (-6)$.

98. Multiply: $-\frac{2}{3} \cdot \frac{5}{6}$.

99. Divide: $\frac{42}{-7}$.

Solve.

100. $x + 25 = -51.4$

101. $\frac{2}{3}x = 18$

102. $0.5m - 13 = 17 + 2.5m$

103. A consultant charges $80 an hour. How many hours did the consultant work in order to make $42,600?

104. If you add two fifths of a number to the number itself, you get 56. What is the number?

INTRODUCTION These developmental units are meant to provide extra instruction for students who have difficulty with any of Sections 1.2, 1.3, 1.5, or 1.6. ❖

Developmental Units

OBJECTIVES

After finishing Section A.1, you should be able to:

a Add any two of the numbers 0, 1, 2, 3, 4, 5, 6, 7, 8, 9.

b Find certain sums of three numbers such as 1 + 7 + 9.

A.1 Basic Addition

a Basic addition can be explained by counting. The sum

$$3 + 4$$

can be found by counting out a set of 3 objects and a separate set of 4 objects, putting them together, and counting all the objects.

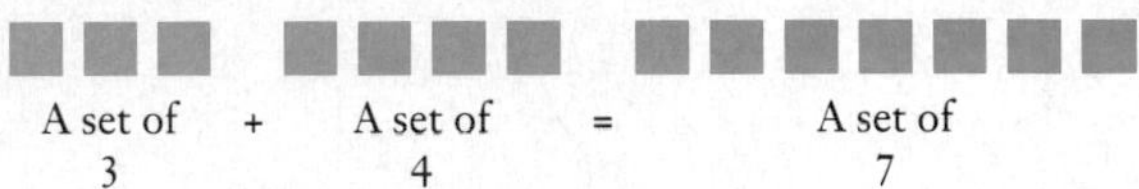

The numbers to be added are called **addends.**

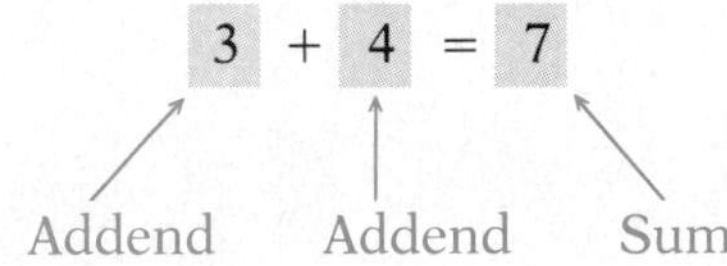

▶ **EXAMPLES** Add. Think of putting sets of objects together. Use fingers or counters if necessary.

1. 5 + 6 = 11

$$\begin{array}{r} 5 \\ +\ 6 \\ \hline 11 \end{array}$$

2. 8 + 5 = 13

$$\begin{array}{r} 8 \\ +\ 5 \\ \hline 13 \end{array}$$ ◀

We can also do these problems by counting up from one of the numbers. For example, in Example 1 start at 5 and count up 6 times: 6, 7, 8, 9, 10, 11.

DO EXERCISES 1–6.

Add; think of joining sets.

1. 4 + 5
9

2. 3 + 4
7

3. $\begin{array}{r} 9 \\ +\ 5 \\ \hline \end{array}$
14

4. $\begin{array}{r} 8 \\ +\ 8 \\ \hline \end{array}$
16

5. $\begin{array}{r} 9 \\ +\ 7 \\ \hline \end{array}$
16

6. $\begin{array}{r} 7 \\ +\ 9 \\ \hline \end{array}$
16

What happens when we add 0? Think of a set of 5 objects. If we add 0 objects to it, we still have 5 objects. Similarly, if we have a set with 0 objects in it and add 5 objects to it, we have a set with 5 objects. Thus

$$5 + 0 = 5 \quad \text{and} \quad 0 + 5 = 5.$$

Adding 0 to a number doesn't change it.

The first printed use of the + symbol was in a book by a German, Johann Widmann, in 1498.

ANSWERS ON PAGE A-7

▶ **EXAMPLES** Add.

3. $0 + 9 = 9$

$$\begin{array}{r} 0 \\ +\,9 \\ \hline 9 \end{array}$$

4. $0 + 0 = 0$

$$\begin{array}{r} 0 \\ +\,0 \\ \hline 0 \end{array}$$

5. $97 + 0 = 97$

$$\begin{array}{r} 97 \\ +\;\;0 \\ \hline 97 \end{array}$$

◀

DO EXERCISES 7–12.

Your objective for this part of the section is to be able to add any of the numbers 0, 1, 2, 3, 4, 5, 6, 7, 8, 9. Adding 0 is easy. The rest of the sums are listed in this table.

+	1	2	3	4	5	6	7	8	9
1	2	3	4	5	6	7	8	9	10
2	3	4	5	6	7	8	9	10	11
3	4	5	6	7	8	9	10	11	12
4	5	6	7	8	9	10	11	12	13
5	6	7	8	9	10	11	12	13	14
6	7	8	9	10	11	12	13	14	15
7	8	9	10	11	12	13	14	15	16
8	9	10	11	12	13	14	15	16	17
9	10	11	12	13	14	15	16	17	18

6 + 7 = 13
Find 6 at the left,
7 at the top.

7 + 6 = 13
Find 7 at the left,
6 at the top.

Memorize the table by saying it to yourself over and over or by using flash cards.

It is very important that you *memorize* the basic addition facts! If you do not, you will always have trouble with addition.

Note the following.

$3 + 4 = 7$ $\quad$ $7 + 6 = 13$ $\quad$ $7 + 2 = 9$
$4 + 3 = 7$ $\quad$ $6 + 7 = 13$ $\quad$ $2 + 7 = 9$

We can add whole numbers in any order. This is the *commutative law of addition.* You need to learn only about half the table above, that is, the shaded part.

DO EXERCISE 13.

Add.

7. $8 + 0$

8

8. $0 + 8$

8

9. $\begin{array}{r} 7 \\ +\,0 \\ \hline \end{array}$

7

10. $\begin{array}{r} 46 \\ +\;\;0 \\ \hline \end{array}$

46

11. $0 + 13$

13

12. $58 + 0$

58

13. Complete this table. Use fingers or counters if necessary.

+	6	5	7	4	9
7	13	12	14	11	16
9	15	14	16	13	18
5	11	10	12	9	14
8	14	13	15	12	17
4	10	9	11	8	13

ANSWERS ON PAGE A-7

b Certain Sums of Three Numbers

To add 3 + 5 + 4, we can add 3 and 5, then 4:

$$3 + 5 + 4 \rightarrow 8 + 4 \rightarrow 12.$$

We can also add 5 and 4, then 3:

$$3 + 5 + 4 \rightarrow 3 + 9 \rightarrow 12.$$

Either way we get 12.

▶ **EXAMPLE 6** Add from the top mentally.

$$\begin{array}{r} 1 \\ 7 \\ +\ 9 \\ \hline \end{array}$$

We first add 1 and 7, getting 8. Then we add 8 and 9, getting 17.

$$\begin{array}{r} 1 \\ 7 \\ +\ 9 \\ \hline \end{array} \quad 1 + 7 \longrightarrow 8, \quad 8 + 9 \longrightarrow 17$$ ◀

▶ **EXAMPLE 7** Add from the top mentally.

$$\begin{array}{r} 2 \\ 4 \\ +\ 8 \\ \hline \end{array} \quad 2 + 4 \longrightarrow 6, \quad 6 + 8 \longrightarrow 14$$ ◀

DO EXERCISES 14–17.

Add from the top mentally.

14. $\begin{array}{r} 1 \\ 6 \\ +\ 9 \\ \hline \end{array}$

16

15. $\begin{array}{r} 2 \\ 3 \\ +\ 4 \\ \hline \end{array}$

9

16. $\begin{array}{r} 6 \\ 1 \\ +\ 4 \\ \hline \end{array}$

11

17. $\begin{array}{r} 5 \\ 2 \\ +\ 8 \\ \hline \end{array}$

15

ANSWERS ON PAGE A-7

NAME SECTION DATE

EXERCISE SET A.1

a Add. Try to do these mentally. If you have trouble, think of putting objects together. Use fingers or counters if necessary.

1. $9 + 2$ **2.** $8 + 9$ **3.** $8 + 7$ **4.** $6 + 7$

5. $7 + 8$ **6.** $9 + 5$ **7.** $5 + 7$ **8.** $7 + 5$

9. $7 + 6$ **10.** $5 + 6$ **11.** $7 + 9$ **12.** $9 + 8$

13. $9 + 7$ **14.** $8 + 4$ **15.** $9 + 1$ **16.** $8 + 2$

17. $3 + 8$ **18.** $0 + 7$ **19.** $4 + 3$ **20.** $2 + 9$

21. $4 + 4$ **22.** $0 + 0$ **23.** $3 + 0$ **24.** $9 + 9$

25. $8 + 6$ **26.** $0 + 9$ **27.** $3 + 7$ **28.** $6 + 8$

29. $2 + 2$ **30.** $7 + 7$ **31.** $6 + 5$ **32.** $2 + 0$

33. $7 + 8$ **34.** $7 + 9$ **35.** $8 + 8$ **36.** $8 + 1$

37. $8 + 3$ **38.** $5 + 8$ **39.** $5 + 9$ **40.** $4 + 1$

41. $4 + 7$ **42.** $6 + 1$

ANSWERS

1. 11
2. 17
3. 15
4. 13
5. 15
6. 14
7. 12
8. 12
9. 13
10. 11
11. 16
12. 17
13. 16
14. 12
15. 10
16. 10
17. 11
18. 7
19. 7
20. 11
21. 8
22. 0
23. 3
24. 18
25. 14
26. 9
27. 10
28. 14
29. 4
30. 14
31. 11
32. 2
33. 15
34. 16
35. 16
36. 9
37. 11
38. 13
39. 14
40. 5
41. 11
42. 7

ANSWERS

43. 13
44. 14
45. 12
46. 6
47. 10
48. 12
49. 10
50. 8
51. 2
52. 9
53. 13
54. 8
55. 10
56. 9
57. 10
58. 6
59. 13
60. 9
61. 8
62. 11
63. 16
64. 7
65. 12
66. 12
67. 8
68. 4
69. 7
70. 10
71. 15
72. 13
73. 4
74. 6
75. 8
76. 8
77. 15
78. 12
79. 13
80. 10
81. 12
82. 17
83. 13
84. 10
85. 16
86. 16
87. 17

43. 6 + 7 **44.** 7 + 7 **45.** 3 + 9 **46.** 6 + 0 **47.** 6 + 4

48. 9 + 3 **49.** 5 + 5 **50.** 5 + 3 **51.** 1 + 1 **52.** 4 + 5

53. 9 + 4 **54.** 0 + 8 **55.** 4 + 6 **56.** 2 + 7 **57.** 3 + 7

58. 3 + 3 **59.** 5 + 8 **60.** 3 + 6 **61.** 4 + 4 **62.** 4 + 7

63. 8 + 8 **64.** 5 + 2 **65.** 4 + 8 **66.** 6 + 6 **67.** 3 + 5

68. 0 + 4 **69.** 3 + 4 **70.** 2 + 8 **71.** 6 + 9 **72.** 4 + 9

b Add from the top mentally.

73. 1 + 1 + 2 **74.** 1 + 2 + 3 **75.** 1 + 4 + 3 **76.** 1 + 3 + 4 **77.** 1 + 6 + 8

78. 1 + 8 + 3 **79.** 1 + 7 + 5 **80.** 3 + 2 + 5 **81.** 4 + 3 + 5 **82.** 1 + 7 + 9

83. 5 + 2 + 6 **84.** 4 + 5 + 1 **85.** 1 + 9 + 6 **86.** 1 + 8 + 7 **87.** 1 + 8 + 8

A.2 More on Addition

a Addition (No Carrying)

To add larger numbers, we can add the ones first, then the tens, then the hundreds, and so on.

▶ **EXAMPLE 1** Add: 5722 + 3234.

```
  5 7 2 2    Add ones.
+ 3 2 3 4
---------
        6

  5 7 2 2    Add tens.
+ 3 2 3 4
---------
      5 6

  5 7 2 2    Add hundreds.
+ 3 2 3 4
---------
    9 5 6

  5 7 2 2    Add thousands.
+ 3 2 3 4
---------
  8 9 5 6
```

This is for explanation.

```
  5 7 2 2
+ 3 2 3 4
---------
  8 9 5 6
```

You should write only this.

◀

DO EXERCISES 1–4.

b Addition (With Carrying)

Carrying Tens

▶ **EXAMPLE 2** Add: 18 + 27.

```
  1 8    Add ones.        Think:    8
+ 2 7                             + 7
-----                             ---
    ?                             1 5   1 ten and 5 ones

  1
  1 8    Write 5 in the ones column.
+ 2 7    Write 1 for a reminder above the tens.
-----    This is called carrying.
    5

  1
  1 8    Add tens.
+ 2 7
-----
  4 5
```

OBJECTIVES

After finishing Section A.2, you should be able to:

a Add two whole numbers when carrying is not necessary.

b Add two whole numbers when carrying is necessary.

Add.

1.
```
  2 4
+ 3 5
-----
   59
```

2.
```
  3 4 6
+ 2 0 3
-------
    549
```

3.
```
  8 3 2 7
+ 1 6 5 2
---------
     9979
```

4.
```
  3 4 6 1
+ 2 0 3 5
---------
     5496
```

ANSWERS ON PAGE A-7

Add.

5.
```
  1 9
+ 3 7
-----
   56
```

6.
```
  4 6
+ 3 9
-----
   85
```

Add.

7.
```
  3 4 1
+ 4 8 8
-------
    829
```

8.
```
  7 3 0
+ 2 9 6
-------
   1026
```

ANSWERS ON PAGE A-7

We can use money to help explain Example 2.

```
   1  8¢ ——→ 1 dime and 8 pennies
 + 2  7¢ ——→ 2 dimes and 7 pennies
     15¢     We first add the pennies.
```

```
   1 dime
   1 8
 + 2 7            We exchange ten pennies for a dime.
 -----
     5 pennies
```

```
   1
   1 8     We now add the dimes. The result is
 + 2 7     4 dimes and 5 pennies.
 -----
   4 5
```

DO EXERCISES 5 AND 6.

Carrying Hundreds

▶ **EXAMPLE 3** Add: 256 + 391.

```
    2  5  6     Add ones.
 +  3  9  1
 ----------
          7
```

```
    1
    2  5  6     Add tens. We get 14 tens. Write 4 in the tens
 +  3  9  1     column and a 1 above the hundreds.
 ----------
       4  7
```

> The carrying here is like exchanging 14 dimes for a 1 dollar bill and 4 dimes.

```
    1
    2  5  6     Add hundreds.
 +  3  9  1
 ----------
    6  4  7
```

DO EXERCISES 7 AND 8.

Carrying Thousands

▶ **EXAMPLE 4** Add: 4803 + 3792.

$$\begin{array}{r} 4\ 8\ 0\ 3 \\ +\ 3\ 7\ 9\ 2 \\ \hline 5 \end{array}$$ Add ones.

$$\begin{array}{r} 4\ 8\ 0\ 3 \\ +\ 3\ 7\ 9\ 2 \\ \hline 9\ 5 \end{array}$$ Add tens.

$$\begin{array}{r} ^{1} \\ 4\ 8\ 0\ 3 \\ +\ 3\ 7\ 9\ 2 \\ \hline 5\ 9\ 5 \end{array}$$ Add hundreds. Write 5 in the hundreds column and 1 above the thousands.

$$\begin{array}{r} ^{1} \\ 4\ 8\ 0\ 3 \\ +\ 3\ 7\ 9\ 2 \\ \hline 8\ 5\ 9\ 5 \end{array}$$ Add thousands. ◀

DO EXERCISE 9.

Combined Carrying

▶ **EXAMPLE 5** Add: 5767 + 4993.

$$\begin{array}{r} ^{1} \\ 5\ 7\ 6\ 7 \\ +\ 4\ 9\ 9\ 3 \\ \hline 0 \end{array}$$ Add ones. Write 0 in the ones column and 1 above the tens.

$$\begin{array}{r} ^{1}\ ^{1} \\ 5\ 7\ 6\ 7 \\ +\ 4\ 9\ 9\ 3 \\ \hline 6\ 0 \end{array}$$ Add tens. Write 6 in the tens column and 1 above the hundreds.

$$\begin{array}{r} ^{1}\ ^{1}\ ^{1} \\ 5\ 7\ 6\ 7 \\ +\ 4\ 9\ 9\ 3 \\ \hline 7\ 6\ 0 \end{array}$$ Add hundreds. Write 7 in the hundreds column and 1 above the thousands.

$$\begin{array}{r} ^{1}\ ^{1}\ ^{1} \\ 5\ 7\ 6\ 7 \\ +\ 4\ 9\ 9\ 3 \\ \hline 1\ 0\ 7\ 6\ 0 \end{array}$$ Add thousands. ◀

DO EXERCISES 10 AND 11.

9. $$\begin{array}{r} 7\ 8\ 5\ 0 \\ +\ 4\ 8\ 4\ 8 \\ \hline \end{array}$$
12,698

Add.

10. $$\begin{array}{r} 7\ 9\ 8\ 9 \\ +\ 5\ 6\ 7\ 2 \\ \hline \end{array}$$
13,661

11. $$\begin{array}{r} 5\ 6{,}7\ 8\ 9 \\ +\ 1\ 4{,}5\ 3\ 9 \\ \hline \end{array}$$
71,328

To the student: If you had trouble with Section 1.2 and have studied developmental units A.1 and A.2 (or parts of them), you should go back and work through Section 1.2.

ANSWERS ON PAGE A-7

❖ SIDELIGHTS

A Number Puzzle

Place the numbers 3, 4, 5, 6, 7, 8, 9 in the boxes so the sum along any line will be 18.

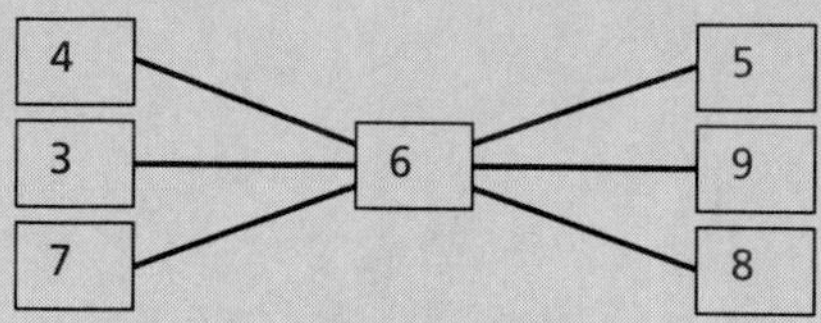

NAME SECTION DATE

EXERCISE SET A.2

a Add.

1. $\begin{array}{r} 14 \\ +\ 5 \\ \hline \end{array}$

2. $\begin{array}{r} 23 \\ +16 \\ \hline \end{array}$

3. $\begin{array}{r} 34 \\ +61 \\ \hline \end{array}$

4. $\begin{array}{r} 54 \\ +35 \\ \hline \end{array}$

5. $\begin{array}{r} 67 \\ +20 \\ \hline \end{array}$

6. $\begin{array}{r} 78 \\ +\ 1 \\ \hline \end{array}$

7. $\begin{array}{r} 308 \\ +541 \\ \hline \end{array}$

8. $\begin{array}{r} 496 \\ +503 \\ \hline \end{array}$

9. $\begin{array}{r} 700 \\ +200 \\ \hline \end{array}$

10. $\begin{array}{r} 801 \\ +\ 67 \\ \hline \end{array}$

11. $\begin{array}{r} 765 \\ +110 \\ \hline \end{array}$

12. $\begin{array}{r} 666 \\ +333 \\ \hline \end{array}$

13. $\begin{array}{r} 523 \\ +325 \\ \hline \end{array}$

14. $\begin{array}{r} 747 \\ +130 \\ \hline \end{array}$

15. $\begin{array}{r} 789 \\ +900 \\ \hline \end{array}$

16. $\begin{array}{r} 8250 \\ +9430 \\ \hline \end{array}$

17. $\begin{array}{r} 6552 \\ +4321 \\ \hline \end{array}$

18. $\begin{array}{r} 3406 \\ +1293 \\ \hline \end{array}$

19. $\begin{array}{r} 7225 \\ +2522 \\ \hline \end{array}$

20. $\begin{array}{r} 7340 \\ +3527 \\ \hline \end{array}$

21. $\begin{array}{r} 4825 \\ +5070 \\ \hline \end{array}$

22. $\begin{array}{r} 2073 \\ +1925 \\ \hline \end{array}$

23. $\begin{array}{r} 7340 \\ +2658 \\ \hline \end{array}$

24. $\begin{array}{r} 9111 \\ +9111 \\ \hline \end{array}$

25. $\begin{array}{r} 2345 \\ +5432 \\ \hline \end{array}$

26. $\begin{array}{r} 7889 \\ +9000 \\ \hline \end{array}$

27. $\begin{array}{r} 9986 \\ +9013 \\ \hline \end{array}$

28. $\begin{array}{r} 52{,}433 \\ +12{,}056 \\ \hline \end{array}$

29. $\begin{array}{r} 43{,}723 \\ +56{,}276 \\ \hline \end{array}$

30. $\begin{array}{r} 51{,}670 \\ +26{,}107 \\ \hline \end{array}$

ANSWERS

1. 19
2. 39
3. 95
4. 89
5. 87
6. 79
7. 849
8. 999
9. 900
10. 868
11. 875
12. 999
13. 848
14. 877
15. 1689
16. 17,680
17. 10,873
18. 4699
19. 9747
20. 10,867
21. 9895
22. 3998
23. 9998
24. 18,222
25. 7777
26. 16,889
27. 18,999
28. 64,489
29. 99,999
30. 77,777

ANSWERS

31. 46

32. 26

33. 55

34. 101

35. 1643

36. 1010

37. 846

38. 628

39. 1204

40. 607

41. 10,000

42. 1010

43. 1110

44. 1227

45. 1111

46. 1717

47. 10,138

48. 6554

49. 6111

50. 8427

51. 9890

52. 11,612

53. 11,125

54. 15,543

55. 16,774

56. 68,675

57. 34,437

58. 166,444

59. 101,315

60. 49,449

61. 45, 56, 67, 78, 89

62. 1 6 15 20 15 6 1; 1 7 21 35 35 21 7 1

b Add.

31. 38 + 8

32. 17 + 9

33. 17 + 38

34. 95 + 6

35. 862 + 781

36. 999 + 11

37. 355 + 491

38. 280 + 348

39. 814 + 390

40. 274 + 333

41. 9990 + 10

42. 999 + 11

43. 999 + 111

44. 839 + 388

45. 909 + 202

46. 808 + 909

47. 8718 + 1420

48. 3854 + 2700

49. 4828 + 1283

50. 6995 + 1432

51. 9889 + 1

52. 6889 + 4723

53. 9128 + 1997

54. 8898 + 6645

55. 9989 + 6785

56. 46,889 + 21,786

57. 23,448 + 10,989

58. 67,658 + 98,786

59. 77,548 + 23,767

60. 44,684 + 4,765

SYNTHESIS

61. Find the next five terms of this sequence.

1, 12, 23, 34, ___, ___, ___, ___, ___

62. Look for a pattern and complete two more rows of this number "triangle."

```
       1
      1 1
     1 2 1
    1 3 3 1
   1 4 6 4 1
 1 5 10 10 5 1
```

A.3 Low-Stress Addition

a If you are having trouble with the addition method studied in Chapter 1, you might try the following. We first need some new notation.

Former notation		*New notation*	
7 +8 15	→	7 8 1 5	Write the answer in small numerals.
6 +3 9	→	6 3 9	

DO EXERCISES 1–4.

To add in columns, we add only the ones digits. Let's do the addition 8 + 7 + 9 + 3.

Step 1.

8
7
1 5
9
+3

8 + 7 = 15. Write 15 using the new notation.

Step 2.

8
7
1 5
9
1 4
+3

5 + 9 = 14. Write 14 below the 9 using the new notation.

Step 3.

8
7
1 5
9
1 4
+3
7

4 + 3 = 7. Write 7 below the 3 using the new notation.

Step 4.

8
7
1 5
9
1 4
+3
7
7

Write the final 7.

Step 5.

8
7
1 5
9
1 4
+ 3
7
2 7

Add the tens.

DO EXERCISES 5 AND 6.

For larger numbers, we add the columns using the new method, and then carry the way we did before.

OBJECTIVE

After finishing Section A.3, you should be able to:

a Add using the low-stress method.

Add using the new notation.

1. 8
+ 6
8
$_1 6_4$

2. 8
+ 3
8
$_1 3_1$

3. 4
+ 3
4
3_7

4. 5
+ 0
5
0_5

Add using the new method.

5. 9
7
5
6
+ 4
31

6. 7
7
7
6
+ 5
32

ANSWERS ON PAGE A-7

Add using the new method.

7.
```
   2  2
   3  1
   6  9
   8  5
 + 4  2
 ------
    249
```

8.
```
   6  5
   4  3
   9  9
 + 1  2
 ------
    219
```

Add using the new method.

9.
```
   2  9  8
   1  4  7
   2  8  3
   9  7  1
 + 4  0  5
 ---------
      2104
```

10.
```
   1  0  2  7
   6  0  3  1
   4  5  6  7
 + 3  2  1  9
 ------------
       14,844
```

ANSWERS ON PAGE A-7

▶ **EXAMPLE 1** Add: 66 + 34 + 28 + 90 + 12.

Step 1.
```
   2
   6   6
   3   4
      10
   2   8
       8
   9   0
       8
 + 1   2
      10
 -------
       0
```
Add the ones using the new method: 20. Carry the 2 above the tens.

Step 2.
```
    2
    6    6
    8
    3    4
   11   10
    2    8
    3    8
    9    0
   12    8
 +  1    2
    3   10
 ---------
   23    0
```
Add the tens using the new new method.

◀

DO EXERCISES 7 AND 8.

▶ **EXAMPLE 2** Add: 886 + 972 + 912 + 363 + 428.

Step 1. Add ones.
```
      2
   8  8  6
   9  7  2
         8
   9  1  2
        10
   3  6  3
         3
 + 4  2  8
        11
 ---------
         1
```

Step 2. Add tens.
```
   2  2
   8  8  6
     10
   9  7  2
      7  8
   9  1  2
      8 10
   3  6  3
     14  3
 + 4  2  8
      6 11
 ---------
      6  1
```

Step 3. Add hundreds.
```
    2    2
    8    8  6
   10   10
    9    7  2
    9    7  8
    9    1  2
   18    8 10
    3    6  3
   11   14  3
 +  4    2  8
    5    6 11
 ------------
   35    6  1
```

◀

DO EXERCISES 9 AND 10.

NAME SECTION DATE

EXERCISE SET A.3

a Add using the new method.

1.
$$\begin{array}{r} 5 \\ 3 \\ 8 \\ 0 \\ +1 \\ \hline \end{array}$$

2.
$$\begin{array}{r} 1 \\ 3 \\ 9 \\ 6 \\ +0 \\ \hline \end{array}$$

3.
$$\begin{array}{rr} 5 & 6 \\ 2 & 8 \\ 4 & 9 \\ 8 & 5 \\ +9 & 0 \\ \hline \end{array}$$

4.
$$\begin{array}{rr} 9 & 0 \\ 1 & 3 \\ 2 & 6 \\ 1 & 2 \\ +3 & 0 \\ \hline \end{array}$$

5.
$$\begin{array}{rrr} 3 & 8 & 8 \\ 4 & 4 & 1 \\ 9 & 9 & 9 \\ 9 & 9 & 0 \\ +9 & 0 & 0 \\ \hline \end{array}$$

6.
$$\begin{array}{rrr} 5 & 5 & 5 \\ 6 & 6 & 6 \\ 6 & 6 & 0 \\ 6 & 0 & 0 \\ +1 & 1 & 1 \\ \hline \end{array}$$

7.
$$\begin{array}{rrr} 1 & 6 & 5 \\ 2 & 0 & 3 \\ 4 & 4 & 2 \\ +3 & 5 & 7 \\ \hline \end{array}$$

8.
$$\begin{array}{rrr} 2 & 7 & 6 \\ 1 & 1 & 1 \\ 4 & 6 & 8 \\ +2 & 3 & 7 \\ \hline \end{array}$$

9.
$$\begin{array}{rrrr} 3 & 7 & 8 & 8 \\ 1 & 2 & 7 & 7 \\ 9 & 1 & 3 & 2 \\ +4 & 5 & 0 & 2 \\ \hline \end{array}$$

10.
$$\begin{array}{rrrr} 6 & 6 & 1 & 0 \\ 5 & 8 & 9 & 3 \\ 2 & 0 & 0 & 4 \\ +8 & 5 & 6 & 1 \\ \hline \end{array}$$

ANSWERS

1. 17

2. 19

3. 308

4. 171

5. 3718

6. 2592

7. 1167

8. 1092

9. 18,699

10. 23,068

ANSWERS

11. 15,085

12. 86,425

13. 2740

14. 10,983

15. 1243

16. 8497

17. 2798

18. 1681

19. 169,351

20. $40,000

11.

	1	2	3	4	5
		2	3	4	5
			3	4	5
				4	5
+					5

12.

	7	7	7	7	9
		7	7	7	9
			7	7	9
				7	9
+					9

Add using the new method.

13. 2345 + 345 + 45 + 5

14. 9888 + 988 + 98 + 9

15. 1110 + 110 + 11 + 12

16. 7677 + 677 + 67 + 76

17. 678 + 567 + 899 + 654

18. 238 + 832 + 328 + 283

SYNTHESIS

19. Attendance for the first three games of a World Series between the New York Yankees and the Los Angeles Dodgers was 56,691, 55,992, and 56,668. Find the total attendance for the first three games.

20. A person received yearly salaries of $12,500, $13,200, and $14,300. What was the total income for the three years?

S.1 Basic Subtraction

a Subtraction can be explained by taking away part of a set.

▶ **EXAMPLE 1** Subtract: 7 − 3.

We can do this by counting out 7 objects and then taking away 3 of them. Then we count the number that remain.

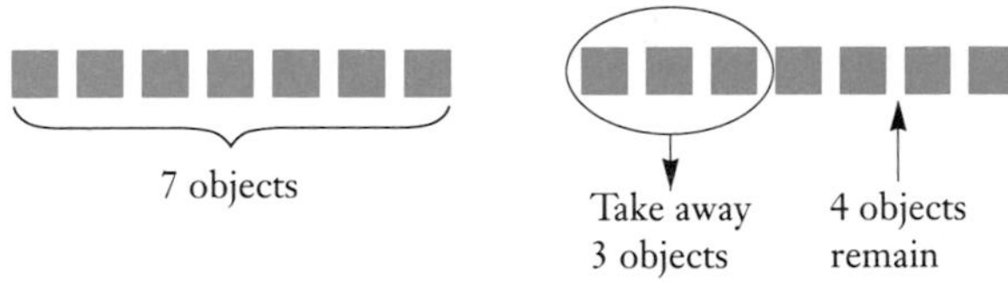

$7 - 3 = 4$ This number is called the **difference.**

We could also do this mentally by counting down: 6, 5, 4. ◀

▶ **EXAMPLES** Subtract. Think of "take away" or "how much more."

2. $11 - 6 = 5$

$$\begin{array}{r} 11 \\ -\ 6 \\ \hline 5 \end{array}$$

3. $17 - 9 = 8$

$$\begin{array}{r} 17 \\ -\ 9 \\ \hline 8 \end{array}$$

◀

DO EXERCISES 1–4.

Your objective for this part of the section is to be able to subtract numbers that are no larger than 18. In Section A.1 you memorized an addition table. That table will enable you to subtract also. First, let's recall how addition and subtraction are related.

An addition:

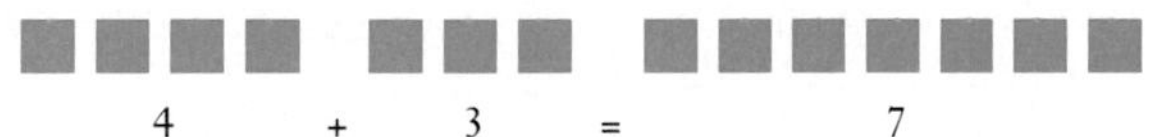

Two related subtractions:

A.

B.

7 – 4 = 3

OBJECTIVE

After finishing Section S.1, you should be able to:

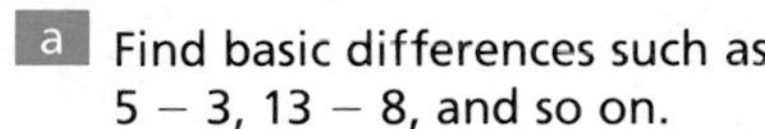

a Find basic differences such as 5 − 3, 13 − 8, and so on.

Subtract.

1. $10 - 6$ 4

2. $11 - 4$ 7

3. $\begin{array}{r} 16 \\ -\ 8 \\ \hline 8 \end{array}$

4. $\begin{array}{r} 10 \\ -\ 7 \\ \hline 3 \end{array}$

ANSWERS ON PAGE A-7

For each addition fact, write two subtraction facts.

5. $8 + 4 = 12$ $12 - 8 = 4$; $12 - 4 = 8$

6. $6 + 7 = 13$ $13 - 6 = 7$; $13 - 7 = 6$

Subtract. Try to do these mentally.

7. $14 - 6$ 8

8. $12 - 5$ 7

9. $13 - 4$ 9

10. $11 - 7$ 4

ANSWERS ON PAGE A-7

Since we know that

$$4 + 3 = 7, \quad \text{A basic addition fact}$$

we also know the two subtraction facts

$$7 - 3 = 4 \quad \text{and} \quad 7 - 4 = 3.$$

▶ **EXAMPLE 4** For the following addition fact, write two subtraction facts: $8 + 9 = 17$.

a) $17 - 9 = 8$ We can get this by moving the 9: $8 + 9 = 17$

b) $17 - 8 = 9$ We can get this by moving the 8: $8 + 9 = 17$ ◀

DO EXERCISES 5 AND 6.

Now let's look at the addition table you memorized. (Remember, you memorized only half of it.)

+	1	2	3	4	5	6	7	8	9
1	2	3	4	5	6	7	8	9	10
2	3	4	5	6	7	8	9	10	11
3	4	5	6	7	8	9	10	11	12
4	5	6	7	8	9	10	11	12	13
5	6	7	8	9	10	11	12	13	14
6	7	8	9	10	11	12	13	14	15
7	8	9	10	11	12	13	14	15	16
8	9	10	11	12	13	14	15	16	17
9	10	11	12	13	14	15	16	17	18

It is *very* important that you have the basic subtraction facts *memorized*. If you do not, you will always have trouble with subtraction.

▶ **EXAMPLE 5** Find $15 - 7$.

We memorized that $7 + 8 = 15$. Right away we know the related subtraction, $15 - 7 = 8$. In the table we find 7 on the left. We go across to 15. Then we go up. There we find the difference of $15 - 7$, which is 8.

To find $15 - 7$, we ask ourselves "7 plus what number is 15?"

$$7 + \square = 15$$

◀

DO EXERCISES 7–10.

NAME SECTION DATE

EXERCISE SET S.1

a Subtract. Try to do these mentally.

1. $7 - 3$
2. $3 - 2$
3. $4 - 1$
4. $2 - 0$
5. $3 - 3$
6. $8 - 8$
7. $5 - 0$
8. $6 - 3$
9. $7 - 6$
10. $9 - 8$
11. $10 - 3$
12. $10 - 5$
13. $\begin{array}{r} 7 \\ -\ 0 \\ \hline \end{array}$
14. $\begin{array}{r} 9 \\ -\ 0 \\ \hline \end{array}$
15. $\begin{array}{r} 8 \\ -\ 8 \\ \hline \end{array}$
16. $\begin{array}{r} 7 \\ -\ 7 \\ \hline \end{array}$
17. $\begin{array}{r} 8 \\ -\ 3 \\ \hline \end{array}$
18. $\begin{array}{r} 5 \\ -\ 2 \\ \hline \end{array}$
19. $\begin{array}{r} 16 \\ -\ 8 \\ \hline \end{array}$
20. $\begin{array}{r} 17 \\ -\ 9 \\ \hline \end{array}$
21. $\begin{array}{r} 12 \\ -\ 6 \\ \hline \end{array}$
22. $\begin{array}{r} 13 \\ -\ 8 \\ \hline \end{array}$
23. $\begin{array}{r} 11 \\ -\ 4 \\ \hline \end{array}$
24. $\begin{array}{r} 12 \\ -\ 9 \\ \hline \end{array}$

ANSWERS

1. 4
2. 1
3. 3
4. 2
5. 0
6. 0
7. 5
8. 3
9. 1
10. 1
11. 7
12. 5
13. 7
14. 9
15. 0
16. 0
17. 5
18. 3
19. 8
20. 8
21. 6
22. 5
23. 7
24. 3

ANSWERS

25. 7
26. 9
27. 6
28. 6
29. 2
30. 2
31. 0
32. 0
33. 4
34. 4
35. 5
36. 4
37. 4
38. 5
39. 9
40. 9
41. 7
42. 9
43. 5
44. 6
45. 6
46. 2
47. 6
48. 7
49. 1
50. 8
51.

25. $\begin{array}{r} 14 \\ -\ 7 \\ \hline \end{array}$ **26.** $\begin{array}{r} 18 \\ -\ 9 \\ \hline \end{array}$ **27.** $\begin{array}{r} 13 \\ -\ 7 \\ \hline \end{array}$

28. $\begin{array}{r} 15 \\ -\ 9 \\ \hline \end{array}$ **29.** $\begin{array}{r} 8 \\ -\ 6 \\ \hline \end{array}$ **30.** $\begin{array}{r} 9 \\ -\ 7 \\ \hline \end{array}$

Subtract.

31. 6 − 6 **32.** 7 − 7 **33.** 11 − 7 **34.** 12 − 8

35. 5 − 0 **36.** 4 − 0 **37.** 13 − 9 **38.** 14 − 9

39. 11 − 2 **40.** 12 − 3 **41.** 16 − 9 **42.** 18 − 9

43. 11 − 6 **44.** 11 − 5 **45.** 10 − 4 **46.** 10 − 8

47. 14 − 8 **48.** 15 − 8 **49.** 9 − 8 **50.** 10 − 2

Fill the 3-gal bucket and pour it into the 5-gal bucket. Fill the 3-gal bucket again, and pour into the 5-gal until full; one gallon remains.

SYNTHESIS

51. A parent sends a child to a river with two buckets. One bucket measures 3 gal and the other 5 gal. How can the child return with exactly 1 gal?

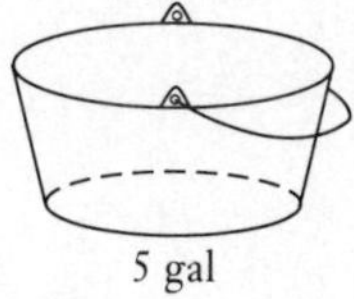
5 gal

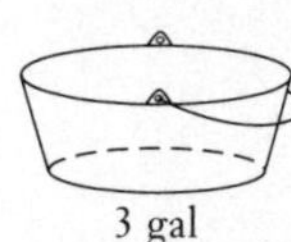
3 gal

S.2 More on Subtraction

a Subtraction (No Borrowing)

To subtract larger numbers, we can subtract the ones first, then the tens, then the hundreds, and so on.

▶ **EXAMPLE 1** Subtract: 5787 − 3214.

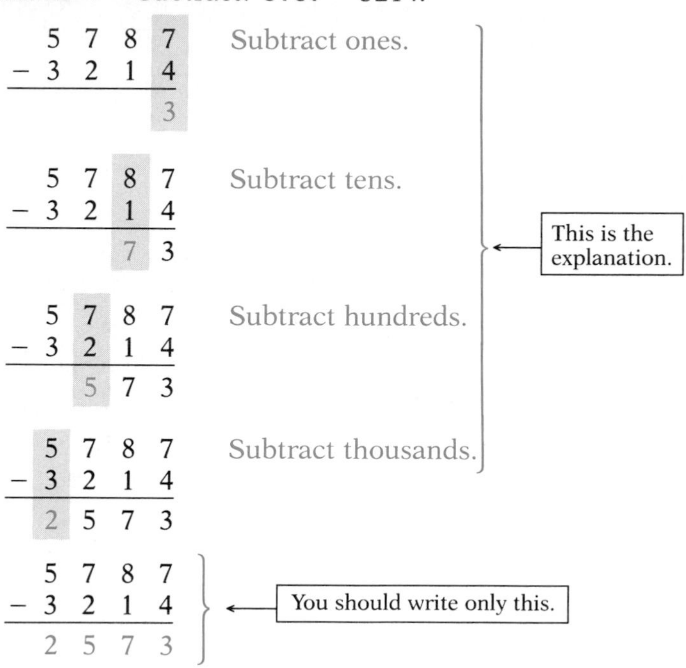

◀

DO EXERCISES 1–4.

b Subtraction (With Borrowing)

We now consider subtraction when borrowing is necessary.

Borrowing from the Tens Place

▶ **EXAMPLE 2** Subtract: 37 − 18.

$$\begin{array}{r} 3\ 7 \\ -1\ 8 \\ \hline ? \end{array}$$

Try to subtract ones.

Think: 7 − 8 is not a whole number!

$$\begin{array}{r} {\scriptstyle 2\ 17} \\ \not{3}\ 7 \\ -1\ 8 \end{array}$$

Borrow a ten. Write 2 above the tens column and 17 above the ones.

$$\begin{array}{r} {\scriptstyle 2\ 17} \\ \not{3}\ 7 \\ -1\ 8 \\ \hline 9 \end{array}$$

Subtract ones.

The borrowing here is like exchanging 3 dimes and 7 pennies for 2 dimes and 17 pennies.

OBJECTIVES

After finishing Section S.2, you should be able to:

a Subtract one whole number from another when borrowing is not necessary.

b Subtract one whole number from another when borrowing is necessary.

Subtract.

1. $\begin{array}{r} 7\ 8 \\ -6\ 4 \\ \hline \end{array}$

14

2. $\begin{array}{r} 2\ 9 \\ -\ \ 9 \\ \hline \end{array}$

20

3. $\begin{array}{r} 5\ 4\ 2 \\ -3\ 0\ 1 \\ \hline \end{array}$

241

4. $\begin{array}{r} 6\ 8\ 9\ 6 \\ -4\ 8\ 7\ 1 \\ \hline \end{array}$

2025

ANSWERS ON PAGE A-7

Subtract.

5.
$$\begin{array}{r} 46 \\ -29 \\ \hline \end{array}$$

17

6.
$$\begin{array}{r} 74 \\ -38 \\ \hline \end{array}$$

36

Subtract.

7.
$$\begin{array}{r} 646 \\ -192 \\ \hline \end{array}$$

454

8.
$$\begin{array}{r} 733 \\ -483 \\ \hline \end{array}$$

250

ANSWERS ON PAGE A-7

$$\begin{array}{rr} \overset{2}{\cancel{3}} & \overset{17}{\cancel{7}} \\ -\ 1 & 8 \\ \hline 1 & 9 \end{array}$$

Subtract tens.

$$\begin{array}{rr} \overset{2}{\cancel{3}} & \overset{17}{\cancel{7}} \\ -\ 1 & 8 \\ \hline 1 & 9 \end{array}$$

You should write only this. ◀

DO EXERCISES 5 AND 6.

Borrowing Hundreds

▶ **EXAMPLE 3** Subtract: 538 − 275.

$$\begin{array}{rrr} 5 & 3 & 8 \\ -2 & 7 & 5 \\ \hline & & 3 \end{array}$$

Subtract ones.

$$\begin{array}{rrr} 5 & 3 & 8 \\ -2 & 7 & 5 \\ \hline & ? & 3 \end{array}$$

Try to subtract tens: 30 − 70 is not a whole number.

$$\begin{array}{rrr} \overset{4}{\cancel{5}} & \overset{13}{\cancel{3}} & 8 \\ -2 & 7 & 5 \\ \hline & & 3 \end{array}$$

Borrow a hundred. Write 4 above the hundreds column and 13 above the tens.

> The borrowing is like exchanging 5 dollars, 3 dimes, and 8 pennies for 4 dollars, 13 dimes, and 8 pennies.

$$\begin{array}{rrr} \overset{4}{\cancel{5}} & \overset{13}{\cancel{3}} & 8 \\ -\ 2 & 7 & 5 \\ \hline & 6 & 3 \end{array}$$

Subtract tens.

$$\begin{array}{rrr} \overset{4}{\cancel{5}} & \overset{13}{\cancel{3}} & 8 \\ -\ 2 & 7 & 5 \\ \hline 2 & 6 & 3 \end{array}$$

Subtract hundreds.

$$\begin{array}{rrr} \overset{4}{\cancel{5}} & \overset{13}{\cancel{3}} & 8 \\ -\ 2 & 7 & 5 \\ \hline 2 & 6 & 3 \end{array}$$

You should write only this. ◀

DO EXERCISES 7 AND 8.

Borrowing More Than Once

Sometimes we have to borrow more than once.

▶ **EXAMPLE 4** Subtract: 672 − 394.

```
   6 12
 6 7̸ 2̸
−3 9 4
------
     8
```

Borrowing tens to subtract ones

```
  16
 5 6̸ 12
 6̸ 7̸ 2̸
−3 9 4
------
 2 7 8
```

Borrowing hundreds to subtract tens ◀

DO EXERCISES 9 AND 10.

▶ **EXAMPLE 5** Subtract: 6357 − 1769.

```
     4 17
 6 3 5̸ 7̸
−1 7 6 9
--------
       8
```

We cannot subtract 9 from 7. We borrow a ten.

```
    14
   2 4̸ 17
 6 3̸ 5̸ 7̸
−1 7 6 9
--------
     8 8
```

We cannot subtract 60 from 40. We borrow a hundred.

```
  12 14
 5 2̸ 4̸ 17
 6̸ 3̸ 5̸ 7̸
−1 7 6 9
--------
 4 5 8 8
```

We cannot subtract 700 from 200. We borrow a thousand. ◀

We can always check by adding the answer to the number being subtracted.

▶ **EXAMPLE 6** Subtract: 8341 − 2673. Check by adding.

```
  12 13
 7 2̸ 3̸ 11
 8̸ 3̸ 4̸ 1̸
−2 6 7 3
--------
 5 6 6 8
```

We check by adding 5668 and 2673.

Check:

```
  1 1 1
  5 6 6 8
+ 2 6 7 3
---------
  8 3 4 1
```

◀

DO EXERCISES 11 AND 12.

Zeros in Subtraction

Before subtracting note the following:

50 is 5 tens;

70 is 7 tens.

Subtract.

9. 563 − 187

376

10. 733 − 488

245

Subtract. Check by adding.

11. 4236 − 1679

2557

12. 7541 − 3867

3674

Complete.

13. 80 = 8 tens

14. 60 = 6 tens

15. 300 = 30 tens

16. 900 = 90 tens

ANSWERS ON PAGE A-7

Complete.

17. 5000 = 500 tens

18. 9000 = 900 tens

19. 5380 = 538 tens

20. 6770 = 677 tens

Subtract.

21. 60 − 18 = 42

22. 480 − 256 = 224

Subtract.

23. 602 − 464 = 138

24. 408 − 364 = 44

Subtract.

25. 4006 − 1238 = 2768

26. 9001 − 7804 = 1197

Subtract.

27. 3000 − 1754 = 1246

28. 8017 − 3289 = 4728

ANSWERS ON PAGE A-7

Then

100 is 10 tens;
200 is 20 tens.

DO EXERCISES 13–16 ON THE PRECEDING PAGE.

Also,

230 is 2 hundreds + 3 tens
or 20 tens + 3 tens
or 23 tens.

Similarly,

1000 is 100 tens;
2000 is 200 tens;
4670 is 467 tens.

DO EXERCISES 17–20.

▶ **EXAMPLE 7** Subtract: 50 − 37.

```
 4 10
 5̸ 0̸     We have 5 tens.
−3 7     We keep 4 of them in the tens column.
 1 3     We put 1 ten, or 10 ones, with the ones.
```

DO EXERCISES 21 AND 22.

▶ **EXAMPLE 8** Subtract: 803 − 547.

```
  7 9 13
  8̶ 0̸ 3̸     We have 8 hundreds, or 80 tens.
−5 4 7      We keep 79 tens.
 2 5 6      We put 1 ten, or 10 ones, with the ones.
```

DO EXERCISES 23 AND 24.

▶ **EXAMPLE 9** Subtract: 9003 − 2789.

```
  8 9 9 13
  9̶ 0̶ 0̸ 3̸     We have 9 thousands, or 900 tens.
−2 7 8 9     We keep 899 tens.
 6 2 1 4     We put 1 ten, or 10 ones, with the ones.
```

DO EXERCISES 25 AND 26.

▶ **EXAMPLES** Subtract.

10.
```
  4 9 9 10
  5̶ 0̶ 0̶ 0̸
−2 8 6 1
 2 1 3 9
```

11.
```
      10
  4 9 0̸ 13
  5̶ 0̶ 1̸ 3̸
−1 8 5 7
 3 1 5 6
```

DO EXERCISES 27 AND 28.

NAME SECTION DATE

EXERCISE SET S.2

a Subtract.

1. $15 - 4$
2. $26 - 13$
3. $64 - 31$
4. $55 - 34$
5. $67 - 20$
6. $78 - 11$
7. $548 - 301$
8. $596 - 403$
9. $700 - 200$
10. $867 - 101$
11. $765 - 111$
12. $666 - 333$
13. $525 - 323$
14. $747 - 130$
15. $988 - 700$
16. $9450 - 8230$
17. $6552 - 4321$
18. $3496 - 1235$
19. $7525 - 2522$
20. $7547 - 3421$
21. $5875 - 2111$
22. $16,843 - 4,321$
23. $38,695 - 37,004$
24. $23,707 - 11,607$
25. $67,899 - 66,673$
26. $99,999 - 1$
27. $56,780 - 56,770$
28. $42,111 - 32,010$
29. $77,654 - 66,611$
30. $23,456 - 12,345$

b Subtract.

31. $-34 - 16$
32. $-86 - 47$
33. $-93 - 28$
34. $-42 - 13$
35. $-86 - 78$
36. $-98 - 89$
37. $-625 - 317$
38. $-726 - 409$

ANSWERS

1. 11
2. 13
3. 33
4. 21
5. 47
6. 67
7. 247
8. 193
9. 500
10. 766
11. 654
12. 333
13. 202
14. 617
15. 288
16. 1220
17. 2231
18. 2261
19. 5003
20. 4126
21. 3764
22. 12,522
23. 1691
24. 12,100
25. 1226
26. 99,998
27. 10
28. 10,101
29. 11,043
30. 11,111
31. 18
32. 39
33. 65
34. 29
35. 8
36. 9
37. 308
38. 317

ANSWERS

39. 126
40. 617
41. 214
42. 89
43. 4402
44. 3555
45. 1503
46. 5889
47. 2387
48. 3649
49. 3832
50. 8144
51. 7750
52. 10,445
53. 33,793
54. 26,869
55. 16
56. 13
57. 16
58. 17
59. 86
60. 281
61. 455
62. 389
63. 571
64. 6148
65. 2200
66. 2113
67. 3748
68. 4313
69. 1063
70. 2418
71. 5206
72. 1459
73. 305
74. 4455

39. $\begin{array}{r} 735 \\ -609 \\ \hline \end{array}$ **40.** $\begin{array}{r} 853 \\ -236 \\ \hline \end{array}$ **41.** $\begin{array}{r} 961 \\ -747 \\ \hline \end{array}$ **42.** $\begin{array}{r} 787 \\ -698 \\ \hline \end{array}$

43. $\begin{array}{r} 6769 \\ -2367 \\ \hline \end{array}$ **44.** $\begin{array}{r} 6431 \\ -2876 \\ \hline \end{array}$ **45.** $\begin{array}{r} 3982 \\ -2479 \\ \hline \end{array}$ **46.** $\begin{array}{r} 7654 \\ -1765 \\ \hline \end{array}$

47. $\begin{array}{r} 5246 \\ -2859 \\ \hline \end{array}$ **48.** $\begin{array}{r} 6328 \\ -2679 \\ \hline \end{array}$ **49.** $\begin{array}{r} 7641 \\ -3809 \\ \hline \end{array}$ **50.** $\begin{array}{r} 8743 \\ -\ \ 599 \\ \hline \end{array}$

51. $\begin{array}{r} 12{,}647 \\ -\ \ 4{,}897 \\ \hline \end{array}$ **52.** $\begin{array}{r} 16{,}222 \\ -\ \ 5{,}777 \\ \hline \end{array}$ **53.** $\begin{array}{r} 46{,}781 \\ -12{,}988 \\ \hline \end{array}$ **54.** $\begin{array}{r} 75{,}654 \\ -48{,}785 \\ \hline \end{array}$

55. $\begin{array}{r} 40 \\ -24 \\ \hline \end{array}$ **56.** $\begin{array}{r} 50 \\ -37 \\ \hline \end{array}$ **57.** $\begin{array}{r} 70 \\ -54 \\ \hline \end{array}$ **58.** $\begin{array}{r} 90 \\ -73 \\ \hline \end{array}$

59. $\begin{array}{r} 140 \\ -\ \ 54 \\ \hline \end{array}$ **60.** $\begin{array}{r} 470 \\ -189 \\ \hline \end{array}$ **61.** $\begin{array}{r} 690 \\ -235 \\ \hline \end{array}$ **62.** $\begin{array}{r} 803 \\ -414 \\ \hline \end{array}$

63. $\begin{array}{r} 703 \\ -132 \\ \hline \end{array}$ **64.** $\begin{array}{r} 6406 \\ -\ \ 258 \\ \hline \end{array}$ **65.** $\begin{array}{r} 2309 \\ -\ \ 109 \\ \hline \end{array}$ **66.** $\begin{array}{r} 3406 \\ -1293 \\ \hline \end{array}$

67. $\begin{array}{r} 6807 \\ -3059 \\ \hline \end{array}$ **68.** $\begin{array}{r} 7340 \\ -3027 \\ \hline \end{array}$ **69.** $\begin{array}{r} 4037 \\ -2974 \\ \hline \end{array}$ **70.** $\begin{array}{r} 4007 \\ -1589 \\ \hline \end{array}$

71. $\begin{array}{r} 8000 \\ -2794 \\ \hline \end{array}$ **72.** $\begin{array}{r} 8002 \\ -6543 \\ \hline \end{array}$ **73.** $\begin{array}{r} 38{,}000 \\ -37{,}695 \\ \hline \end{array}$ **74.** $\begin{array}{r} 16{,}043 \\ -11{,}588 \\ \hline \end{array}$

S.3 Low-Stress Subtraction

a If you are having trouble with subtraction, you might try the following method. Let's do the subtraction 64 − 28. Leave a space between the numbers.

```
  6 4
  [ ]
− 2 8
```

We then try to subtract ones, but we cannot subtract 8 from 4, so we need to borrow a ten. We write this:

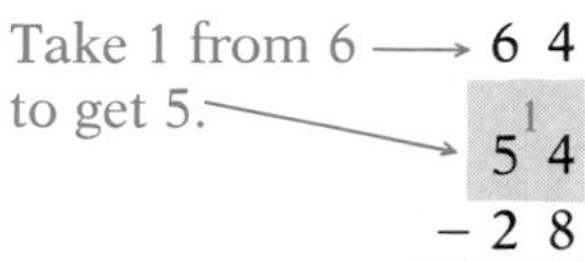


```
  6 4
  5 ¹4     1 ten comes to the ones column.
− 2 8
```

Now we subtract.

```
  6  4
  5 ¹4     Subtract ones.
− 2  8
     6
```

```
  6  4
  5 ¹4     Subtract tens.
− 2  8
  3  6
```

DO EXERCISES 1 AND 2.

Using this method, we do *all* the borrowing first.

▶ **EXAMPLE 1** Subtract: 534 − 278.

We cannot subtract 8 from 4 so we borrow a ten.

```
Instead of 3 tens, we          5  3  4
now have 2 tens.          →       2 ¹4     1 ten comes to the
                             − 2  7  8     ones column.
```

```
Instead of 5 hundreds,    →    5  3  4
we now have 4 hundreds.   →    4 ¹2 ¹4     We now have 12 in
                             − 2  7  8     the tens column.
```

Now we subtract.

```
  5  3  4
  4 ¹2 ¹4     Subtract ones.
− 2  7  8
        6
```

OBJECTIVE

After finishing Section S.3, you should be able to:

a Subtract using the low-stress method.

Subtract using the new method.

1.
```
  7   3
− 4   6
```
27

2.
```
  4   5
− 1   9
```
26

ANSWERS ON PAGE A-7

Subtract using the new low-stress method.

3.
```
  5  3  8

- 1  7  9
---------
       359
```

Subtract using the low-stress method.

4.
```
  7  5  3  8

- 4  2  7  5
------------
        3263
```

5.
```
  9  2  3  1

- 2  8  7  7
------------
        6354
```

6.
```
  6  0  0  0

- 3  4  4  8
------------
        2552
```

7.
```
  7  0  4  4

- 3  6  9  4
------------
        3350
```

ANSWERS ON PAGE A-7

```
    5  3  4
    4 ¹2 ¹4
  - 2  7  8
  ---------
       5  6
```
Subtract tens.

```
    5  3  4
    4 ¹2 ¹4
  - 2  7  8
  ---------
    2  5  6
```
Subtract hundreds.

DO EXERCISE 3.

▶ **EXAMPLE 2** Subtract: 6256 − 2548.

```
    6  2  5  6
    5 ¹2  4 ¹6
  - 2  5  4  8
  ------------
    3  7  0  8
```
Remember to do all the borrowing first.

▶ **EXAMPLE 3** Subtract: 8341 − 2876.

```
    8  3  4  1
    7 ¹2 ¹3 ¹1
  - 2  8  7  6
  ------------
    5  4  6  5
```

▶ **EXAMPLE 4** Subtract: 5000 − 2861.

```
    5  0  0  0
    4  9  9 ¹0
  - 2  8  6  1
  ------------
    2  1  3  9
```

▶ **EXAMPLE 5** Subtract: 5013 − 1857.

```
    5  0  1  3
    4  9 ¹0 ¹3
  - 1  8  5  7
  ------------
    3  1  5  6
```

DO EXERCISES 4–7.

NAME SECTION DATE

EXERCISE SET S.3

a Subtract using the low-stress method.

1. $\begin{array}{r} 86 \\ -47 \\ \hline \end{array}$

2. $\begin{array}{r} 94 \\ -29 \\ \hline \end{array}$

3. $\begin{array}{r} 35 \\ -17 \\ \hline \end{array}$

4. $\begin{array}{r} 42 \\ -18 \\ \hline \end{array}$

5. $\begin{array}{r} 409 \\ -329 \\ \hline \end{array}$

6. $\begin{array}{r} 801 \\ -493 \\ \hline \end{array}$

7. $\begin{array}{r} 962 \\ -848 \\ \hline \end{array}$

8. $\begin{array}{r} 878 \\ -794 \\ \hline \end{array}$

9. $\begin{array}{r} 7431 \\ -3876 \\ \hline \end{array}$

10. $\begin{array}{r} 7679 \\ -2367 \\ \hline \end{array}$

11. $\begin{array}{r} 8654 \\ -1765 \\ \hline \end{array}$

12. $\begin{array}{r} 6555 \\ -2677 \\ \hline \end{array}$

13. $\begin{array}{r} 5037 \\ -2974 \\ \hline \end{array}$

14. $\begin{array}{r} 3406 \\ -2193 \\ \hline \end{array}$

15. $\begin{array}{r} 8000 \\ -5443 \\ \hline \end{array}$

16. $\begin{array}{r} 7000 \\ -3899 \\ \hline \end{array}$

17. $\begin{array}{r} 16,233 \\ -6,666 \\ \hline \end{array}$

18. $\begin{array}{r} 12,547 \\ -4,058 \\ \hline \end{array}$

19. $\begin{array}{r} 46,178 \\ -13,999 \\ \hline \end{array}$

20. $\begin{array}{r} 76,650 \\ -49,406 \\ \hline \end{array}$

ANSWERS

1. 39
2. 65
3. 18
4. 24
5. 80
6. 308
7. 114
8. 84
9. 3555
10. 5312
11. 6889
12. 3878
13. 2063
14. 1213
15. 2557
16. 3101
17. 9567
18. 8489
19. 32,179
20. 27,244

ANSWERS

21. 2326
22. 1157
23. 2689
24. 2428
25. 7306
26. 7662
27. 986
28. 2804
29. 27,617
30. 31,113
31. 23,903
32. 17,057
33. 9778
34. 12,448
35. a) 8976 ft
b) 2720 m

21. $\begin{array}{r} 5003 \\ -2677 \\ \hline \end{array}$

22. $\begin{array}{r} 8900 \\ -7743 \\ \hline \end{array}$

23. $\begin{array}{r} 9998 \\ -7309 \\ \hline \end{array}$

24. $\begin{array}{r} 8911 \\ -6483 \\ \hline \end{array}$

25. $\begin{array}{r} 7600 \\ -\ \ 294 \\ \hline \end{array}$

26. $\begin{array}{r} 8008 \\ -\ \ 346 \\ \hline \end{array}$

27. $\begin{array}{r} 1000 \\ -\ \ \ 14 \\ \hline \end{array}$

28. $\begin{array}{r} 3002 \\ -\ \ 198 \\ \hline \end{array}$

29. $\begin{array}{r} 30,504 \\ -\ \ 2887 \\ \hline \end{array}$

30. $\begin{array}{r} 40,023 \\ -\ \ 8910 \\ \hline \end{array}$

31. $\begin{array}{r} 98,716 \\ -74,813 \\ \hline \end{array}$

32. $\begin{array}{r} 84,263 \\ -67,206 \\ \hline \end{array}$

33. $\begin{array}{r} 45,666 \\ -35,888 \\ \hline \end{array}$

34. $\begin{array}{r} 56,781 \\ -44,333 \\ \hline \end{array}$

SYNTHESIS

35. The elevation of Denver, Colorado, is 5280 ft (1600 m). The elevation of Long's Peak, near Denver, is 14,256 ft (4320 m).

a) How much higher, in feet, is Long's Peak than Denver?

b) How much higher, in meters, is Long's Peak than Denver?

M.1 Basic Multiplication

a To multiply, we start with two numbers, called **factors,** and get a third number, called a **product.** Multiplication can be explained by counting. The product 3×5 can be found by counting out 3 sets of 5 objects each, joining them (in a rectangular array if desired), and counting all the objects.

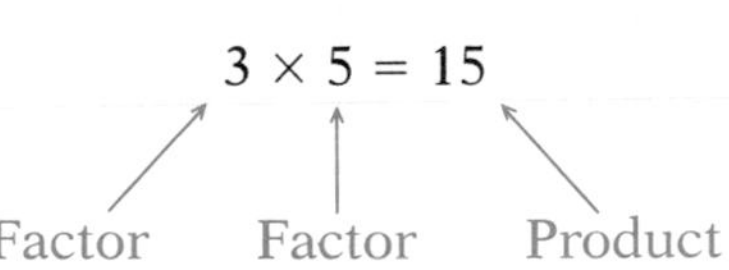

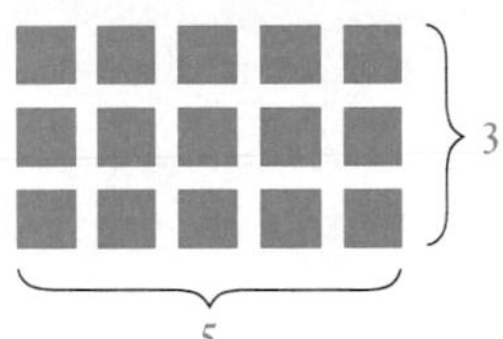

We can also think of multiplication as repeated addition.

$$3 \times 5 = 5 + 5 + 5 = 15$$

→ 3 addends of 5

▶ **EXAMPLES** Multiply. If you have trouble, think either of putting sets together in a rectangular array or of repeated addition.

1. $5 \times 6 = 30$

$$\begin{array}{r} 6 \\ \times\ 5 \\ \hline 30 \end{array}$$

2. $8 \times 4 = 32$

$$\begin{array}{r} 4 \\ \times\ 8 \\ \hline 32 \end{array}$$

◀

DO EXERCISES 1–4.

How do we multiply by 0? Consider $4 \cdot 0$. Using repeated addition, we see that

$$4 \cdot 0 = 0 + 0 + 0 + 0 = 0.$$

→ 4 addends of 0

We can also think of this using sets. That is, $4 \cdot 0$ is 4 sets with 0 objects in them, so the total is 0.

Consider $0 \cdot 4$. Using repeated addition, this is 0 addends of 4, which is 0. Using sets, this is 0 sets with 4 objects, which is 0.

In general,

> **Multiplying by 0 gives 0.**

▶ **EXAMPLES** Multiply.

3. $13 \times 0 = 0$

$$\begin{array}{r} 0 \\ \times\ 13 \\ \hline 0 \end{array}$$

4. $0 \cdot 11 = 0$

$$\begin{array}{r} 11 \\ \times\ 0 \\ \hline 0 \end{array}$$

5. $0 \cdot 0 = 0$

$$\begin{array}{r} 0 \\ \times\ 0 \\ \hline 0 \end{array}$$

◀

DO EXERCISES 5 AND 6.

OBJECTIVE

After finishing Section M.1, you should be able to:

a Multiply any two of the numbers 0, 1, 2, 3, 4, 5, 6, 7, 8, 9.

Multiply. Think of joining sets in a rectangular array or of repeated addition.

1. $7 \cdot 8$ (The dot "·" means the same as "×".) 56

2.

$$\begin{array}{r} 9 \\ \times\ 4 \\ \hline 36 \end{array}$$

3. $4 \cdot 7$ 28

4.

$$\begin{array}{r} 7 \\ \times\ 6 \\ \hline 42 \end{array}$$

Multiply.

5. $8 \cdot 0$ 0

6.

$$\begin{array}{r} 17 \\ \times\ 0 \\ \hline 0 \end{array}$$

ANSWERS ON PAGE A-7

Multiply.

7. $8 \cdot 1$ 8

8.
$$\begin{array}{r} 23 \\ \times\ 1 \\ \hline 23 \end{array}$$

9. Complete this table.

×	6	7	8	9
5	30	35	40	45
6	36	42	48	54
7	42	49	56	63
8	48	56	64	72
9	54	63	72	81

ANSWERS ON PAGE A-7

How do we multiply by 1? Consider $5 \cdot 1$. Using repeated addition, we see that

$$5 \cdot 1 = \underbrace{1 + 1 + 1 + 1 + 1}_{} = 5.$$

→5 addends of 1

We can also think of this using sets. That is, $5 \cdot 1$ is 5 sets with 1 object in each set, so the total is 5.

Consider $1 \cdot 5$. Using repeated addition, this is 1 addend of 5, which is 5. Using sets, this is 1 set of 5 objects, which is 5.

In general,

Multiplying a number by 1 doesn't change it.

This is a very important property.

► **EXAMPLES** Multiply.

6. $13 \cdot 1 = 13$

$$\begin{array}{r} 1 \\ \times\ 13 \\ \hline 13 \end{array}$$

7. $1 \cdot 7 = 7$

$$\begin{array}{r} 7 \\ \times\ 1 \\ \hline 7 \end{array}$$

8. $1 \cdot 1 = 1$

$$\begin{array}{r} 1 \\ \times\ 1 \\ \hline 1 \end{array}$$ ◄

DO EXERCISES 7 AND 8.

Your objective for this section is to be able to multiply any of the numbers 0, 1, 2, 3, 4, 5, 6, 7, 8, 9. Multiplying by 0 and 1 is easy. The rest of the products are listed in the table below.

×	2	3	4	5	6	7	8	9
2	4	6	8	10	12	14	16	18
3	6	9	12	15	18	21	24	27
4	8	12	16	20	24	28	32	36
5	10	15	20	25	30	35	40	45
6	12	18	24	30	36	42	48	54
7	14	21	28	35	42	49	56	63
8	16	24	32	40	48	56	64	72
9	18	27	36	45	54	63	72	81

$5 \times 7 = 35$
Find 5 at the left, and 7 at the top.

It is *very* important that you have the basic multiplication facts *memorized.* If you do not, you will always have trouble with multiplication.

The commutative law says that we can multiply numbers in any order. Thus you need to learn only about half the table.

DO EXERCISE 9.

NAME SECTION DATE

EXERCISE SET M.1

a Multiply. Try to do these mentally.

1. 3×4
2. 1×3
3. 6×4
4. 6×0
5. 7×1
6. 0×2
7. 4×6
8. 2×2
9. 10×1
10. 6×5
11. 1×10
12. 5×2
13. 2×5
14. 9×7
15. 3×7
16. 9×6
17. 2×6
18. 7×0
19. 9×8
20. 18×1
21. 8×9
22. 1×8
23. 0×4
24. 8×0
25. 4×7
26. 3×8
27. 1×7
28. 5×9
29. 0×5
30. 2×9
31. 6×7
32. 7×1
33. 0×7
34. 5×7
35. 1×9
36. 9×5
37. 5×8
38. 0×0
39. 8×5
40. 2×8

ANSWERS

1. 12
2. 3
3. 24
4. 0
5. 7
6. 0
7. 24
8. 4
9. 10
10. 30
11. 10
12. 10
13. 10
14. 63
15. 21
16. 54
17. 12
18. 0
19. 72
20. 18
21. 72
22. 8
23. 0
24. 0
25. 28
26. 24
27. 7
28. 45
29. 0
30. 18
31. 42
32. 7
33. 0
34. 35
35. 9
36. 45
37. 40
38. 0
39. 40
40. 16

ANSWERS

41. 16
42. 25
43. 64
44. 81
45. 1
46. 0
47. 9
48. 4
49. 36
50. 8
51. 48
52. 0
53. 27
54. 18
55. 45
56. 0
57. 42
58. 10
59. 0
60. 48
61. 54
62. 0
63. 25
64. 72
65. 49
66. 15
67. 27
68. 0
69. 8
70. 9
71. 2
72. 5
73. 32
74. 6
75. 15
76. 6
77. 8
78. 8
79. 20
80. 20
81. 64
82. 16
83. 10
84. 0

41. $4 \cdot 4$ **42.** $5 \cdot 5$ **43.** $8 \cdot 8$ **44.** $9 \cdot 9$

45. $1 \cdot 1$ **46.** $0 \cdot 0$ **47.** $3 \cdot 3$ **48.** $2 \cdot 2$

49. $6 \cdot 6$ **50.** $1 \cdot 8$ **51.** $8 \cdot 6$ **52.** $0 \cdot 1$

53. $3 \cdot 9$ **54.** $2 \cdot 9$ **55.** $9 \cdot 5$ **56.** $6 \cdot 0$

57. $6 \cdot 7$ **58.** $10 \cdot 1$ **59.** $0 \cdot 10$ **60.** $6 \cdot 8$

61. $9 \cdot 6$ **62.** $8 \cdot 0$ **63.** $5 \cdot 5$ **64.** $9 \cdot 8$

65. $7 \cdot 7$ **66.** $3 \cdot 5$ **67.** $9 \cdot 3$ **68.** $0 \cdot 2$

69. $1 \cdot 8$ **70.** $1 \cdot 9$ **71.** $2 \cdot 1$ **72.** $5 \cdot 1$

73. $8 \cdot 4$ **74.** $3 \cdot 2$ **75.** $5 \cdot 3$ **76.** $1 \cdot 6$

77. $4 \cdot 2$ **78.** $2 \cdot 4$ **79.** $4 \cdot 5$ **80.** $5 \cdot 4$

81. $8 \cdot 8$ **82.** $4 \cdot 4$ **83.** $5 \cdot 2$ **84.** $8 \cdot 0$

M.2 More on Multiplication

a Multiplying Multiples of 10, 100, and 1000

Here we consider multiplication by multiples of 10, 100, and 1000. These are numbers such as 10, 20, 30, 100, 400, 1000, and 7000.

Multiplying by a Multiple of 10

We know that

$$50 = 5 \text{ tens}, \quad 340 = 34 \text{ tens}, \quad \text{and} \quad 2340 = 234 \text{ tens}.$$

Turning this around, we see that to multiply by 10, all we need to do is write a 0 on the end.

To multiply by 10, write 0 on the end.

▶ **EXAMPLES** Multiply.

1. $10 \cdot 6 = 60$
2. $10 \cdot 47 = 470$
3. $10 \cdot 583 = 5830$ ◀

DO EXERCISES 1–5.

Let's find $4 \cdot 90$. This is $4 \cdot (9 \text{ tens})$, or 36 tens. This is the same as multiplying the 4 and 9 and writing a 0 on the end. Thus, $4 \cdot 90 = 360$.

▶ **EXAMPLES** Multiply.

4. $5 \cdot 70 = 350$ ← $5 \cdot 7$, then write a 0
5. $8 \cdot 80 = 640$
6. $5 \cdot 60 = 300$ ◀

DO EXERCISES 6 AND 7.

Multiplying by a Multiple of 100

Note the following:

$$300 = 3 \text{ hundreds}, \quad 4700 = 47 \text{ hundreds}, \quad 56{,}800 = 568 \text{ hundreds}.$$

Turning this around, we see that to multiply by 100, all we need to do is write two 0's on the end.

To multiply by 100, write 00 on the end.

OBJECTIVES

After finishing Section M.2, you should be able to:

a Multiply multiples of 10, 100, and 1000.
b Multiply larger numbers by 0, 1, 2, 3, 4, 5, 6, 7, 8, and 9.
c Multiply by multiples of 10, 100, and 1000.

Multiply.

1. $10 \cdot 7$
   70
2. $10 \cdot 45$
   450
3. $10 \cdot 273$
   2730
4. $10 \cdot 10$
   100
5. $10 \cdot 100$
   1000

Multiply.

6. 70×8
   560
7. 60×6
   360

ANSWERS ON PAGE A-7

Multiply.

8. 100 · 7

700

9. 100 · 23

2300

10. 100 · 723

72,300

11. 100 · 100

10,000

12. 100 · 1000

100,000

Multiply.

13. 700 × 8

5600

14. 400 × 4

1600

Multiply.

15. 1000 · 9

9000

16. 1000 · 852

852,000

17. 1000 · 10

10,000

18. 1000 · 100

100,000

19. 1000 · 1000 1,000,000

ANSWERS ON PAGE A-7

▶ **EXAMPLES** Multiply.

7. 100 · 6 = 600

8. 100 · 39 = 3900

9. 100 · 448 = 44,800 ◀

DO EXERCISES 8–12.

Let's find 4 · 900. This is 4 · (9 hundreds). If we use addition, this is

9 hundreds + 9 hundreds + 9 hundreds + 9 hundreds,
or 36 hundreds,

which is the same as multiplying 4 and 9 and writing two 0's on the end. Thus, 4 · 900 = 3600.

▶ **EXAMPLES** Multiply.

10. 6 · 800 = 4800 ← 6 · 8, then write 00

11. 9 · 700 = 6300

12. 5 · 500 = 2500 ◀

DO EXERCISES 13 AND 14.

Multiplying by a Multiple of 1000

Note the following:

6000 = 6 thousands and 19,000 = 19 thousands.

Turning this around, we see that to multiply by 1000, all we need to do is write three 0's on the end.

To multiply by 1000, write 000 on the end.

▶ **EXAMPLES** Multiply.

13. 1000 · 8 = 8000

14. 1000 · 13 = 13,000

15. 1000 · 567 = 567,000 ◀

DO EXERCISES 15–19.

Multiplying Multiples by Multiples

Let's multiply 50 and 30. This is 50 · (3 tens), or 150 tens, or 1500. This is the same as multiplying 5 and 3 and writing two 0's at the end.

To multiply multiples of tens, hundreds, thousands, and so on:
1. **Multiply the one-digit numbers.**
2. **Count the number of zeros.**
3. **Write that many 0's on the end.**

▶ **EXAMPLES** Multiply.

16. $80 \times 60 = 4800$ — 80: 1 zero at end; 60: 1 zero at end. $6 \cdot 8$, then write 00

17. $800 \times 60 = 48{,}000$ — 800: 2 zeros at end; 60: 1 zero at end. $6 \cdot 8$, then write 000

18. $800 \times 600 = 480{,}000$ — 800: 2 zeros at end; 600: 2 zeros at end. $6 \cdot 8$, then write 0,000

19. $800 \times 50 = 40{,}000$ — 800: 2 zeros at end; 50: 1 zero at end. $5 \cdot 8$, then write 000 ◀

DO EXERCISES 20–23.

b Multiplying Larger Numbers

The product 2×13 can be represented as

$$\begin{aligned} 2 \times (1 \text{ ten} + 3) &= (1 \text{ ten} + 3) + (1 \text{ ten} + 3) \\ &= 2 \text{ tens} + 6. \end{aligned}$$

We are doubling 13. We can do this by doubling the tens, and doubling the ones. Thus,

$$2 \times 13 = 26.$$

The product 3×24 can be represented as

$$\begin{aligned} 3 \times (2 \text{ tens} + 4) &= (2 \text{ tens} + 4) + (2 \text{ tens} + 4) + (2 \text{ tens} + 4) \\ &= 6 \text{ tens} + 12 = 6 \text{ tens} + 1 \text{ ten} + 2 \\ &= 7 \text{ tens} + 2 \\ &= 72. \end{aligned}$$

We multiply the 4 ones by 3, getting 12
We multiply the 2 tens by 3, getting +60
Then we add: 72

▶ **EXAMPLE 20** Multiply: 3×24.

```
   2 4
 ×   3
   1 2 ← Multiply the 4 ones by 3.
   6 0 ← Multiply the 2 tens by 3.
   7 2 ← Add.
```
◀

DO EXERCISES 24–26.

▶ **EXAMPLE 21** Multiply: 5×734.

```
    7 3 4
 ×      5
      2 0 ← Multiply ones by 5.
    1 5 0 ← Multiply tens by 5.
  3 5 0 0 ← Multiply hundreds by 5.
  3 6 7 0 ← Add.
```
◀

DO EXERCISES 27 AND 28.

Multiply.

20. 9000×6 — 54,000

21. 80×70 — 5600

22. 800×70 — 56,000

23. 600×30 — 18,000

Multiply.

24. 14×2 — 28

25. 58×2 — 116

26. 37×4 — 148

Multiply.

27. 823×6 — 4938

28. 1348×5 — 6740

ANSWERS ON PAGE A-7

Multiply using the short form.

29. $\begin{array}{r} 58 \\ \times\ 2 \\ \hline \end{array}$ 116

30. $\begin{array}{r} 37 \\ \times\ 4 \\ \hline \end{array}$ 148

31. $\begin{array}{r} 823 \\ \times\ 6 \\ \hline \end{array}$ 4938

32. $\begin{array}{r} 1348 \\ \times\ 5 \\ \hline \end{array}$ 6740

Multiply.

33. $\begin{array}{r} 746 \\ \times\ 8 \\ \hline \end{array}$ 5968

34. $\begin{array}{r} 746 \\ \times\ 80 \\ \hline \end{array}$ 59,680

35. $\begin{array}{r} 746 \\ \times 800 \\ \hline \end{array}$ 596,800

ANSWERS ON PAGE A-7

▶ **EXAMPLE 22** Multiply: 5 × 734.

$\begin{array}{r} {}^{2} \\ 7\,3\,4 \\ \times\quad 5 \\ \hline 0 \end{array}$ Multiply ones by 5. Write 0 in the ones column and 2 above the tens.

$\begin{array}{r} {}^{1\,2} \\ 7\,3\,4 \\ \times\quad 5 \\ \hline 7\,0 \end{array}$ Multiply tens by 5 and add 2 tens: 5 · (3 tens) = 15 tens, and 15 tens + 2 tens = 17 tens. Write 7 in the tens column and 1 above the hundreds.

$\begin{array}{r} {}^{1\,2} \\ 7\,3\,4 \\ \times\quad 5 \\ \hline 3\,6\,7\,0 \end{array}$ Multiply hundreds by 5 and add 1 hundred: 5 · (7 hundreds) = 35 hundreds, and 35 hundreds + 1 hundred = 36 hundreds.

$\begin{array}{r} {}^{1\,2} \\ 7\,3\,4 \\ \times\quad 5 \\ \hline 3\,6\,7\,0 \end{array}$ You should write only this.

Avoid writing the reminders unless necessary. ◀

DO EXERCISES 29–32.

C Multiplication by Multiples of 10, 100, and 1000

To multiply 327 by 50, we multiply by 10 (write a 0), and then multiply 327 by 5.

$$\begin{array}{r} 3\,2\,7 \\ \times\quad 5\,0 \\ \hline 1\,6{,}3\,5\,0 \end{array}$$

Write a 0, then 5 · 327.

▶ **EXAMPLE 23** Multiply: 400 × 289.

$\begin{array}{r} 2\,8\,9 \\ \times 4\,0\,0 \\ \hline 0\,0 \end{array}$ Write two 0's.

$\begin{array}{r} 2\,8\,9 \\ \times\quad 4\,0\,0 \\ \hline 1\,1\,5{,}6\,0\,0 \end{array}$ Multiply 4 and 289: $\begin{array}{r} {}^{3\,3} \\ 2\,8\,9 \\ \times\quad 4 \\ \hline 1\,1\,5\,6 \end{array}$

$\begin{array}{r} {}^{3\,3} \\ 2\,8\,9 \\ \times\quad 4\,0\,0 \\ \hline 1\,1\,5{,}6\,0\,0 \end{array}$ You should write only this. ◀

DO EXERCISES 33–35.

NAME SECTION DATE

EXERCISE SET M.2

a Multiply.

1. 10×8
2. 10×6
3. 10×9
4. 9×10
5. 7×10
6. 20×8
7. 30×7
8. 45×10
9. 78×10
10. 50×9
11. 80×7
12. 90×4
13. 100×8
14. 100×3
15. 100×9
16. 100×10
17. 67×100
18. 321×100
19. 3457×100
20. 400×3
21. 700×7
22. 500×8
23. 600×7
24. 100×100
25. 1000×7
26. 1000×9
27. 1000×2
28. 457×1000
29. 7888×1000
30. 6769×1000
31. 2000×9
32. 5000×4
33. 6000×8
34. 8000×2
35. 3000×2
36. 1000×1000
37. 40×30
38. 70×70
39. 20×10
40. 80×50

ANSWERS

1. 80
2. 60
3. 90
4. 90
5. 70
6. 160
7. 210
8. 450
9. 780
10. 450
11. 560
12. 360
13. 800
14. 300
15. 900
16. 1000
17. 6700
18. 32,100
19. 345,700
20. 1200
21. 4900
22. 4000
23. 4200
24. 10,000
25. 7000
26. 9000
27. 2000
28. 457,000
29. 7,888,000
30. 6,769,000
31. 18,000
32. 20,000
33. 48,000
34. 16,000
35. 6000
36. 1,000,000
37. 1200
38. 4900
39. 200
40. 4000

ANSWERS

41. 2500
42. 12,000
43. 49,000
44. 6000
45. 63,000
46. 120,000
47. 150,000
48. 800,000
49. 120,000
50. 24,000,000
51. 16,000,000
52. 80,000
53. 147
54. 444
55. 2965
56. 4872
57. 6293
58. 3460
59. 16,236
60. 29,376
61. 30,924
62. 13,508
63. 87,554
64. 195,384
65. 3480
66. 2790
67. 3360
68. 7020
69. 20,760
70. 36,150
71. 6840
72. 10,680
73. 358,800
74. 109,800
75. 583,800
76. 299,700
77. 11,346,000
78. 23,390,000
79. 61,092,000
80. 73,032,000

41. 50×50

42. 400×30

43. 700×70

44. 200×30

45. 700×90

46. 400×300

47. 500×300

48. 4000×200

49. 6000×20

50. 6000×4000

51. 4000×4000

52. 8000×10

b Multiply.

53. 49×3

54. 74×6

55. 593×5

56. 609×8

57. 899×7

58. 865×4

59. 8118×2

60. 3264×9

61. 7731×4

62. 6754×2

63. $43{,}777 \times 2$

64. $32{,}564 \times 6$

c Multiply.

65. 58×60

66. 93×30

67. 42×80

68. 78×90

69. 346×60

70. 723×50

71. 342×20

72. 267×40

73. 897×400

74. 366×300

75. 834×700

76. 333×900

77. 5673×2000

78. 4678×5000

79. 6788×9000

80. 9129×8000

M.3 Low-Stress Multiplication

a If you are having trouble with the multiplication method we have been using, you might try the following. We again need new notation.

Regular notation	*New notation*
$7 \times 6 = 42$	$7 \times 6 = 4_2$
$4 \times 5 = 20$	$4 \times 5 = 2_0$
$2 \times 3 = 6$	$2 \times 3 = 0_6$

DO EXERCISES 1–3.

To do more complicated problems, we multiply from left to right and record answers using the new notation.

▶ **EXAMPLE 1** Multiply 6×357. Use the new notation.

```
   3 5 7
×      6
 1
   8
```

Multiply 300 by 6: $6 \times 300 = 1800$. Write this as shown using 1_8. Note that 1 is in the thousands place and 8 is below it in the hundreds place.

```
   3 5 7
×      6
 1 3
   8 0
```

Multiply 50×6: $6 \times 50 = 300$. Write this as shown by the 3_0. Note that 3 is in the hundreds place and 0 is below it in the tens place.

```
   3 5 7
×      6
 1 3 4
   8 0 2
```

Multiply 7 by 6: $6 \times 7 = 42$. Write this as shown by the 4_2. Note that 4 is in the tens place and 2 is below it in the ones place.

```
   3 5 7
×      6
 1 3 4
   8 0 2
 2 1 4 2
```

Note that an advantage of this method is that no carrying (in multiplication) is necessary.

Add.

DO EXERCISES 4 AND 5.

OBJECTIVE

After finishing Section M.3, you should be able to:

a Multiply using the low-stress method.

Multiply using the new notation.

1. 4×8 3_2

2. 5×6 3_0

3. 3×3 0_9

Multiply using the new method.

4.
```
  6 8
×   3
```
204

5.
```
  6 8 3
×     7
```
4781

ANSWERS ON PAGE A-7

Multiply using the new method.

6.
$$\begin{array}{r} 687 \\ \times\ \ 52 \\ \hline \end{array}$$

35,724

7.
$$\begin{array}{r} 619 \\ \times 432 \\ \hline \end{array}$$

267,408

ANSWERS ON PAGE A-7

▶ **EXAMPLE 2** Multiply: 347 × 896.

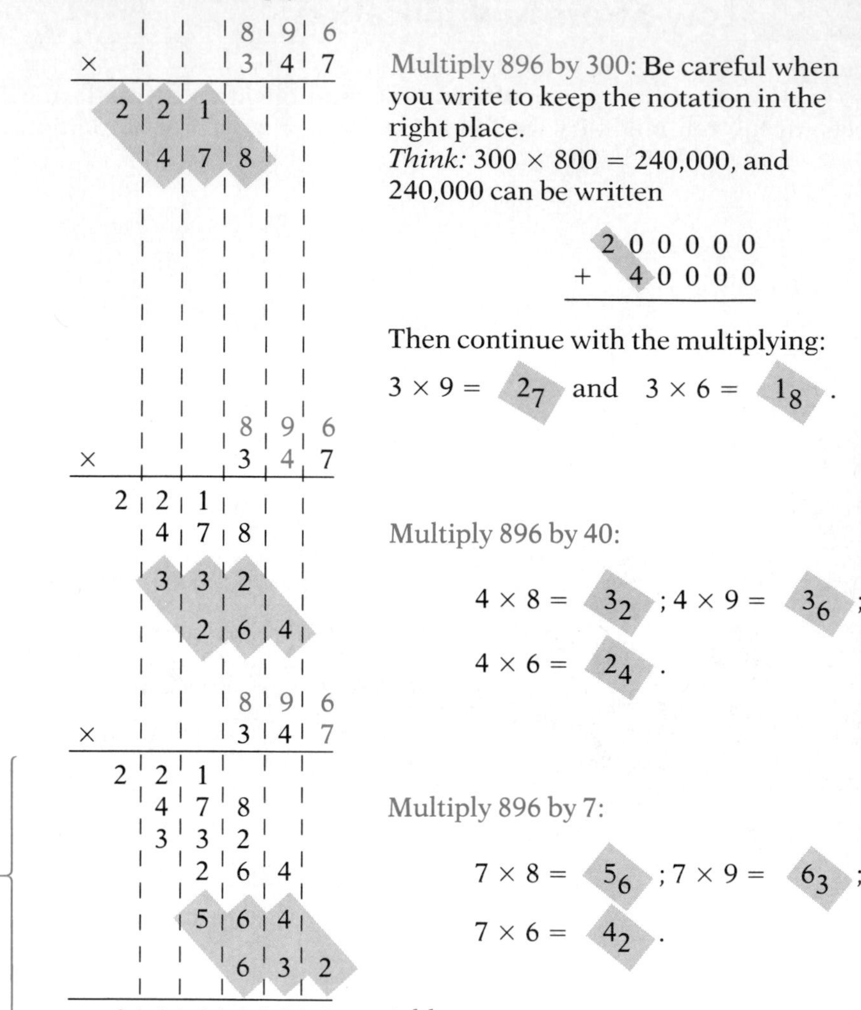

DO EXERCISES 6 AND 7.

NAME SECTION DATE

EXERCISE SET M.3

a Multiply using the new method.

1. 9×4
2. 8×5
3. 4×2
4. 2×5

5. 65×8
6. 87×4
7. 94×6
8. 76×9

9. 509×3
10. 806×7
11. 432×5
12. 617×2

13. 3195×6
14. 3642×8
15. 9229×7
16. 7867×4

17. 53×90
18. 78×60
19. 65×48
20. 87×34

21. 640×72
22. 666×66
23. 444×33
24. 509×88

ANSWERS

1. 36
2. 40
3. 8
4. 10
5. 520
6. 348
7. 564
8. 684
9. 1527
10. 5642
11. 2160
12. 1234
13. 19,170
14. 29,136
15. 64,603
16. 31,468
17. 4770
18. 4680
19. 3120
20. 2958
21. 46,080
22. 43,956
23. 14,652
24. 44,792

ANSWERS

25. 310,490

26. 577,280

27. 224,900

28. 272,190

29. 263,898

30. 673,017

31. 269,496

32. 197,136

33. 1,177,848

34. 1,227,858

35. 5,446,770

36. 20,336,745

37. 32,681,565

38. 38,714,037

39. 548,186,862

40. 694,891,776

25. $\begin{array}{r} 610 \\ \times 509 \\ \hline \end{array}$

26. $\begin{array}{r} 820 \\ \times 704 \\ \hline \end{array}$

27. $\begin{array}{r} 346 \\ \times 650 \\ \hline \end{array}$

28. $\begin{array}{r} 633 \\ \times 430 \\ \hline \end{array}$

29. $\begin{array}{r} 543 \\ \times 486 \\ \hline \end{array}$

30. $\begin{array}{r} 789 \\ \times 853 \\ \hline \end{array}$

31. $\begin{array}{r} 591 \\ \times 456 \\ \hline \end{array}$

32. $\begin{array}{r} 444 \\ \times 444 \\ \hline \end{array}$

33. $\begin{array}{r} 7182 \\ \times \quad 164 \\ \hline \end{array}$

34. $\begin{array}{r} 5433 \\ \times \quad 226 \\ \hline \end{array}$

35. $\begin{array}{r} 9814 \\ \times \quad 555 \\ \hline \end{array}$

36. $\begin{array}{r} 5489 \\ \times 3705 \\ \hline \end{array}$

37. $\begin{array}{r} 7539 \\ \times 4335 \\ \hline \end{array}$

38. $\begin{array}{r} 9039 \\ \times 4283 \\ \hline \end{array}$

39. $\begin{array}{r} 76,413 \\ \times \quad 7174 \\ \hline \end{array}$

40. $\begin{array}{r} 84,352 \\ \times \quad 8238 \\ \hline \end{array}$

D.1 Basic Division

a Division can be explained by arranging a set of objects in a rectangular array. This can be done in two ways.

▶ **EXAMPLE 1** Divide: $18 \div 6$.

Method 1. We can do this division by taking 18 objects and determining how many rows, each with 6 objects, we can arrange the objects into.

Since there are 3 rows of 6 objects, we have

$$18 \div 6 = 3.$$

Method 2. We can also arrange the objects into 6 rows and determine how many objects are in each row.

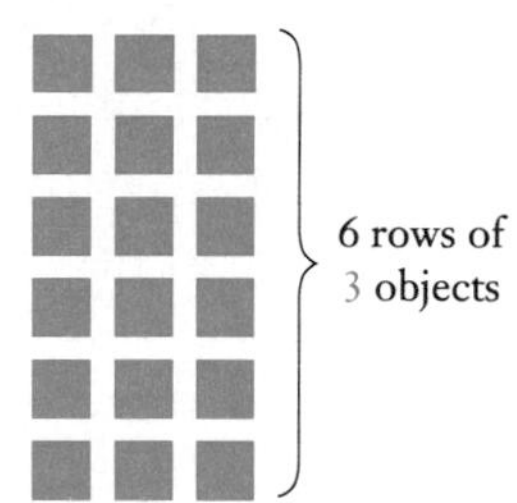

Since there are 3 objects in each of the 6 rows, we have

$$18 \div 6 = 3.$$ ◀

We can also use fractional notation for division. That is, a/b means $a \div b$.

▶ **EXAMPLES** Divide.

2. $36 \div 9 = 4$

3. $\frac{42}{7} = 6$

4. $\frac{24}{3} = 8$ ◀

DO EXERCISES 1–4.

OBJECTIVE

After finishing Section D.1, you should be able to:

a Find basic quotients such as $20 \div 5$, $56 \div 7$, and so on.

Divide.

1. $24 \div 6$ 4

2. $64 \div 8$ 8

3. $\frac{63}{7}$ 9

4. $\frac{27}{9}$ 3

ANSWERS ON PAGE A-7

For each multiplication fact, write two division facts.

5. $6 \cdot 2 = 12$

$12 \div 2 = 6$;
$12 \div 6 = 2$

6. $7 \times 6 = 42$

$42 \div 6 = 7$;
$42 \div 7 = 6$

ANSWERS ON PAGE A-7

Your objective for this part of the section is to be able to divide numbers that are no larger than 81. In Section M.1 you memorized a multiplication table. That table will enable you to divide as well. First, let's recall how multiplication and division are related.

A multiplication:

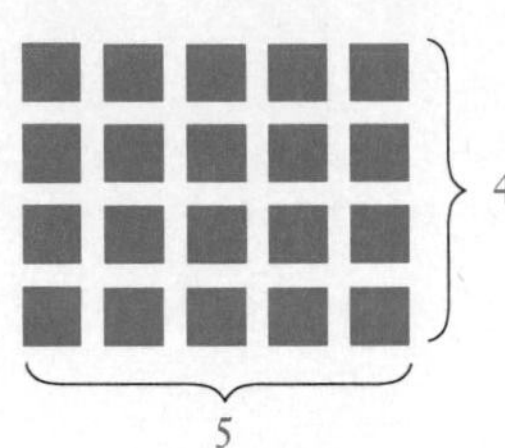

$5 \cdot 4 = 20.$

Two related divisions:

A.

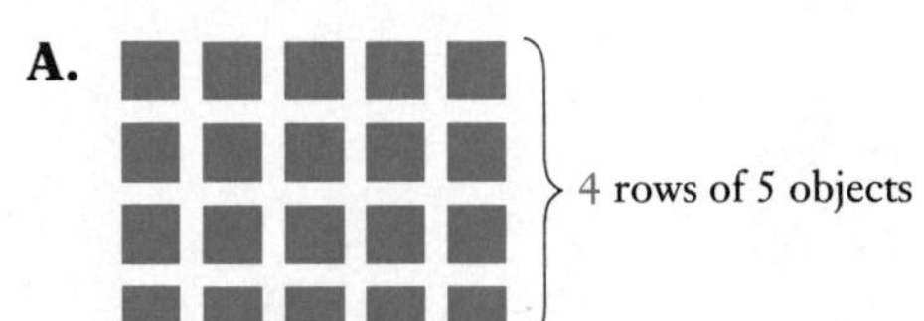

$20 \div 5 = 4.$

B.

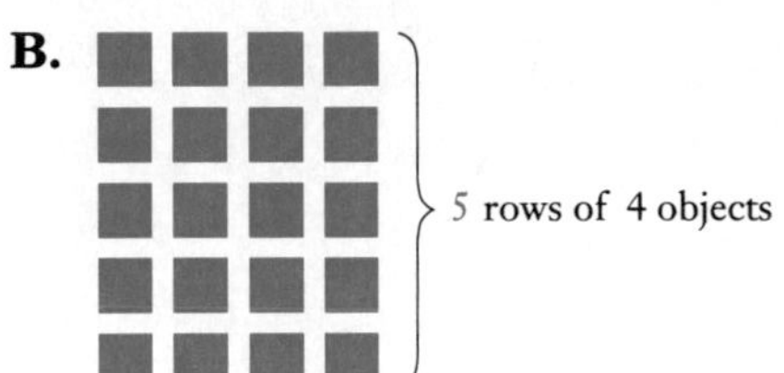

$20 \div 4 = 5.$

Since we know that

$5 \cdot 4 = 20,$ A basic multiplication fact

we also know the two division facts

$20 \div 5 = 4$ and $20 \div 4 = 5.$

▶ **EXAMPLE 5** From $7 \cdot 8 = 56$, write two division facts.

a) $56 \div 7 = 8$ We can get this by moving the 7: $7 \cdot 8 = 56$

b) $56 \div 8 = 7$ We can get this by moving the 8: $7 \cdot 8 = 56$ ◀

DO EXERCISES 5 AND 6.

Now let's look at the multiplication table you memorized. You can use it in reverse as a division table.

×	2	3	4	5	6	7	8	9
2	4	6	8	10	12	14	16	18
3	6	9	12	15	18	21	24	27
4	8	12	16	20	24	28	32	36
5	10	15	20	25	30	35	40	45
6	12	18	24	30	36	42	48	54
7	14	21	28	35	42	49	56	63
8	16	24	32	40	48	56	64	72
9	18	27	36	45	54	63	72	81

It is *very* important that you have the basic division facts *memorized*. If you do not, you will always have trouble with division.

▶ **EXAMPLE 6** Find $35 \div 7$.

You memorized that $7 \cdot 5 = 35$. Thus you know the related division $35 \div 7 = 5$. In the table we find 7 on the left. We go across to 35. Then we go up. There we find the quotient $35 \div 7$, which is 5.

To find $35 \div 7$, we ask ourselves "7 times what number is 35?"

$$7 \cdot \square = 35$$ ◀

DO EXERCISES 7–10.

Division by 1

Note that

$$3 \div 1 = 3 \quad \text{because} \quad 3 = 3 \cdot 1; \qquad \frac{14}{1} = 14 \quad \text{because} \quad 14 = 14 \cdot 1.$$

Any number divided by 1 is that same number.

▶ **EXAMPLES** Divide.

7. $\frac{8}{1} = 8$

8. $6 \div 1 = 6$

9. $34 \div 1 = 34$ ◀

DO EXERCISES 11–13.

Division by Zero

Why can't we divide by 0? Suppose the number 4 could be divided by 0. Then if $\square$ were the answer,

$$4 \div 0 = \square \quad \text{and this would mean} \quad 4 = \square \cdot 0 = 0. \quad \text{False!}$$

Suppose 12 could be divided by 0. If $\square$ were the answer,

$$12 \div 0 = \square \quad \text{and this would mean} \quad 12 = \square \cdot 0 = 0. \quad \text{False!}$$

Thus, $a \div 0$ would be some number $\square$ such that $a = 0 \cdot \square = 0$. So the only possible number that could be divided by 0 would be 0 itself.

Divide.

7. $28 \div 4$ 7

8. $81 \div 9$ 9

9. $\frac{16}{2}$ 8

10. $\frac{54}{6}$ 9

Divide.

11. $6 \div 1$ 6

12. $\frac{13}{1}$ 13

13. $1 \div 1$ 1

ANSWERS ON PAGE A-7

Divide, if possible. If not possible, write "not defined."

14. $\frac{8}{4}$ 2

15. $\frac{5}{0}$ Not defined

16. $\frac{0}{5}$ 0

17. $\frac{0}{0}$ Not defined

18. $12 \div 0$ Not defined

19. $100 \div 10$ 10

20. $\frac{5}{3-3}$ Not defined

21. $\frac{8-8}{4}$ 0

Divide.

22. $23 \div 23$ 1

23. $\frac{67}{67}$ 1

24. $\frac{41}{41}$ 1

25. $17 \div 17$ 1

26. $17 \div 1$ 17

27. $\frac{54}{54}$ 1

ANSWERS ON PAGE A-7

But such a division would give us any number we wish, for

$$\left.\begin{array}{lll} 0 \div 0 = 8 & \text{because} & 0 = 8 \cdot 0; \\ 0 \div 0 = 3 & \text{because} & 0 = 3 \cdot 0; \\ 0 \div 0 = 7 & \text{because} & 0 = 7 \cdot 0. \end{array}\right\} \text{ All true!}$$

We avoid the preceding difficulties by agreeing to exclude division by 0.

Division by 0 is not defined. (We agree not to divide by 0.)

Dividing Zero by Other Numbers

Note that

$0 \div 3 = 0$ because $0 = 0 \cdot 3$; $\quad \frac{0}{12} = 0$ because $0 = 0 \cdot 12$.

Zero divided by any number greater than 0 is 0.

▶ **EXAMPLES** Divide.

10. $0 \div 8 = 0$

11. $0 \div 22 = 0$

12. $\frac{0}{9} = 0$ ◀

DO EXERCISES 14–21.

Division of a Number by Itself

Note that

$3 \div 3 = 1$ because $3 = 1 \cdot 3$; $\quad \frac{34}{34} = 1$ because $34 = 1 \cdot 34$.

Any number greater than 0 divided by itself is 1.

▶ **EXAMPLES** Divide.

13. $8 \div 8 = 1$

14. $27 \div 27 = 1$

15. $\frac{32}{32} = 1$ ◀

DO EXERCISES 22–27.

D.2 Building Further Division Skills

a Division by "Guess, Multiply, and Subtract"

To understand the process of division, we use a method known as "guess, multiply, and subtract."

▶ **EXAMPLE 1** Divide 275 ÷ 4. Use "guess, multiply, and subtract."

We *guess* a partial quotient of 35. We could guess *any* number, say 4, 16, or 30. We *multiply* and *subtract* as follows:

```
      3 5 ←— Partial quotient
  4 )2 7 5
    1 4 0 ←— 35 · 4
    1 3 5 ←— Remainder
```

Next, look at 135 and *guess* another partial quotient, say 20. Then *multiply* and *subtract:*

```
      2 0 ←— Second partial quotient
      3 5
  4 )2 7 5
    1 4 0
    1 3 5
      8 0 ←— 20 · 4
      5 5 ←— Remainder
```

Next, look at 55 and *guess* another partial quotient, say 13. Then *multiply* and *subtract:*

```
      1 3 ←— Third partial quotient
      2 0
      3 5
  4 )2 7 5
    1 4 0
    1 3 5
      8 0
      5 5
      5 2 ←— 13 · 4
        3 ←— Remainder is less than 4
```

OBJECTIVES

After finishing Section D.2, you should be able to:

a Divide using the "guess, multiply, and subtract" method.

b Divide by estimating multiples of thousands, hundreds, tens, and ones.

Divide using the "guess, multiply, and subtract" method.

1. $6\overline{)454}$ 75 R 4

2. $32\overline{)747}$ 23 R 11

ANSWERS ON PAGE A-7

Divide using the "guess, multiply, and subtract" method.

3. $7\overline{)6789}$ 969 R 6

4. $64\overline{)3012}$ 47 R 4

ANSWERS ON PAGE A-7

Since we cannot subtract any more 4's, the division is finished. We add our partial quotients.

```
      6 8 ←— Quotient (sum of guesses)
      1 3
      2 0
      3 5
  4 )2 7 5
    1 4 0
    1 3 5
      8 0
      5 5
      5 2
        3
```

Check: $275 = (4 \times 68) + 3$

275	272 + 3
	275

The answer is 68 R 3. This tells us that with 275 objects, we could make 68 rows of 4 and have 3 left over. ◀

The partial quotients (guesses) can be made in any manner so long as subtraction is possible.

DO EXERCISES 1 AND 2 ON THE PRECEDING PAGE.

▶ **EXAMPLE 2** Divide: 1506 ÷ 32.

```
          4 7 ←— Quotient (sum of guesses)
          2 0 ⎫
            2 ⎬ ←— Guesses
          2 0 ⎪
            5 ⎭
  3 2 )1 5 0 6
        1 6 0 ←— 5 · 32
      1 3 4 6
        6 4 0 ←— 20 · 32
        7 0 6
          6 4 ←— 2 · 32
        6 4 2
        6 4 0 ←— 20 · 32
            2 ←— Remainder
```

The answer is 47 R 2. ◀

Remember, you can *guess any partial quotient* so long as subtraction is possible.

DO EXERCISES 3 AND 4.

b Division by Estimating Multiples

Let's refine the guessing we have used to divide in Section D.1. We guess multiples of 10, 100, and 1000, and so on.

▶ **EXAMPLE 3** Divide: $435 \div 6$.

a) Are there any 1000's (that is, $6 \cdot 1000$'s) in 435? No, $6 \cdot 1000 = 6000$, and 6000 is larger than 435.

b) Are there any 100's (that is, $6 \cdot 100$'s) in 435? No, $6 \cdot 100 = 600$, and 600 is larger than 435.

c) Are there any 10's (that is, $6 \cdot 10$'s) in 435? Yes, $6 \cdot 10 = 60$. To find how many, we find products of 6 and multiples of 10.

Try to do this mentally.

$$\begin{aligned} 6 \cdot 10 &= 60 \\ 6 \cdot 20 &= 120 \\ 6 \cdot 30 &= 180 \\ 6 \cdot 40 &= 240 \\ 6 \cdot 50 &= 300 \\ 6 \cdot 60 &= 360 \\ 6 \cdot 70 &= 420 \\ 6 \cdot 80 &= 480 \end{aligned}$$

← 435 is here; there are 70 sixes in 435.

$$\begin{array}{r} 70 \\ 6\,\overline{)\,435} \\ \underline{420} \\ 15 \end{array}$$

d) Now go to the ones place in the remainder. Are there any 6's in the ones place of the quotient?

$$\begin{aligned} 6 \cdot 1 &= 6 \\ 6 \cdot 2 &= 12 \\ 6 \cdot 3 &= 18 \end{aligned}$$

← 15 is here, so there are 2 sixes in 15.

$$\begin{array}{r} \underline{72} \\ 2 \\ 70 \\ 6\,\overline{)\,435} \\ \underline{420} \\ 15 \\ \underline{12} \\ 3 \end{array}$$

The answer is 72 R 3. ◀

DO EXERCISE 5.

▶ **EXAMPLE 4** Divide: $7643 \div 3$.

a) Are there any $3 \cdot 1000$'s in 7643? Yes, $3 \cdot 1000 = 3000$. To find how many, we find products of 3 and multiples of 1000.

$$\begin{aligned} 3 \cdot 1000 &= 3000 \\ 3 \cdot 2000 &= 6000 \\ 3 \cdot 3000 &= 9000 \end{aligned}$$

← 7643 is here, so there are 2000 threes in 7643.

$$\begin{array}{r} 2000 \\ 3\,\overline{)\,7643} \\ \underline{6000} \\ 1643 \end{array}$$

Divide.

5. $4\,\overline{)\,385}$ 96 R 1

ANSWER ON PAGE A-7

Divide.

6. 7)8 8 4 6   1263 R 5

ANSWER ON PAGE A-7

b) Now go to the hundreds place in the first remainder. Are there any $3 \cdot 100$'s in the hundreds place of the quotient?

$3 \cdot 100 = 300$
$3 \cdot 200 = 600$
$3 \cdot 300 = 900$
$3 \cdot 400 = 1200$
$3 \cdot 500 = 1500$
$3 \cdot 600 = 1800$ ← 1643

```
      5 0 0
    2 0 0 0
 3 )7 6 4 3
    6 0 0 0
    -------
    1 6 4 3
    1 5 0 0
    -------
      1 4 3
```

c) Now go to the tens place in the second remainder. Are there any $3 \cdot 10$'s in the tens place of the quotient?

$3 \cdot 10 = 30$
$3 \cdot 20 = 60$
$3 \cdot 30 = 90$
$3 \cdot 40 = 120$
$3 \cdot 50 = 150$ ← 143

```
        4 0
      5 0 0
    2 0 0 0
 3 )7 6 4 3
    6 0 0 0
    -------
    1 6 4 3
    1 5 0 0
    -------
      1 4 3
      1 2 0
    -------
        2 3
```

d) Now go to the ones place in the third remainder. Are there any 3's in the ones place of the quotient?

$3 \cdot 1 = 3$
$3 \cdot 2 = 6$
$3 \cdot 3 = 9$
$3 \cdot 4 = 12$
$3 \cdot 5 = 15$
$3 \cdot 6 = 18$
$3 \cdot 7 = 21$ ← 23
$3 \cdot 8 = 24$

```
    2 5 4 7
    -------
          7
        4 0
      5 0 0
    2 0 0 0
 3 )7 6 4 3
    6 0 0 0
    -------
    1 6 4 3
    1 5 0 0
    -------
      1 4 3
      1 2 0
    -------
        2 3
        2 1
    -------
          2
```

The answer is 2547 R 2.

DO EXERCISE 6.

A Short Form

Here is a shorter way to write Example 4.

Instead of this,

```
      2 5 4 7
      -------
            7
          4 0
        5 0 0
      2 0 0 0
    ---------
  3 )7 6 4 3
     6 0 0 0
     -------
     1 6 4 3
     1 5 0 0
     -------
       1 4 3
       1 2 0
     -------
         2 3
         2 1
     -------
           2
```

Short form

we write this.

```
      2 5 4 7
    ---------
  3 )7 6 4 3
     6 0 0 0
     -------
     1 6 4 3
     1 5 0 0
     -------
       1 4 3
       1 2 0
     -------
         2 3
         2 1
     -------
           2
```

We write a 2 above the thousand digit in the dividend to record 2000. We write a 5 to record 500. We write a 4 to record 40. We write a 7 to record 7.

DO EXERCISES 7 AND 8.

▶ **EXAMPLE 5** Divide 2637 ÷ 41. Use the short form.

```
              6
         --------
  4 1 )2 6 3 7
       2 4 6 0
       -------
         1 7 7
```

```
              6 4
         --------
  4 1 )2 6 3 7
       2 4 6 0
       -------
         1 7 7
         1 6 4
       -------
           1 3
```

The answer is 64 R 13. ◀

DO EXERCISES 9 AND 10.

Divide using the short form.

7. $2\overline{)648}$ 324

8. $9\overline{)3758}$ 417 R 5

Divide.

9. $11\overline{)415}$ 37 R 8

10. $46\overline{)1075}$ 23 R 17

ANSWERS ON PAGE A-7

❖ SIDELIGHTS

Calculator Corner: Number Patterns

Look for a pattern in the following. We can use a calculator for the calculations.

$$\begin{aligned} 3 \times 5 &= 15 \\ 33 \times 5 &= 165 \\ 333 \times 5 &= 1665 \\ 3333 \times 5 &= 16{,}665 \end{aligned}$$

Do you see a pattern? If so, find $33{,}333 \times 5$ without the use of your calculator.

EXERCISES

In each of the following, do the first four calculations using a calculator. Look for a pattern. Then do the last calculation without your calculator.

1.
4×6 24
44×6 264
444×6 2664
4444×6 26,664
$44{,}444 \times 6$ 266,664

2.
9×9 81
99×89 8811
999×889 888,111
9999×8889 88,881,111
$99{,}999 \times 88{,}889$ 8,888,811,111

3.
77×78 6006
777×78 60,606
7777×78 606,606
$77{,}777 \times 78$ 6,066,606
$777{,}777 \times 78$ 60,666,606

4.
$1 \cdot 13 \cdot 76{,}923$ 999,999
$2 \cdot 13 \cdot 76{,}923$ 1,999,998
$3 \cdot 13 \cdot 76{,}923$ 2,999,997
$4 \cdot 13 \cdot 76{,}923$ 3,999,996
$5 \cdot 13 \cdot 76{,}923$ 4,999,995

NAME SECTION DATE

EXERCISE SET D.2

a Divide using the "guess, multiply, and subtract" method.

1. $4\overline{)277}$ **2.** $2\overline{)399}$ **3.** $8\overline{)737}$ **4.** $6\overline{)831}$

5. $5\overline{)8619}$ **6.** $3\overline{)8775}$ **7.** $9\overline{)7777}$ **8.** $8\overline{)4179}$

9. $7\overline{)3691}$ **10.** $2\overline{)5794}$ **11.** $20\overline{)875}$ **12.** $30\overline{)987}$

13. $21\overline{)999}$ **14.** $23\overline{)975}$ **15.** $85\overline{)7757}$ **16.** $54\overline{)2821}$

17. $111\overline{)3219}$ **18.** $102\overline{)5612}$

19. $346\overline{)78,910}$ **20.** $781\overline{)15,999}$

ANSWERS

1. 69 R 1
2. 199 R 1
3. 92 R 1
4. 138 R 3
5. 1723 R 4
6. 2925
7. 864 R 1
8. 522 R 3
9. 527 R 2
10. 2897
11. 43 R 15
12. 32 R 27
13. 47 R 12
14. 42 R 9
15. 91 R 22
16. 52 R 13
17. 29
18. 55 R 2
19. 228 R 22
20. 20 R 379

ANSWERS

21. 21

22. 118

23. 91 R 1

24. 321 R 2

25. 964 R 3

26. 679 R 5

27. 1328 R 2

28. 27 R 8

29. 23

30. 321 R 18

31. 224 R 10

32. 27 R 60

33. 87 R 1

34. 75

35. 22 R 85

36. 49 R 46

37. 383 R 91

38. 74 R 440

39. 2437 × 6

b Divide.

21. $5\overline{)105}$ **22.** $6\overline{)708}$ **23.** $9\overline{)820}$ **24.** $3\overline{)965}$

25. $5\overline{)4823}$ **26.** $8\overline{)5437}$ **27.** $7\overline{)9298}$ **28.** $41\overline{)1115}$

29. $46\overline{)1058}$ **30.** $24\overline{)7722}$ **31.** $38\overline{)8522}$ **32.** $81\overline{)2247}$

33. $72\overline{)6265}$ **34.** $74\overline{)5550}$

35. $94\overline{)2153}$ **36.** $82\overline{)4064}$

37. $117\overline{)44{,}902}$ **38.** $740\overline{)55{,}200}$

SYNTHESIS

39. Place the digits 2, 3, 4, 6, and 7 correctly in the blanks.

$$\begin{array}{r} (\_)(\_)(\_)(\_) \\ \times \quad (\_) \\ \hline 1\ 4{,}\ 6\ 2\ 2 \end{array}$$

Tables

TABLE 1

SQUARE ROOTS

N	$\sqrt{N}$	N	$\sqrt{N}$	N	$\sqrt{N}$	N	$\sqrt{N}$
2	1.414	27	5.196	52	7.211	77	8.775
3	1.732	28	5.292	53	7.280	78	8.832
4	2	29	5.385	54	7.348	79	8.888
5	2.236	30	5.477	55	7.416	80	8.944
6	2.449	31	5.568	56	7.483	81	9
7	2.646	32	5.657	57	7.550	82	9.055
8	2.828	33	5.745	58	7.616	83	9.110
9	3	34	5.831	59	7.681	84	9.165
10	3.162	35	5.916	60	7.746	85	9.220
11	3.317	36	6	61	7.810	86	9.274
12	3.464	37	6.083	62	7.874	87	9.327
13	3.606	38	6.164	63	7.937	88	9.381
14	3.742	39	6.245	64	8	89	9.434
15	3.873	40	6.325	65	8.062	90	9.487
16	4	41	6.403	66	8.124	91	9.539
17	4.123	42	6.481	67	8.185	92	9.592
18	4.243	43	6.557	68	8.246	93	9.644
19	4.359	44	6.633	69	8.307	94	9.695
20	4.472	45	6.708	70	8.367	95	9.747
21	4.583	46	6.782	71	8.426	96	9.798
22	4.690	47	6.856	72	8.485	97	9.849
23	4.796	48	6.928	73	8.544	98	9.899
24	4.899	49	7	74	8.602	99	9.950
25	5	50	7.071	75	8.660	100	10
26	5.099	51	7.141	76	8.718		

Answers

ARGIN EXERCISE ANSWERS

IAPTER 1

argin Exercises, Section 1.1, pp. 3-6

3 thousands + 7 hundreds + 2 tens + 8 ones
3 ten thousands + 6 thousands + 2 hundreds + 2 tens +
ones **3.** 3 thousands + 2 tens + 1 one
2 thousands + 9 ones **5.** 5 thousands + 7 hundreds
5689 **7.** 87,128 **8.** 9003 **9.** Fifty-seven
. Twenty-nine **11.** Eighty-eight **12.** Two hundred
ur **13.** Nineteen thousand, two hundred four **14.** One
illion, seven hundred nineteen thousand, two hundred
ur **15.** Twenty-two billion, three hundred one million,
ven hundred nineteen thousand, two hundred four
. 213,105,329 **17.** 2 ten thousands **18.** 2 hundred
ousands **19.** 2 millions **20.** 2 ten millions **21.** 4
. 7 **23.** 9 **24.** 0

argin Exercises, Section 1.2, pp. 9-12

4 + 6 = 10 **2.** \$15 + \$13 = \$28 **3.** 40 mi + 50 mi = 90 mi **4.** 5 ft + 7 ft = 12 ft **5.** 50 in^2 + 60 in^2 = 110 in^2 **6.** 200 mi^2 + 400 mi^2 = 600 mi^2 **7.** 10 gal + 18 gal = 28 gal **8.** 3000 tons + 7000 tons = 10,000 tons **9.** 9745 **10.** 13,465 **11.** 27 **12.** 27 **13.** 38 **14.** 61 **15.** 27,474

Margin Exercises, Section 1.3, pp. 15-18

1. 16 oz − 5 oz = 11 oz **2.** 400 acres − 100 acres = 300 acres **3.** 7 = 2 + 5 **4.** 17 = 9 + 8 **5.** 5 = 13 − 8 and 8 = 13 − 5 **6.** 11 = 14 − 3 and 3 = 14 − 11 **7.** 15 + □ = 32; □ = 32 − 15 **8.** 10 + □ = 23; □ = 23 − 10 **9.** 3801 **10.** 6328 **11.** 4747 **12.** 56 **13.** 205 **14.** 658 **15.** 2851 **16.** 1546

Margin Exercises, Section 1.4, pp. 21-24

1. 40 **2.** 50 **3.** 70 **4.** 100 **5.** 40 **6.** 80 **7.** 90 **8.** 140 **9.** 470 **10.** 240 **11.** 290 **12.** 600 **13.** 800 **14.** 800 **15.** 9300 **16.** 8000 **17.** 8000 **18.** 19,000 **19.** 69,000 **20.** 200 **21.** 1800 **22.** 5000 **23.** 11,000 **24.** $<$ **25.** $>$

26. > **27.** < **28.** < **29.** >

Margin Exercises, Section 1.5, pp. 27-32

1. $8 \times 4 = 32$ mi **2.** $10 \cdot 75 = 750$ mL **3.** $12 \cdot 20 = 240$ **4.** $4 \cdot 6 = 24$ yd^2 **5.** 1035 **6.** 3024 **7.** 46,252 **8.** 205,065 **9.** 144,432 **10.** 287,232 **11.** 14,075,720 **12.** 169,920 **13.** 17,345,600 **14. (a)** 1081; **(b)** 1081; (c) same **15.** 40 **16.** 15 **17.** 210,000; 160,000

Margin Exercises, Section 1.6, pp. 35-42

1. $112 \div 14 = \square$ **2.** $112 \div 8 = \square$ **3.** $15 = 5 \cdot 3$ **4.** $72 = 9 \cdot 8$ **5.** $6 = 12 \div 2$; $2 = 12 \div 6$ **6.** $6 = 42 \div 7$; $7 = 42 \div 6$ **7.** 6; $9 \cdot 6 = 54$ **8.** 6 R 7; $6 \cdot 9 = 54$, $54 + 7 = 61$ **9.** 4 R 5; $4 \cdot 12 = 48$, $48 + 5 = 53$ **10.** 6 R 13; $24 \cdot 6 = 144$, $144 + 13 = 157$ **11.** 59 R 3 **12.** 1475 R 5 **13.** 1015 **14.** 134 **15.** 63 R 12 **16.** 807 R 4 **17.** 1088 **18.** 360 R 4 **19.** 800 R 47

Margin Exercises, Section 1.7, pp. 45-48

1. 7 **2.** 5 **3.** No **4.** Yes **5.** 5 **6.** 10 **7.** 5 **8.** 22 **9.** 22,490 **10.** 9022 **11.** 570 **12.** 3661 **13.** 8 **14.** 45 **15.** 77 **16.** 3311 **17.** 6114 **18.** 8 **19.** 16 **20.** 644 **21.** 96 **22.** 94

Margin Exercises, Section 1.8, pp. 51-58

1. 659 **2.** 98 gal **3.** 158,578 sq mi **4.** 549 kWh **5.** \$521 **6.** 2260 gal **7.** 103° C **8.** \$1560 **9.** 2800 **10.** \$5184 **11.** 100,375 sq mi **12.** 4320 min **13.** 275 cartons; 5 left over **14.** 45 gal **15.** 70 min **16.** 106

Margin Exercises, Section 1.9, pp. 63-66

1. 5^4 **2.** 5^5 **3.** 10^2 **4.** 10^4 **5.** 10,000 **6.** 100 **7.** 512 **8.** 32 **9.** 51 **10.** 30 **11.** 584 **12.** 84 **13.** 4; 1 **14.** 52; 52 **15.** 29 **16.** 1880 **17.** 305 **18.** 93 **19.** 100; 52 **20.** 1880 **21.** 305 **22.** 93 **23.** 46 **24.** 4

CHAPTER 2

Margin Exercises, Section 2.1, pp. 75-80

1. 1, 2, 3, 6 **2.** 1, 2, 4, 8 **3.** 1, 2, 5, 10 **4.** 1, 2, 4, 8, 16, 32 **5.** $5 = 1 \cdot 5$, $45 = 9 \cdot 5$, $100 = 20 \cdot 5$ **6.** $10 = 1 \cdot 10$, $60 = 6 \cdot 10$, $110 = 11 \cdot 10$ **7.** 5, 10, 15, 20, 25, 30, 35, 40, 45, 50 **8.** Yes **9.** Yes **10.** No **11.** 13, 19, 41 are prime; 4, 6, 8 are composite; 1 is neither **12.** $2 \cdot 3$ **13.** $2 \cdot 2 \cdot 3$ **14.** $3 \cdot 3 \cdot 5$ **15.** $2 \cdot 7 \cdot 7$ **16.** $2 \cdot 3 \cdot 3 \cdot 7$ **17.** $2 \cdot 2 \cdot 2 \cdot 2 \cdot 3 \cdot 3$

Margin Exercises, Section 2.2, pp. 83-86

1. Yes **2.** No **3.** Yes **4.** No **5.** No **6.** Yes **7.** No **8.** Yes **9.** Yes **10.** No **11.** Yes **12.** No **13.** Yes **14.** No **15.** No **16.** Yes **17.** No **18.** Yes **19.** Yes **20.** No **21.** Yes **22.** No **23.** No **24.** Yes **25.** No **26.** Yes **27.** No **28.** Yes **29.** Yes **30.** No **31.** No **32.** Yes

Margin Exercises, Section 2.3, pp. 89-92

1. 1 numerator; 6 denominator **2.** 5 numerator; 7 denominator **3.** 22 numerator; 3 denominator **4.** $\frac{1}{2}$ **5.** $\frac{1}{3}$ **6.** $\frac{1}{3}$ **7.** $\frac{1}{6}$ **8.** $\frac{5}{8}$ **9.** $\frac{2}{3}$ **10.** $\frac{3}{4}$ **11.** $\frac{4}{6}$ **12.** $\frac{4}{3}$ **13.** $\frac{5}{5}$ **14.** $\frac{5}{4}$ **15.** $\frac{7}{4}$ **16.** $\frac{2}{5}$ **17.** $\frac{2}{3}$ **18.** $\frac{2}{6}, \frac{4}{6}$ **19.** 1 **20.** 1 **21.** 1 **22.** 1 **23.** 1 **24.** 1 **25.** 0 **26.** 0 **27.** 0 **28.** 0 **29.** 8 **30.** 10 **31.** 346 **32.** 1

Margin Exercises, Section 2.4, pp. 95-98

1. $\frac{2}{3}$ **2.** $\frac{5}{8}$ **3.** $\frac{10}{3}$ **4.** $\frac{33}{8}$ **5.** $\frac{46}{5}$ **6.** $\frac{15}{56}$ **7.** $\frac{32}{15}$ **8.** $\frac{3}{100}$ **9.** $\frac{14}{3}$

10.

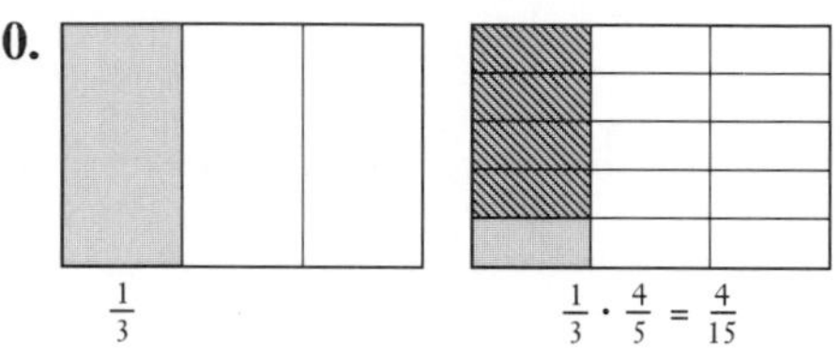

11. $\frac{3}{8}$ **12.** $\frac{63}{100}$ sq cm **13.** $\frac{3}{56}$

Margin Exercises, Section 2.5, pp. 101-104

1. $\frac{8}{16}$ **2.** $\frac{30}{50}$ **3.** $\frac{52}{100}$ **4.** $\frac{200}{75}$ **5.** $\frac{12}{9}$ **6.** $\frac{18}{24}$ **7.** $\frac{90}{100}$ **8.** $\frac{9}{45}$ **9.** $\frac{56}{49}$ **10.** $\frac{1}{4}$ **11.** $\frac{5}{6}$ **12.** 5 **13.** $\frac{4}{3}$ **14.** $\frac{7}{8}$ **15.** $\frac{89}{78}$ **16.** $\frac{8}{7}$ **17.** $\frac{1}{4}$ **18.** = **19.** $\neq$

Margin Exercises, Section 2.6, pp. 107-108

1. $\frac{7}{12}$ **2.** $\frac{1}{3}$ **3.** 6 **4.** $\frac{5}{2}$ **5.** 10 lb

Margin Exercises, Section 2.7, pp. 113-116

1. $\frac{5}{2}$ **2.** $\frac{7}{10}$ **3.** $\frac{1}{9}$ **4.** 5 **5.** $\frac{8}{7}$ **6.** $\frac{8}{3}$ **7.** $\frac{1}{10}$ **8.** 100 **9.** 1 **10.** $\frac{14}{15}$ **11.** $\frac{4}{5}$ **12.** 32 **13.** 320 **14.** 200 gal

CHAPTER 3

Margin Exercises, Section 3.1, pp. 127-132

1. 45 **2.** 40 **3.** 30 **4.** 24 **5.** 10 **6.** 80 **7.** 40 **8.** 360 **9.** 2520 **10.** 18 **11.** 24 **12.** 36 **13.** 210 **14.** 2520 **15.** 3780

Margin Exercises, Section 3.2, pp. 135-138

1. $\frac{4}{5}$ **2.** 1 **3.** $\frac{1}{2}$ **4.** $\frac{3}{4}$ **5.** $\frac{5}{6}$ **6.** $\frac{29}{24}$ **7.** $\frac{5}{9}$ **8.** $\frac{413}{1000}$ **9.** $\frac{759}{1000}$ **10.** $\frac{197}{210}$ **11.** $\frac{11}{10}$ lb

Margin Exercises, Section 3.3, pp. 141-144

1. $\frac{1}{2}$ **2.** $\frac{3}{8}$ **3.** $\frac{1}{2}$ **4.** $\frac{1}{12}$ **5.** $\frac{13}{18}$ **6.** $\frac{1}{2}$ **7.** $<$ **8.** $>$ **9.** $>$ **10.** $>$ **11.** $<$ **12.** $\frac{1}{6}$ **13.** $\frac{11}{40}$ **14.** $\frac{11}{20}$ cup

Margin Exercises, Section 3.4, pp. 147-150

1. $1\frac{2}{3}$ **2.** $8\frac{3}{4}$ **3.** $12\frac{2}{3}$ **4.** $\frac{22}{5}$ **5.** $\frac{61}{10}$ **6.** $\frac{29}{6}$ **7.** $\frac{37}{4}$ **8.** $\frac{62}{3}$ **9.** $2\frac{1}{3}$ **10.** $1\frac{1}{10}$ **11.** $18\frac{1}{3}$ **12.** $807\frac{2}{3}$ **13.** $134\frac{23}{45}$

Margin Exercises, Section 3.5, pp. 153-156

1. $7\frac{2}{5}$ **2.** $12\frac{1}{10}$ **3.** $13\frac{7}{12}$ **4.** $1\frac{1}{2}$ **5.** $3\frac{1}{6}$ **6.** $3\frac{1}{3}$ **7.** $3\frac{2}{3}$ **8.** $17\frac{1}{12}$ yd **9.** $3\frac{3}{4}$ m **10.** $23\frac{1}{4}$ gal

Margin Exercises, Section 3.6, pp. 161-164

1. 20 **2.** $1\frac{7}{8}$ **3.** $12\frac{4}{5}$ **4.** $8\frac{1}{3}$ **5.** 16 **6.** $7\frac{3}{7}$ **7.** $1\frac{7}{8}$ **8.** $\frac{7}{10}$ **9.** $227\frac{1}{2}$ mi **10.** 20 **11.** $240\frac{3}{4}$ sq ft

CHAPTER 4

Margin Exercises, Section 4.1, pp. 177-180

1. Twenty-seven and three tenths **2.** Two and four thousand five hundred thirty-three ten thousandths **3.** Two hundred forty-five and eighty-nine hundredths **4.** Thirty-one thousand, four hundred seventy-nine and seven hundred sixty-four thousandths **5.** Four thousand, two hundred seventeen and $\frac{56}{100}$ dollars **6.** Thirteen and $\frac{98}{100}$ dollars **7.** $\frac{896}{1000}$ **8.** $\frac{2378}{100}$ **9.** $\frac{56,789}{10,000}$ **10.** $\frac{19}{10}$ **11.** 7.43 **12.** 0.406 **13.** 6.7089 **14.** 0.9

Margin Exercises, Section 4.2, pp. 183-184

1. 2.04 **2.** 0.06 **3.** 0.58 **4.** 1 **5.** 0.8989 **6.** 21.05 **7.** 2.8 **8.** 13.9 **9.** 234.4 **10.** 7.0 **11.** 0.64 **12.** 7.83 **13.** 34.68 **14.** 0.03 **15.** 0.943 **16.** 8.004 **17.** 43.112 **18.** 37.401 **19.** 7459.355 **20.** 7459.35 **21.** 7459.4 **22.** 7459 **23.** 7460 **24.** 7500 **25.** 7000

Margin Exercises, Section 4.3, pp. 187-190

1. 10.917 **2.** 34.2079 **3.** 4.969 **4.** 3.5617 **5.** 9.40544 **6.** 912.67 **7.** 2514.773 **8.** 10.754 **9.** 0.339 **10.** 0.5345 **11.** 0.5172 **12.** 7.36992 **13.** 1194.22 **14.** 4.9911 **15.** 38.534 **16.** 14.164

Margin Exercises, Section 4.4, pp. 195-196

1. 137.5 gal **2.** $0.43 per pound

CHAPTER 5

Margin Exercises, Section 5.1, pp. 209-214

1. 348.2 **2.** 34.82 **3.** 3.482 **4.** 0.3482 **5.** 529.48 **6.** 5.0594 **7.** 34.2906 **8.** 8415 **9.** 0.056 **10.** 34,590.6 **11.** 3.45906 **12.** 0.00073 **13.** 730 **14.** 1569¢ **15.** 17¢ **16.** $0.35 **17.** $5.77 **18.** 17,000,000 **19.** 5,100,000,000

Margin Exercises, Section 5.2, pp. 217-222

1. 0.6 **2.** 1.5 **3.** 0.47 **4.** 0.32 **5.** 3.75 **6.** 0.25 **7. (a)** 375; **(b)** 15 **8.** 4.9 **9.** 12.8 **10.** 15.625 **11.** 12.78 **12.** 0.001278 **13.** 0.09847 **14.** 67.832 **15.** 0.78314 **16.** 1105.6 **17.** 0.2426 **18.** 728.44

Margin Exercises, Section 5.3, pp. 227-230

1. 0.8 **2.** 0.45 **3.** 0.275 **4.** 1.32 **5.** 0.4 **6.** 0.375 **7.** $0.1\overline{6}$ **8.** $0.\overline{6}$ **9.** $0.\overline{45}$ **10.** $1.\overline{09}$ **11.** $0.\overline{428571}$ **12.** 0.7; 0.67; 0.667 **13.** 0.8; 0.81; 0.808 **14.** 6.2; 6.25; 6.245 **15.** 0.72 **16.** 0.552

Margin Exercises, Section 5.4, pp. 233-234

1. b) **2.** a) **3.** d) **4.** b) **5.** a) **6.** d) **7.** b) **8.** c) **9.** b) **10.** b) **11.** c) **12.** a) **13.** c) **14.** c)

Margin Exercises, Section 5.5, pp. 237-240

1. $37.28 **2.** $368.75 **3.** 96.52 sq cm **4.** $0.35 **5.** 28.6 miles per gallon

CHAPTER 6

Margin Exercises, Section 6.1, pp. 253-256

1. 4 and 2, 10 and 5, 16 and 8; answers may vary **2.** 60 and 40, 12 and 8, 24 and 16; answers may vary **3.** $\frac{21.1}{182.5}$ **4.** $\frac{107}{365}$ **5.** $\frac{7}{11}$ **6.** $\frac{3.4}{0.189}$ **7.** $\frac{4}{3}$ **8.** $\frac{3}{4}$ **9.** $\frac{2964}{11,400}$ **10.** $\frac{4}{7\frac{2}{3}}$ **11.** Yes **12.** No **13.** No **14.** 14 **15.** $11\frac{1}{4}$ **16.** 10.5 **17.** 9 **18.** 10.8

Margin Exercises, Section 6.2, pp. 259-260

1. 5 km/h **2.** 12 km/h **3.** 0.3 km/h **4.** 1100 m/sec **5.** 4 m/sec **6.** 14.5 m/sec **7.** 250 ft/sec **8.** 2 gal/day **9.** 20.64 ¢/oz **10.** Can A

Margin Exercises, Section 6.3, pp. 265-268

1. 3360 km **2.** $110.50 **3.** 2.93 **4.** 21 cm **5.** 1620 **6.** 140 oz

CHAPTER 7

Margin Exercises, Section 7.1, pp. 279-282

1. $\frac{70}{100}$; $70\times\frac{1}{100}$; 70×0.01 **2.** $\frac{23.4}{100}$; $23.4\times\frac{1}{100}$; 23.4×0.01
3. $\frac{100}{100}$; $100\times\frac{1}{100}$; 100×0.01 **4.** 0.34 **5.** 0.789 **6.** 0.1208
7. 0.021 **8.** 24% **9.** 347% **10.** 100% **11.** 40%
12. 38%

Margin Exercises, Section 7.2, pp. 285-288

1. 25% **2.** 87.5%, or $87\frac{1}{2}\%$ **3.** $66.\overline{6}\%$, or $66\frac{2}{3}\%$
4. $83.\overline{3}\%$, or $83\frac{1}{3}\%$ **5.** 57% **6.** 76% **7.** $\frac{3}{5}$ **8.** $\frac{13}{400}$ **9.** $\frac{2}{3}$

10.

$\frac{1}{5}$	$\frac{5}{6}$	$\frac{3}{8}$
0.2	$0.83\overline{3}$	0.375
20%	$83.\overline{3}\%$ or $83\frac{1}{3}\%$	$37\frac{1}{2}\%$

Margin Exercises, Section 7.3, pp. 291-294

1. $12\%\times50=a$ **2.** $a=40\%\times60$ **3.** $45=20\%\times t$
4. $120\%\times y=60$ **5.** $16=n\times40$ **6.** $b\times84=10.5$ **7.** 6
8. \$35.20 **9.** 225 **10.** \$50 **11.** 40% **12.** 12.5%

Margin Exercises, Section 7.4, pp. 297-300

1. $\frac{12}{100}=\frac{a}{50}$ **2.** $\frac{40}{100}=\frac{a}{60}$ **3.** $\frac{20}{100}=\frac{45}{b}$ **4.** $\frac{120}{100}=\frac{60}{b}$ **5.** $\frac{n}{100}=\frac{16}{40}$
6. $\frac{n}{100}=\frac{10.5}{84}$ **7.** 6 **8.** 35.2 **9.** \$225 **10.** 50 **11.** 40%
12. 12.5%

Margin Exercises, Section 7.5, pp. 303-308

1. 44% **2.** 5 lb **3.** 9% **4.** 4% **5.** \$10,682

Margin Exercises, Section 7.6, pp. 313-314

1. \$22.14; \$391.09 **2.** 6% **3.** \$420

Margin Exercises, Section 7.7, pp. 317-320

1. \$5628 **2.** 25% **3.** \$1675 **4.** \$33.60; \$106.40

Margin Exercises, Section 7.8, pp. 323-326

1. \$602 **2.** \$451.50 **3.** \$56 **4.** \$2464.20 **5.** \$2101.25

CHAPTER 8

Margin Exercises, Section 8.1, pp. 337-340

1. 75 **2.** 54.9 **3.** 81 **4.** 19.4 **5.** 143.1 yd/game
6. 28 mpg **7.** 2.54 **8.** 94 **9.** 17 **10.** 17 **11.** 91 **12.** 17
13. 67.5 **14.** 45 **15.** 34; 67 **16.** 13; 24; 27; 28; 67; 89
17. **(a)** 82; **(b)** 81; **(c)** 74, 86, 96, 67, 82

Margin Exercises, Section 8.2, pp. 343-348

1. 2 P.M. on Sunday **2.** \$1.016 **3.** From 8 A.M. to 5 P.M. **4.** Dallas, Denver, and San Francisco
5. \$140-\$190 **6.** None **7.** 600 **8.** Iraq and Israel
9. 8.625×50, or 431.25 **10.** Brazil, France, USSR, US
11. USSR and US **12.** India and France
13.

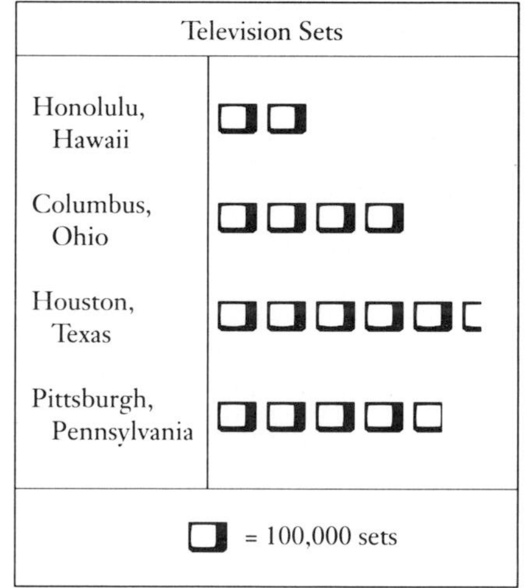

Margin Exercises, Section 8.3, pp. 353-358

1. Boredom **2.** Work or military service **3.** About 5%
4. Approximately \$17,000 **5.** Approximately \$14,000
6. Approximately \$6000
7.

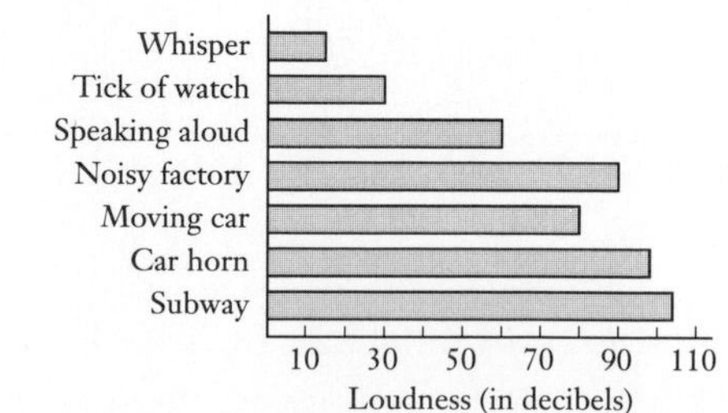

8. Sixth week **9.** Second week **10.** About 225
11. 1980 and 1982 **12.** Approximately \$50 **13.** 1985

14.

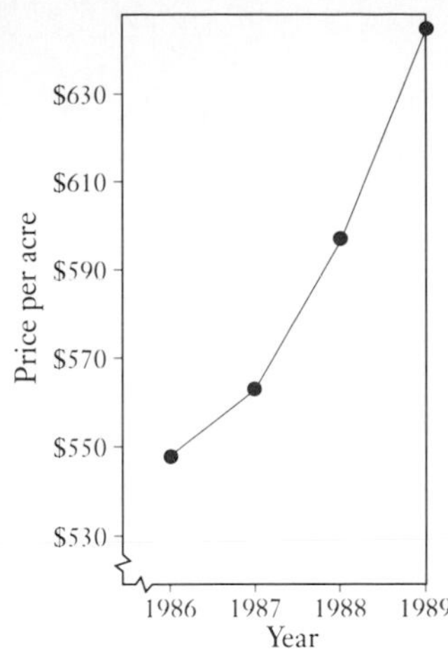

Margin Exercises, Section 8.4, pp. 363-366

1. \$540 **2.** 38% **3.** 6 to 1 **4.** **(a)** 173°; **(b)** 108°; **(c)** 29°; **(d)** 50°; **(e)**

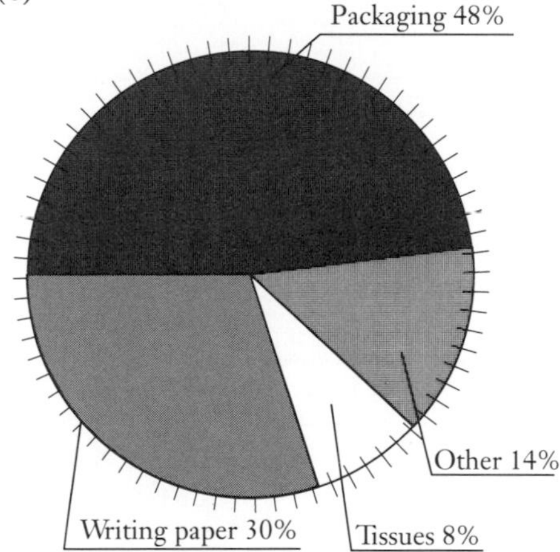

CHAPTER 9

Margin Exercises, Section 9.1, pp. 383-386

1. 2 **2.** 3 **3.** $1\frac{1}{2}$ **4.** $2\frac{1}{2}$ **5.** 288 **6.** 43.5 **7.** 6 **8.** $1\frac{5}{12}$ **9.** 8 **10.** 33 **11.** $11\frac{2}{3}$, or $11.\overline{6}$ **12.** 5 **13.** 31,680

Margin Exercises, Section 9.2, pp. 389-394

1. cm **2.** km **3.** mm **4.** m **5.** cm **6.** m **7.** 23,000 **8.** 400 **9.** 178 **10.** 9040 **11.** 7.814 **12.** 781.4 **13.** 0.967 **14.** 8,900,000 **15.** 6.78 **16.** 97.4 **17.** 0.1 **18.** 8.451 **19.** 90.909 **20.** 804.5 **21.** 1995.273

Margin Exercises, Section 9.3, pp. 397-398

1. 26 cm **2.** 46 mm **3.** 12 cm **4.** 17.5 yd **5.** 32 km **6.** 40 km **7.** 21 yd **8.** 31.2 km **9.** 70 ft; \$346.50

Margin Exercises, Section 9.4, pp. 401-402

1. 8 cm^2 **2.** 56 km^2 **3.** $18\frac{3}{8}$ yd^2 **4.** 144 km^2 **5.** 118.81 m^2 **6.** $12\frac{1}{4}$ yd^2 **7.** 659.75 m^2

Margin Exercises, Section 9.5, pp. 405-408

1. 43.8 cm^2 **2.** 12.375 km^2 **3.** 96 m^2 **4.** 18.7 cm^2 **5.** 100 m^2 **6.** 88 cm^2 **7.** 54 m^2

Margin Exercises, Section 9.6, pp. 411-414

1. 12 km **2.** 5 ft **3.** 62.8 **4.** 88 m **5.** 20.096 cm **6.** $78\frac{4}{7}$ km^2 **7.** 339.62 cm^2 **8.** The pizza pan by 13.04 in^2

Margin Exercises, Section 9.7, pp. 419-422

1. 81 **2.** 100 **3.** 121 **4.** 144 **5.** 169 **6.** 196 **7.** 225 **8.** 256 **9.** 289 **10.** 324 **11.** 625 **12.** 400 **13.** 3 **14.** 4 **15.** 11 **16.** 10 **17.** 9 **18.** 8 **19.** 18 **20.** 20 **21.** 15 **22.** 13 **23.** 1 **24.** 0 **25.** 2.236 **26.** 8.832 **27.** 12.961 **28.** $c = \sqrt{41}$; $c \approx 6.403$ **29.** $a = \sqrt{75}$; $a \approx 8.660$ **30.** $b = \sqrt{120}$; $b \approx 10.954$ **31.** $a = \sqrt{175}$; $a \approx 13.229$ **32.** $\sqrt{424}$ ft $\approx$ 20.591 ft

CHAPTER 10

Margin Exercises, Section 10.1, pp. 437-440

1. 12 cm^3 **2.** 38.4 m^3 **3.** 128 ft^3 **4.** 40 **5.** 20 **6.** mL **7.** mL **8.** L **9.** L **10.** 970 **11.** 8.99 **12.** 4.8 **13.** **(a)** 118.28 mL; **(b)** 0.11828 L **14.** \$1.60

Margin Exercises, Section 10.2, pp. 443-444

1. 785 ft^3 **2.** 67,914 m^3 **3.** $91{,}989\frac{1}{3}$ ft^3 **4.** 38.77272 cm^3 **5.** 1695.6 m^3 **6.** 528 in^3

Margin Exercises, Section 10.3, pp. 447-450

1. 80 **2.** 8640 **3.** 32,000 **4.** kg **5.** kg **6.** mg **7.** g **8.** t **9.** 6200 **10.** 0.3048 **11.** 77 **12.** 234.4 **13.** 6700 **14.** 7200 **15.** 1461 **16.** 1440 **17.** 168

Margin Exercises, Section 10.4, pp. 455-456

1. 80° C **2.** 0° C **3.** –20° C **4.** 80° F **5.** 100° F **6.** 50° F **7.** 176° F **8.** 95° F **9.** 35° C **10.** 45° C

Margin Exercises, Section 10.5, pp. 459-460

1. 9 **2.** 45 **3.** 2880 **4.** 2.5 **5.** 3200 **6.** 1,000,000 **7.** 100 **8.** 28,800 **9.** 0.043 **10.** 0.678

CHAPTER 11

Margin Exercises, Section 11.1, pp. 471-476

1. 8, −5 **2.** 134, −76 **3.** −10, 148 **4.** −137, 389

5. 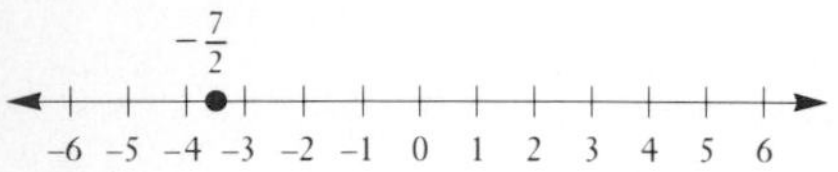

6. 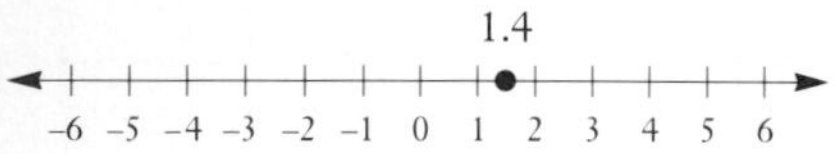

7. 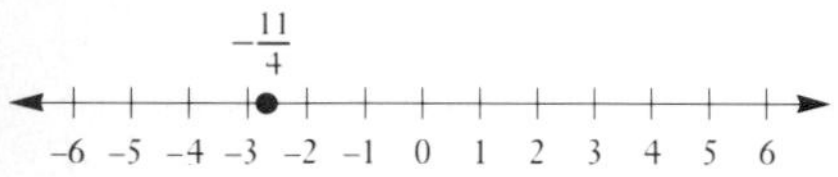

8. −0.375 **9.** $-0.\overline{54}$ **10.** $1.\overline{3}$ **11.** < **12.** < **13.** > **14.** > **15.** > **16.** < **17.** < **18.** > **19.** 8 **20.** 0 **21.** 9 **22.** $\frac{2}{3}$ **23.** 5.6

Margin Exercises, Section 11.2, pp. 479-482

1. −8 **2.** −3 **3.** −8 **4.** 4 **5.** 0 **6.** −2 **7.** −11 **8.** −12 **9.** 2 **10.** −4 **11.** −2 **12.** 0 **13.** −22 **14.** 3 **15.** 0.53 **16.** 2.3 **17.** −7.7 **18.** −6.2 **19.** $-\frac{2}{9}$ **20.** $-\frac{19}{20}$ **21.** −58 **22.** −56 **23.** −14 **24.** −12 **25.** 4 **26.** −8.7 **27.** 7.74 **28.** $\frac{8}{9}$ **29.** 0 **30.** −12 **31.** −14, 14 **32.** −1, 1 **33.** 19, −19 **34.** 1.6, −1.6 **35.** $-\frac{2}{3}, \frac{2}{3}$ **36.** $\frac{9}{8}, -\frac{9}{8}$ **37.** 4 **38.** 13.4 **39.** 0 **40.** $-\frac{1}{4}$

Margin Exercises, Section 11.3, pp. 485-486

1. −10 **2.** 3 **3.** −5 **4.** −2 **5.** −11 **6.** 4 **7.** −2 **8.** −6 **9.** −16 **10.** 7.1 **11.** 3 **12.** 0 **13.** $\frac{3}{2}$ **14.** −8 **15.** 7 **16.** −3 **17.** −23.3 **18.** 0 **19.** −9 **20.** 17 **21.** 12.7

Margin Exercises, Section 11.4, pp. 489-490

1. $2 \cdot 10 = 20$; $1 \cdot 10 = 10$; $0 \cdot 10 = 0$; $-1 \cdot 10 = -10$; $-2 \cdot 10 = -20$; $-3 \cdot 10 = -30$ **2.** −18 **3.** −100 **4.** −80 **5.** $-\frac{5}{9}$ **6.** −30.033 **7.** $-\frac{7}{10}$ **8.** $1 \cdot (-10) = -10$; $0 \cdot (-10) = 0$; $-1 \cdot (-10) = 10$; $-2 \cdot (-10) = 20$; $-3 \cdot (-10) = 30$ **9.** 12 **10.** 32 **11.** 35 **12.** $\frac{20}{63}$ **13.** $\frac{2}{3}$ **14.** 13.455 **15.** −30 **16.** 30 **17.** −32 **18.** $-\frac{8}{3}$ **19.** −30 **20.** −30.75 **21.** $-\frac{5}{3}$ **22.** 120 **23.** −120 **24.** 6

Margin Exercises, Section 11.5, pp. 493-496

1. −2 **2.** 5 **3.** −3 **4.** 9 **5.** −6 **6.** $-\frac{30}{7}$ **7.** Undefined **8.** 0 **9.** $\frac{3}{2}$ **10.** $-\frac{4}{5}$ **11.** $-\frac{1}{3}$ **12.** −5 **13.** $\frac{1}{5.78}$ **14.** $-\frac{3}{2}$

15.

Number	Opposite	Reciprocal
$\frac{2}{3}$	$-\frac{2}{3}$	$\frac{3}{2}$
$-\frac{5}{4}$	$\frac{5}{4}$	$-\frac{4}{5}$
0	0	Undefined
1	−1	1
−4.5	4.5	$-\frac{1}{4.5}$

16. $\frac{4}{7} \cdot \left(-\frac{5}{3}\right)$ **17.** $5 \cdot \left(-\frac{1}{8}\right)$ **18.** $-10 \cdot \left(\frac{1}{7}\right)$ **19.** $-\frac{2}{3} \cdot \frac{7}{4}$ **20.** $-5 \cdot \left(\frac{1}{7}\right)$ **21.** $-\frac{20}{21}$ **22.** $-\frac{12}{5}$ **23.** $\frac{16}{7}$ **24.** −7 **25.** −1237 **26.** 381

CHAPTER 12

Margin Exercises, Section 12.1, pp. 507-512

1. 64 **2.** 28 **3.** 60 **4.** 25 **5.** 16

6.

	$1 \cdot x$	x
$x = 3$	3	3
$x = -6$	−6	−6
$x = 4.8$	4.8	4.8

7.

	$2x$	$5x$
$x = 2$	4	10
$x = -6$	−12	−30
$x = 4.8$	9.6	24

8. 36; 36 **9.** 90; 90 **10.** 64; 64 **11.** 14; 14 **12.** 30; 30 **13.** 12; 12 **14.** $5x$; $-4y$; 3 **15.** $-4y$; $-2x$; $3z$ **16.** $3x - 15$ **17.** $5x - 5y + 20$ **18.** $-2x + 6$ **19.** $-5x + 10y - 20z$ **20.** $6(x - 2)$ **21.** $3(x - 2y + 3)$ **22.** $2(8a - 18b + 21)$ **23.** $-4(3x - 8y + 4z)$ **24.** $3x$ **25.** $6x$ **26.** $-8x$ **27.** $0.59x$ **28.** $3x + 3y$ **29.** $-4x - 5y - 7$

Margin Exercises, Section 12.2, pp. 515-516

1. −5 **2.** 13.2 **3.** −6.5

Margin Exercises, Section 12.3, pp. 519-520

1. $-\frac{4}{5}$ **2.** $-\frac{7}{4}$ **3.** 8 **4.** −18

Margin Exercises, Section 12.4, pp. 523-528

1. 5 **2.** 4 **3.** 4 **4.** 39 **5.** $-\frac{3}{2}$ **6.** −4.3 **7.** −3 **8.** 800 **9.** 1 **10.** 2 **11.** 2 **12.** $\frac{17}{2}$ **13.** $\frac{8}{3}$ **14.** −4.3

Margin Exercises, Section 12.5, pp. 531-536

1. $x - 12$ **2.** $y + 12$, or $12 + y$ **3.** $m - 4$ **4.** $\frac{1}{2} \cdot p$ **5.** $6 + 8x$, or $8x + 6$ **6.** $a - b$ **7.** $59\%x$, or $0.59x$ **8.** $xy - 200$ **9.** $p + q$ **10.** 28 in., 30 in. **11.** 5 **12.** 240.9 mi **13.** Width: 9 ft; length: 12 ft

DEVELOPMENTAL UNITS

Margin Exercises, Section A.1, pp. 552-554

1. 9 **2.** 7 **3.** 14 **4.** 16 **5.** 16 **6.** 16 **7.** 8 **8.** 8 **9.** 7 **10.** 46 **11.** 13 **12.** 58

13.

+	6	5	7	4	9
7	13	12	14	11	16
9	15	14	16	13	18
5	11	10	12	9	14
8	14	13	15	12	17
4	10	9	11	8	13

14. 16 **15.** 9 **16.** 11 **17.** 15

Margin Exercises, Section A.2, pp. 557-560

1. 59 **2.** 549 **3.** 9979 **4.** 5496 **5.** 56 **6.** 85 **7.** 829 **8.** 1026 **9.** 12,698 **10.** 13,661 **11.** 71,328

Margin Exercises, Section A.3, pp. 563-564

1. $\overset{8}{{}_{1}6_{4}}$ **2.** $\overset{8}{{}_{1}3_{1}}$ **3.** $\overset{4}{3_{7}}$ **4.** $\overset{5}{0_{5}}$ **5.** 31 **6.** 32 **7.** 249 **8.** 219 **9.** 2104 **10.** 14,844

Margin Exercise, Section S.1, pp. 567-568

1. 4 **2.** 7 **3.** 8 **4.** 3 **5.** $12 - 8 = 4$; $12 - 4 = 8$ **6.** $13 - 6 = 7$; $13 - 7 = 6$ **7.** 8 **8.** 7 **9.** 9 **10.** 4

Margin Exercise, Section S.2, pp. 571-574

1. 14 **2.** 20 **3.** 241 **4.** 2025 **5.** 17 **6.** 36 **7.** 454 **8.** 250 **9.** 376 **10.** 245 **11.** 2557 **12.** 3674 **13.** 8 **14.** 6 **15.** 30 **16.** 90 **17.** 500 **18.** 900 **19.** 538 **20.** 677 **21.** 42 **22.** 224 **23.** 138 **24.** 44 **25.** 2768 **26.** 1197 **27.** 1246 **28.** 4728

Margin Exercise, Section S.3, pp. 577-578

1. 27 **2.** 26 **3.** 359 **4.** 3263 **5.** 6354 **6.** 2552 **7.** 3350

Margin Exercises, Section M.1, pp. 581-582

1. 56 **2.** 36 **3.** 28 **4.** 42 **5.** 0 **6.** 0 **7.** 8 **8.** 23

9.

×	6	7	8	9
5	30	35	40	45
6	36	42	48	54
7	42	49	56	63
8	48	56	64	72
9	54	63	72	81

Margin Exercises, Section M.2, pp. 585-588

1. 70 **2.** 450 **3.** 2730 **4.** 100 **5.** 1000 **6.** 560 **7.** 360 **8.** 700 **9.** 2300 **10.** 72,300 **11.** 10,000 **12.** 100,000 **13.** 5600 **14.** 1600 **15.** 9000 **16.** 852,000 **17.** 10,000 **18.** 100,000 **19.** 1,000,000 **20.** 54,000 **21.** 5600 **22.** 56,000 **23.** 18,000 **24.** 28 **25.** 116 **26.** 148 **27.** 4938 **28.** 6740 **29.** 116 **30.** 148 **31.** 4938 **32.** 6740 **33.** 5968 **34.** 59,680 **35.** 596,800

Margin Exercises, Section M.3, pp. 591-598

1. 3_2 **2.** 3_0 **3.** 0_9 **4.** 204 **5.** 4781 **6.** 35,724 **7.** 267,408

Margin Exercises, Section D.1, pp. 595-598

1. 4 **2.** 8 **3.** 9 **4.** 3 **5.** $12 \div 2 = 6$; $12 \div 6 = 2$ **6.** $42 \div 6 = 7$; $42 \div 7 = 6$ **7.** 7 **8.** 9 **9.** 8 **10.** 9 **11.** 6 **12.** 13 **13.** 1 **14.** 2 **15.** Not defined **16.** 0 **17.** Not defined **18.** Not defined **19.** 10 **20.** Not defined **21.** 0 **22.** 1 **23.** 1 **24.** 1 **25.** 1 **26.** 17 **27.** 1

Margin Exercises, Section D.2, pp. 601-606

1. 75 R 4 **2.** 23 R 11 **3.** 969 R 6 **4.** 47 R 4 **5.** 96 R 1 **6.** 1263 R 5 **7.** 324 **8.** 417 R 5 **9.** 37 R 8 **10.** 23 R 17

EXERCISE SET AND TEST ANSWERS

Book Diagnostic Pretest, p. xxiii

1. 1807 **2.** 29 **3.** 15,087 **4.** 21 glasses, 2 oz left over **5.** $\frac{3}{10}$ **6.** $\frac{3}{2}$ **7.** $2 \cdot 2 \cdot 2 \cdot 2 \cdot 3 \cdot 3$ **8.** $\frac{1}{2}$ cup **9.** $\frac{13}{24}$ **10.** $15\frac{2}{5}$ **11.** $1\frac{3}{5}$ m **12.** 30 **13.** 0.0001 **14.** 25.6 **15.** 39.815 **16.** 186.9 mi **17.** 0.03 **18.** $2.\overline{3}$ **19.** 2.4 **20.** \$119.95 **21.** 4.5 **22.** 12 **23.** 10.0 ¢/oz **24.** 12.5% **25.** 0.0135 **26.** 24% **27.** \$19.55 **28.** Average: $24.\overline{3}$; median: 25;

mode: 25 **29.** 18.5 **30.** 85 **31.** 72 **32.** 0.00004 **33.** 36π cm^2; 12π cm **34.** 7.5 ft^2; 11 ft **35.** 120 **36.** 0.005 **37.** 18 **38.** 2.5 **39.** 3052.08 cm^3 **40.** 4.2 **41.** $-0.\overline{4}$ **42.** -0.1 **43.** $-\frac{1}{12}$ **44.** -3 **45.** $\frac{4}{5}$ **46.** \$25.56 **47.** 8

CHAPTER 1

Pretest: Chapter 1, p. 2

1. Three million, seventy-eight thousand, fifty-nine **2.** 6 thousands + 9 hundreds + 8 tens + 7 ones **3.** 2,047,398,589 **4.** 6 ten thousands **5.** 956,000 **6.** 60,000 **7.** 10,216 **8.** 4108 **9.** 22,976 **10.** 503 R 11 **11.** < **12.** > **13.** 5542 **14.** 22 **15.** 34 **16.** 25 **17.** 12 lb **18.** 126 **19.** 22,216,100 **20.** 2292 sq ft **21.** 25 **22.** 64 **23.** 0 **24.** 0

Exercise Set 1.1, p. 7

1. 5 thousands + 7 hundreds + 4 tens + 2 ones **3.** 2 ten thousands + 7 thousands + 3 hundreds + 4 tens + 2 ones **5.** 9 thousands + 1 ten **7.** 2 thousands + 3 hundreds **9.** 2475 **11.** 68,939 **13.** 7304 **15.** 1009 **17.** Seventy-seven **19.** Eighty-eight thousand **21.** One hundred twenty-three thousand, seven hundred sixty-five **23.** Seven million, seven hundred fifty-four thousand, two hundred eleven **25.** Two hundred forty-four million, eight hundred thirty-nine thousand, seven hundred seventy-two **27.** One million, nine hundred fifty-four thousand, one hundred sixteen **29.** 2,233,812 **31.** 8,000,000,000 **33.** 217,503 **35.** 2,173,638 **37.** 206,658,000 **39.** 5 thousands **41.** 5 hundreds **43.** 3 **45.** 0 **47.** All 9's as digits. Answers may vary. For an 8-digit readout, it would be 99,999,999.

Exercise Set 1.2, p. 13

1. 3 yards + 6 yards = 9 yards **3.** \$23 + \$31 = \$54 **5.** 387 **7.** 4998 **9.** 164 **11.** 1000 **13.** 1110 **15.** 1010 **17.** 1201 **19.** 847 **21.** 10,139 **23.** 6608 **25.** 16,784 **27.** 34,432 **29.** 101,310 **31.** 100,111 **33.** 28 **35.** 25 **37.** 35 **39.** 87 **41.** 230 **43.** 130 **45.** 149 **47.** 169 **49.** 842 **51.** 11,679 **53.** 22,654 **55.** 12,765,097 **57.** 7992 **59.** $1 + 99 = 100$, $2 + 98 = 100$, and so on, ..., $49 + 51 = 100$. Then $49 \cdot 100 = 4900$ and $4900 + 50 + 100 = 5050$.

Exercise Set 1.3, p. 19

1. $2400 - 800 = \square$ **3.** $10 = 3 + 7$ **5.** $13 = 5 + 8$ **7.** $23 = 14 + 9$ **9.** $43 = 27 + 16$ **11.** $6 = 15 - 9$; $9 = 15 - 6$ **13.** $8 = 15 - 7$; $7 = 15 - 8$ **15.** $17 = 23 - 6$; $6 = 23 - 17$ **17.** $23 = 32 - 9$; $9 = 32 - 23$ **19.** $190 + \square = 220$; $\square = 220 - 190$ **21.** 12 **23.** 44 **25.** 533 **27.** 1126 **29.** 39 **31.** 298 **33.** 226 **35.** 234 **37.** 5382 **39.** 1493 **41.** 2187 **43.** 3831 **45.** 7748 **47.** 33,794 **49.** 56 **51.** 36 **53.** 84 **55.** 454 **57.** 771 **59.** 2191 **61.** 3749 **63.** 1053 **65.** 4206 **67.** 10,305 **69.** 7 ten thousands

Exercise Set 1.4, p. 25

1. 50 **3.** 70 **5.** 730 **7.** 900 **9.** 100 **11.** 1000 **13.** 3600 **15.** 2900 **17.** 6000 **19.** 8000 **21.** 45,000 **23.** 373,000 **25.** 17,600 **27.** 5130 **29.** 220; incorrect **31.** 890; incorrect **33.** 17,500 **35.** 5200 **37.** 1600 **39.** 1500 **41.** 31,000 **43.** 69,000 **45.** < **47.** > **49.** < **51.** > **53.** > **55.** > **57.** 86,754 **59.** 30,411 **61.** 69,594

Sidelight: Number Patterns: Magic Squares, p. 32

1. First row: 8, 3, 4; second row: 1, 5, 9; third row: 6, 7, 2 **2.** First row: 1, 12, 14, 7; second row: 4, 15, 9, 6; third row: 13, 2, 8, 11; fourth row: 16, 5, 3, 10 **3.** 78 should be 79 **4.** 18 should be 17

Exercise Set 1.5, p. 33

1. $32 \cdot \$10 = \$320$ **3.** $8 \cdot 8 = 64$ **5.** $3 \cdot 6 = 18$ ft^2 **7.** 870 **9.** 2,340,000 **11.** 520 **13.** 564 **15.** 1527 **17.** 64,603 **19.** 4770 **21.** 3120 **23.** 46,080 **25.** 14,652 **27.** 207,672 **29.** 503,076 **31.** 166,260 **33.** 11,794,332 **35.** 20,723,872 **37.** 362,128 **39.** 20,064,048 **41.** 25,236,000 **43.** 302,220 **45.** 49,101,136 **47.** 30,525 **49.** 298,738 **51.** $50 \cdot 70 = 3500$ **53.** $30 \cdot 30 = 900$ **55.** $900 \cdot 300 = 270{,}000$ **57.** $400 \cdot 200 = 80{,}000$ **59.** $6000 \cdot 5000 = 30{,}000{,}000$ **61.** $8000 \cdot 6000 = 48{,}000{,}000$ **63.** 4370 **65.** 2350; 2300; 2000

Exercise Set 1.6, p. 43

1. $176 \div 4 = \square$ **3.** $\$184{,}000 \div \$23{,}000 = \square$ **5.** $24 = 3 \cdot 8$ **7.** $22 = 1 \cdot 22$ **9.** $54 = 9 \cdot 6$ **11.** $37 = 37 \cdot 1$ **13.** $9 = 45 \div 5$; $5 = 45 \div 9$ **15.** $37 = 37 \div 1$; $1 = 37 \div 37$ **17.** $8 = 64 \div 8$ **19.** $11 = 66 \div 6$; $6 = 66 \div 11$ **21.** 55 R 2 **23.** 108 **25.** 307 **27.** 906 R 3 **29.** 74 R 1 **31.** 92 R 2 **33.** 1703 **35.** 987 R 5 **37.** 52 R 52 **39.** 29 R 5 **41.** 40 R 12 **43.** 90 R 22 **45.** 29 **47.** 105 R 3 **49.** 507 R 1 **51.** 1007 R 1 **53.** 23 **55.** 107 R 1 **57.** 370 **59.** 609 R 15 **61.** 304 **63.** 3508 R 219 **65.** 8070 **67.** 7 thousands + 8 hundreds + 8 tens + 2 ones **69.** 9, with 2 left over (The butts from the first seven cigarettes are recycled.)

Exercise Set 1.7, p. 49

1. 14 **3.** 0 **5.** 29 **7.** 0 **9.** 8 **11.** 14 **13.** 1035 **15.** 25 **17.** 450 **19.** 90,900 **21.** 32 **23.** 143 **25.** 79 **27.** 45 **29.** 324 **31.** 743 **33.** 37 **35.** 66 **37.** 15 **39.** 48 **41.** 175 **43.** 335 **45.** 104 **47.** 45 **49.** 4056 **51.** 17,603 **53.** 18,252 **55.** 205 **57.** 55 **59.** $6 = 48 \div 8$; $8 = 48 \div 6$ **61.** >

Exercise Set 1.8, p. 59

1. 5693 **3.** 449 m **5.** 2995 cubic centimeters **7.** 100 cubic centimeters **9.** 304,000 **11.** \$91 **13.** 665 cal **15.** 3600 sec **17.** 2808 sq ft **19.** 7815 mi **21.** \$1638 **23.** 5130 sq yd **25.** 38 **27.** 15 **29.** \$27 **31.** 38 bags; 11 kg left over **33.** 16 **35.** 11 in.; 770 mi **37.** 480 **39.** 525 min, or 8 hr 45 min **41.** 186,000 mi

Sidelight: Palindrome Numbers, p. 66

1. 11,011 **2.** 5115

Exercise Set 1.9, p. 67

1. 3^4 **3.** 5^2 **5.** 7^5 **7.** 10^3 **9.** 49 **11.** 729 **13.** 20,736 **15.** 121 **17.** 22 **19.** 20 **21.** 100 **23.** 1 **25.** 49 **27.** 27 **29.** 434 **31.** 41 **33.** 88 **35.** 4 **37.** 303 **39.** 20 **41.** 70 **43.** 295 **45.** 32 **47.** 906 **49.** 62 **51.** 102 **53.** 110 **55.** 7 **57.** 544 **59.** 708 **61.** 24; $1+5\cdot(4+3)=36$ **63.** 7; $12\div(4+2)\cdot 3-2=4$

Summary and Review: Chapter 1, p. 69

1. [1.1a] 2 thousands + 7 hundreds + 9 tens + 3 ones **2.** [1.1c] Two million, seven hundred eighty-one thousand, four hundred twenty-seven **3.** [1.1e] 7 ten thousands **4.** [1.1d] \$2,626,100,000,000 **5.** [1.2b] 5979 **6.** [1.2b] 66,024 **7.** [1.2b] 22,098 **8.** [1.2b] 98,921 **9.** [1.3d] 1153 **10.** [1.3d] 1147 **11.** [1.3d] 2274 **12.** [1.3d] 17,757 **13.** [1.3d] 444 **14.** [1.3d] 4766 **15.** [1.5b] 420,000 **16.** [1.5b] 6,276,800 **17.** [1.5b] 684 **18.** [1.5b] 44,758 **19.** [1.5b] 3404 **20.** [1.5b] 506,748 **21.** [1.5b] 27,589 **22.** [1.5b] 3,456,000 **23.** [1.6c] 5 **24.** [1.6c] 12 R 3 **25.** [1.6c] 80 **26.** [1.6c] 207 R 2 **27.** [1.6c] 384 R 1 **28.** [1.6c] 4 R 46 **29.** [1.6c] 54 **30.** [1.6c] 452 **31.** [1.6c] 5008 **32.** [1.6c] 4389 **33.** [1.7b] 45 **34.** [1.7b] 546 **35.** [1.7b] 8 **36.** [1.8a] 1982 **37.** [1.8a] 2785 bu **38.** [1.8a] \$19,748 **39.** [1.8a] 2825 cal **40.** [1.8a] \$501 **41.** [1.8a] 152 beakers, 17 L left over **42.** [1.8a] 10 **43.** [1.4a] 345,800 **44.** [1.4a] 345,760 **45.** [1.4a] 346,000 **46.** [1.4b] $41,300+19,700=61,000$ **47.** [1.4b] $38,700-24,500=14,200$ **48.** [1.5c] $400\cdot 700=280,000$ **49.** [1.4c] $>$ **50.** [1.4c] $<$ **51.** [1.9a] 8^3 **52.** [1.9b] 16 **53.** [1.9b] 36 **54.** [1.9c] 65 **55.** [1.9c] 233 **56.** [1.9c] 56 **57.** [1.9c] 32 **58.** [1.9d] 260

Test: Chapter 1, p. 71

1. [1.1a] 8 thousands + 8 hundreds + 4 tens + 3 ones **2.** [1.1c] Thirty-eight million, four hundred three thousand, two hundred seventy-seven **3.** [1.1e] 5 **4.** [1.2b] 9989 **5.** [1.2b] 63,791 **6.** [1.2b] 34 **7.** [1.2b] 10,515 **8.** [1.3d] 3630 **9.** [1.3d] 1039 **10.** [1.3d] 6848 **11.** [1.3d] 5175 **12.** [1.5b] 41,112 **13.** [1.5b] 5,325,600 **14.** [1.5b] 2405 **15.** [1.5b] 534,264 **16.** [1.6c] 3 R 3 **17.** [1.6c] 70 **18.** [1.6c] 97 **19.** [1.6c] 805 R 8 **20.** [1.8a] 1955 **21.** [1.8a] 92 packages, 3 left over **22.** [1.8a] 18 **23.** [1.8a] 120,000 sq m **24.** [1.8a] 1808 lb **25.** [1.8a] 20 **26.** [1.8a] 305 sq mi **27.** [1.8a] 56 **28.** [1.8a] 66,444 sq mi **29.** [1.8a] \$271 **30.** [1.7b] 46 **31.** [1.7b] 13 **32.** [1.7b] 14 **33.** [1.4a] 35,000 **34.** [1.4a] 34,580 **35.** [1.4a] 34,600 **36.** [1.4b] $23,600+54,700=78,300$ **37.** [1.4b] $54,800-23,600=31,200$ **38.** [1.5c] $800\cdot 500=400,000$ **39.** [1.4c] $>$ **40.** [1.4c] $<$ **41.** [1.9a] 12^4 **42.** [1.9b] 343 **43.** [1.9b] 8 **44.** [1.9c] 64 **45.** [1.9c] 96 **46.** [1.9c] 2 **47.** [1.9d] 216 **48.** [1.9c] 18

CHAPTER 2

Pretest: Chapter 2, p. 74

1. Prime **2.** $2\cdot 2\cdot 5\cdot 7$ **3.** No **4.** Yes **5.** 1 **6.** 68 **7.** 0 **8.** $\frac{1}{4}$ **9.** $\frac{6}{5}$ **10.** 20 **11.** $\frac{5}{4}$ **12.** $\frac{8}{7}$ **13.** $\frac{1}{11}$ **14.** 24 **15.** $\frac{3}{4}$ **16.** 30 **17.** $\neq$ **18.** \$36 **19.** $\frac{1}{24}$ m

Sidelight: Factors and Sums, p. 80

First row: 48, 90, 432, 63; second row: 7, 18, 36, 14, 12, 6, 21, 11; third row: 9, 2, 2, 10, 8, 10, 21; fourth row: 29, 19, 42

Exercise Set 2.1, p. 81

1. 1, 2, 4, 8, 16 **3.** 1, 2, 3, 6, 9, 18, 27, 54 **5.** 1, 2, 4 **7.** 1, 7 **9.** 1 **11.** 1, 2, 7, 14, 49, 98 **13.** 4, 8, 12, 16, 20, 24, 28, 32, 36, 40 **15.** 20, 40, 60, 80, 100, 120, 140, 160, 180, 200 **17.** 3, 6, 9, 12, 15, 18, 21, 24, 27, 30 **19.** 12, 24, 36, 48, 60, 72, 84, 96, 108, 120 **21.** 10, 20, 30, 40, 50, 60, 70, 80, 90, 100 **23.** 9, 18, 27, 36, 45, 54, 63, 72, 81, 90 **25.** No **27.** Yes **29.** Yes **31.** No **33.** No **35.** Neither **37.** Composite **39.** Prime **41.** Prime **43.** $2\cdot 2\cdot 2$ **45.** $2\cdot 7$ **47.** $2\cdot 11$ **49.** $5\cdot 5$ **51.** $2\cdot 5\cdot 5$ **53.** $13\cdot 13$ **55.** $2\cdot 2\cdot 5\cdot 5$ **57.** $5\cdot 7$ **59.** $2\cdot 2\cdot 2\cdot 3\cdot 3$ **61.** $7\cdot 11$ **63.** $2\cdot 2\cdot 2\cdot 2\cdot 7$ **65.** $2\cdot 2\cdot 3\cdot 5\cdot 5$ **67.** 26 **69.** 0 **71.** A rectangular array of 6 rows of 9 objects each, or 9 rows of 6 objects each

Exercise Set 2.2, p. 87

1. 46; 300; 224; 36; 45,270; 4444 **3.** 300; 224; 36; 4444 **5.** 300; 36; 45,270 **7.** 36; 711; 45,270 **9.** 75; 324; 42; 501; 3009; 2001 **11.** 200; 75; 2345; 55,555 **13.** 324 **15.** 200 **17.** 138 **19.** \$680 **21.** $2\cdot 2\cdot 2\cdot 3\cdot 5\cdot 5\cdot 13$ **23.** $2\cdot 2\cdot 3\cdot 3\cdot 7\cdot 11$

Exercise Set 2.3, p. 93

1. 3 numerator; 4 denominator **3.** 11 numerator; 20 denominator **5.** $\frac{2}{4}$ **7.** $\frac{1}{8}$ **9.** $\frac{2}{3}$ **11.** $\frac{3}{4}$ **13.** $\frac{4}{8}$ **15.** $\frac{6}{12}$ **17.** $\frac{5}{8}$ **19.** $\frac{3}{5}$ **21.** 0 **23.** 5 **25.** 1 **27.** 1 **29.** 0 **31.** 1 **33.** 1 **35.** 1 **37.** 1 **39.** 8 **41.** 34,560 **43.** 35,000 **45.** 201 min **47.** $\frac{1200}{2700}; \frac{540}{2700}; \frac{360}{2700}; \frac{600}{2700}$

Exercise Set 2.4, p. 99

1. $\frac{3}{5}$ **3.** $\frac{5}{6}$ **5.** $\frac{8}{11}$ **7.** $\frac{70}{9}$ **9.** $\frac{2}{5}$ **11.** $\frac{6}{5}$ **13.** $\frac{21}{4}$ **15.** $\frac{85}{6}$ **17.** $\frac{1}{6}$ **19.** $\frac{1}{40}$ **21.** $\frac{2}{15}$ **23.** $\frac{4}{15}$ **25.** $\frac{9}{16}$ **27.** $\frac{14}{39}$ **29.** $\frac{7}{100}$ **31.** $\frac{49}{64}$ **33.** $\frac{1}{1000}$ **35.** $\frac{182}{285}$ **37.** $\frac{12}{25}$ sq m **39.** $\frac{7}{16}$ L **41.** $\frac{3}{8}$ cup **43.** $\frac{230}{1000}$ **45.** 204 **47.** 3001

Exercise Set 2.5, p. 105

1. $\frac{5}{10}$ **3.** $\frac{36}{48}$ **5.** $\frac{27}{30}$ **7.** $\frac{28}{32}$ **9.** $\frac{20}{48}$ **11.** $\frac{51}{54}$ **13.** $\frac{75}{45}$ **15.** $\frac{42}{132}$ **17.** $\frac{1}{2}$ **19.** $\frac{3}{4}$ **21.** $\frac{1}{5}$ **23.** 3 **25.** $\frac{3}{4}$ **27.** $\frac{7}{8}$ **29.** $\frac{6}{5}$ **31.** $\frac{1}{3}$ **33.** 6 **35.** $\frac{1}{3}$ **37.** = **39.** ≠ **41.** = **43.** ≠ **45.** = **47.** ≠ **49.** = **51.** ≠ **53.** 4992 sq ft **55.** $\frac{2}{5}$ **57.** No; $\frac{92}{564} \neq \frac{84}{634}$ because $92 \cdot 634 \neq 564 \cdot 84$

Exercise Set 2.6, p. 109

1. $\frac{1}{3}$ **3.** $\frac{1}{8}$ **5.** $\frac{1}{10}$ **7.** $\frac{1}{6}$ **9.** $\frac{27}{10}$ **11.** $\frac{14}{9}$ **13.** 1 **15.** 1 **17.** 1 **19.** 1 **21.** 2 **23.** 5 **25.** 9 **27.** 9 **29.** $\frac{26}{5}$ **31.** $\frac{98}{5}$ **33.** 60 **35.** 30 **37.** $\frac{1}{5}$ **39.** $\frac{9}{25}$ **41.** $\frac{11}{40}$ **43.** $\frac{5}{14}$ **45.** \$9 **47.** 625 **49.** $\frac{1}{3}$ cup **51.** \$1600 **53.** 160 mi **55.** \$3375 for food; \$2700 for housing; \$1350 for clothing; \$1500 for savings; \$3375 for taxes; \$1200 for other **57.** 35 **59.** 4673

Exercise Set 2.7, p. 117

1. $\frac{6}{5}$ **3.** $\frac{1}{6}$ **5.** 6 **7.** $\frac{3}{10}$ **9.** $\frac{4}{5}$ **11.** $\frac{4}{15}$ **13.** 4 **15.** 2 **17.** $\frac{1}{8}$ **19.** $\frac{3}{7}$ **21.** 8 **23.** 35 **25.** 1 **27.** $\frac{2}{3}$ **29.** $\frac{9}{4}$ **31.** 144 **33.** 75 **35.** 2 **37.** $\frac{3}{5}$ **39.** 315 **41.** $\frac{1}{10}$ m **43.** 32 **45.** 24 **47.** 16 L **49.** 288 km; 108 km

Summary and Review: Chapter 2, p. 119

1. [2.1d] $2 \cdot 5 \cdot 7$ **2.** [2.1d] $2 \cdot 3 \cdot 5$ **3.** [2.1d] $5 \cdot 3 \cdot 3$ **4.** [2.1d] $2 \cdot 3 \cdot 5 \cdot 5$ **5.** [2.2a] No **6.** [2.2a] No **7.** [2.2a] Yes **8.** [2.2a] No **9.** [2.1c] Prime **10.** [2.3a] Numerator: 2; denominator: 7 **11.** [2.3b] $\frac{3}{5}$ **12.** [2.3b] $\frac{3}{8}$ **13.** [2.3c] 0 **14.** [2.3c] 1 **15.** [2.3c] 48 **16.** [2.5b] 6 **17.** [2.5b] $\frac{2}{3}$ **18.** [2.5b] $\frac{1}{4}$ **19.** [2.3c] 1 **20.** [2.3c] 0 **21.** [2.5b] $\frac{2}{5}$ **22.** [2.3c] 18 **23.** [2.5b] 4 **24.** [2.5b] $\frac{1}{3}$ **25.** [2.5c] ≠ **26.** [2.5c] = **27.** [2.5c] ≠ **28.** [2.5c] = **29.** [2.6a] $\frac{3}{2}$ **30.** [2.6a] 56 **31.** [2.6a] $\frac{5}{2}$ **32.** [2.6a] 24 **33.** [2.6a] $\frac{2}{3}$ **34.** [2.6a] $\frac{1}{14}$ **35.** [2.6a] $\frac{2}{3}$ **36.** [2.6a] $\frac{1}{22}$ **37.** [2.7a] $\frac{5}{4}$ **38.** [2.7a] $\frac{1}{3}$ **39.** [2.7a] 9 **40.** [2.7a] $\frac{36}{47}$ **41.** [2.7b] 2 **42.** [2.7b] $\frac{9}{2}$ **43.** [2.7b] $\frac{11}{6}$ **44.** [2.7b] $\frac{1}{4}$ **45.** [2.7b] 300 **46.** [2.7b] $\frac{9}{4}$ **47.** [2.7b] 1 **48.** [2.7b] $\frac{4}{9}$ **49.** [2.7c] $\frac{3}{10}$ **50.** [2.7c] 240 **51.** [2.7d] 160 km **52.** [2.6b] $\frac{2}{5}$ cup **53.** [2.6b] \$6 **54.** [2.7d] 18 **55.** [1.7b] 24 **56.** [1.7b] 469 **57.** [1.8a] 1118 mi **58.** [1.8a] \$512 **59.** [1.6c] 408 R 9 **60.** [1.3d] 3607

Test: Chapter 2, pp. 121

1. [2.1d] $2 \cdot 3 \cdot 3$ **2.** [2.1d] $2 \cdot 2 \cdot 3 \cdot 5$ **3.** [2.2a] Yes **4.** [2.2a] No **5.** [2.3a] 4 numerator; 9 denominator **6.** [2.3b] $\frac{3}{4}$ **7.** [2.3c] 26 **8.** [2.3c] 1 **9.** [2.3c] 0 **10.** [2.5b] $\frac{1}{2}$ **11.** [2.5b] 6 **12.** [2.5b] $\frac{1}{14}$ **13.** [2.5c] = **14.** [2.5c] ≠ **15.** [2.6a] 32 **16.** [2.6a] $\frac{3}{2}$ **17.** [2.6a] $\frac{5}{2}$ **18.** [2.6a] $\frac{1}{10}$ **19.** [2.7a] $\frac{8}{5}$ **20.** [2.7a] 4 **21.** [2.7a] $\frac{1}{18}$ **22.** [2.7b] $\frac{3}{10}$ **23.** [2.7b] $\frac{8}{5}$ **24.** [2.7b] 18 **25.** [2.7c] 64 **26.** [2.7c] $\frac{7}{4}$ **27.** [2.6b] 28 lb **28.** [2.7d] $\frac{3}{40}$ m **29.** [1.7b] 1805 **30.** [1.7b] 101 **31.** [1.8a] 3635 mi **32.** [1.6c] 380 R 7 **33.** [1.3d] 4434

Cumulative Review: Chapters 1-2, p. 123

1. [1.1d] 584,017,800 **2.** [1.1c] Five million, three hundred eighty thousand, six hundred twenty-one **3.** [1.1e] 0 **4.** [1.2b] 17,797 **5.** [1.2b] 8866 **6.** [1.3d] 4946 **7.** [1.3d] 1425 **8.** [1.5b] 16,767 **9.** [1.5b] 8,266,500 **10.** [2.6a] $\frac{3}{20}$ **11.** [2.6a] $\frac{1}{6}$ **12.** [1.6c] 241 R 1 **13.** [1.6c] 62 **14.** [2.7b] $\frac{3}{50}$ **15.** [2.7b] $\frac{16}{45}$ **16.** [1.4a] 428,000 **17.** [1.4a] 5300 **18.** [1.4b] $749,600 + 301,400 = 1,051,000$ **19.** [1.5c] $700 \times 500 = 350,000$ **20.** [1.4c] > **21.** [2.5c] ≠ **22.** [1.9b] 81 **23.** [1.9c] 36 **24.** [1.9d] 2 **25.** [2.1a] 1, 2, 4, 7, 14, 28 **26.** [2.1d] $28 = 2 \cdot 2 \cdot 7$ **27.** [2.1c] Composite **28.** [2.2a] Yes **29.** [2.2a] No **30.** [2.3c] 35 **31.** [2.5b] 7 **32.** [2.5b] $\frac{2}{7}$ **33.** [2.3c] 0 **34.** [1.7b] 37 **35.** [2.7c] $\frac{3}{2}$ **36.** [1.7b] 3 **37.** [1.7b] 24 **38.** [1.8a] 32,119 **39.** [1.8a] 11,719 **40.** [1.8a] \$75 **41.** [2.6b] 3 cups **42.** [2.7d] 8 days **43.** [1.8a] Westside **44.** [1.8a, 2.6b] Yes; \$150 **45.** [2.3b] $\frac{3}{6}$, or $\frac{1}{2}$

CHAPTER 3

Pretest: Chapter 3, p. 126

1. 120 **2.** < **3.** $\frac{61}{8}$ **4.** $5\frac{1}{2}$ **5.** $399\frac{1}{12}$ **6.** $11\frac{31}{60}$ **7.** $6\frac{1}{6}$ **8.** $13\frac{3}{5}$ **9.** $21\frac{2}{3}$ **10.** 6 **11.** $1\frac{2}{3}$ **12.** $\frac{2}{9}$ **13.** $21\frac{1}{4}$ lb **14.** $4\frac{1}{4}$ cu ft **15.** $351\frac{1}{5}$ km **16.** $22\frac{1}{2}$ cups

Sidelight: Application of LCMs: Planet Orbits, p. 132

1. Every 60 yr **2.** Every 420 yr **3.** Every 420 yr

Exercise Set 3.1, p. 133

1. 4 **3.** 50 **5.** 40 **7.** 54 **9.** 150 **11.** 120 **13.** 72 **15.** 420 **17.** 144 **19.** 288 **21.** 30 **23.** 105 **25.** 72 **27.** 60 **29.** 36 **31.** 24 **33.** 48 **35.** 50 **37.** 143 **39.** 420 **41.** 378 **43.** 810 **45.** 250 **47.** $\frac{4}{3}$ **49.** 24 in. **51.** 70,200

Exercise Set 3.2, p. 139

1. 1 **3.** $\frac{3}{4}$ **5.** $\frac{3}{2}$ **7.** $\frac{7}{24}$ **9.** $\frac{3}{2}$ **11.** $\frac{19}{24}$ **13.** $\frac{9}{10}$ **15.** $\frac{29}{18}$ **17.** $\frac{31}{100}$ **19.** $\frac{41}{60}$ **21.** $\frac{189}{100}$ **23.** $\frac{7}{8}$ **25.** $\frac{13}{24}$ **27.** $\frac{17}{24}$ **29.** $\frac{3}{4}$ **31.** $\frac{437}{500}$ **33.** $\frac{53}{40}$ **35.** $\frac{391}{144}$ **37.** $\frac{3}{4}$ lb **39.** $\frac{23}{12}$ mi **41.** $\frac{4}{5}$ L; $\frac{8}{5}$ L; $\frac{2}{5}$ L **43.** $\frac{173}{100}$ cm **45.** $\frac{4}{9}$

Exercise Set 3.3, p. 145

1. $\frac{2}{3}$ **3.** $\frac{3}{4}$ **5.** $\frac{5}{8}$ **7.** $\frac{1}{24}$ **9.** $\frac{1}{2}$ **11.** $\frac{9}{14}$ **13.** $\frac{3}{5}$ **15.** $\frac{7}{10}$ **17.** $\frac{17}{60}$ **19.** $\frac{53}{100}$ **21.** $\frac{26}{75}$ **23.** $\frac{9}{100}$ **25.** $\frac{13}{24}$ **27.** $\frac{1}{10}$ **29.** $\frac{1}{24}$ **31.** $\frac{13}{16}$ **33.** $\frac{31}{75}$ **35.** $\frac{13}{75}$ **37.** < **39.** > **41.** < **43.** < **45.** > **47.** > **49.** < **51.** $\frac{1}{15}$ **53.** $\frac{2}{15}$ **55.** $\frac{1}{15}$ **57.** $\frac{1}{4}$ **59.** $\frac{3}{2}$ **61.** Rice **63.** $\frac{19}{24}$ **65.** $\frac{145}{144}$

Exercise Set 3.4, p. 151

1. $\frac{17}{3}$ **3.** $\frac{25}{4}$ **5.** $\frac{81}{8}$ **7.** $\frac{51}{10}$ **9.** $\frac{103}{5}$ **11.** $\frac{59}{6}$ **13.** $\frac{73}{10}$ **15.** $\frac{13}{8}$ **17.** $\frac{51}{4}$ **19.** $\frac{43}{10}$ **21.** $\frac{203}{100}$ **23.** $\frac{200}{3}$ **25.** $\frac{279}{50}$ **27.** $1\frac{3}{5}$ **29.** $4\frac{2}{3}$ **31.** $4\frac{1}{2}$ **33.** $5\frac{7}{10}$ **35.** $7\frac{4}{7}$ **37.** $7\frac{1}{2}$ **39.** $11\frac{1}{2}$ **41.** $1\frac{1}{2}$ **43.** $7\frac{57}{100}$ **45.** $43\frac{1}{8}$ **47.** $108\frac{5}{8}$ **49.** $906\frac{3}{7}$ **51.** $40\frac{4}{7}$ **53.** $55\frac{1}{51}$ **55.** 18 **57.** $\frac{1}{4}$ **59.** $8\frac{2}{3}$ **61.** $52\frac{2}{7}$

Exercise Set 3.5, p. 157

1. $6\frac{1}{2}$ **3.** $2\frac{11}{12}$ **5.** $14\frac{7}{12}$ **7.** $12\frac{1}{10}$ **9.** $16\frac{5}{24}$ **11.** $21\frac{1}{2}$ **13.** $27\frac{7}{8}$ **15.** $27\frac{13}{24}$ **17.** $1\frac{3}{5}$ **19.** $4\frac{1}{10}$ **21.** $21\frac{17}{24}$ **23.** $12\frac{1}{4}$ **25.** $15\frac{3}{8}$ **27.** $7\frac{5}{12}$ **29.** $13\frac{3}{8}$ **31.** $11\frac{5}{18}$ **33.** $5\frac{14}{15}$ lb **35.** $17\frac{11}{20}$ cm **37.** $18\frac{4}{5}$ cm **39.** $95\frac{1}{5}$ km **41.** 39 in. **43.** $\$103\frac{3}{8}$ **45.** $3\frac{1}{6}$ gal **47.** $188\frac{3}{20}$ cm **49.** $1\frac{1}{2}$ gal **51.** $28\frac{3}{4}$ yd **53.** $7\frac{3}{8}$ ft **55.** $1\frac{9}{16}$ in. **57.** $\frac{1}{10}$ **59.** $8\frac{7}{12}$

Exercise Set 3.6, p. 165

1. $22\frac{2}{3}$ **3.** $2\frac{5}{12}$ **5.** $8\frac{1}{6}$ **7.** $9\frac{31}{40}$ **9.** $24\frac{91}{100}$ **11.** $209\frac{1}{10}$ **13.** $6\frac{1}{4}$ **15.** $1\frac{1}{5}$ **17.** $3\frac{9}{16}$ **19.** $1\frac{1}{8}$ **21.** $1\frac{8}{43}$ **23.** $\frac{9}{40}$ **25.** 700 **27.** 7 oz **29.** $343\frac{3}{4}$ lb **31.** $1\frac{1}{4}$ lb opossum meat, $1\frac{1}{4}$ tsp salt, $\frac{2}{3}$ tsp black pepper, $\frac{1}{3}$ cup flour, $\frac{1}{4}$ cup water, 2 medium sweet potatoes, 1 tbsp sugar; $7\frac{1}{2}$ lb opossum meat, $7\frac{1}{2}$ tsp salt, 4 tsp black pepper, 2 cups flour, $1\frac{1}{2}$ cups water, 12 medium sweet potatoes, 6 tbsp sugar **33.** 68° F **35.** $14\frac{2}{5}$ min **37.** 15 mpg **39.** 4 cu ft **41.** 24 lb **43.** $35\frac{115}{256}$ sq in. **45.** $59{,}538\frac{1}{8}$ sq m **47.** 1,429,017 **49.** 588 **51.** $35\frac{57}{64}$ **53.** $\frac{4}{9}$ **55.** $\frac{9}{5}$, or $1\frac{4}{5}$

Summary and Review: Chapter 3, p. 169

1. [3.1a] 36 **2.** [3.1a] 90 **3.** [3.1a] 30 **4.** [3.2b] $\frac{63}{40}$ **5.** [3.2b] $\frac{19}{48}$ **6.** [3.2b] $\frac{29}{15}$ **7.** [3.2b] $\frac{7}{16}$ **8.** [3.3a] $\frac{1}{3}$ **9.** [3.3a] $\frac{1}{8}$ **10.** [3.3a] $\frac{5}{27}$ **11.** [3.3a] $\frac{11}{18}$ **12.** [3.3b] > **13.** [3.3b] > **14.** [3.3c] $\frac{19}{40}$ **15.** [3.3c] $\frac{2}{5}$ **16.** [3.4a] $\frac{15}{2}$ **17.** [3.4a] $\frac{67}{8}$ **18.** [3.4a] $\frac{13}{3}$ **19.** [3.4a] $\frac{75}{7}$ **20.** [3.4b] $2\frac{1}{3}$ **21.** [3.4b] $6\frac{3}{4}$ **22.** [3.4b] $12\frac{3}{5}$ **23.** [3.4b] $3\frac{1}{2}$ **24.** [3.4b] $877\frac{1}{3}$ **25.** [3.4b] $456\frac{5}{23}$ **26.** [3.5a] $10\frac{2}{5}$ **27.** [3.5a] $11\frac{11}{15}$ **28.** [3.5a] $10\frac{2}{3}$ **29.** [3.5a] $8\frac{1}{4}$ **30.** [3.5b] $7\frac{7}{9}$ **31.** [3.5b] $4\frac{11}{15}$ **32.** [3.5b] $4\frac{3}{20}$ **33.** [3.5b] $13\frac{3}{8}$ **34.** [3.6a] 16 **35.** [3.6a] $3\frac{1}{2}$ **36.** [3.6a] $2\frac{21}{50}$ **37.** [3.6a] 6 **38.** [3.6b] 12 **39.** [3.6b] $1\frac{7}{17}$ **40.** [3.6b] $\frac{1}{8}$ **41.** [3.6b] $\frac{9}{10}$ **42.** [3.6c] 15 **43.** [3.5c] $\$70\frac{3}{8}$ **44.** [3.6c] 30 cups **45.** [3.5c] $8\frac{3}{8}$ cups **46.** [2.6a] $\frac{6}{5}$ **47.** [2.7b] $\frac{3}{2}$ **48.** [1.5b] 708,048 **49.** [1.8a] 17 days

Test: Chapter 3, p. 171

1. [3.1a] 48 **2.** [3.2a] 3 **3.** [3.2b] $\frac{37}{24}$ **4.** [3.2b] $\frac{79}{100}$ **5.** [3.3a] $\frac{1}{3}$ **6.** [3.3a] $\frac{1}{12}$ **7.** [3.3a] $\frac{1}{12}$ **8.** [3.3b] > **9.** [3.3c] $\frac{1}{4}$ **10.** [3.4a] $\frac{7}{2}$ **11.** [3.4a] $\frac{79}{8}$ **12.** [3.4b] $4\frac{1}{2}$ **13.** [3.4b] $8\frac{2}{9}$

14. [3.4b] $162\frac{7}{11}$ **15.** [3.5a] $14\frac{1}{5}$ **16.** [3.5a] $14\frac{5}{12}$ **17.** [3.5b] $4\frac{7}{24}$ **18.** [3.5b] $6\frac{1}{6}$ **19.** [3.6a] 39 **20.** [3.6a] $4\frac{1}{2}$ **21.** [3.6a] $5\frac{5}{6}$ **22.** [3.6b] 6 **23.** [3.6b] 2 **24.** [3.6b] $\frac{1}{36}$ **25.** [3.6c] $17\frac{1}{2}$ cups **26.** [3.6c] 80 **27.** [3.5c] $160\frac{5}{12}$ kg **28.** [3.5c] $2\frac{1}{2}$ in. **29.** [1.5b] 346,636 **30.** [2.7b] $\frac{8}{5}$ **31.** [2.6a] $\frac{10}{9}$ **32.** [1.8a] 535 bottles, 10 oz left over

Cumulative Review: Chapters 1-3, p. 173

1. [1.1e] 5 **2.** [1.1a] 6 thousands + 7 tens + 5 ones **3.** [1.1c] Twenty-nine thousand, five hundred **4.** [1.2b] 899 **5.** [1.2b] 8982 **6.** [3.2b] $\frac{5}{12}$ **7.** [3.5a] $8\frac{1}{4}$ **8.** [1.3d] 5124 **9.** [1.3d] 4518 **10.** [3.3a] $\frac{5}{12}$ **11.** [3.5b] $1\frac{1}{6}$ **12.** [1.5b] 5004 **13.** [1.5b] 293,232 **14.** [2.6a] $\frac{3}{2}$ **15.** [2.6a] 15 **16.** [3.6a] $7\frac{1}{3}$ **17.** [1.6c] 715 **18.** [1.6c] 56 R 11 **19.** [3.4b] $56\frac{11}{45}$ **20.** [2.7b] $\frac{4}{7}$ **21.** [3.6b] $7\frac{1}{3}$ **22.** [1.4a] 38,500 **23.** [3.1a] 72 **24.** [2.2a] No **25.** [2.1a] 1, 2, 4, 8, 16 **26.** [2.3b] $\frac{1}{4}$ **27.** [3.3b] > **28.** [3.3b] < **29.** [2.5b] $\frac{4}{5}$ **30.** [2.5b] 32 **31.** [3.4a] $\frac{37}{8}$ **32.** [3.4b] $5\frac{2}{3}$ **33.** [1.7b] 93 **34.** [3.3c] $\frac{5}{9}$ **35.** [2.7c] $\frac{12}{7}$ **36.** [1.7b] 905 **37.** [1.8a] \$235 **38.** [1.8a] \$108 **39.** [1.8a] 297 sq ft **40.** [1.8a] 31 **41.** [2.6b] $\frac{2}{5}$ tsp **42.** [3.6c] 39 lb **43.** [3.6c] 16 **44.** [3.2c] $\frac{33}{20}$ km

CHAPTER 4

Pretest: Chapter 4, p. 176

1. Two and three hundred forty-seven thousandths **2.** Three thousand, two hundred sixty-four and $\frac{78}{100}$ dollars **3.** $\frac{21}{100}$ **4.** $\frac{5408}{1000}$ **5.** 3.79 **6.** 0.0539 **7.** 3.2 **8.** 0.099 **9.** 0.562 **10.** 21.0 **11.** 21.04 **12.** 21.045 **13.** 607.219 **14.** 0.8684 **15.** 113.664 **16.** 6.7437 **17.** 91.732 **18.** 39.0901 **19.** 7.9951 **20.** 252.9937 **21.** 3.27 **22.** 6345.2117 **23.** \$285.95 **24.** 1081.6

Exercise Set 4.1, p. 181

1. Twenty-three and two tenths **3.** One hundred thirty-five and eighty-seven hundredths **5.** Thirty-four and eight hundred ninety-one thousandths **7.** Three hundred twenty-six and $\frac{48}{100}$ dollars **9.** Zero and $\frac{67}{100}$ dollar **11.** $\frac{68}{10}$ **13.** $\frac{17}{100}$ **15.** $\frac{146}{100}$ **17.** $\frac{2046}{10}$ **19.** $\frac{3142}{1000}$ **21.** $\frac{4603}{100}$ **23.** $\frac{13}{100{,}000}$ **25.** $\frac{20{,}003}{1000}$ **27.** $\frac{10{,}008}{10{,}000}$ **29.** $\frac{45{,}672}{10}$ **31.** 0.8 **33.** 0.92 **35.** 9.3 **37.** 8.89 **39.** 250.8 **41.** 3.798 **43.** 0.0078 **45.** 0.56788 **47.** 21.73 **49.** 0.66 **51.** 34.17 **53.** 0.376193 **55.** 6170 **57.** 6000 **59.** 4.909

Exercise Set 4.2, p. 185

1. 0.58 **3.** 0.111 **5.** 0.001 **7.** 235.07 **9.** 0.4545 **11.** $\frac{4}{100}$ **13.** 0.78 **15.** 0.84384 **17.** 1.9 **19.** 0.1 **21.** 0.2 **23.** 0.6 **25.** 2.7 **27.** 13.4 **29.** 123.7 **31.** 0.89 **33.** 0.67 **35.** 0.42 **37.** 1.44 **39.** 3.58 **41.** 0.01 **43.** 0.325 **45.** 0.667 **47.** 17.002 **49.** 0.001 **51.** 10.101 **53.** 0.116 **55.** 300 **57.** 283.136 **59.** 283 **61.** 34.5439 **63.** 34.54 **65.** 35 **67.** 830 **69.** 182 **71.** 6.78346 **73.** 99.99999

Sidelight: A Number Pattern, p. 190

1. 21 **2.** 36 **3.** 78 **4.** 5050

Exercise Set 4.3, p. 191

1. 334.37 **3.** 1576.215 **5.** 132.560 **7.** 64.413 **9.** 50.0248 **11.** 0.835 **13.** 771.967 **15.** 20.8649 **17.** 227.4680 **19.** 8754.8221 **21.** 1.3 **23.** 49.02 **25.** 45.61 **27.** 85.921 **29.** 2.4975 **31.** 3.397 **33.** 8.85 **35.** 3.37 **37.** 1.045 **39.** 3.703 **41.** 0.9902 **43.** 99.66 **45.** 4.88 **47.** 12.608 **49.** 2546.973 **51.** 44.001 **53.** 2.491 **55.** 32.7386 **57.** 1.6666 **59.** 2344.90886 **61.** 199.897 **63.** 19.251 **65.** 384.68 **67.** 582.97 **69.** 35,000 **71.** 345.8

Exercise Set 4.4, p. 197

1. 118.5 gal **3.** \$3.01 **5.** 102.8° F **7.** 22,691.5 **9.** \$6.59 **11.** \$984.89 **13.** 3.4 yr **15.** 18.09 min **17. (a)** 2438 mi; **(b)** 3900.8 km **19.** 225.8 mi **21.** \$1171.74 **23.** 15.8 million **25.** 62.2 million **27.** \$84.70 **29.** 78.1 cm **31.** 2.31 cm **33.** 1.4° F **35.** No **37.** 6335 **39.** 2803

Summary and Review: Chapter 4, p. 201

1. [4.1a] Three and forty-seven hundredths **2.** [4.1a] Thirty-one thousandths **3.** [4.1a] Five hundred ninety-seven and $\frac{25}{100}$ dollars **4.** [4.1a] Zero and $\frac{98}{100}$ dollars **5.** [4.1b] $\frac{9}{100}$ **6.** [4.1b] $\frac{4561}{1000}$ **7.** [4.1b] $\frac{89}{1000}$ **8.** [4.1b] $\frac{30{,}227}{10{,}000}$ **9.** [4.1c] 0.034 **10.** [4.1c] 4.2603 **11.** [4.1c] 27.91 **12.** [4.1c] 0.006 **13.** [4.2a] 0.034 **14.** [4.2a] 0.91 **15.** [4.2a] 0.741 **16.** [4.2a] 1.041 **17.** [4.2b] 17.4 **18.** [4.2b] 17.43 **19.** [4.2b] 17.429 **20.** [4.2b] 4.272 **21.** [4.2b] 4.27 **22.** [4.2b] 4.3 **23.** [4.3a] 574.519 **24.** [4.3a] 0.6838 **25.** [4.3a] 499.829 **26.** [4.3a] 62.6932 **27.** [4.3a] 229.1 **28.** [4.3a] 45.601 **29.** [4.3b] 29.2092 **30.** [4.3b] 790.29 **31.** [4.3b] 29.148 **32.** [4.3b] 70.7891 **33.** [4.3b] 685.0519 **34.** [4.3b] 7.9953 **35.** [4.3c] 496.2795 **36.** [4.3c] 4.9911 **37.** [4.4a] 11.16 **38.** [4.4a] 3.5 yr **39.** [4.4a] 6365.1 bu

40. [4.4a] \$5888.74 **41.** [1.2b] 14,605 **42.** [3.2b] $\frac{49}{30}$ **43.** [1.3d] 3389 **44.** [3.3a] $\frac{1}{30}$

Test: Chapter 4, p. 203

1. [4.1a] Two and thirty-four hundredths **2.** [4.1a] One thousand, two hundred thirty-four and $\frac{78}{100}$ dollars **3.** [4.1b] $\frac{91}{100}$ **4.** [4.1b] $\frac{2769}{1000}$ **5.** [4.1c] 0.74 **6.** [4.1c] 3.7047 **7.** [4.2a] 0.162 **8.** [4.2a] 0.9 **9.** [4.2a] 0.078 **10.** [4.2b] 5.7 **11.** [4.2b] 5.68 **12.** [4.2b] 5.678 **13.** [4.3a] 405.219 **14.** [4.3a] 0.7902 **15.** [4.3a] 186.5 **16.** [4.3a] 1033.23 **17.** [4.3b] 48.357 **18.** [4.3b] 19.0901 **19.** [4.3b] 1.9946 **20.** [4.3b] 152.8934 **21.** [4.3c] 8.982 **22.** [4.3c] 3365.6597 **23.** [4.4a] \$3627.65 **24.** [4.4a] 10.57 sec **25.** [3.2b] $\frac{19}{18}$ **26.** [1.2b] 13,652 **27.** [3.3a] $\frac{11}{18}$ **28.** [1.3d] 2155

Cumulative Review: Chapters 1-4, p. 205

1. [1.1a] 1 ten thousand + 2 thousands + 7 hundreds + 5 tens + 8 ones **2.** [4.1a] Eight hundred two and $\frac{53}{100}$ dollars **3.** [4.1b] $\frac{1009}{100}$ **4.** [3.4a] $\frac{27}{8}$ **5.** [4.1c] 0.035 **6.** [2.1a] 1, 2, 3, 6, 11, 22, 33, 66 **7.** [2.1d] $2 \cdot 3 \cdot 11$ **8.** [3.1a] 140 **9.** [1.4a] 7000 **10.** [4.2b] 6962.47 **11.** [3.5a] $6\frac{2}{9}$ **12.** [4.3a] 235.397 **13.** [1.2b] 5495 **14.** [3.2b] $\frac{1}{2}$ **15.** [1.3d] 826 **16.** [4.3b] 8446.53 **17.** [3.3a] $\frac{1}{72}$ **18.** [3.5b] $3\frac{2}{5}$ **19.** [1.5b] 182,820 **20.** [2.6a] $\frac{2}{7}$ **21.** [3.6a] $13\frac{7}{11}$ **22.** [2.6a] $1\frac{1}{2}$ **23.** [3.6b] $1\frac{1}{2}$ **24.** [2.7b] $\frac{48}{35}$ **25.** [1.6c] 38 **26.** [3.4b] $205\frac{20}{21}$ **27.** [3.3b] < **28.** [4.2a] > **29.** [2.5c] = **30.** [4.2a] < **31.** [4.3c] 0.795 **32.** [2.7c] $\frac{25}{8}$ **33.** [1.7b] 121 **34.** [3.3c] $\frac{7}{45}$ **35.** [4.3c] 4.985 **36.** [2.7c] $\frac{6}{5}$ **37.** [1.8a] 807,643 **38.** [1.8a] 40,800 **39.** [4.4a] 7.1 billion **40.** [4.4a] \$204.54 **41.** [3.6c] $1\frac{1}{8}$ cups **42.** [3.5c] $1\frac{7}{8}$ yards **43.** [3.2c] $\frac{7}{24}$ **44.** [1.8a] 1536 **45.** [1.9c, 2.6a, 3.3a, 3.5b] $\frac{9}{32}$ **46.** [4.1c, 4.3a] 17.887

CHAPTER 5

Pretest: Chapter 5, p. 208

1. 38.54 **2.** 0.6179 **3.** 46.3 **4.** 435.4724 **5.** 3.672 **6.** 0.42735 **7.** 0.32456 **8.** 739.62 **9.** 3.625 **10.** 1.32 **11.** 0.48 **12.** 30.4 **13.** 0.38 **14.** 0.57698 **15.** 75,689 **16.** 0.00004653 **17.** 84.26 **18.** 224 **19.** 3.5 **20.** 1.4 **21.** 1.4375 **22.** 13.25 **23.** 2.75 **24.** $0.\overline{7}$ **25.** $4.\overline{142857}$ **26.** 4.1 **27.** 4.14 **28.** 4.143 **29.** \$89.70 **30.** 2.17 km **31.** \$3397.71 **32.** 7496¢ **33.** \$135.49 **34.** 48,600,000,000,000 **35.** 38.66475 **36.** 1548.8836 **37.** 49.34375 **38.** 58.17

Sidelight: Calculator Corner: Number Patterns, p. 214

1. 9, 1089, 110889, 11108889, 1111088889 **2.** 54, 6534, 665334, 66653334, 6666533334 **3.** 111, 1221, 12321, 123321, 1233321 **4.** 111, 222, 333, 444, 555 **5.** 88, 888, 8888, 88888, 888888 **6.** 9, 98, 987, 9876, 98765 **7.** 48, 408, 4008, 40008, 400008 **8.** 24, 2904, 295704, 29623704, 2962903704 **9.** 1, 121, 12321, 1234321, 123454321 **10.** 81, 9801, 998001, 99980001, 9999800001 **11.** 6006, 60606, 606606, 6066606, 60666606 **12.** 999999, 1999998, 2999997, 3999996, 4999995

Exercise Set 5.1, p. 215

1. 60.2 **3.** 6.72 **5.** 0.252 **7.** 0.522 **9.** 237.6 **11.** 783,686.852 **13.** 780 **15.** 8.923 **17.** 0.09768 **19.** 0.782 **21.** 521.6 **23.** 3.2472 **25.** 897.6 **27.** 322.07 **29.** 55.68 **31.** 3487.5 **33.** 50.0004 **35.** 114.42902 **37.** 789 **39.** 13.284 **41.** 90.72 **43.** 0.0028728 **45.** 0.72523 **47.** 1.872115 **49.** 45,678 **51.** 4567.8 **53.** 2888¢ **55.** 66¢ **57.** \$0.34 **59.** \$34.45 **61.** 2,830,000,000,000 **63.** $11\frac{1}{5}$ **65.** 342 **67.** 10^{21}

Exercise Set 5.2, p.223

1. 2.99 **3.** 23.78 **5.** 7.48 **7.** 7.2 **9.** 1.143 **11.** 4.041 **13.** 0.07 **15.** 70 **17.** 20 **19.** 0.4 **21.** 0.41 **23.** 8.5 **25.** 9.3 **27.** 0.625 **29.** 0.26 **31.** 15.625 **33.** 2.34 **35.** 0.47 **37.** 0.2134567 **39.** 21.34567 **41.** 1023.7 **43.** 56,780 **45.** 9.3 **47.** 0.0090678 **49.** 2107 **51.** 303.003 **53.** 446.208 **55.** 24.14 **57.** 13.0072 **59.** 19.3204 **61.** 96.13 **63.** 10.49 **65.** 911.13 **67.** 205 **69.** $15\frac{1}{8}$ **71.** $\frac{6}{7}$

Exercise Set 5.3, p. 231

1. 0.6 **3.** 0.325 **5.** 0.2 **7.** 0.85 **9.** 0.475 **11.** 0.975 **13.** 0.52 **15.** 20.016 **17.** 0.25 **19.** 0.575 **21.** 0.72 **23.** 1.1875 **25.** $0.2\overline{6}$ **27.** $0.\overline{3}$ **29.** $1.\overline{3}$ **31.** $1.1\overline{6}$ **33.** $0.\overline{571428}$ **35.** $0.91\overline{6}$ **37.** 0.3; 0.27; 0.267 **39.** 0.3; 0.33; 0.333 **41.** 1.3; 1.33; 1.333 **43.** 1.2; 1.17; 1.167 **45.** 0.6; 0.57; 0.571 **47.** 0.9; 0.92; 0.917 **49.** 11.06 **51.** $417.51\overline{6}$ **53.** 0.20425 **55.** 21 **57.** $3\frac{2}{5}$ **59.** 325 **61.** $0.\overline{142857}$ **63.** $0.\overline{428571}$ **65.** $0.\overline{714285}$ **67.** $0.\overline{1}$ **69.** $0.\overline{001}$ **71.** $0.\overline{012345679}$ **73.** $1.\overline{2345678901}$ **75.** $\frac{2}{3}$, $\frac{5}{7}$, $\frac{15}{19}$, $\frac{11}{13}$, $\frac{17}{20}$, $\frac{13}{15}$

Exercise Set 5.4, p. 235

1. d **3.** c **5.** a **7.** c **9.** 1.6 **11.** 6 **13.** 60 **15.** 2.3 **17.** 180 **19.** a **21.** c **23.** b **25.** b **27.** $2 \cdot 2 \cdot 3 \cdot 3 \cdot 3$ **29.** $\frac{5}{16}$ **31.** Yes

Exercise Set 5.5, p. 241

1. \$230.86 **3.** \$30.07 **5.** 62.5 mi **7.** \$5.95 **9.** 250,205.04 sq ft **11.** \$53.28 **13.** \$139.36 **15.** \$465.78 **17.** 887.4 km **19.** \$57.35 **21.** 20.2 mpg **23.** 11.9752 cu ft **25.** \$10 **27.** 1032 cal **29.** 14.5 mpg **31.** 0.307 **33.** \$8.70 **35.** \$4.15 **37.** \$394.03 **39.** 6020.48 sq m **41.** \$196,987.20 **43.** $13\frac{7}{12}$ **45.** $34\frac{11}{12}$ **47.** **(a)** \$13.38; **(b)** \$14.49; **(c)** a; **(d)** \$13.78; **(e)** d

Summary and Review: Chapter 5, p. 245

1. [5.1a] 12.96 **2.** [5.1a] 0.14442 **3.** [5.1a] 4.3 **4.** [5.1a] 0.2784 **5.** [5.1a] 1.073 **6.** [5.1a] 0.2184 **7.** [5.1a] 0.02468 **8.** [5.1a] 24,680 **9.** [5.2a] 7.5 **10.** [5.2a] 3.2 **11.** [5.2a] 0.45 **12.** [5.2a] 45.2 **13.** [5.2a] 1.6 **14.** [5.2a] 1.022 **15.** [5.2a] 2.763 **16.** [5.2a] 0.2763 **17.** [5.2a] 1.274 **18.** [5.2a] 1389.2 **19.** [5.2b] 6.95 **20.** [5.2b] 42.54 **21.** [5.4a] 272 **22.** [5.4a] 4 **23.** [5.4a] 216 **24.** [5.4a] \$125 **25.** [5.3a] 2.6 **26.** [5.3a] 1.28 **27.** [5.3a] 2.75 **28.** [5.3a] 3.25 **29.** [5.3a] $1.1\overline{6}$ **30.** [5.3a] $1.\overline{54}$ **31.** [5.3b] 1.5 **32.** [5.3b] 1.55 **33.** [5.3b] 1.545 **34.** [5.5a] \$239.80 **35.** [5.5a] 82.67 km **36.** [5.5a] \$32.59 **37.** [5.5a] 24.36 cups; 104.4 cups **38.** [5.5a] \$8.98 **39.** [5.1b] \$82.73 **40.** [5.1b] \$4.87 **41.** [5.1b] 2493¢ **42.** [5.1b] 986¢ **43.** [5.1b] 3,400,000,000 **44.** [5.1b] 1,200,000 **45.** [5.2c] 1.8045 **46.** [5.2c] 57.1449 **47.** [5.3c] 15.6375 **48.** [5.3c] $41.537\overline{3}$ **49.** [3.6a] $43\frac{3}{4}$ **50.** [3.6b] $3\frac{3}{4}$ **51.** [3.5a] $19\frac{4}{5}$ **52.** [3.5b] $6\frac{3}{5}$ **53.** [2.5b] $\frac{1}{2}$ **54.** [2.1d] $2 \cdot 2 \cdot 2 \cdot 2 \cdot 2 \cdot 2 \cdot 3$

Test: Chapter 5, p. 247

1. [5.1a] 8 **2.** [5.1a] 0.03 **3.** [5.1a] 3.7 **4.** [5.1a] 0.2079 **5.** [5.1a] 1.088 **6.** [5.1a] 0.31824 **7.** [5.1a] 0.21345 **8.** [5.1a] 739.62 **9.** [5.2a] 4.75 **10.** [5.2a] 0.44 **11.** [5.2a] 0.24 **12.** [5.2a] 30.4 **13.** [5.2a] 0.19 **14.** [5.2a] 0.34689 **15.** [5.2a] 34,689 **16.** [5.2a] 0.0000123 **17.** [5.2b] 84.26 **18.** [5.4a] 198 **19.** [5.4a] 4 **20.** [5.3a] 1.6 **21.** [5.3a] 0.88 **22.** [5.3a] 5.25 **23.** [5.3a] 0.75 **24.** [5.3a] $1.\overline{2}$ **25.** [5.3a] $2.\overline{142857}$ **26.** [5.3b] 2.1 **27.** [5.3b] 2.14 **28.** [5.3b] 2.143 **29.** [5.5a] \$119.70 **30.** [5.5a] 2.37 km **31.** [5.5a] \$1675.50 **32.** [5.1b] 8795¢ **33.** [5.1b] \$9.49 **34.** [5.1b] 38,700,000,000,000 **35.** [5.2c] 40.0065 **36.** [5.2c] 384.8464 **37.** [5.3c] 302.4 **38.** [5.3c] $52.339\overline{4}$ **39.** [3.5b] $26\frac{1}{2}$ **40.** [3.5a] $2\frac{11}{16}$ **41.** [3.6b] $1\frac{1}{8}$ **42.** [3.6b] 14 **43.** [2.5b] $\frac{11}{18}$ **44.** [2.1d] $2 \cdot 2 \cdot 2 \cdot 3 \cdot 3 \cdot 5$

Cumulative Review: Chapters 1-5 , p. 249

1. [3.4a] $\frac{20}{9}$ **2.** [4.1b] $\frac{3052}{1000}$ **3.** [5.3a] 1.4 **4.** [5.3a] $0.\overline{54}$ **5.** [2.1c] Prime **6.** [2.2a] Yes **7.** [1.9c] 1754 **8.** [5.2c] 4.364 **9.** [4.2b] 584.90 **10.** [5.3b] 218.56 **11.** [5.4a] 160 **12.** [5.4a] 4 **13.** [1.5c] 12,800,000 **14.** [5.4a] 6 **15.** [3.5a] $6\frac{1}{20}$ **16.** [1.2b] 139,116 **17.** [3.2b] $\frac{31}{18}$ **18.** [4.3a] 145.953 **19.** [1.3d] 710,137 **20.** [4.3b] 13.097 **21.** [3.5b] $\frac{5}{7}$ **22.** [3.3a] $\frac{1}{110}$ **23.** [2.6a] $\frac{1}{6}$ **24.** [1.5b] 5,317,200 **25.** [5.1a] 4.78 **26.** [5.1a] 0.0279431 **27.** [5.2a] 2.122 **28.** [1.6c] 1843 **29.** [5.2a] 13862.1 **30.** [2.7b] $\frac{5}{6}$ **31.** [4.3c] 0.78 **32.** [1.7b] 28 **33.** [5.2b] 8.62 **34.** [1.7b] 367,251 **35.** [3.3c] $\frac{1}{18}$ **36.** [2.7c] $\frac{1}{2}$ **37.** [1.8a] 11,222 **38.** [2.7d] \$500 **39.** [1.8a] 86,400 **40.** [2.6b] \$2400 **41.** [4.4a] \$258.77 **42.** [3.5c] $6\frac{1}{2}$ lb **43.** [3.2c] 2 lb **44.** [5.5a] 467.28 **45.** [3.6c] 144 **46.** [5.5a] \$0.91

CHAPTER 6

Pretest: Chapter 6, p. 252

1. $\frac{35}{43}$ **2.** $\frac{0.079}{1.043}$ **3.** 22.5 **4.** 0.75 **5.** 25.5 miles per gallon **6.** 0.075 bushels per minute **7.** 5.79 cents per ounce **8.** Brand B **9.** 1944 km **10.** 22 **11.** 12 min **12.** 393.75 miles

Exercise Set 6.1, p. 257

1. $\frac{4}{5}$ **3.** $\frac{0.4}{12}$ **5.** $\frac{2}{12}$ **7.** $\frac{2.7}{13.1}$; $\frac{13.1}{2.7}$ **9.** No **11.** Yes **13.** 45 **15.** 12 **17.** 10 **19.** 20 **21.** 5 **23.** 18 **25.** 22 **27.** 28 **29.** $9\frac{1}{3}$ **31.** $2\frac{8}{9}$ **33.** 0.06 **35.** 5 **37.** 1 **39.** 1 **41.** 14 **43.** $2\frac{3}{16}$ **45.** = **47.** 50 **49.** 14.5 **51.** $\frac{13,339,000}{145,304,000} \approx 0.09$; $\frac{145,304,000}{13,339,000} \approx 10.89$

Exercise Set 6.2, p. 261

1. 40 km/h **3.** 11 m/sec **5.** 152 yd/day **7.** 25 km/hr; 0.04 hr/km **9.** 0.623 gal/sq ft **11.** 2.5 servings/lb **13.** 186,000 mi/sec **15.** 2.3 km/h **17.** 560 mi/hr **19.** \$9.50/yd **21.** 21.46 ¢/oz **23.** \$1.08/lb **25.** A **27.** B **29.** B **31.** B **33.** Eight 16-oz bottles **35.** 1.7 million **37.** 67,819 **39.** Approx. 11.54 m/sec; 0.08666 sec/m

Sidelight: An Application of Ratio: State Lottery Profits, p. 268

1. New York **2.** Vermont **3.** $618.3 million **4.** $3.6042 billion **5.** 0.1706 **6.** $461.3 million **7.** $74.3 million **8.** $13.67 per person **9.** $34.81 per person **10.** $45.08 **11.** Illinois

Exercise Set 6.3, p. 269

1. 702 km **3.** $84.60 **5.** 3.57 **7.** 954 **9.** 12 lb **11.** 322 **13.** 120 lb **15.** 1980 **17.** 58.1 mi **19.** $7\frac{1}{3}$ in **21.** No; the money will be gone in 24 weeks; $133.33 more **23.** 2150

Summary and Review: Chapter 6, p. 271

1. [6.1a] $\frac{47}{84}$ **2.** [6.1a] $\frac{46}{1.27}$ **3.** [6.1a] $\frac{83}{100}$ **4.** [6.1a] $\frac{0.72}{197}$ **5.** [6.1a] $\frac{5200}{1070}$ **6.** [6.1c] 32 **7.** [6.1c] $\frac{1}{40}$ **8.** [6.1c] 7 **9.** [6.1c] 24 **10.** [6.2a] 23.54 km/h **11.** [6.2a] 0.638 gal/sq ft **12.** [6.2a] $25.36/kg **13.** [6.2a] 0.72 servings/lb **14.** [6.2b] 5.7 cents/oz **15.** [6.2b] 5.45 cents/oz **16.** [6.2b] B **17.** [6.2b] B **18.** [6.3a] $4.45 **19.** [6.3a] 351 **20.** [6.3a] 832 km **21.** [6.3a] 27 acres **22.** [6.3a] 2,392,000 kg **23.** [6.3a] 6 in. **24.** [6.3a] 2622 **25.** [4.4a] $1630.40 **26.** [2.5c] = **27.** [2.5c] ≠ **28.** [5.1a] 10,672.74 **29.** [5.2a] 45.5

Test: Chapter 6, p. 273

1. [6.1a] $\frac{85}{97}$ **2.** [6.1a] $\frac{0.34}{124}$ **3.** [6.1c] 12 **4.** [6.1c] 360 **5.** [6.2a] 0.625 m/sec **6.** [6.2a] $1\frac{1}{3}$ servings/lb **7.** [6.2b] 19.39 cents/oz **8.** [6.2b] B **9.** [6.3a] 1512 km **10.** [6.3a] 44 **11.** [6.3a] 4.8 min **12.** [6.3a] 525 mi **13.** [4.4a] 25.8 million lb **14.** [2.5c] ≠ **15.** [5.1a] 17,324.14 **16.** [5.2a] 0.9944

Cumulative Review: Chapters 1-6, p. 275

1. [4.3a] 513.996 **2.** [3.5a] $6\frac{3}{4}$ **3.** [3.2b] $\frac{7}{20}$ **4.** [4.3b] 30.491 **5.** [4.3b] 72.912 **6.** [3.3a] $\frac{7}{60}$ **7.** [5.1a] 222.076 **8.** [5.1a] 567.8 **9.** [3.6a] 3 **10.** [5.2a] 43 **11.** [1.6c] 899 **12.** [2.7b] $\frac{3}{2}$ **13.** [1.1b] 3 ten thousands + 7 tens + 4 ones **14.** [4.1a] One hundred twenty and seven hundredths **15.** [4.2a] 0.7 **16.** [4.2a] 0.8 **17.** [2.1d] $2 \cdot 2 \cdot 2 \cdot 2 \cdot 3 \cdot 3$ **18.** [3.1a] 140 **19.** [2.3b] $\frac{5}{8}$ **20.** [2.5b] $\frac{5}{8}$ **21.** [5.3c] 5.718 **22.** [5.3c] 0.179 **23.** [6.1a] $\frac{0.3}{15}$ **24.** [6.1b] Yes **25.** [6.2a] 55 m/sec **26.** [6.2b] The 14-oz can **27.** [6.1c] 30.24 **28.** [5.2b] 26.4375 **29.** [2.7c] $\frac{8}{9}$ **30.** [6.1c] 128 **31.** [4.3c] 33.34 **32.** [3.3c] $\frac{76}{175}$ **33.** [5.5a] 42.2025 km **34.** [6.3a] 7 min **35.** [4.4a] 976.9 mi **36.** [1.8a] 476,986 **37.** [2.4c] $39 **38.** [2.7d] 12 **39.** [3.5c] $2\frac{1}{4}$ cups **40.** [5.5a] 132 **41.** [6.3a] 60 mph **42.** [6.2b] The 12-oz bag

CHAPTER 7

Pretest: Chapter 7, p. 278

1. 0.87 **2.** 53.7% **3.** 75% **4.** $\frac{37}{100}$ **5.** $x = 60\% \times 75$; 45 **6.** $\frac{n}{100} = \frac{35}{50}$; 70% **7.** 90 lb **8.** 20% **9.** $14.30; $300.30 **10.** $5152 **11.** $112.50 discount; $337.50 sale price **12.** $99.60 **13.** $20 **14.** $7128.60

Sidelight: Calculator Corner: Finding Whole-Number Remainders in Division, p. 282

1. 28 R 2 **2.** 116 R 3 **3.** 74 R 10 **4.** 415 R 3

Exercise Set 7.1, p. 283

1. $\frac{90}{100}$; $90 \times \frac{1}{100}$; 90×0.01 **3.** $\frac{12.5}{100}$; $12.5 \times \frac{1}{100}$; 12.5×0.01 **5.** 0.67 **7.** 0.456 **9.** 0.5901 **11.** 0.1 **13.** 0.01 **15.** 2 **17.** 0.001 **19.** 0.0009 **21.** 0.0018 **23.** 0.2319 **25.** 0.9 **27.** 0.108 **29.** 0.458 **31.** 47% **33.** 3% **35.** 100% **37.** 33.4% **39.** 75% **41.** 40% **43.** 0.6% **45.** 1.7% **47.** 27.18% **49.** 2.39% **51.** 2.5% **53.** 24% **55.** $33\frac{1}{3}$ **57.** $0.\overline{6}$ **59.** Multiply by 100

Sidelight: Applications of Ratio and Percent: The Price-Earnings Ratio and Stock Yields, p. 288

1. 6.0, 8.1% **2.** 11.9, 1.8% **3.** 4.7, 3.2% **4.** 9.4, 6.8%

Exercise Set 7.2, p. 289

1. 41% **3.** 1% **5.** 20% **7.** 30% **9.** 50% **11.** 62.5% **13.** 40% **15.** $66.\overline{6}\%$, or $66\frac{2}{3}\%$ **17.** $16.\overline{6}\%$, or $16\frac{2}{3}\%$ **19.** 16% **21.** 5% **23.** 34% **25.** 36% **27.** $\frac{4}{5}$ **29.** $\frac{5}{8}$ **31.** $\frac{1}{3}$ **33.** $\frac{1}{6}$ **35.** $\frac{29}{400}$ **37.** $\frac{1}{125}$ **39.** $\frac{7}{20}$

41.

$\frac{1}{8}$	$\frac{1}{6}$	$\frac{1}{5}$	$\frac{1}{4}$	$\frac{1}{3}$	$\frac{3}{8}$	$\frac{2}{5}$
0.125	$0.1\overline{6}$	0.2	0.25	$0.\overline{3}$	0.375	0.4
$12\frac{1}{2}\%$ or 12.5%	$16\frac{2}{3}\%$ or $16.\overline{6}\%$	20%	25%	$33\frac{1}{3}\%$ or $33.\overline{3}\%$	$37\frac{1}{2}\%$ or 37.5%	40%

$\frac{1}{2}$	$\frac{3}{5}$	$\frac{5}{8}$	$\frac{2}{3}$	$\frac{3}{4}$	$\frac{4}{5}$	$\frac{5}{6}$
0.5	0.6	0.625	$0.\overline{6}$	0.75	0.8	$0.8\overline{3}$
50%	60%	$62\frac{1}{2}\%$ or 62.5%	$66\frac{2}{3}\%$ or $66.\overline{6}\%$	75%	80%	$83\frac{1}{3}\%$ or $83.\overline{3}\%$

$\frac{7}{8}$	$\frac{1}{1}$
0.875	1
$87\frac{1}{2}\%$ or 87.5%	100%

43. 5 **45.** 18.75 **47.** $5.\overline{405}\%$

Exercise Set 7.3, p. 295

1. $y = 41\% \times 89$ **3.** $89 = a \times 99$ **5.** $13 = 25\% \times y$ **7.** 90 **9.** 45 **11.** $15 **13.** 1.05 **15.** 24% **17.** 200% **19.** 50% **21.** 125% **23.** 40 **25.** $40 **27.** 88 **29.** 20 **31.** 6.25 **33.** $846.60 **35.** $\frac{9}{100}$ **37.** 0.89 **39.** $880 (Can vary); $843.20

Exercise Set 7.4, p. 301

1. $\frac{82}{100} = \frac{a}{74}$ **3.** $\frac{n}{100} = \frac{4.3}{5.9}$ **5.** $\frac{25}{100} = \frac{14}{b}$ **7.** $42 **9.** 440 **11.** 80 **13.** 2.88 **15.** 25% **17.** 102% **19.** 25% **21.** 93.75% **23.** $72 **25.** 90 **27.** 88 **29.** 20 **31.** 25 **33.** $780.20 **35.** $1134 (Can vary); $1118.64

Exercise Set 7.5, p. 309

1. 27; 133 **3.** 536; 264 **5.** 32.5%; 67.5% **7.** 20.4 mL; 659.6 mL **9.** 25% **11.** 45%; $37\frac{1}{2}\%$; $17\frac{1}{2}\%$ **13.** 8% **15.** 20% **17.** $9030 **19.** $8400 **21.** 5.2832 billion; about 5.3677 billion; about 5.4536 billion **23.** $12,500 **25.** 166; 156; 146; 136; 122 **27.** Neither; they are the same **29.** About 5 ft, 6 in.

Exercise Set 7.6, p. 315

1. $20.46; $268.46 **3.** $11.40; $201.35 **5.** 5% **7.** 4% **9.** $2800 **11.** $800 **13.** $33.25 **15.** 5.6% **17.** 37 **19.** $1.\overline{18}$ **21.** $5214.72

Sidelight: An Application: Water Loss, p. 320

1. 120 lb **2.** 1.2 lb **3.** 9.6 lb **4.** 12 lb **5.** 24 lb

Exercise Set 7.7, p. 321

1. $3690 **3.** 5% **5.** $980 **7.** $6860 **9.** $519.80 **11.** $30; $270 **13.** $3; $2 **15.** $12.50; $112.50 **17.** 40%; $360 **19.** $387; $30\frac{6}{17}\%$ **21.** $0.\overline{5}$ **23.** $0.91\overline{6}$ **25.** $5924.25; $73,065.75

Exercise Set 7.8, p. 327

1. $26 **3.** $248 **5.** $150.50 **7.** $120.83 **9.** $413.60 **11.** $484 **13.** $236.75 **15.** $466.56 **17.** $2184.05 **19.** 4.5 **21.** $33\frac{1}{3}$ **23.** $1434.53

Summary and Review: Chapter 7, p. 329

1. [7.1c] 48.3% **2.** [7.1c] 36% **3.** [7.2a] 37.5% **4.** [7.2a] $33.\overline{3}\%$, or $33\frac{1}{3}\%$ **5.** [7.1b] 0.735 **6.** [7.1b] 0.065 **7.** [7.2b] $\frac{6}{25}$ **8.** [7.2b] $\frac{63}{1000}$ **9.** [7.3a, b] $30.6 = x\% \times 90$; 34% **10.** [7.3a, b] $63 = 84\% \times n$; 75 **11.** [7.3a, b] $y = 38\frac{1}{2}\% \times 168$; 64.68 **12.** [7.4a, b] $\frac{24}{100} = \frac{16.8}{b}$; 70 **13.** [7.4a, b] $\frac{n}{100} = \frac{22.2}{30}$; 74% **14.** [7.4a, b] $\frac{38\frac{1}{2}}{100} = \frac{a}{168}$; 64.68 **15.** [7.5a] $182 **16.** [7.5b] 12% **17.** [7.5b] 82,400 **18.** [7.5b] 20% **19.** [7.5a] 168 **20.** [7.6a] $14.40 **21.** [7.6a] 5% **22.** [7.7a] 11% **23.** [7.7b] $42; $308 **24.** [7.7b] $42.70; $262.30 **25.** [7.7a] $29.40 **26.** [7.8a] $9.60 **27.** [7.8a] $21.60 **28.** [7.8a] $31.90 **29.** [7.8a] $15.25 **30.** [7.8b] $224.72 **31.** [7.8b] $188.16 **32.** [6.1c] $18\frac{2}{3}$ **33.** [5.2b] 64 **34.** [5.2b] 7.6123 **35.** [6.1c] 42 **36.** [5.3a] $3.\overline{6}$ **37.** [5.3a] $1.\overline{571428}$ **38.** [3.4b] $3\frac{2}{3}$ **39.** [3.4b] $17\frac{2}{7}$

Test: Chapter 7, p. 331

1. [7.1b] 0.89 **2.** [7.1c] 67.4% **3.** [7.2a] 87.5% **4.** [7.2b] $\frac{13}{20}$ **5.** [7.3a, b] $m = 40\% \times 55$; 22 **6.** [7.4a, b] $\frac{n}{100} = \frac{65}{80}$; 81.25% **7.** [7.5a] 50 lb **8.** [7.5b] 20% **9.** [7.6a] $16.20; $340.20 **10.** [7.7a] $630 **11.** [7.7b] $40; $160 **12.** [7.8a] $8.52 **13.** [7.8a] $4.30 **14.** [7.8b] $127.69 **15.** [5.2b] 222 **16.** [6.1c] 16 **17.** [5.3a] $1.41\overline{6}$ **18.** [3.4b] $3\frac{21}{44}$

Cumulative Review: Chapters 1-7, p. 333

1. [4.1b] $\frac{91}{1000}$ **2.** [5.3a] $2.1\overline{6}$ **3.** [7.1b] 0.03 **4.** [7.2a] 112.5% **5.** [6.1a] $\frac{10}{1}$ **6.** [6.2a] $23\frac{1}{3}$ km/h **7.** [3.3b] $<$ **8.** [2.5c] $<$ **9.** [1.4b] 296,200 **10.** [1.4b] 50,000 **11.** [1.9d] 13 **12.** [5.2c] 1.5 **13.** [3.5a] $3\frac{1}{30}$ **14.** [4.3a] 49.74 **15.** [1.2b] 515,150 **16.** [4.3b] 0.02 **17.** [3.5b] $\frac{2}{3}$ **18.** [3.3a] $\frac{2}{63}$ **19.** [2.6a] $\frac{1}{6}$ **20.** [1.5b] 853,142,400

21. [5.1a] 1.38036 **22.** [3.6b] $1\frac{1}{2}$ **23.** [5.2a] 12.25 **24.** [1.6c, 3.4b] $123\frac{1}{3}$ **25.** [1.7b] 95 **26.** [4.3c] 8.13 **27.** [2.7c] 9 **28.** [3.3c] $\frac{1}{12}$ **29.** [6.1c] 40 **30.** [6.1c] $8\frac{8}{21}$ **31.** [4.4a] $6878.84 **32.** [5.5a] $11.50 **33.** [6.2b] 7.5 cents per ounce **34.** [6.3a] 608 km **35.** [7.5b] 30% increase **36.** [7.6a] $5.67 **37.** [1.8a] 2210 **38.** [3.6c] 5 **39.** [3.2c] $1\frac{1}{2}$ km **40.** [6.3a] 60 mi **41.** [7.8a, b] Bank A **42.** [3.6b, 7.5b] 12.5% increase

CHAPTER 8

Pretest: Chapter 8, p. 336

1. **(a)** 51; **(b)** 51.5; **(c)** 46, 50, 53, 55 **2.** **(a)** 3; **(b)** 3; **(c)** 5, 4, 3, 2, 1 **3.** **(a)** 12.75; **(b)** 17; **(c)** 4 **4.** 55 km/h **5.** 76 **6.**

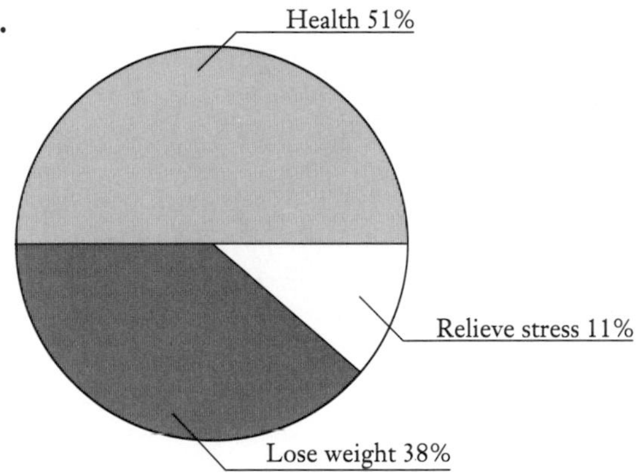

7. **(a)** $298; **(b)** $172; **(c)** $134 **8.**

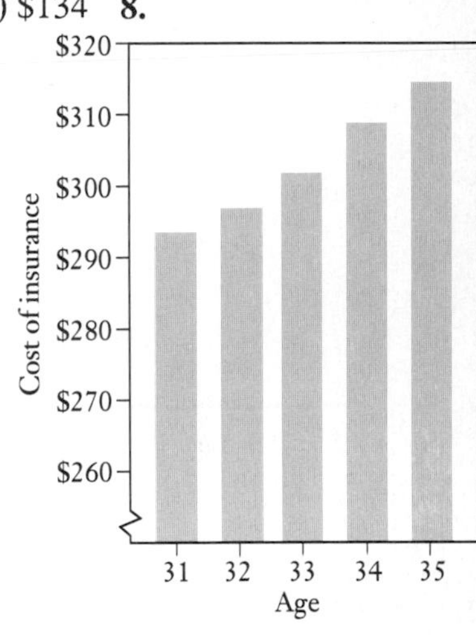

9. 1988 **10.** 19,000

Exercise Set 8.1, p. 341

1. Average: 12; median: 13.5; mode: 15 **3.** Average: 20; median: 20; mode: 5, 10, 15, 20, 25, 30, 35 **5.** Average: 5.2; median: 5.7; mode: 7.4 **7.** Average: 239.5; median: 234; mode: 234 **9.** Average: 256.25 lb; median: 257.5 lb; mode: 260 lb **11.** Average: 40°; median: 40°; mode: 43°, 40°, 23°, 38°, 54°, 35°, 47°, **13.** 29 mpg **15.** 2.75 **17.** Average: $9.75; median: $9.79; mode: $9.79 **19.** 90 **21.** $18,460 **23.** $\frac{4}{9}$ **25.** 1.999396 **27.** 171

Sidelight: A Problem-Solving Extra, p.348

1. 520,000 **2.** 261,800 **3.** 401,900

Exercise Set 8.2, p. 349

1. 4 in. **3.** 94 **5.** 21 **7.** Not given in table **9.** 2740 **11.** 294 **13.** Calisthenics **15.** Aerobic dance **17.** 660 **19.** Moderate walking **21.** 121 **23.** 1987 **25.** 1983 and 1984 **27.** 7000 **29.** 1986 **31.** 3 **33.** 16 **35.** 13 **37.** Out **39.**

US Automobile Registration

1960
1965
1970
1975
1980
1985

[412CNG] = 10,000,000 cars

41. 27,859.5 sq mi

Exercise Set 8.3, p. 359

1. Los Angeles **3.** 22 **5.** 2 **7.** Syracuse **9.** Approximately \$270 **11.** San Francisco **13.** Approximately \$50

15.

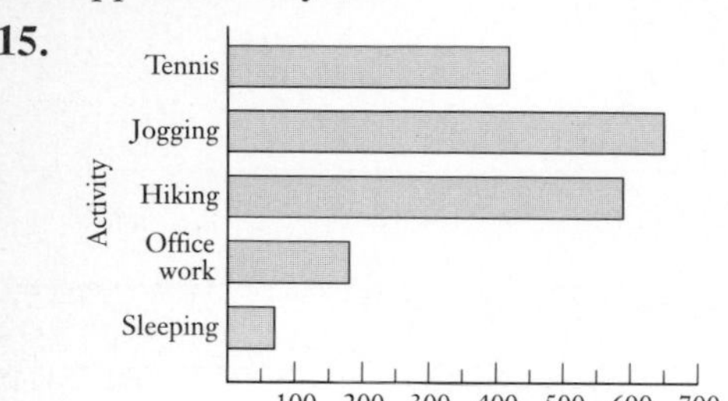

17. 3:00 P.M. **19.** Approximately 15° **21.** Between 5:00 P.M. and 6:00 P.M. **23.** 1991 **25.** Approximately \$17 million **27.** Approximately \$1.5 million

29.

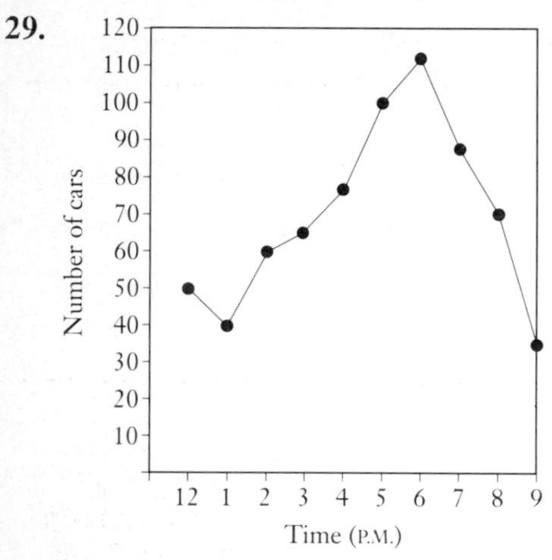

31. Between 8 P.M. and 9 P.M. **33.** 18% **35.** $66.\overline{6}\%$, or $66\frac{2}{3}\%$

Exercise Set 8.4, p. 367

1. 3.7% **3.** 270 **5.** 6.8% **7.** Gas purchased **9.** 4¢ **11.** 16¢ **13.**

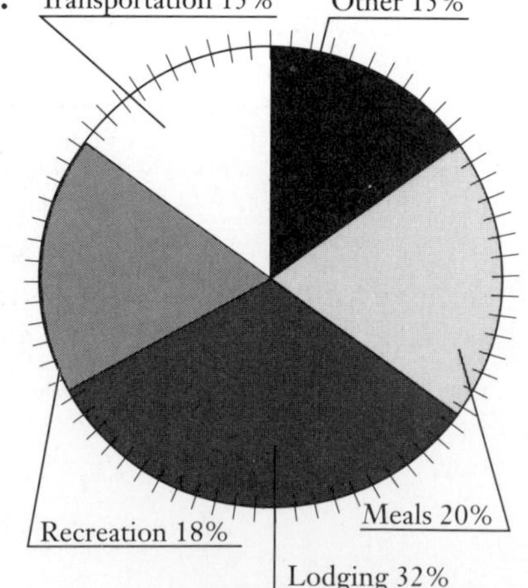

15. 25% **17.** $\frac{3}{4}$

Summary and Review: Chapter 8, p. 369

1. [8.1a] 38.5 **2.** [8.1a] 13.4 **3.** [8.1a] 1.55 **4.** [8.1a] 1840 **5.** [8.1a] $\$16.\overline{6}$ **6.** [8.1a] $321.\overline{6}$ **7.** [8.1b] 38.5 **8.** [8.1b] 14 **9.** [8.1b] 1.8 **10.** [8.1b] 1900 **11.** [8.1b] \$17 **12.** [8.1b] 375 **13.** [8.1c] 26 **14.** [8.1c] 11; 17 **15.** [8.1c] 0.2 **16.** [8.1c] 700; 800 **17.** [8.1c] \$17 **18.** [8.1c] 20 **19.** [8.1a, b] \$110.5; \$107 **20.** [8.1a] $66.1\overline{6}°$ **21.** [8.1a] 96 **22.** [8.2a] Boston **23.** [8.2a] \$0.50 **24.** [8.2a] \$2.18 **25.** [8.2a] \$70,716 **26.** [8.2a] Washington, D.C. **27.** [8.2a] Canton, Ohio **28.** [8.2b] USSR **29.** [8.2b] 91 **30.** [8.2b] US **31.** [8.2b] 103 **32.** [8.3a] \$6.25 **33.** [8.3a] Hong Kong **34.** [8.3a] New York City **35.** [8.3a] Dallas **36.** [8.3a] \$4.50 **37.** [8.3a] Sidney **38.** [8.3c] Under 20 **39.** [8.3c] Approximately 12 **40.** [8.3c] Approximately 13 per 100 drivers **41.** [8.3c] Between 45 and 74 **42.** [8.3c] Approximately 11 per 100 drivers **43.** [8.3c] Under 20 **44.** [8.4a] 30% **45.** [8.4a] Outlying suburbs **46.** [8.4a] 64% **47.** [8.4a] 7.5 **48.** [6.3a] 12,600 mi **49.** [7.5a] 215.28 million **50.** [7.3a, 7.4a] $222.\overline{2}\%$, or $222\frac{2}{9}\%$ **51.** [7.3a, 7.4a] 50% **52.** [2.7b] $\frac{9}{10}$ **53.** [2.7b] $\frac{5}{12}$

Test: Chapter 8, p. 373

1. [8.1a] 50 **2.** [8.1a] 3 **3.** [8.1a] 15.5 **4.** [8.1b, c] Median: 50.5; mode: 45, 49, 52, 54 **5.** [8.1b, c] Median: 3; mode: 1, 2, 3, 4, 5 **6.** [8.1b, c] Median: 17.5; mode: 17, 18 **7.** [8.1a] 58 km/h **8.** [8.1a] 76 **9.** [8.2a] Hiking with 20-lb load **10.** [8.2a] Hiking with 10-lb load **11.** [8.2a] Walking 3.5 mph **12.** [8.2a] Fitness walking **13.** [8.3b]

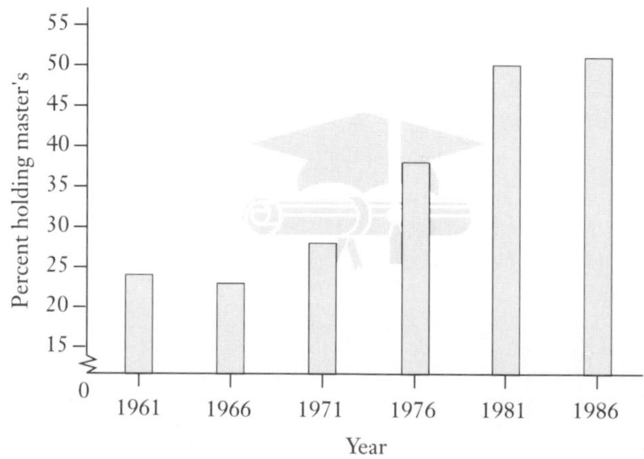

14. [8.2b] 25 **15.** [8.2b] Carew **16.** [8.2b] 145 **17.** [8.2b] Garr **18.** [8.2b] 10 **19.** [8.3c] Increasing **20.** [8.3c] 1970 to 1975 **21.** [8.3c] \$1 billion **22.** [8.3c] \$2 billion

23. [8.4b]

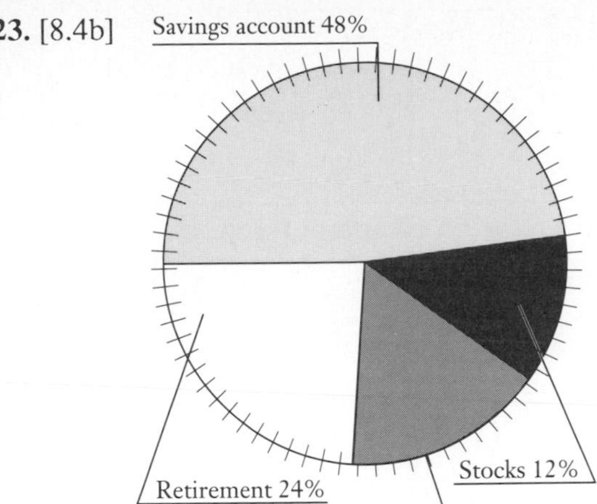

24. [2.7b] $\frac{25}{4}$ **25.** [7.3a, 7.4a] 68 **26.** [7.5a] 15,600 **27.** [6.3a] 340

Cumulative Review: Chapters 1-8, p. 377

1. [1.1e] 5 hundreds **2.** [1.9c] 128 **3.** [2.1a] 1, 2, 3, 4, 5, 6, 10, 12, 15, 20, 30, 60 **4.** [4.2b] 52.0 **5.** [3.4a] $\frac{33}{10}$ **6.** [5.1b] \$2.10 **7.** [5.1b] \$3,250,000,000 **8.** [6.1b] No **9.** [3.5a] $6\frac{7}{10}$ **10.** [4.3a] 44.6351 **11.** [3.3a] $\frac{1}{3}$ **12.** [4.3b] 325.43 **13.** [3.6a] 15 **14.** [1.5b] 2,740,320 **15.** [2.7b] $\frac{9}{10}$ **16.** [1.6c, 3.4b] $4361\frac{1}{2}$ **17.** [6.1c] $9\frac{3}{5}$ **18.** [2.7c] $\frac{3}{4}$ **19.** [5.2b] 6.8 **20.** [1.7b] 15,312 **21.** [6.2b] 5.5 cents per ounce **22.** [7.5a] 3324 **23.** [4.4a] 6.2 pounds **24.** [3.6c] $\frac{1}{4}$ yd **25.** [1.8a] 2572 billion **26.** [2.4c] $\frac{3}{8}$ cup **27.** [3.3d] $\frac{1}{4}$ **28.** [6.3a] 1122 **29.** [5.5a] \$9.55 **30.** [7.7a] 7% **31.** [8.4a] 56% **32.** [8.4a] 94% **33.** [8.4a] 15 **34.** [8.3b]

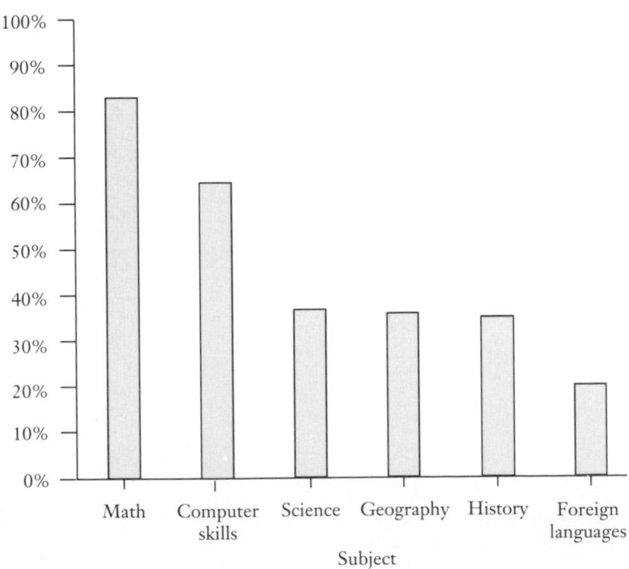

35. [8.1c] 5 **36.** [8.1b] 62 **37.** [7.5b, 8.1a] 12% decrease

CHAPTER 9

Pretest: Chapter 9, p. 382

1. 96 **2.** $\frac{5}{12}$ **3.** 8460 **4.** 0.92 **5.** 131 mm **6.** 100 ft^2 **7.** 22 cm^2 **8.** $32\frac{1}{2}$ ft^2 **9.** 4 m^2 **10.** 9.6 m **11.** 30.144 m **12.** 72.3456 m^2 **13.** 92 in^2 **14.** 9 **15.** 9.849 **16.** $c = 20$ **17.** $b = \sqrt{45}$; $b \approx 6.708$

Sidelight: Applications of Baseball, p. 386

1. (a) 0.367; (b) 185; (c) 1.000; **2.** (a) 0.756; (b) 0.557; (c) 4.000 **3.** (a) 333; (b) 4; (c) 2129 **4.** (a) 12,364; (b) 1051; (c) yes **5.** About 96,281; about 138,551

Exercise Set 9.1, p. 387

1. 12 **3.** $\frac{1}{12}$ **5.** 5280 **7.** 468 **9.** 7 **11.** $1\frac{1}{2}$ **13.** 15,840 **15.** $\frac{1}{4}$ **17.** $3\frac{1}{3}$ **19.** 52,800 **21.** $1\frac{1}{2}$ **23.** 1 **25.** 110 **27.** 2 **29.** 300 **31.** 30 **33.** $\frac{1}{36}$ **35.** 126,720 **37.** 238.7 **39.** 23,870 **41.** 0.0041 in.

Exercise Set 9.2, p. 395

1. **(a)** 1000; **(b)** 0.001 **3.** **(a)** 10; **(b)** 0.1 **5.** **(a)** 0.01; **(b)** 100 **7.** 6700 **9.** 0.98 **11.** 8.921 **13.** 0.05666 **15.** 566,600 **17.** 4.77 **19.** 688 **21.** 0.1 **23.** 100,000 **25.** 142 **27.** 0.82 **29.** 450 **31.** 0.000024 **33.** 0.688 **35.** 230 **37.** 3.92 **39.** 100 **41.** 6.6 **43.** 88.495 **45.** 1.75 **47.** 0.234 **49.** 13.85 **51.** 4

Exercise Set 9.3, p. 399

1. 17 mm **3.** 15.25 cm **5.** 13 m **7.** 30 ft **9.** 79.14 cm **11.** 88 ft **13.** 182 mm **15.** 826 m; \$1197.70 **17.** 99 cm **19.** **(a)** 14; **(b)** \$33.60; **(c)** 39 m; **(d)** \$33.15; **(e)** \$76.70 **21.** 0.561 **23.** 961

Exercise Set 9.4, p. 403

1. 15 km^2 **3.** 1.4 cm^2 **5.** 6.25 mm^2 **7.** 8100 ft^2 **9.** 50 ft^2 **11.** 169.883 cm^2 **13.** 484 ft^2 **15.** 3237.61 km^2 **17.** 1197 m^2 **19.** 630.36 m^2 **21.** **(a)** 24.75 m^2; **(b)** \$207.90 **23.** 107.5 mm^2 **25.** 45.2% **27.** 55%

Exercise Set 9.5, p. 409

1. 32 cm^2 **3.** 36 m^2 **5.** 51 ft^2 **7.** 64 m^2 **9.** 45.5 mm^2 **11.** 8.05 cm^2 **13.** 297 cm^2 **15.** 7 m^2 **17.** $55\frac{1}{8}$ ft^2 **19.** 675 cm^2 **21.** 10,816 in^2 **23.** 852.04 m^2 **25.** $\frac{37}{400}$ **27.** 137.5%

Exercise Set 9.6, p. 415

1. 14 cm **3.** $1\frac{1}{2}$ in. **5.** 16 ft **7.** 0.7 cm **9.** 44 cm **11.** $4\frac{5}{7}$ in. **13.** 100.48 ft **15.** 4.396 cm **17.** 154 cm^2 **19.** $1\frac{43}{56}$ in^2 **21.** 803.84 ft^2 **23.** 1.5386 cm^2 **25.** 3 cm; 18.84 cm; 28.26 cm^2 **27.** 151,976 km^2 **29.** 3.454 m **31.** 2.5 cm; 1.25 cm; 4.90625 cm^2 **33.** 65.94 m^2 **35.** 45.68 ft **37.** 26.84 yd **39.** 45.7 yd **41.** 100.48 m^2 **43.** 6.9972 cm^2 **45.** 48.8886 cm^2 **47.** 87.5% **49.** $66.\overline{6}$% **51.** 3.142 **53.** $3d$; πd; circumference of one ball, since $\pi > 3$

Exercise Set 9.7, p. 423

1. 10 **3.** 15 **5.** 25 **7.** 22 **9.** 23 **11.** 100 **13.** 6.928 **15.** 2.828 **17.** 4.243 **19.** 2.449 **21.** 3.162 **23.** 8.660 **25.** 14 **27.** 13.528 **29.** $c = \sqrt{34}$; $c \approx 5.831$ **31.** $c = \sqrt{98}$; $c \approx 9.899$ **33.** $a = 5$ **35.** $b = 8$ **37.** $c = 13$ **39.** $b = 24$ **41.** $a = \sqrt{399}$; $a \approx 19.975$ **43.** $b = \sqrt{224}$; $b \approx 14.967$ **45.** $c = 13$ **47.** $\sqrt{250}$ m $\approx$ 15.811 m **49.** $\sqrt{8450}$ ft $\approx$ 91.924 ft **51.** $h = \sqrt{500}$ ft; $h \approx 22.361$ ft **53.** 0.456 **55.** 1.23 **57.** The areas are the same.

Summary and Review: Chapter 9, p. 427

1. [9.1a] $2\frac{2}{3}$ **2.** [9.1a] 30 **3.** [9.2a] 0.03 **4.** [9.2a] 0.004 **5.** [9.1a] 72 **6.** [9.2a] 400,000 **7.** [9.1a] $1\frac{1}{6}$ **8.** [9.2a] 0.15 **9.** [9.3a] 23 m **10.** [9.3a] 4.4 m **11.** [9.3b] 228 ft **12.** [9.4a] 81 ft^2 **13.** [9.4a] 12.6 cm^2 **14.** [9.5a] 60 cm^2 **15.** [9.5a] 35 mm^2 **16.** [9.5a] 22.5 m^2 **17.** [9.5a] 27.5 cm^2 **18.** [9.5a] 88 m^2 **19.** [9.5a] 126 in^2 **20.** [9.4b] 840 ft^2 **21.** [9.6a] 8 m **22.** [9.6a] $\frac{14}{11}$ in., or $1\frac{3}{11}$ in. **23.** [9.6a] 14 ft **24.** [9.6a] 20 cm **25.** [9.6b] 50.24 m **26.** [9.6b] 8 in. **27.** [9.6c] 200.96 m^2 **28.** [9.6c] $5\frac{1}{11}$ in^2 **29.** [9.6d] 1038.555 ft^2 **30.** [9.7a] 8 **31.** [9.7c] $c = \sqrt{850}$; $c \approx 29.155$ **32.** [9.7c] $b = \sqrt{51}$; $b \approx 7.141$ **33.** [9.7c] $c = \sqrt{89}$ ft; $c \approx 9.434$ ft **34.** [9.7c] $a = \sqrt{76}$ cm; $a \approx 8.718$ **35.** [7.1c] 47% **36.** [7.2a] 92% **37.** [7.1b] 0.567 **38.** [7.2b] $\frac{73}{100}$

Test: Chapter 9, p. 431

1. [9.1a] 48 **2.** [9.1a] $\frac{1}{3}$ **3.** [9.2a] 6000 **4.** [9.2a] 0.87 **5.** [9.3a] 32.82 cm **6.** [9.4a] 625 m^2 **7.** [9.5a] 25 cm^2 **8.** [9.5a] 12 m^2 **9.** [9.5a] 18 ft^2 **10.** [9.6a] $\frac{1}{4}$ in. **11.** [9.6a] 9 cm **12.** [9.6b] $\frac{11}{14}$ in. **13.** [9.6c] 254.34 cm^2 **14.** [9.6d] 103.815 km^2 **15.** [9.7a] 15 **16.** [9.7b] 9.327 **17.** [9.7c] $c = 40$ **18.** [9.7c] $b = \sqrt{60}$; $b \approx 7.746$ **19.** [9.7c] $c = \sqrt{2}$; $c \approx 1.414$ **20.** [9.7c] $b = \sqrt{51}$; $b \approx 7.141$ **21.** [7.1c] 93% **22.** [7.2a] 81.25% **23.** [7.1b] 0.932 **24.** [7.2b] $\frac{1}{3}$

Cumulative Review: Chapters 1-9, p. 433

1. [1.5b] 50,854,100 **2.** [2.4b] $\frac{1}{12}$ **3.** [5.3c] 15.2 **4.** [3.6b] $\frac{2}{3}$ **5.** [5.2a] 35.6 **6.** [3.5b] $\frac{5}{6}$ **7.** [2.2a] Yes **8.** [2.2a] No **9.** [2.1d] $3 \cdot 3 \cdot 11$ **10.** [3.1a] 245 **11.** [5.3b] 35.8 **12.** [4.1a] One hundred three and sixty-four thousandths **13.** [8.1a, b] $17.8\overline{3}$, 17.5 **14.** [7.1c] 8% **15.** [7.2a] 60% **16.** [9.7a] 11 **17.** [9.7b] 5.39 **18.** [9.1a] 6 **19.** [9.3a] 14.3 m **20.** [9.5a] 297.5 yd^2 **21.** [5.2b] 150.5 **22.** [1.7b] 19,248 **23.** [2.7c] $\frac{15}{2}$ **24.** [3.3c] $\frac{2}{35}$ **25.** [8.2a] 5.5 hr **26.** [8.2a] 13.8 hr **27.** [8.2a] 2.5 hr **28.** [8.2a] 9 hr **29.** [7.7b] $11\frac{9}{11}$% **30.** [1.8a] 165,000,000,000 lb **31.** [9.4b, 9.6d] 248.64 in^2 **32.** [6.3a] $8\frac{3}{4}$ lb **33.** [7.5a] Yes **34.** [3.2c] $1\frac{1}{4}$ hr **35.** [4.4a] 255.8 mi **36.** [2.4c] \$4100 **37.** [7.5b, 9.4b] $41\frac{2}{3}$% increase

CHAPTER 10

Pretest: Chapter 10, p. 436

1. 2.304 **2.** 2400 **3.** 80 **4.** 8800 **5.** 4800 **6.** 0.62 **7.** 3.4 **8.** 420 **9.** 384 **10.** 64 **11.** 2560 **12.** 24 **13.** 25°C **14.** 98.6°F **15.** 144 **16.** 2,000,000 **17.** 1000 g **18.** 160 cm^3 **19.** 1256 ft^3 **20.** $33,493.\overline{3}$ yd^3 **21.** 150.72 cm^3

Exercise Set 10.1, p. 441

1. 768 cm^3 **3.** 45 cm^3 **5.** 75 m^3 **7.** $357\frac{1}{2}$ yd^3 **9.** 1000; 1000 **11.** 87,000 **13.** 0.049 **15.** 0.000401 **17.** 78,100 **19.** 320 **21.** 128 **23.** 32 **25.** 500 mL **27.** 125 mL **29.** 5832 yd^3 **31.** \$39 **33.** 1000 **35.** 57,480 in^3; 33.3 ft^3

Exercise Set 10.2, p. 445

1. 803.84 in^3 **3.** 353.25 cm^3 **5.** 41,580,000 yd^3 **7.** $4,186,666\frac{2}{3}$ in^3 **9.** 124.72 m^3 **11.** $1437\frac{1}{3}$ km^3 **13.** 113,982 ft^3 **15.** 24.64 cm^3 **17.** 33,880 m^3 **19.** 367.38 m^3 **21.** 113.0 m^3 **23.** 137,188,693,333.33 km^3 **25.** $6.9\overline{7}$ ft^3; $52\frac{1}{3}$ gal

Exercise Set 10.3, p. 451

1. 2000 **3.** 3 **5.** 64 **7.** 7000 **9.** 0.1 **11.** 6 **13.** 1000 **15.** 10 **17.** $\frac{1}{100}$, or 0.01 **19.** 1000 **21.** 10 **23.** 234,000 **25.** 5.2 **27.** 6.7 **29.** 0.0502 **31.** 6.78 **33.** 6.9 **35.** 800,000 **37.** 1000 **39.** 0.0034 **41.** 24 **43.** 60 **45.** $365\frac{1}{4}$ **47.** 336 **49.** 8.2 **51.** 16 **53.** 125 **55.** 0.4535 **57.** 1000 **59.** 0.125 mg **61.** 4 **63.** 8 mL **65.** $\frac{2}{3}$ oz **67.** About 31.7 yr

Exercise Set 10.4, p. 457

1. 80°*C* **3.** 60°*C* **5.** 30°*C* **7.** −10°*C* **9.** 190°*F* **11.** 140°*F* **13.** 10°*F* **15.** 40°*F* **17.** 77°*F* **19.** 104°*F* **21.** 5432°*F* **23.** 30°*C* **25.** 55°*C* **27.** 37°*C* **29.** 234 **31.** 336 **33.** 260.6°*F*

Exercise Set 10.5, p. 461

1. 144 **3.** 640 **5.** $\frac{1}{144}$ **7.** 198 **9.** 396 **11.** 12,800 **13.** 27,878,400 **15.** 5 **17.** 1 **19.** 214.75 ft² **21.** 20,000,000 **23.** 140 **25.** 23.456 **27.** 0.12 **29.** 2500 **31.** 1.875 ft² **33.** $13\frac{1}{3}$ ft² **35.** 42.05915 cm² **37.** 4.2 **39.** $\frac{1}{640}$, or 0.0015625

Summary and Review: Chapter 10, p. 463

1. [10.1a] 93.6 m³ **2.** [10.1a] 193.2 cm³ **3.** [10.3a] 112 **4.** [10.3b] 0.004 **5.** [10.3c] $\frac{4}{15}$ **6.** [10.1b] 0.464 **7.** [10.3c] 180 **8.** [10.3b] 4700 **9.** [10.3a] 16,140 **10.** [10.1b] 830 **11.** [10.3c] $\frac{1}{4}$ **12.** [10.3b] 0.04 **13.** [10.3b] 200 **14.** [10.3b] 30 **15.** [10.1b] 0.06 **16.** [10.3a] 1600 **17.** [10.1b] 400 **18.** [10.3a] $1\frac{1}{4}$ **19.** [10.3c] 50 **20.** [10.1b] 160 **21.** [10.1b] 7.5 **22.** [10.1b] 13.5 **23.** [10.4b] 80.6°*F* **24.** [10.4b] 20°*C* **25.** [10.2a] 31,400 ft³ **26.** [10.2b] $33.49\overline{3}$ cm³ **27.** [10.2c] 4.71 in³ **28.** [10.5a] 36 **29.** [10.5b] 300,000 **30.** [10.5a] 14.375 **31.** [10.5b] 0.06 **32.** [10.1c] 18.72 L **33.** [7.8a] $39.58 **34.** [1.9b] 27 **35.** [1.9b] 22.09 **36.** [1.9b] 103.823 **37.** [1.9b] $\frac{1}{16}$ **38.** [9.1a] 13,200 **39.** [9.1a] 4 **40.** [9.2a] 45.68 **41.** [9.2a] 45,680

Test: Chapter 10, p. 465

1. [10.1a] 84 cm³ **2.** [10.1b] 3.08 **3.** [10.1b] 240 **4.** [10.3a] 64 **5.** [10.3a] 8220 **6.** [10.3b] 3800 **7.** [10.3b] 0.4325 **8.** [10.3b] 2.2 **9.** [10.3c] 300 **10.** [10.3c] 360 **11.** [10.1b] 32 **12.** [10.1b] 1280 **13.** [10.1b] 40 **14.** [10.4b] 35°*C* **15.** [10.4b] 138.2°*F* **16.** [10.5a] 1728 **17.** [10.5b] 0.0003 **18.** [10.1a] 420 in³ **19.** [10.2a] 1177.5 ft³ **20.** [10.2b] $4186.\overline{6}$ yd³ **21.** [10.2c] 113.04 cm³ **22.** [7.8a] $880 **23.** [1.9b] 1000 **24.** [1.9b] $\frac{1}{16}$ **25.** [1.9b] 9.8596 **26.** [1.9b] 0.00001 **27.** [9.1a] 42 **28.** [9.1a] 250 **29.** [9.2a] 2300 **30.** [9.2a] 3400

Cumulative Review: Chapters 1-10, p. 467

1. [3.5a] $4\frac{1}{6}$ **2.** [5.3c] 49.2 **3.** [4.3b] 87.52 **4.** [1.6c] 1234 **5.** [1.9d] 2 **6.** [1.9c] 1565 **7.** [4.1b] $\frac{1209}{1000}$ **8.** [7.2b] $\frac{17}{100}$ **9.** [3.3b] < **10.** [2.5c] = **11.** [10.3a] $\frac{3}{8}$ **12.** [10.4b] 59 **13.** [10.1b] 87 **14.** [10.3c] $\frac{3}{20}$ **15.** [10.5a] 27 **16.** [9.2a] 0.17 **17.** [9.3a, 9.5a] 380 cm; 5500 cm² **18.** [9.3a, 9.5a] 32.3 ft, 56.55 ft² **19.** [8.3c] 20 lb **20.** [8.3c] 1984 **21.** [6.1c] $14\frac{2}{5}$ **22.** [6.1c] $4\frac{2}{7}$ **23.** [5.2b] 113.4 **24.** [3.3c] $\frac{1}{8}$ **25.** [8.1a] 99 **26.** [1.8a] 528 million **27.** [10.2b] 4187 cm³ **28.** [7.8a] $24 **29.** [9.7c] 17 m **30.** [7.6a] 6% **31.** [3.5c] $2\frac{1}{8}$ yd **32.** [5.5a] $21.82 **33.** [6.2b] The 8-qt box **34.** [2.6b] $\frac{7}{20}$ km **35.** [9.1a, 10.1a] 278 ft³

CHAPTER 11

Pretest: Chapter 11, p. 470

1. > **2.** > **3.** > **4.** < **5.** −0.625 **6.** $-0.\overline{6}$ **7.** $-0.\overline{90}$ **8.** 12 **9.** 2.3 **10.** 0 **11.** −5.4 **12.** $\frac{2}{3}$ **13.** −17 **14.** 38.6 **15.** $-\frac{17}{15}$ **16.** −5 **17.** 63 **18.** $-\frac{5}{12}$ **19.** −98 **20.** 8 **21.** 24 **22.** 26

Sidelight: An Application: Leap Years, p. 476

1. Yes **2.** No **3.** Yes **4.** No **5.** No **6.** No **7.** No **8.** Yes

Exercise Set 11.1, p. 477

1. 5, −12 **3.** −1286; 29,028 **5.** 750, −125 **7.** 20, −150, 300

9.

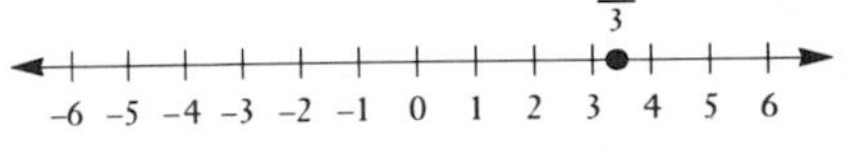

11.

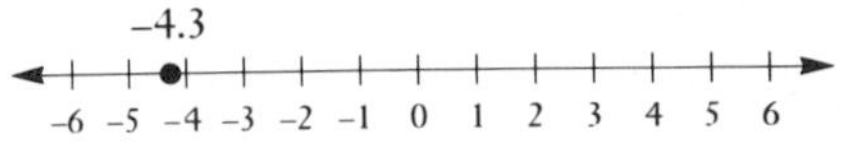

13. −0.625 **15.** $-1.\overline{6}$ **17.** $-1.1\overline{6}$ **19.** −0.875 **21.** $-0.\dot{3}5$ **23.** > **25.** > **27.** > **29.** < **31.** > **33.** > **35.** < **37.** > **39.** > **41.** < **43.** 7 **45.** 11 **47.** 4 **49.** 325 **51.** $\frac{10}{7}$

53. 14.8 **55.** $2 \cdot 2 \cdot 2 \cdot 2 \cdot 2 \cdot 2 \cdot 3$ **57.** 96 **59.** < **61.** $-\frac{5}{6}$, $-\frac{3}{4}$, $-\frac{2}{3}$, $\frac{1}{6}$, $\frac{3}{8}$, $\frac{1}{2}$

Exercise Set 11.2, p. 483

1. −7 **3.** −4 **5.** 0 **7.** −8 **9.** −7 **11.** −27 **13.** 0 **15.** −42 **17.** 0 **19.** 0 **21.** 3 **23.** −9 **25.** 7 **27.** 0 **29.** 45 **31.** −1.8 **33.** −8.1 **35.** $-\frac{1}{5}$ **37.** $-\frac{8}{7}$ **39.** $-\frac{3}{8}$ **41.** $-\frac{29}{35}$ **43.** $-\frac{11}{15}$ **45.** −6.3 **47.** $\frac{7}{16}$ **49.** 39 **51.** 50 **53.** −1093 **55.** −24 **57.** 26.9 **59.** −9 **61.** $\frac{14}{3}$ **63.** −65 **65.** $\frac{5}{3}$ **67.** 14 **69.** −10 **71.** All positive **73.** Positive **75.** Negative

Exercise Set 11.3, p. 487

1. −4 **3.** −7 **5.** −6 **7.** 0 **9.** −4 **11.** −7 **13.** −6 **15.** 0 **17.** 11 **19.** −14 **21.** 5 **23.** −7 **25.** −5 **27.** −3 **29.** −23 **31.** −68 **33.** −73 **35.** 116 **37.** −2.8 **39.** $-\frac{1}{4}$ **41.** $\frac{1}{12}$ **43.** $-\frac{17}{12}$ **45.** $\frac{1}{8}$ **47.** 19.9 **49.** −9 **51.** −0.01 **53.** −2.7 **55.** −3.53 **57.** $-\frac{1}{2}$ **59.** $\frac{6}{7}$ **61.** $-\frac{41}{30}$ **63.** $-\frac{1}{156}$ **65.** 37 **67.** −62 **69.** 6 **71.** 107 **73.** 219 **75.** 96.6 cm^2 **77.** 116 m

Exercise Set 11.4, p. 491

1. −16 **3.** −24 **5.** −72 **7.** 16 **9.** 42 **11.** −120 **13.** −238 **15.** 1200 **17.** 98 **19.** −12.4 **21.** 24 **23.** 21.7 **25.** $-\frac{2}{5}$ **27.** $\frac{1}{12}$ **29.** −17.01 **31.** $-\frac{5}{12}$ **33.** 420 **35.** $\frac{2}{7}$ **37.** −60 **39.** 150 **41.** $-\frac{2}{45}$ **43.** 1911 **45.** 50.4 **47.** $\frac{10}{189}$ **49.** −960 **51.** 17.64 **53.** $-\frac{5}{784}$ **55.** −756 **57.** −720 **59.** −30, 240 **61.** $2 \cdot 2 \cdot 2 \cdot 2 \cdot 2 \cdot 2 \cdot 2 \cdot 2 \cdot 2 \cdot 3 \cdot 3$ **63.** $33\frac{1}{3}\%$, or $33.\overline{3}\%$ **65.** 72 **67.** 1944 **69.** −17 **71. (a)** One must be negative and one must be positive; **(b)** Either or both must be zero; **(c)** Both must be negative or both must be positive

Exercise Set 11.5, p. 497

1. −6 **3.** −13 **5.** −2 **7.** 4 **9.** −8 **11.** 2 **13.** −12 **15.** −8 **17.** Undefined **19.** −9 **21.** $-\frac{7}{15}$ **23.** $\frac{1}{13}$ **25.** $-\frac{9}{8}$ **27.** $\frac{5}{3}$ **29.** $\frac{9}{14}$ **31.** $\frac{9}{64}$ **33.** −2 **35.** $\frac{11}{13}$ **37.** −16.2 **39.** Undefined **41.** −7 **43.** −7 **45.** −334 **47.** 14 **49.** 1880 **51.** 12 **53.** 8 **55.** −86 **57.** 37 **59.** −1 **61.** −10 **63.** 25 **65.** −7988 **67.** −3000 **69.** 60 **71.** 1 **73.** 10 **75.** $-\frac{13}{45}$ **77.** $-\frac{4}{3}$

Summary and Review: Chapter 11, p. 499

1. [11.1e] 38 **2.** [11.1e] 7.3 **3.** [11.1e] $\frac{5}{2}$ **4.** [11.1e] −0.2 **5.** [11.1c] −1.25 **6.** [11.1c] $-0.8\overline{3}$ **7.** [11.1c] $-0.41\overline{6}$ **8.** [11.1c] $-0.\overline{27}$

9. [11.1b]

−2.5

−6 −5 −4 −3 −2 −1 0 1 2 3 4 5 6

10. [11.1b]

$\frac{8}{9}$

−6 −5 −4 −3 −2 −1 0 1 2 3 4 5 6

11. [11.1d] < **12.** [11.1d] > **13.** [11.1d] > **14.** [11.1d] < **15.** [11.2b] −3.8 **16.** [11.2b] $\frac{3}{4}$ **17.** [11.2b] 34 **18.** [11.2b] 5 **19.** [11.5b] $\frac{8}{3}$ **20.** [11.5b] $-\frac{1}{7}$ **21.** [11.5b] −10 **22.** [11.2a] −3 **23.** [11.2a] $-\frac{7}{12}$ **24.** [11.2a] −4 **25.** [11.2a] −5 **26.** [11.3a] 4 **27.** [11.3a] $-\frac{7}{5}$ **28.** [11.3a] −7.9 **29.** [11.4a] 54 **30.** [11.4a] −9.18 **31.** [11.4a] $-\frac{2}{7}$ **32.** [11.4a] −210 **33.** [11.5a] −7 **34.** [11.5c] −3 **35.** [11.5c] $\frac{3}{4}$ **36.** [11.5d] 40.4 **37.** [11.5d] −62 **38.** [11.5d] −5 **39.** [9.4a] 210 cm^2 **40.** [3.1a] 270 **41.** [2.1d] $2 \cdot 2 \cdot 2 \cdot 3 \cdot 3 \cdot 3 \cdot 3$ **42.** [7.3b] 36% **43.** [11.1e, 11.3a] $-\frac{5}{8}$ **44.** [11.1e, 11.5d] −2.1

Test: Chapter 11, p. 501

1. [11.1d] < **2.** [11.1d] > **3.** [11.1d] > **4.** [11.1d] < **5.** [11.1c] −0.125 **6.** [11.1c] $-0.\overline{4}$ **7.** [11.1c] $-0.\overline{18}$ **8.** [11.1e] 7 **9.** [11.1e] $\frac{9}{4}$ **10.** [11.1e] −2.7 **11.** [11.2b] $-\frac{2}{3}$ **12.** [11.2b] 1.4 **13.** [11.2b] 8 **14.** [11.5b] $-\frac{1}{2}$ **15.** [11.5b] $\frac{7}{4}$ **16.** [11.3a] 7.8 **17.** [11.2a] −8 **18.** [11.2a] $\frac{7}{40}$ **19.** [11.3a] 10 **20.** [11.3a] −2.5 **21.** [11.3a] $\frac{7}{8}$ **22.** [11.4a] −48 **23.** [11.4a] $\frac{3}{16}$ **24.** [11.5a] −9 **25.** [11.5c] $\frac{3}{4}$ **26.** [11.5c] −9.728 **27.** [11.5d] 109 **28.** [9.4a] 55.8 ft^2 **29.** [7.3b] 48% **30.** [2.1d] $2 \cdot 2 \cdot 2 \cdot 5 \cdot 7$ **31.** [3.1a] 240 **32.** [11.1e, 11.3a] 15

Cumulative Review: Chapters 1-11, p. 503

1. [7.1b] 0.263 **2.** [11.1c] $-0.\overline{45}$ **3.** [10.3b] 834 **4.** [10.5b] 0.0275 **5.** [11.1e] 4.5 **6.** [11.3a] −11 **7.** [6.2a] 12.5 m/sec **8.** [9.6a, b, c] 35 mi, 220 mi, 3850 mi^2 **9.** [9.7a] 15 **10.** [9.7b] 8.31 **11.** [11.4a] −10 **12.** [11.5a] 3 **13.** [11.2a] 8

14. [8.2a]

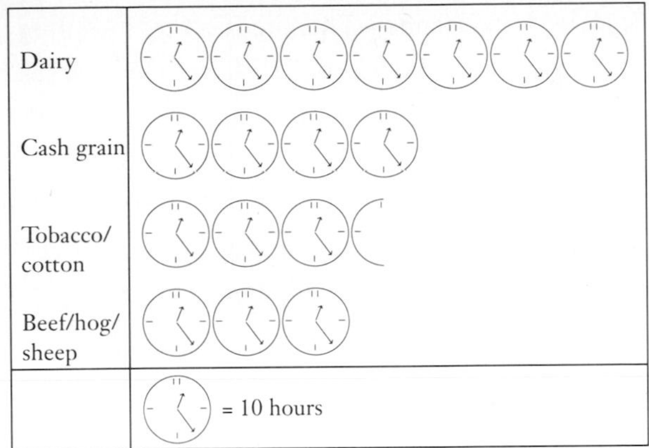

15. [5.1a] 0.01485 **16.** [11.3a] −3 **17.** [11.3a] $-\frac{3}{22}$ **18.** [11.4a] $-\frac{1}{24}$ **19.** [3.5b] $1\frac{5}{6}$ **20.** [11.5c] $-\frac{1}{4}$ **21.** [11.5d] −57.5 **22.** [1.5b] 9,640,500,000 **23.** [11.2a] 32.49 **24.** [11.5d] Undefined **25.** [7.3b] 87.5% **26.** [7.3b] 32 **27.** [4.4a] $133.05 **28.** [10.2a] 307.72 cm^3 **29.** [8.1a] $44.8\overline{3}°$ **30.** [7.5a] 78 **31.** [9.4b] 6902.5 m^2 **32.** [3.5c] $2\frac{11}{12}$ cups **33.** [1.8a] $239,441 **34.** [5.5a] 4.55 km

CHAPTER 12

Pretest: Chapter 12, p. 506

1. $\frac{5}{16}$ **2.** $78\%x$, or $0.78x$ **3.** $9z-18$ **4.** $-4a-2b+10c$ **5.** $4(x-3)$ **6.** $3(2y-3z-6)$ **7.** $-3x$ **8.** $2x+2y+18$ **9.** −7 **10.** −1 **11.** 2 **12.** 3 **13.** Width: 34 m; length: 39 m **14.** $650

Exercise Set 12.1, p. 513

1. 42 **3.** 3 **5.** 1 **7.** 6 **9.** 240; 240 **11.** 160; 160 **13.** $2b+10$ **15.** $7-7t$ **17.** $30x+12$ **19.** $7x+28+42y$ **21.** $-7y+14$ **23.** $45x+54y-72$ **25.** $-4x+12y+8z$ **27.** $-3.72x+9.92y-3.41$ **29.** $2(x+2)$ **31.** $5(6+y)$ **33.** $7(2x+3y)$ **35.** $5(x+2+3y)$ **37.** $8(x-3)$ **39.** $4(8-y)$ **41.** $2(4x+5y-11)$ **43.** $6(3x-2y+1)$ **45.** $19a$ **47.** $9a$ **49.** $8x+9z$ **51.** $-19a+88$ **53.** $4t+6y-4$ **55.** $-8x$ **57.** $5n$ **59.** $-16y$ **61.** $17a-12b-1$ **63.** $4x+2y$ **65.** $7x+y$ **67.** $0.8x+0.5y$

Exercise Set 12.2, p. 517

1. 4 **3.** −20 **5.** −14 **7.** −18 **9.** 15 **11.** −14 **13.** 2 **15.** 20 **17.** −6 **19.** $\frac{7}{3}$ **21.** $-\frac{7}{4}$ **23.** $\frac{41}{24}$ **25.** $-\frac{1}{20}$ **27.** 5.1 **29.** 12.4 **31.** −5 **33.** $1\frac{5}{6}$ **35.** $-\frac{10}{21}$ **37.** −11 **39.** $-\frac{5}{12}$ **41.** 342.246 **43.** $-\frac{26}{15}$ **45.** −10 **47.** All real numbers **49.** $-\frac{5}{17}$ **51.** 13, −13

Exercise Set 12.3, p. 521

1. 6 **3.** 9 **5.** 12 **7.** −40 **9.** 5 **11.** −7 **13.** −6 **15.** 6 **17.** −63 **19.** 36 **21.** −21 **23.** $-\frac{3}{5}$ **25.** $-\frac{3}{2}$ **27.** $\frac{9}{2}$ **29.** 7 **31.** −7 **33.** 8 **35.** 15.9 **37.** 62.8 ft; 20 ft; 314 ft^2 **39.** 8000 ft^3 **41.** −8655 **43.** No solution **45.** No solution

Exercise Set 12.4, p. 529

1. 5 **3.** 8 **5.** 10 **7.** 14 **9.** −8 **11.** −8 **13.** −7 **15.** 15 **17.** 6 **19.** 4 **21.** 6 **23.** −3 **25.** 1 **27.** −20 **29.** 6 **31.** 7 **33.** 2 **35.** 5 **37.** 2 **39.** 10 **41.** 4 **43.** 0 **45.** −1 **47.** $-\frac{4}{3}$ **49.** $\frac{2}{5}$ **51.** −2 **53.** −4 **55.** $\frac{4}{5}$ **57.** −6.5 **59.** < **61.** 2 **63.** −2

Exercise Set 12.5, p. 537

1. $b+6$, or $6+b$ **3.** $c-9$ **5.** $6+q$, or $q+6$ **7.** $b+a$, or $a+b$ **9.** $y-x$ **11.** $x+w$, or $w+x$ **13.** $r+s$, or $s+r$ **15.** $2x$ **17.** $5t$ **19.** $97\%x$, or $0.97x$ **21.** 52 **23.** 29 **25.** $1.99 **27.** 19 **29.** −10 **31.** 20 m, 40 m, 120 m **33.** Width: 100 ft; length: 160 ft; area: 16,000 ft^2 **35.** Length: 27.9 cm; width: 21.6 cm **37.** 450.5 mi **39.** 20 **41.** Length: 12 cm; width: 9 cm **43.** 5 half dollars, 10 quarters, 20 dimes, 60 nickels

Summary and Review: Chapter 12, p. 539

1. [12.2a] −22 **2.** [12.3a] 7 **3.** [12.3a] −192 **4.** [12.2a] 1 **5.** [12.3a] $-\frac{7}{3}$ **6.** [12.2a] 25 **7.** [12.2a] $\frac{1}{2}$ **8.** [12.3a] $-\frac{15}{64}$ **9.** [12.2a] 9.99 **10.** [12.4a] −8 **11.** [12.4b] −5 **12.** [12.4b] $-\frac{1}{3}$ **13.** [12.4a] 4 **14.** [12.4b] 3 **15.** [12.4b] 4 **16.** [12.4b] 16 **17.** [12.5a] $19\%x$, or $0.19x$ **18.** [12.5b] $591 **19.** [12.5b] 27 **20.** [12.5b] 3 m, 5 m **21.** [12.5b] 9 **22.** [12.5b] Width: 11 cm; length: 17 cm **23.** [12.5b] 1800.4 **24.** [9.6a, b, c] 40 ft; 125.6 ft; 1256 ft^2 **25.** [10.1a] 1702 cm^3 **26.** [11.2a] −45 **27.** [11.4a] −60 **28.** [12.5b] Amazon: 6437 km; Nile: 6671 km **29.** [11.1e; 12.4a] 23, −23 **30.** [11.1e, 12.3a] 20; −20

Test: Chapter 12, p. 541

1. [12.2a] 8 **2.** [12.2a] 26 **3.** [12.3a] −6 **4.** [12.3a] 49 **5.** [12.4b] −12 **6.** [12.4a] 2 **7.** [12.4a] −8 **8.** [12.2a] $-\frac{7}{20}$ **9.** [12.4b] 2.5 **10.** [12.5b] Width: 7 cm; Length: 11 cm **11.** [12.5b] 6 **12.** [12.5a] $x-9$ **13.** [11.4a] 180

14. [11.2a] $-\frac{2}{9}$ **15.** [9.6a, b, c] 140 yd; 439.6 yd; 15,386 yd^2 **16.** [10.1a] 1320 ft^3 **17.** [11.1e, 12.4a] 15, −15 **18.** [12.5b] 60

Cumulative Review: Chapters 1-12, p. 543

1. [1.1e] 7 **2.** [1.1a] 7 thousands + 4 hundreds + 5 ones **3.** [1.2b] 1012 **4.** [1.2b] 21,085 **5.** [3.2b] $\frac{5}{26}$ **6.** [3.5a] $5\frac{7}{9}$ **7.** [4.3a] 493.971 **8.** [4.3a] 802.876 **9.** [1.3d] 152 **10.** [1.3d] 674 **11.** [3.3a] $\frac{5}{24}$ **12.** [3.5b] $2\frac{17}{24}$ **13.** [4.3b] 19.9973 **14.** [4.3b] 34.241 **15.** [1.5b] 4752 **16.** [1.5b] 266,287 **17.** [3.6a] $4\frac{1}{12}$ **18.** [2.6a] $\frac{6}{5}$ **19.** [2.6a] 10 **20.** [5.1a] 259.084 **21.** [1.6c] 573 **22.** [1.6c] 56 R 10 **23.** [3.4b] $56\frac{5}{17}$ **24.** [2.7b] $\frac{3}{2}$ **25.** [3.6b] $\frac{7}{90}$ **26.** [5.2a] 39 **27.** [1.4a] 68,000 **28.** [4.2b] 0.428 **29.** [5.3b] 21.84 **30.** [2.2a] Yes **31.** [2.1a] 1, 3, 5, 15 **32.** [3.1a] 800 **33.** [2.5b] $\frac{7}{10}$ **34.** [2.5b] 55 **35.** [3.4b] $3\frac{3}{5}$ **36.** [2.5c] $\neq$ **37.** [3.3b] $<$ **38.** [4.2a] 1.001 **39.** [1.4c] $>$ **40.** [2.3b] $\frac{3}{5}$ **41.** [4.1c] 0.037 **42.** [5.3a] 0.52 **43.** [5.3a] $0.\overline{8}$ **44.** [7.1b] 0.07 **45.** [4.1b] $\frac{463}{100}$ **46.** [3.4a] $\frac{29}{4}$ **47.** [7.2b] $\frac{2}{5}$ **48.** [7.2a] 85% **49.** [7.1c] 150% **50.** [1.7b] 555 **51.** [5.2b] 64 **52.** [2.7c] $\frac{5}{4}$ **53.** [6.1c] $76\frac{1}{2}$, or 76.5 **54.** [1.8a] $675 **55.** [1.8a] 65 min **56.** [3.5c] $\$25\frac{3}{4}$ **57.** [4.4a] 485.9 mi **58.** [1.8a] $8100 **59.** [1.8a] $423 **60.** [6.3a] $\frac{3}{10}$ km **61.** [5.5a] $84.96 **62.** [6.3a] 13 gal **63.** [6.2b] 17 cents/oz **64.** [7.8a] $480 **65.** [7.7a] 7% **66.** [7.5b] 30,160 **67.** [8.1a, b, c] 28; 26; 18 **68.** [1.9b] 324 **69.** [1.9b] 400 **70.** [9.7a] 3 **71.** [9.7a] 11 **72.** [9.7b] 4.472 **73.** [9.7c] $c = \sqrt{50}$ ft; $c \approx 7.071$ ft **74.** [9.1a] 12 **75.** [9.2a] 428 **76.** [10.3c] 72 **77.** [10.3b] 20 **78.** [10.3a] 80 **79.** [10.3b] 0.08 **80.** [10.1b] 8.19 **81.** [10.1b] 5 **82.** [9.3a, 9.4a] 25.6 m; 25.75 m^2 **83.** [9.5a] 25 in^2 **84.** [9.5a] 128.65 yd^2 **85.** [9.5a] 61.6 cm^2 **86.** [9.6a, b, c] 20.8 in.; 65.312 in.; 339.6224 in^2 **87.** [10.1a] 52.9 m^3 **88.** [10.2a] 803.84 cm^3 **89.** [10.2c] $267.94\overline{6}$ cm^3 **90.** [10.2b] $267.94\overline{6}$ cm^3 **91.** [1.9c] 238 **92.** [1.9c] 172 **93.** [11.1e,11.4a] 3 **94.** [11.2a] 14 **95.** [11.3a] $\frac{1}{3}$ **96.** [11.4a] 30 **97.** [11.4a] $-\frac{2}{7}$ **98.** [11.5a] −8 **99.** [12.4a] −5 **100.** [12.3a] 4 **101.** [12.4b] −8 **102.** [12.5a] $y + 17$ **103.** [12.5a] $38\%x$, or $0.38x$ **104.** [12.5b] 39 **105.** [12.5b] 36

Final Examination, p. 547

1. [1.1a] 8 thousands + 3 hundreds + 4 tens + 5 ones **2.** [1.1e] 7 **3.** [4.3a] 215.6177 **4.** [3.5a] $8\frac{3}{4}$ **5.** [3.2b] $\frac{5}{3}$ **6.** [1.2b] 759 **7.** [1.2b] 21,562 **8.** [4.3a] 158.8837 **9.** [1.3d] 5561 **10.** [1.3d] 5937 **11.** [4.3b] 30.392 **12.** [3.5b] $\frac{5}{8}$ **13.** [3.3a] $\frac{1}{12}$ **14.** [4.3b] 99.16 **15.** [3.6a] $24\frac{3}{8}$ **16.** [2.6a] $\frac{2}{3}$ **17.** [2.6a] 6 **18.** [1.5b] 5814 **19.** [1.5b] 234,906 **20.** [5.1a] 226.327 **21.** [1.6c] 517 R 1 **22.** [3.4b] $517\frac{1}{8}$ **23.** [1.6c] 197 **24.** [2.7b] $\frac{2}{3}$ **25.** [3.6b] $1\frac{19}{66}$ **26.** [5.2a] 48 **27.** [1.4a] 43,000 **28.** [4.2b] 6.79 **29.** [5.3b] 7.384 **30.** [2.2a] Yes **31.** [2.1a] 1, 2, 4, 8 **32.** [3.1a] 230 **33.** [2.5b] $\frac{3}{2}$ **34.** [2.5b] 10 **35.** [3.4b] $7\frac{2}{3}$ **36.** [2.5c] $=$ **37.** [3.3b] $<$ **38.** [4.2a] 0.9 **39.** [1.4c] $<$ **40.** [2.3b] $\frac{1}{3}$ **41.** [7.1b] 0.499 **42.** [5.3a] 0.24 **43.** [5.3a] $0.\overline{27}$ **44.** [4.1c] 7.86 **45.** [3.4a] $\frac{23}{4}$ **46.** [7.2b] $\frac{37}{100}$ **47.** [4.1b] $\frac{897}{1000}$ **48.** [7.1c] 77% **49.** [7.2a] 96% **50.** [6.1c] 3.84, or $3\frac{21}{25}$ **51.** [3.3c] $\frac{1}{25}$ **52.** [1.7b] 25 **53.** [4.3c] 245.7 **54.** [7.5b] 5% **55.** [4.4a] $242.60 **56.** [2.7d] 80 **57.** [1.8a] $348 **58.** [3.5c] $3\frac{3}{4}$ m **59.** [1.8a] $1311 **60.** [1.8a] $39 **61.** [1.8a] 13 mpg **62.** [5.5a] $16.99 **63.** [6.3a] 540 km **64.** [6.2b] $3.98 **65.** [7.5a] 39 **66.** [7.8a] $60 **67.** [8.1a, b, c] $\$15.1\overline{6}$; $12; $12 **68.** [8.4a] **(a)** Blue; **(b)** 1050 **69.** [1.9b] 625 **70.** [1.9b] 256 **71.** [9.7a] 7 **72.** [9.7a] 25 **73.** [9.7b] 4.899 **74.** [9.7c] $a = \sqrt{85}$ ft; $a \approx 9.220$ ft **75.** [9.1a] 5 **76.** [9.2a] 2.371 **77.** [10.1b] 5000 **78.** [10.3a] 14,000 **79.** [10.3c] 1440 **80.** [10.3b] 5340 **81.** [10.3b] 7.54 **82.** [10.1b] 5 **83.** [9.3a, 9.4a] 24.8 m; 26.88 m^2 **84.** [9.5a] 38.775 ft^2 **85.** [9.5a] 153 m^2 **86.** [9.5a] 216 dm^2 **87.** [9.6a, b, c] 4.3 yd; 27.004 yd; 58.0586 yd^2 **88.** [10.1a] 68.921 ft^3 **89.** [10.2a] 314,000 m^3 **90.** [10.2b] $4186.\overline{6}$ m^3 **91.** [10.2c] $104,666.\overline{6}$ m^3 **92.** [1.9c] 383 **93.** [1.9c] 99 **94.** [11.1e] 32 **95.** [11.3a] −22 **96.** [11.2a] −22 **97.** [11.4a] 30 **98.** [11.4a] $-\frac{5}{9}$ **99.** [11.5a] −6 **100.** [12.2a] −76.4 **101.** [12.3a] 27 **102.** [12.4b] −15 **103.** [12.5b] 532.5 **104.** [12.5b] 40

DEVELOPMENTAL UNITS

Exercise Set A.1, p. 555

1. 11 **2.** 17 **3.** 15 **4.** 13 **5.** 15 **6.** 14 **7.** 12 **8.** 12 **9.** 13 **10.** 11 **11.** 16 **12.** 17 **13.** 16 **14.** 12 **15.** 10 **16.** 10 **17.** 11 **18.** 7 **19.** 7 **20.** 11 **21.** 8 **22.** 0 **23.** 3

24. 18 **25.** 14 **26.** 9 **27.** 10 **28.** 14 **29.** 4 **30.** 14 **31.** 11 **32.** 2 **33.** 15 **34.** 16 **35.** 16 **36.** 9 **37.** 11 **38.** 13 **39.** 14 **40.** 5 **41.** 11 **42.** 7 **43.** 13 **44.** 14 **45.** 12 **46.** 6 **47.** 10 **48.** 12 **49.** 10 **50.** 8 **51.** 2 **52.** 9 **53.** 13 **54.** 8 **55.** 10 **56.** 9 **57.** 10 **58.** 6 **59.** 13 **60.** 9 **61.** 8 **62.** 11 **63.** 16 **64.** 7 **65.** 12 **66.** 12 **67.** 8 **68.** 4 **69.** 7 **70.** 10 **71.** 15 **72.** 13 **73.** 4 **74.** 6 **75.** 8 **76.** 8 **77.** 15 **78.** 12 **79.** 13 **80.** 10 **81.** 12 **82.** 17 **83.** 13 **84.** 10 **85.** 16 **86.** 16 **87.** 17

Sidelight: A Number Puzzle, p. 560

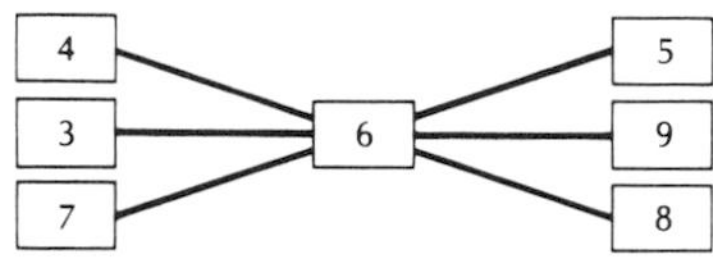

Exercise Set A.2, p. 561

1. 19 **2.** 39 **3.** 95 **4.** 89 **5.** 87 **6.** 79 **7.** 849 **8.** 999 **9.** 900 **10.** 868 **11.** 875 **12.** 999 **13.** 848 **14.** 877 **15.** 1689 **16.** 17,680 **17.** 10,873 **18.** 4699 **19.** 9747 **20.** 10,867 **21.** 9895 **22.** 3998 **23.** 9998 **24.** 18,222 **25.** 7777 **26.** 16,889 **27.** 18,999 **28.** 64,489 **29.** 99,999 **30.** 77,777 **31.** 46 **32.** 26 **33.** 55 **34.** 101 **35.** 1643 **36.** 1010 **37.** 846 **38.** 628 **39.** 1204 **40.** 607 **41.** 10,000 **42.** 1010 **43.** 1110 **44.** 1227 **45.** 1111 **46.** 1717 **47.** 10,138 **48.** 6554 **49.** 6111 **50.** 8427 **51.** 9890 **52.** 11,612 **53.** 11,125 **54.** 15,543 **55.** 16,774 **56.** 68,675 **57.** 34,437 **58.** 166,444 **59.** 101,315 **60.** 49,449 **61.** 45, 56, 67, 78, 89

62. 1 6 15 20 15 6 1;

1 7 21 35 35 21 7 1

Exercise Set A.3, p. 565

1. 17 **2.** 19 **3.** 308 **4.** 171 **5.** 3718 **6.** 2592 **7.** 1167 **8.** 1092 **9.** 18,699 **10.** 23,068 **11.** 15,085 **12.** 86,425 **13.** 2740 **14.** 10,983 **15.** 1243 **16.** 8497 **17.** 2798 **18.** 1681 **19.** 169,351 **20.** $40,000

Exercise Set S.1, p. 569

1. 4 **2.** 1 **3.** 3 **4.** 2 **5.** 0 **6.** 0 **7.** 5 **8.** 3 **9.** 1 **10.** 1 **11.** 7 **12.** 5 **13.** 7 **14.** 9 **15.** 0 **16.** 0 **17.** 5 **18.** 3 **19.** 8 **20.** 8 **21.** 6 **22.** 5 **23.** 7 **24.** 3 **25.** 7 **26.** 9 **27.** 6 **28.** 6 **29.** 2 **30.** 2 **31.** 0 **32.** 0 **33.** 4 **34.** 4 **35.** 5 **36.** 4 **37.** 4 **38.** 5 **39.** 9 **40.** 9 **41.** 7 **42.** 9 **43.** 5 **44.** 6 **45.** 6 **46.** 2 **47.** 6 **48.** 7 **49.** 1 **50.** 8 **51.** Fill the 3-gal bucket and pour it into the 5-gal bucket. Fill the 3-gal bucket again, and pour into the 5-gal until full; one gal remains.

Exercise Set S.2, p. 575

1. 11 **2.** 13 **3.** 33 **4.** 21 **5.** 47 **6.** 67 **7.** 247 **8.** 193 **9.** 500 **10.** 766 **11.** 654 **12.** 333 **13.** 202 **14.** 617 **15.** 288 **16.** 1220 **17.** 2231 **18.** 2261 **19.** 5003 **20.** 4126 **21.** 3764 **22.** 12,522 **23.** 1691 **24.** 12,100 **25.** 1226 **26.** 99,998 **27.** 10 **28.** 10,101 **29.** 11,043 **30.** 11,111 **31.** 18 **32.** 39 **33.** 65 **34.** 29 **35.** 8 **36.** 9 **37.** 308 **38.** 317 **39.** 126 **40.** 617 **41.** 214 **42.** 89 **43.** 4402 **44.** 3555 **45.** 1503 **46.** 5889 **47.** 2387 **48.** 3649 **49.** 3832 **50.** 8144 **51.** 7750 **52.** 10,445 **53.** 33,793 **54.** 26,869 **55.** 16 **56.** 13 **57.** 16 **58.** 17 **59.** 86 **60.** 281 **61.** 455 **62.** 389 **63.** 571 **64.** 6148 **65.** 2200 **66.** 2113 **67.** 3748 **68.** 4313 **69.** 1063 **70.** 2418 **71.** 5206 **72.** 1459 **73.** 305 **74.** 4455

Exercise Set S.3, p. 579

1. 39 **2.** 65 **3.** 18 **4.** 24 **5.** 80 **6.** 308 **7.** 114 **8.** 84 **9.** 3555 **10.** 5312 **11.** 6889 **12.** 3878 **13.** 2063 **14.** 1213 **15.** 2557 **16.** 3101 **17.** 9567 **18.** 8489 **19.** 32,179 **20.** 27,244 **21.** 2326 **22.** 1157 **23.** 2689 **24.** 2428 **25.** 7306 **26.** 7662 **27.** 986 **28.** 2804 **29.** 27,617 **30.** 31,113 **31.** 23,903 **32.** 17,057 **33.** 9778 **34.** 12,448 **35.** **(a)** 8976 ft; **(b)** 2720 m

Exercise Set M.1, p. 583

1. 12 **2.** 3 **3.** 24 **4.** 0 **5.** 7 **6.** 0 **7.** 24 **8.** 4 **9.** 10 **10.** 30 **11.** 10 **12.** 10 **13.** 10 **14.** 63 **15.** 21 **16.** 54 **17.** 12 **18.** 0 **19.** 72 **20.** 18 **21.** 72 **22.** 8 **23.** 0 **24.** 0 **25.** 28 **26.** 24 **27.** 7 **28.** 45 **29.** 0 **30.** 18 **31.** 42 **32.** 7 **33.** 0 **34.** 35 **35.** 9 **36.** 45 **37.** 40 **38.** 0 **39.** 40 **40.** 16 **41.** 16 **42.** 25 **43.** 64 **44.** 81 **45.** 1 **46.** 0 **47.** 9 **48.** 4 **49.** 36 **50.** 8 **51.** 48 **52.** 0 **53.** 27 **54.** 18 **55.** 45 **56.** 0 **57.** 42 **58.** 10 **59.** 0 **60.** 48 **61.** 54 **62.** 0 **63.** 25 **64.** 72 **65.** 49 **66.** 15 **67.** 27 **68.** 0 **69.** 8 **70.** 9 **71.** 2 **72.** 5 **73.** 32 **74.** 6 **75.** 15 **76.** 6 **77.** 8 **78.** 8 **79.** 20 **80.** 20 **81.** 64 **82.** 16 **83.** 10 **84.** 0

Exercise Set M.2, p. 589

1. 80 **2.** 60 **3.** 90 **4.** 90 **5.** 70 **6.** 160 **7.** 210 **8.** 450 **9.** 780 **10.** 450 **11.** 560 **12.** 360 **13.** 800 **14.** 300 **15.** 900 **16.** 1000 **17.** 6700 **18.** 32,100 **19.** 345,700 **20.** 1200 **21.** 4900 **22.** 4000 **23.** 4200 **24.** 10,000 **25.** 7000 **26.** 9000 **27.** 2000 **28.** 457,000 **29.** 7,888,000 **30.** 6,769,000 **31.** 18,000 **32.** 20,000 **33.** 48,000 **34.** 16,000 **35.** 6000 **36.** 1,000,000 **37.** 1200 **38.** 4900 **39.** 200 **40.** 4000 **41.** 2500 **42.** 12,000 **43.** 49,000 **44.** 6000 **45.** 63,000 **46.** 120,000 **47.** 150,000 **48.** 800,000 **49.** 120,000 **50.** 24,000,000 **51.** 16,000,000 **52.** 80,000 **53.** 147 **54.** 444 **55.** 2965 **56.** 4872 **57.** 6293 **58.** 3460 **59.** 16,236 **60.** 29,376 **61.** 30,924 **62.** 13,508 **63.** 87,554 **64.** 195,384 **65.** 3480 **66.** 2790 **67.** 3360 **68.** 7020 **69.** 20,760 **70.** 36,150 **71.** 6840 **72.** 10,680 **73.** 358,800 **74.** 109,800 **75.** 583,800 **76.** 299,700 **77.** 11,346,000 **78.** 23,390,000 **79.** 61,092,000 **80.** 73,032,000

Exercise Set M.3, p. 593

1. 36 **2.** 40 **3.** 8 **4.** 10 **5.** 520 **6.** 348 **7.** 564 **8.** 684 **9.** 1527 **10.** 5642 **11.** 2160 **12.** 1234 **13.** 19,170 **14.** 29,136 **15.** 64,603 **16.** 31,468 **17.** 4770 **18.** 4680 **19.** 3120 **20.** 2958 **21.** 46,080 **22.** 43,956 **23.** 14,652 **24.** 44,792 **25.** 310,490 **26.** 577,280 **27.** 224,900 **28.** 272,190 **29.** 263,898 **30.** 673,017 **31.** 269,496 **32.** 197,136 **33.** 1,177,848 **34.** 1,227,858

35. 5,446,770 **36.** 20,336,745 **37.** 32,681,565 **38.** 38,714,037 **39.** 548,186,862 **40.** 694,891,776

Exercise Set D.1, p. 599

1. 3 **2.** 8 **3.** 4 **4.** 6 **5.** 1 **6.** 32 **7.** 9 **8.** 5 **9.** 7 **10.** 5 **11.** 37 **12.** 29 **13.** 5 **14.** 9 **15.** 4 **16.** 5 **17.** 6 **18.** 9 **19.** 5 **20.** 9 **21.** 8 **22.** 9 **23.** 6 **24.** 7 **25.** 3 **26.** 2 **27.** 9 **28.** 6 **29.** 2 **30.** 3 **31.** 7 **32.** 2 **33.** 8 **34.** 4 **35.** 7 **36.** 3 **37.** 2 **38.** 3 **39.** 6 **40.** 6 **41.** 1 **42.** 4 **43.** 6 **44.** 3 **45.** 7 **46.** 9 **47.** 1 **48.** 0 **49.** Not defined **50.** Not defined **51.** 7 **52.** 9 **53.** 8 **54.** 7 **55.** 1 **56.** 5 **57.** 0 **58.** 1 **59.** 0 **60.** 3 **61.** 1 **62.** 4 **63.** 7 **64.** 1 **65.** 6 **66.** 0 **67.** 1 **68.** 5 **69.** 2 **70.** 8 **71.** 0 **72.** 4 **73.** 0 **74.** 8 **75.** 3 **76.** 8 **77.** Not defined **78.** Not defined **79.** 4 **80.** 1 **81.** Never stops. Never get rid of 4's.

Sidelight: Calculator Corner: Number Patterns, p. 606

1. 24; 264; 2664; 26,664; 266,664 **2.** 81; 8811; 888,111; 88,881,111; 8,888,811,111 **3.** 6006; 60,606; 606,606; 6,066,606; 60,666,606 **4.** 999,999; 1,999,999; 2,999,997; 3,999,996; 4,999,995

Exercise Set D.2, p. 607

1. 69 R 1 **2.** 199 R 1 **3.** 92 R 1 **4.** 138 R 3 **5.** 1723 R 4 **6.** 2925 **7.** 864 R 1 **8.** 552 R 3 **9.** 527 R 2 **10.** 2897 **11.** 43 R 15 **12.** 32 R 27 **13.** 47 R 12 **14.** 42 R 9 **15.** 91 R 22 **16.** 52 R 13 **17.** 29 **18.** 55 R 2 **19.** 228 R 22 **20.** 20 R 379 **21.** 21 **22.** 118 **23.** 91 R 1 **24.** 321 R 2 **25.** 964 R 3 **26.** 679 R 5 **27.** 1328 R 2 **28.** 27 R 8 **29.** 23 **30.** 321 R 18 **31.** 224 R 10 **32.** 27 R 60 **33.** 87 R 1 **34.** 75 **35.** 22 R 85 **36.** 49 R 46 **37.** 383 R 91 **38.** 74 R 440 **39.** 2437×6

Index